Natural hazards in El Salvador

Edited by

William I. Rose
Geological Engineering & Sciences
Michigan Technological University
Houghton, Michigan 49931
USA

Julian J. Bommer
Department of Civil & Environmental Engineering
Imperial College London
South Kensington Campus
London SW7 2AZ
UK

Dina L. López
Department of Geological Sciences
Ohio University
316 Clippinger Laboratories
Athens, Ohio 45701
USA

Michael J. Carr
Rutgers University
New Brunswick, New Jersey 08901
USA

Jon J. Major
U.S. Geological Survey
Cascades Volcano Observatory
1300 SE Cardinal Court
Vancouver, Washington 98683
USA

THE
GEOLOGICAL
SOCIETY
OF AMERICA

Special Paper 375

3300 Penrose Place, P.O. Box 9140 ▪ Boulder, Colorado 80301-9140 USA

2004

Published by The Geological Society of America, Inc.
3300 Penrose Place, P.O. Box 9140, Boulder, Colorado 80301
www.geosociety.org

Printed in U.S.A.

GSA Books Science Editor: Abhijit Basu

Library of Congress Cataloging-in-Publication Data

Natural hazards in El Salvador / William I. Rose … [et al.].
    p. cm. — (Special paper ; 375)
  Includes bibliographic references and index.
  ISBN 0-8137-2375-2 (softcover)
    1. Natural disasters--El Salvador. 2. Hazardous geographic environments--El Salvador. 3. Emergency management--El Salvador. I. Rose, William I. (William Ingersol) II. Geological Society of America. III. Special papers (Geological Society of America) ; 375.

GB5011.22.N38 2004
551.2'097284—dc22

                                                    2003049451

Cover: Oblique aerial view to the south of Las Colinas landslide shortly after it occurred on Saturday, January 13, 2001. Source area is located at the top of the forested slope. The landslide killed approximately 585 people. (Photograph by Manuel Orellana, courtesy of *El Diario de Hoy*.) See "The Las Colinas landslide, Santa Tecla: A highly destructive flowslide triggered by the January 13, 2001, El Salvador earthquake," by S.G. Evans and A.L. Bent, p. 25–37.

10 9 8 7 6 5 4 3 2 1

Contents

Geological Society of America
Special Paper 375
2004

Natural hazards and risk mitigation in El Salvador: An introduction

William I. Rose
Geological Engineering & Sciences, Michigan Technological University, Houghton, Michigan 49931, USA

Julian J. Bommer
Department of Civil & Environmental Engineering, Imperial College London, South Kensington Campus, London SW7 2AZ, UK

Ciro A. Sandoval
Department of Humanities, Michigan Technological University, Houghton, Michigan 49931, USA

ABSTRACT

This volume brings together papers on current research into natural hazards in El Salvador and efforts to mitigate their impact on the society. It recognizes the need and potential for such work around the world and especially in many developing countries. In El Salvador, researchers on geological hazards obtain very frequent experience with real hazards and, through their work, aim to help develop strategies to mitigate the terrible suffering and monetary cost that is associated with their impact.

Keywords: natural hazards, El Salvador, risk mitigation, earthquakes, volcanoes, landslides.

INTRODUCTION

Much of the work in this volume is collaborations of foreign scientists with local counterparts, with the aim of providing training and exchanging experiences. The volume is likely to be widely read in El Salvador, and also we hope by scientists and engineers from outside El Salvador—people who are doing similar work in other countries or people who might consider working in El Salvador itself. In this brief introduction we provide a thumbnail sketch of El Salvador and its environment, plus an overview of the natural hazards to which it is subjected and the social factors that both exacerbate the impact of natural hazards and present obstacles to effective risk mitigation programs.

THE COUNTRY OF EL SALVADOR

El Salvador is the smallest country of Central America, with just 21,040 km². Its population is 6.4 million (2001) resulting in an average density of ~304 per km², the region's highest. According to the World Bank, the country has a per capita income of just over $2,000, higher than Guatemala, Honduras, and Nicaragua (http://www.worldbank.org/data). However, this income level masks a stark inequality between a small wealthy elite and a large and very poor majority (Barry, 1991).

Population density is amplified in the southwestern third of El Salvador, where 3/4 of the population is now settled in the area west of Lake Ilopango and south of the city of Santa Ana (Rosa and Barry, 1995). The people live on a rugged topography made up of young volcanoes and eroded older ones. Settlement patterns were originally dictated by fertility and water availability on the volcanic slopes in the Great Interior Valley. Coffee cultivation in the end of the nineteenth century changed land use and ownership when productive land was put in control of a few families, and left most of the people without their own land. During the last two decades the migration from the northern and eastern parts of the country has been driven by persistent rural poverty and by the fratricidal war from 1980 to 1992. The last two decades has also seen a huge exodus from El Salvador with more than a million Salvadorans having moved abroad, primarily to the United States. The mass emigrations have also had a marked effect on the national economy, giving rise to a situation where U.S. dollars, sent back by Salvadorans living in the United States, reached a level of US$1.75 billion in the year 2000. This is six times greater than the value of coffee exports and almost three times greater than the net foreign exchange generated by the garment assembly industry, which has mushroomed in recent years. The U.S. dollar is now the official currency of El Salvador.

The topography and geology of El Salvador provides many advantages, including significant hydrological and geothermal power, scenic landscapes for tourism, and rich tropical soils. El Salvador borders the Middle America Trench, the active subduction boundary and seismic zone between the Cocos and Caribbean plates (Dewey et al., this volume, Chapter 27). As a consequence of plate subduction, El Salvador is bisected by the volcanic front, a linear belt of active volcanoes and accompanying seismic zone. Thus, El Salvador faces high rates of both subduction zone and upper-crustal earthquakes (White et al., this volume, Chapter 28), explosive eruptions, and landslides—three of the most destructive geological hazards. Floods are also important. The affirmation by the historian William Durant, that civilization exists by geological consent subject to changes without prior notice, is definitely applicable to this tiny country. The shadowy ambiguity of a volcano—beauty, fertility, energy, and its inevitable danger—is an important symbol for El Salvador, and it colors everyone's hazard perception.

RECENT EVENTS

Recent history has reminded Salvadorians of many of the natural hazards in their otherwise delightful natural surroundings. In addition to geological hazards, El Salvador is located in the subtropical hurricane zone and is visited by both Atlantic and Pacific cyclones. In November 1998, Hurricane Mitch killed 240 people, displaced 85,000, and cost the country an estimated $388 million, equivalent to 3% of the gross domestic product (GDP) (Mowforth, 2001; Araniva de Gonzalez, this volume, Chapter 34). In early 2001, a series of earthquakes and associated landslides killed 1259 people, destroyed or damaged more than 300,000 houses, and caused about $1.6 billion in damages (Bommer et al., 2002). This was the fifth large earthquake in 50 yr, consistent with the pattern of El Salvador being hit by a destructive earthquake once per decade on average. The last earthquake before this was the upper-crustal event that struck San Salvador in October 1986, leaving 1500 dead and causing damages equivalent to 31% of the GDP in the same year. Similarly destructive earthquakes occurred in 1951 and 1965.

Volcanoes have been kinder in recent years, although landslides triggered in loose deposits on their slopes have had a terrible impact. A notorious example was the mudslide triggered on the slopes of the San Salvador volcano by heavy rainfall in September 1982, burying 500 people and leaving another 2400 homeless (CEPRODE, 1994). Although the morphology of the volcanic front resembles both Guatemala and Nicaragua, historic volcanic activity in El Salvador is sparse, compared to either of its neighbors. Prehistory was markedly different, however (Sheets, this volume, Chapter 8), and this longer time perspective integrates more information and more accurately portrays volcanic risk. The three largest cities (San Salvador, Santa Ana,

San Miguel) are each located on the flanks of potentially active volcanoes (with the same names). In the center of El Salvador is Ilopango Caldera, an active volcano (Mann et al., Chapter 12; Richer et al., Chapter 13; López et al., Chapter 14: all this volume) masquerading as a beautiful lake, which has devastated El Salvador four times in the past 56,000 yr, the last at ~A.D. 429 (Dull, this volume, Chapter 18). The country's major seaport, Acajutla, through which 45% of the external trade passes, is located on a huge debris avalanche from Santa Ana volcano (Siebert et al., this volume, Chapter 2). San Salvador is on top of thick deposits of the Tierra Blanca Joven tephra of Ilopango (10 km east; Rolo et al., this volume, Chapter 5) and intercalated mudflows and fall deposits of San Salvador volcano (7 km west; Sofield, this volume, Chapter 11). Both of these volcanoes are likely to be active again. Although the impact on the population was limited, the most recent eruption of San Salvador volcano, in 1917, covered part of the northern side of the volcano with lava, creating a rough area that remains unpopulated.

As if it were not enough for this small country to have the geographical misfortune to be prone to the full spectrum of natural hazards, its recent history has been severely marked by disaster from human conflict. The hugely uneven distribution of the country's wealth led to a civil war from 1980 to 1992 that left 75,000 people dead, the vast majority of them noncombatants, and cost more than $2 billion. The decade of peace that has followed the signing of agreements between the guerrillas of the Farabundo Marti para la Liberacion Nacional (FMLN) and the El Salvadorian government has not brought the stability and prosperity that was expected (Europa World Yearbook, 2000). Poverty, particularly in rural areas, has not been eliminated, and most Salvadorians feel that their economic situation is precarious or worse. United Nations statistics on human development paint a worrying picture (http://www.deasrrollohumano.org.sv). A particular problem that has arisen in the postwar period is a massive increase in crime, particularly violent crime, which has created a situation of great insecurity for most of the population. A Gallup poll taken in February 2002 (http://www.cidgallup.com), just one year after the earthquakes, reveals that the two overriding preoccupations of Salvadorians were unemployment and delinquency. Environmental degradation is another very serious problem and a concern for Salvadorians (PRISMA, 1995).

RISK MITIGATION EFFORTS

Until recently, El Salvador's government has not made a large commitment to natural hazards mitigation. International researchers did not study the geology of El Salvador much, because of war and an environment perceived as unsafe. During the 1980s, the war was the obvious obstacle to any risk mitigation programs. In the last decade, the relative lack of concerted efforts in hazards mitigation likely reflects the severity of more immediately demanding problems such as water quality, sanitation, various health issues, etc. This is despite the fact that 50% of the population reported, in the Gallup poll mentioned previously, that they or a family member had been adversely affected by the earthquakes in 2001.

The combined effect of Hurricane Mitch and the 2001 earthquakes, however, has had an impact. In response to these, the Salvadorian government has taken the bold step in the past two years to establish a new government agency, Servicio Nacional de Estudios Territoriales (SNET), in the Ministerio de Ambiente and Recursos Naturales. The new agency aims to deal with the mitigation of natural hazards, and it will try to consolidate and strengthen infrastructure. Work on hazard characterization involves technical aspects such as monitoring of earthquakes, volcanoes, meteorology/hydrology, and slopes. But education of the public about risk is also badly needed. Geoscience education in El Salvador is minimal—the only university degree program in the whole country is one in physics with an option in geophysics. Training in earthquake-resistant structural design is also limited (López et al., this volume, Chapter 23). Although no studies document this, there seems to be an unusual sophistication amongst Salvadorians about risk perception. This comes from the continual experience with war and natural hazards in recent years. Few in the population are without friends and relatives that were impacted. This has taught them to compare low-level risks with some sophistication. As we know from our experience in developed countries, the public, government officials, and even insurance companies are quite ignorant about low-level risk and there can be difficulty in discussing these issues rationally. The cultural, social and historical context of each country makes education about natural hazards challenging. It is not enough to know the technical aspects.

There are many questions to consider when thinking about mitigating the risks from natural hazards in a developing country. The poorest Salvadorans live in the areas with highest risk (on steep slopes and along rivers). Most of these natural events cannot be prevented or even significantly modified by man's effort. Recent history has made Salvadorans more aware of seismic hazards than volcanic ones, but this may well be misleading. How can education effectively address this? How does a mitigation effort produce value added from expenditures that are very badly needed for other purposes as well? What should be the goal of a risk mitigation effort? How are low-probability but apocalyptic hazards addressed? How can the work by foreign scientists and engineers in El Salvador best reinforce local infrastructure? What kinds of technology transfer are most effective? How can local administrative authorities in El Salvador be strengthened in order to enable them to impose land use control and building regulations that can control the impact of future events?

CONCLUSION

The first volume devoted to geological studies in El Salvador is now published. It puts together information that can be used by Salvadorans for many purposes, such as land use, risk management, and education. We hope that this volume will also increase interest in international intercultural communications and science about natural hazards. The volume is complementary to studies that have looked at natural hazards and local capacities for dealing with their impact in the region (Ordoñez et al., 1999; Trujillo et al., 2000), providing a specific focus on El Salvador and adding more detailed and up-to-date technical information. We hope that others may be inspired to produce similar compilations for other countries in Central and South America that share similar problems in terms of natural hazards and developing economies.

We believe that the high and growing costs of natural hazards can be cut by working together in a way that benefits everyone—giving scientists from abroad access to the most dramatic and current natural events and providing a vehicle for strengthening local knowledge and capability.

REFERENCES CITED

Araniva de Gonzalez, A.E., 2004, Social perspectives on hazards and disasters in El Salvador, *in* Rose, W.I., et al., eds., Natural hazards in El Salvador: Boulder, Colorado, Geological Society of America Special Paper 375, p. 453–459 (this volume).
Barry, T., 1991, Central America inside out: New York, Grove Widenfeld, 501 p.
Bommer, J., Benito, M.B., Ciudad-Real, M., Lemoine, A., López-Menjívar, M.A., Madariaga, R., Mankelow, J., Méndez de Hasbun, P., Murphy, W., Nieto-Lovo, M., Rodríguez-Pineda, C.E., and Rosa, H., 2002, The El Salvador earthquakes of January and February 2001: Context, characteristics, and implications for seismic risk: Soil Dynamics and Earthquake Engineering, v. 22, no. 5, p. 398–418.
CEPRODE (Centro de Protección para Desastres), 1994, Caracterización de los desastres en El Salvador: tipología y vulnerabilidad socioeconómica. San Salvador: Centro de Protección para Desastres, 61 p.
Dewey, J.W., White, R.A., and Hernández, D.A., 2004, Seismicity and tectonics of El Salvador, *in* Rose, W.I., et al., eds., Natural hazards in El Salvador: Boulder, Colorado, Geological Society of America Special Paper 375, p. 363–378 (this volume).
Dull, R., 2004, Lessons from the mud, lessons from the Maya: Paleoecological records of the Tierra Blanca Joven eruption, *in* Rose, W.I., et al., eds., Natural hazards in El Salvador: Boulder, Colorado, Geological Society of America Special Paper 375, p. 237–244 (this volume).
Europa World Yearbook, 2000, International organizations, El Salvador, Volume 1 (41st edition): London, Europa Publications, p. 1334–1350.
López, D.L., Ransom, L., Pérez, N.M., Hernández, P., and Monterosa, J., 2004, Dynamics of diffuse degassing at Ilopango Caldera, El Salvador, *in* Rose, W.I., et al., eds., Natural hazards in El Salvador: Boulder, Colorado, Geological Society of America Special Paper 375, p. 191–202 (this volume).
López, M., Bommer, J.J., and Pinho, R., 2004, Seismic hazard assessments, seismic design codes, and earthquake engineering in El Salvador, *in* Rose W.I., et al., eds., Natural hazards in El Salvador: Boulder, Colorado, Geological Society of America Special Paper 375, p. 301–320 (this volume).
Mann, C.P., Stix, J., Vallance, J.W., and Richer, M., 2004, Subaqueous intracaldera volcanism, Ilopango Caldera, El Salvador, Central America, *in* Rose, W.I., et al., eds., Natural hazards in El Salvador: Boulder, Colorado, Geological Society of America Special Paper 375, p. 159–174 (this volume).
Mowforth, M., 2001, Storm warnings: Hurricanes George and Mitch and the lessons for development: London, The Catholic Institute for International Relations, 84 p.
Ordoñez, A., Trujllo, M., and Hernández, R., 1999, Mapeo de riesgos y vulnerabilidad en Centroamérica y México: Estudio de capacidades locales para trabajar en situaciones de emergencia: Managua, Nicaragua, Oxfam, 100 p.
PRISMA, 1995, El Salvador: Dinámica de Degradación Ambiental, San Salvador: Programa Salvadoreño de Investigación Sobre Desarrollo y Medio Ambiente (PRISMA), 44 p.
Richer, M., Mann, C.P., and Stix, J., 2004, Mafic magma injection triggering eruption at Ilopango Caldera, El Salvador, Central America, *in* Rose, W.I., et al., eds., Natural hazards in El Salvador: Boulder, Colorado, Geological Society of America Special Paper 375, p. 175–189 (this volume).

Rolo, R., Bommer, J.J., Houghton, B.F., Vallance, J.W., Berdousis, P., Mavrommati, C., and Murphy, W., 2004, Geologic and engineering characterization of Tierra Blanca pyroclastic ash deposits, *in* Rose, W.I., et al., eds.,
Natural hazards in El Salvador: Boulder, Colorado, Geological Society of
America Special Paper 375, p. 55–67 (this volume).

Rosa, H., and Barry, D., 1995, Población, territorio y medio ambiente en El
Salvador: San Salvador, Programa Salvadoreño de Investigación Sobre
Desarrollo y Medio Ambiente (PRISMA) Boletín, no. 11, 16 p.

Sheets, P., 2004, Apocalypse then: Social science approaches to volcanism,
people, and cultures in the Zapotitán Valley, El Salvador, *in* Rose, W.I., et
al., eds., Natural hazards in El Salvador: Boulder, Colorado, Geological
Society of America Special Paper 375, p. 109–120 (this volume).

Siebert, L., Kimberly, P., and Pullinger, C.R., 2004, The voluminous Acajutla
debris avalanche from Santa Ana volcano, western El Salvador, and comparison with other Central American edifice-failure events, *in* Rose, W.I.,
et al., eds., Natural hazards in El Salvador: Boulder, Colorado, Geological
Society of America Special Paper 375, p. 5–23 (this volume).

Sofield, D., 2004, Eruptive History and Volcanic Hazards of Volcán San Salvador, *in* Rose, W.I., et al., eds., Natural hazards in El Salvador: Boulder,
Colorado, Geological Society of America Special Paper 375, p. 147–158
(this volume).

Trujillo, M., Ordóñez, A., and Hernández, C., 2000, Risk-mapping and local
capacities: Lessons from Mexico and Central America. Oxfam Working
Papers: Oxford, England, Oxfam, 78 p.

White, R.A., Ligorría, J.P., and Cifuentes, I.L., 2004, Seismic history of the
Middle America subduction zone along El Salvador, Guatemala, and Chi-
apas, Mexico, 1526–2000, *in* Rose, W.I., et al., eds., Natural hazards in
El Salvador: Boulder, Colorado, Geological Society of America Special
Paper 375, p. 379–396 (this volume).

Manuscript Accepted by the Society June 16, 2003

Geological Society of America
Special Paper 375
2004

The voluminous *Acajutla debris avalanche from Santa Ana volcano, western El Salvador, and comparison with other Central American edifice-failure events*

Lee Siebert*
Paul Kimberly
Smithsonian Institution, Global Volcanism Program, Washington, D.C. 20013-7012, USA

Carlos R. Pullinger
Servicio Nacional de Estudios Territoriales (SNET), San Salvador, El Salvador

ABSTRACT

Collapse of Santa Ana volcano during the late Pleistocene produced the voluminous and extremely mobile Acajutla debris avalanche, which traveled ~50 km south into the Pacific Ocean, forming the broad Acajutla Peninsula. The subaerial deposit covers ~390 km^2; inclusion of a possible additional ~150 km^2 submarine component gives an estimated volume of 16 ± 5 km^3. Hummocks are present to beyond the coastline but are most prominent in four clusters corresponding to the location of buried bedrock ridges. Bulking in distal portions incorporated accessory Tertiary-to-Quaternary volcaniclastic rocks and ignimbrites. Modern Santa Ana volcano was constructed within the collapse scarp, visible only on its northwest side, following an apparent transition in eruptive style. More than 286,000 people, the country's main port, and important agricultural land now overlie the Acajutla debris-avalanche deposit, which is one of only a few in Central America to exceed 10 km^3 in size. Because major edifice failures are high-impact, low-frequency events, the probability of a future Acajutla-scale collapse is very low. However, a collapse even an order of magnitude smaller in volume from modern Santa Ana volcano would impact heavily populated areas. The Acajutla failure was perpendicular to a NW-trending fissure system cutting across Santa Ana volcano, which may also influence future failure orientations. The current structure of Santa Ana volcano suggests that future collapses are most likely to the southwest, but the possibility of northward failures cannot be excluded.

Keywords: Santa Ana volcano, edifice failure, debris avalanche, volcanic hazards.

INTRODUCTION

Large-scale volcanic edifice collapse is a common process at volcanoes in a wide variety of tectonic settings. Although steep-sided continental-margin stratovolcanoes are particularly susceptible to edifice failure, which has consequently been identified at many volcanoes in México and Central America, little attention has been devoted to this topic in El Salvador. We report here on a

voluminous volcaniclastic deposit that extends from the Santa Ana volcanic complex in the main volcanic front of western El Salvador to the Pacific Ocean. The avalanche extended the otherwise linear Pacific Ocean shoreline, forming a broad 20 km wide delta, and a significant submarine component extends offshore.

The subaerial component of this deposit was for many years mapped both as part of the regional Pliocene Bálsamo Formation and overlying Quaternary alluvial deposits (Weber and Weisemann, 1978). Pullinger (1998) noted the debris-avalanche origin of this deposit during a study of the Santa Ana volcanic

*E-mail: siebert@volcano.si.edu

Siebert, L., Kimberly, P., and Pullinger, C.R., 2004, The voluminous Acajutla debris avalanche from Santa Ana volcano, western El Salvador, and comparison with other Central American edifice-failure events, *in* Rose, W.I., Bommer, J.J., López, D.L., Carr, M.J., and Major, J.J., eds., Natural hazards in El Salvador: Boulder, Colorado, Geological Society of America Special Paper 375, p. 5–23. For permission to copy, contact editing@geosociety.org. © 2004 Geological Society of America

complex. More detailed mapping during the present study was undertaken to further delineate the extent of the deposit and its characteristics, to determine the impact of the edifice-failure event on the eruptive style and subsequent constructional history of the Santa Ana volcanic complex, and to assess the volcanic hazard implications of this voluminous edifice failure, one of the largest in Central America. Seismically induced small-volume slope failures have caused extensive devastation in El Salvador (see papers in this volume by Evans and Bent, Chapter 3; Jibson et al., Chapter 6; and Konagai et al., Chapter 4); this study underscores the potential hazards in El Salvador from landslides several orders of magnitude larger in volume.

REGIONAL GEOLOGY

El Salvador occupies the west-central portion of the Chortis block of the Caribbean plate. A roughly E-W–trending chain of volcanoes extends from the México/Guatemala border across El Salvador to the Gulf of Fonseca, ~150 km inboard of the Middle American Trench (Fig. 1). Crustal thickness drops from ~50 km at either end to ~32–37 km over a broad portion of the center of the Central American arc. Sinistral movement of the plate-bounding Motagua-Polochic fault system has caused extensional faulting in southeastern Guatemala and western El Salvador (Burkhart and Self, 1985), and in contrast to the extremities of the arc, volcanoes in El Salvador and Nicaragua are not constructed over topographic highs, but rise in isolation above broad down-dropped crustal blocks.

Santa Ana volcano lies adjacent to a plate segment boundary proposed by Stoiber and Carr (1973), although later work (Burbach et al., 1984) questioned micro-segmentation of the descending Cocos plate. The Santa Ana volcanic complex in western El Salvador is situated at the southern margin of the Median Trough that transects the country and lies between San Salvador volcano and the E-W–trending cluster of small stratovolcanoes and underlying silicic calderas known as the Sierra de Apaneca or Ahuachapán Range (Fig. 1).

The 165 km³ (Carr, 1984) Santa Ana volcanic complex (Fig. 2) forms the largest and highest stratovolcano in El Salvador. The broad flanks of the 2381 m high basaltic-to-andesitic volcano extend north to the city of Santa Ana and south to Sonsonate and are home to more than a million inhabitants, 15% of the population of El Salvador (Pullinger, 1998). Basement rocks underlying the Santa Ana complex (Fig. 3) are older than ~200,000 yr (Pullinger, 1998) and exposed southward toward the coast. They consist of the Miocene-Pliocene Bálsamo Formation and the basal part of the Quaternary San Salvador Formation (Weber and Weisemann, 1978; Baxter, 1984). Three broad benches and a series of nested craters, the youngest of which contains an acidic lake, form the 1.5 km wide, flat-topped summit of the volcano. The lake and surrounding crater walls are fumarolically active, and the crater has been the source of phreatomagmatic eruptions in historical time, most recently in 1920. A 30 km long, NW-SE–trending fissure system that transects the

volcano from its summit to lower flanks (Fig. 1) has been active since the late Pleistocene and is the locus of vents ranging from phreatomagmatic craters on the NW end to historically active cinder cones on the SE (Pullinger, 1998).

The dramatic unvegetated cone of Izalco volcano was constructed since 1770 on the southern flank of Santa Ana (Carr and Pontier, 1981). Frequent strombolian eruptions visible from the coast caused it to be called the "Lighthouse of the Pacific." Coatepeque caldera partially truncates the eastern side of ancestral Santa Ana volcano. The 7 × 10 km wide caldera, now largely filled by a scenic lake, was formed during two major late-Pleistocene rhyodacitic pyroclastic-flow-producing eruptions. The Arce pyroclastic-flow deposit, distinguishable by its large biotite content, has been dated using the high-precision ^{40}Ar/^{39}Ar method at 72,000 ± 2000 yr B.P. (Rose et al., 1999). The Congo eruption was dated using a high-sensitivity ^{14}C procedure at 56,900 +2800 −2100 yr B.P. (Rose et al., 1999). Post-caldera basaltic cinder cones and lava flows were erupted along the caldera rim, and a half dozen rhyodacitic lava domes were emplaced along a NE-SW line on the caldera floor near the lake margins.

ACAJUTLA DEBRIS–AVALANCHE DEPOSIT

Distribution and Morphology

The Acajutla debris-avalanche deposit (Fig. 4) is broadly exposed SSW of Santa Ana volcano to the coast, where it forms the Acajutla Peninsula. The deposit is named after the peninsula and the overlying coastal city of Acajutla. In proximal areas, the deposit is overlain by lava flows and tephra deposits from modern Santa Ana and Izalco volcanoes that extend as far as 18 km from Santa Ana volcano, but exposures of the avalanche deposit have been found to within less than 10 km of the current summit of Santa Ana. On the east, the deposit is bounded by the rugged hills of the Bálsamo Range, and on the west it laps onto erosionally dissected pyroclastic-flow deposits of the basal San Salvador Formation from the Sierra de Apaneca.

The indistinct margins of the deposit on the NE and western sides are approximately inferred from scattered outcrops, deflection of drainages, and topographic smoothing of dissected valleys by infill of the deposit. The medial portion of the deposit south of the mantling lava flows from Santa Ana volcano contains abundant hummocks prominent on topographic maps. Hummocks in this area, ~25–30 km from the source, reach to ~60 m high and are typically ~100–300 m in longest dimension. Some apparent hummocks are instead kipukas of Bálsamo Formation rocks in line with NE-SW–trending Bálsamo ridges to the east (Fig. 5). Hummock density is highest between ~15 and 30 km from the volcano, but hummocks are found to the current shoreline and offshore.

The NE boundary of the avalanche deposit in the flat-lying area north of the Bálsamo Range is uncertain because the deposit is largely obscured by overlying younger lava flows. The avalanche deposit was exposed in a now-filled utility trench under construction in 2002 about one km southeast of the village of

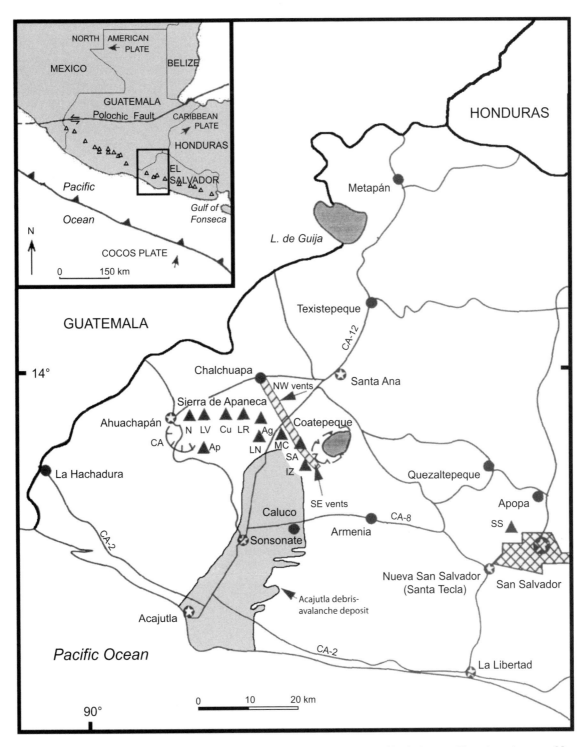

Figure 1. Location map. Triangles denote Quaternary volcanoes: CA—Concepción de Ataco caldera, Ap—Apaneca, N—La Ninfas, LV—Laguna Verde, Cu—Cuyanausol, LR—Las Ranas, Ag—El Aguila, LN—Los Naranjos, MC—Malacara, SA—Santa Ana, IZ—Izalco, SS—San Salvador. Inset map of northern Central America shows tectonic setting; teeth are on upthrust plate; open triangles represent volcanoes. Light-gray shaded area marks approximate subaerial extent of Acajutla debris-avalanche deposit. After Pullinger (1998).

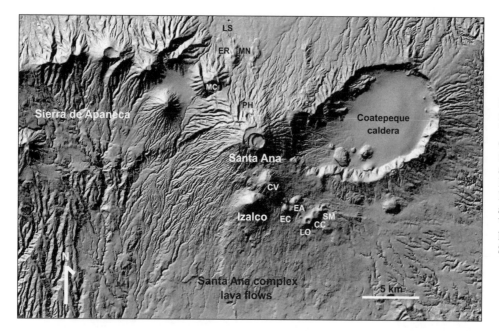

Figure 2. Shaded-relief digital elevation model view of the Santa Ana massif. Vents of the NW-trending fissure zone cutting across the volcano and extending ~7 km north of the top of the image are labeled from north to south: LS—Laguna Seca–El Hoyo, ER—El Retiro, MN—Montañita, MC—Malacara, PH—Plan de Hoyo, CV—Cerro Verde, EC—El Conejal, EA—El Astillero, LO—La Olla, CC—Cerro Chino, SM—San Marcelino.

Modern Santa Ana volcano	Magmatic and phreatomagmatic tephra-fall deposits and lava flows from modern Santa Ana volcano, flank vents, and Izalco volcano; upper San Salvador Formation (late-Pleistocene to Holocene).
Acajutla debris-avalanche	Acajutla debris-avalanche deposit (late Pleistocene), containing material from ancestral Santa Ana volcano and accessory material from the Bálsamo and basal San Salvador Formations.
Congo	Congo pyroclastic-fall and -flow deposits from caldera-forming eruption at Coatepeque, K-Ar dated 56,900 B.P.
Arce	Arce pyroclastic-fall and -flow deposits from caldera-forming eruption at Coatepeque, K-Ar dated 72,000 B.P.
Ancestral Santa Ana volcano	Pyroclastic-fall and block-and-ash flow deposits and lava flows from ancestral Santa Ana volcano; member S2 of San Salvador Formation (late Pleistocene).
San Salvador Fm.	Silicic pyroclastic-flow and volcaniclastic deposits and basaltic-to-andesitic effusive rocks of the Sierra de Apaneca; members S1 and S2 of the basal San Salvador Formation (late Pleistocene).
Bálsamo Fm.	Mafic-to-intermediate volcaniclastic and localized effusive rocks of the basal Bálsamo Formation (Miocene-Pliocene).

Figure 3. Generalized stratigraphic column of the Santa Ana area, modified from Weber and Weisemann (1978) and Baxter (1984). K-Ar dates are from Rose et al. (1999).

Caluco (Fig. 4); this is near the inferred eastern margin of the deposit. Two exposures of the Arce pyroclastic-flow deposit from Coatepeque volcano dated at 72,000 yr B.P. (Rose et al., 1999) are present nearby at the Hacienda Comalapa 2.5 km south of Caluco and along the Río Chiquihuat south of the Hacienda Victoria 3 km SE of Caluco, but Arce deposits are not exposed farther west and are considered to underlie the avalanche deposit. Small conical hills resembling debris-avalanche hummocks are exposed ~10 km east of the inferred deposit margin ~3 km north of the Sonsonate–San Salvador highway (CA-8 on Fig. 1) on both sides of the road to Coatepeque caldera. Excavations have

shown that these hills, however, are Mayan archaeological sites (Sheets, 1983).

Farther south, the eastern margin of the Acajutla deposit is buried by flat-lying sediments from Bálsamo Range river valleys that were likely initially the site of ephemeral avalanche-dammed lakes. Although these tributary valleys intersect the deposit at an angle oblique to the avalanche travel direction, the buried avalanche deposit is arbitrarily inferred to extend halfway up the flat-lying portion of these valleys due to lateral spreading of the avalanche by analogy with the 1980 Mount St. Helens debris avalanche, which extended at least 3.5 km up Coldwater Creek

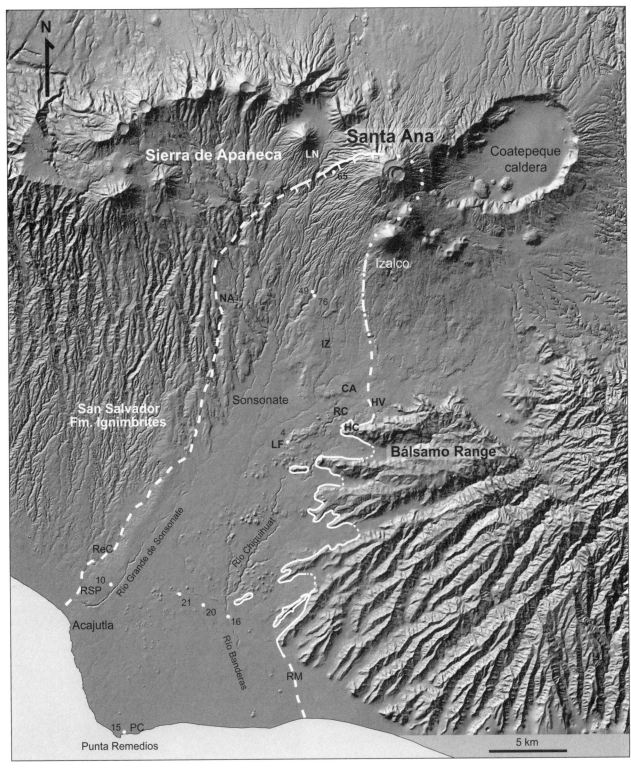

Figure 4. Shaded-relief digital elevation model of western El Salvador showing extent of Acajutla debris-avalanche deposit. Dashed and solid lines mark inferred extent of Acajutla debris-avalanche deposit, hachured line marks exposed margin of avalanche source area, dots indicate inferred buried margin. Dashed-dot line lies on crest of arcuate escarpment south of Izalco volcano. CA—Caluco, HC—Hacienda Comalapa, HV—Hacienda Victoria, IZ—town of Izalco, LF—maximum extent of lava flows from modern Santa Ana volcano, LN—Los Naranjos, NA—Nahuizalco, PC—Playa Los Cóbonos, RC—Río Chiquihuat, ReC—Río el Coyol, RM—Río Mandinga, RSP—Río San Pedro. Numbers mark locations of samples used for sedimentological or geochemical analyses.

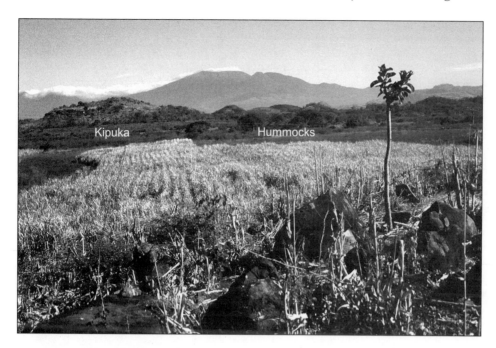

Figure 5. Coarse blocks litter the hummock surface in the foreground, and a hummock cluster can be seen to the right of the larger hill at the left, a kipuka of Bálsamo Formation material that served as a barrier to avalanche transport. Santa Ana volcano (center skyline) lies 30 km away.

in a direction 140° oblique to that of the avalanche travel path (Glicken, 1986). The southeasternmost margin of the deposit is considered to lie near the Río Mandinga, whose southwestward trajectory is deflected to the SE after the river extends beyond the confining Bálsamo ridges.

On the western side, the Acajutla avalanche deposit underlies the city of Sonsonate and is exposed in quarries on the western side of the city, where it directly overlies ignimbrite deposits containing abundant black, glassy blocks. Farther south, the avalanche deposit feathers out just west of the Río San Pedro (Fig. 4). North of here the avalanche deposit laps onto ignimbrite deposits forming steep-walled gorges along the Río el Coyol, ~3 km north of Highway 2 (Fig. 1), and SW-trending topography here oblique to the roughly N-S–trending ridges below the Sierra de Apaneca is considered to be underlain by the avalanche deposit. The Río Grande de Sonsonate (also known as the Río Sensunapán) occupies the topographic inflection point below where the western margin of the avalanche lapped onto the ignimbrite ridges.

Submarine Extent

The Acajutla Peninsula forms the largest topographic irregularity along a 900 km long stretch of the Pacific coast between the Gulf of Tehuantepec off the coast of Oaxaca, México, and the Gulf of Fonseca at the southeastern tip of El Salvador. The ~20 km wide peninsula extends up to 7 km off the former coastline. Vigorous longshore drift prevents the formation of deltas off the Salvadoran coast; thus, a large delta is not likely to have existed prior to the collapse, and the pre-collapse shoreline would likely have paralleled the current coastline. The tip of the asymmetrical peninsula lies west of the axis of the avalanche deposit,

whose western side has been eroded by longshore currents forming vertical cliffs ~15 m high at the port of Acajutla (Fig. 6). Semi-indurated, normally graded beach deposits lap onto the avalanche deposit at the tip of the peninsula at Punta Remedios (Fig. 4), and post-avalanche deposition has smoothed the shoreline to the east.

A 1:25,000-scale bathymetric map prepared to identify shipping channels for Acajutla, the country's largest port, delineates submarine topography of the western part of the deposit. The map extends eastward to cover an area beyond the southernmost subaerial extent of the deposit at Punta Remedios (Fig. 4).

Figure 6. Irregular coastal cliffs at the city of Acajutla expose the debris-avalanche deposit and contrast with the low, linear, beach-fringed shoreline in the distance west of the deposit.

Small rocky shoals and closed submarine contours indicate that hummocky topography continues offshore. The abundant map notations of "muchas rocas" ("many rocks") and "no sonedos" ("no soundings") distinguish the rocky avalanche deposit from offshore areas on either side. Irregular terrain with hummocks ranging to ~5 m in height is present to ~5 km offshore at the southernmost extent of detailed bathymetry.

The submarine extent of the avalanche deposit is inferred from the inflection point of otherwise shore-parallel bathymetric contours. The 35 m contour deflects, while the 40 m contour does not, suggesting that the avalanche deposit extends to a depth of just over 35 m ~5 km SW of the tip of the Acajutla Peninsula. Although comparable bathymetry is not available to the east, a 1:250,000 map shows 6-, 10- and 20-fathom contours. A pronounced deflection of the 20-fathom (37 m) submarine contour south of the mouth of Río Mandinga (Fig. 4) may correspond to the eastern deposit margin. If this inference is correct, the submarine component of the Acajutla debris-avalanche deposit extends up to 11 km from the eastern shoreline, and the currently submarine component of the deposit would be nearly 200 km^2.

The lack of detailed bathymetry leaves remaining uncertainty regarding the eastern submarine margin of the avalanche. Although detailed bathymetric contours on the western side suggest that the deposit extends to ~35 m depth at its terminus WSW of Punta Remedios, it is possible that the deflected 20-fathom bathymetric contour is underlain by beach-sand deposits banked up against the avalanche deposit margin by longshore currents, and thus the avalanche may not have reached this depth along its full terminus. Using a more conservative extrapolated 15-fathom contour to reflect the deposit margin results in an estimated currently submarine area of ~150 km^2. This implies that ~215 km^2 of the avalanche deposit lies beyond the former shoreline and that the terminus of the deposit lies a minimum of 46 and a maximum of 50 km from the volcano.

Impact of an avalanche of this magnitude into the Pacific would almost certainly have produced a tsunami, although tsunami deposits this old would likely not be preserved. The largest slope-failure-related tsunamis have occurred when the basal failure plane extended below sea level and the edifice failure itself had a submarine component, such as at Ritter Island in Melanesia in 1888 (Johnson, 1987) and Oshima-Oshima Island off Hokkaido in 1742 (Satake and Kato, 2001). This was not the case at Santa Ana; however, significant tsunamis elsewhere have resulted from the sudden impact of debris avalanches into the sea (Siebert et al., 1987).

Textural Characteristics

Despite the massive size of the Acajutla debris-avalanche deposit, few natural cross-sectional exposures are available. Scattered aggregate quarries provided ephemeral exposures critical to determining the textural characteristics of the deposit (Fig. 7). Block-facies material (Fig. 7A) is abundant in avalanche hummocks, where quarries expose spectacular sections of color-mottled units of the former edifice transported

relatively intact from their source on the volcano. The best-exposed outcrop was at Cerro Jícaro, ~20 km from the volcano (sample #4 on Figure 4), where quarrying operations revealed a colorful assemblage of slightly sheared and faulted units from Santa Ana volcano. Individual clasts include andesitic, basaltic andesite, and olivine basaltic lithologies (Pullinger, 1998). The juxtaposition of multicolored units highlighted evidence of normal faulting and horst-and-graben formation on the scale of up to a few meters offset. Large clasts up to 6 m in diameter were exposed, and jigsaw blocks were present. Some individual units could be traced several tens of meters across the quarry wall.

A quarry in another hummock south of Highway 2 (sample #21, Fig. 4) ~30 km from the volcano exposed large segments of essentially intact bedded ash-and-lapilli layers transported without significant disruption (Fig. 7B). Tephra layers up to several tens of cm thick were offset only slightly along normal faults and appeared almost as pristine as an outcrop on the flanks of the volcano; smaller segments of tephra packages were found on the coast. A wide variety of relatively undisturbed segments of the Santa Ana edifice were observed in this hummock, including interbedded tephra layers, massive and brecciated lava flows, and a 20 × 30 m segment of a basaltic (52.4% SiO_2) block-and-ash-flow deposit with abundant juvenile clasts.

Mixed-facies or transitional material in intrahummock areas, within which block-facies material was transported, constitutes the volumetrically dominant component of the deposit. This material consists of a homogenized heterolithologic matrix containing clasts from Santa Ana volcano mixed at the margins of the deposit with significant amounts of accessory material from the Bálsamo Formation on the east and Quaternary ignimbrites on the west. Ripped-up soil clasts are also present (Fig. 7C). Mixed-facies material is exposed only in stream banks of rivers superposed on the surface of the avalanche deposit. Fieldwork in 1998 benefited from heavy rains associated with Hurricane Mitch that scoured riverbeds and produced unusually pristine exposures of mixed-facies material not readily apparent during a later visit. This revealed large segments of block-facies material transported within the mixed facies that could easily be overlooked when streambeds and banks are more vegetated and obscured. Clastic dikes (Fig. 7D) of mixed-facies material intruding into megaclasts of block-facies material are common. These clastic dikes are planar to irregular in form and can be traced for many meters into block-facies material several tens of meters in dimension. Block-facies material transported within the mixed facies ranged up to several tens of meters in size in medial portions of the avalanche deposit along the Río Chiquihuat ESE of Sonsonate. A wave-cut terrace beyond the farthest subaerial extent of the avalanche displays an extensive lag deposit of course boulders up to several m in size from which fine-grained matrix has been removed.

Sedimentological Characteristics

Preliminary study of the sedimentological properties of the Acajutla debris-avalanche deposit shows similarities with those of other debris-avalanche deposits. Size analyses of lahar and

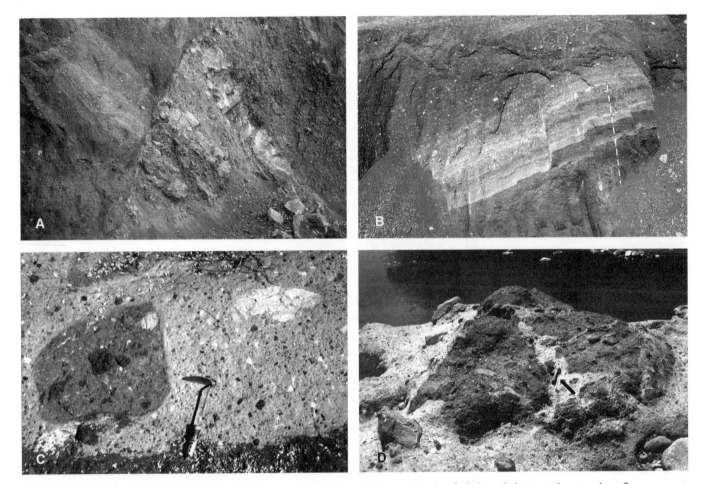

Figure 7. Textural characteristics. A: Block facies in quarry 30 km from Santa Ana, showing fault-bounded contact between lava-flow segments (right) and tephra layers (left). Large block at lower right is ~1 m in diameter. B: Bedded lapilli-and-ash layers offset along small normal faults at same quarry, scale bar is 2 m long. C: Light-colored soil clast (upper right) and small block-facies clast (left) transported within mixed facies at Río Banderas, 32 km from the volcano; trowel is 26 cm long. D: Light-colored clastic dikes intrude block-facies megaclasts at same location as C. Hammer (see arrow) is 31 cm long.

debris-avalanche deposits suffer from underrepresentation of clasts at the coarse end of the size spectrum. The coarse fraction is sometimes incorporated by field point counting (Major and Voight, 1986) or photographic analysis (Glicken, 1986, 1996; Siebert et al., 1989) of coarser clasts, but even these techniques underrepresent very coarse clasts. We analyzed data from lahars at Mount St. Helens (Major and Voight, 1986), where both techniques were used for samples from the same deposits, to investigate the degree of variability of the two procedures. Calculation of the average median size (Md_ϕ) and sorting (σ_ϕ) parameters for 20 samples using the sieving technique only and for 13 samples supplemented by coarse-clast data show that the two procedures had more of an effect on median grain size than sorting, but that both parameters showed relatively small effects. The average of Md_ϕ and σ_ϕ for each technique varied by only 0.10 ø. The coarse-end shift that would be expected using the point-count analysis increased to 0.48 ø when three outlier samples used for the sieving-only procedure were discarded because they were signifi-

cantly coarser than any of the samples used for coarse-tail point counting; however, the median sorting differential in this case decreased to only 0.03. Although these data are not necessarily representative for lahar and debris-avalanche deposits worldwide, they provide perspective for comparison of data using the two techniques.

Size-fraction histograms (Fig. 8) of block-facies material from the Acajutla deposit (Table 1) do not show as pronounced a bimodal distribution as those from Mount St. Helens (Glicken, 1986, 1996) and Augustine (Siebert et al., 1989), although they lie within the range of Mount St. Helens and other Cascade Range samples (Siebert et al., 1989). This reflects in part the broadening of the histogram distribution using the sieving-only procedure (Major and Voight, 1986). Preliminary data suggest that mixed- or lahar-facies material displays significantly higher silt and clay contents. Sorting and size parameters (Inman, 1952) of the Acajutla deposit are compared with those of other avalanche deposits around the world and with pyroclastic-flow and

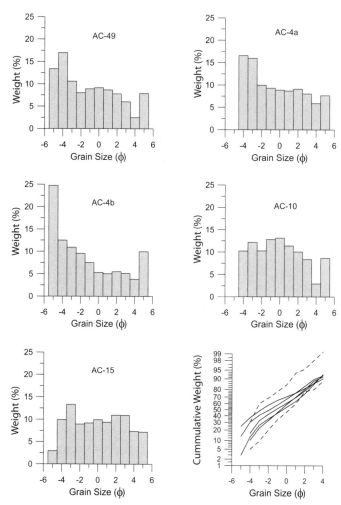

Figure 8. Grain-size plots of debris-avalanche samples; bottom axis is size fraction in ø units. Histogram bar to right of 4ø includes silt- and clay-sized fractions. Distance from source: AC-49, 10 km; AC-4a and 4b, 19 km; AC-10, 33 km; AC-15, 42 km. Normal probability plot (lower right) compares Acajutla samples (solid lines) with range of Mount St. Helens (1980) samples (dashed lines) from Glicken (1986).

TABLE 1. GRAIN-SIZE DISTRIBUTION OF
DEBRIS-AVALANCHE DEPOSIT SAMPLES

Sample #	Gravel %	Sand %	Mud* %	Md_ϕ[†]	σ_ϕ[§]	Distance[#] (km)
AC-49	57.9	34.2	7.9	−1.89	3.46	10
AC-4a	51.9	40.5	7.6	−1.20	3.38	19
AC-4b	66.2	24.8	10	−2.81	3.96	19
AC-10	45.5	45.9	8.6	−0.66	2.90	33
AC-15	44.3	48.5	7.2	−0.43	3.32	42

*% mud column includes silt and clay.
[†]Md_ϕ = Inman median diameter.
[§]σ_ϕ = Inman sorting coefficient.
[#]Distance is from Santa Ana volcano.

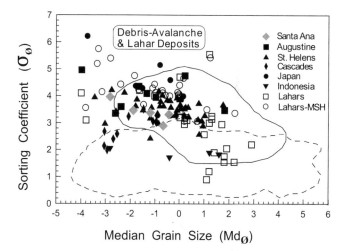

Figure 9. Size parameters of debris-avalanche deposits (solid symbols) compared with those of lahars (open symbols): Santa Ana data (this study, gray large diamond symbols); Augustine (Siebert et al., 1989); Mount St. Helens (Glicken, 1986, 1996); Cascade Range and Indonesia (Siebert et al., 1989); Japan (Murai, 1961); lahar data (Walker, 1971); Mount St. Helens southwest flank lahars (Major and Voight, 1986). Augustine and Mount St. Helens avalanche data corrected for coarse fraction; other data by sieving only. Solid and dashed lines outline, respectively, the pyroclastic-flow and plinian fall fields of Walker (1971).

pyroclastic-fall fields established by Walker (1971). The Acajutla samples are coarser and more poorly sorted than pyroclastic-fall deposits and lie at the coarse end or outside of the field of pyroclastic-flow deposits (Fig. 9). As with other debris-avalanche deposits, there is considerable overlap with the size and sorting characteristics of lahar deposits.

Source of the Acajutla Debris Avalanche

The geometry of the Acajutla debris-avalanche deposit allows three possible source areas—the Sierra de Apaneca, Santa Ana volcano, and Coatepeque caldera. The small volume of individual volcanoes forming the Sierra de Apaneca massif precludes the Apaneca Range as a source of the massive Acajutla debris avalanche and requires an origin from the Santa Ana–Coatepeque complex.

Debris-avalanche clasts, including eight from one hummock at Cerro Jícaro (sample number 4 on Fig. 4) were analyzed chemically (Table 2) and compared with analyses of rocks from Santa Ana and Coatepeque volcanoes. The most common rock type was basaltic andesite, but basaltic, andesitic, and dacitic rocks were also present (Fig. 10), all with high TiO_2 values. Both Santa Ana and debris-avalanche deposit samples show relatively high TiO_2 values dissimilar to most rocks from the adjacent, more silicic Coatepeque caldera (Fig. 11), and the chemistry of Acajutla avalanche deposit samples is consistent with a source from Santa Ana volcano (Pullinger, 1998).

Although modern Santa Ana volcano has buried most of the avalanche scarp, part of the failure scarp is exposed on the NW side (Fig. 12). An arcuate ridge displays a steep south-fac-

14   *L. Siebert, P. Kimberly, and C.R. Pullinger*

TABLE 2. MAJOR AND TRACE-ELEMENT GEOCHEMISTRY

Sample #	AC-1	AC-2	AC-3	AC-4	AC-5	AC-6	AC-7	AC-8	AC-15	AC-20	AC-21	AC-65	AC-76
SiO_2	54.09	51.80	65.45	52.86	53.76	65.35	54.29	53.82	52.80	53.05	51.33	52.43	60.36
TiO_2	1.15	1.14	0.76	1.17	1.17	0.74	0.92	0.89	1.03	0.94	1.03	0.95	0.78
Al_2O_3	19.54	19.99	16.71	18.30	17.68	16.07	20.99	20.55	19.40	19.79	20.23	18.89	16.23
Fe_2O_3	n.a.	n.a.	n.a.	n.a.	n.a.	n.a.	n.a.	n.a.	5.83	3.48	2.35	3.39	2.57
FeO	8.53	9.61	5.63	10.61	10.46	5.59	8.51	8.23	3.42	5.23	7.03	6.29	3.68
MnO	0.19	0.19	0.12	0.20	0.18	0.12	0.17	0.18	0.20	0.18	0.19	0.18	0.15
MgO	3.33	3.95	1.52	4.37	4.11	1.55	2.66	2.8	2.87	3.03	3.33	3.99	1.22
CaO	8.27	8.86	3.80	8.49	8.15	3.83	8.98	8.92	8.21	8.54	9.56	8.47	3.22
Na_2O	4.08	3.61	3.66	3.29	3.38	4.02	3.62	3.79	4.09	3.96	3.55	3.58	3.18
K_2O	1.17	1.03	3.81	1.83	1.87	3.81	0.84	1.01	1.39	1.20	1.13	1.45	3.00
P_2O_5	0.30	0.26	0.32	0.37	0.34	0.33	0.27	0.26	0.29	0.28	0.26	0.29	0.32
LOI	n.a.	n.a.	n.a.	n.a.	n.a.	n.a.	n.a.	n.a.	0.32	−0.08	−0.18	−0.27	4.81
Sum	100.65	100.44	101.78	101.49	101.10	101.41	101.25	100.45	99.85	99.6	99.82	99.64	99.52
Trace-element data (XRF ppm)													
V	n.a.	n.a.	n.a.	n.a.	n.a.	n.a.	n.a.	n.a.	147	153	174	175	76
Cr	n.a.	n.a.	n.a.	n.a.	n.a.	n.a.	n.a.	n.a.	<3<	<3<	<3<	<3<	<3<
Co	n.a.	n.a.	n.a.	n.a.	n.a.	n.a.	n.a.	n.a.	30	30	34	37	19
Ni	n.a.	n.a.	n.a.	n.a.	n.a.	n.a.	n.a.	n.a.	7	10	9	13	8
Cu	n.a.	n.a.	n.a.	n.a.	n.a.	n.a.	n.a.	n.a.	82	79	103	145	44
Zn	n.a.	n.a.	n.a.	n.a.	n.a.	n.a.	n.a.	n.a.	77	88	84	86	105
Rb	n.a.	n.a.	n.a.	n.a.	n.a.	n.a.	n.a.	n.a.	29	21	27	34	105
Sr	n.a.	n.a.	n.a.	n.a.	n.a.	n.a.	n.a.	n.a.	605	518	652	535	271
Y	n.a.	n.a.	n.a.	n.a.	n.a.	n.a.	n.a.	n.a.	26	24	24	25	43
Zr	n.a.	n.a.	n.a.	n.a.	n.a.	n.a.	n.a.	n.a.	123	109	108	134	317
Ba	n.a.	n.a.	n.a.	n.a.	n.a.	n.a.	n.a.	n.a.	550	495	429	478	1032
Nb	n.a.	n.a.	n.a.	n.a.	n.a.	n.a.	n.a.	n.a.	8	4	<2<	4	11

Note: Samples AC-1 to AC-8 are XRF analyses by C.P. at Michigan Technological University (normalized values listed in Pullinger, 1988); remaining data are Smithsonian Institution XRF analyses from this study. Sample locations: AC-1 to AC-8, Cerro Jícaro (location AC-4); AC-15, AC-20, and AC-21, debris-avalanche deposit (AC-21 is a juvenile bomb from a pyroclastic-flow deposit within an avalanche hummock); AC-76, pumice sample from a dacitic airfall deposit overlying debris-avalanche deposit; AC-65, lava flow within the Acajutla avalanche headwall scarp.

ing wall that truncates lower-angle, northwest-dipping lavas and pyroclastic deposits. Bedded scoria deposits from modern Santa Ana drape the top of the scarp, whose unburied portion is up to ~200 m high. The opposite northern flanks of the scarp are cut by deeply dissected and more widely spaced valleys, in contrast to the shallower, closely spaced valleys of the modern Santa Ana edifice. Stream valleys to the south of the scarp are deflected by the scarp, and valleys on either side have distinctly different orientations.

This scarp, exposed for a more than 5 km length before truncating the Plan de Hoyo depression (Fig. 12) and disappearing beneath the modern Santa Ana edifice, is interpreted to be a remnant of the original horseshoe-shaped failure scarp of the Acajutla avalanche (Pullinger, 1998). The Quebrada la Periquera cuts across the scarp, but a subtle change in slope across the river to the SW along trend of the scarp suggests that it lies beneath deposits from Los Naranjos volcano. Although much of the ridge is blanketed by coffee plantations, the road between Sonsonate and the city of Santa Ana traverses a 500 m wide portion of the scarp, allowing inspection of the mostly vegetated failure headwall. Farther to the east the failure scarp is buried by the modern

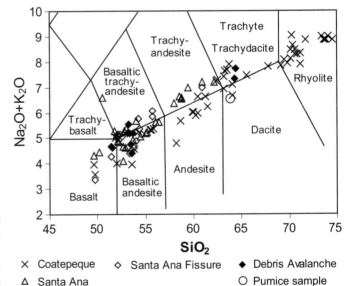

Figure 10. TAS plot (Le Bas et al., 1986) of avalanche deposit samples (solid symbols) and those of the Santa Ana volcanic complex. Data sources listed in Table 2.

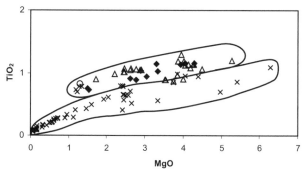

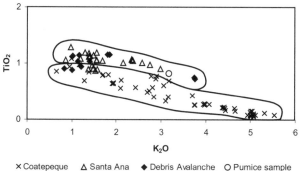

Figure 11. K$_2$O and MgO vs. TiO$_2$. Solid lines outline high and low Ti fields. Data sources listed in Table 2.

× Coatepeque △ Santa Ana ◆ Debris Avalanche ○ Pumice sample

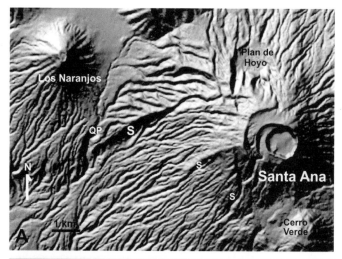

Figure 12. A: Digital elevation model of Santa Ana complex showing avalanche scarp headwall (labeled by large letter "S" at the middle left), partially overtopped by modern Santa Ana volcano. Smaller letters "s" at lower right mark inferred failure scarp on modern Santa Ana volcano. The Quebrada la Periquera (QP) flows south from the plain east of Los Naranjos volcano and cuts across the Acajutla avalanche scarp. B: Vertical arboreal strips of coffee plantations delineate part of the 5 km long headwall scarp, which descends diagonally across the middle of the photo to the left, as viewed from the flanks of modern Santa Ana volcano, within the avalanche caldera. Conical Los Naranjos volcano (center) and other peaks of the Sierra de Apaneca form the horizon.

Santa Ana edifice. Sudden increases in barranca depths on the northern side of Santa Ana volcano and a subtle change in slope west of the rim of Coatepeque caldera, however, suggest the location of the buried scarp.

Although the precise location of the eastern side of the scarp is not known, if the buried scarp is symmetrical around the center of the modern Santa Ana summit crater complex, its width there would be ~4 km and extend to 6 km SW of the current location of Izalco volcano. An intriguing semi-arcuate N-S–trending escarpment extends from ~500 to 1000 m elevation at the SW foot of Izalco volcano (Figs. 2 and 4). Although oriented partially obliquely to the inferred location of the buried eastern failure scarp, it has a different trend than the fissure system extending across the Santa Ana complex. It further contrasts in having a vertical offset up to ~100 m, raising the possibility that it is a remnant of a failure scarp, either from the Acajutla or another collapse event. If the former, the horseshoe-shaped caldera left by the Acajutla collapse would have dimensions of ~6 × 11 km. This is comparable to other collapse scarps related to avalanches of volumes exceeding 10 km³, such as the 6 × 10 km wide scarp at Barú in Panamá (see below) and the 8.5 × 13 km one at Raung volcano eastern Java (Siebert, 2002a). If this escarpment is part of the eastern failure scarp at Santa Ana, the avalanche caldera would have an unusual asymmetrical orientation, with a much longer eastern side. This configuration would be expected, however, where one side of the failure

plane intersected a topographic high beyond the volcano, such as occurred at Pacaya volcano in Guatemala, whose failure scarp walls are also asymmetrical.

Volume

The subaerial extent of the Acajutla avalanche deposit is shown in Figure 4. The subaerial deposit below the present-day 800 m contour (including the originally submarine area of the 64 km² Acajutla Peninsula) covers ~390 km². Inclusion of the more conservative 15-fathom estimate for the offshore component gives a total estimated area of ~540 km². Lack of exposures precludes direct measurements of the thickness of the deposit, which is the most problematical variable in determining deposit volumes. Comparison of present contours of the Acajutla Peninsula and offshore bathymetry with inferred N22°W-trending shore-parallel pre-avalanche contours, however, allows estimation of the thickness of the distal part of the avalanche, including its submarine component.

The avalanche deposit is thus inferred to be ~40–50 m thick over a broad area near the axis of the deposit along the former shoreline, 35 km from Santa Ana, and to average slightly under 30 m thick along the entire former shoreline. The southernmost tip of the deposit, at Punta Remedios, is adjacent to the inferred 30 m pre-collapse submarine contour, indicating a deposit thickness here of ~40 m along the axis of the deposit, a figure that would decrease laterally. Similarly, the deposit is estimated to have ranged from ~20 to 30 m thick within a few km of the shoreline and to be ~20 m thick in hummocky terrain ~5 km offshore at the furthermost extent of detailed bathymetry. The average estimated thickness of the submarine deposit within an ~50 km^2 area covered by detailed bathymetry was 17 m, and 15 m may be a representative thickness for the entire currently submarine component of the deposit. It should be noted that some of the offshore volume inferred from bathymetry may represent beach sands transported on top of the submarine avalanche deposit by longshore currents.

The thickness of the subaerial component of the deposit is even less constrained, but it would have been substantially more than that of the submarine component. For comparison, the order of magnitude smaller 1980 Mount St. Helens avalanche deposit ranged from ~10 m thick in its distal portion to as much as 195 m thick proximally. The deposit exceeded its average thickness of just under 40 m for ~2/3 of its extent (Glicken, 1986). A similar ratio at Acajutla would place the average thickness near the former shoreline. Thus, using the known average thickness of 30 ± 10 m at the former shoreline as a proxy for the average thickness of the entire Acajutla deposit would give an estimated volume of 16 ± 5 km^3, placing it among the largest known avalanche deposits from a continental volcano.

Despite large uncertainties about the dimensions of the collapse scarp, evidence suggests that an avalanche of this magnitude could have originated from ancestral Santa Ana volcano. The exposed headwall scarp, much of which has been buried by growth of the modern volcano, is only ~200 m high, but elsewhere headwall scarps often exceed 1 km in height. Even the more conservative inferred source area dimensions at Santa Ana are comparable to those of other failure scarps of avalanches exceeding 10 km^3 in volume, such as Meru volcano in Tanzania and Socompa volcano in Chile. Comparison of avalanche-volume/edifice-volume ratios places present-day Santa Ana within the range of other collapse events worldwide (Siebert, 1996), and the ancestral edifice would have had an even larger volume. Furthermore, dilation of the avalanche and incorporation of nontrivial amounts of exotic material during transport implies that the in situ collapse volume would have been substantially less than 16 ± 5 km^3.

Age of Acajutla Avalanche

Morphological evidence suggests a late-Quaternary age for the Acajutla debris avalanche. Sufficient time has elapsed for erosion to modify the symmetry of the Acajutla Peninsula, eroding cliffs on the western side and redistributing material eastward, where the shoreline is more subdued. Wave planation has created a several-hundred-m-wide terrace, prominent at low tide, offshore from Punta Remedios. Several post-collapse drainages, such as the Río Chiquihuat/Río Banderas, have been established across the surface of the deposit; however, the largest of these, the Río Grande de Sonsonate, remains confined to the western deposit margin.

The age of the Acajutla debris avalanche is limited by the rhyolitic Arce and Congo tephra-fall and pyroclastic-flow deposits associated with the creation of the stratigraphically older Coatepeque caldera. The voluminous pyroclastic-flow deposits are not found overlying the Acajutla debris-avalanche deposit, although the Arce pyroclastic-flow deposit is more than 8 m thick immediately adjacent to the avalanche deposit. Neither the Arce nor Congo fall deposits were found overlying the avalanche deposit, although their isopachs imply that potential thicknesses of up to 2–3 m would be found in the proximal part of the avalanche deposit (CEL, 1992). The ~57,000 yr B.P. age of the Congo Formation (Rose et al., 1999) thus currently provides the best upper limit on the age of the Acajutla avalanche.

The lower age limit of the avalanche is not well constrained. The avalanche predates the formation of modern Santa Ana volcano. Overlying lavas from modern Santa Ana volcano are not dated, but soil profiles on the avalanche deposit suggest a limiting age of ~10,000 yr (Vallance, 1997, personal commun.). Thicker soil profiles seen elsewhere imply an age of several tens of thousands of years. The age of the Acajutla debris avalanche is thus considered to be very latest Pleistocene, younger than ~57,000 yr. This provides a crude upper limit on the age of the modern Santa Ana edifice, which was constructed within the collapse scarp and partially overtops it. The collapse also appears to predate formation of Los Naranjos volcano, located immediately NW of the Santa Ana collapse scarp (Fig. 12). The debris apron from Los Naranjos volcano appears to overlie the lower SW part of the scarp. This small Izalco-sized volcano was mapped as Holocene by Weber and Weisemann (1978), but its age is not precisely known.

Possible Younger Avalanche and Associated Eruption

Evidence seen to date implies emplacement of the Acajutla debris avalanche in a single instantaneous failure event. Detailed studies of avalanche deposits at other volcanoes initially assumed to be a single event have however shown them to have originated during multiple failures, and we do not exclude this possibility at Santa Ana. Examination of recently obtained digital elevation model (DEM) images suggests the possibility of an edifice-collapse event from modern Santa Ana volcano that could in proximal areas overlie the Acajutla debris-avalanche deposit. An indistinct 1.6 × 2 km horseshoe-shaped depression cuts the upper

SW flank of modern Santa Ana volcano (Fig. 12). The SW rim of the outer Santa Ana crater intersects and truncates this shallow depression, which could originate from a smaller edifice failure that preceded construction of the current summit crater complex of Santa Ana.

The most proximal observed exposure of a debris-avalanche deposit at Santa Ana lies ~10 km SW of the current summit. Erosional dissection of the modern volcano is not sufficiently advanced to expose outcrops with multiple avalanche deposits, but if more than one major edifice failure occurred, this proximal debris-avalanche deposit is likely from the younger collapse event. A light-brown dacitic pumice-fall deposit (SiO_2 63.8%) directly overlies the debris-avalanche deposit here. The pumice-fall deposit is ~15–20 cm thick and contains vitric, sparsely crystalline pumice clasts up to 4 cm. At more proximal outcrops this contact is buried beneath thick tephra layers from modern Santa Ana volcano; more distally, only soil was seen overlying the avalanche deposit.

The total alkali/silica ratio chemistry of the dacitic pumice is similar to mafic pumice clasts of the Arce fall deposit from Coatepeque volcano (Fig. 10). The chemistry of the pumice suggests an affinity with high-TiO_2 rocks from Santa Ana volcano, although some overlap exists with andesitic rocks from Coatepeque caldera (Fig. 11). The Arce andesitic pumices, in contrast, are dense black and gray clasts that do not form a distinct tephra layer but are interspersed in the dominant white rhyodacitic pumice deposit. Other eruptions from Coatepeque caldera are distinctly smaller in volume than the caldera-forming rhyodacitic eruptions, and given the distance from Coatepeque (18 km; twice the distance of the outcrop from Santa Ana volcano), the clast size of the light-brown pumices is more consistent with an origin from Santa Ana volcano. Although this would be the most silicic in situ rock analyzed from Santa Ana volcano, dacitic rocks occur within the Acajutla debris-avalanche deposit. A Sierra de Apaneca origin is also less likely because no large pumice-producing eruptions postdate the 57,000 yr B.P. Congo pumice-fall deposit from Coatepeque, which precedes the avalanche.

We thus tentatively attribute the andesitic pumice-fall deposit to Santa Ana volcano. Although the pumice-fall deposit directly overlies the avalanche deposit without evidence of erosion, their synchroneity is not known with certainty. At several locations, pumice fragments distinct from the glassy mafic ignimbrite clasts of Sierra de Apaneca provenance are found within mixed-facies material of the Acajutla debris-avalanche deposit. Further work is needed to determine the provenance of these pumice fragments and to address questions of multiple edifice collapses and synchroneity of the pumice-fall deposit with a collapse event.

DISCUSSION

Causes of Failure

A wide variety of factors contribute to edifice failure (McGuire, 1996; Siebert et al., 1987; van Wyk de Vries and Francis, 1997). Extensive hydrothermal alteration, which facilitates edi-

fice instability (López and Williams, 1993; Reid et al., 2001), has affected the flat-lying summit crater area of the modern Santa Ana edifice, although Reid et al. noted that alteration in areas of gentle slope high on the volcano has a lesser impact on edifice instability. Although areas of hydrothermal alteration are present in block-facies material in the Acajutla debris-avalanche deposit, the extent of alteration of the ancestral volcano is not currently known.

The inferred SW-facing orientation of the failure scarp at Santa Ana is oblique-to-perpendicular to the NW-SE–trending fissure system cutting through the Santa Ana complex. Faulting along this trend continues southward into the Bálsamo Range (Pullinger, 1998). The variable morphology and degree of dissection of the volcanoes constructed along this fissure suggest that it has been active for much of the history of the Santa Ana complex. Failure perpendicular to dike trends within volcanoes has been noted at many other volcanoes (Siebert et al., 1987; Ellsworth and Voight, 1995), and extension and elevated pore pressures produced by dike emplacement may have been a factor contributing to the Santa Ana collapse.

Avalanche Emplacement Mechanisms

Block-facies material containing segments of the ancestral Santa Ana edifice is found throughout the deposit all the way to the coast. The number of block-facies hummocks decreases significantly beyond ~30 km from the volcano, but four distinct hummock clusters occur in the medial part of the deposit (Fig. 13). Each hummock cluster correlates with an extension of major linear Bálsamo Formation ridges. Exposures of kipukas of Bálsamo material indicate that these ridges, now largely buried by the avalanche deposit, would have extended well beyond their current margins into the central part of the valley. They acted as natural barriers to avalanche movement and caused rapid deceleration of the avalanche and stranding of block-facies hummocks on the proximal side of the ridges (Fig. 13). Similar stranding of rafted hummocks around large kipukas of basement material was observed at Raung volcano in Indonesia (Siebert, 2002a), due to both avalanche deceleration and grounding of block-facies material when host mixed-facies material thinned as it swept around the natural barriers.

The incorporation of exotic materials during transport indicates that significant bulking of the avalanche occurred during emplacement. The western margin of the deposit contains large numbers of clasts of mafic ignimbrites over which the avalanche traveled. Likewise, the eastern margin of the deposit contains rounded clasts from the adjacent Bálsamo Formation that in some cases form a major component of distal hummocks. Ripped-up soil clasts from the pre-failure valley floor were also incorporated into the avalanche. The amount of exotic material increases toward the margins of the deposit, but ignimbrite clasts, accretionary lapilli, and soil clasts are also present at the Río Banderas in the center of the deposit.

The volumetrically dominant mixed-facies portion of the deposit may be transitional into lahar facies. Preliminary data

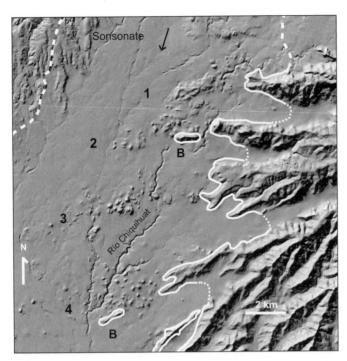

Figure 13. Hummock clusters in the medial portion of the debris-avalanche deposit are numbered 1–4; letters "B" indicate kipukas of Bálsamo Formation rocks along trend with Bálsamo ridges. Arrow at top marks generalized flow direction of avalanche from Santa Ana volcano (~15 km NW). Digital elevation model is based on 10 m contours and does not show small hummocks.

suggest that mixed- or lahar-facies material displays high clay contents (derived either from the ancestral edifice or from incorporated accessory material), which could place it in the cohesive debris flow range (Vallance and Scott, 1997). Several lines of evidence, however, suggest that a significant portion of this component of the avalanche was not water saturated. The Río Banderas streambed at the Highway 2 bridge (Fig. 1) displays abundant block-facies material up to tens of meters wide with smooth to very irregular margins within the mixed-facies material. Clast clusters and breccia schlieren, which would likely have been dispersed in a water-saturated environment, were seen to be abundant here at the time of fresh exposures created by Hurricane Mitch erosion, and clast clusters up to 8 m wide containing clasts up to ~80 cm were also observed in a 25 m deep steep-walled gorge cut in mixed-facies material in the Río Sensunapán (Río Grande de Sonsonate) south of the coastal highway.

Both melting of glacial icecaps and hydrothermal alteration producing clay mineralization can contribute to avalanche/lahar transformation and enhanced mobility (Carrasco-Núñez et al., 1993; Vallance and Scott, 1997). Glaciation would not have been a factor at Santa Ana volcano, and the extent to which hydrothermal alteration had affected ancestral Santa Ana is not currently known. The vast bulk of the distal portion of the avalanche deposit is obscured beneath vegetation or farmlands. This inhib-

its representative inspection of this important component of the deposit, and although water-saturated domains may have been present within the avalanche, their extent remains speculative.

Avalanche Mobility

The extremely mobile Acajutla avalanche traveled 46–50 km from the summit of Santa Ana to ~35 m below sea level off the coastline of the Acajutla Peninsula. At some volcanoes, the height of the pre-failure edifice can be estimated by restoring contours connecting both sides of the failure scarp, but this is precluded at Santa Ana, where only one side of the failure scarp is preserved. The vertical differential between the toe of the avalanche deposit and the current summit is ~2420 m, which results in an H/L ratio of 0.05. Although the geometry of the failure scarp suggests that the summit of ancestral Santa Ana was higher than that of the modern summit, this differential probably would not have exceeded a few hundred meters, which would have little effect on the H/L ratio. Edifice elevations in Central America show a distinct pattern, decreasing dramatically from the extremities of the arc toward the center. This has been attributed to systematic variations in crustal thickness, basement elevations, and magma densities (Carr, 1984), and to hydrostatic magma source/lithostatic pressure relationships producing an upper limit on cone growth (Rose et al., 1977). The height of the modern Santa Ana edifice, as with other Salvadoran volcanoes, is comparable to that of higher-elevation volcanoes at either end of the arc. This, coupled with the additional factors affecting volcano growth in this part of the arc, suggests that the ancestral volcano would not have risen significantly above the current elevation.

The 0.05 Acajutla H/L ratio is the lowest currently known in Central America. This makes it one of the most mobile avalanches known globally, comparable to the 0.05 H/L ratios of slope-failure events at Nevado del Colima in México (Stoopes and Sheridan, 1992) and Peteroa volcano in Chile (Moreno, 1991) and the 0.04 H/L ratio of the Raung avalanche in Indonesia (Siebert, 2002a). Their mobility (also calculated using current edifice heights) was enhanced by large volume (each exceeded 10 km³), a possible lahar-facies component, and emplacement down valleys bounded by mountainous terrain that inhibited lateral expansion.

Evolution of Santa Ana Volcano

Inspection of rocks in the avalanche headwall scarp traversed by a 500 m long section of the Sonsonate–Santa Ana highway shows lava flows and tephra layers from ancestral Santa Ana interspersed with at least five block-and-ash flow deposits containing prismatically jointed juvenile blocks. The pyroclastic-flow deposits, some of which exceed 10 m in thickness, may have originated during periods of strong explosive activity and/or the collapse of lava flow fronts on steep slopes, as is typical at explosively active, steep-sided stratovolcanoes like Colima in México and Fuego volcano in Guatemala. A >20 m thick pyroclastic-flow deposit with abundant juvenile blocks was also found in a hummock of the Acajutla debris avalanche. Block-and-ash flow deposits similar to those exposed in the collapse scarp of ancestral

Santa Ana volcano and in the avalanche deposit have not been identified from modern Santa Ana volcano, which in contrast has produced lava flows and magmatic/phreatomagmatic explosive eruptions without significant pyroclastic flows. The occurrence of block-and-ash-flow deposits of possible lava flow front collapse origin at distances beyond the current base of the volcano further implies that the ancestral Santa Ana volcano may have been higher and had steeper flanks than the modern edifice and that the collapse modified both the volcano's structure and eruptive style. Cone collapse elsewhere has marked a shift in eruptive style and geochemistry, such as seen at Tata Sabaya in Bolivia (de Silva et al., 1993) and Parinacota (Wörner et al., 1988) in Chile, and further work to investigate the extent to which this has occurred at Santa Ana is warranted.

The following stages have been recognized in the evolution of Santa Ana volcano:

1. Growth of ancestral Santa Ana produced an edifice that had a larger footprint on the west and south sides than the modern volcano. A large depression (Plan de Hoyo) formed north of the summit along the trend of the "Chalchuapa Graben" (CEL, 1992) and was partly filled on its south side by the ancestral summit cone, which may have been higher than the current summit.

2. Major edifice collapse in a SW direction truncated the summit and the Plan de Hoyo depression, leaving a massive avalanche caldera perhaps as large as 6.5 × 11 km. The resulting Acajutla debris avalanche traveled ~50 km into the Pacific Ocean.

3. Growth of modern Santa Ana volcano filled in much of the avalanche scarp and overtopped it on the north and east sides. At a time when the modern volcano was near its present elevation, the SW side apparently collapsed, perhaps in association with an explosive eruption of dacitic pumice, leaving a 1.6 × 2 km wide horseshoe-shaped depression (Fig. 12).

4. Ejecta from growth of the summit cone complex largely filled this depression, and the earliest cone overtopped the scarp headwall. The SW rim of this cone was breached during construction of the inner summit cones, and renewed regional faulting across the SW flank failure scarp on modern Santa Ana was obscured to the north by near-vent ejecta. Eruptions continued into historical time from the central vent complex and the NW-SE fissure system.

Hazard Implications

Hazards from large-scale edifice collapse are among the most severe of potential events at volcanoes. The high velocity and widespread extent of volcanic debris avalanches preclude hazard mitigation after the onset of the event. More than 286,000 people currently live on top of the Acajutla debris-avalanche deposit. Topography indicates that any major southward-directed failure would place large areas currently underlain by the Acajutla deposit at risk. Because the reconstructed modern Santa Ana edifice is of smaller volume than the ancestral volcano, a future failure of modern Santa Ana to the south would probably be of lesser extent than the Acajutla debris avalanche. Comparison with H/L ratios typical of other large slope failures (Siebert

et al., 1987) suggests, however, that even an order of magnitude smaller failure (such as the one inferred from modern Santa Ana volcano) would likely be sufficiently mobile to impact the city of Sonsonate. A major northward-directed failure perpendicular to the fissure-system cutting across Santa Ana volcano could place the city of Santa Ana itself (only ~15 km to the north) within the area of potential impact. Failure in this direction could also produce an avalanche that would sweep into Lake Coatepeque. As occurred at Spirit Lake at Mount St. Helens in 1980, tsunamis with runup of hundreds of meters are possible, potentially overtopping the caldera wall. Although the buttressing effect of the remnant ancestral Santa Ana volcano preserved on the north and east may preferentially promote failure to the south, tectonic factors discussed below suggest that the possibility of failure to the north should not be excluded.

Catastrophic edifice failure is, fortunately, a low-frequency event. Hazard planning for high-impact, low-frequency events is problematic. Modern Santa Ana volcano has grown to a point where it is of sufficient size to produce a major collapse. Massive edifice failure on the scale of the Acajutla event would be truly catastrophic were it to occur again, but such an event may not occur for periods of time far in excess of time frames considered by hazard planners. Land-use zoning that precludes habitation of areas potentially affected by a future collapse of Santa Ana is not a viable option. Consequently, a focus on the identification of short-term precursors to collapse, accompanied by education of public officials and residents regarding potential hazards, may be the only practical course of action.

COMPARISON WITH OTHER CENTRAL AMERICAN EDIFICE-FAILURE EVENTS

Santa Ana is one of only a few volcanoes currently known to have undergone large-scale edifice collapse in El Salvador. Another collapse occurred at San Vicente volcano, east of Ilopango caldera (Major et al., 2004, this volume, Chapter 7). The Tecoluca debris avalanche and associated lahar from San Vicente volcano traveled ~25 km SE to the Río Lempa. One of us (C.P.) has observed another avalanche deposit of uncertain provenance near Lolotique in eastern El Salvador. These Salvadoran examples are among the more than two dozen similar large-volume edifice-collapse events in Central America (Table 3).

In order to evaluate the frequency of avalanches the size of the Acajutla debris avalanche, we assessed the occurrence of other avalanche deposits in Central America. The Acajutla debris avalanche is one of several in Central America with volumes in excess of 10 km^3. The exposed volume of the Escuintla debris-avalanche deposit from collapse of ancestral Meseta volcano at Volcán de Fuego in Guatemala is ~9 km^3 (Vallance et al., 1995); inclusion of the inferred now-buried proximal portion gives a total volume of ~15 km^3 (Siebert et al., 1994).

An even larger edifice-collapse event took place at Volcán Barú in Panamá (IRHE, 1987), leaving a 6 × 10 km wide avalanche caldera breached to the west and producing the massive

TABLE 3. CENTRAL AMERICAN DEBRIS AVALANCHES (≥0.1 km³)

Volcano*	Avalanche	Direction	Length† (km)	H/L§	Area (km²)	Volume# (km³)	References
GUATEMALA							
Tacaná?	Río Las Majadas	NNE	–	–	–	–	Mercado and Rose, 1992; this study
Tajumulco?	–	South	–	–	–	–	Vallance et al., 1995
Siete Orejas?	–	South	–	–	–	[>1]	Vallance et al., 1995
Almolonga	Cerro Quemado	NW	6	0.13	13	0.1	Conway et al., 1992
Acatenango	La Democracia	SSW	42	0.09	210	5	Vallance et al., 1995; Siebert et al., 1994; Basset, 1996
Fuego	Escuintla	SSE	50	0.08	440	15	Vallance et al., 1995; Siebert et al., 1994
Pacaya	Río Metapa	SSW	25	0.10	55	>1	Vallance et al., 1995; Siebert et al., 1994
Tecuamburro	Miraflores	SE	15	0.12	–	[>1]	Duffield et al., 1989
Ixtahuan?	Los Achiotes	SE	15?	0.07	–	–	Reynolds, 1987; Siebert et al., 1994
EL SALVADOR							
Santa Ana	Acajutla	SW	48	0.04	540	16 ± 5	Pullinger, 1998; this study
Santa Ana?	Modern Santa Ana	SW	–	–	–	–	This study
San Vicente	Tecoluca	SE	24	0.09	–	[>1]	Major et al., 2004; this study
Unknown source	Lolotique	North?	–	–	–	–	Pullinger, 2002, pers. commun.
NICARAGUA							
San Cristóbal	El Chonco	NW	–	–	–	[<1]	Van Wyk de Vries and Borgia, 1996
Mombacho	Las Isletas	NE	>12	0.11	45	(1)	Ui, 1972; Van Wyk de Vries and Francis, 1997; this study
Mombacho	South Crater	South	12	0.11	–	(1)	Ui, 1972; Van Wyk de Vries and Francis, 1997
Mombacho	–	SE	>10	0.13	–	–	Van Wyk de Vries and Francis, 1997
COSTA RICA							
Cerro Cacao	Quebrada Grande	SE	16	0.07	36–52	1.7 ± 0.3	Alvarado and Vega, 2002
Rincón de la Vieja	Azufrales	–	9	0.12	11–18	0.3 ± 0.1	Alvarado and Vega, 2002
Miravalles	Fortuna	SW	19	0.11	92–127	7 ± 1	Alvarado and Vega, 2002
Miravalles	Upala	NE	13	0.08	>20	–	Alvarado and Vega, 2002
Miravalles	Mesas	SW?	6	–	3	0.1	Alvarado and Vega, 2002
Tenorio	Tierras Morenas	South	17	0.07	80–118	2.0 ± 0.4	Alvarado and Carr, 1993
Platanar	Chocosuela	NW?	20	0.10	–	–	Alvarado and Carr, 1993
Barva?	Valle Central	South	–	–	–	–	Alvarado and Vega, 2002
Irazú	Prusia	–	5.5	0.15	5.2	1.4	Alvarado and Vega, 2002
Irazú	Río Costa Rica	–	–	–	5.5	–	Alvarado and Vega, 2002
Turrialba	Angostura	ESE	15	0.07	15–28	1.1 ± 0.3	Soto, 1988; Alvarado and Vega, 2002
Turrialba	Santa Rosa	–	5	0.07	15–28	–	Alvarado and Vega, 2002
PANAMÁ							
Cerro Colorado	Río Chiriquí Viejo	South	–	–	–	(>10)	IRHE, 1987; this study
Barú	Hato del Volcán	WSW	–	–	–	(25)	IRHE, 1987; this study
Barú ?	–	SW?	–	–	–	–	This study

*Volcano names followed by a "?" represent edifice-collapse events inferred from source area or deposit morphology on maps or aerial photographs, but without field identification of deposits.

†Length is distance from summit or headwall scarp to terminus of debris-avalanche deposit.

§H/L is ratio between vertical drop and travel length; in some cases may include lahar facies of edifice-collapse event.

#Volume data in parentheses are source area volumes; figures in square brackets are order-of-magnitude volume assessments.

Hato del Volcán debris-avalanche deposit, which extends beyond the Pan-American Highway. Restoration of inferred pre-failure contours suggests that as much as 30 km³ may have collapsed (Siebert, 2002b). About 5 km³ of this volume remained within the horseshoe-shaped caldera in the form of large toreva blocks, distinct from the post-collapse lava dome complex that later grew to a height above the caldera rim. The debris-avalanche deposit from Barú volcano overlaps another avalanche deposit from Pleistocene Cerro Colorado volcano and may have formed during late-Holocene time. The basal sediment layer cored from a lake characterized by Behling (2000) as a volcanic crater (but which is more likely a pond formed within a closed depression on the surface of the avalanche deposit) was dated at 2860 ± 50 yr B.P. and may provide a limiting age for the collapse. Cerro Colorado, located immediately NW of Barú, has a 7 km wide horseshoe-shaped caldera (now partially filled by post-collapse lava domes) whose area is comparable to that of the Barú caldera and whose original volume thus also likely exceeded 10 km³.

Examination of Quaternary edifice failures in Central America the size of the Acajutla deposit suggests that although such catastrophic collapses are not unique, their probability is very low. Failures one to two orders of magnitude smaller in volume, although still of low frequency, are more common (Table 3) and would be sufficiently mobile to impact heavily populated areas. Hazards associated with Central American slope failures also include associated eruptions and tsunamis. In addition to the possible collapse-related explosive eruption at Santa Ana, the collapse of Cerro Quemado volcano in Guatemala triggered a lateral blast (Conway et al., 1992) an order of magnitude smaller than at Mount St. Helens in 1980. Pyroclastic surges accompanying collapse of Pacaya volcano in Guatemala reached as far as 10 km (Vallance et al., 1995). The Acajutla avalanche is the only Pacific coastal plain avalanche in Central America known to have reached the sea and likely to have produced a tsunami. Lacustrine tsunamis, however, would probably have occurred during collapses of Mombacho volcano into Lake Nicaragua and could also occur at Lake Coatepeque during a northward collapse of Santa Ana.

Regional tectonic factors have influenced failure directions of volcanoes on both margins of the Caribbean plate. In contrast to West Indies arc slope failures, which occur largely in a direction opposite that of the ocean-floor trench, most of the Central American slope failures occur toward the trench side of the arc (Siebert, 2002b). These preferred failure directions have been attributed to the construction of volcanoes on regional basement slopes. In the West Indies, this reflects the higher slope of islands toward the deep backarc Grenada Basin (Boudon et al., 2002). In Guatemala, this has been attributed to the construction of volcanoes over inclined basements at the trenchward margin of the arc (Vallance et al., 1995). A similar topographic effect is seen in the Cofre de Perote–Citlaltépetl Range in the eastern Mexican Volcanic Belt. All slope failures in this massif, constructed above the margin of the Altiplano where it drops dramatically toward the coastal plain, have occurred to the east (Carrasco-Núñez et al., 2002). In contrast, the central and eastern sides of the Chortis block of the Caribbean plate in El Salvador and Nicaragua are characterized by extensional grabens, where the basement topographic effect is less pronounced, and collapses here have had more variable orientations. Future collapses at Santa Ana to the north, although perhaps less likely, thus should not be ruled out.

CONCLUSIONS

A voluminous volcaniclastic deposit extending from the volcanic front in western El Salvador to the Pacific Ocean was formed during the largest slope-failure event known in El Salvador. The geometry of the debris-avalanche deposit allows three possible source areas, but the geochemical affinity of rocks within the debris-avalanche deposit with those from Santa Ana volcano indicates its origin. The 16 ± 5 km³ avalanche extended the former shoreline by 7 km, forming the Acajutla Peninsula. The subaerial portion of the deposit covers an area of ~390 km², and bathymetry implies an additional submarine component of 150 km². The collapse is younger than the ~57,000 yr B.P. deposits associated with the formation of Coatepeque caldera.

Avalanche hummocks are most abundant in the medial portion of the deposit but are found to the coast and offshore. Significant bulking of the deposit during emplacement incorporated material from the Bálsamo Formation on the east and from Tertiary mafic ignimbrites on the west. Stratigraphy in the collapse scarp wall suggests that the massive edifice collapse marked a transition in the eruptive style at Santa Ana.

The collapse occurred perpendicular to a NW-SE–trending fissure system cutting across Santa Ana, and extension and elevated pore pressures related to dike intrusion may have contributed to the collapse. In contrast to volcanoes at the western and eastern ends of Central America that were constructed over inclined basement substrates and failed preferentially in the direction of regional slope, Santa Ana is a free-standing edifice overlying the Median Trough of El Salvador, and future failure to the north, although perhaps less likely, is also possible.

ACKNOWLEDGMENTS

We thank Bill Rose for his long-term promotion of collaborative research projects in El Salvador and his support of field sessions for this project in 1998 and 2002 through his National Science Foundation grant. Jim Vallance initially identified the debris-avalanche origin of the Acajutla deposit in the field and provided a tentative soil-thickness age for the deposit. Tim Rose and Tim Gooding of the Smithsonian Institution provided help with sample preparation for geochemical analyses, and Heather Njo performed XRF analyses. Tom Jorstad and Bill Boykins of the Smithsonian's sedimentology laboratory provided guidance for size analyses. Steve Schilling of the U.S. Geological Survey developed the DEM files used in several of the figures. Jon Major, Nancy Riggs, and Jim Vallance provided valuable comments in review.

REFERENCES CITED

Alvarado, G.E., and Carr, M., 1993, The Platanar-Aguas Zarcas volcanic centers, Costa Rica: Spatial-temporal association of Quaternary calc-alkaline and alkaline volcanism: Bulletin of Volcanology, v. 55, p. 443–453.

Alvarado, G.E., and Vega, E., 2002, Los grandes deslizamientos tipo debris avalanche en Costa Rica: VIII Seminario Nacional de Geotecnia, III Encuentro Centroamericano de Geotecnistas, Costa Rica, p. 48–61.

Basset, T., 1996, Histoire éruptive et évaluation des aléas du volcano Acatenango (Guatemala) [Ph.D. thesis]: University of Genève, published in Terre & Environment, Section des Sciences de la Terre, Université de Genève, v. 3, p. 1–240 and appendices.

Baxter, S., 1984, Lexico estratigrafico de El Salvador: San Salvador: Comisión Ejectiva Hidroeléctrica del Rio Lempa (CEL), 108 p.

Behling, H., 2000, A 2860-year high-resolution pollen and charcoal record from the Cordillera de Talamanca in Panama; a history of human and volcanic forest disturbance: Holocene, v. 10, p. 387–393.

Boudon, G., Le Friant, A., Deplus, C., Komorowski, J.-C., and Semet, M.P., 2002, Montagne Pelee 1902–2002, Explosive volcanism in subduction zones, St. Pierre, Martinique, May 12–16, 2002, Abstracts, p. 66.

Burbach, G.V., Frohlich, C., Pennington, W.D., and Matumoto, T., 1984, Seismicity and tectonics of the subducted Cocos plate: Journal of Geophysical Research, v. 89, p. 7719–7735.

Burkhart, B., and Self, S., 1985, Extension and rotation of crustal blocks in the northern Central America and effect on the volcanic arc: Geology, v. 13, p. 22–26.

Carr, M.J., 1984, Symmetrical and segmented variation of physical and geochemical characteristics of the Central America volcanic front: Journal of Volcanology and Geothermal Research, v. 20, p. 231–252.

Carr, M.J., and Pontier, N.K., 1981, Evolution of a young parasitic cone towards a mature central vent: Izalco and Santa Ana volcanoes in El Salvador, Central America: Journal of Volcanology and Geothermal Research, v. 11, p. 277–292.

Carrasco-Núñez, G., Vallance, J.W., and Rose, W.I., 1993, A voluminous avalanche-induced lahar from Citlaltépetl volcano, Mexico: Implications for hazard assessment: Journal of Volcanology and Geothermal Research, v. 59, p. 35–46.

Carrasco-Núñez, G., Díaz-Castellón, G., Hubbard, R., Siebert, L., Sheridan, M.A., Lozano, L., Zimbelman, D., Watters, R., and Rodríguez, S., 2002, Flank collapse events at the Citlaltépetl-Cofre de Perote Range, eastern Mexican Volcanic Belt: Montagne Pelee 1902–2002, Explosive Volcanism in Subduction Zones, St. Pierre, Martinique, May 12–16, 2002, Abstracts, p. 66.

CEL (Comisión Ejectiva Hidroeléctrica del Rio Lempa), 1992, Desarrollo de los Recursos Geotérmicos del Area Centro-Occidental de El Salvador. Prefactibilidad Geotérmica del Area de Coatepeque. Reconocimiento Geotérmico. Informe Final. Geotermica Italiana. Unpublished internal report.

Conway, F.M., Vallance, J.W., Rose, W.I., Johns, G.W., and Paniagua, S., 1992, Cerro Quemado, Guatemala: The volcanic history and hazards of an exogenous volcanic dome complex: Journal of Volcanology and Geothermal Research, v. 52, p. 303–323.

de Silva, S.L., Davidson, J.P., Croudace, I.W., and Escobar, A., 1993, Volcanological and petrological evolution of Volcan Tata Sabaya, SW Bolivia: Journal of Volcanology and Geothermal Research, v. 55, p. 305–335.

Duffield, W.A., Heiken, G.H., Wohletz, K.H., Maassen, L.W., Dengo, G., and Mckee, E.H., 1989, Geology and geothermal potential of the Tecuamburro volcano area of Guatemala: Geothermal Resources Council Transactions, v. 13, p. 125–131.

Ellsworth, D., and Voight, B., 1995, Dike intrusion as a trigger for large earthquakes and the failure of volcano flanks: Journal of Geophysical Research, v. 100, p. 6005–6024.

Evans, S., and Bent, A.L., 2004, The Las Colinas landslide, Santa Tecla: A highly destructive flowslide triggered by the January 13, 2001, El Salvador earthquake, in Rose, W.I., et al., eds., Natural hazards in El Salvador: Boulder, Colorado, Geological Society of America Special Paper 375, p. 25–37 (this volume).

Glicken, H., 1986, Rockslide–debris avalanche of May 18, 1980, Mount St. Helens Volcano [Ph.D. thesis]: Santa Barbara, University of California, 303 p.

Glicken, H., 1996, Rockslide–debris avalanche of May 18, 1980, Mount St. Helens volcano, Washington: U.S. Geological Survey Open-File Report 96-677, 90 p. (currently available in electronic version only at the Cascades Volcano Observatory Web site: http://vulcan.wr.usgs.gov/Projects/Glicken/framework.html) (June 2003).

IRHE, 1987, Final report on the reconnaissance study of geothermal resources in the Republic of Panama: Instituto de Recursos Hidraulicos y Electrificacion (IRHE), Quito, Ecuador: IRHE-DIB-OLADE, 72 p.

Inman, D.L., 1952, Measures describing the size distribution of sediments: Journal of Sedimentary Petrology, v. 22, p. 125–145.

Jibson, R.W., Crone, A.J., Harp, E.L., Baum, R.L., Major, J.J., Pullinger, C.R., Escobar, C.D., Martínez, M., and Smith, M.E., 2004, Landslides triggered by the 13 January and 13 February 2001 earthquakes in El Salvador, in Rose, W.I., et al., eds., Natural hazards in El Salvador: Boulder, Colorado, Geological Society of America Special Paper 375, p. 69–88 (this volume).

Johnson, R.W., 1987, Large-scale volcanic cone collapse: The 1888 slope failure of Ritter volcano, and other examples from Papua New Guinea: Bulletin of Volcanology, v. 49, p. 669–679.

Konagai, K., Johansson, J., Mayorca, P., Uzuoka, R., Yamamoto, T., Miyajima, M., Pulido, N., Sassa, K., Fukuoka, H., and Duran, F., 2004, Las Colinas landslide: Rapid and long-traveling soil flow caused by the January 13, 2001, El Salvador earthquake, in Rose, W.I., et al., eds., Natural hazards in El Salvador: Boulder, Colorado, Geological Society of America Special Paper 375, p. 39–53 (this volume).

Le Bas, M.J., Le Maitre, R.W., Streckinson, A., and Zanettin, B., 1986, A chemical classification of volcanic rocks based on total alkali-silica diagram: Journal of Petrology, v. 77, p. 24–37.

López, D.L., and Williams, S.N., 1993, Catastrophic volcanic collapse: Relation to hydrothermal processes: Science, v. 260, p. 1794–1796.

Major, J.J., and Voight, B., 1986, Sedimentology and clast orientations of the 18 May 1980 southwest-flank lahars, Mount St. Helens, Washington: Journal of Sedimentary Petrology, v. 56, p. 691–705.

Major, J.J., Schilling, S.P., Pullinger, C.R., and Escobar, C.D., 2004, Debris-flow hazards at San Salvador, San Vicente, and San Miguel volcanoes, El Salvador, in Rose, W.I., et al., eds., Natural hazards in El Salvador: Boulder, Colorado, Geological Society of America Special Paper 375, p. 89–108 (this volume).

McGuire, W.J., 1996, Volcano instability: A review of contemporary themes, in McGuire, W.J., et al., eds., Volcano instability on the Earth and other planets: Geological Society [London] Special Publication 110, p. 1–23.

Mercado, R., and Rose, W.I., 1992, Reconocimiento geologico y evaluacion preliminar de peligrosidad del volcán Tacaná, Guatemala/Mexico: Geofisica Internacional, v. 31, p. 205–237.

Moreno, H., 1991, The southern Andes volcanoes (33°–41°30′S), Chile: 6th Congreso Geológico Chileno–Guía de Excursión, PC-3, 26 p.

Murai, I., 1961, A study of the textural characteristics of pyroclastic flow deposits in Japan: Bulletin of the Earthquake Research Institute, Tokyo University, v. 39, p. 133–248.

Pullinger, C., 1998, Evolution of the Santa Ana volcanic complex, El Salvador [M.S. thesis]: Houghton, Michigan Technological University, 151 p.

Reid, M.E., Sisson, T.W., and Brien, D.L., 2001, Volcano collapse promoted by hydrothermal alteration and edifice shape, Mount Rainier, Washington: Geology, v. 29, p. 779–781.

Reynolds, J.H., 1987, Timing and sources of Neogene and Quaternary volcanism in south-central Guatemala: Journal of Volcanology and Geothermal Research, v. 33, p. 9–22.

Rose, W.I., Jr., Grant, N.K., Hahn, G.A., Lange, I.M., Powell, J.L., Easter, J., and DeGraff, J.M., 1977, The evolution of Santa Maria volcano, Guatemala: Journal of Geology, v. 85, p. 63–87.

Rose, W.I., Conway, F.M., Pullinger, C.R., Deino, A., and McIntosh, W.C., 1999, An improved framework for late Quaternary silicic eruptions in northern Central America: Bulletin of Volcanology, v. 61, p. 106–120.

Satake, K., and Kato, Y., 2001, The 1741 Oshima-Oshima eruption: Extent and volume of submarine debris avalanche: Geophysical Research Letters, v. 28, p. 427–430.

Sheets, P., editor, 1983, Archeology and volcanism in Central America: The Zapotitan Valley of El Salvador: Austin, University of Texas Press, 307 p.

Siebert, L., 1996, Hazards of large volcanic debris avalanches and associated eruptive phenomena, in Scarpa, R., and Tilling, R.I., eds., Monitoring and mitigation of volcano hazards: Berlin, Springer-Verlag, p. 541–572.

Siebert, L., 2002a, Landslides resulting from structural failure of volcanoes, in Evans, S.G., and De Graff, J.V., eds., Catastrophic landslides: Effects, occurrence, and mechanisms: Boulder, Colorado, Geological Society of America Reviews in Engineering Geology 15, p. 209–235.

Siebert, L., 2002b, Large-scale edifice failures at the western margin of the Caribbean plate: Montagne Pelee 1902–2002, Explosive Volcanism in Subduction Zones, St. Pierre, Martinique, May 12–16, 2002, Abstracts, p. 71.

Siebert, L., Glicken, H., and Ui, T., 1987, Volcanic hazards from Bezymianny- and Bandai-type eruptions: Bulletin of Volcanology, v. 49, p. 435–459.

Siebert, L., Glicken, H., and Kienle, J., 1989, Debris avalanches and lateral blasts at Mount St. Augustine volcano, Alaska: National Geographic Research, v. 5, p. 232–249.

Siebert, L., Vallance, J.W., and Rose, W.I., 1994, Quaternary edifice failures at volcanoes in the Guatemalan highlands [abs.]: Eos (Transactions, American Geophysical Union), v. 75, p. 367.

Soto, G., 1988, Estructuras volcano-tectónicas del volcán Turrialba, Costa Rica, América Central: Actas V Congresso Geologial Chileno, 8–12 Agosoto, 1988, Santiago, v. III-I, p. 163–175.

Stoiber, R.E., and Carr, M.J., 1973, Quaternary volcanic segmentation of Central America: Bulletin of Volcanology, v. 37, p. 304–325.

Stoopes, G.R., and Sheridan, M.F., 1992, Giant debris avalanches from the Colima Volcanic Complex, Mexico: Implications for long-runout landslides (>100 km) and hazard assessment: Geology, v. 20, p. 299–302.

Ui, T., 1972, Recent volcanism in Masaya-Granada area, Nicaragua: Bulletin of Volcanology, v. 36, p. 174–190.

Vallance, J.W., and Scott, W.E., 1997, The Osceola Mudflow from Mount Rainier; sedimentology and hazard implications of a huge clay-rich debris flow: Geological Society of America Bulletin, v. 109, p. 143–163.

Vallance, J.W., Siebert, L., Rose, W.I., Jr., Girón, J.R., and Banks, N.G., 1995, Edifice collapse and related hazards in Guatemala, *in* Ida, Y., and Voight, B., eds., Models of magmatic processes and volcanic eruptions: Journal of Volcanology and Geothermal Research, v. 66, p. 337–355.

van Wyk de Vries, B., and Borgia, A., 1996, The role of basement in volcano deformation, *in* McGuire, W.J., et al., eds., Volcano instability on the Earth and other planets: Geological Society [London] Special Publication 110, p. 95–110.

van Wyk de Vries, B., and Francis, P.W., 1997, Catastrophic collapse at stratovolcanoes induced by gradual volcano spreading: Nature, v. 387, p. 387–390.

Walker, G.P.L., 1971, Grain-size characteristics of pyroclastic deposits: Journal of Geology, v. 79, p. 696–714.

Weber, H.S., and Weisemann, G., 1978, Geologische karte der Republic El Salvador/Mitelamerica. 1:100,000 herausgegeben von der Bunderstalt für Geowissenschaften und Rohstoffe, Hannover, 6 sheets.

Wörner, G., Harmon, R.S., Davidson, J., Moorbath, S., Turner, D.L., McMillan, N., Nye, C., Lopez-Escobar, L., and Moreno, H., 1988, The Nevados de Payachata volcanic group (18°S/69°W, N. Chile): Geological, geochemical and isotopic observations: Bulletin of Volcanology, v. 50, p. 287–303.

MANUSCRIPT ACCEPTED BY THE SOCIETY JUNE 16, 2003

Geological Society of America
Special Paper 375
2004

The Las Colinas landslide, Santa Tecla: A highly destructive flowslide triggered by the January 13, 2001, El Salvador earthquake

Stephen G. Evans*

Department of Earth Sciences, University of Waterloo, 200 University Avenue West, Waterloo, Ontario N2L 3G1, Canada

Allison L. Bent

*National Earthquake Hazards Program, Geological Survey of Canada,
7 Observatory Crescent, Ottawa, Ontario K1A 0Y3, Canada*

ABSTRACT

The Las Colinas landslide was one of thousands of landslides triggered by the January 13th El Salvador earthquake (M_w 7.6) in early 2001. The landslide was highly destructive. It led to the death of ~585 people when it swept into a residential area of Santa Tecla, a suburb of San Salvador. The landslide originated from the top of a steep escarpment and involved pyroclastic deposits (silty sands and sandy silts) interbedded with paleosol horizons. The initial volume of the landslide was only ~130,000 m^3. The runout distance of the landslide, which developed into a rapid flowslide, was 735 m over a vertical distance of 166 m giving a H/L ratio of 0.23. The flowslide ran its final 460 m over a slope of only 3°. The flowslide debris was mainly dry but may have been partially saturated. It is postulated that strong seismic shaking amplified by topographic effects led to tensile stripping of the initial failure mass, which then lost strength very rapidly as it moved downslope and disintegrated into cohesionless debris. Urban topography consisting of buildings and streets may have inhibited debris spreading and channelized debris resulting in a long runout. The Las Colinas flowslide illustrates that runout behavior determines the landslide hazard at the base of the source slope and raises the question of landslide risk at the base of the Balsamo Escarpment, where existing residential developments are located within the runout distance of similar flowslide events that could occur in the future.

Keywords: earthquake-triggered landslide, Las Colinas landslide, flowslide, slope stability, runout, deaths, casualties.

INTRODUCTION

On January 13, 2001, a major earthquake (M_w 7.6) struck southern El Salvador and triggered thousands of landslides (Jibson and Crone, 2001; Bommer et al., 2002a). The most devastating occurred in the Las Colinas area of Santa Tecla (Fig. 1), a western suburb of the capital San Salvador, at an epicentral distance of ~105 km (Fig. 2; Bent and Evans, this volume, Chapter 29). The landslide buried or damaged more than 300 houses at the foot of the steep source slope (Fig. 1), and the death toll from the slide is estimated at ~585 persons (Comité de Emergencia Nacional [El Salvador]).

This paper reports initial results of our reconnaissance field observations at Las Colinas on January 24–29, 2001, which were carried out before rehabilitation of the site was completed. Our objectives are to document and describe the Las Colinas landslide, to contribute to the understanding of the event and thus contribute to the mitigation of landslide risk in similar settings in El Salvador and elsewhere in Central America.

*sgevans@uwaterloo.ca

Evans, S.G., and Bent, A.L., 2004, The Las Colinas landslide, Santa Tecla: A highly destructive flowslide triggered by the January 13, 2001, El Salvador earthquake, *in* Rose, W.I., Bommer, J.J., López, D.L., Carr, M.J., and Major, J.J., eds., Natural hazards in El Salvador: Boulder, Colorado, Geological Society of America Special Paper 375, p. 25–37. For permission to copy, contact editing@geosociety.org. © 2004 Geological Society of America

Figure 1. Oblique aerial view toward the southwest of the Las Colinas flowslide taken on January 13, 2001. Source area is at the top of the steep escarpment of the Cordillera Balsamo (indicated at A). Note heavy vegetation on the slopes of the escarpment. Distal limit of debris (east lobe), arrowed at B, is 735 m from the head of the landslide and 166 m below it. Elevation at the head of the slide is ~1075 m above sea level (masl). Photograph courtesy of *La Prensa Grafica*.

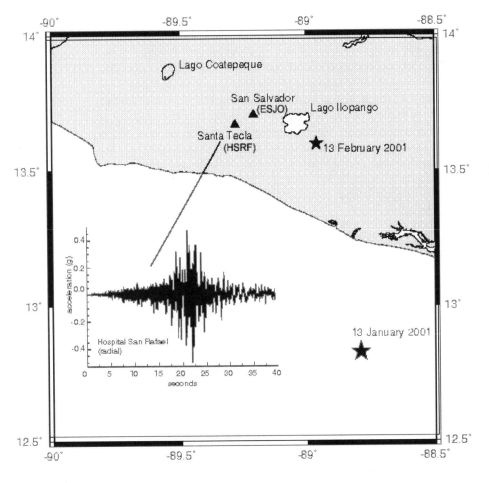

Figure 2. Location map of epicenters (stars) of January 13 and February 13, 2001, earthquakes. Size of stars proportional to earthquake magnitude. Strong-motion stations at Santa Tecla (HSRF) and San Salvador (ESJO) are shown by triangles. 40 s of the accelerogram for January 13 earthquake from HSRF is also shown. Peak radial acceleration is 0.50 g. HSRF is located 1.4 km from the source of the Las Colinas flowslide.

GEOLOGY OF SLOPES IN THE VICINITY OF SAN SALVADOR

San Salvador is located in the vicinity of several recently active volcanoes. The city, and its Santa Tecla suburb, skirt the lower eastern and southern slopes of San Salvador volcano (Fig. 3), which last erupted in 1917 (Simkin and Siebert, 1994). The surficial geology of the San Salvador area is dominated by the primary and reworked products of Upper Tertiary to Holocene volcanism (Schmidt-Thomé, 1975; Weber, 1978). The stratigraphic sequence in the San Salvador area consists, from the oldest to the youngest, of the Balsamo, Cuscatlán, and San Salvador Formations that are largely composed of pyroclastic deposits and associated volcaniclastics (Table 1). The units vary in thickness, reflecting the irregular topographic surface on which they were deposited.

The so-called Tierra Blanca occurs within the upper part of the San Salvador Formation and blankets the surface of the area (Schmidt-Thomé, 1975); it reaches up to 50 m in thickness in the eastern part of the San Salvador area. While the geotechnics of the Tierra Blanca have been extensively investigated by Bommer and his coworkers (e.g., Bommer et al., 1998; Bommer et al., 2002b), geotechnical characterization of the older pyroclastic deposits of the Cuscatlán and Balsamo Formations has not been carried out. Some important features of the geotechnical behavior of the Tierra Blanca deposits, which may apply to similar deposits in the older sequences, are (1) the contribution of negative pore pressures and a light cementation to the cohesive element of shear strength, (2) their strain-softening behavior when this cohesive component is destroyed either by saturation or seismic shaking, and (3) the high frictional component of the material in its residual state ($\phi \approx 35°$). When disaggregated, the Tierra Blanca can be classified as a silty sand or a sandy silt.

HISTORIC EARTHQUAKES AND EARTHQUAKE-TRIGGERED LANDSLIDES IN SAN SALVADOR

San Salvador is subject to frequent strong earthquakes. The city has been largely destroyed on at least eight occasions since 1576, with other earthquakes causing extensive damage (Lomnitz and Schulz, 1966). Eleven major earthquakes with magnitudes greater than 6 have occurred in El Salvador since 1857; these earthquakes have caused widespread landsliding in the San Salvador area and have been concentrated in areas of pyroclastic deposits (Rymer and White, 1989; Bommer and Rodriguez, 2002).

As recently as 1986, a major earthquake triggered many landslides in the San Salvador area. The 1986 El Salvador earthquake (M_s 5.4) was a shallow crustal event that triggered several hundred landslides that killed over 200 people. The epicenter was located just to the south of the San Salvador city limits, and landsliding was largely confined to the Tierra Blanca deposits of the San Salvador Formation. Importantly, Rymer (1987) and Rymer and White (1989) describe a highly mobile flowslide, which they termed a "rapid soil flow," that in broad outline showed similar behavior to the flowslides we observed in the vicinity of Santa

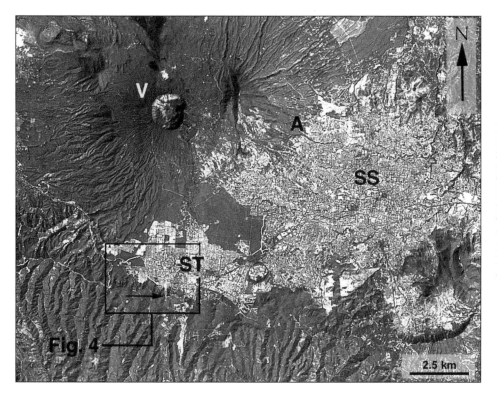

Figure 3. LANDSAT image of the San Salvador area on February 7, 2001. The main urban mass of San Salvador (SS) is located to the east and southeast of the Volcan de San Salvador (V), which last erupted in 1917. Dark-toned lava flows noted north of the vent date from this eruption. Santa Tecla (ST) is a western suburb of San Salvador. Area of Figure 4 is outlined and the Las Colinas landslide is arrowed. A is impact area of debris flow triggered by heavy rains in 1982 that resulted in ~500 deaths.

TABLE 1: STRATIGRAPHY OF VOLCANIC MATERIALS IN VICINITY OF SAN SALVADOR

Formation	Age	Stratigraphic members	Remarks
San Salvador	Pleisto-Holocene	SS1—white acidic pumice ash (Tierra Blanca) up to 50 m thick produced from within Lake Ilopango depression. SS2—brown pyroclastic and volcanic epiclastic deposits (Tobas Color Café) up to 25 m thick. SS3—andesitic and basaltic effusive rocks, locally scoria, from La Laguna crater partially interbedded in SS2.	Soil layers are present in the formation. Tierra Blanca widely distributed throughout region. Thickest in the east where pre-eruption valleys are buried. Black soil has developed on Tierra Blanca.
Cuscatlán	Plio-Pleistocene	C1—acidic to intermediate-acidic effusive rocks, partly of the same age, partly older than C2. C2—Acidic pyroclastic and volcanic epiclastic rocks (locally ignimbritic and welded tuff deposits up to 25 m thick) up to 80 m thick.	Brown loamified soil horizons are present in the Cuscatlán. Pyroclastic rocks consist of fine-grain, light brown to whitish-yellow pumice, tuff, and locally dark-gray ash material.
Bálsamo	Mio-Pliocene	B1—andesitic basaltic effusive rocks up to 30 m thick, partly of the same age as B2. B2—Volcanic epiclastic and basic pyroclastic rocks (intercalated with andesitic lava flows) up to 100 m thick.	Thick red loam layers (paleosols) occur at several horizons in the formation.

Note: Modified after Schmidt-Thomé (1975) and Weber (1978).

Tecla in 2001. The 1986 flowslide originated in the failure of volcanic tuff at the top of a slope, near a ridge crest; initial failure volume was estimated to be 8000 m³, and the flowslide ran out a distance of 440 m over a vertical distance of only ~85 m to yield a travel angle, defined as $\tan^{-1}(H/L)$ where H is the height of the travel path (the elevation difference between the head of the landslide scarp and the distal limit of the debris) and L is its horizontal length, of only 11°. In contrast to the Las Colinas flowslide, the initial failure mass of the 1986 event was apparently saturated. The source of the saturation was thought to be a perched water table within the source volcaniclastic sequence.

Heavy rains have also triggered major landslides in El Salvador in the historical period. In September 1982 ~500 people were killed in San Salvador when a large rainfall-triggered debris flow originating on the steep eastern flank of San Salvador volcano struck the Montebello area on the northern outskirts of the city (Fig. 3; Finnson et al., 1996; Kiernan and Ledru, 1996).

THE 2001 EL SALVADOR EARTHQUAKES

Two major earthquakes struck El Salvador in early 2001 (Fig. 2). The first earthquake (M_w 7.6) occurred at 11:33 local time on Saturday, January 13, at a depth of 60 km within the subducting Cocos plate (Bent et al., 2001; Bommer et al., 2002a; Bent and Evans, 2004). The epicenter is located ~105 km south-southeast of the Las Colinas landslide site. Strong-motion data recorded at the Hospital San Rafael (HSRF), located 1.4 km to northeast of the landslide source, indicate a duration of shaking of ~43 s (Bent and Evans, 2004). A peak radial acceleration of 0.50 g was recorded at HSRF, compared to 0.38 g recorded at the San Salvador station (ESJO), located at a similar epicentral distance, 9 km to the east (Fig. 2).

The January earthquake triggered many landslides in addition to the Las Colinas landslide (Jibson and Crone, 2001; Bent et al., 2001; Bommer et al., 2002a; Jibson et al., 2004). The earthquake occurred in El Salvador's dry season (November–April), following a dryer than average wet season, which usually extends from May to October. The mean annual rainfall in San Salvador is 1800–2200 mm (Rymer and White, 1989).

A second major earthquake (M_w 6.6) occurred on February 13, 2001. This earthquake was a shallow crustal event within the Caribbean plate at a depth of 13 km. The epicenter was located 40 km east-southeast of San Salvador (Fig. 2). This earthquake also triggered widespread landsliding (Baum et al., 2001; Bommer et al., 2002a; Jibson et al., 2004).

THE LAS COLINAS LANDSLIDE

Overview

The Las Colinas landslide (Fig. 1) was one of two flowslides to occur on a steep north-facing escarpment (Figs. 3 and 4) that marks the northern edge of the Cordillera Balsamo and one of many we observed to have occurred from ridgetop sites throughout the cordillera. The escarpment (hereafter referred to as the Balsamo Escarpment) is developed in pyroclastic materials of the San Salvador, Cuscatlán, and Balsamo Formations, underlain by resistant flow and ignimbritic units. The crest of the escarpment, running between 1065 and 1150 m above sea level (masl), marks the divide between drainage flowing southwest into the Pacific and drainage flowing into the interior valley of El Salvador. The head of the landslide is at elevation 1075 masl. According to witnesses, the landslide occurred suddenly and extremely rapidly during or immediately after the seismic shaking of the January

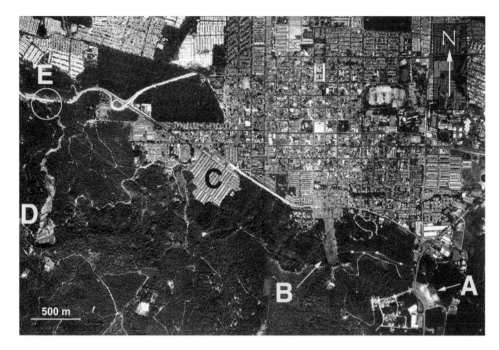

Figure 4. Aerial photograph mosaic of the steep north-facing escarpment of the Cordillera Balsamo in the vicinity of Santa Tecla (El Salvador government photographs flown on January 16, 2001). A—Construction site where geological section is exposed (see text for discussion). B—Las Colinas landslide. C—New residential development of Pinares de Suiza. D—Source area and track of Las Chorros flowslide, which traveled 1.3 km downslope and struck the Pan-American Highway at E.

13th earthquake. An army officer interviewed by the authors reported that survivors estimated the duration of the landslide to be on the order of 45 s, indicating an average velocity of ~16 m/s for a runout of 735 m.

The Las Colinas landslide (Fig. 5) is classified as a flowslide (Hutchinson, 1992; Hungr et al., 2001). Flowslides are "characterized by a suddenness of failure, following some disturbance, and a rapid and extensive runout, commonly over very gentle or horizontal ground, which comes to rest abruptly. Materials involved are generally cohesionless, weakly cemented, and are invariably of high porosity" (Hutchinson, 1992, p. 827). Hungr et al. (2001) further define a flowslide as involving "liquefaction of material originating from the landslide source" (p. 226).

A profile of the landslide was surveyed during our field investigations (Fig. 6A). Three phases of the landslide are distinguished: initial failure of the source slope at the top of the slope above Las Colinas at 1075 masl (Fig. 6A), downslope movement of the disintegrating failed mass down a steep slope to 937 masl

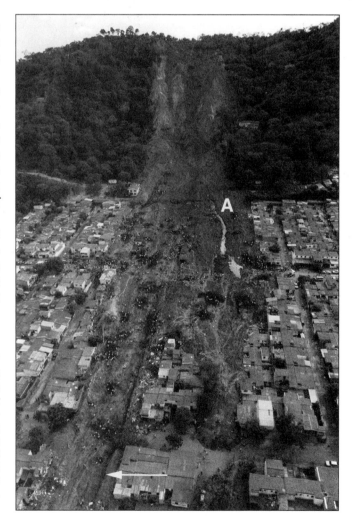

Figure 5. Oblique aerial view to the south of Las Colinas landslide shortly after it occurred on Saturday, January 13, 2001. Source area is located at the top of the forested slope. Water issuing from a burst water main is evident at the base of the slope (A). The view shows that flowslide debris covered houses (mid-center of photograph) before splitting into two tongues that were channeled by streets oriented parallel to the direction of flow (lower-center of photograph). Levees are noted at the margins of the east debris tongue (arrowed at bottom left). Lateral margins of the main body of the debris also appear to be constrained by orientation of buildings. Clusters of rescuers are seen digging in debris. (Photograph by Manuel Orellana, courtesy of *El Diario de Hoy*.)

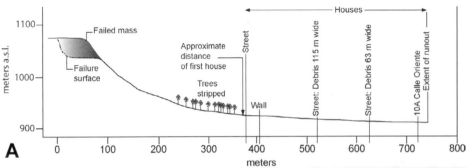

Figure 6. A: Profile of Las Colinas flowslide surveyed with laser range finder and Abney level on January 27, 2001. B: Aerial view, toward the southwest, of Las Colinas flowslide site after extensive remediation of site and after much of the landslide debris had been cleared away. Note proximity of adjacent development to the foot of Balsamo Escarpment (photograph January 26, 2001).

(a vertical distance of 138 m), and runout of the debris into the residential area where it traveled a further horizontal distance of 460 m over a slope of only 3° (Figs. 6A and 6B). The extensive low-angle runout is an aspect of the Las Colinas flowslide path geometry that poses a problem in the geotechnical interpretation of the event and thus in the landslide hazard assessment of areas at the foot of the Balsamo Escarpment.

Initial Failure

A profile (Fig. 7) constructed from pre-event 1:5000 topographic maps indicates that the failed slope at Las Colinas consisted of two segments: a steep upper segment from 1076 to 975 masl, which slopes at ~32°, and a lower-angle slope thought to correspond to a colluvial wedge sloping at ~15°. The steepness of the upper part of the slope is indicative of its considerable shear strength. The failure mass originated from the upper part of the steep slope near the Finca Gloria coffee farm and retrogressed into the flat surface at the top of the slope. Examination of pre-earthquake aerial photographs show the slope to be well vegetated with dense tropical forest, although the top of the failed slope had been cleared for coffee cultivation. Evidence for previous failure along the escarpment

is ambiguous; whereas there are no scars visible in the tropical forest, the sharp indented, alcove-like features along the crest of the escarpment (Fig. 4) may represent vegetated scars of previous failures.

The shape of the failure surface is sketched out in Figure 7 on the basis of field observations in January 2001. The base of the initial failure mass is located ~48 m above the base of the slope, 37.5 m below the top of the slope. It is also noted that the failure surface is strongly asymmetrical, almost chair-shaped in profile, with a steep backscarp and a horizontal to gently northward sloping basal component. The steep backscarp is suggestive of tensional failure; indeed, a series of tension cracks were noted up to 25 m behind the backscarp. In plan the scar is semicircular in shape and is 100 m wide at its widest point. The volume of the failed mass is estimated at 130,000 m³. Total retrogression is on the order of 10 m.

Exposures in the landslide scar (Fig. 8) show that the landslide involved interlayered brown loam-like pyroclastic materials underlain by a coarser, white to yellow-white pumiceous tephra layer that is visible at the base of the scar (Fig. 8). The boundary between the Cuscatlán Formation and the overlying San Salvador Formation is thought to correspond to a thin dark coarse tephra unit that drapes preexisting topography (Fig. 8).

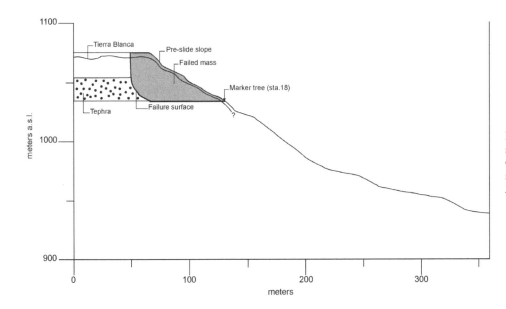

Figure 7. Profile of source slope constructed from 1:5000 topographic maps. Geology of source area and failure surface based on field examination in January 2001.

Figure 8. Scar of Las Colinas landslide (photograph January 27, 2001). Stratigraphy exposed is interpreted as follows: 5—Cuscatlán Formation white to yellow-white pumice at base of scar (discussed in text); 4—pyroclastic deposits interbedded with paleosol horizons (Cuscatlán Formation); 3—weathered pyroclastic deposits (Cuscatlán Formation); 2—Tobas Color Café pyroclastic deposits (San Salvador Formation); 1—just below the top of the scarp, thin white San Salvador Formation Tierra Blanca tephra layer that also drapes over slope to the right. This layer, thought to be the Tierra Blanca Joven of Dull et al. (2001) dating from the fourth century A.D., is overlain by organic soil.

Above this the Tobas Color Café unit (unit SS2 in Table 1) is inferred to be present and above this is the thin Tierra Blanca tephra layer (Fig. 8). The San Salvador Formation is underlain by pyroclastic units of the Plio-Pleistocene Cuscatlán Formation, which are interlayered with paleosols. A thick upper tephra unit, brown in color, is underlain by a series of darker brown soils interlayered with pyroclastics, which in turn are underlain by a thick (~3–4 m) white to yellow-white pumiceous tephra visible at the base of the exposure (Fig. 8). It is probable that the underlying materials that form most of the slope below the landslide scar are Balsamo Formation rocks that consist of resistant ignimbritic and lava flow units (Weber, 1978). Although geological controls on the shape and location of the failure surface are not known precisely and cannot be established without drilling, reconciliation of the surveyed profile with exposures in the scar strongly suggests that the basal shear surface of the initial failure is near the base of the lower white to yellow-white tephra unit within the Cuscatlán Formation.

The broad sequence in the landslide scarp was exposed in more detail in a freshly cut slope at a construction site along the east side of Highway CA-4, ~1 km to the southeast of the Las Colinas landslide (loc. A, Fig. 4). Here the lower yellow-white pumice layer was underlain by a paleosol that acted as an aquitard so that the top of the paleosol was quite wet in the floor of the excavation. In sampling the materials in the cut slope exposure, the cohesive nature of the materials in the sequence was evident as was the lack of very coarse fragments in the pyroclastic units.

Notably, there was widespread evidence of very strong shaking along the crest of the escarpment in the vicinity of the head of the landslide (cf. Jibson et al., 2004). This evidence included snapped trees, dramatic structural damage to metal structures at Finca Gloria, as well as widespread cracking and fissuring along the axis of the ridge.

Downslope Movement

In moving down the steep upper slope below 1037 masl, the mass swept the forest cover before it and entrained a small thickness (estimated as <1 m) of the residual soil cover down to 975 masl. In this part of the flowslide the trees directly adjacent to the path were stripped of branches and bark (Figs. 6A, 9), suggesting that at a path distance (defined as the horizontal distance from the head of the landslides) of ~300 m, the flowslide was already moving very rapidly and was accompanied by a fast-moving airborne dust cloud. The presence of a pronounced depression at the base of the slope, visible in Figure 5 below the steep segment, suggests that an unknown but limited volume of the colluvial wedge was entrained in the flowslide.

Runout into the Residential Area

The fully developed flowslide struck the first houses at 927 masl, 110 m below the base of the initial failure at a path distance of 375 m (Fig. 6A). The mass then covered a number of houses for a distance of 200 m before splitting into two (east and west) tongues at a path distance of 575 m (Figs. 6 and 7). The west lobe was channeled down a street parallel to the direction of movement and halted at a path distance of 625 m. The east lobe, also channeled down a street, was more mobile and traveled a further distance of 125 m to a distal limit equivalent to a path distance of 735 m, 135 m below the base of the failure (Figs. 5 and 6). In Figure 5 levees are visible at the lateral margins of the east tongue indicating plug-type flow. Thus, the flowslide crossed four streets in a swath ranging from 115 m to 63 m in width (Figs. 6A, B).

The depth of the flow was evident from mudlines on houses remaining along the lateral margin of the debris (Figs. 10 and 11). Between the first and second street encountered, the flow debris was ~5 m deep, sufficient to bury most two-story homes in this area. Between the second and third street the debris thinned to ~3 m, sufficient to bury single-story homes. In the distal part of the flowslide, between the third and fourth street, flow debris was as much as 2 m in depth. Debris penetrated lower floors and destroyed walls of homes up to its distal limit (Fig. 11).

An estimate of the landslide volume based on debris area and thickness is 150,000 m³, a volume that is ~20% more than the initial volume estimate. This order of bulking of the original volume is typical for landslides (cf. Pierson, 1998).

Figure 9. Flowslide deposits in the foreground and condition of trees adjacent to path suggesting damage by a rapidly moving airborne dust cloud ("stripped trees" area in Figure 6A) at a path distance of ~300 m. Scar is visible at top left. House behind trees is at 962 masl (photograph January 28, 2001).

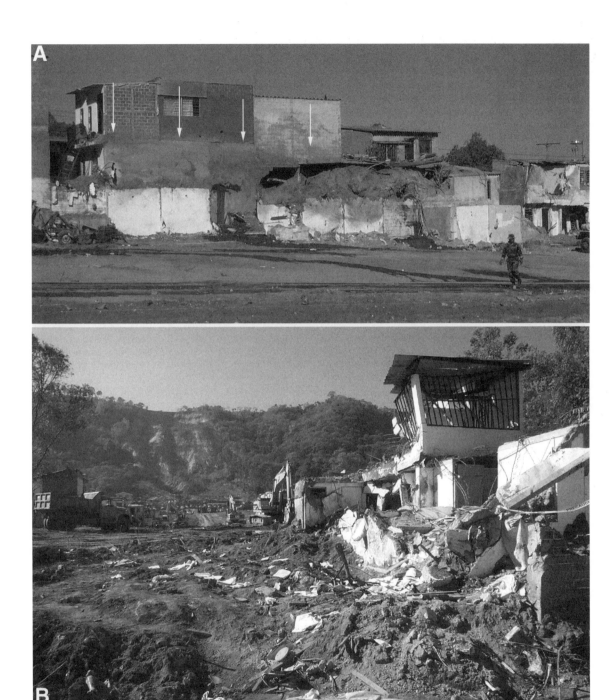

Figure 10. Flowslide debris and damage in Las Colinas I. A: Mudline on residential buildings along the western margin of the landslide after debris had been cleared away (photograph January 28, 2001). Height of mudline (vertical arrows) indicates debris depth of ~5 m. Location is between first and second streets in Figure 6A. B: Landslide debris and damage at distal limit of flowslide. Scar is visible in background at a horizontal distance of 735 m. Note texture of landslide debris (photograph January 28, 2001).

Figure 11. Flowslide debris and damage in Las Colinas II. Ground views toward landslide source from east lobe of flowslide. A: View toward landslide source shortly after it occurred on January 13, 2001. Camera location is between third and fourth streets (see Figure 6A). Note texture of debris, the levees that mark the margins of the flow, and destruction of houses at left (photograph courtesy of El Diario de Hoy). B: View of swath of damage from the third street after debris had been removed. Debris mudline is 3.2 m above street level at right, indicated by white arrow (photograph January 24, 2001).

Debris Characteristics

The flowslide debris (Figs. 10 and 11) consists primarily of brown weathered pyroclastic material of the San Salvador and Cuscatlán Formation source materials. We note that pumice fragments were relatively rare in the debris. Grain-size analysis of ten samples of the debris (Fig. 12) indicates the material is mostly finer than sand size (<2 mm). The percentage of fine sand or finer (<0.1 mm) ranges from 40% to 60%, and silt-size particles or finer ranges from 20% to 40%. These data confirm the visual observation that the texture of the debris was pre-dominantly fine sand to sandy silt (Figs. 10 and 11). Particles coarser than 2 mm consisted of white tephra particles mixed in with the finer brown debris.

Role of Seepage in Initial Failure and Degree of Saturation of Debris

Oblique aerial photographs taken on the afternoon of January 13 (Figs. 1 and 5) and our field observations in late January 2001 show no evidence of seepage in the landslide scar and no subsequent erosion by running water of the slope below.

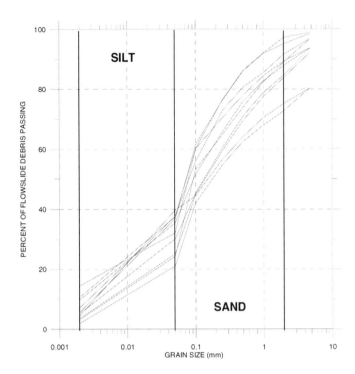

Figure 12. Grain-size analysis of ten samples of flowslide debris collected from the Las Colinas site in January 2001.

Although the possibility exists that elevated pore-water pressures could have developed in the colluvial wedge at the base of the slope as a result of loading by the impact of the debris from above, no evidence was noted for seepage in this zone during our field investigations. It is therefore concluded that the source material for the flowslide was generally dry (or at most partially saturated) during initial failure and subsequent motion. El Salvador Armed Forces personnel interviewed at the site reported that the debris was generally dry and this was confirmed by an examination of newspaper photographs taken in the immediate aftermath of the flowslide (Fig. 10B).

Some newspaper photographs taken immediately after the landslide, however, show the presence of wet debris in part of the landslide. We conclude that this wet debris was limited to an area adjacent to a broken water main on the western margin of the landslide (Fig. 5).

MOBILITY OF THE LAS COLINAS FLOWSLIDE

Landslide mobility can be measured as the ratio of height of landslide path (H) to length of landslide path (L) (Fig. 13). While the physical significance of this ratio is uncertain (e.g., Legros, 2002), it remains a useful index of landslide mobility (Corominas, 1996) and allows a dimensionless comparison of mobility with other landslides. Based on our field measurement of the landslide profile, the Las Colinas flowslide shows H =

166, L = 735, and H/L = 0.23. The equivalent travel angle is 12.7°.

Comparing this mobility with landslides of similar volume, 130,000 m³ (e.g., Finlay et al., 1999), it is seen that the Las Colinas landslide is highly mobile for its volume (Fig. 13). The data for Las Colinas plot near the lower 95% confidence limit calculated by Corominas (1996), and the flowslide shares mobility characteristics with some highly mobile flowslides in coal mine waste rock and the Chalk of northwest Europe (Hutchinson, 2002).

The length of the path (L) is 735 m and is scaled to its initial volume (V) by $L = 15V^{1/3}$, a scaling expression almost identical to that obtained for volcanic landslides (Legros, 2002). The flowslide debris covered an area of ~2.8 ha. This yields an average thickness of 4.7 m, which corresponds generally to that observed (Figs. 10 and 11). In terms of deposit area–source volume relationships, the area of the deposit (A) scales to the initial volume (V) as $A = 11.6V^{2/3}$. These scaling relationships may be useful in future hazard assessment at the base of the Balsamo Escarpment.

DISCUSSION

The destruction at Las Colinas was caused by a relatively small but extremely rapidly moving earthquake-triggered flowslide that exhibited an unusually long runout. These types of landslides commonly occur in the pyroclastic deposits of Central America as a result of earthquake triggering. The rapidly moving landslide debris was primarily dry.

The Las Colinas flowslide, as well as other observed events in the Cordillera Balsamo, originated in the uppermost part of steep slopes that culminate in a ridge or escarpment crest. This fact, together with the extensive evidence of major shaking, suggests that topographic amplification of seismic shaking (e.g., Ashford et al., 1997), probably augmented by amplification due to the low density of the pyroclastic materials, is an important factor in the genesis of the flowslide.

In some respects, the initial failure of the Las Colinas flowslide is similar to that of the Fatalak landslide triggered by the 1990 Manjil earthquake in Iran (Ishihara et al., 1992). In the Fatalak landslide it was speculated (Ishihara et al., 1992) that initial failure was due to what we term here "tensile stripping" generated by high-intensity seismic shaking of ridgetop or slope-top materials, where tensile forces initiate sliding. Similar tensile stresses were found to occur in a numerical analysis (Bommer et al., 2002b) of a hypothetical Tierra Blanca slope subject to the Santa Tecla accelerogram in Figure 2 (Fig. 6 in Bommer et al., 2002b). In that analysis failure was dependent upon the formation of a seismically induced tension crack.

In the absence of obvious evidence for the existence of positive pore-water pressures, the mechanism of strength loss, or brittleness (Bishop, 1973), and the source of the brittleness implied in the behavior of the Las Colinas flowslide remain unexplained. A number of alternatives present themselves: *(a)* brittleness due to the destruction of suction forces in the partially saturated soil by seismic-shaking–induced rupture of

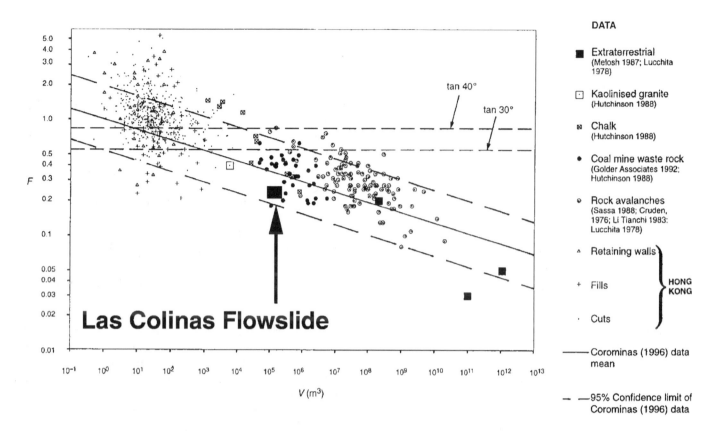

Figure 13. Comparative mobility of Las Colinas flowslide shown on plot of F (F = H/L [defined in text]) versus log landslide volume (V) for various landslides (for references see Finlay et al., 1999). Regression line and 95% confidence limits calculated by Corominas (1996) are shown (modified after Finlay et al., 1999, Figure 4).

menisci, *(b)* brittleness due to a loss of cohesion through the destruction of cementation in the pyroclastic materials by seismic shaking and, *(c)* generation of excess pore-water pressure in a thin zone at the base of the flowing mass. In the case of the Las Colinas landslide, the evidence for water contributing to high pore-water pressures is ambiguous and raises the question as to whether excess pore-air pressures, generated by the strain-induced collapse of a porous pyroclastic fabric, may have also acted to reduce the frictional resistance of the partially saturated flowslide mass.

With reference to flowslide debris behavior, it is probable that lateral spreading of the debris, which would have resulted in a lower runout length along the axis of movement, was inhibited because of the channelization of the debris by streets aligned parallel to the movement direction, as is evident in Figure 5, thus enhancing the runout distance.

Since the high-velocity landslide struck a densely populated urban neighborhood (Figs. 1 and 5), the Las Colinas flowslide was highly destructive. The casualties at Las Colinas and the geometry of the landslide deposit indicate a loss rate of at least 211 persons/ha for the event.

CONCLUSION

The 2001 Las Colinas flowslide is a tragic reminder that earthquake-triggered landslides are a major hazard in El Salvador and in other parts of Central America. Steep slopes composed of pyroclastic materials are particularly prone to failure during earthquake shaking. The source areas for these landslides are commonly located at ridge crests or ridgetops, and it is speculated that initial failure may result from tensile stripping. Failures frequently generate mobile flowslides that move very rapidly over gentle slopes and achieve long runout distances. Landslide hazard in these areas is therefore determined by the runout behavior of landslides at the base of their source slopes. The mechanism of long runout at Las Colinas remains undetermined. It is concluded, however, that the long runout of the flowslide is in part explained by channelization by urban topography wherein streets parallel to movement channeled the debris and enhanced travel distance by inhibiting the lateral spreading of the debris. This has important implications for landslide hazard assessment at the base of the Balsamo Escarpment and other urban areas in similar settings in Central America. The San Salvador area is a densely populated urban complex that is subject to frequent geologic hazard events. Population living near the base of steep slopes is therefore vulnerable to high loss rates as a result of future earthquake-triggered flowslides.

ACKNOWLEDGMENTS

Fieldwork in El Salvador was facilitated by Mauricio Rosales Rivera, El Salvador Ambassador to Canada. Funding was provided by Emergency Preparedness Canada and Natural Resources Canada. The cooperation of the El Salvador Ministerio de Medio Ambiente y Recursos Naturales, in particular Ernesto Lopez Zepeda, is gratefully acknowledged. The assistance of the El Salvador Armed Forces is also acknowledged. We would also like to thank *El Diario de Hoy* (especially Marina León de Menjívar) and *La Prensa Grafica* for providing the invaluable photographs of the Las Colinas landslide and the permission to use them here. Michelle Poirier (Natural Resources Canada) processed the LANDSAT image in Figure 3. S.G.E. is also thankful to Professor K. Ishihara for bringing the Iran landslide, discussed in the text, to his attention. Enthusiastic critical reviews by Jon Major, Réjean Couture, Scott Burns, and Kevin Schmidt greatly improved the original manuscript.

REFERENCES CITED

Ashford, S.A., Sitar, N., Lysmer, J., and Deng, N., 1997, Topographic effects on the seismic response of steep slopes: Bulletin of the Seismological Society of America, v. 87, p. 701–709.
Baum, R.L., Crone, A.J., Escobar, D., Harp, E.L., Major, J.J., Martinez, M., Pullinger, C., and Smith, M.E., 2001, Assessment of landslide hazards resulting from the February 13, 2001, El Salvador Earthquake: U.S. Geological Survey Open-File Report 01-119, 20 p.
Bent, A.L., and Evans, S.G., 2004, The M$_w$ 7.6 El Salvador Earthquake of 13 January 2001 and implications for seismic hazard in El Salvador, *in* Rose, W.I., et al., eds., Natural hazards in El Salvador: Geological Society of America Special Paper 375, p. 397–404 (this volume).
Bent, A.L., Evans, S.G., Lau, D.T., and Law, K.T., 2001, The M$_w$ 7.6 El Salvador earthquake of 13 January 2001 and implications for seismic hazard in Central America: Proceedings, International Conference on Seismic Risk in the Caribbean region, Dominican Republic, July 2001.
Bishop, A.W., 1973, The stability of tips and spoil heaps: Quarterly Journal of Engineering Geology, v. 6, p. 335–376.
Bommer, J.J., and Rodriguez, C.E., 2002, Earthquake-induced landslides in Central America: Engineering Geology, v. 63, p. 189–220.
Bommer, J., Rolo, R., and Méndez, P., 1998, Propiedades mécanicas de la Tierra Blanca y la inestabilidad de Taludes: Revista ASIA, no. 128, p. 15–21.
Bommer, J.J., Benito, M.B., Ciudad-Real, M., Lemoine, A., López-Menjívar, M.A., Madariaga, R., Mankelow, J., Méndez de Hasbun, P., Murphy, W., Nieto-Lovo, M., Rodríguez-Pineda, C.E., and Rosa, H., 2002a, The El Salvador earthquakes of January and February 2001: Context, characteristics, and implications for seismic risk: Soil Dynamics and Earthquake Engineering, v. 22, p. 389–418.
Bommer, J.J., Rolo, R., Mitroulia, A., and Berdousis, P., 2002b, Geotechnical properties and seismic slope stability of volcanic soils: 12th European Conference on Earthquake Engineering, Paper 695, 10 p.
Corominas, J., 1996, The angle of reach as a mobility index for small and large landslides: Canadian Geotechnical Journal, v. 33, p. 260–271.
Cruden, D.M., 1976, Major rock slides in the Rockies: Canadian Geotechnical Journal, v. 13, no. 8, p. 8–20.
Dull, R.A., Southon, J.R., and Sheets, P., 2001, Volcanism, ecology and culture: A reassessment of the Volcán Ilopango TBJ eruption in the southern Maya realm: Latin American Antiquity, v. 12, p. 25–44.

Finlay, P.J., Mostyn, G.R., and Fell, R., 1999, Landslide risk assessment: prediction of travel distance: Canadian Geotechnical Journal, v. 36, p. 556–562.
Finnson, H., Bäcklin, C., and Bodare, A., 1996, Landslide hazard at the San Salvador volcano, El Salvador, *in* Senneset, K., ed., Landslides, Proceedings of the 7th International Symposium on Landslides, Trondheim, Norway: Rotterdam, Balkema, v. 1, p. 215–220.
Golder Associates, 1992, Travel characteristics of debris from dump failures in mountainous terrain, Contract 23440-0-918/01-X8G: Interim Report to the Department of Supply and Services, Canada.
Hungr, O., Evans, S.G., Bovis, M.J., and Hutchinson, J.N., 2001, A review of the classification of landslides of the flow type: Environmental and Engineering Geoscience, v. 7, p. 221–238.
Hutchinson, J.N., 1988, General report: morphological and geotechnical parameters of landslides in relation to geology and hydrogeology, *in* Bonnard, C., ed., Landslides, Proceedings of the 5th International Symposium on Landslides, Lausanne, Switzerland: Rotterdam, The Netherlands, A.A. Balkema, v. 1, p. 3–35.
Hutchinson, J.N., 1992, Flow slides from natural slopes and waste tips: Proceedings, III Simposio Nacional sobre Taludes y Laderas Inestables, La Coruña, v. 3, p. 827–841.
Hutchinson, J.N., 2002, Chalk flows from the coastal cliffs of northwest Europe, *in* Evans, S.G., and DeGraff, J.V., eds., Catastrophic landslides: Effects, Occurrence, and mechanisms: Boulder, Colorado, Geological Society of America Reviews in Engineering Geology XV, p. 257–302.
Ishihara, K., Haeri, S.M., Moinfar, A.A., Towhata, I., and Tsujino, S., 1992, Geotechnical aspects of the June 20, 1990, Manji earthquake in Iran: Soils and Foundations, v. 32, p. 61–78.
Jibson, R.W., and Crone, A.J., 2001, Observations and recommendations regarding landslide hazards related to the January 13, 2001, M-7.6 El Salvador earthquake: U.S. Geological Survey Open-File Report 01-0141, 19 p.
Jibson, R.W., Crone, A.J., Harp, E.L., Baum, R.L., Major, J.J., Pullinger, C.R., Escobar, C.D., Martínez, M., and Smith, M.E., 2004, Landslides triggered by the 13 January and 13 February 2001 earthquakes in El Salvador, *in* Rose, W.I., et al., eds., Natural hazards in El Salvador: Boulder, Colorado, Geological Society of America Special Paper 375, p. 69–88 (this volume).
Kiernan, S.H., and Ledru, O., 1996, Remedial measures against landslide hazards at San Salvador volcano, El Salvador [M.S. thesis]: Stockholm, Royal Institute of Technology, 100 p. and appendices.
Legros, F., 2002, The mobility of long-runout landslides: Engineering Geology, v. 63, p. 301–321.
Li Tianchi, C., 1983, A mathematical model for predicting the extent of a major rock fall: Zeitschrift für Geomorphologie, N.F., v. 27, no. 24, p. 473–482.
Lomnitz, C., and Schulz, R., 1966, The San Salvador earthquake of May 3, 1965: Bulletin of the Seismological Society of America, v. 56, p. 561–575.
Lucchita, B.K., 1978, A large landslide on Mars: Geological Society of America Bulletin, v. 89. p. 1601–1609.
Melosh, H.J., 1987, The mechanics of large rock avalanches: Geological Society of America Reviews in Engineering Geology VII, p. 41–49.
Pierson, T.C., 1998, An empirical model for estimating travel times for wet volcanic flows: Bulletin of Volcanology, v. 60, p. 98–109.
Rymer, M.J., 1987, The San Salvador earthquake of October 10, 1986—Geologic aspects: Earthquake Spectra, v. 3, p. 435–463.
Rymer, M.J., and White, R.A., 1989, Hazards in El Salvador from earthquake-induced landslides, *in* Brabb, E., and Harrod, B., eds., Landslides: Extent and economic significance: Rotterdam, Balkema, p. 105–109.
Sassa, K., 1988, Special lecture: geotechnical model for the motion of landslides, *in* Bonnard, C., ed., Landslides, Proceedings of the 5th International Symposium on Landslides, Lausanne, Switzerland: Rotterdam, The Netherlands, A.A. Balkema, v. 1, p. 37–55.
Schmidt-Thomé, M., 1975, The geology in the San Salvador area (El Salvador, Central America), a basis for city development and planning: Geologischen Jahrbuch, v. B13, p. 207–228.
Simkin, T., and Siebert, L., 1994, Volcanoes of the world (2nd edition): Tucson, Arizona, Geoscience Press, 349 p.
Weber, H.S., compiler, 1978, Mapa Geologico de la Republica de El Salvador. Hoja San Salvador: scale 1:100,000, Bundesanstalt fur Geowissenschaften und Rohstoffe, Hannover, Germany.

MANUSCRIPT ACCEPTED BY THE SOCIETY JUNE 16, 2003

Geological Society of America
Special Paper 375
2004

Las Colinas landslide:
Rapid and long-traveling soil flow caused by
the January 13, 2001, El Salvador earthquake

Kazuo Konagai*
Jorgen Johansson
Paola Mayorca
Institute of Industrial Science, University of Tokyo, 4-6-1 Komaba, Meguro-ku, Tokyo 153-8505, Japan

Ryosuke Uzuoka
Department of Civil Engineering, Tohoku University, Japan

Tetsuro Yamamoto
Department of Civil Engineering, Yamaguchi University, Japan

Masakatsu Miyajima
Department of Civil Engineering, Kanazawa University, Japan

Nelson Pulido
Earthquake Disaster Mitigation Research Center, Kobe, Japan

Kyoji Sassa
Hiroshi Fukuoka
Disaster Prevention Research Center, Kyoto University, Japan

Freddy Duran
Department of Civil Engineering, Kyoto University, Japan

ABSTRACT

Two devastating earthquakes struck El Salvador within a month. The first quake of January 13, 2001, which was centered off El Salvador's southern coast, damaged or destroyed nearly 108,000 houses and killed at least 944 people. A considerable amount of soil (~200,000 m³) was fluidized on a mountain ridge rising south behind the Las Colinas area of Nueva San Salvador (Santa Tecla). The average slope was at most ~13°, and yet the fluidized soil flowed ~400 m across the residential area, destroying many houses and killing more than 700 people. This report outlines the findings obtained through reconnaissance by a mission dispatched by the Japan Society of Civil Engineers and the laboratory tests that followed it.

Keywords: long-traveling soil flow, grain crushing, microtremor.

*Konagai@iis.u-tokyo.ac.jp

Konagai, K., Johansson, J., Mayorca, P., Uzuoka, R., Yamamoto, T., Miyajima, M., Pulido, N., Sassa, K., Fukuoka, H., and Duran, F., 2004, Las Colinas landslide: Rapid and long-traveling soil flow caused by the January 13, 2001, El Salvador earthquake, *in* Rose, W.I., Bommer, J.J., López, D.L., Carr, M.J., and Major, J.J., eds., Natural hazards in El Salvador: Boulder, Colorado, Geological Society of America Special Paper 375, p. 39–53. For permission to copy, contact editing@geosociety.org. © 2004 Geological Society of America.

K. Konagai et al.

INTRODUCTION

El Salvador is one of the smallest and most crowded nations in Central America. It extends ~240 km westward from the Gulf of Fonseca to the border with Guatemala. Two devastating earthquakes struck this country within a month. The first quake of January 13, 2001, which was centered off El Salvador's southern coast, damaged or destroyed nearly 108,000 houses and killed at least 944 people. One of the most spectacular aspects of the January 13 earthquake was the damage inflicted by landslides. Among them, the Las Colinas landslide was the most tragic. A considerable amount of soil (~200,000 m^3) was fluidized on a mountain ridge rising south behind the Las Colinas area of Nueva San Salvador (Santa Tecla, La Libertad), which destroyed many houses. About 500 fatalities due to this landslide were reported at the time of our investigation (see La Libertad in Table 1, Comité de Emergencia Nacional, January 30, 2001). The number of deaths, however, continued to rise as additional information was acquired and ultimately reached 747 (Sassa et al., 2002).

On January 19, 2001, the Japan Society of Civil Engineers (JSCE) decided to dispatch an investigation team to El Salvador (see Konagai et al., 2001a, 2001b, 2001c, 2002). Though JSCE is composed of researchers having a wide range of expertise, the reconnaissance team, which consisted of eight experts, had little chance to thoroughly investigate every civil engineering specialty represented during their short stay (February 1–6) in El Salvador. The strategy of the JSCE team was to make a reconnaissance investigation focusing mostly on landslide-induced damages. This report outlines the important findings of our investigation of the Las Colinas landslide and discusses possible preventive measures for reducing loss of human lives and social disruptions from future landslides.

OVERVIEW OF LAS COLINAS LANDSLIDE

Landslides caused by the January 13 earthquake were found nationwide. The statistics from the National Emergency Committee, COEN (Table 1), show that the majority of the hundreds of landslides triggered by the earthquake occurred in the departments of La Libertad, La Paz, Sonsonate, Santa Ana, Usulután, San Salvador, and San Miguel. The Ministry of Environment and Natural Resources, MARN (Ministerio de Medio Ambiente y Recursos Naturales) built a geographical information system (GIS) database of earthquake-induced landslides immediately after the January 13 event. The database describes landslides in terms of their locations, scales, and instabilities. It includes information not only from MARN, but also from COEN and the Ministry of Public Works (MOP, Ministerio de Obras Públicas). Figures 1 and 2 show locations of registered landslides on topographical and geological maps, respectively. The landslides are found mostly on mountainsides of volcanoes such as Santa Ana, San Vicente, and Tecapa or in pyroclastic and/or epiclastic deposits such as Tierra Blanca and Tobas Color Cafes (Weber et al., 1978).

Among all, the Las Colinas landslide in La Libertad was the largest. A considerable amount of soil (~200,000 m^3) slipped down the slope rising behind a new residential district of Nueva San Salvador, destroying hundreds of houses and killing more than 700 people. Balsamo Ridge rises south behind the city, and the slope failure took place on its northern side (Fig. 3). Other

TABLE 1. DAMAGE STATISTICS

Department	Dead	Injured	Damaged public buildings	Damaged houses	Collapsed houses	Buried houses	Landslides
La Libertad	685*	2183	48	14558	15723	687	20
La Paz	44	147	272	25076	17996	0	75
Sonsonate	48	1295	38	17773	10501	0	82
Santa Ana	47	327	5	13925	4823	0	27
Cuscatlán	20	43	47	4762	4282	0	17
Usulután	27	786	335	30716	29293	0	38
San Salvador	24	391	76	12836	10372	0	133
San Miguel	19	43	23	10624	2902	0	26
San Vicente	29	81	40	17292	5218	0	4
La Unión	1	8	98	2136	268	0	1
Ahuachapán	0	247	60	18540	6553	0	12
Cabañas	0	7	31	1153	309	0	4
Morazán	0	3	35	94	5	0	1
Chalatenango	0	4	47	307	16	1	5
TOTAL	944	5565	1155	169692	108261	688	445

Note: Damage statistics are from Comité de Emergencia Nacional, January 30, 2001.
*The majority of these fatalities are the result of the Las Colinas landslide. The number of deaths at Las Colinas, however, continued to rise after our visit, and ultimately reached 747 (Sassa et al., 2002).

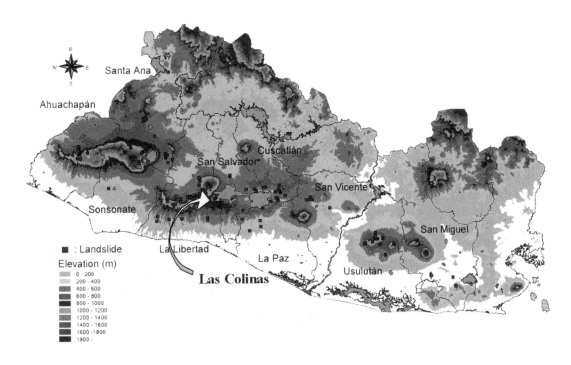

Figure 1. Landslide locations plotted on a topographical map. (Topographical map: MOP [Ministerio de Obras Publicas], Instituto Geografico Nacional, 1984.)

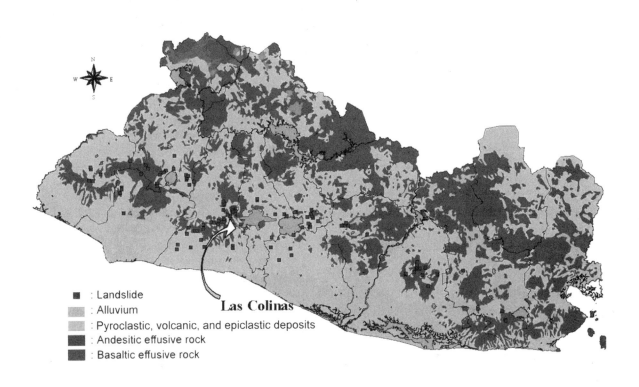

Figure 2. Landslide locations plotted on a geological map. (Geological map: Guzman, 2000.)

Figure 3. Bird's eye view of the Las Colinas landslide. White color of Zone 3 is the result of applied disinfectant.

small landslides and cracks were also found along the mountain ridge. A flat area that slopes gently southwards is located just behind the top of the landslide scarp, and two unfinished brick houses, which were supposed to become a school, occupied the flat. A coffee plantation spreads along the ridge and irrigation of the coffee plants keeps the soil locally moistened.

The Las Colinas sliding surface was surveyed using a laser-based theodolite (Laser Ace 300) connected to a portable computer. The theodolite has a built-in digital compass, and with its laser beam it determines the azimuth, dip angle, and horizontal distance to a point. The surveyed slope was plotted with the map software SURFER (Fig. 4). The horizontal distances from the top

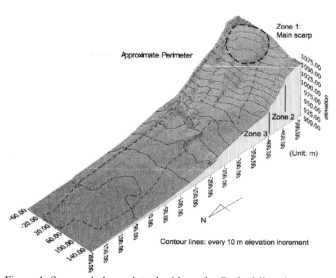

Figure 4. Surveyed slope plotted with surfer. Dashed line shows approximate perimeter of landslide.

end of the scar to the toe of the slope and to the farthest reach of the soil mass are 480 m and 700 m, respectively. The elevations of the head scarp and the toe of the slide are 1080 m and 920 m, respectively. Therefore, the landslide descended 160 m over a horizontal distance of 700 m. The average inclination from the top of the source area to the toe of the deposit is ~13°.

The exposed failure surface can be divided roughly into three zones (Fig. 4). The uppermost zone (Zone 1) is the source area, a hollow ~150 m in diameter. The head scarp is displaced 20–30 m from the original face of the ridge. The soil mass fitted into this hollow was estimated to be ~150,000 m³. North beyond the hollow is the steepest slope (Zone 2), which becomes gradually gentle as it approaches the toe. In Zone 3, the depositional area, the landslide debris had been removed by the time of our visit, and the cleared area had been disinfected (Fig. 3).

To estimate the total volume of the landslide mass, we scanned a 1:5000-scale topographic map (MOP, 1970) of the source area on the ridge and compared it with the surveyed configuration matching both original and surveyed origins (GPS: 13.665907° N, 89.286890° W, 920 m above sea level). The total slide volume was estimated to be ~200,000 m³, a little larger than the soil mass fitted in the uppermost hollow. This suggests that there were other secondary sources of soil mass added to the landslide in Zone 2.

DETAILED FEATURES OF THE SLOPE

Source Area (Zone 1)

Photographs (Fig. 5) of the soil layers on the west half of the uppermost scar were taken on January 14 and February 4, 2001. Dark- and light-colored stripes appearing on the scarp show a stratified soil profile having a top lapilli tuff layer 2 m thick (see Fig. 6) overlying a pair of differently colored (ocher and white) pumice

Figure 5. West slope of main scarp (photographed by José Antonio Rivas). A: January 14. B: February 4.

Figure 6. A piece of lapilli tuff. Hammer is 327 mm long.

Figure 7. Cracks and joints on the exposed intact pumice. Hammer is 327 mm long.

layers 11.9 m thick. Bedding planes separate the lapilli tuff into sublayers of varying thickness (10–25 cm). The dark brown color of the soil indicates that it was moist, and this pair of photos shows that the dark-colored stripes had thinned in the weeks following the earthquake. Therefore, the exposed soils were drying. This fact suggests that the intact soils along the slip surface were wet before the event. White and/or ocher-colored pumice soils appear beneath the dark and wet soils. They were also moist and weakly cemented. The pumice was easily broken into small pieces or into powder just by rubbing it together. Broken fragments of this pumice had subangular to angular shapes. Some pieces were medium-sized (1 mm × 4 mm × 7 mm), and some were fairly large (reaching 12 mm × 25 mm × 35 mm). As shown in Figure 7, cracks and joints were observed even on intact sections of pumice.

The main scarp showed that the soil there was deeply gouged. To define the shape of the scarp, we marked 43 points along its perimeter using a GPS receiver (Fig. 8). Behind the scarp, exten-

sive cracks ran almost straight from west to east (see photographs in Fig. 8). Crack openings (12 visible cracks; Table 2 and Fig. 8) were measured perpendicular to the scarp. The total opening of 1.25 m across the 22 m distance of the measured line amounts to about a 5% average strain within the soil behind the scar. Exten-

Cracks appearing on the top terrace behind the scar (**Zone 1**): Crack map shown below.

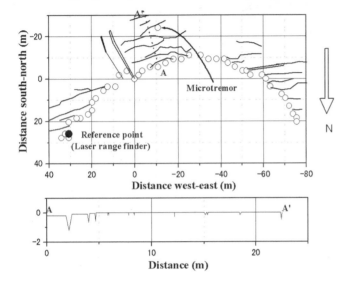

Figure 8. Perimeter of scar, Las Colinas Landslide.

sive cracks were found along the ridge crest even in areas that did not slide; these cracks can certainly cause further slides.

Limited evidence for liquefaction was observed near the top of the slope. A pair of photos (Fig. 9) taken on January 14, the day after the earthquake, show that edges of cracks were lightly dusted with fine sand. This fine sediment may be evidence that the earthquake caused some soils to liquefy even though field observation did not suggest that the groundwater level was very high.

At the time of our investigation, some downward streaks of brown soil remained on the bottom of the upper hollow (arrow in Fig. 10), and broken pieces of a mortar-block fence were caught on the soil mass remaining on the outer edge of the hollow (arrow in Fig. 11). These features may be evidence that most of the initial 150,000 m³ soil mass had been thrown forward and then flowed down the lower steep slope. Along the observed streaks, a number of broken pieces of lapilli tuff and some uprooted trees were found.

Steep Sliding Surface (Zone 2)

Figure 12 shows a complete view of the lower steep sliding surface (Zone 2). As shown in Figure 13, much of this slope was covered with a thin film of mud that included porous fragments of pumice. The mud film was noticeably stiff after it dried completely.

TABLE 2. CRACK OPENINGS ALONG LINE A–A' IN FIGURE 8

No.	Distance from A (cm)	Vertical offset (cm)	Opening (cm)	Cumulative opening (cm)	Depth (cm)
1	180–240	10	60	60	122
2	390–410	7	20	80	65
3	460–471	5	11	91	60
4	580–581	0	1	92	20
5	590–593	0	3	95	20
6	780–782	0	2	97	20
7	833–834	0	1	98	20
8	1215–1217	0	2	100	30
9	500–1502	0	2	102	20
10	1525–1535	0	10	112	20
11	1841–1859	0	8	120	50
12	2234–2249	0	5	125	20

The film covered trees, bamboo, and other plants that were still well rooted but had been overridden by the landslide. This fact suggests that most of the original sediment in Zone 2 was not scraped off, and it is consistent with the result from the three-dimensional survey of the slope configuration. This fact also suggests that the

Figure 9. Crack at the top terrace (photographed on January 14, 2001, by José Antonio Rivas, Geotérmica Salvadoreña).

Figure 10. Lower part of the hollow: A piece of mortar block wall (within the circle) is caught on the small amount of the soil mass stopped at the lower edge of the hollow. The arrow shows downward streaks of brown soil remained on the bottom of the upper hollow.

bottom surface of the landslide mass was wet and may have liquefied (Sassa, 2000), a phenomenon possibly related to some weak tephra layers spreading over less permeable paleosols.

Deposition Area (Zone 3)

The landslide mass surged across a small ravine entering from the east and splashed up a small bank. We observed splash marks 6–8 m high on some trees on this bank (Fig. 14). The mobilized soil mass overran or inundated many houses and caused more than 700 deaths. A complete view of Zone 3, taken from the toe of the exposed slip surface, is shown in Figure 15A. By the time of our investigation, the landslide mass and destroyed houses had been removed, and the cleared area had been disinfected.

There were many splash marks of mud remaining on house walls, trees, etc., at the time of our visit. In general, the splash marks were higher near the toe of the slope than in the middle or distal reaches of the landslide deposition zone. The highest splash was found ~8 m above ground level on a tree trunk near the toe of the head slope. Figure 15B shows the walls of a dwelling on the

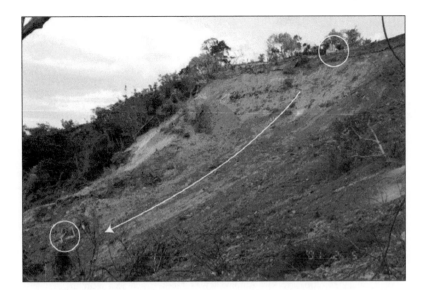

Figure 11. Fallen block wall.

Figure 12. Steep slope (Zone 2).

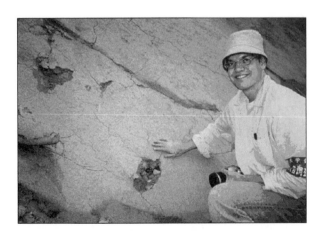

Figure 13. Thin mud film.

eastern perimeter of the landslide deposit. Mud spatters on the wall follow a parabolic shape. The parabola, having a peak height of 4.5 m, drops downward and reaches the ground after about a 5 m horizontal run. This geometry suggests that the time, t, needed for the splashes to reach the ground from their peak height Δh (4.5 m) was ~1 s ($\Delta h = g \cdot t^2/2$). During this time, the splashes ran ~5 m horizontally, indicating a horizontal speed of 5 m/s. Because the estimated velocity is for the eastern edge of the landslide mass, the main stream of debris might have flowed faster despite moving over, around, and through dwellings standing close together.

SOIL TESTS

We hypothesize that crushing of fragile, porous soil grains caused a buildup of pore-water pressure that drastically decreased the friction angle among the grains and allowed the landslide to travel an unusually long distance. To test this hypothesis, we conducted in situ strength tests and ring shear tests on samples taken back to the laboratory.

In Situ Test

The standard penetration test (SPT; ASTM D1586, Eurocode 7) is the most widely used test for assessing in situ soil strength. The purpose of this test is to obtain a representative soil sample using a split-barrel sampler and to measure the resistance of the soil to penetration of the sampler. The sampler is driven into the ground at the base of a borehole with a free-falling 63.5 kg trip hammer dropping a distance of 760 mm. During the test, a count is made of the number of blows required to drive the sampler through an initial 150 mm, and a subsequent 300 mm, test zone. The blow count for

Figure 14. Photograph from the toe of the slope looking downhill. The landslide mass seems to have splashed up a small bank (white arrow). Splash marks were found 6–8 m high on tree trunks (above the frame of the photo).

Figure 15. Photographs from Zone 3. A: Complete view of Zone 3 (February 4, 2001). B: House wall spotted with mud splashes.

the full 300 mm test zone is termed the N_s value, and is an indication of the relative density of sands and gravels. Although we desired to conduct an SPT at Balsamo Ridge, we instead conducted portable cone penetration tests (CPT; Japan Geotechnical Society, 1995) because of the limited amount of time we had to spend in the field.

The cone penetration test consists of a pair of rods coupled upright with a knocking head. A cone is attached to the bottom end of the rods, and a hammer is dropped along a guide groove cut straight on the surface of the upper rod (Fig. 16). The procedure for conducting a CPT is as follows:

1. Assemble the equipment.

2. Hold the equipment upright on a testing point.

3. Confirm that the hammer is never stuck in the groove and drops smoothly along the guide rod.

4. Drop the hammer from the height 50 cm above the knocking head. When the cone is driven ~10 cm deep, record the number of blows, N, and the exact depth, d, reached by this driving. N_d is then defined as:

$$N_d = 10\frac{N}{d}$$

5. When the penetrometer sinks into the soil under its own weight, record the depth that the cone has sunk. In this case, N_d is set at 0.

6. Stop the test when, after ten blows, the depth has not reached 2 cm.

7. Observe the soil when pulling out the equipment.

A positive correlation exists between N_s values obtained from a standard penetration test and N_d values obtained from a portable dynamic penetration test (Japan Geotechnical Society, 1995).

$$N_s = \frac{N_d}{1.5}$$

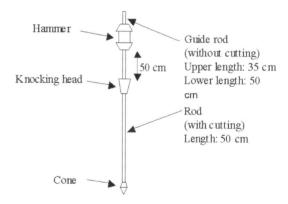

Figure 16. Cone penetration test equipment.

Cone penetration tests were performed at three sites near the Las Colinas landslide on Balsamo Ridge to assess in situ soil strength. Because we intended to sound the strength of the pumice layer, which was considered to be the material most responsible for the initiation of the slide, we tested a site (loc. C1, Fig. 17) near

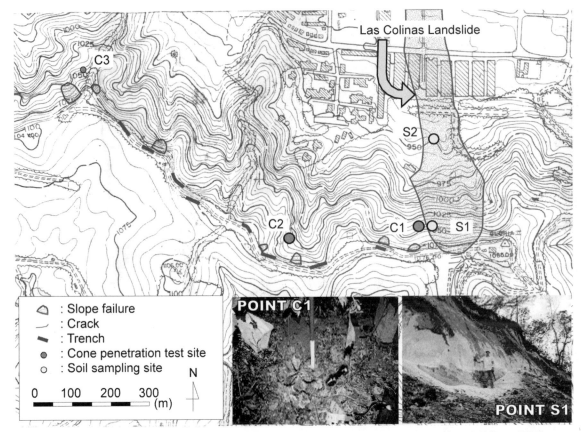

Figure 17. In situ tests and sampling sites in Las Colinas. (Topography: Ministerio de Obras Públicas, 1970).

the toe of the source area on Balsamo Ridge immediately above the pumice layer. Variations of the estimated N_s values with depth at the three locations tested are shown in Figure 18. In general, the equivalent N_s values at point C1 are considerably lower than those from other locations. Extremely low values approaching zero were found at depths of 1.2 m and 2.5 m. These extremely small N_s values suggest the presence of two weak layers, which may be the pumice and/or fragmental volcanic deposits thought to be the layers most responsible for the landslide, judging from their depths.

Ring Shear Test

Because wet pumice was considered to have been responsible for the initiation of the slide, pumice samples were taken (loc. S1, Fig. 17) to examine material crushability. The crushability of soil grains is closely related to pore-water-pressure buildup during rapid shear. We used a ring shear apparatus (DPRI-5; Sassa, 1996; Fig. 19) to simulate the rapid shear of an undrained sample. Outer and inner diameters of the cylindrical shear box were 120 and 80 mm, respectively. A maximum shearing velocity of 10 cm/s can be realized within this box.

The concept of the ring shear landslide simulation test is illustrated in Figure 20 (Sassa, 2000). The ratio of shear to normal stresses along a sliding surface is simulated in the interior of

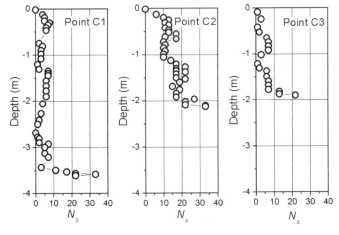

Figure 18. Equivalent N_s values at three sites. $N_s = N_d/1.5$, where N_d is the number of blows when the penetration depth of 10 cm is reached.

the shear box, and the necessary physical quantities of a soil sample in shear (normal stress, torque, pore-water pressure, shearing velocity, and the apparent friction angle) are monitored in real time. A cross section of the shear box is shown in Figure 21 (see Sassa, 1996, 2000, for further details).

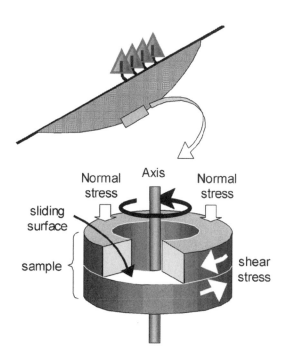

Figure 20. Schematic drawing of ring shear landslide simulation test (Sassa, 2000).

Figure 19. Ring shear apparatus.

Two ring shear tests were performed on the landslide soil (pumice) having the mechanical properties listed in Table 3. Samples having different degrees of saturation were prepared: one fully saturated, another 80% saturated. All samples were normally consolidated under a stress state corresponding to a slip surface inclined 20 degrees at a depth of 13 m. Shearing stresses were applied subsequently to undrained samples after consolidation at a loading and unloading rate of 0.098 kPa/s. Samples failed and began rotating when the stress state reached the failure line. Shearing was continued until an equivalent of 15–20 m displacement was achieved.

Figure 22A shows the effective stress paths for the two tests. As soon as the first test sample (degree of saturation, $Sr = 100\%$) began shearing, pore-water pressure appeared to increase, and this increase was followed by dilation of the pumice. When the residual failure line (RFL) was reached, the apparent friction angle of the sample decreased suddenly. The drop in shear resistance is plotted in Figure 22B with respect to shear displacement. When the sheared pumice sample had

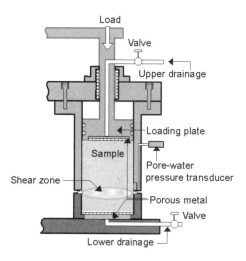

Figure 21. Cross section (right half) of undrained ring shear box (Sassa, 2000).

TABLE 3. MECHANICAL PROPERTIES OF SAMPLE

Property	Pumice
Mean grain size, D_{50} (mm)	0.91
Effective grain size, D_{10} (mm)	0.25
Uniformity coefficient, U_c	5.0
Specific gravity, G_s	2.28

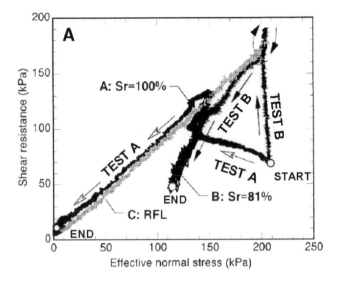

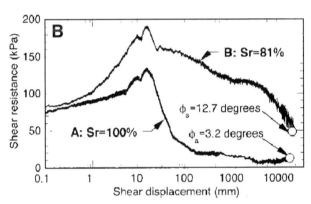

slipped 20 m, it achieved an apparent friction angle of ~3.2°, which suggests that sliding-surface liquefaction occurred (Sassa, 1996). The average slope gradient of the landslide measured in the field, however, was ~13° over the entire travel distance, much higher than the 3.2° achieved in the ring shear test using saturated soil.

Owing to the differences in apparent friction angles estimated in the field and measured on saturated samples in the laboratory, we conducted a second test at a smaller saturation value (80%). In this test, the stress path rose until it reached the RFL (Fig. 22A), at which point the apparent friction angle began dropping along the RFL. The stress path deviated from the RFL as it dropped, however, presumably because the measurement sensor did not keep good track of pore-water pressure of the unsaturated pumice sample. The apparent friction angle was still falling even after the sample had slipped 20 m, though the apparent friction angle of 12.7° was closer to the average slope gradient measured in the field. Two possible reasons are considered to explain the differences between the field measurements and the laboratory test results. First, the slope gradient estimated in the field was obtained by assuming that the potential energy of the initial landslide mass was exhausted completely through friction on the slip surface. However, some considerable part of the energy must have been used to destroy dwellings as well as to deform the landslide mass. Second, the degree of saturation of the mass at failure may have been lower than those tested in the ring shear tests because the earthquake took place in the middle of the dry season in El Salvador. Nevertheless, the ring shear test of the water-saturated pumice soil shows clearly that the apparent friction angle can drop to as low as a few degrees after slipping short distances. These results indicate that the volcanic soil responsible for this landslide is particularly susceptible to failure during rainy seasons.

MICROTREMOR MEASUREMENT

Analysis of microtremors provided evidence of topographic amplification of ground motion. Such amplification probably contributed to initiation of the landslide. Microtremors were measured at both the top terrace behind the uppermost scar of the Las Colinas landslide (Figs. 3 and 4) and at the toe of the ridge. Fourier spectra of three components of the microtremor as well

Figure 22. Ring shear test results. A: Effective stress path. The residual failure line (RFL; C line) was obtained after TEST A by the following procedure: (1) The drainage valve on the top of the shear box was opened, and the pore pressure built up in the specimen was released. (2) Normal stress was then gradually decreased at a constant rate of 0.2 kPa/s to zero, keeping the constant shearing speed of 0.2 mm/s. B: Shear strength vs. shear displacement.

as ratios of these spectra (horizontal component/vertical component: H/V ratios hereafter) are shown in Figures 23 and 24. Thin lines correspond to four separate windows of 40.96 s opening, and the thick lines show the averages. At the top of the ridge, two peaks at 0.6 and 1.1 Hz (on the H/V diagram) are distinguished for the EW and NS components, while at the toe of the ridge, a 0.75 Hz component in the NS direction is predominant whereas no clear peak is observed for the EW component.

In order to evaluate the topographical effect of strong-motion amplification at Santa Tecla, the microtremors at the top and toe of Balsamo Ridge were compared. The Fourier spectra of the three components at the top of the ridge were divided by the corresponding components at the toe of the ridge. The

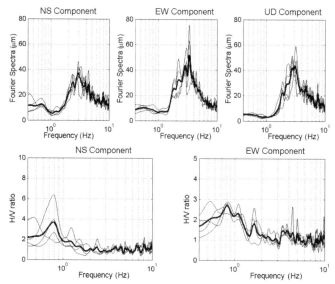

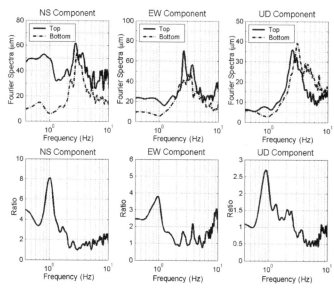

Figure 23. Fourier spectra of microtremors at the top of Balsamo Ridge. See text.

Figure 25. Ratios of Fourier spectra (top of the ridge / toe of the ridge). See text.

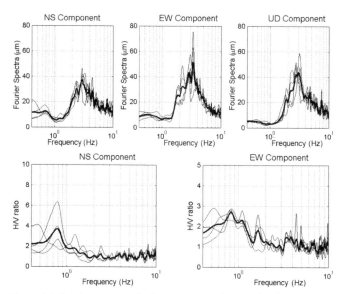

Figure 24. Fourier spectra of microtremors at the toe of Balsamo Ridge. See text.

results are shown in Figure 25. The three upper graphs depict the average of the Fourier spectra at the top and the toe of the ridge. The three lower graphs show the ratios of the average Fourier spectra (top of the ridge/toe of the ridge). The spectral ratios of all NS, EW, and even UD (vertical) components show clear peaks at ~1 Hz. The peak value of 8 is highest in the NS direction, the direction normal to the rim of the ridge, probably reflecting the topographical effect.

Figure 26 shows spectral acceleration and velocity for 5% damping for the NS component of strong ground motion recorded during the January 13 earthquake at Hospital San Rafael, ~1 km away from the Las Colinas landslide site. If we assume that Balsamo Ridge can be described as a single-degree-of-freedom system having a natural frequency of 1 Hz and a damping ratio of 5%–6% (8–10 times amplification), then peak accelerations and velocities of ~400–500 cm/s^2 and 90 cm/s may have been reached on this ridge during the January 13 earthquake.

SUMMARY AND RECOMMENDATIONS

The January 13 earthquake triggered hundreds of landslides in El Salvador, and most of those landslides were initiated at volcanoes and/or in thick volcanic sediments. Among the landslides triggered, the Las Colinas landslide was the most tragic. A large mass of soil (~150,000 m^3) was thrown off the rim of a mountain ridge rising behind the Las Colinas area of Nueva San Salvador. This amount of soil is not unusually large when compared with the 32–36 million m^3 landslide that failed at Mount Ontake, Japan, during the 1984 West Nagano earthquake (U.S. Geological Survey, 1999). The Las Colinas landslide, however, was devastating in terms of damage because the landslide mass surged ~400 m across a residential district, destroyed many houses, and killed more than 700 people.

Extensive cracks remaining along Balsamo Ridge suggest a serious threat of further slides. The city of Nueva San Salvador established and controlled potential hazard zones along the toe of Balsamo Ridge: a red zone having the highest risk and a yellow zone of slightly less risk. Within a week of the earthquake, ~14,000 residents were evacuated from these zones due to the

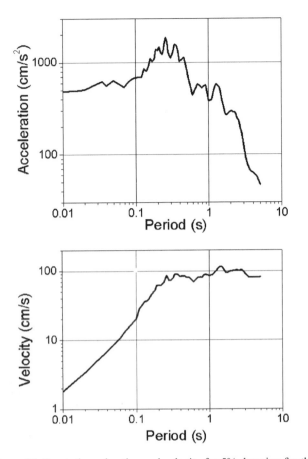

Figure 26. Spectral acceleration and velocity for 5% damping for the NS component of strong ground motion recorded during the January 13 earthquake at Hospital San Rafael, Las Colinas.

menace of possible landslides, and ~4000 refugees remained living in tents as late as February 3, 2001.

Countermeasures are necessary and should be taken because even flat-lying areas away from sudden changes in slope angle may still be within the possible reach of future landslides.

Landslides can range in size from small movements of loose debris to massive collapses of entire summits. For short- to medium-length slopes, the following measures will be effective for assessing and mitigating landslide hazards.

Short-Term Measures

1. Establish crack maps.
2. Cover or fill soil cracks to prevent water infiltration.
3. Continuously monitor crack openings.

Long-Term Measures

1. Remove partially detached soil masses.
2. Install preventive drainage works.

3. Anchor slopes.
4. Reinforce slopes with concrete crib arrangements.
5. Install piles.

Extremely large slope failures, however, are very difficult to mitigate. We strongly recommend, therefore, that governmental officials develop and enforce land-use building ordinances that regulate construction in areas susceptible to landslides and debris flows. Effective land-use ordinances for areas like Nueva San Salvador require knowledge of landslide travel distances and the structure of volcanic tuff and pumice deposits that underlie potential source areas. Such information allows scientists and engineers not only to simulate the possible travel distances and velocities of future landslides, but it is also useful for designing possible alarm systems.

Public education about landslide hazards will be also effective. If local people and governmental officials are familiar with the land around them, they may be able to detect early signs of landslide activity. The ring shear test of the water-saturated pumice soil presented here clearly shows that the apparent friction angle can drop down as low as a few degrees after slipping short distances. This indicates that the volcanic soil involved in the Las Colinas landslide is particularly susceptible to failure during intense rainfall. Any changes in patterns of water drainage should be watched carefully. Early signs of possible landslide activity include steadily developing cracks, spallation of rock fragments, or progressive tilting of trees. Such features are easy to observe, and they are pieces of information that the local people and authorities can use to prepare for a possible evacuation.

ACKNOWLEDGMENTS

The authors, members of the JSCE reconnaissance team, are indebted to the many Salvadorian and Japanese people who devotedly offered the team members every convenience during their investigation. The team was briefed on the entire scope of the earthquake-related damage by Jose Antonio Rodriguez, General Manager, Geothermal of El Salvador. Rodriguez, together with Salvador Handel, Geothermal of El Salvador, kindly arranged the team's reconnaissance trip and coordinated the schedules of specialists and officials at the Ministry of Environment, Ministry of Public Works, Ministry of Housing, and other organizations including Programa Salvadoreño de Investigación sobre Desarrollo y Medio Ambiente (PRISMA), Centro de Investigaciones Geotecnicas (CIG), and Universidad Centroamericana "Jose Simeon Canas" (UCA). Saburo Yuzawa, Ambassador, Japanese Embassy in El Salvador, kindly negotiated with ministers for every possible convenience that the Salvadorian counterparts could provide. The El Salvador military provided helicopter transport that allowed wide coverage of extensive areas affected by the earthquake. Yuzawa also coordinated an assembly for discussing the findings of the JSCE team. Lastly, all the members of the JSCE reconnaissance team would like to express their sincere sympathy to the people affected by the two devastating earthquakes in El Salvador.

REFERENCES CITED

Comité de Emergencia Nacional, 2001, http://www.gobernacion.gob.sv/Webcoen/COEN-SISNAE.htm (June 2003).

Guzman, P.A., 2000, Atlas de El Salvador: El Salvador Instituto Geografico Nacional, http://www.libroslatinos.com/cgi-bin/libros/CA21578?location=lca (June 2003).

Japan Geotechnical Society, 1995, Portable cone penetration test, Soil Investigation Method, p. 221–226 (in Japanese).

Konagai, K., Johansson, J., Mayorca, P., Uzuoka, R., Yamamoto, T., Miyajima, M., Pulido, N., and Duran, F., 2001a, Japan Society of Civil Engineers reconnaissance team provisional report on the January 13, 2001 earthquake occurred off the coast of El Salvador: http://www.jsce.or.jp/e/index.html (June 2003).

Konagai, K., Johansson, J., Mayorca, P., Uzuoka, R., Yamamoto, T., Miyajima, M., Pulido, N., and Duran, F., 2001b, Japan Society of Civil Engineers reconnaissance team report on the January 13, 2001 earthquake occurred off the coast of El Salvador: http://www.jsce.or.jp/e/index.html (June 2003).

Konagai, K., Johansson, J., Mayorca, P., Uzuoka, R., Yamamoto, T., Miyajima, M., Pulido, N., and Duran, F., 2001c, Las Colinas landslide caused by the January 13, 2001 earthquake: JSCE Symposium of Earthquake Engineering, v. 26, no. 1, p. 29–32.

Konagai, K., Johansson, J., Mayorca, P., Uzuoka, R., Yamamoto, T., Miyajima, M., Pulido, N., Sassa, K., Fukuoka, H., and Duran, F., 2002, Las Colinas landslide caused by the January 13, 2001 Off the Coast of El Salvador Earthquake: Journal of Japan Association for Earthquake Engineering, v. 2, no. 1, p. 1–15.

MOP (Ministerio de Obras Públicas), Instituto Geográfico Nacional, 1970, Topographic map, scale 1:5000. Nueva San Salvador, Topographic map, scale 1: 5000, http://clearinghouse.cnr.gob.sv/ign/metadatos/NSSalvador50k.html (June 2003).

MOP (Ministerio de Obras Públicas), Instituto Geográfico Nacional, 1984, Topographical map: Clearing House de datos geograficos de El Salvador, scale 1:50,000, http://clearinghouse.cnr.gob.sv/ign/invign50k.htm, (June 2003).

Sassa, K., 1996, Prediction of earthquake induced landslides, Special Lecture, *in* Senneset, K., ed., Landslides, Glissements de terrain, Proceedings of the 7th International Symposium on Landslides, Trondheim, Norway: Rotterdam, Balkema, v. 1, p. 115–132.

Sassa, K., 2000, Mechanism of flows in granular soils: Invited paper, Proceedings of GeoEng2000, Melbourne, v. 1, p. 1671–1702.

Sassa, K., Fukuoka, H., and Wang, G., 2002, Activities of "APERIF: Areal Prediction of Earthquake and Rain Induced Rapid and Long Traveling Flow Phenomena," *in* Fukuoka, H., ed., Proceedings of the Kansai Branch Symposium, Japan Landslide Society, ISBN 4-9900618-9-6 C3051, p. 61–78 (in Japanese).

U.S. Geological Survey (USGS), 1999, Earthquake and rainfall triggers landslide and lahar, Mt. Ontake Volcano, Japan, Sept. 14, 1984: http://volcanoes.usgs.gov/Hazards/What/Lahars/OntakeLahar.html http://www.consrv.ca.gov/cgs/rghm/ap/index.htm (June 2003).

Weber, H.S., Wiesemann, G., Lorenz, W., and Schmidt-Thome, M., 1978, Mapa geólogico de la República de El Salvador/América Central: Bundesanstalt füer Geowissenshaften und Rohstoffe, Hannover, Germany, scale 1:100,000, 6 sheets.

MANUSCRIPT ACCEPTED BY THE SOCIETY JUNE 16, 2003

Geological Society of America
Special Paper 375
2004

Geologic and engineering characterization of Tierra Blanca pyroclastic ash deposits

Reinaldo Rolo
Julian J. Bommer*
Department of Civil & Environmental Engineering, Imperial College London,
South Kensington Campus, London SW7 2AZ, UK

Bruce F. Houghton
Department of Geology and Geophysics, SOEST-University of Hawaii at Manoa,
1680 East-West Road, Honolulu, Hawaii 96822, USA

James W. Vallance*
Department of Civil Engineering, McGill University,
817 Sherbrooke St. West, Montréal, Québec, H3A 2K6, Canada

Panagiotis Berdousis
Christina Mavrommati
Imperial College London, South Kensington Campus, London SW7 2AZ, UK

William Murphy
School of Earth Sciences, University of Leeds, Leeds LS2 9JT, UK

ABSTRACT

The pyroclastic deposits, known as Tierra Blanca Joven, underlie most of metropolitan San Salvador and other areas surrounding Lake Ilopango. The Tierra Blanca Joven deposits are products of a complex sequence of pyroclastic flows and falls that occurred during the A.D. 430 eruption of Ilopango Caldera. Very fine, compact white ash-lapilli predominates in both flow and fall units. Laboratory tests carried out on high-quality, undisturbed Tierra Blanca Joven samples show negative pore-water pressures and weak cementation. They also reveal how the strength and compressibility of these sediments can change significantly when the suction and bonding are lost upon soaking or remolding. Thick Tierra Blanca Joven deposits contribute to landslide risk during heavy rainfalls and strong earthquakes in the region.

Keywords: volcanic deposition, pyroclastic deposits, Lake Ilopango Caldera, Tierra Blanca, chemical composition, mechanical and index properties, strength and deformation characteristics.

*Corresponding author: Bommer, Department of Civil & Environmental Engineering, Imperial College London, South Kensington Campus, London SW7 2AZ, UK; j.bommer@imperial.ac.uk. Present address, Vallance: U.S. Geological Survey, 1300 SE Cardinal Court, Bldg. 10, Suite 100, Vancouver, Washington, USA.

Rolo, R., Bommer, J.J., Houghton, B.F., Vallance, J.W., Berdousis, P., Mavrommati, C., Murphy, W., 2004, Geologic and engineering characterization of Tierra Blanca pyroclastic ash deposits, *in* Rose, W.I., Bommer, J.J., López, D.L., Carr, M.J., and Major, J.J., eds., Natural hazards in El Salvador: Boulder, Colorado, Geological Society of America Special Paper 375, p. 55–67. For permission to copy, contact editing@geosociety.org. © 2004 Geological Society of America.

INTRODUCTION

The young pyroclastic Tierra Blanca deposits outcrop over the area occupied by the capital city of San Salvador (Fig. 1). Many other important population centers in El Salvador, including Santa Tecla and San Vicente, also occupy sites blanketed with several meters of silicic ash and pumice belonging to the Tierra Blanca deposits.

In this paper we examine the geologic and geotechnical character of the Tierra Blanca soils from the perspective of their contribution to slope instability and erosion in El Salvador.

Tierra Blanca soils generally form a competent foundation material for buildings, but it is highly susceptible to erosion by runoff from the surrounding hills during the intense tropical rainfalls that are common between June and September. This process leads to the formation of narrow, deeply incised stream channels, locally known as barrancas. Many of these channels have been filled to provide land for development during the rapid and uncontrolled expansion of San Salvador in recent decades. The fill material generally consists of Tierra Blanca and frequently is inadequately compacted, leading to additional problems of erosion during heavy precipitation (Schmidt-Thomé, 1975). Pronounced settlements of these fills under earthquake shaking also occur (Chierazzi, 1987), sometimes leading to severe structural damage, as was the case at the Bachillerato building of the Colegio Guadalupano in San Salvador in the 1986 earthquake (Ledbetter and Bommer, 1987).

Tierra Blanca material has also been observed to be susceptible to liquefaction but this hazard is limited because of the deep water table in most urban areas (>35 m). Nonetheless, liquefaction and lateral spreading of Tierra Blanca did cause damage on the shores of Lake Ilopango in the earthquakes of 3 May 1965 and 13 January 2001 (Rosenblueth and Prince, 1966; Bommer et al., 2002; Orense et al., 2002).

The main engineering problem posed by the Tierra Blanca Joven deposit, however, is its susceptibility to landslides during heavy rainfall and earthquake ground shaking. Earthquakes in northern Central America invariably trigger large numbers of landslides in volcanic soils such as the Tierra Blanca Joven sequence (Bommer and Rodríguez, 2002). The geographi-cal extension of the areas affected by landslides triggered by earthquakes in the Central American region is not exceptional, but the number of landslides is generally greater by an order of magnitude or more than the numbers observed in earthquakes of similar size elsewhere in the world (Rymer, 1987). For example, the number of landslides triggered by the 1976 Guatemalan earthquake lies far above the count from other earthquakes in the database of Keefer (1984), just as the 1986 San Salvador earthquake stands out in the same way in the database of Rodríguez et al. (1999). These earthquake-induced landslides cause considerable disruption and damage. More than 200 people died as a direct result of landslides in the 1986 San Salvador earthquake, and more than three-quarters of the death toll in the 13 January 2001 earthquake was directly attributable to landslides.

Rymer and White (1989) have identified the areas of highest landslide hazard in El Salvador as being those where poorly consolidated volcanic tuffs such as the Tierra Blanca units are encountered. These hazardous areas include the coastal cordillera, the interior valley, and particularly the volcanic chain that runs through the valley, which coincide with the most densely populated regions of El Salvador, resulting from settlement patterns driven by coffee cultivation. The slides triggered on slopes formed within Tierra Blanca are often small—less than 50 m³ (Rymer, 1987)—particularly in near-vertical road cuts and on the banks of ravines (barrancas). However, their impact is severe as a result of the high population density and the tendency of rural migrants to the cities to settle at precarious sites. This problem is well illustrated in Figure 2, showing the comparison of homelessness resulting from numerous earthquakes to population growth.

As a part of a reliable landslide hazard assessment program for El Salvador, correlations are needed to relate observed slope failure patterns and the engineering-geology characteristics of the affected soils. The development of effective landslide risk mitigation measures, especially if these are to include slope stabilization, requires knowledge of the mechanical behavior of the soil. The objective of this paper is to provide data that contribute to the geological and geotechnical characterization of the Tierra Blanca Joven so as to improve understanding and mitigation of landslide hazards in El Salvador and neighboring countries.

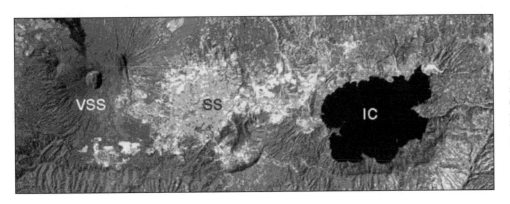

Figure 1. 1986 LANDSAT image showing the city of San Salvador (SS). Ilopango Lake occupies Ilopango Caldera (IC), the source of the Tierra Blanca tephras. VSS—Volcán San Salvador.

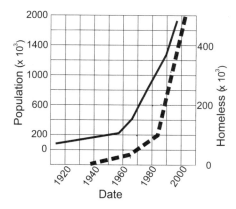

Figure 2. Population estimates (left, finer line) for San Salvador since 1917 and number of people left homeless (right, bolder line) as a result of the 1936, 1965, 1986, and 2001 El Salvador earthquakes.

GEOLOGICAL ORIGIN AND CHARACTER OF THE TIERRA BLANCA JOVEN TUFF SEQUENCE

Geographic and Geologic Setting

San Salvador (metropolitan population ~1.75 million) is the capital of El Salvador and the second largest city in Central America. San Salvador has been severely damaged by earthquakes and accompanying landslides 15 times since 1700 (Lardé-Larin, 1948; Harlow et al., 1993). The city is partially built on an active cone volcano, Volcán San Salvador to the west (Fig. 1), last active in 1917, and on Terra Blanca Joven tephras and ignimbrite, products of the much larger Ilopango Caldera to the east (Fig. 1), last active in 1889–90.

Ilopango Caldera is a part of the Central American volcanic front located in a graben system that bisects El Salvador (Meyer-Abich, 1956; Stoiber and Carr, 1973). This large (~100 km²) volcanic center is the most recently active, and perhaps the most dangerous, caldera in Central America. About 25 km to the WNW and ESE are the Volcán San Salvador (Boquerón) and San Vicente (Chichontepec) volcano complexes. Between Volcán San Salvador and Lake Ilopango (which occupies the Ilopango Caldera) a cluster of andesitic domes called Cerro San Jacinto rise ~500 m above the surrounding topography. All these are separate volcanic systems. Ilopango Caldera is dangerous because of its history of large, violent, explosive eruptions and its proximity to San Salvador (see Fig. 1). In the last 75,000 years, Ilopango Caldera has been the source of four exceptionally large and violent explosive Tierra Blanca eruptions that produced widespread tephra layers and ignimbrites.

Older Pyroclastic Units

There are at least four Tierra Blanca (TB) pyroclastic sequences (Hart and Steen-McIntyre, 1983), each separated

from the others by a brown to reddish-brown paleosol several meters thick. These sequences are informally called TB4, TB3, TB2, and TBJ (Tierra Blanca Joven), from oldest to youngest. All of the TB units have Ilopango Caldera as their source. Rose et al. (1999) obtained a date of 56,900 ± 2800 years B.P. on an ash layer that underlies the oldest of these (TB4). Thus, TB4 is younger than ca. 60 ka. Well-developed B horizons on each of the TB units but TBJ have prominent reddish brown color, clay development, and ped structures that suggest ~10,000 years of weathering (coloration, clay formation, and ped development below the organic layer are relative indications of age). The dacitic to rhyolitic Terra Blanca units alternate with weathered decimeter- to meter-thick andesitic ashes derived from the adjacent composite volcanoes.

The Terra Blanca Joven Eruption

The most recent of the Tierra Blanca eruptions has a calibrated age of A.D. 430 ± 20 (Dull et al., 2001). During this event, pyroclastic flows from Ilopango spread throughout and totally devastated the central Salvadoran highlands, and swept down steep valleys onto the Pacific coastal plain, all areas that were then populated by early Mayan civilization. Now, these areas are again densely populated. The flows extended 40 km from Lake Ilopango. In addition, ash fell across much of Central America, southern Mexico, and nearby parts of the Pacific Ocean and Gulf of Mexico (Fig. 3). The areal extent, thickness, and landslide susceptibility of TBJ make it one of the most hazardous geologic units in San Salvador and the surrounding areas.

Distribution and Character of TBJ Units

The deposits of the TBJ eruption are striking in their diversity, both in terms of the distribution and thickness of individual phases and in their material properties. Thin sheet-like products of sedimentation of ash and lapilli from the high eruption plumes alternate with valley-confined but much thicker pyroclastic-flow units. Some units are thin, well-sorted and highly permeable pumice-lapilli beds, others are thin to thick, low-permeability fine ashes, and yet others are thick, very poorly sorted mixtures of very fine ash- to block-sized particles. This diversity of volcanological and engineering properties creates difficulty in assessing the landslide hazard associated with the TBJ tephra. The total volume of these eruptives is variously estimated to range between 20 km³ and 50 km³ magma volume.

A general TBJ sequence (Fig. 4) from bottom to top comprises a few centimeters to decimeters of basal plinian-fall deposits, a sequence of pyroclastic flows and falls that varies from a few meters to a few tens of meters thick, a wet and dry fall sequence that ranges from tens of centimeters to a few meters thick, and a thick (meters to tens of meters) pyroclastic flow sequence with abundant fine ash becoming more common toward the top.

From the perspective of engineering properties, the lower and upper pyroclastic flow-and-fall sequences are of paramount

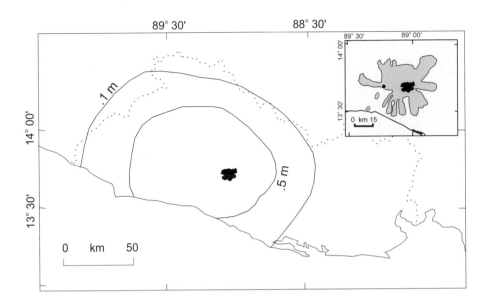

Figure 3. Extent and overall thickness of the ash deposits resulting from the most recent pyroclastic eruption (TBJ) of Ilopango Caldera. Inset shows approximate extent of pyroclastic flows (modified from Hart and Steen-McIntyre, 1983). Dot on inset shows approximate location where TBJ samples were recovered for laboratory testing.

Figure 4. Photograph of the Tierra Blanca Joven (TBJ) sequence showing the upper ignimbrite, the intermediate wet and dry fall deposits, the lower pyroclastic flow-and-fall sequence, the basal plinian fall deposit, and the underlying paleosol.

importance as these units are volumetrically largest and most likely to control slope failures. Basal fall deposits comprise fine to coarse ash and lapilli and are volumetrically insignificant (in San Salvador, generally only a few centimeters thickness of pumice lapilli); see Figure 4.

The lower pyroclastic unit comprises flow and fall facies of energetic pyroclastic eruptions. Lower pyroclastic flow units are rich in fine ash and cumulatively up to tens of meters thick in valleys near the caldera (Fig. 4) and up to several meters thick on intervening ridges. Intercalated between pyroclastic flows and at the top of the sequence are fine and very fine ash fall deposits derived from pyroclastic flows that are planar bedded and rich in well-formed concentrically banded accretionary lapilli (ash "hailstones"). In the vicinity of San Salvador, fine ash layers at the top of the sequence have cumulative thickness between 1 and 2 meters, thinning rapidly to the west.

Intermediate fall deposits comprise coarse ash to fine lapilli beds that were emplaced dry and alternating very fine ash beds, dominated by irregular-shaped, accretionary lapilli, and whose syndepositional deformation suggests wet emplacement. This unit ranges up to 3 m thick on the southwestern rim of the caldera and is 0.5 to 2 m thick in San Salvador.

The upper, main ignimbrite comprises multiple layers of very poorly sorted ash to block-sized pumice and dense "wall-rock" particles. Individual beds become thinner and much finer grained in the upper part of this unit. Intercalated fine ash beds (coignimbrite ash) with accretionary lapilli become increasingly common up-section. The ignimbrite has thickness maxima in the valley bottoms of 10 to 30 m. This unit is of major significance from a perspective of geotechnical investigations in San Salvador, not only because it is the most superficial but also because of its relatively significant thickness. The upper ignimbrite and associated coignimbrite fine ash beds have a broad radial distribution about the Ilopango Caldera consisting of fans or tongues of ignimbrite filling pre-TBJ drainages. The deposit becomes finer grained and thins to a meter-scale veneer on the interfluves.

Lateral Changes within the TBJ Units

From an engineering perspective, depositional variability in the TBJ deposits creates two distinct slope instability conditions. In paleovalleys radiating outward from Ilopango, thick ignimbrite forms fans up to several tens of meters thick, and slope failures may occur purely within this material. On paleo-interfluves pyroclastic flow units thin rapidly and the total TBJ sequence is only a few meters thick. For this geologic condition, slope failures often extend well below TBJ to affect most or all of the older Tierra Blanca succession.

Chemical Composition

Qualitative scanning electron microscope analyses by Mavrommati (2000) revealed that silicon (Si) is the dominant element in TB samples. The silicon content is at least an order of magnitude larger than most of the other components. Aluminum (Al), common in volcanic soils, is the second most abundant element followed by calcium (Ca), oxygen (O), and magnesium (Mg). This is consistent with the dominant component of the TBJ tephras being rhyolitic volcanic glass (74% SiO_2, Mann et al., this volume, Chapter 12). The chief minerals in these samples are plagioclase feldspar and subordinate hornblende (amphibole) and hypersthene. Other minerals include ilmenite, magnetite, and amorphous iron oxides/hydroxides. The latter may be a gel. These results agree with petrographic and mineralogical analyses by Amaya Dubón and Hayem Brevé (2000).

GEOTECHNICAL CHARACTERIZATION OF TIERRA BLANCA MATERIALS

The properties of samples of Tierra Blanca deposits presented herein are based primarily on a series of laboratory experiments carried out at Imperial College in London on three block samples recovered from around San Salvador (Rolo, 1998; Mavrommati, 2000; Berdousis, 2001). Additional information is taken from a report published in an engineering journal in El Salvador (Guzmán Urbina and Melara, 1996) and from some studies carried out by the Universidad Centroamericana "José Simeón Cañas" in San Salvador (e.g., Amaya Dubón and Hayem Brevé, 2000), in collaboration with whom this work has been developed.

Soil Sample Extraction

Block samples were recovered on three separate visits to El Salvador, from different locations around San Salvador, as shown in Table 1. Each of the samples is from the fine-grained top of the upper pyroclastic-flow sequence of the Tierra Blanca Joven deposit. These locations were chosen because they were of easy access and also because they had deposits thought to be representative of TBJ products. Coarse-grained facies of the upper pyroclastic sequence were not chosen because of the difficulty of collecting, and running tests on, such coarse-grained deposits.

The samples, whose dimensions were ~200 × 250 × 300 mm, were extracted using the technique described by Clayton et al. (1995), whereby a pit is first excavated and then the sample dug out before being wrapped in successive layers of burlap and wax; the first layer applied was plastic wrapping film. The samples were finally sealed in a layer of wax and then extracted directly into purpose-built plywood boxes, which were sealed with screws and then with insulating tape.

All three samples were extracted during the latter part of the dry season in El Salvador, which lasts from October to April or May, and therefore are likely to have had moisture contents close to the minimum values attained. Furthermore, the samples were extracted from depths of between 1.5 and 2.0 m, generally not more than 1 m in from a cut, this location being of relevance to slope stability problems as well as allowing the sample extraction to be carried out within a few hours. In all three cases the samples were stored, fully sealed, in a relatively stable environment until testing began in May or June of the same year as when the samples were extracted.

Index Properties

The index properties of Tierra Blanca as obtained from the block samples are summarized in Table 2.

TABLE 1. SAMPLES OF TIERRA BLANCA

Date of extraction	Location	Coordinates	
		N°	W°
April 1998	Car park, Universidad Centroamericana, San Salvador (Rolo, 1998)	13°40'52.1"	89°14'13.0"
April 2000	Los Pinares, Cantón El Carmen, Calle El Carmen, San Antonio Abad (Mavrommati, 2000)	13°42'53.0"	89°14'38.8"
February 2001	Cumbres de Cuscutlán, Santa Elena (Berdousis, 2001)	13°40'00.0"	89°14'51.6"

TABLE 2. INDEX PROPERTIES OF TIERRA BLANCA

Reference		Moisture content w(%)	Specific gravity G_s	Void ratio e	Porosity n(%)	Degree of saturation S_r (%)	Unit weight (kN/m³)
Berdousis (2001)		7.95	2.44	0.86	46.1	22.63	13.91
Mavrommati (2000)		9.92	2.50	1.07	51.7	23.44	13.02
Amaya Dubón and Hayem Brevé (2000)	San Antonio	20.88	2.26	0.79	44.0	59.81	14.98
	Bosques de Cuscatlán	29.48	2.29	1.12	53.0	60.11	13.70
Guzmán Urbina and Melara (1996)	minimum	16.70	2.43	0.78	47.5	43.00	10.80
	maximum	28.10	2.50	1.24	62.3	82.50	12.90
Rolo (1998)		7.63	N.D.*	0.97	49.2	21.80	13.70

*Not determined.

The soil typically has moisture contents that vary from 7.5 to 29.5% reflected by degrees of saturation of ~23%, but that can be up to 80%. These variations arise because the samples were extracted during different seasons; i.e., the wetter, more saturated, specimens examined in other studies were obtained during the rainy season.

The particle size distributions of the materials above are presented in Figure 5. According to these curves, Tierra Blanca can be classified as silty sand or sandy silt, although the proportion of fine particles varies considerably. However, Tierra Blanca is an extremely heterogeneous soil, and its index properties seem to depend quite significantly on the location from which the soil samples were extracted.

Suction Measurements

The water table in the San Salvador area is well below ground level. It can be deeper than 35 m even during the wet season. Negative pore-water tensions (suctions), therefore, develop in the desiccated sediments above the water table.

Soil suction (matrix suction) has been measured in block samples of Tierra Blanca using the filter paper method (Chandler and Gutiérrez, 1986; Crilly and Chandler, 1993) and the Imperial College suction probe (Ridley and Burland, 1994, 1995); see Table 3.

The samples contain different moisture contents, w (%), and this is partly reflected in the variations in negative pore-water pressures mentioned above. Tierra Blanca saturates at a

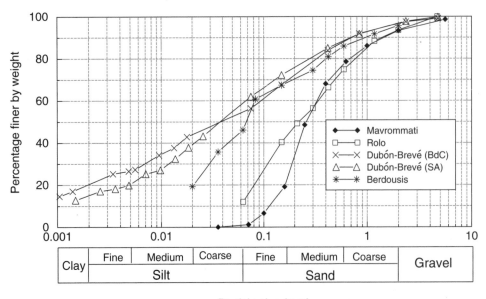

Figure 5. Grain size distribution of Tierra Blanca samples.

TABLE 3. SUCTION MEASUREMENTS OF TIERRA BLANCA (in kN/m²)

Reference	w(%)	Filter paper	Suction probe
Mavrommati (2000)	10.0	508	536
Amaya Dubón and Hayem Brevé (2000)	36.0	11	N.D.*
Berdousis (2001)	7.9	N.D.*	796
Rolo (1998)	7.6	335	456
*Not determined.			

moisture content of ~40%, which explains the low value of suction recorded by Amaya Dubón and Hayem Brevé (2000). It is also important to notice that even though the moisture contents of Rolo's (1998) samples were slightly lower than those of Mavrommati (2000), the suction measured by the latter was somewhat higher (even though both materials have similar grading curves). It is not clear what the explanation for this may be, but it could be related to differences in the soil particles (e.g., size, mineralogy), the soil structure, or even in the chemistry of the pore water. Also the filter paper measurements by Rolo (1998) seem to be unrealistically low when compared to those from the suction probe and to the ones obtained by Mavrommati (2000). In the authors' experience, there are experimental difficulties with the filter paper method, particularly when used on granular materials, which may well lead to such low values of suction. The large difference between the results of Rolo (1998) and Berdousis (2001), in spite of the similar values of moisture content, can be attributed to a larger proportion of finer particles in the sample tested by the latter.

Microfabric of Tierra Blanca

Small samples of Tierra Blanca were impregnated with epoxy resin and placed under the scanning electron microscope by Mavrommati (2000). Typical micrographs can be seen in Figure 6.

The open nature of the material is immediately apparent from these pictures. The grains are loosely packed, having large voids between them. No strong links among them can be observed, which suggests a metastable structure. The particles also seem to have predominantly angular shapes. Another dominant feature is the presence of vesicular pumice, a feature typical of silicic pyroclastic deposits. Recent measurements carried out by the third author at the University of Hawaii on samples from other TBJ subunits show that the mean vesicularity/porosity of the particles ranges from 53 to 71%.

In general the pores and particles appear to have a random distribution. The difference in the shade of the different areas in Figure 6 is related to the atomic mass of the material infilling the space, i.e., void spaces filled with air appear black, dense iron-titanium oxides appear bright white.

Double Odometer Tests

Burland (1990) has shown that natural materials may posses a structure such that their normal compression curve lies to the right of their intrinsic or reconstituted counterpart. Strength due to postdepositional processes such as ageing, leaching, and chemical bonding or cementation (Mitchell, 1992; Leroueil and Vaughan, 1990) causes this particular structure. As explained by Burland (1965), suction can have an effect similar to that of the cementation.

In order to study the individual effects of suction and cementation on the compressibility of TBJ, the double odometer test described by Jennings and Knight (1957) has been used. The idea is to utilize a set of intact and reconstituted samples at the natural and saturated moisture contents that can be subjected to consolidation tests. Rolo (1998) and Mavrommati (2000) have tested high-quality (block) samples of Tierra Blanca (Fig. 7). The lower

Figure 6. Scanning electron micrographs of two samples of Tierra Blanca.

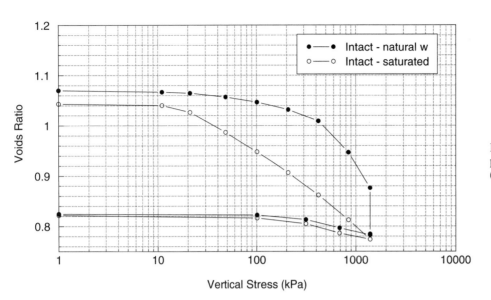

Figure 7. Compressibility of intact samples of TB at natural moisture content (w) and saturated.

curve in Figure 7 represents the saturated, i.e., zero suction, intact samples (soaked at zero vertical stress), whereas the higher line shows the compression curve of the intact material at its natural moisture content. This latter sample was soaked at a vertical total stress of 1386 kPa.

Figure 8 shows similar results for reconstituted samples formed under comparable initial conditions to that of the intact specimens (i.e., at natural moisture content and saturated). The results show similar features to those explained in Jennings and Burland (1962). The unsaturated specimens present a stronger structure that enables them to reach higher stress states at the same density as that of the saturated samples. Moreover, the unsaturated

samples collapse abruptly upon soaking, almost reaching the saturated compression line and following it upon unloading.

The collapse potential (how large a deformation can be expected) of a soil upon inundation can be estimated from the double odometer test by means of the following relationship:

$$CP(\%) = \frac{e_n - e_s}{e_n} * 100 \tag{1}$$

where e_n and e_s are the void ratios of the material at the natural moisture content and in the saturated state at any given vertical stress.

Collapse potential, CP (see Fig. 9), of the undisturbed and remolded materials are similar at low stresses, but they diverge

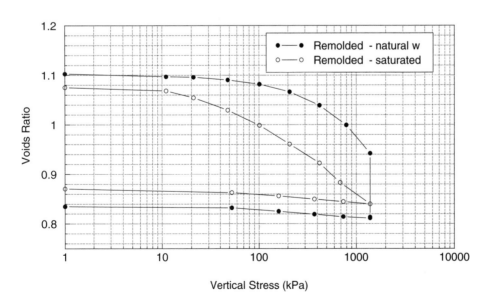

Figure 8. Compressibility of remolded samples at natural moisture content (w) and in the saturated state.

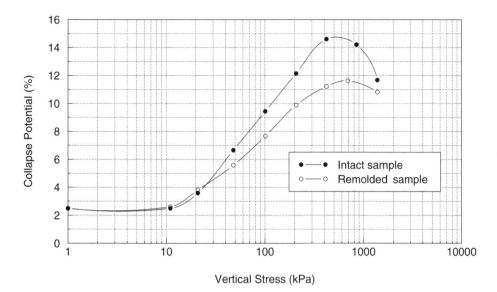

Figure 9. Collapse potential obtained from double odometer tests.

somewhat as the stress level increases. The additional strength of the intact material may arise from cementation at interparticle contacts.

The maximum collapse potential observed for the undisturbed sample is ~15% at 421 kPa. At a more typical level of stress for these deposits (~100–200 kPa), the CP is ~9–12%. The volume losses upon saturation occur suddenly and could cause the large number of soil slumps occurring during heavy rainfalls in El Salvador. After reaching the maximum value of 15%, the collapse potential starts to decrease as the larger stresses reduce the effects of suction and cementation on structure. Furthermore, at large stresses particle breakage may take place as the soil starts to reach its normal compression line (Coop and Lee, 1993).

Shear Strength of Tierra Blanca

Direct shear box tests have been carried out on undisturbed and reconstituted samples of Tierra Blanca materials (Mavrommati, 2000; Berdousis, 2001). A similar approach to that of the double odometer tests was adopted, whereby saturated and natural samples were tested so as to compare the isolated effects of suction and cementation. Representative failure envelopes from these tests are shown in Figure 10. The undisturbed sample at natural moisture content presented an angle of shearing resistance of $\phi' = 39°$ with an apparent cohesion, c, of 30 kPa (Fig. 10). These strength parameters rapidly decreased when an equivalent saturated specimen was sheared, showing $\phi' = 34°$ and c = 0 kPa. The remolded saturated sample showed strength parameters similar to the undisturbed saturated sample ($\phi' = 35°$ and c = 0 kPa). The remolded sample at natural moisture content revealed a slightly higher ϕ' of 36° with a cohesion intercept of c = 10 kPa (Fig. 10). The higher values of the angle of shear resistance and apparent cohesion compared to the remolded sample at natural moisture content (Table 4) suggest some effects of natural structure and

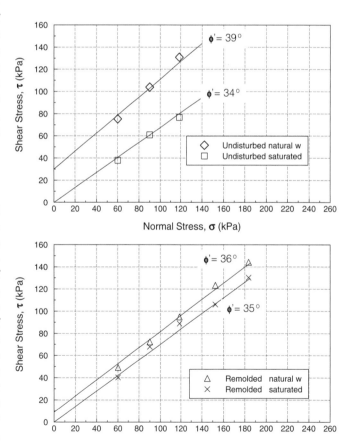

Figure 10. Failure envelopes for intact and remolded samples at natural moisture content and saturated.

cementation. It is evident that the cementing agent in the natural, undisturbed sample contributes quite significantly to shear strength of the material by increasing its angle of shearing resistance. This cementation appears to be rather weak, however, and its effects are lost upon both wetting and remolding.

Negative pore-water pressures appear to cause some of the apparent cohesion observed in the intact sample (with a cohesion 20 kPa larger than that of the remolded sample at natural moisture content). The soil suction contributes to this component of strength rather than to the angle of shearing resistance.

Figure 11 presents representative stress–strain curves for the four cases discussed before (i.e., intact and reconstituted samples at natural moisture content and saturated). The intact sample at natural moisture content shows a clear peak stress ratio and a stiffer response than the other samples. The saturated and remolded samples do not show a peak. The response of the intact specimen is brittle and the post-rupture strength converges, at large deformations, with that of the intact and remolded saturated samples. A slight tendency to dilate accompanied the increased strength of the intact sample (suggesting some kind of particle interlocking), whereas a net compressive deformation occurred in the other specimens.

The shear strength of the intact and remolded saturated samples converge at large deformations, but the normalized strength of the reconstituted unsaturated sample is slightly higher than that of the other specimens at large deformations (Fig. 11). The apparent cohesion generated by the negative pore-water pressures is responsible, in part, for the greater shear strength of this latter sample. All the normalized strengths would be expected to converge at larger deformations, however.

Amaya Dubón and Hayem Brevé (2000) report somewhat different values for strength parameters for TBJ. They found that samples obtained from the San Antonio Abad area had c = 6 kPa, $\phi' = 35°$, whereas samples from Cumbres de Cuscatlán showed c = 1.2 kPa and $\phi' = 30°$. Nevertheless, their Tierra Blanca samples contained significantly larger proportions of fine particles as well as higher moisture contents (Table 2 and Fig. 5), which may explain the lower values of ϕ' and c. Bernal Riosalido (2002) found from triaxial tests on undisturbed samples that c = 70–90 kPa and $\phi' = 34°$, but his samples also had a large proportion of fine particles (40 to 75%).

Two additional direct shear tests carried out at higher normal stresses (400 and 500 kN/m²) on undisturbed, unsaturated samples showed that increasing the stress level precluded increased shear strength (see Figure 12). Any potential curvature of the failure envelope at the higher stress level reflecting nonlinearity would have to be unrealistically marked to explain the decay in the angle of shearing resistance observed at the highest stresses. For a curved failure envelope to reasonably fit the experimental data shown in Figure 12, the value of ϕ' toward the higher stress region would have to be ~27°, which seems too low compared to the reconstituted saturated samples.

Another explanation for the lower value of ϕ' at higher stresses could be that since there was not a clear peak in the

TABLE 4. SUMMARY OF STRENGTH PARAMETERS
OF TIERRA BLANCA

	c (kPa)	Φ' (°)
Intact samples, natural moisture content		
This study (vertical stress <400 kPa)	30	39
This study (vertical stress ≥400 kPa)	0	35
Amaya Dubón & Hayem Brevé (2000)		
San Antonio	6	35
Cumbres de Cuscatlán	1.2	30
Bernal Riosalido (2002)	70–90	34
Intact samples, saturated	0	34
Remolded samples, natural moisture content	10	36
Remolded samples, saturated	0	35

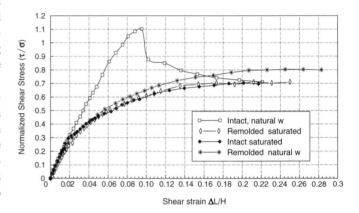

Figure 11. Stress–strain curves for intact and remolded samples.

stress–strain response during the tests at higher normal stresses, the effects of any cementation or possible strength owing to negative pore-water pressures may have been suppressed, resulting in a failure envelope similar to that of the reconstituted saturated samples. However, further tests are required at the intermediate stress level to clarify this point.

One important observation from the above results is the remarkably uniform strength characteristics displayed by the Tierra Blanca samples, despite their apparent significant heterogeneity in other respects. Bommer et al. (2001) have shown that the strength values presented herein appear to reflect in situ conditions, and the variation observed due to loss of suction and cementation do indeed induce instability in typical slopes. However, we do not recommend that engineering analyses and design be conducted without site-specific geotechnical investigation.

DISCUSSION AND CONCLUSIONS

The higher values of shear strength observed in the intact samples at natural moisture content (compared to the recon-

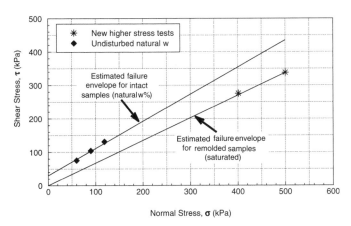

Figure 12. Drained shear strength of undisturbed samples at natural moisture content at the higher 400 and 500 kPa normal stresses. Also showing the failure envelopes of the intact and reconstituted samples at natural moisture content and saturated, respectively.

stituted samples) can be attributed to natural structure and soil fabric. Negative pore-water pressures also make a small contribution to the shear strength, as they generate an apparent cohesion intercept on the Mohr-Coulomb failure envelope. However, the reduction in the angle of shearing resistance upon inundation and an increased cohesive component of strength in the natural undisturbed sample seem to indicate that TB could have some weak interparticle bonding or cementation.

Cementation has been observed in many natural geomaterials, including soft and stiff clays, sands, residual soils, and weak or weathered rocks (Leroueil and Vaughan, 1990). This cementation or bonding may arise from a single agent or a combination of different agents such as dissolution and precipitation of silica, iron oxides, carbonates, organic matter, recrystallization of minerals during weathering, cold-welding of interparticle contacts in sands due to high pressures, and modification of the adsorbed water and interparticle forces in clayey soils (Leroueil and Vaughan, 1990; Clough et al., 1981). In the case of Tierra Blanca, any cementing agent at interparticle contacts is more likely to have developed due to chemical processes rather than mechanical ones, which generally require long periods of time and large stresses to develop. It could be possible that silica viscous agents act as a cementing vehicle in TB as a result of chemical weathering and aging. It is, however, very difficult to evaluate with certainty the nature of the cementing agent because it cannot be easily identified with techniques such as petrologic thin sections, scanning electron microscope images, or X-ray diffraction, and normal chemical analyses would only reveal the overall material composition. The presence of silica cementing agents has been noticed in different soils by Mitchell and Solymar (1984), Leroueil and Vaughan (1990), Belloni and Morris (1991), and Butenuth et al. (1995), who have also recognized their contribution to the strength and stability of geomaterials.

The effects of various degrees of cementation have been studied by several authors in both natural and reconstituted materials (Leroueil and Vaughan, 1990; Clough et al., 1981; Lade and Overton, 1989; Mitchell and Solymar, 1984; Cuccovillo and Coop, 1999). These studies corroborate the observations from the TB experiments, with cementation providing additional strength and stiffness to the soil mass, giving it the capacity to reach stress states that would otherwise be unsustainable. Near-vertical slopes observed in Central America and other parts of the world present a clear example of the effects of cementation and suction within the soil matrix (Bommer and Rodríguez, 2002; Sitar, 1990; Sitar and Clough, 1983).

Various authors have studied volcanic soils from other regions of the world (see Bommer et al., 2001). Yamanouchi and Murata (1973) studied sediments (so-called "Shirasu") originating from Pleistocene volcanic eruptions that are widely distributed in the southern Kyushu region of Japan. O'Rourke and Crespo (1988) show some results from a study of the properties of a volcaniclastic formation that is found in the Andes of Ecuador and Colombia, known as Cangahua. González de Vallejo et al. (1981) presented a study of a volcanic soil located in the northeastern part of Tenerife, Spain. Belloni and Morris (1991) performed a study on the behavior of volcanic debris from the northeast side of Ecuador in South America that represents the final phases of the explosive activity of volcanoes. In all these studies it was observed that volcanic deposits have a metastable, open structure and relatively low moisture contents, which lead to negative pore-water pressures within the soils. Additionally, rather weak cementation seems to be present in most volcanic deposits reviewed. Most authors agree that these two factors play an important role in the stability of slopes formed by these deposits and that they have invariably shown a tendency to fail during heavy rainfall and seismic activity.

However, slope failures in volcanic ashes in other regions of the world do not seem to occur with the same frequency as those observed in El Salvador, or at least the impact is not as intense. Possible factors contributing to this may be the high seismic activity and marked periods of intense rainfall in El Salvador. Another important factor contributing to the greater hazard posed by the TB sediments arises due to the concentration of population in the areas where these deposits are widespread (e.g., in San Salvador) and to the sometimes inadequate construction practices adopted.

The present work has highlighted various features that appear to contribute to the strength of Tierra Blanca. Nevertheless, further research is necessary to study landslide hazard in Central America due to rainfall and seismic activity. In situ measurements of soil suction, assessment of slope geomorphology, more sophisticated laboratory tests to measure shear strength and stiffness, and detailed static and dynamic slope stability analyses would help to identify the potential variations in the strength of this soil and the effects of such variations in the stability of these deposits. This information would ultimately result in safer and more economical engineering design.

ACKNOWLEDGMENTS

The authors are indebted to many individuals for their assistance with different parts of the work. Patricia Méndez de Hasbun, Arturo Escalante, Carlos Amaya Dubón, Enrique Hayem, José Mauricio Cepeda, and Joseph Mankelow all assisted with the recovery of the soil samples from El Salvador. Steve Ackerley, Kieran Dineen, Andrew Ridley, Julio Colmenares, Mónica Melgarejo, Graham Keefe, Rehne Yang, Nick Buenfeld, and Geoff Fowler all contributed in one way or another to the series of laboratory tests. We are also grateful to Michael de Freitas, Matthew Coop, Richard J. Chandler, Christine Butenuth, Sarada K. Sarma, and Ken Walsh for sharing their experience and insights at different stages of the work. Funding for parts of this work was generously provided by the Commission of European Communities, the Royal Academy of Engineering, and the Natural Environment Research Council. Houghton and Vallance were supported by a grant from the U.S. National Science Foundation, grant number EAR 9903291.

REFERENCES CITED

Amaya Dubón, C.A., and Hayem Brevé, E.A., 2000, Introducción al estudio de los suelos parcialmente saturados e inicio de la caracterización de la Tierra Blanca del AMSS [B.S. thesis]: San Salvador, Universidad Centroamericana, 366 p.

Belloni, L., and Morris, D., 1991, Earthquake-induced shallow slides in volcanic debris soils: Géotechnique, v. 41, no. 4, p. 539–551.

Berdousis, P., 2001, Engineering characterisation of a volcanic soil from Central America [M.S. thesis]: London, Imperial College, University of London, 205 p.

Bernal Riosalido, A., 2002, Estabilidad de taludes en terremotos: El deslizamiento de Las Colinas, El Salvador, en el terremoto de 13 de Enero de 2001, *in* Proceedings, Segundo Congreso Iberoamericano de Ingeniería Sísmica, Madrid: Asociación Española de Ingeniería Sísmica, p. 781–791.

Bommer, J.J., and Rodríguez, C.E., 2002, Earthquake-induced landslides in Central America: Engineering Geology, v. 63, no. 3/4, p. 189–220.

Bommer, J.J., Rolo, R., Mitroulia, A., and Berdousis, P., 2001, Geotechnical properties and seismic slope stability of volcanic soils, *in* Proceedings, 12th European Conference on Earthquake Engineering, London: Oxford, UK, Elsevier, Paper no. 695.

Bommer, J.J., Benito, M.B., Ciudad-Real, M., Lemoine, A., López-Menjivar, M.A., Madariaga, R., Mankelow, J., Méndez de Hasbun, P., Murphy, W., Nieto-Lavo, M., Rodríguez-Pineda, C.E., and Rosa, H., 2002, The El Salvador earthquakes of January and February 2001—Context, characteristics and implications for seismic risk: Soil Dynamics and Earthquake Engineering, v. 22, no. 5, p. 389–418.

Burland, J.B., 1965, Some aspects of the mechanical behaviour of partly saturated soils, *in* Aitchison, G.D., ed., Symposium-in-print, Moisture Equilibrium and Moisture Changes in Soils Beneath Covered Areas: Sydney, Australia, Butterworths, p. 270–278.

Burland, J.B., 1990, On the compressibility and shear strength of natural soils: Géotechnique, v. 40, no. 3, p. 329–378.

Butenuth, C., Frey, M.L., de Freitas, M.H., Passas, N., and Forero-Duenas, C.A., 1995, Silica gels—A possible explanation for slope failures in certain rocks, *in* Maund, J.G., and Eddleston, M., eds., Proceedings, Engineering Group of the Geological Society of London: Coventry, 31st Annual Conference Engineering Group Geological Society, p. 185–191.

Chandler, R.J., and Gutiérrez, C.I., 1986, The filter paper method of suction measurement: Géotechnique, v. 36, p. 265–268.

Chierazzi, R., 1987, The San Salvador earthquake of October 10, 1986—Geotechnical effects: Earthquake Spectra, v. 3, no. 3, p. 483–489.

Clayton, C.R.I., Matthews, M.C., and Simons, N.E., 1995, Site investigation (2nd edition): Oxford, Blackwell Science, 584 p.

Clough, G.W., Sitar, N., and Bachus, R.C., 1981, Cemented sands under static loading: Journal of the Geotechnical Engineering Division, American Society of Civil Engineers, v. 107, no. GT6, p. 799–818.

Coop, M.R., and Lee, I.K., 1993, The behaviour of granular soils at elevated stresses, *in* Houlsby, G.T., and Schofield, A.N., eds., Proceedings, Wroth Memorial Symposium: Predictive Soil Mechanics: London, Thomas Telford, p. 186–198.

Crilly, M.S., and Chandler, R.J., 1993, A method of determining the state of desiccation in clay soils: Waterford, UK, Building Research Establishment (BRE) information paper, 4/93 (IP4/93), 4 p.

Cuccovillo, T., and Coop, M.R., 1999, On the mechanics of structured sands: Géotechnique, v. 49, no. 6, p. 741–760.

Dull, R.A., Southon, J.R., and Sheets, P., 2001, Volcanism, ecology, and culture—A reassessment of the Volcan Ilapongo TBJ eruption in the southern Maya realm: Latin American Antiquity, no. 12, p. 25–44.

González de Vallejo, L.I., Jiménez Salas, J.A., and Leguey Jiménez, S., 1981, Engineering geology of the tropical volcanic soils of La Laguna, Tenerife: Engineering Geology, v. 17, no. 1–2, p. 1–17.

Guzmán Urbina, M.A., and Melara, E., 1996, Propiedades ingenieriles del suelo del área metropolitana de San Salvador, El Salvador: Revista ASIA—Asociación Salvadoreña de Ingenieros y Arquitectos, San Salvador, El Salvador, no. 122, p. 14–22.

Harlow, D.H., White, R.A., Rymer, M.J., and Alvarez, G.S., 1993, The San Salvador Earthquake of 10 October 1986 and its historical context: Bulletin of the Seismological Society of America, v. 83, p. 1143–1154.

Hart, W.J., and Steen-McIntyre, V., 1983, Terra Blanca tephra, *in* Sheets, P.D., ed., Archeology and volcanism in Central America: Austin, University of Texas Press, p. 14–34.

Jennings, J.E.B., and Burland, J.B., 1962, Limitations to the use of effective stresses in partly saturated soils: Géotechnique, v. 12, p. 125–144.

Jennings, J.E.B., and Knight, K., 1957, The prediction of total heave from the double oedometer test: Transport Society, African Institute of Civil Engineers, no. 7, p. 285–291.

Keefer, D.K., 1984, Landslides caused by earthquakes: Geological Society of America Bulletin, v. 95, p. 406–421.

Lade, P.V., and Overton, D.D., 1989, Cementation effects in frictional materials: Journal of Geotechnical Engineering, American Society of Civil Engineers, v. 115, no. GT10, p. 1373–1377.

Lardé-Larin, J., 1948, Génesis del volcán del Playón: Volcanológica Salvadoreña, Revista del Ministerio de Cultura, v. 7, no. 24, p. 101–111.

Ledbetter, S.R., and Bommer, J.J., 1987, The San Salvador earthquake of 10 October 1986: A field report by Earthquake Engineering Field Investigation Team, London, The Institution of Civil Engineers, 99 p.

Leroueil, S., and Vaughan, P.R., 1990, The general and congruent effects of structure in natural soils and weak rocks: Géotechnique, v. 40, no. 3, p. 467–488.

Mavrommati, Z.C., 2000, Seismic behaviour of slopes in an unsaturated volcanic soil [M.S. thesis]: London, Imperial College, University of London, 116 p.

Meyer-Abich, H., 1956, Los volcanes activos de Guatemala y El Salvador, Anales del Servicio Geológico Nacional de El Salvador, no. 3, p. 67–70.

Mitchell, J.K., 1992, Fundamentals of soil behaviour: New York, Wiley, 437 p.

Mitchell, J.K., and Solymar, Z.V., 1984, Time-dependent strength gain in freshly deposited or densified sand: Journal of the Geotechnical Division, American Society of Civil Engineers, v. 110, no. GT11, p. 1559–1576.

Orense, R., Vargas-Monge, W., and Cepeda, J., 2002, Geotechnical aspects of the January 13, 2001 El Salvador earthquake: Soil and Foundations, v. 42, no. 4, p. 57–68.

O'Rourke, T.D., and Crespo, E., 1988, Cemented volcanic soils: Journal of Geotechnical Engineering, American Society of Civil Engineers, v. 114, no. 10, p. 1126–1147.

Ridley, A.M., and Burland, J.B., 1994, A new instrument for the measurement of soil moisture suction: Géotechnique, v. 43, no. 2, p. 321–324.

Ridley, A.M., and Burland, J.B., 1995, A pore pressure probe for the in situ measurement of soil suction, *in* Craig, C., ed., Advances in site investigation practice: London, Thomas Telford, p. 510–520.

Rodríguez, C.E., Bommer, J.J., and Chandler, R.J., 1999, Earthquake-induced landslides, 1980–1997: Soil Dynamics and Earthquake Engineering, v. 18, p. 325–346.

Rolo, R., 1998, Elements of seismic hazard in El Salvador and Central America [M.S. thesis]: London, Imperial College, University of London, 205 p.

Rose, W.I., Conway, F.M., Pullinger, C.R., Deino, A., and McIntosh, W.C., 1999, An improved age framework for late Quaternary silicic eruptions in northern Central America: Bulletin of Volcanology, v. 61, p. 106–120.

Rosenblueth, E., and Prince, J., 1966, El temblor de San Salvador, 3 de Mayo 1965: Ingeniería (Journal by the Instituto de Ingeniería of the Universidad Nacional de México), v. 36, p. 31–58.

Rymer, M.J., 1987, The San Salvador earthquake of October 10, 1986—Geologic aspects: Earthquake Spectra, v. 3, no. 3, p. 435–463.

Rymer, M.J., and White, R.A., 1989, Hazards in El Salvador from earthquake-induced landslides, *in* Brabb, E.E., and Harrod, B.L., eds., Landslides—Extent and economic significance: Rotterdam, Balkema, p. 105–109.

Schmidt-Thomé, M., 1975, The geology in the San Salvador area (El Salvador, Central America), a basis for city development and planning: Geologisches Jahrbuch, v. 13, p. 207–228.

Sitar, N., 1990, Seismic response of steep slopes in weakly cemented sands and gravels, *in* Duncan, J.M., ed., Proceedings, Memorial H. Bolton Seed symposium: Vancouver, Canada, v. 2, p. 67–82.

Sitar, N., and Clough, G.W., 1983, Seismic response of steep slopes in cemented soils: Journal of Geotechnical Engineering, American Society of Civil Engineers, v. 109, no. GT2, p. 210–227.

Stoiber, R.E., and Carr, M.J., 1973, Quaternary volcanic and tectonic segmentation of Central America: Bulletin of Volcanology, v. 37, p. 304–325.

Yamanouchi, T., and Murata, H., 1973, Brittle failure of a volcanic ash soil "Shirasu," *in* Proceedings, 8th International Conference on Soil Mechanics and Foundation Engineering, Moscow: USSR National Society for Soil Mechanics and Foundation Engineering, v. 1, p. 495–500.

MANUSCRIPT ACCEPTED BY THE SOCIETY JUNE 16, 2003

Geological Society of America
Special Paper 375
2004

Landslides triggered by the 13 January and 13 February 2001 earthquakes in El Salvador

Randall W. Jibson*
Anthony J. Crone
Edwin L. Harp
Rex L. Baum
U.S. Geological Survey, Box 25046, MS 966, Denver Federal Center, Denver, Colorado 80225, USA

Jon J. Major
U.S. Geological Survey, 1300 SE Cardinal Court, Suite 100, Vancouver, Washington 98683, USA

Carlos R. Pullinger
C. Demetrio Escobar
Mauricio Martínez
*Servicio Nacional de Estudios Territoriales, Km 5 1/2 Carretera a Santa Tecla,
Avenida Las Mercedes, Edificio ISTA, San Salvador, El Salvador*

Mark E. Smith
U.S. Geological Survey, Box 25046, MS 415, Denver Federal Center, Denver, Colorado 80225, USA

ABSTRACT

During a one-month period in early 2001, El Salvador experienced two devastating earthquakes. On 13 January, a M-7.7 earthquake centered ~40 km off the southern coast in the Pacific Ocean caused widespread damage and fatalities throughout much of the country. The earthquake triggered thousands of landslides that were broadly scattered across the southern half of the country. The most damaging landslide, a rapidly moving mass of ~130,000 m³, occurred in the Las Colinas neighborhood of Santa Tecla, where ~585 people were killed. Another large landslide (~750,000 m³) near the city of San Vicente blocked the Pan-American Highway for several weeks. One month later, on 13 February, a M-6.6 earthquake occurred ~40 km east-southeast of San Salvador and triggered additional thousands of landslides in the area east of Lake Ilopango. The landslides were concentrated in a 2500 km² area and were particularly abundant in areas underlain by thick deposits of poorly consolidated, late Pleistocene and Holocene Tierra Blanca rhyolitic tephras erupted from Ilopango caldera. Most of the triggered landslides were relatively small, shallow failures, but two large landslides occurred that blocked the El Desagüe River and the Jiboa River.

The two earthquakes triggered similar types of landslides, but the distribution of triggered landslides differed because of different earthquake source parameters. The large-magnitude, deep, offshore earthquake triggered broadly scattered landslides over a large region, whereas the shallow, moderate-magnitude earthquake centered within the country triggered a much smaller, denser concentration of landslides. These results are significant in the context of seismic-hazard mitigation for various earthquake scenarios.

Keywords: seismic hazards, landslides, El Salvador, landslide dams.

*jibson@usgs.gov

Jibson, R.W., Crone, A.J., Harp, E.L., Baum, R.L., Major, J.J., Pullinger, C.R., Escobar, C.D., Martínez, M., and Smith, M.E., 2004, Landslides triggered by the 13 January and 13 February 2001 earthquakes in El Salvador, *in* Rose, W.I., Bommer, J.J., López, D.L., Carr, M.J., and Major, J.J., eds., Natural hazards in El Salvador: Boulder, Colorado, Geological Society of America Special Paper 375, p. 69–88. For permission to copy, contact editing@geosociety.org. © 2004 Geological Society of America

INTRODUCTION

During a one-month period in January and February 2001, El Salvador experienced two major earthquakes that caused widespread damage and fatalities throughout much of the country. On 13 January 2001, a moment-magnitude (**M**)-7.7 earthquake occurred ~40 km off the southern coast of El Salvador at a depth of ~60 km beneath the Pacific Ocean (Fig. 1). Exactly one month later, on 13 February 2001, a **M**-6.6 earthquake occurred ~40 km east-southeast of San Salvador at a depth of 15 km. Both earthquakes triggered thousands of landslides over a large part of the country. These landslides, most of which were fairly shallow (<5 m), caused most of the earthquake fatalities, destroyed hundreds of structures, blocked roads, and dammed or clogged streams and rivers. Remarkably, this widespread landslide damage was triggered during El Salvador's dry season following a below-average rainfall year (Evans and Bent, this volume, Chapter 3); had the earthquakes occurred during the wet season, landslide damage probably would have been considerably worse.

Landslides are a common occurrence in the geologic setting of El Salvador, a country situated in a very active tectonic environment where the Cocos plate is being subducted beneath the Caribbean plate (Bommer et al., 1996). Rising magma created by the subduction process has produced a chain of active volcanoes throughout El Salvador, which have erupted episodically and deposited widespread, poorly consolidated tephra in many parts of the country. The combination of active tectonism, steep topography, relatively young, weak volcanic rocks, a warm, humid, subtropical climate with heavy rains, and relatively frequent moderate and large earthquakes all contribute to an environment prone to landsliding.

Some tephra deposits that are widespread in El Salvador (described subsequently) are particularly susceptible to landsliding during earthquakes. In fact, Bommer et al. (2002) found that, for a given earthquake magnitude, both the numbers of seismically triggered landslides and the areas affected by those landslides tend to be greater in the geologic environment of El Salvador than in different geologic, geomorphic, and climatic environments. Historical records show that El Salvador has suffered damage from earthquake-triggered landslides in at least ten earthquakes since 1857 (Rymer and White, 1989).

Soon after the 2001 earthquakes, the U.S. Geological Survey (USGS) responded to requests for technical assistance by the government of El Salvador and dispatched teams of scientists to El Salvador to work with Salvadoran colleagues in making rapid hazard assessments and mitigation recommendations. Bommer et al. (2002) likewise investigated and documented post-earthquake effects, including landslides. Preliminary reconnaissance reports and hazard assessments by the USGS teams were published soon after the earthquakes (Baum et al., 2001; Harp and Vallance, 2001; Jibson and Crone, 2001). This paper draws on these early reports to briefly describe the source parameters of the two earthquakes, document major landslides

triggered by the earthquakes, and qualitatively compare and contrast the landslide types and distributions triggered by the two earthquakes. Because we conducted rapid reconnaissance and provided on-the-spot hazard assessments, our investigations were brief and did not permit a comprehensive assessment of landslide effects or detailed investigations of individual landslides. However, the observations and interpretations in this paper provide an important documentation of the types of landslides triggered, the environments in which they occurred, and some of their devastating effects.

Terminology in this paper follows Varnes's (1978) classification system, which classifies slope movements by type of movement and type of material (Table 1). Also following Varnes, we use the term landslide generically to describe all varieties of slope movements; thus, landslide can refer to slope movements such as rock falls, debris flows, slumps, etc.

THE 13 JANUARY 2001 EARTHQUAKE

The earthquake of 13 January 2001 occurred at 11:33 a.m. local time and had a **M** of 7.7. The earthquake was located ~40 km off the coast of El Salvador beneath the Pacific Ocean (13.049° N, 88.660° W) at an estimated focal depth of 60 km (Fig. 1). The focal mechanism indicates normal faulting in the overriding Caribbean plate (U.S. Geological Survey, NEIC, 2001a).

Strong earthquake shaking affected a large part of El Salvador. Most of the southern two-thirds of the country had ground accelerations of at least 0.1 g, and several strong-motion stations in south-central El Salvador recorded peak accelerations of 0.3 g or greater. The maximum recorded shaking occurred in La Libertad along the southern coast (Fig. 2), where an acceleration of 1.11 g was recorded (Bommer et al., 2002).

The earthquake affected more than 1.3 million people and caused 844 deaths and 4723 injuries; more than 277,000 houses were damaged or destroyed (U.S. Agency for International Development, 2001). Landslides accounted for most of the earthquake damage and fatalities.

Overview of Landslides Triggered by the Earthquake

The 13 January earthquake triggered widespread damaging landslides in the southern half of El Salvador. Landslides also were reported as far away as west-central Guatemala at epicentral distances of ~350 km (Bommer et al., 2002). Although we were unable to accurately locate a landslide limit across the entire region, we estimated a limit (Fig. 1) enclosing an area of ~25,000 km² (including parts of Guatemala not shown) based on aerial and ground reconnaissance and landslides reported by Bommer et al. (2002). Within this landslide limit is an area of more concentrated landslides encompassing ~5000 km² (Fig. 1). The most significant concentrations of landslides occurred on slopes around the periphery of San Salvador, in the Cordillera Bálsamo region west and south of San Salvador, in areas around Lake Ilopango and Lake Coatepeque, and on steep flanks of

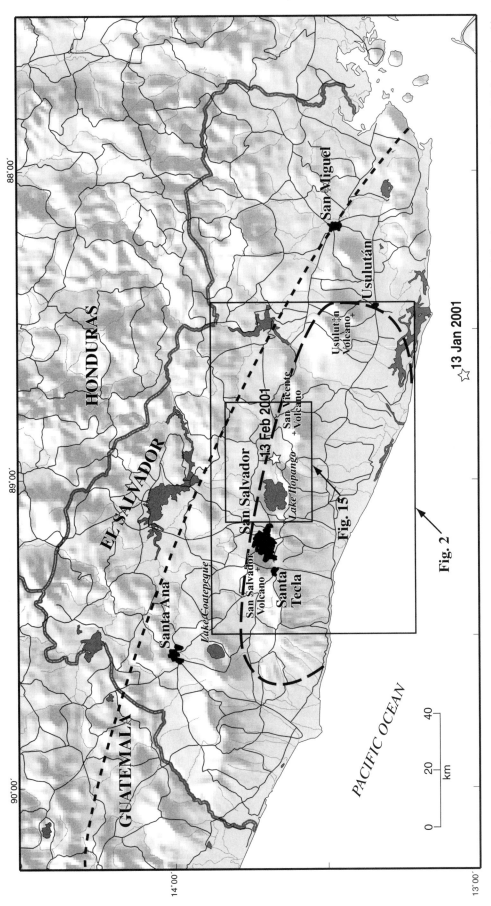

Figure 1. Map of El Salvador showing epicenters (stars) of 13 January (M-7.7) and 13 February (M-6.6) earthquakes; short-dashed line is approximate limit of landslides triggered by 13 January earthquake; long-dashed line outlines area of concentrated landslides from the 13 January earthquake; black lines are roads; gray lines are rivers and streams; large rectangle outlines area of Figure 2; small rectangle outlines area of Figure 15.

TABLE 1. LANDSLIDE CLASSIFICATION SYSTEM

Type of movement		Bedrock	Type of material	
			Engineering soils	
			Primarily coarse	Primarily fine
Falls		Rock fall	Debris fall	Earth fall
Topples		Rock topple	Debris topple	Earth topple
Slides	Rotational	Rock slump	Debris slump	Earth slump
	Translational	Rock slide	Debris slide	Earth slide
Lateral spreads		Rock spread	Debris spread	Earth spread
Flows		Rock flow	Debris flow	Earth flow
Complex		Combination of two or more principal types of movement		

Note: Modified from Varnes (1978).

some volcanoes in the southern part of the country, particularly Usulután volcano (Figs. 1 and 2).

Most of the triggered landslides were relatively small (tens to hundreds of cubic meters), shallow (<5 m) falls and slides in surficial rock and debris. Landslide concentrations were greatest where two types of Pleistocene and Holocene volcanic rocks crop out: (1) relatively soft, weak pyroclastic deposits and (2) solid, indurated rocks that originated as lava flows. The largest number of landslides occurred in pyroclastic deposits; these slides tended to be highly disrupted masses of rock and earth that fell and slid into jumbled piles of landslide debris. Earthquake-triggered landslides in this type of material have been documented in many previous earthquakes in El Salvador (Rymer and White, 1989). Landslides originating in the harder lava flows were volumetrically small but very hazardous; they consisted primarily of boulders as large as several tens of cubic meters that were shaken loose from steep outcrops and then bounced and rolled down steep slopes. These boulders caused great damage when they hit buildings, vehicles, or people.

The January earthquake also caused two large, deep (tens of meters) landslides. The most damaging was the Las Colinas landslide in Santa Tecla, a western suburb of the capital city of San Salvador (Fig. 2). The Las Colinas landslide had an estimated volume of 130,000 m³ (Evans and Bent, this volume, Chapter 3) and caused ~585 fatalities when it slid off the north slope of Bálsamo Ridge; the landslide had an abnormally long runout distance and destroyed everything in its path in this densely populated neighborhood. A second large landslide near San Vicente (Fig. 2), which we estimate had a volume of ~750,000 m³, buried a few hundred meters of the Pan-American Highway under tens of meters of debris.

The January earthquake also triggered liquefaction and associated lateral spreading along coastal areas from La Libertad eastward to the mouth of the Lempa River and along riverbanks east of Usulután (Fig. 2). Because our mission focused on assessing landslide hazards, we did not investigate liquefaction areas in detail but merely noted some locations where liquefaction effects were observed.

Significant Landslides and Areas of Landslide Activity

Cordillera Bálsamo

The January earthquake triggered widespread landsliding in the Cordillera Bálsamo, a broad, deeply dissected upland area southwest of San Salvador. Bálsamo Ridge defines the northern boundary of the Cordillera Bálsamo, and the ridge separates the cordillera to the south from a broad flat valley the north, which is occupied by Santa Tecla (Fig. 2). Cordillera Bálsamo is underlain by the Bálsamo Formation (Weber et al., 1978), which consists of volcanic breccias, lavas, and other well-indurated volcanic rocks. In the area near San Salvador, the top of the Bálsamo Formation is marked by weathered soil that contains sufficient clay and fine-grained material to act as a groundwater barrier that locally perches water in porous, overlying, young volcanic deposits. Such perched groundwater conditions probably contribute to seismic instability because high pore-water pressures can develop in saturated layers during earthquake shaking (Harp et al., 1984).

In central El Salvador, the Bálsamo Formation is overlain by a sequence of latest Pleistocene and Holocene rhyolitic and andesitic tephras erupted from San Salvador Volcano and Ilopango caldera (Rose et al., 1999). Four named deposits were erupted from the Ilopango caldera, an 8 × 11 km depression that is now filled by Lake Ilopango (Fig. 1). The four deposits, named Tierra Blanca Joven (TBJ) and TB2–TB4, erupted from the caldera since 56.9 ka (Rose et al., 1999). The youngest unit, TBJ, was deposited about A.D. 430 (Dull et al., 2001). The oldest unit (TB4) buried the preexisting landscape locally to depths of 2 m or more; TB4 is overlain by younger, interstratified tephra from Ilopango caldera and San Salvador Volcano along Bálsamo Ridge. The thickness of the TB4 and younger deposits can vary considerably over short distances because they were deposited on and flowed over preexisting topography; the deposits are commonly thicker in ancient valleys compared to adjacent ridges. The generalized stratigraphy of post–Bálsamo Formation deposits in the Bálsamo Ridge area is shown in Table 2.

The Ilopango Tierra Blanca tephras are particularly prone to landsliding (Rymer and White, 1989; Bommer and Rodríguez,

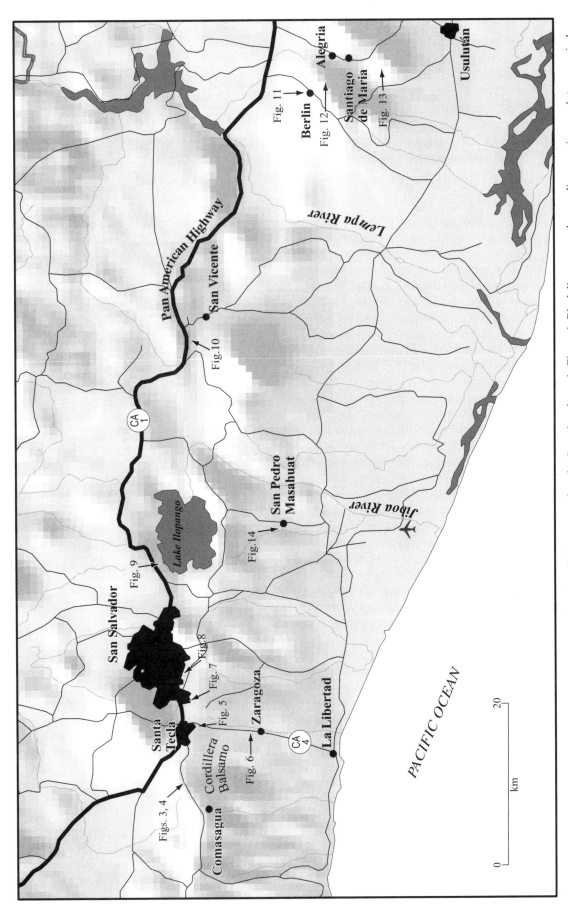

Figure 2. Map showing area of greatest landslide concentration from 13 January earthquake. Location shown in Figure 1. Black lines are roads; gray lines are rivers and streams; airplane symbol shows location of international airport. Locations of photographs shown in Figures 3–14 shown by arrows.

TABLE 2. GENERALIZED STRATIGRAPHY OF LASTEST
QUATERNARY VOLCANIC DEPOSITS OF BALSAMO RIDGE

Depth (m)	Deposit
0–2	Stratified, loose, basaltic tephra from San Salvador Volcano.
2–4	Massive, tan to orange-brown, weathered ash.
4–6	Interstratified, dark, loose basaltic tephra, brown weathered massive ash, and loose volcanic lapilli.
6+	TB4 volcanic ash, white, soft to loose, well sorted.

Figure 3. Shattered ridge near Comasagua (see Fig. 2 for location).

2002; Konagai et al., this volume, Chapter 4; Rolo et al., this volume, Chapter 5). Weak cementation and generally negative pore-water pressures provide strength for these deposits to be stable under most conditions; however, when saturated by heavy rainfall or shaken during strong earthquakes, they can lose strength and collapse (Bommer and Rodríguez, 2002; Bommer et al., 2002; Rolo et al., this volume). Our reconnaissance indicated that the distribution of abundant earthquake-induced landslides in the Cordillera Bálsamo generally coincided with the presence of relatively thick deposits of TB4 and overlying pyroclastic sediment.

Our aerial and ground reconnaissance showed that the earthquake triggered thousands of landslides throughout much of the Cordillera Bálsamo, most of which were shallow (<5 m thick), disrupted earth or debris slides and rock falls. Keefer (1984) classified these types of earthquake-triggered landslides as "disrupted slides and falls" because they tend to involve materials that are unconsolidated, very weakly cemented, or highly fractured, and the triggered landslides disaggregate and form deposits of highly disrupted material. Much of the Cordillera Bálsamo is blanketed with fine, weakly cemented tephra that collapsed and disaggregated during the strong earthquake shaking. Collapse and disaggregation of tephra deposits left thick accumulations of loose, fine debris on the ground surface. Many steep-walled ridges in this region were thoroughly shattered by amplified shaking of the weak surficial deposits. Fissuring and incipient landsliding along the edges of such ridges were widespread.

The road from Santa Tecla to Comasagua traverses the Cordillera Bálsamo (Fig. 2). Along most of its length, this road follows the crest of a narrow, steep-sided ridge where ground shaking was amplified; the ridge is thoroughly shattered along most of its length (Fig. 3). The shattering, in turn, led to widespread slope failures that produced large volumes of landslide material. The ground surface along much of the ridge was fissured, road cuts had failed extensively, and nearly all man-made structures were damaged or destroyed. The shaking left the road in very poor condition; disaggregated tephra covered the road with as much as 0.5 m of fine, powdery dust, and loose sediment continued to accumulate from the failed road cuts (Fig. 4). At the western edge of Santa Tecla, the road crossed a large, deep debris slide similar in size and scale to the Las Colinas landslide, which is described below.

Figure 4. Road-cut failures near Comasagua (see Fig. 2 for location).

Las Colinas Landslide

By far the most devastating landslide triggered by the January earthquake was the Las Colinas landslide (GPS: 13° 39.662′ N, 89° 17.188′ W) that cascaded off the steep northern flank of Bálsamo Ridge. The landslide destroyed hundreds of houses and killed ~585 people, the largest loss of life in one location caused by this earthquake. Detailed descriptions and analyses of the Las Colinas landslide are the subject of two other papers in this volume (Evans and Bent, Chapter 3; Konagai et al., Chapter 4), and so we provide only a brief overview.

The Las Colinas landslide, a rapid earth flow having an estimated volume of 130,000 m³, originated at an altitude of 1075 m on Bálsamo Ridge and traveled 735 m northward through the Las Colinas neighborhood of Santa Tecla (Fig. 5); the landslide dropped ~165 m from source to terminus (Evans and Bent, this volume). Evans and Bent (this volume) estimate

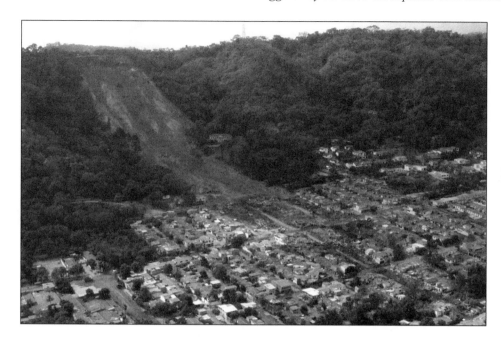

Figure 5. Las Colinas landslide (view to southwest) (see Fig. 2 for location).

that the overall landslide velocity was ~15 m/s and that it may have reached velocities as great as 30 m/s in some places. The near-vertical scarp at the landslide headwall exposed the upper 25–30 m of material involved in the landslide, which appeared to be only somewhat moist at the time of the landslide (Evans and Bent, this volume). Deposits of the Cuscatlán Formation and the Tierra Blanca units of the San Salvador Formation, typical of deposits covering Bálsamo Ridge and the Cordillera Bálsamo (Weber et al., 1978), were exposed in the main scarp (Evans and Bent, this volume).

A strong-motion seismometer located on the valley bottom in Santa Tecla, 1.4 km from the Las Colinas landslide source, recorded a peak ground acceleration of 0.5 g (Bommer et al., 2002), a very high level of shaking considering the site is ~100 km from the earthquake source. Evidence along the crest of Bálsamo Ridge, however, indicated that shaking there probably was even greater. We saw trees snapped off, boulders thrown vertically and then laterally from their sockets, and deep fissures along the edge of the ridge, all of which suggest shaking at or above 1 g. Such amplification of shaking is common along the edges of steep ridges (Ashford et al., 1997). This very strong shaking, along with the thick, loose to poorly consolidated, volcaniclastic deposits, the steep topography on the northern flank of Bálsamo Ridge, and possibly the presence of a relatively impermeable soil at the top of the Bálsamo Formation were contributing factors to the Las Colinas landslide.

Zaragoza

In and around Zaragoza (Fig. 2), rock falls occurred where fracturing in lava flows formed large boulders that were susceptible to being shaken loose during earthquakes. A road cut on highway CA-4 north of Zaragoza (GPS: 13° 36.176′ N, 89° 17.161′

W) produced one such damaging rock fall that had a volume of ~2500 m³ (Fig. 6). One house was undermined in the initial rock fall, and several additional houses were perched precariously along the newly formed scarp. The rock below these houses was extensively fractured and dilated by earthquake shaking and posed significant risk of failing in aftershocks, heavy rainfall, or future large earthquakes.

Santa Elena

Rock falls involving large boulders occurred in the Santa Elena neighborhood in the southwestern part of San Salvador (Fig. 2). A steep ridge that has several outcrops of massive, hard lava flows extends along the southern edge of Santa Elena (GPS: 13° 39.639′ N, 89° 15.711′ W), and these rock outcrops produced rock falls during the earthquake. Fractures in the outcrops create rock blocks 1–20 m³ in volume that can be shaken loose during earthquakes; the blocks then roll and bounce down the 30°–40° slope at high speed. Several such boulders are strewn through the forest all the way to the base of the slope, indicating that this process has occurred previously. During the 13 January earthquake, a 5 m³ boulder broke from the rock outcrop and rolled downslope rapidly enough to knock down a 0.5 m diameter tree and travel to within 30 m of the houses at the base of the slope (Fig. 7). Had this boulder continued downslope, it would have severely damaged any structure it hit.

The hazard from rock falls in Santa Elena varies with position along the ridge. The hazard is greatest on the western end of the ridge and lessens somewhat eastward because the rock-fall sources are higher on the slope on the eastern end (boulders released higher on the slope are more likely to be slowed and stopped by the thick forest covering the slope). Although heavy rainfall can also trigger rock falls in this area, the greatest

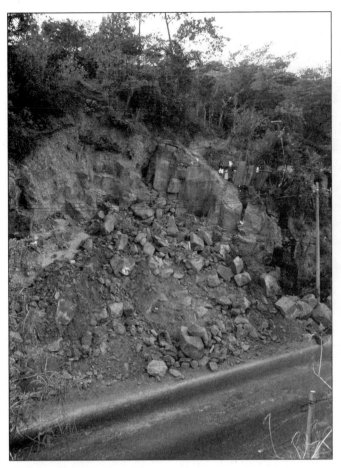

Figure 6. Rock fall from road cut north of Zaragoza (see Fig. 2 for location).

Figure 7. Boulder shaken loose by 13 January earthquake that came to rest near houses in Santa Elena (see Fig. 2 for location). Measuring stick is 2 m long. Note scars on tree at right where the boulder tore bark off the trunk.

hazard is from rocks shaken loose in moderate or large earthquakes. Houses in this area are built directly against the base of the slope, and houses in the uppermost part of this neighborhood are at significant risk from rock-fall hazards.

La Cima

The earthquake cracked the ground surface and overlying structures in La Cima, a large housing development built on cut-and-fill pads on the southern outskirts of San Salvador (Fig. 2; GPS: 13° 39.817′ N, 89° 12.965′ W). During the earthquake, cracks as wide as 10 cm formed in some parts of La Cima. The cracks extended through road pavement and through several houses, causing serious damage (Fig. 8). The cracks appeared to have formed at or near the boundary of the cut slope and the fill slope, a correlation that has been observed in earthquakes elsewhere. Cracking along this boundary results from either differential settlement or an impedance contrast created when seismic waves encounter the boundary between stiffer intact material and looser fill (Stewart et al., 1996). Renewed cracking or failure of the steep edge of the fill in future earthquakes or rainy periods is possible.

Lake Ilopango and Lake Coatepeque

Steep slopes cut by streams in the area around Lake Ilopango (Figs. 1, 2) produced hundreds of debris falls (Fig. 9) having volumes as great as several thousand cubic meters (Harp and Vallance, 2001). Slopes in this area are blanketed with thick deposits of weak, rhyolitic Tierra Blanca tephra.

The earthquake also triggered dense concentrations of debris falls in rhyolitic pumice deposits in the area around Lake Coatepeque (Fig. 1). Tephra layers in this area, however, are not as thick as those around Lake Ilopango, and so the number and volumes of landslide were smaller on average.

Las Leonas Landslide

The January earthquake triggered a large debris slide (Fig. 10) from a high, south-facing road cut on the Pan-American Highway at Las Leonas (GPS: 13° 39.963′ N, 88° 48.738′ W), ~4 km northwest of San Vicente (Fig. 2). The landslide buried several cars and killed 12 people. The roadway at the landslide location was a divided four-lane highway that traversed the

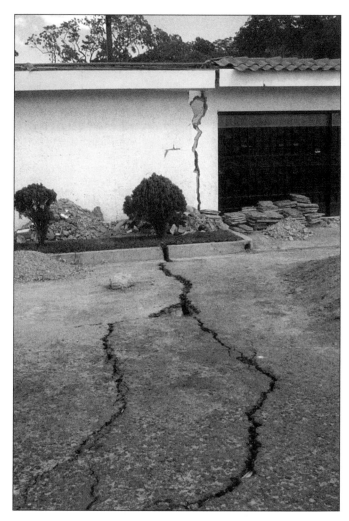

Figure 8. Cracks at the cut-fill boundary in La Cima (see Fig. 2 for location).

Figure 9. Landslides in Tierra Blanca ash deposits near Lake Ilopango (see Fig. 2 for location).

slopes on the northern flank of the Istepeque River valley. Our field observations indicated that the slide was ~250 m wide at its base, 120 m high, and 25 m thick, yielding an estimated volume of 750,000 m³.

The main scarp of the landslide formed along the fault plane of a south-dipping normal fault that bounds a graben that forms the valley. This fault created a zone of weakened, sheared tephritic material, which facilitated slope failure during earthquake shaking. We inspected the slope above the crown of the landslide and did not see open cracks or other evidence of impending failure of material above the scarp, which suggests that the failure location was controlled strongly by the preexisting fault-plane surface in the slope.

The formation of a large landslide at this location can be attributed primarily to the zone of weakness along the fault plane that defined a large, deep mass of weak volcanic rock that failed during strong shaking. Also, the toe of the slope was oversteep-

ened by a road cut, which, in turn, was destabilized by people excavating rocks from the cut to sell as building aggregate.

Usulután Region

The region around Usulután (Figs. 1, 2) experienced widespread but sparsely scattered landslides, primarily rock falls from road cuts and rock and debris slides from gully walls on the flanks of volcanoes.

At a geothermal plant near Berlín (Fig. 2), a rock fall of several hundred cubic meters occurred on the steep slope behind one of the wells (GPS: 13° 30.947′ N, 88° 30.740′ W). The slope that failed is the scarp of an older, larger landslide. Several boulders from the rock fall stopped within a few meters of the well equipment; the biggest single boulder was ~5 m³ (Fig. 11). Another landslide headed just below the road to the well, and the road had cracks parallel to the edge of the slope.

Several road cuts produced rock falls and rock slides along roads linking Berlín, Alegría, and Santiago de Maria (Fig. 2). The larger rock falls had volumes of several thousand cubic meters (Fig. 12). At least one slope in this area also produced rock falls during Hurricane Mitch in 1998, which shows that identical types of landslides in this area can be triggered by heavy rainfall or earthquake shaking. Strong shaking dilated fractures 10 cm or more in some of the rock-fall scars, and many precariously perched boulders were ready to fall.

Near the town of Alegría, a small lake basin in an old volcanic crater has steep walls tens of meters high. The earthquake triggered several rock falls from the crater walls, which had volumes as large as a few thousand cubic meters (GPS: 13° 29.563′ N, 88° 29.523′ W), and some rock falls temporarily closed the road around the lake.

Volcanoes in the southern part of the country shed numerous rock and debris slides from the walls of preexisting gullies. These failures had volumes of several cubic meters to several thousand

Figure 10. Las Leonas landslide blocking the Pan-American Highway near San Vicente (see Figure 2 for location).

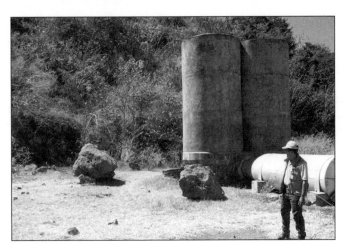

Figure 11. Boulders from rock fall above geothermal well near Berlín (see Fig. 2 for location).

cubic meters. Usulután Volcano had perhaps the highest concentration of landslides (Fig. 13) in this region.

San Pedro Masahuat

Many steep walls of deeply incised gullies around San Pedro Masahuat (Fig. 2) failed during the earthquake. San Pedro Masahuat is built on a thick (tens of meters) deposit of white, powdery volcanic ash, and deep gorges are eroded into the ash on the edge of town. These gorges have near-vertical walls 70–80 m high and are 50 m or more wide. The earthquake triggered massive failures along walls of the gorges; blocks and slabs of ash hundreds to thousands of cubic meters in volume collapsed and cascaded

to the bases of the steep slopes (Fig. 14). Numerous cracks and fissures parallel to the edge of the gorge opened as far as 30 m from the edge. These cracks typically had several centimeters of displacement downward and toward the bluff edge and thus defined incipient landslides that likely will fail in future rainy seasons or earthquakes.

THE 13 FEBRUARY 2001 EARTHQUAKE

Exactly one month after the devastating 13 January earthquake, a second major earthquake struck El Salvador. The 13 February 2001 earthquake occurred at 8:22 a.m. local time and had a moment magnitude (**M**) of 6.6. The earthquake was located ~40 km east-southeast of San Salvador in the area east of Lake Ilopango (13.671° N, 88.938° W; Figs. 1, 15) at an estimated focal depth of 15 km. The focal mechanism indicated strike-slip crustal faulting in the overriding Caribbean plate (U.S. Geological Survey, NEIC, 2001b).

Owing to the smaller magnitude and shallower depth of the February earthquake, the effects and damage were more localized than those in the January event. Nevertheless, the earthquake still affected more than 1.5 million people and caused 315 deaths and 3399 injuries. About 57,000 houses were damaged or destroyed by strong shaking or ground failure (U.S. Agency for International Development, 2001).

Overview of Landslides Triggered by the Earthquake

The February earthquake triggered thousands of landslides in an area of perhaps 2500 km². We could not establish an accurate landslide limit, because to distinguish landslides triggered

Figure 12. Landslide on road from Berlín to Alegría (see Fig. 2 for location).

Figure 13. Rock and debris slides in gullies on the slopes of Usulután Volcano (see Fig. 2 for location).

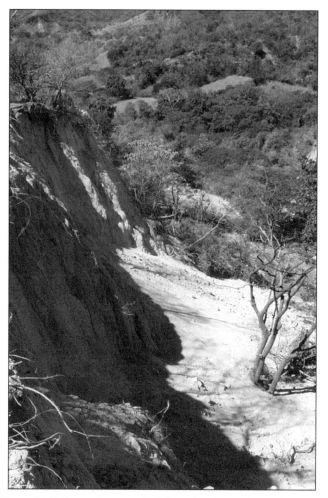

Figure 14. Failures of near-vertical slopes in ash near San Pedro Masahuat (see Fig. 2 for location).

in the February earthquake from those triggered in January was generally impossible outside the immediate epicentral area. The most concentrated landsliding occurred in an area of ~1000 km² (Fig. 15). Thick deposits of late Pleistocene and Holocene rhyolitic tephras that are particularly susceptible to failure during seismic shaking (Rymer and White, 1989; Rolo et al., this volume, Chapter 5) underlie the epicentral area (Weber et al., 1978). These deposits erode easily and formed steep-sided valleys that produced thousands of relatively small (<1000 m³), shallow (<5 m) falls and slides. In places, these landslides were so highly concentrated that they coalesced into nearly continuous failures along canyon walls (Fig. 16).

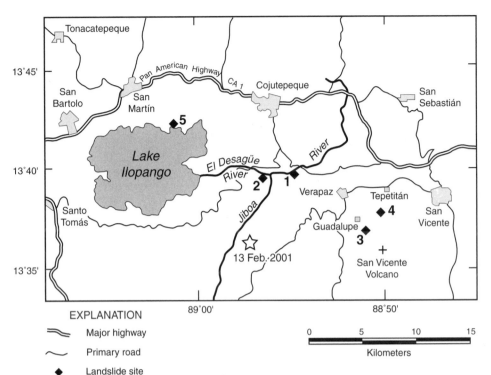

Figure 15. Map showing area of greatest concentration of landslides in the 13 February earthquake. Figure 1 shows location. Numbered symbols correspond to significant landslide locations: 1—Jiboa River landslide (photo in Figure 19); 2—El Desagüe River landslide (photos in Figs. 17 and 18); 3—Quebrada del Muerto rock slide (photo in Fig. 20); 4—Quebrada El Blanco debris slides (photo in Fig. 21); 5—San Agustín–Quebrada El Chaguite area.

Figure 16. Coalescing failures in ash deposits east of Lake Ilopango (see Fig. 1 for location of Lake Ilopango).

The earthquake also triggered large landslides in river valleys east of Lake Ilopango and at San Vicente Volcano (Fig. 1). Particularly large landslides occurred in the drainages of the El Desagüe and Jiboa Rivers near Lake Ilopango (Fig. 15); these slides dammed the rivers and impounded lakes, which posed a potentially significant downstream threat. Several large landslides also occurred on the northern slopes of San Vicente Volcano. Prominent failures included a large rock slide in the upper part of the Quebrada Del Muerto and numerous failures along the

steep valley walls in the upper part of the Quebrada El Blanco and adjacent drainages (Fig. 15).

Earthquake-induced liquefaction and lateral spreading caused localized damage around the northeastern shore of Lake Ilopango. The greatest damage from liquefaction and lateral spreading occurred in San Agustín, ~10 km northwest of the earthquake epicenter (Fig. 15), where fissures and slump scarps formed on a populated alluvial fan. Buildings that spanned the fissures and scarps suffered considerable damage.

Significant Landslides and Areas of Landslide Activity

Jiboa River Landslide

About 2 km upstream from the confluence of the El Desagüe and Jiboa Rivers (Fig. 15), a landslide having an estimated volume of 12 million m³ dammed the Jiboa River (Fig. 17). The Jiboa River landslide consists primarily of rhyolitic tephra and pyroclastic debris of the Cuscatlán Formation (Weber et al., 1978), and most of the material is poorly compacted and very porous. The landslide dam is ~700 m long (parallel to the river drainage), 250 m wide (across the valley), and 60–70 m thick.

The dam impounded a lake that was ~25 m deep, 400 m long, and filling at a rate of less than 0.1 m³/s (<8600 m³/d) by mid-March 2001. The lake-surface altitude at that time was ~475 m. If filled completely, the lake would have had an estimated maximum depth of 60 m, a length of ~2 km, and a volume of ~7 million m³.

The surface of the landslide deposit was hummocky and irregular. Because of the surface irregularity, two possible breach

Figure 17. Landslide dam on the Jiboa River (view southwest). The impounded lake behind the landslide is at the lower left (see Fig. 15 for location).

points existed on the landslide dam: one along the northern margin of the dam at an altitude of ~510 m and another along the southern margin of the dam at an altitude of ~515 m. (All altitudes were estimated from field measurements made with barometric altimeters and hand-held GPS receivers, which have considerable inaccuracy.) On the basis of our estimates of the potential maximum lake volume and stream-gage records of average flow rates for the Jiboa River, we projected that, without engineering intervention, the lake could fill and overtop the dam within several months.

Initial hazard assessment of landslide dam. The Jiboa River landslide and the lake it impounded presented a significant threat to people, property, and infrastructure downstream. If the lake filled to capacity, it would have overtopped the dam and generated a potentially devastating flood. On the basis of a case study of a similar landslide dam on the Pisque River in Ecuador (Asanza et al., 1992), we speculated that a catastrophic flood could develop quickly and that the lake could drain in several hours if it overtopped the dam. To quantify the threat by overtopping of the landslide dam, we estimated a range of values for peak discharges at the point of breach and the time it would take to reach peak discharge. We used a model that predicts flood discharges after failures of natural dams (Walder and O'Connor, 1997). The model is based on the kinematics of breach formation and the hydraulics of flow through the breach. Parameters that must be estimated for the model include drop in lake level (d), water volume released (V), and breach erosion rate (k). Walder and O'Connor (1997) show that d is typically 50%–100% of the dam height and that k generally ranges from 10 to 100 m/h. From the projected hypsometry of the maximum lake impoundment, we determined that ~95% of the lake volume (~6.7 million m^3) could empty if a breach eroded the dam by 40 m (~66% of the dam height). Therefore, by setting $d = 40$ m, $V = 6.7$ million m^3, and using bounding values of k, we estimated that the maximum

discharge at the breach could range from ~1000 to 10,000 m^3/s and that peak discharge could occur in less than one hour. Discharges of the Jiboa River were recorded between 1961 and 1996 at the Montecristo gaging station, 20 km downstream from the dam (División de Meteorológico y Hidrológico, 2001). The greatest peak discharge recorded during this period was 642 m^3/s. Therefore, the range of estimated peak discharges caused by overtopping of the landslide dam is ~1.5 to 15 times greater than the maximum discharge recorded along the Jiboa River valley.

A catastrophic flood on the Jiboa River caused by failure of the landslide dam would follow a complex flow path and possibly increase in magnitude downstream. Below the dam, the flood would flow violently down a narrow valley to the confluence with the El Desagüe River, 2 km downstream. At the confluence, some floodwater would flow westward up the El Desagüe River toward the outlet of Lake Ilopango; the remainder would turn southward and flow down the Jiboa River. Floodwater that surged up the El Desagüe River would likely overtop and erode part of the El Desagüe landslide dam (described subsequently). Erosion of that landslide dam would release some of the water it impounds, but the maximum volume of water contributed from that impoundment (~500,000 m^3) is less than 10% of the potential maximum volume released from the Jiboa River impoundment (~7 million m^3). These estimates assume a worst case in which all of the impounded waters are released, which is probably unlikely.

Discharge records from gaging stations along the upper and middle reaches of the Jiboa River can be used to roughly estimate the travel times of flood peaks through the parts of the valley downstream from the landslide dam (División de Meteorológico y Hidrológico, 2001). The upstream gage, Jiboa River at San Ramon, was situated at the site of the Jiboa River landslide, and the most downstream gage, Jiboa River at Montecristo, was situated ~20 km downstream from the landslide. Three flood peaks (all less than 200 m^3/s) that occurred during the common period of operation for these gages (1972–80) traveled the 20 km from the site of the landslide dam to the Montecristo gage in 2–4 h. Furthermore, the flood hydrographs showed little attenuation between the San Ramon and Montecristo gages because this section of the Jiboa River is incised into a bedrock canyon.

Directly downstream from the Montecristo gage site, the Jiboa River valley widens abruptly, and the river flows onto the broad coastal plain. In the confined, narrow bedrock valley a flood would retain much of its velocity, energy, and erosive power, but in the valley's broad lower reaches a flood would spread across the coastal plain dissipating much of its energy and attenuating its peak while inundating large areas of flood plain. A catastrophic flood of 1000 to 10,000 m^3/s would inundate much of the river's flood plain and could travel downstream faster than the smaller, seasonal floods.

The impact of a catastrophic flood on the lower reaches of the Jiboa River is difficult to assess accurately because of the complex effects of water flowing up, and then back down, the El Desagüe River valley, the subtle relief of the coastal plain, a lack of accurate topographic data along the flood reach, and the uncertainty in the timing and magnitude of the flood peak. Nevertheless, such a flood in the lower Jiboa River valley would likely affect settlements on the lower flood plain, the coastal highway bridge that spans the river, and perhaps partly inundate the international airport, which lies adjacent to the Jiboa River flood plain (Fig. 2).

Mitigation of landslide-flood hazard. To avoid a potentially catastrophic flood along the Jiboa River caused by uncontrolled overtopping of the landslide dam, the government of El Salvador excavated a spillway across the landslide to reduce the maximum volume of the impounded lake and to provide some control on the timing of water release. The north breach point was selected as the most advantageous site for spillway excavation for several reasons: (1) The north breach point coincided with an existing roadway, providing relatively easy access for heavy equipment to construct a spillway. (2) A spillway on the northern side of the landslide dam required less excavation than a spillway along the southern margin: ~50–60 m long versus ~700 m long. (3) The northern margin of the landslide dam provided a safer work site. The southern margin of the dam lies directly beneath the steep (~70°), 200 m high headscarp of the landslide, where rock falls that occurred nearly continuously would pose a constant hazard to workers. (4) Two breach points existed along the northern margin of the landslide dam: an upstream (eastern) point adjacent to the roadway, and a downstream point several hundred meters to the west. If the lake overtopped the upstream breach point before completion of the spillway, the floodwater would flow into and inundate a

large tributary valley (Quebrada Seca) of the Jiboa River, but the downstream breach point would prevent this water from immediately flowing down the main Jiboa River channel.

The spillway that was eventually excavated across the Jiboa River landslide dam was 20 m deep, ~100 m long, and ~15 m wide at the bottom of the cut (Fig. 18). Lake water began flowing over the spillway on 1 July 2001. Soon after water began flowing over the dam, a 2–3 m deep gully eroded through the spillway, but the channel rapidly stabilized, and no significant additional erosion has occurred. Reasons for this stabilization are unclear and may involve various factors: (1) Precipitation and surface runoff in El Salvador subsequent to emplacement of the landslide have been below normal, which possibly contributed to the low initial outflow over the spillway and a consequent lack of rapid erosion and catastrophic flooding. (2) Even without significant inflow, once the lake began to spill over the landslide, it could have emptied catastrophically if rapid erosion persisted. The fact that the erosion stabilized suggests that a resistant lens of material may exist that impeded erosion. (3) The dam may be porous enough that seepage reduces the head available for flow over the spillway.

As of January 2003, a lake averaging ~15 m deep having an estimated volume of 2.5 million m³ remained behind the landslide dam, which could be stable for a number of years. However, the erodible nature of the material composing the landslide dam makes it questionable whether the lake will remain indefinitely or whether a rapid breach of the spillway may yet occur in average or above-average water years when outflow over the spillway will be greater.

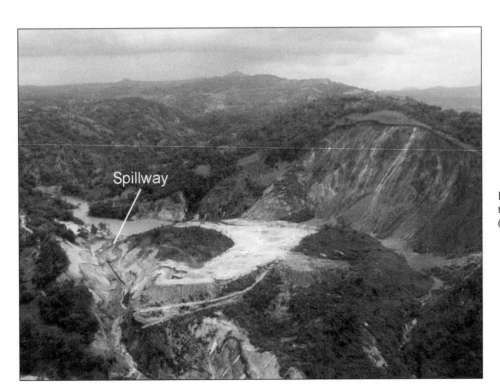

Figure 18. Spillway excavated through the toe of Jiboa River landslide dam (see Fig. 15 for location).

The spillway was excavated to reduce the magnitude of a flood caused by overtopping of the dam and to provide some control over the timing of water release. A more difficult problem is the possibility of a dam failure induced by piping, in which lake water percolates through the dam and erodes conduits through which the lake water can drain. As impoundment depth increased, the potential for piping increased because the impounded water exerted increasing pressure on the dam. During and after spillway construction the dam was monitored visually for seepage indicative of piping and potentially decreasing dam stability. No seepage had been observed as of January 2003.

El Desagüe River Landslide

The earthquake triggered a second large landslide in the valley of the El Desagüe River (Fig. 15). The El Desagüe River drains eastward out of Lake Ilopango and is a tributary to the Jiboa River. Approximately 6.5 km east of the outlet of Lake Ilopango, a landslide consisting of coarse, bouldery debris and having an estimated volume of 1.5 million m^3 dammed the El Desagüe River and impounded a shallow lake (Fig. 19). The landslide originated from the south wall of the valley and dammed the river ~100 m upstream from its confluence with the Jiboa River. About 70% of the landslide deposit consists of andesitic breccia of the Bálsamo Formation; the remainder of the deposit consists of rhyolitic pyroclastic debris from the Cuscatlán Formation (Weber et al., 1978). The andesite breccia contains clasts ranging in size from coarse sand to boulders as large as 3 m in diameter.

Flow in the El Desagüe River consists mainly of outflow from Lake Ilopango. Despite the large size of Lake Ilopango, the amount of water that drains down the El Desagüe River is modest. Streamflow records from 1961 to 1973 indicate that the average annual discharge in the El Desagüe River ranges from 0.01 to 0.53 m^3/s, and typical maximum flow during the rainy season is ~2–5 m^3/s (División de Meteorológico y Hidrológico, 2001). The maximum peak discharge reported during the period of record was 7.48 m^3/s.

The lake impounded by the El Desagüe landslide is ~1.5 km long and contains an estimated 500,000 m^3 of water. The lake's maximum depth is estimated at 5 m on the basis of heights of treetops exposed near the landslide dam (Fig. 19). Soon after the earthquake, a spillway channel was excavated by hand across the toe of the landslide to allow the lake to drain. In mid-March 2001, the flow in this 3 m wide, 2 m deep channel was ~0.2 m^3/s, and it was not eroding the spillway. Pebble- to cobble-sized clasts in the landslide debris effectively armored the spillway and prevented erosion under ambient low-flow conditions. Since the earthquake, drainage across the spillway has maintained the lake at a nearly constant volume. Higher rainy season discharges since the earthquake have not significantly modified the channel armor along the spillway; the armor has minimized spillway erosion and prevented an erosional breach of the landslide dam. The degree of channel armoring, the shallowness of the lake, and the typically modest discharges along the El Desagüe River suggest that failure of this dam by overtopping is unlikely. Even

Figure 19. Slump/rock avalanche that dammed the El Desagüe River near its confluence with the Jiboa River (see Fig. 15 for location). A shallow lake upstream from the landslide (left side of photo) is draining through a spillway that was hand-dug across the toe of the slide.

discharges as great as the maximum recorded discharge on the El Desagüe River are unlikely to significantly disrupt this channel armor and cause a catastrophic breach of the landslide dam.

The most significant event that could affect the lake impounded behind the El Desagüe landslide would be a catastrophic dam-break flood resulting from a failure of the Jiboa River landslide dam. Such a flood would flow up the El Desagüe River valley owing to the T-shaped configuration of the confluence of the El Desagüe and Jiboa Rivers (Fig. 15). That type of flood could flow across and erode the landslide dam, which could then release the water impounded behind the El Desagüe dam catastrophically. However, the volume of water impounded behind the El Desagüe dam is small compared to the volume of water that could potentially be released by a catastrophic failure of the Jiboa River landslide dam. Although release of the water impounded in the El Desagüe River would add to the flood volume downstream, its release under the scenario described above would likely cause little additional downstream damage.

Las Leonas Landslide

The 13 February earthquake triggered failure of the slope above the 13 January scarp of the Las Leonas landslide, and again the Pan-American Highway was blocked by landslide debris (Fig. 2). Removal of debris from the 13 January landslide that had blocked the Pan-American Highway had been completed by ~7 February, and the highway had been reopened to traffic. After this second major failure, the government of El Salvador rerouted the highway around the landslide to avoid future instability problems.

San Vicente Volcano

The earthquake triggered landslides in several deeply incised drainages on the northern flank of San Vicente Volcano (Fig. 15). A large rock slide occurred in the upper part of the Quebrada Del Muerto drainage on the volcano's northwest flank. On the northern flank, and lower on the volcano's slopes, extensive failures occurred along steep walls of three drainages that are deeply incised into an apron of coarse, unconsolidated debris-flow, pyroclastic, and alluvial deposits. These failures coalesced into debris slides that filled channel bottoms with loose deposits. Potential remobilization and transport of that loose sediment as debris flows or by sediment-laden floods poses a significant hazard to communities downstream, particularly in the Quebrada El Blanco, because the town of Tepetitán is located on the alluvial fan of this drainage.

   Quebrada Del Muerto rock slide. The Quebrada Del Muerto rock slide has an estimated volume of 200,000 m³ and is composed mainly of large blocks of lithified andesite. The rock slide occurred high on the northwest flank of San Vicente Volcano at the head of the channel of Quebrada Del Muerto (Fig. 15). The channel passes directly east of the village of Guadalupe, and because this drainage is so close to the village, community officials were concerned that the rock slide could remobilize as debris flows and endanger the village. Debris within the rock slide consists primarily of large boulders, some as much as several meters in diameter (Fig. 20). The large size and indurated character of the rock fragments and the minimal drainage area above the deposit make it unlikely that this landslide debris will remobilize into significant debris flows that could threaten downstream settlements or structures, even during heavy rainfall.

   Quebrada El Blanco debris slides. Low on the northern slope of San Vicente Volcano, numerous coalescing debris slides were triggered by the earthquake along the 40 to 50 m deep incised drainage of Quebrada El Blanco (Fig. 15). Unconsolidated pyroclastic, epiclastic, and debris-flow deposits in the steep valley walls failed during the earthquake and produced numerous debris slides that were typically 1–3 m thick; we estimate their cumulative volume to be ~250,000 m³ (Fig. 21).

   Debris from these slides clogged the channel bottom predominantly with silty sand. This sediment could remobilize as, or be entrained by, debris flows during rainy seasons. Large debris flows triggered in Quebrada El Blanco could extend down to the broad alluvial fan formed near the base of the channel on which the small town of Tepetitán is located. About 1 km upslope from Tepetitán, discharges in Quebrada El Blanco are confined to a channel that is incised several meters into older alluvial deposits. However, closer to the town the drainage emerges onto a broad open slope that is cut by shallow channels less than 2 m deep. This slope is the depositional surface of previous debris flows, and the active channels of the fan have routed flows near and toward Tepetitán in the recent past. The most recent debris flow that reached the town occurred in 1934, caused several fatalities, and buried parts of the old village under ~60 cm of sediment (Major et al., this volume, Chapter 7). After the 1934 event most of the town was relocated to a higher, safer site, but over the years the original site has been resettled.

   The shallow channels and flat topography near Tepetitán make it difficult to determine the exact paths future debris flows

Figure 20. View toward source of the Quebrada del Muerto rock slide on the northwest flank of San Vicente Volcano (see Fig. 15 for location). Largest boulders in foreground are about 2 m across.

Figure 21. View to the northwest down the upper reaches of Quebrada El Blanco channel showing coalescing debris slides from the valley walls (see Fig. 15 for location). The valley bottom is clogged with landslide debris that could mobilize into debris flows during heavy rainfall.

will take. Shallow channels near Tepetitán could get clogged with debris and divert flows westward, away from the town, or northeastward toward the town. The channels could also remain open and guide the flows toward the town. Debris-flow simulation models (Major et al., this volume, Chapter 7) suggest that debris flows having volumes greater than 100,000 m³ could potentially reach the town.

San Agustín Delta

Liquefaction and consequent lateral spreading along the northeastern shore of Lake Ilopango during the 13 February earthquake damaged or destroyed many homes in the village of San Agustín (Fig. 15). The village is situated on an active alluvial fan of Quebrada El Chaguite, which drains ~10 km² of deeply dissected terrain between San Agustín and the northern edge of the Ilopango caldera rim. Lateral spreading on this fan locally formed fissures and small scarps in parts of the village and in other areas adjacent to the lakeshore (Fig. 22). The water level in Lake Ilopango is a major control on the local depth of the water table; therefore, sites close to the shoreline were more prone to liquefaction and lateral spreading than sites more distant from the lake or at higher altitudes.

DISCUSSION

The occurrence of two major earthquakes only one month apart had devastating consequences for El Salvador. Both earthquakes caused significant damage and fatalities, but each had different source characteristics that resulted in different damage patterns. These earthquakes typify two fairly common hazard scenarios that are anticipated in seismically active regions: a large-magnitude, distant earthquake and a moderate-magnitude, proximal earthquake. Comparison of landslide patterns from these two events helps us understand what to expect in such seismic scenarios.

The 13 January earthquake was a large-magnitude (**M**-7.7), moderately deep (60 km) and distant (40 km offshore) event. The resulting landslide pattern was generally what might be expected from such an event: landslides were scattered over a very broad region in the southern half of El Salvador and Guatemala. Landslide concentrations appeared to relate more to localized areas of high susceptibility (steep, weak slopes) than to localization of strong shaking. One possible exception to this was the Las Colinas landslide, which occurred in an area that appears to have experienced an abnormally high intensity of ground shaking compared to surrounding areas.

The 13 February earthquake was a moderate-magnitude (**M**-6.6), shallow (15 km) event located near the center of the country. Landslides triggered by this earthquake were more densely concentrated in a much smaller area near the epicenter. This concentration appeared to relate closely to a limited area of most intense shaking.

Figure 23 shows the approximate areas affected by landsliding in the two earthquakes on a plot of landslide areas versus

Figure 22. Lateral-spread landslide on the northeastern shore of Lake Ilopango (see Fig. 15 for location).

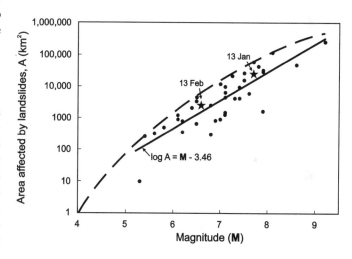

Figure 23. Area affected by landslides from worldwide earthquakes (modified from Keefer, 2002). Dashed line is upper bound; solid line is best-fit regression line. Stars are the 2001 El Salvador earthquakes, which plot above the regression line but below the upper bound.

magnitudes from 40 worldwide earthquakes (Keefer, 2002). Areas from both earthquakes plot above the mean regression line but are in the main cluster of data. Thus, areas affected by landslides in these earthquakes are above the worldwide average but are well within the upper bound of worldwide data. Using the same data set, Keefer (1984) also plotted maximum epicentral distance to landslides versus magnitude; his data indicate that the 13 January earthquake (**M**-7.7) could have triggered landslides as far as ~330 km from the epicenter. Although their precise locations are unknown, landslides from the January earthquake were reported in west-central Guatemala (Bommer et al., 2002) at or slightly beyond that distance. These observations are consistent with conclusions of other studies that suggest that earthquakes in El Salvador tend to trigger landslides over larger than average areas (Bommer and Rodríguez, 2002; Bommer et al., 2002; Rodríguez et al., 1999).

Types of landslides triggered by the two earthquakes were broadly similar and were consistent with observations from earthquakes worldwide: The large majority of triggered landslides were shallow (<5 m), disrupted falls and slides in rock and debris (Keefer, 1984, 1999; Rodríguez et al., 1999). Each earthquake also triggered localized liquefaction and lateral spreading as well as a few large, deep landslides that caused great damage and, in the case of Las Colinas, great loss of life.

A major factor in the distribution of landslides associated with both earthquakes was the presence of thick deposits of poorly consolidated Holocene and late Pleistocene tephra. These materials are particularly susceptible to failure during seismic shaking owing to their low tensile strength and their tendency to erode readily and form deeply incised, steep-walled gullies (Bommer and Rodríguez, 2002; Rolo et al., this volume, Chapter 5). Such materials have been observed in other geologically similar areas to be particularly susceptible to failure during earthquakes (Bommer and Rodríguez, 2002; Harp et al., 1981). Seismic shaking can cause these types of materials to collapse, disaggregate, and fill gully bottoms with large masses of loose debris that are then remobilized during wet periods and transported through the alluvial system (Keefer, 1999). This can lead to dramatically increased sedimentation in subsequent rainy seasons and possibly to the generation of debris flows during intense or prolonged rainfall.

The January earthquake also triggered damaging rock falls from hard, fractured lava flows. When such rock falls are triggered on steep slopes, boulders can bounce and roll at high speeds for significant distances and do great damage to anything they hit. Potential sources of rock falls are generally easy to identify because the hard lava flows tend to crop out prominently.

The earthquakes occurred in the dry season when soil moisture and groundwater levels were at annually low levels. Had these earthquakes occurred during the wet season, the number, size, and areal extent of landslides likely would have been far greater. For example, the 1906 San Francisco, California, earthquake (M = 8.2) occurred near the end of the rainy season in an above-average rainfall year; thousands of landslides were triggered over a very broad region including the Santa Cruz Mountains, which extend along the fault-rupture zone (Lawson, 1908). By contrast, the 1989 Loma Prieta, California, earthquake (M = 7.1), centered in the Santa Cruz Mountains, occurred at the end of the dry season after ~5 yr of sustained drought. This earthquake produced far fewer landslides, many of which moved a very limited distance or did not develop completely (Keefer and Manson, 1998). Schuster et al. (1998) and Keefer and Manson (1998) concluded that the significant difference in landslide triggering from these two earthquakes is attributable not just to the magnitude difference but to soil-moisture conditions that relate to seasonal rainfall.

SUMMARY AND CONCLUSIONS

In early 2001, El Salvador experienced two major earthquakes in a period of just one month. The 13 January **M**-7.7 earthquake occurred at a depth of 60 km and was located ~40 offshore. The 13 February **M**-6.6 earthquake occurred at a depth of 15 km and was located near the center of the country. Both earthquakes triggered thousands of landslides that caused most of the damage and fatalities. In both earthquakes, the large majority of landslides were shallow slides and falls in rock and debris, but both earthquakes also triggered a few large, deep landslides that were very damaging.

Landslides from the 13 January earthquake were broadly scattered over the southern half of El Salvador and Guatemala extending across an area of perhaps 25,000 km^2, somewhat above average for earthquakes in this magnitude range. Areas of greatest landslide concentration included the Cordillera Bálsamo, an upland area of deeply dissected terrain composed of young, weak volcaniclastic deposits; areas around Lake Ilopango; and gullies on the flanks of volcanoes in the Usulután region. The most devastating landslide triggered by the earthquake was the Las Colinas landslide from the north slope of Bálsamo Ridge. This 130,000 m^3 slide detached from a steep ridge above Santa Tecla and flowed 735 m into the densely populated Las Colinas neighborhood, killing ~585 people. Another large slide (~750,000 m^3) blocked the Pan-American Highway near San Vicente.

The 13 February 2001 earthquake in central El Salvador triggered thousands of landslides in the area between Lake Ilopango and the city of San Vicente. In much of this area, strong ground shaking caused extensive failures on steep walls of deeply incised valleys. Two large landslides also were triggered that blocked two major rivers in the area: the El Desagüe and Jiboa Rivers. The El Desagüe River landslide impounded a small, shallow lake of ~500,000 m^3; lake level quickly stabilized because the low-volume outflow from the lake was channeled through a small, hand-dug spillway. The Jiboa River landslide impounded a lake that could have reached a maximum volume of ~7 million m^3, which poses a more significant hazard. A 20 m deep spillway was excavated through the landslide dam to prevent the lake from reaching its projected capacity; although some erosion of the spillway has occurred, its channel appears to have stabilized, at least temporarily.

The distributions of triggered landslides from these earthquakes were consistent with what has been observed in other earthquakes in the region (Bommer et al., 2002; Rymer and White, 1989) and is intuitively expected. The large, moderately deep and distant earthquake of 13 January triggered broadly scattered landslides across a large area encompassing the southern parts of El Salvador and Guatemala. Because moderately high levels of ground shaking occurred throughout much of this area, landslide concentrations were related primarily to areas of very high susceptibility owing to steep slopes in young, weak volcanic material. The shallow, moderate-magnitude, onshore earthquake of 13 February triggered landslides over a much smaller area, about one-tenth the size of that affected in the January earthquake. Landslides were densely concentrated in the epicentral area of highest-intensity ground shaking. In both earthquakes, the areas most heavily affected by landslides were underlain by thick deposits of unconsolidated late Pleistocene and Holocene volcanic tephra.

Because El Salvador is located along a major plate boundary, it will continue to experience earthquakes in the future. The earthquakes of January–February 2001 demonstrated that El Salvador must be prepared for a broad spectrum of seismic-hazard scenarios, ranging from large, distant earthquakes to small and moderate, proximal earthquakes. While these two types of earthquakes produce different areal distributions of hazards, they are both capable of causing enormous damage and need to be factored into seismic-hazard planning throughout El Salvador.

ACKNOWLEDGMENTS

Field investigations in El Salvador were facilitated through the assistance and support of the U.S. Agency for International Development (USAID) and the El Salvadoran Ministerio de Medio Ambiente y Recursos Naturales. Assistance in conducting field investigations during the January earthquake was provided by Luz Antonina Barrios and Arturo Quezada of Geotérmica Salvadoreña (GESAL). Willie Rodriguez (U.S. Geological Survey) assisted with planning and logistics for our field operations. Margo Johnson composed the maps in Figures 1 and 2. Gerald Wieczorek and Robert Jarrett reviewed the manuscript.

REFERENCES CITED

Asanza, M., Plaza-Nieto, G., Yepes, H., Schuster, R.L., and Ribadeneira, S., 1992, Landslide blockage of the Pisque River, northern Ecuador, *in* Bell, D.H., ed., Landslides, Glissements de Terrain, Proceedings of the Sixth International Symposium, Christchurch, New Zealand, 10–14 February, 1992: Rotterdam, A.A. Balkema, p. 1229–1234.

Ashford, S.A., Sitar, N., Lysmer, J., and Deng, N., 1997, Topographic effects on the seismic response of steep slopes: Bulletin of the Seismological Society of America, v. 87, p. 701–709.

Baum, R.L., Crone, A.J., Escobar, D., Harp, E.L., Major, J.J., Martinez, M., Pullinger, C., and Smith, M.E., 2001, Assessment of landslide hazards resulting from the February 13, 2001, El Salvador earthquake: U.S. Geological Survey Open-File Report 01-119, 20 p.

Bommer, J.J., and Rodríguez, C.E., 2002, Earthquake-induced landslides in Central America: Engineering Geology, v. 63, p. 189–220.

Bommer, J.J., Hernandez, D.A., Navarrete, J.A., and Salazar, W.M., 1996, Seismic hazard assessments for El Salvador: Geofísica Internacional, v. 35, p. 227–244.

Bommer, J.J., Benito, M.B., Ciudad-Real, M., Lemoine, A., López-Menjívar, M.A., Madariaga, R., Mankelow, J., Méndez de Hasbun, P., Murphy, W., Nieto-Lovo, M., Rodríguez-Pineda, C.E., and Rosa, H., 2002, The El Salvador earthquakes of January and February 2001: Context, characteristics, and implications for seismic risk: Soil Dynamics and Earthquake Engineering, v. 22, p. 389–418.

División de Meteorológico y Hidrológico, 2001, Dirección general de recursos naturales renovables—Anuario Hidrológico 1961–1996: Ministerio de Agricultura y Ganadería, El Salvador.

Dull, R.A., Southton, J.R., and Sheets, P., 2001, Volcanism, ecology, and culture: A reassessment of the Volcán Ilopango TBJ eruption in the southern Mayan realm: Latin American Antiquity, v. 12, p. 25–44.

Evans, S.G., and Bent, A.L., 2004, The Las Colinas landslide, Santa Tecla, El Salvador: A highly destructive flowslide triggered by the 13 January 2001

earthquake, *in* Rose, W.I., et al., eds., Natural hazards in El Salvador: Boulder, Colorado, Geological Society of America Special Paper 375, p. 25–37 (this volume).

Harp, E.L., and Vallance, J., 2001, Landslide hazards in El Salvador triggered by the 13 January 2001 earthquake: U.S. Geological Survey report to U.S. Agency for International Development, 6 p.

Harp, E.L., Wilson, R.C., and Wieczorek, G.F., 1981, Landslides from the February 4, 1976, Guatemala earthquake: U.S. Geological Survey Professional Paper 1204-A, 35 p.

Harp, E.L., Sarmiento, J., and Cranswick, E., 1984, Seismic-induced pore-water pressure records from the Mammoth Lakes, California, earthquake sequence of 25 to 27 May 1980: Bulletin of the Seismological Society of America, v. 74, p. 1381–1393.

Jibson, R.W., and Crone, A.J., 2001, Observations and recommendations regarding landslide hazards related to the January 13, 2001, M-7.6 El Salvador earthquake: U.S. Geological Survey Open-File Report 01-141, 19 p.

Keefer, D.K., 1984, Landslides caused by earthquakes: Geological Society of America Bulletin, v. 95, p. 406–421.

Keefer, D.K., 1999, Earthquake-induced landslides and their effects on alluvial fans: Journal of Sedimentary Research, v. 69, p. 84–104.

Keefer, D.K., 2002, Investigating landslide caused by earthquakes—A historical review: Surveys in Geophysics, v. 23, p. 473–510.

Keefer, D.K., and Manson, M.W., 1998, Regional distribution and characteristics of landslides generated by the earthquake, *in* Keefer, D.K., ed., The Loma Prieta, California, Earthquake of October 17, 1989—Landslides: U.S. Geological Survey Professional Paper 1551-C, p. C7–C32.

Konagai, K., Johansson, J., Mayorca, P., Uzuoka, R., Yamamoto, T., Miyajima, M., Pulido, N., Sassa, K., Fukuoka, H., and Duran, F., 2004, Las Colinas landslide: Rapid and long-traveling soil flow caused by the January 13, 2001, El Salvador earthquake, *in* Rose, W.I., et al., eds., Natural hazards in El Salvador: Boulder, Colorado, Geological Society of America Special Paper 375, p. 39–53 (this volume).

Lawson, A.C., 1908, The California Earthquake of April 18, 1906—Report of the State Earthquake Investigation Committee, Volume 1: Washington, D.C., Carnegie Institution, 451 p.

Major, J.J., Schilling, S.P., Pullinger, C.R., and Escobar, C.D., 2004, Debris-flow hazards at San Salvador, San Vicente, and San Miguel volcanoes, El Salvador, *in* Rose, W.I., et al., eds., Natural hazards in El Salvador: Boulder, Colorado, Geological Society of America Special Paper 375, p. 89–108 (this volume).

Rodríguez, C.E., Bommer, J.J., and Chandler, R.J., 1999, Earthquake-induced landslides: 1980–1997: Soil Dynamics and Earthquake Engineering, v. 18, p. 325–346.

Rolo, R., Bommer, J.J., Houghton, B.F., Vallance, J.W., Berdousis, P., Mavrommati, C., and Murphy, W., 2004, Geologic and engineering characterization of Tierra Blanca pyroclastic ash deposits, *in* Rose, W.I., et al., eds., Natural hazards in El Salvador: Boulder, Colorado, Geological Society of America Special Paper 375, p. 55–67 (this volume).

Rose, W.I., Conway, F.M., Pullinger, C.R., Deino, A., and McIntosh, W.C., 1999, An improved age framework for late Quaternary silicic eruptions in northern Central America: Bulletin of Volcanology, v. 61, p. 106–120.

Rymer, M.J., and White, R.A., 1989, Hazards in El Salvador from earthquake-induced landslides, *in* Brabb, E.E., and Harrod, B.L., eds., Landslides: Extent and economic significance: Rotterdam, A.A. Balkema, p. 105–109.

Schuster, R.L., Wieczorek, G.F., and Hope, D.G., 1998, Landslide dams in Santa Cruz County, California, resulting from the earthquake, *in* Keefer, D.K., ed., The Loma Prieta, California, earthquake of October 17, 1989—Landslides: U.S. Geological Survey Professional Paper 1551-C, p. C51–C70.

Stewart, J.P., Seed, R.B., and Bray, J.D., 1996, Incidents of ground failure from the Northridge earthquake: Bulletin of the Seismological Society of America, v. 86, no. 1B, p. S300–S318.

U.S. Agency for International Development, 2001, El Salvador—Earthquake II Fact Sheet #8 (FY 2001) 15 March 2001: www.usaid.gov/hum_response/ofda/elsalveq2_fs8_fy01.html (last visited 15 Jan. 2003).

U.S. Geological Survey, NEIC (National Earthquake Information Center), 2001a, Earthquake Bulletin: http://neic.usgs.gov/neis/bulletin/01_EVENTS/010113173329/ (last visited 23 Sept. 2003).

U.S. Geological Survey, NEIC (National Earthquake Information Center), 2001b, Earthquake Bulletin: http://neic.usgs.gov/neis/bulletin/01_EVENTS/010213142205/ (last visited 23 Sept. 2003).

Varnes, D.J., 1978, Slope movement types and processes, *in* Schuster, R.L., and Krizek, R.J., eds., Landslides—Analysis and control: Transportation

Research Board, National Academy of Science, Special Report 176, p. 11–33.

Walder, J.S., and O'Connor, J.E., 1997, Methods for predicting peak discharge of floods caused by failure of natural and constructed earthen dams: Water Resources Research, v. 33, p. 2337–2348.

Weber, H.S., Wiesemann, G., Lorenz, W., and Schmidt-Thome, M., 1978, Mapa geológico de la República de El Salvador/América Central: Bunde-sanstalt fuer Geowissenschaften und Rohstoffe, Hannover, Germany, scale 1:100,000, 6 sheets.

MANUSCRIPT ACCEPTED BY THE SOCIETY JUNE 16, 2003

Geological Society of America
Special Paper 375
2004

Debris-flow hazards at San Salvador, San Vicente, and San Miguel volcanoes, El Salvador

Jon J. Major*
Steven P. Schilling
U.S. Geological Survey, Cascades Volcano Observatory, 1300 SE Cardinal Court, Vancouver, Washington 98683, USA

Carlos R. Pullinger
C. Demetrio Escobar
Servicio Nacional de Estudios Territoriales, San Salvador, El Salvador

ABSTRACT

Volcanic debris flows (lahars) in El Salvador pose a significant risk to tens of thousands of people as well as to property and important infrastructure. Major cities and nearly a third of the country's population are located near San Salvador, San Vicente, and San Miguel volcanoes. Debris flows traveling as little as 4 km from source at these volcanoes put hundreds to thousands of lives, property, and infrastructure at risk.

We used a statistically based model that relates debris-flow volume to cross-sectional and planimetric inundation areas to evaluate spatial patterns of inundation from a suite of debris flows ranging in volume from 100,000 m³ to as large as 100 million m³ and examined prehistoric deposits and a limited number of historical events at these volcanoes to estimate probable frequencies of recurrence. Our analyses show that zones of greatest debris-flow hazard generally are focused within 10 km of the summits of the volcanoes. For typical debris-flow velocities (3–10 m/s), these hazard areas can be inundated within a few minutes to a few tens of minutes after the onset of a debris flow. Our analyses of debris-flow recurrence at these volcanoes suggest that debris flows with volumes of 100,000 m³ to as large as 500,000 m³ have probable return periods broadly in the range of ~10 to 100 yr. Debris flows having volumes less than 100,000 m³ probably recur more frequently, especially at San Miguel volcano.

Despite the limited extents of the hazard zones portrayed in our analyses, even the smallest debris flows could be devastating. Urban and agricultural expansions have encroached onto the flanks of the volcanoes, and debris-flow–hazard zones extend well into areas that are settled densely or used for agriculture. Therefore, people living, working, or recreating along channels that drain the volcanoes must learn to recognize potentially hazardous conditions, be aware of the extents of debris-flow–hazard zones, and be prepared to evacuate to safer ground when hazardous conditions develop.

Keywords: debris flow, volcano, landslide, hazard mapping, GIS, El Salvador.

*jjmajor@usgs.gov

First Page Tag Line: Major, J.J., Schilling, S.P., Pullinger, C.R., and Escobar, C.D., 2004, Debris-flow hazards at San Salvador, San Vicente, and San Miguel volcanoes, El Salvador, in Rose, W.I., Bommer, J.J., López, D.L., Carr, M.J., and Major, J.J., eds., Natural hazards in El Salvador: Boulder, Colorado, Geological Society of America Special Paper 375, p. 89–108. For permission to copy, contact editing@geosociety.org. © 2004 Geological Society of America

INTRODUCTION

El Salvador is a small (21,000 km²) Central American country, about the size of the state of Massachusetts (USA), subject to recurring natural hazards that include volcanic eruptions, earthquakes, hurricanes, floods, landslides, debris flows, fires, and tsunamis. The country is particularly vulnerable to hazardous events along a chain of major volcanoes and calderas, which spans ~200 km (Fig. 1), owing to the densities of population and infrastructure that encroach on or near several volcanoes. More than 40% of the country's population resides in metropolitan areas that lie at the bases of four volcanoes: Santa Ana, San Salvador, San Vicente, and San Miguel (Fig. 1). In addition to urban centers, numerous small villages and coffee plantations are located on or around the flanks of all of El Salvador's volcanoes, and the Pan-American highway crosses the lowermost slopes of at least two volcanoes. The proximity of dense populations and major transportation routes to several volcanoes increases the likelihood of disastrous consequences from even small volcano-related events.

Populations situated on or close to volcanoes are at risk not only from future eruptions, but also from events unrelated to eruptions, such as landslides and debris flows (watery flows of mud, rock, and debris—also known as lahars when they occur on a volcano) triggered by heavy rainfalls, earthquakes, or releases of water stored in lakes and reservoirs. The catastrophe that occurred in 1998 at Casita volcano, Nicaragua (Scott, 2000; Kerle and van Wyk de Vries, 2001), where torrential rainfall from Hurricane Mitch triggered a landslide that transformed into a rapidly moving debris flow that destroyed two villages and killed more than 2000 people (Scott, 2000), highlights the risk posed by noneruption events to the populace near volcanoes. In El Salvador, earthquakes and torrential rainfalls have triggered similar landslides at several volcanoes, and some of those landslides also have transformed into debris flows that inundated populated areas downstream.

Here, we describe the hazards of landslides and debris flows in general and discuss potential hazards from future debris flows at San Salvador, San Vicente, and San Miguel volcanoes in particular. We show, in accompanying hazard zonation maps, which areas around these three volcanoes are likely to be at risk from future debris flows. We also estimate the probabilities of occurrence of debris flows of varying volumes in an effort to link frequencies and magnitudes of events.

LANDSLIDE, DEBRIS AVALANCHE, AND DEBRIS FLOW

Slope failures at volcanoes occur in many forms, but commonly they consist of various forms of falls, slides, or flows (Varnes, 1978). Falls consist of detached fragments of rock or debris that bounce or roll downslope and accumulate in fragmental piles. Slides consist chiefly of masses of material that rotate or translate along relatively distinct surfaces. Flows are characterized by slope failures that take on a fluid-like motion. Large (>10⁶ m³) slope failures that move particularly rapidly and exhibit some combination of slide- and flow-like character commonly are referred

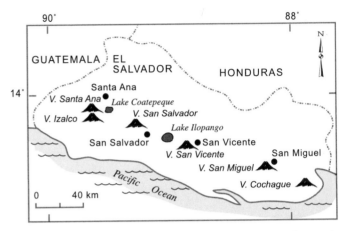

Figure 1. Location of major cities and significant Quaternary volcanoes in El Salvador. Circles indicate major cities, triangles indicate major volcanoes. Lake Coatepeque and Lake Ilopango are large silicic calderas. Base map modified from the CIA World Factbook (http://www.cia.gov).

to as debris avalanches (Varnes, 1978). These various forms of slope failure are referred to generically as landslides. In this paper, however, we distinguish debris flows from the generic category of landslides, and we refer to slope failures as debris avalanches where appropriate to highlight a particular style of landslide.

Slope failures at volcanoes, as elsewhere, are triggered by a variety of processes at a variety of scales. Earthquakes, torrential rains, or steam explosions commonly can trigger landslides up to a few tens of meters deep and perhaps to a few tens of millions of m³ in volume (e.g., Endo et al., 1989; McGuire, 1996; Baum et al., 2001; Crone et al., 2001). Landslides of these sizes typically travel up to several kilometers from their sources, unless they transform into debris flows that can travel farther. Nevertheless, they can leave deposits up to tens of meters thick and destroy everything in their paths. Particularly strong earthquakes, magma rising upward through a volcano, or warm, acidic groundwater circulating through cracks and porous zones deep inside a volcano can trigger larger landslides resulting from deeper-seated failures (e.g., Voight et al., 1983; Siebert et al., 1987; López and Williams, 1993; McGuire, 1996; Day, 1996; Elsworth and Voight, 1996; Watters et al., 2000; Reid et al., 2001; Finn et al., 2001). Magma intrusion fractures rock, mechanically weakens a volcanic edifice, and commonly causes destabilizing changes in edifice morphology (e.g., Voight et al., 1983). Circulating thermal waters can alter competent rock to low-strength clay (e.g., López and Williams, 1993; Watters et al., 2000), or be thermally or mechanically pressurized (Day, 1996) and gradually weaken an edifice so that it is susceptible to large landslides (Reid et al., 2001; Finn et al., 2001). Landslides induced by destabilization caused by magmatic and hydrothermal processes can have volumes on the order of 10⁸–10¹⁰ m³, and they can erode deeply and remove large segments of a volcano (Voight et al., 1981; Siebert, 1984; Francis and Self, 1987; Crandell, 1989; McGuire, 1996; Glicken, 1998; Yamagishi, 1996; Siebert et al., this volume, Chapter 2). Landslides of this scale at volcanoes commonly are called debris avalanches (Voight et al., 1981; Glicken, 1998), although volcanic debris avalanches can have smaller volumes (e.g., Kerle and van Wyk de Vries, 2001). Large debris avalanches (>10^7 m^3) can travel tens of kilometers from a volcano, attain speeds in excess of 150 km per hour (e.g., Voight, 1981), leave deposits tens of meters thick on valley floors, and destroy everything in their paths.

Like landslides, debris flows also are triggered by a variety of processes at a variety of scales. In tropical climates such as in El Salvador, debris flows commonly form when rainfall- or earthquake-triggered landslides transform into flowing masses of mud, rock, and water (e.g., Iverson et al., 1997, 2000; Scott et al., 2001). Debris flows at tropical volcanoes also can form by erosion of loose sediment on the slopes of a volcano (e.g., Rodolfo, 1989; Rodolfo and Arguden, 1991; Pierson et al., 1996; Lavigne et al., 2000a) or along river channels in which large amounts of volcanic sediment have been deposited (e.g., Janda et al., 1996; Pierson et al., 1996; Thouret et al., 1998; Suwa and Yamakoshi, 1999; Manville, 2002). The size of a debris flow generally is controlled by the amount of water and sediment available, the characteristics of the sediment, and the mechanism of formation. Large debris avalanches can mobilize tens to thousands of millions of m^3 of material and produce large debris flows; smaller landslides or erosion of loose sediment from volcano slopes or river channels generally produce smaller debris flows.

Debris flows, like floods, can inundate flood plains and flatten or submerge structures in low-lying areas. They can travel many tens of kilometers at speeds of tens of kilometers per hour (e.g., Pierson, 1995; Major et al., 1996; Lavigne et al., 2000a, 2000b) and destroy or damage everything in their paths through burial or impact. Debris flows follow river channels and leave deposits of muddy sand and gravel (terminology after Folk, 1984) commonly to a few meters thick. Debris flows are particularly hazardous because they can travel long distances from a volcano and they affect stream valleys where human settlement usually is greatest. In some instances, debris flows, as well as landslides, can choke a main channel or block a tributary channel and impound a lake behind the blockage (e.g., Umbal and Rodolfo, 1996). Commonly, impounded water overtops a blockage and generates a secondary flood that endangers people and property (e.g., Costa and Schuster, 1988).

Like floods, debris flows recur with frequencies that vary with flow magnitude. Small debris flows on the order of tens to thousands of cubic meters in volume recur most frequently (perhaps annually to every few years), whereas extremely large debris flows on the order of hundreds of millions of cubic meters typically recur on the order of millennia to tens of millennia (e.g., Scott et al., 1995).

Large landslides and debris flows can cause secondary sedimentation problems that persist for years to decades (Major et al., 2000). Stream channels partly or completely choked with sediment may become unstable and undergo rapid and dramatic adjustments in width as streams rework and disperse the sediment down-valley (e.g., Meyer and Janda, 1986; Rodolfo and Arguden,

1991; Scott et al., 1996). Furthermore, when stream channels are choked with sediment, they have less conveyance capacity. As a result, stormflow discharges that previously may have passed unnoticed can pose potentially significant threats to people living in low-lying areas. Although low-lying areas along river valleys are most susceptible to the effects of secondary sedimentation following emplacement of landslides or debris flows, areas on higher ground adjacent to river channels can be threatened by channel migration. Not all landslides and debris flows induce channel instabilities, however. For example, slugs of sediment delivered by relatively small landslides and debris flows (≤50,000 m^3) may simply disperse and disappear along a channel and cause little secondary disturbance (e.g., Sutherland et al., 2002).

PAST EVENTS AT SAN SALVADOR, SAN VICENTE, AND SAN MIGUEL VOLCANOES

San Salvador Volcano

San Salvador volcano rises above El Salvador's capital and largest city, San Salvador (Figs. 2 and 3). The San Salvador metropolitan area has a population of nearly two million, and a population density of several hundreds of people per square kilometer (Urban Land Information System, http://ww2.mcgill.ca/ulis). Approximately 675,000 people reside within 10 km of the summit of the volcano (J.W. Ewert, U.S. Geological Survey, 2003, personal commun.). San Salvador volcano has not erupted for more than 80 yr, but it has a long geologic history of repeated, and sometimes violent, eruptions (Sofield, this volume, Chapter 11). The volcano is composed of remnants of multiple eruptive centers, and these remnants are commonly referred to by several names. The central part of the volcano, which has a gentle

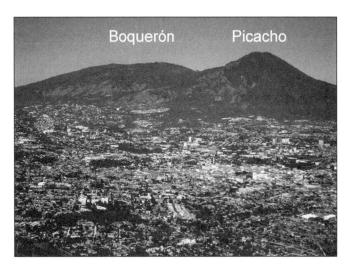

Figure 2. San Salvador volcano viewed from the southeast. The broad edifice of Boquerón is on the left, and steep-sided Picacho is on the right. The San Salvador metropolitan area is in the foreground.

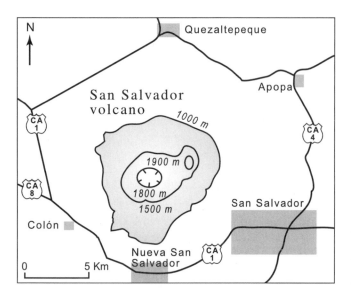

Figure 3. Schematic map showing locations of major communities and transportation routes around San Salvador volcano. CA1 is the Pan-American highway.

profile and contains a large crater, is known as Boquerón, and it rises to an altitude of ~1890 m (Fig. 2). Picacho, the prominent peak of highest elevation (1960 m) to the northeast of the crater (Fig. 2), and Jabalí, a peak to the northwest of the crater, represent remnants of an older, larger edifice.

San Salvador volcano has erupted intermittently for more than 70,000 yr, and historical observations of eruptions date back nearly 500 yr (Sofield, this volume, Chapter 11). Three eruptions have occurred since the early 1500s. The majority of the eruptive products consist mainly of basaltic to andesitic lavas and tephra falls emplaced around the flanks of the volcano. These eruptive products are the result of relatively quiescent to modest explosive activity. However, a distinctive gray, dacitic pumice-fall deposit, known as the G-1 unit (Sofield, this volume, Chapter 11), marks a significant episode of explosive activity at San Salvador volcano and serves as a useful marker horizon. This unit lies between tephra-fall deposits related to eruptions of Ilopango caldera (Fig. 1)—the Tierra Blanca 3 (TB3) and Tierra Blanca 4 (TB4) units—which have been estimated to be ~30,000 to 50,000 years old (Rose et al., 1999; J.W. Vallance, U.S. Geological Survey, 2000, personal commun.). Boquerón exploded violently ~800 yr ago in an eruption that was perhaps similar to, but smaller than, the eruption that produced the G-1 deposit. Pyroclastic-flow and tephra-fall deposits related to this eruption are found at the rim of Boquerón and on the western flank of San Salvador volcano (Sofield, this volume, Chapter 11). The most recent eruptive activity occurred in 1917, when, following an earthquake, small explosions evaporated a small crater lake and then formed a cinder cone inside the Boquerón crater, and a lava flow extruded onto the northwestern flank of the volcano.

In addition to eruptive activity, historical debris flows unrelated to eruptions have occurred at least twice (Table 1). In 1982,

a rainfall-triggered debris flow having a volume of 200,000 to 300,000 m³ originated on the eastern flank of Picacho (Fig. 4), flowed more than 4 km from its source and into the northwestern part of San Salvador city, and killed 300 to 500 people (Finnson et al., 1996). A similar event occurred in 1934 (Fig. 4; C.R. Pullinger, Servicio Nacional de Estudios Territoriales, unpublished data), but because population centers were farther from the flanks of the volcano, it did not cause any fatalities.

San Vicente Volcano

San Vicente volcano is located ~50 km east of San Salvador (Fig. 1). The volcano, which consists of two prominent cones that are composed largely of andesite lava flows (Fig. 5), towers above the city of San Vicente and the towns of Tepetitán, Verapaz, Guadalupe, Zacatecoluca, and Tecoluca (Fig. 6). Minor communities and coffee plantations are dispersed around the flanks of the volcano, and major transportation routes are located near the lowermost southern and eastern flanks (Fig. 6). Nearly 100,000 people live within 10 km of the summit of the volcano (J.W. Ewert, U.S. Geological Survey, 2003, personal commun.).

A volcanic complex, which has produced highly explosive eruptions and also quiescently emplaced lava flows and lava domes, has existed at San Vicente volcano for about two million years (Rotolo et al., 1998). Stratigraphic relations between pyroclastic and epiclastic deposits from San Vicente volcano and tephra deposits from eruptions of Ilopango caldera indicate the most recent major eruption occurred prior to about A.D. 430; paleosol development and deposit weathering suggest it probably occurred long before permanent human habitation of the area (~1200 B.C.; Black, 1983).

Previous studies recognize at least three stages of evolution of San Vicente volcano (Rotolo et al., 1998; Rotolo and Castorina, 1998). The oldest rocks are between one million and two million years old and consist of basalt and andesite lava flows extruded over hundreds of thousands of years. Sometime after about one million years ago, relatively quiescent lava emplacement was interrupted by a phase of explosive activity that produced a sequence of pyroclastic flows, pyroclastic surges, and thick tephra fall. The timing of this explosive activity is unknown; however, an intrasequence pair of thick (~2 m), well-developed paleosols and the construction of the modern volcano after the cessation of this explosive activity suggest that it occurred many tens to perhaps hundreds of thousands of years ago.

Volcaniclastic and epiclastic deposits have formed an apron of sediment that has accumulated at the base of the volcano. On the basis of weathering and soil development, the youngest part of that debris apron abuts the north side of the volcano. On other flanks of the volcano, deposits in the debris apron are probably as much as many tens of thousands of years old; they commonly are extensively weathered, contain decomposed clasts, and are capped by well-developed lateritic soils. The Tierra Blanca Joven (TBJ) deposit, the youngest regional tephra deposit from an eruption of Ilopango caldera, overlies the debris apron around

TABLE 1. HISTORICAL DEBRIS FLOWS AT SAN SALVADOR, SAN VICENTE, AND SAN MIGUEL VOLCANOES

Volcano	Event	Date	Flank	Trigger	Volume* (thousands of m³)	Source
San Salvador	debris flow	1934	E	rainfall	[≤300]	Pullinger (upub. data)
	debris flow	1982	E	rainfall	~200–300	Finnson et al. (1996)
San Vicente	debris flow	1774	NE	rainfall	[≤300]	Brauer et al. (1995)
	debris flow	1912	NNW	rainfall	[≤500]	M. Diaz, SNET, pers. commun. (2002)
	debris flow	1934[†]	N	rainfall	[≤500]	Romano (1997)
	debris flow	1934[†]	NW	rainfall	[≤300]	Romano (1997)
	landslide	1936	many	earthquake	N.D.	Bommer and Rodríguez (2002)
	debris flow	1995	S	rainfall	[≤500]	Pullinger (unpub. data)
	landslide	2001	NW	earthquake	~250	Baum et al. (2001)
	debris flow	2001[†]	NW	rainfall	~100–200	Crone et al. (2001)
	debris flow	2001[†]	N	rainfall	~50	Pullinger (unpub. data)
San Miguel	debris flow	1775		earthquake	N.D.	Feldman (1993)
	debris flow	1945	N	rainfall	<100	Blanco et al. (2002)
	debris flow	1951	N	rainfall	<100	Blanco et al. (2002)
	debris flow	1965	N	rainfall	<100	Blanco et al. (2002)
	debris flow	1975	N	rainfall	<100	Blanco et al. (2002)
	debris flow	1985	N	rainfall	[≤300]	Smithsonian Institution (2002)
	debris flow	1988	E	rainfall	<100	Blanco et al. (2002)
	debris flow	1992	N	rainfall	[≤300]	Smithsonian Institution (2002)
	debris flow	1994	E	rainfall	<100	Blanco et al. (2002)
	debris flow	1999	N	rainfall	<100	Blanco et al. (2002)
	debris flow	2000	N	rainfall	[≤300]	Smithsonian Institution (2002)
	debris flow	2001	NW	rainfall	[≤300]	Smithsonian Institution (2002)

Note: SNET—Servicio Nacional de Estudios Territoriales.

*Volumes in brackets are estimates; actual flow volumes are unknown. On the basis of comparison with inundation simulations shown in figures 14–16, and information on distal runout of the debris flows, probable maximum volumes are estimated. N.D. means no data.

[†]For purposes of estimating probabilities, these flows are considered to be single events because they were triggered by the same rainfall, but on different flanks of the volcano.

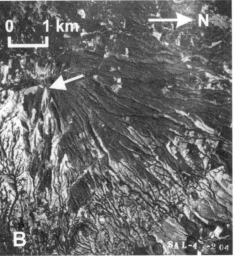

Figure 4. Aerial views of source areas of rainfall-triggered debris flows on the east flank of Picacho (San Salvador volcano). A: Source area, flow track, and depositional area of 1982 debris flow (photograph taken November 1982). B: Source area (arrow) of debris flow that occurred in 1934 (photograph taken in 1949).

Figure 5. San Vicente volcano viewed from the northeast. Note the dual cones of the volcano. The oldest part of the volcanic complex is located to the right, and the youngest cone is on the left, indicating that the focus of volcanic activity has migrated roughly eastward with time. Note the broad apron of volcaniclastic and epiclastic debris that surrounds the volcano. Photograph by Kristal Dorion.

Figure 6. Schematic map showing locations of major communities and transportation routes around San Vicente volcano. CA2 is the Coastal highway; local road to the east of San Vicente volcano connects the towns of San Vicente, Tecoluca, and Zacatecoluca.

the volcano (Fig. 7). The TBJ deposit has a radiocarbon age of about A.D. 430 (Dull et al., 2001).

A particularly noteworthy deposit near San Vicente volcano contains many small hills, known as hummocks, composed of volcanic rock and debris. This deposit, found near and beyond Tecoluca (Figs. 6 and 8), extends at least as far as the Río Lempa, 25 km southeast from the volcano, and represents a debris avalanche that resulted from collapse of a massive segment of the volcano. The age, extent, and volume of the deposit are not known.

Despite a lack of historical eruptive activity, several major debris flows from San Vicente volcano have been reported in the past few centuries (Table 1). Earthquake- and rainfall-triggered landslides and debris flows occurred in 1774, 1912, 1934, 1936, 1995, and 2001 (Brauer et al., 1995; Romano, 1997; Baum et al., 2001; Bommer and Rodríguez, 2002; Crone et al., 2001; L.A. Barrios, Geotérmica Salvadoreña, 2000, personal commun.; unpublished interviews of elderly residents of Tepetitán, 2001; M. Diaz, Servicio Nacional de Estudios Territoriales, 2002, personal commun.). Others may have occurred in historical time, but are not recorded. The 1774 debris flow occurred on the northeast flank of the volcano and affected the village of San Vicente (Fig. 6). The 1912 and 1934 debris flows occurred on the north and northwest flanks of the volcano; those from the north flank flowed more than 6 km from the summit of the volcano and severely damaged the villages of Verapaz and Tepetitán (Fig. 6). Shallow landslides on the south flank of the volcano in 1995 (Fig. 9A) transformed into debris flows that damaged the roadway between Tecoluca and Zacatecoluca (Fig. 6). An earthquake in February 2001 triggered landslides on the north and northwest flanks of the volcano (Fig. 9B), which locally deposited more than 200,000 m³ of sediment and rock debris into headwater channels that drain the

Figure 7. Debris-flow deposits in the sedimentary apron that abuts the north side of San Vicente volcano. The deposits lie beneath the Tierra Blanca Joven (TBJ) tephra-fall deposit from an eruption of Ilopango caldera, which has a radiocarbon date of about A.D. 430. The outcrop is ~4 m tall.

volcano (Baum et al., 2001; Jibson et al., this volume, Chapter 6). In September 2001, heavy rains triggered shallow landslides and debris flows on the north and northwest flanks of the volcano (Crone et al., 2001; C.R. Pullinger, Servicio Nacional de Estudios Territoriales, unpublished data). A debris flow from the northwest flank of the volcano (Fig. 10A) damaged temporary shelters along the outskirts of Guadalupe (Figs. 6 and 10B) and caused two

Figure 8. Deposit related to a Quaternary debris avalanche from San Vicente volcano. The deposit, of unknown age, extends at least 25 km southeast of the volcano. A: Aerial photograph showing hummocky terrain southeast of Tecoluca. B: Poorly sorted, matrix-supported texture of the deposit. Note the light-colored "fragile" clast entrained within the darker-colored matrix of the deposit.

Figure 9. Landslides at San Vicente volcano. A: Shallow landslides on the south flank, triggered by rainfall in 1995, rapidly transformed into debris flows that traveled ~8–10 km. B: An earthquake on February 13, 2001, triggered a relatively shallow landslide on the northwest flank. This slide did not transform into a debris flow, but deposited ~200,000 m^3 of coarse debris in the channel that heads above the village of Guadalupe (cf. Figure 6; Baum et al., 2001; Jibson et al., this volume, Chapter 6).

Figure 10. In September 2001, a rainfall-triggered debris flow occurred on the northwest flank of San Vicente volcano and swept past the town of Guadalupe 5 km downstream (see Crone et al., 2001). A: Oblique aerial view of the source area of the debris flow. B: Oblique aerial view of the debris-flow deposit on the northeast side of the town of Guadalupe. Photographs courtesy of A.J. Crone, U.S. Geological Survey.

fatalities (Crone et al., 2001). On the north flank, the debris flows remained confined to channels and caused no significant damage.

San Miguel Volcano

San Miguel volcano, located in eastern El Salvador (Fig. 1), stands above the city of San Miguel and neighboring villages (Figs. 11, 12) and is one of the country's most active volcanoes. Several small villages and coffee plantations are located on or around the flanks of the volcano, and the Pan-American and

Coastal highways pass the volcano's lowermost northern and southern flanks. Approximately 70,000 people live within 10 km of the summit of the volcano (J.W. Ewert, U.S. Geological Survey, 2003, written commun.). The volcano has erupted at least 29 times since 1699 (Williams and Meyer-Abich, 1955; Simkin and Siebert, 1994). Quiescent emplacement of lava flows and minor explosions that generated modest tephra falls characterize this historical activity (Escobar et al., 1993; Chesner et al., this volume, Chapter 16). Little is known, however, about prehistoric eruptions of the volcano. Chemical analyses of prehistoric lava flows and thin tephra

Figure 11. San Miguel volcano viewed from the west. Deep erosion scars on the flank of the volcano are source areas for debris flows.

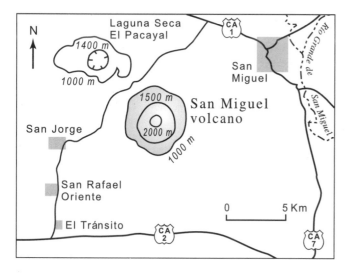

Figure 12. Schematic map showing locations of major communities and transportation routes around San Miguel volcano. CA1 is the Pan-American highway; CA2 is the Coastal highway.

falls from San Miguel volcano indicate that it has a predominantly basaltic character (Chesner et al., this volume, Chapter 16). The chemical composition of eruptive products and a lack of evidence for large cataclysmic eruptions suggest that prehistoric eruptions probably were similar in nature to the historical eruptions. Unlike San Salvador and San Vicente volcanoes, San Miguel volcano apparently does not have a history of violent explosive eruptions.

Historical debris flows have occurred several times at San Miguel volcano (Table 1). Feldman (1993) provides a terse report of a possible earthquake-triggered debris flow in 1775; others report several events within the past 60 yr (Blanco et al., 2002; Smithsonian Institution, 2002). Debris flows triggered by heavy rains originated on the east, north, and northwestern slopes of the volcano at least 11 times since 1945 and primarily damaged road-

ways (Table 1; Blanco et al., 2002; Smithsonian Institution, 2002). Modern debris flows at San Miguel volcano thus far have caused only minor loss of property and primarily have been a short-term inconvenience; however, such minor impacts do not negate the seriousness of the threat posed by debris flows from this volcano.

METHODOLOGY FOR ASSESSING DEBRIS-FLOW–HAZARD ZONATION

Traditionally, assessments of debris-flow hazards at a volcano are based upon detailed mapping of the extents of debris-flow deposits and extrapolation of estimated inundation limits among drainage basins. Any attempt to presently assess the hazards of volcanic debris flows in El Salvador on the basis of such traditional methods would be difficult, however, because details of the debris-flow histories of the volcanoes are not well developed, nor are the extents of debris-flow deposits at the volcanoes well exposed. Therefore, we rely on a simulation model to determine potential inundation areas of future debris flows at San Salvador, San Vicente, and San Miguel volcanoes.

Debris-Flow–Hazard Simulation Model

We used a physically based, statistically constrained simulation model calibrated with data from other volcanoes (Iverson et al., 1998; Schilling, 1998) to estimate potential areas of inundation by debris flows of various volumes from San Salvador, San Vicente, and San Miguel volcanoes. This semiempirical model relies on scaling and statistical relationships among debris-flow volume, cross-sectional area of channel inundation, and planimetric area of inundation. The predictive equations for inundation are given by $A = 0.05V^{2/3}$ and $B = 200V^{2/3}$, where A is the cross-sectional area of inundation, B is the planimetric area of inundation (Fig. 13), and V is debris-flow volume. The proportionality coefficients were obtained from a statistical analysis of a limited number of volcanic debris flows worldwide (Iverson et al., 1998). By using a range of prospective debris-flow volumes to evaluate A and B, a range of inundation areas can be plotted for debris flows of increasing volume and decreasing probability. This mathematically predictive method of defining inundation areas removes the subjectivity and ambiguity commonly associated with traditional mapping techniques.

Embedding the predictive equations of the statistical model in a geographic information system (GIS) computer program automates delineation of debris-flow-hazard boundaries (Schilling, 1998). After coupling the program, called LAHARZ, with digital elevation models (DEMs) of topography, a user selects the stream drainages and debris-flow volumes of interest, and LAHARZ calculates a set of nested debris-flow-inundation zones. Key aspects of program implementation and limitations are highlighted below; detailed descriptions of the program are given in Iverson et al. (1998) and Schilling (1998).

The program distributes a specified volume of debris along a channel beginning at a user-specified starting point. Selection

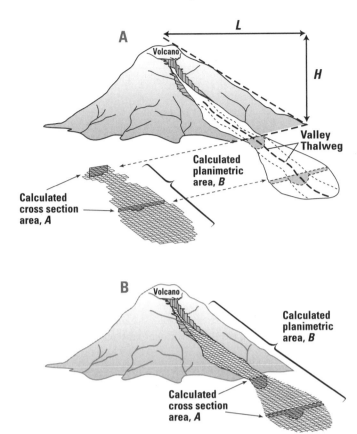

Figure 13. Diagram of the association between dimensions of an idealized debris flow and the cross-sectional and planimetric areas of inundation calculated by LAHARZ for a hypothetical volcano. A: Methodology used in original application of LAHARZ by Iverson et al. (1998), in which the debris-flow source is located on volcano's flank, but calculations of debris-flow inundation begin at the boundary of an energy cone, defined by the ratio H/L. B: Methodology used in this analysis, in which LAHARZ begins calculations of debris-flow inundation at a user-specified source area on the volcano's flank. See text for discussion.

of a starting point is problematic, however, because source areas of debris flows are difficult to predict precisely, and because debris flows can entrain, rather than deposit, sediment on steep slopes (e.g., Scott, 1988; Pierson et al., 1990). Therefore, Iverson et al. (1998) and Schilling (1998) used the concept of an "energy cone" to define a boundary for selecting starting points. Conceptually, an energy cone characterizes the mobility of gravity-driven processes such as debris flows, landslides, lava flows, and pyroclastic flows by comparing horizontal runout distance of an event to the vertical descent from a source region (Malin and Sheridan, 1982; Hayashi and Self, 1992). The dimensions of the cone are defined by an aspect ratio H/L, where H represents vertical descent from the apex of the cone, sited at the summit of a volcano, and L represents horizontal extent (Fig. 13A). Iverson et al. (1998) and Schilling (1998) defined a limiting boundary for debris-flow starting locations as the intersection between an energy cone, used to characterize the mobility of hazardous

processes near a volcano, and local topography; specific starting points for routing debris flows begin where stream channels intersect the energy cone boundary (Fig. 13A). This method for selecting starting points for routing debris flows works well for large-volume flows ($\geq 10^6$ m^3) that travel well beyond the boundary of a typically dimensioned energy cone (e.g., Malin and Sheridan, 1982; Hayashi and Self, 1992; Iverson et al., 1998), but it is not suitable for smaller debris flows that do not travel far beyond the base of an edifice.

In El Salvador, historical volcanic debris flows have had volumes of less than 10^6 m^3 (Table 1). Our initial simulations suggested that debris flows having volumes less than 10^6 m^3 would not travel far beyond the bases of the volcanoes even when started at the boundary of a typically dimensioned energy cone. Therefore, we modified LAHARZ to start debris-flow inundation at any user-specified point along a channel of interest rather than at a cone-defined boundary (Fig. 13B) in order to highlight risks closer to the volcano. In general, the starting points for debris-flow-inundation zones shown herein represent the highest altitude at which the hydrography of a selected stream channel was depicted on the DEM. In reality, debris flows may originate higher in channel headwaters. For a simulation of a large-volume (10^8 m^3) debris avalanche at San Vicente volcano, however, we used an energy cone boundary defined by H/L = 0.2 as the starting point. We selected this value for the energy-cone aspect ratio because it is consistent with an H/L ratio characteristic of the mobility of proximal hazardous phenomena at these volcanoes (see Sofield, this volume, Chapter 11; Chesner et al., this volume, Chapter 16).

LAHARZ calculates inundation areas of debris flows from the embedded statistical equations for each user-specified volume of flow, and then assembles a sequence of grid cells on a DEM that accommodates the calculated areas (Schilling, 1998). Starting at a user-specified initiation cell on a DEM, the program calculates, for each volume of flow, a planimetric inundation area and an inundation-area value that applies to all cross sections along a channel. The program assembles and fills topographic cross sections from DEM grid cells that begin at the starting elevation and advance successively downstream. At each cross section, the program stores the maximum elevations and locations of the cells encountered that satisfy the equation that relates cross-sectional inundation area and volume. The program then advances downstream to the next stream cell, calculates a new cross-sectional fill level and assembles the cells to accommodate this fill, connects the locations of the cells at maximum fill to those encountered during previous calculations upstream, and tracks the cumulative planimetric area of inundation. The program repeats the process of downstream advance and assemblage of cells that accommodate the maximum level of fill until the cumulative planimetric inundation area satisfies the predicted planimetric inundation area.

Although the LAHARZ method of estimating the extent of debris-flow inundation is automated and reproducible, any hazard zone computed by the program should be considered a guideline rather than an absolute definition of a hazardous area. LAHARZ is a statistically based program that utilizes topographic information obtained from a DEM; its output is constrained by the quality of the regression equations and the resolution and accuracy of the DEM employed. Boundaries output by the model are sometimes ragged in detail. This results from construction of a finite number of cross sections defined by a finite number of DEM grid cells (Iverson et al., 1998; Schilling, 1998). In areas of low relief, subtle features on the landscape such as minor swells, elevated berms, or channel levees on nearly planar surfaces can strongly influence the limits of inundation. However, if such features are not faithfully represented in the DEM, they will not influence predicted limits of inundation, especially those of small-volume debris flows. For the simulation of a 10^8 m^3 debris flow at San Vicente volcano we visually smoothed the output from LAHARZ to have a less ragged appearance. All other simulations are presented exactly as output.

Digital Elevation Data

To operate, LAHARZ requires a DEM of the topography covering the area of interest around a volcano. At the time of our analyses, suitable DEMs of the areas of interest were unavailable. Therefore, we produced 1:50,000-scale drainage-enforced DEMs from clear-film separates of the topography, hydrography, and cultural features of appropriate 1:50,000 quadrangle maps obtained from the Instituto Geográfico Nacional in El Salvador. The clear-film separates were scanned, vectorized, tagged, and gridded to produce 10 m horizontal-resolution DEMs. Adjoining quadrangles were edge-matched for consistency.

Volumes of Design Debris Flows

For each volcano, we compute nested hazard zones that depict anticipated inundation by hypothetical debris flows having different volumes. We focus on estimating inundation areas of debris flows similar in size to, or slightly larger than, those that have occurred historically at the volcanoes. Flows of these magnitudes or smaller, triggered mainly by earthquakes or torrential rains, are the most likely to recur again. However, at San Salvador and San Vicente volcanoes, we also consider potential areas of inundation from debris flows that are much larger than those that have occurred historically. At San Salvador volcano, sparse geologic evidence suggests that landslides as large as 2 million m^3 in volume have occurred (Sofield, 1998), so we consider the possibility of a debris flow of that magnitude. At San Vicente volcano, we consider potential areas of inundation from a debris flow spawned by a rare large debris avalanche, the type of event that might be triggered by a very strong earthquake or by magmatic intrusion, because we found geologic evidence for a large debris avalanche at this volcano. At San Miguel volcano, we do not estimate areas of inundation by a debris flow resulting from a large debris avalanche. Throughout its history, San Miguel has erupted mafic, mainly basaltic, lavas and small-volume tephras, and much of that activity has occurred at flank vents ~1000 m below the summit (Chesner et al., this volume, Chapter 16). There is no evidence for a debris avalanche at San Miguel

volcano, and events of this type are particularly rare at basaltic volcanoes. Owing to the nature of volcanism at San Miguel, and to the focus of volcanic activity at vents well below the summit, we consider the possibility of a debris avalanche at San Miguel volcano to be extremely remote.

Regardless of initiation mechanism, we assume that the full volume of a hypothetical debris flow is mobilized instantly. In many instances, rainfall-triggered debris flows result from nearly instantaneous transformation of shallow landslides (e.g., Iverson et al., 1997, 2000), and many documented volcanic debris avalanches have transformed rapidly into debris flows as well (e.g., Scott et al., 2001). Earthquake-triggered landslides also can transform rapidly into debris flows, especially if the failed debris is saturated (e.g., Martinez et al., 1995; Keefer, 1999). Debris flows that result mainly from erosion of channel sediment (e.g., Rodolfo and Arguden, 1991; Pierson et al., 1996) evolve more slowly over longer distances than do those that evolve from landslides, but for modeling purposes we assume that debris flows that might form by sediment erosion are mobilized instantly as well. Although debris flows can increase in volume substantially as they move downslope (e.g., Pierson et al., 1990; Vallance and Scott, 1997; Scott, 2000; Lavigne and Thouret, 2002), we assume that the volume of debris distributed along channels is the volume of debris that was mobilized instantly at the point source. The model does not account for longitudinal entrainment of sediment.

San Salvador Volcano

Debris-flow–hazard zones at San Salvador volcano were constructed by modeling flow volumes ranging from 100,000 m^3 to 2 million m^3 (Table 2). This range of volumes depicts hypothetical debris flows considered typical of the sizes likely to be generated by earthquakes or heavy rainfall. The largest debris flows are most likely to occur on the eastern and western flanks of the volcano, from source areas on Picacho and Jabalí, because these are areas that have steep slopes ($\geq$30 degrees) and the source materials are composed predominantly of rocks and colluvium that are deeply weathered (Sofield, this volume, Chapter 11). Debris flows are least likely to occur on the north and south flanks of the volcano,

TABLE 2. VOLUMES OF DEBRIS FLOWS (IN m^3) USED TO SIMULATE INUNDATION AREAS AT SAN SALVADOR, SAN VICENTE, AND SAN MIGUEL VOLCANOES

Degree of hazard	San Salvador	San Vicente	San Miguel
Low	2 million	100 million*	N.D.
	1 million	1 million	1 million
	500,000	500,000	500,000
	300,000	300,000	300,000
High	100,000	100,000	100,000

Note: N.D.—No data. We did not simulate hazards of debris flows larger than 1 million m^3 at San Miguel volcano.
*Estimate of probable maximum debris flow generated by a large debris avalanche.

where channels head on Boquerón. On those flanks, channel heads have gentle slope gradients (≤ 20 degrees) and source materials are composed predominantly of unweathered or slightly weathered lavas (Sofield, this volume, Chapter 11).

Under particular conditions, debris flows larger than 2 million m³ might occur, particularly from Picacho or Jabalí. Such conditions might include a large earthquake centered beneath or near the peak. In February 2001, a M6.5 earthquake triggered an ~10 million m³ landslide in volcaniclastic material along the Río Jiboa near San Vicente volcano (Baum et al., 2001; Jibson et al., this volume, Chapter 6). That landslide, the largest to occur in El Salvador for at least the past several hundred years (Rymer and White, 1989), formed close to the earthquake epicenter. If a large earthquake occurred beneath or near San Salvador volcano, a landslide composed of weathered rocks and tephras greater than 2 million m³ might fail on Picacho or Jabalí, particularly if the material had high antecedent moisture content. However, we think it is unlikely that such a landslide would mobilize into a debris flow that traverses a single channel. The east flank of Picacho and the west flank of Jabalí are deeply dissected by closely spaced channels, and a debris flow spawned by a large landslide on those flanks would likely be dispersed among several channels, many of which are not interconnected.

Unless San Salvador volcano undergoes extreme deformation, intruding magma is not likely to oversteepen the slopes of Picacho or Jabalí and trigger large landslides. Future magmatic activity is likely to be focused in or near the crater of Boquerón or along radial fracture zones where activity has been focused for the past several centuries (Sofield, this volume, Chapter 11), and both Picacho and Jabalí are significantly separated from the central crater. Intruding magma may, however, generate large earthquakes that might trigger large landslides as noted above.

Although the probability of debris flows is greater on the east and west flanks than on the north and south flanks of San Salvador volcano, we estimate areas of potential debris-flow inundation along channels that head on all flanks of the volcano. We select 2 million m³ as the volume of the largest debris flow that moves along any single channel beneath Picacho and Jabalí. On other flanks of the volcano, we estimate inundation by debris flows having volumes as large as 1 million m³. However, owing to the low-gradient profile of Boquerón and the nature of the source debris on its flanks, any debris flows triggered from that part of the volcano likely would be much smaller than 1 million m³. On the basis of the geology and topography of its edifice, the sizes of the limited number of documented historical debris flows, and the apparent lack of large-volume debris-flow deposits in its stratigraphic record, we surmise that future debris flows triggered at San Salvador volcano by heavy rainfalls and earthquakes most likely will have volumes ≤500,000 m³, and that they probably will occur on Picacho or Jabalí.

San Vicente Volcano

Debris-flow–hazard zones at San Vicente volcano were constructed by modeling flow volumes ranging from 100,000 m³

to 100 million m³ (Table 2). This range of volumes depicts hypothetical debris flows considered typical of volumes generated by earthquakes and torrential rainfalls, as well as a debris avalanche resulting from the failure of a massive segment of the volcano. Landslides at San Vicente volcano triggered by an earthquake in February 2001 had estimated volumes as great as 250,000 m³ (Baum et al., 2001; Jibson et al., this volume, Chapter 6), and rainfall-triggered debris flows in September 2001 had volumes estimated to range from 50,000 m³ to 200,000 m³ (Crone et al., 2001; C.R. Pullinger, Servicio Nacional de Estudios Territoriales, unpublished data).

A debris avalanche on the order of 100 million m³ is considered to be the largest probable event caused by massive flank failure at San Vicente volcano. We make this estimate by drawing upon an analogy to the 1980 eruption of Mount St. Helens (USA), which generated a well-documented debris avalanche. The Mount St. Helens debris avalanche removed ~25% of the volcano's total volume above an altitude where slopes had an average angle of ~30 degrees (Scott et al., 1997). At San Vicente volcano slopes above 1600 m altitude equal or exceed 30 degrees. The volume of San Vicente volcano above this altitude is ~1 billion m³; however, this is the combined volume of the two cones that compose the edifice (see Figs. 5 and 6). Each cone of San Vicente volcano above 1600 m altitude has a volume of ~500 million m³. If we assume that magma would probably intrude one cone rather than both simultaneously, and also assume that a debris avalanche at San Vicente volcano would remove a percentage of the edifice similar to the percentage removed at Mount St. Helens, then the maximum probable volume of a large debris avalanche at San Vicente volcano is slightly more than 100 million m³. For simplification (and to reflect uncertainty in the value), we rounded the modeled volume downward and assumed that the entire avalanche volume would mobilize into a debris flow. San Vicente volcano has collapsed in a massive debris avalanche at least once; however, the volume of that avalanche is unknown.

San Miguel Volcano

Modeled debris flows at San Miguel volcano ranged in volume from 100,000 m³ to 1 million m³ (Table 2). Although a few prehistoric, as well as modern, debris-flow deposits have been identified at this volcano, the volumes of those debris flows are poorly known. We therefore simulated a range of hypothetical flow volumes similar to that used at the other two volcanoes.

DEBRIS-FLOW–HAZARD ZONATION ANALYSIS

Extents of Debris-Flow–Hazard Zones

Potential areas of debris-flow inundation along major channels that drain San Salvador, San Vicente, and San Miguel volcanoes are displayed as nested zones on the basis of the sizes of the debris flows (Figs. 14–16). The locations and sizes of areas affected by future debris flows will depend on the locations of the source areas and nature of the triggering mechanisms, the volume

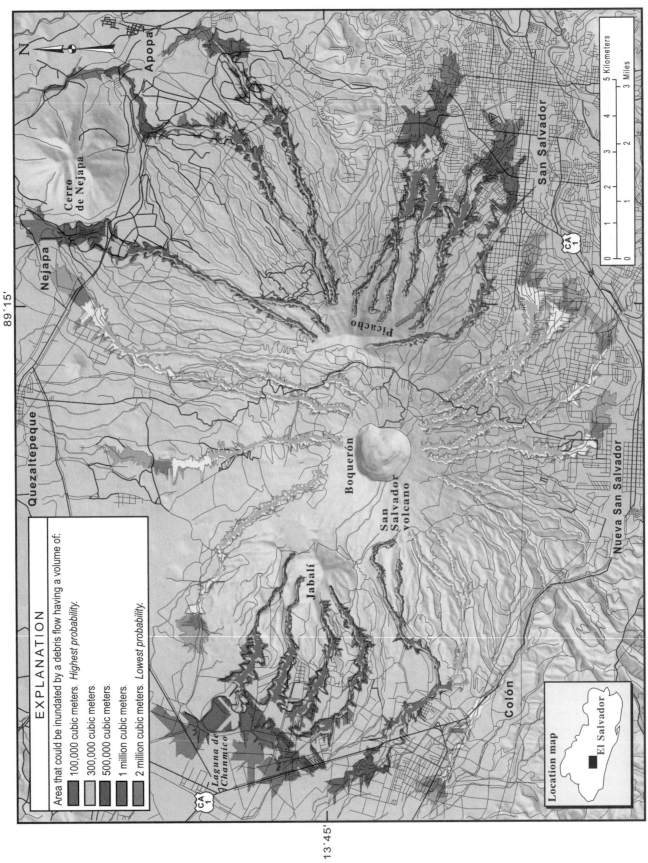

89°15'

N

Quezaltepeque

Nejapa

Apopa

Cerro
de Nejapa

Picacho

San Salvador

Boquerón

San
Salvador
volcano

Jabalí

Laguna de
Chanmico

CA
1

Colón

Nueva San Salvador

CA
1

13°45'

EXPLANATION

Area that could be inundated by a debris flow having a volume of:

100,000 cubic meters. *Highest probability.*

300,000 cubic meters.

500,000 cubic meters.

1 million cubic meters.

2 million cubic meters. *Lowest probability.*

Location map

■ El Salvador

0   1   2   3   4   5 Kilometers
0   1   2   3 Miles

Base maps from El Salvador 1:50,000 scale series: San Salvador quadrangle, 1984 (2357II); San Salvador quadrangle, 1983 (2357III); from best available source. Digital Base Maps from Titan Averstar, Inc. Universal Transverse Mercator projection, Zone 16, Horizontal Datum North American 1927, Vertical Datum Mean Sea Level, Spheroid Clarke 1866.

Figure 14. Debris-flow-hazard zonation for San Salvador volcano. Channels that head on the volcano are subject to debris flows generated by debris avalanches, torrential rains, or earthquakes. Debris-flow–hazard zones are subdivided into five zones on the basis of hypothetical debris-flow volumes. The subdued color tones on the north and south flanks of the volcano indicate that these flanks are much less likely to experience debris flows than are the east and west flanks below Picacho and Jabalí. Drainages without hazard zones depicted were not studied in this analysis. Note the large areas of high population density that may be affected by future debris flows.

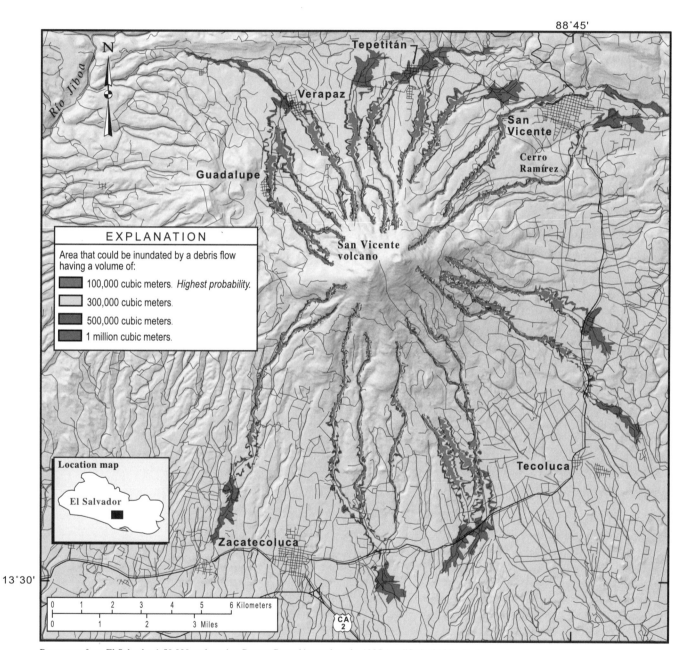

Base maps from El Salvador 1:50,000 scale series: Puente Cuscatlán quadrangle, 1985 (modified) (2456I); Berlín quadrangle, 1985 (2456II); La Herradura quadrangle, 1983 (2456III); San Vicente quadrangle, 1983 (2456IV); from best available source. Digital Base Maps from Titan Averstar, Inc. Universal Transverse Mercator projection, Zone 16, Horizontal Datum North American 1927, Vertical Datum Mean Sea Level, Spheroid Clarke 1866.

Figure 15. Debris-flow-hazard zonation for San Vicente volcano. Channels that head on the volcano are subject to debris flows generated by debris avalanches, torrential rains, or earthquakes. Debris-flow–hazard zones are subdivided into five zones on the basis of hypothetical debris-flow volumes. A: Hazard zones for debris flows ranging in volume from 100,000 m^3 to 1 million m^3. B: Hazard zone for a debris flow triggered by a 100 million m^3 debris avalanche. Drainages without hazard zones depicted were not studied in this analysis.

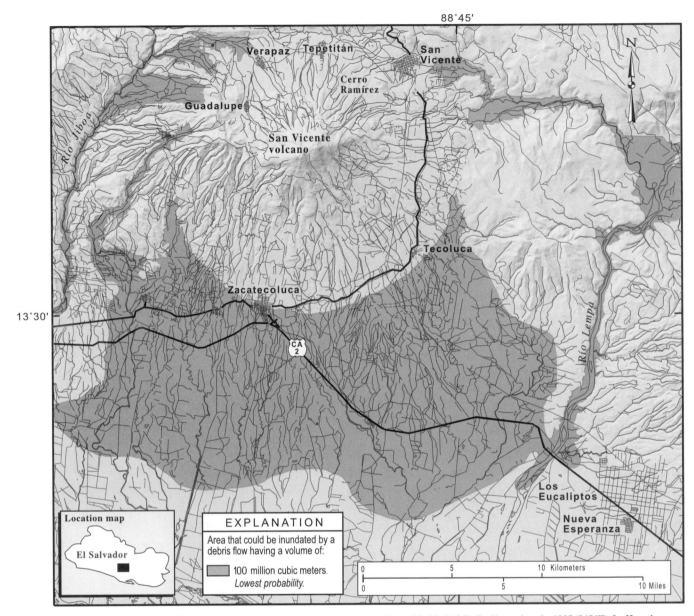

Figure 15 (*continued*).

Base maps from El Salvador 1:50,000 scale series: Puente Cuscatlán quadrangle, 1985 (modified?) (2456I); Berlín quadrangle, 1985 (2456II); La Herradura quadrangle, 1983 (2456III); San Vicente quadrangle, 1983 (2456IV); from best available source. Digital Base Maps from Titan Averstar, Inc. Universal Transverse Mercator projection, Zone 16, Horizontal Datum North American 1927, Vertical Datum Mean Sea Level, Spheroid Clarke 1866.

and character of rock and sediment involved, and the shapes of the channels along which debris flows pass. Although we show sharp boundaries for the debris-flow–hazard zones, the limit of the hazard does not change or end abruptly at these boundaries. Instead, the hazard decreases gradually with distance from the volcano and rapidly with increasing elevation above channel floors. Areas immediately beyond the outer hazard zones should not be regarded as hazard-free, because the limits of inundation can be located only approximately, especially in areas of low relief. Uncertainties about the sources, sizes, and mobilities of

future debris flows, and the statistical nature of the model, preclude locating the inundation boundaries precisely.

The hazard maps do not depict all areas subject to hazardous debris flows from the three volcanoes. For the analyses presented here, we depict zones of potential inundation along prominent channels directed toward populous areas. Other channels for which we have not modeled potential inundation should not be considered as areas devoid of debris-flow hazard.

In general, significant debris-flow–hazard zones lie within ~10 km (plan view) of the summits of the volcanoes (Figs. 14–

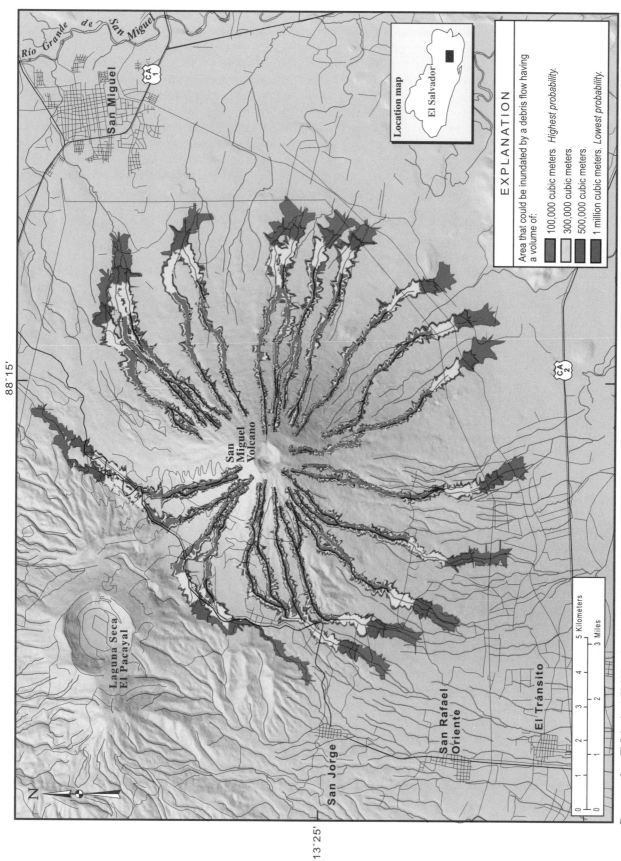

88°15'

13°25'

San Miguel

CA 1

Río Grande de San Miguel

Location map

El Salvador

EXPLANATION

Area that could be inundated by a debris flow having a volume of:

100,000 cubic meters. *Highest probability.*

300,000 cubic meters.

500,000 cubic meters.

1 million cubic meters. *Lowest probability.*

San Miguel Volcano

Laguna Seca El Pacayal

San Jorge

San Rafael Oriente

El Tránsito

CA 2

N

0   1   2   3   4   5 Kilometers
0   1   2   3 Miles

Base maps from El Salvador 1:50,000 scale series: Usulután quadrangle, 1983 (2556III); San Miguel quadrangle, 1983 (?) (2556II); from best available source. Digital Base Maps from Titan Averstar, Inc. Universal Transverse Mercator projection, Zone 16, Horizontal Datum North American 1927, Vertical Datum Mean Sea Level, Spheroid Clarke 1866.

Figure 16. Debris-flow-hazard zonation for San Miguel volcano. Channels that head on the volcano are subject to debris flows generated by torrential rains or earthquakes. Debris-flow–hazard zones are subdivided into four zones on the basis of hypothetical debris-flow volumes. Drainages without hazard zones depicted were not studied in this analysis.

16). Even hypothetical debris flows as large as 2 million m³ at San Salvador volcano extend no more than ~15 km from the Boquerón crater (Fig. 14). The hazard zone for a 100 million m³ debris flow at San Vicente volcano, representing a catastrophic flank failure, extends as much as 25 km from the summit and broadly covers the landscape surrounding the volcano (Fig. 15B).

The extent of debris-flow inundation at the three volcanoes is controlled strongly by local topography. Debris flows will originate in, and flow along, drainages deeply incised on the flanks of each volcano. However, the gradients of most channels decrease abruptly, and channels are incised less deeply, near the base of each edifice. As a result, debris flows will overflow channels, spread, and stop near the bases of the volcanoes. The most distant hazard zones are associated with those channels that remain deeply incised beyond the volcanoes, in which debris flows can remain confined, such as to the southwest and northeast of San Salvador volcano, and to the north and northeast of San Vicente volcano.

Despite the limited extents of the hazard zones portrayed by our analyses, even the smallest debris flows could be devastating. The city of San Salvador and surrounding communities have encroached onto the lower flanks of San Salvador volcano; all major towns near San Vicente volcano are located within 10 km of its summit; and although major towns around San Miguel volcano lie beyond 10 km from the summit, smaller villages, coffee plantations, and important transportation routes are located on its lower flanks and lie within 10 km. At each volcano, the hazard zones of even the smallest debris flows extend well into areas that are settled densely or used for agriculture.

Probabilities of Debris-Flow Occurrences

Small debris flows occur more frequently than large debris flows. Thus, the nested debris-flow–hazard zones depict relative degrees of hazard and show that the likelihood of inundation decreases with distance from a volcano and with increasing elevation above channel beds. Areas immediately adjacent to channel margins are at substantially higher risk of inundation than are areas farther removed. The degree of hazard would be more informative, however, if frequencies of occurrence could be defined for debris flows of various magnitudes, similar to those defined for floods. Debris flows, however, are more episodic, less spatially regular, and less well documented than floods, and accurately quantifying the annual probabilities of inundation for the hazard zones depicted in Figures 14–16 is difficult given the paucity of well-dated events having well-constrained volumes. Nevertheless, from the historical data compiled in Table 1, from analyses of geologic deposits, and from regional occurrences of landslides and debris flows, we can gain some insight into possible recurrence frequencies.

The regional recurrence of small- to moderate-volume (≤500,000 m³) landslides and debris flows triggered by earthquakes provides some insight into the broad recurrence of events of these sizes, although the probabilities of recurrence are not specific to any given site, let alone to the volcanoes of interest.

Hundreds to thousands of earthquake-triggered landslides have occurred throughout El Salvador at least a dozen times between 1857 and 2001 (Rymer and White, 1989; Bommer and Rodríguez, 2002). Volumes of those landslides have ranged from a few hundred m³ to more than 10 million m³, but most had volumes of less than a few to a few tens of thousands of m³ (Rymer and White, 1989; Baum et al., 2001; Jibson and Crone, 2001; Jibson et al., this volume, Chapter 6). Thus, earthquake-induced landslides of small to moderate volumes occur very frequently in El Salvador, on average about once per decade. If such landslides occur in wet materials, they could transform into debris flows of comparable magnitudes. The annual probability of an earthquake-triggered landslide (p ~0.1) possibly bounds the maximum frequency of debris flows at all but San Miguel volcano (discussed below).

San Salvador Volcano

Definitive evidence for debris flows as voluminous as 1 to 2 million m³ is lacking at San Salvador volcano. Sofield (1998), however, reports finding a landslide deposit possibly as voluminous as 2 million m³ on the western slopes of Jabalí (Fig. 14). Debris from the landslide was deposited shortly after an eruption around A.D. 1200. As far as is known, this is the only landslide or debris flow of this magnitude that has occurred since deposition of the G-1 tephra ca. 30 ka to 50 ka. Granted the evidence is circumstantial, but on the basis of an imperfect geologic record the annual probability of a debris flow on the order of 10^6 m³ is apparently less than ~1 in 30,000. Smaller debris flows are more likely to occur; therefore, inundation zones of debris flows having volumes <1 million m³ depict areas of substantially greater hazard.

At least two rainfall-triggered debris flows from Picacho have occurred in the past 70 yr, and they have had volumes on the order of 200,000 to 300,000 m³ (Table 1; Finnson et al., 1996; C.R. Pullinger, Servicio Nacional de Estudios Territoriales, unpublished data). Rainfall that approached or exceeded the greatest intensities, durations, and amounts that have been recorded in over 50 yr in the San Salvador area triggered a debris flow in 1982 (Finnson et al., 1996; Fig. 4). Little is known about another debris flow from an adjacent drainage in 1934 (Fig. 4) other than its occurrence and probable similarity in volume to the 1982 event. On the basis of the historical activity since the last eruption of the volcano, debris flows from Picacho having volumes as large as 300,000 m³ may recur about every 35–40 yr. If we combine the generalized recurrence frequencies of earthquake-triggered landslides across the country with the limited historical data on debris flows at San Salvador volcano, then we suggest that the annual probability of a debris flow from Picacho and perhaps Jabalí having a volume ≤300,000 m³ is bounded, to orders of magnitude, within the range from 1 in 100 to 1 in 10 ($0.01 < p < 0.1$). The annual probability of a debris flow from Boquerón, which is composed predominantly of unweathered lava flows (Sofield, this volume, Chapter 11) and has source slopes having gradients of less than 20 degrees, is less than 1 in 100. To clarify, these are estimates of occurrence probabilities at the volcano, not for any given drainage.

San Vicente Volcano

Debris flows from San Vicente volcano having volumes equal to or exceeding 1 million m³ have an annual probability of occurrence of less than 1 in 10,000 (p < 0.0001), judging by soil development on such deposits. Debris-flow deposits of this magnitude are capped by well-developed soil profiles that are similar in character to those that cap some of the Pleistocene tephra deposits from eruptions of Ilopango caldera (the TB2, TB3, and TB4 units). Those tephra deposits have been estimated to range from ~15,000 to 50,000 yr old (Rose et al., 1999; J.W. Vallance, U.S. Geological Survey, 2001, personal commun.).

Earthquake- and rainfall-triggered debris flows smaller than 1 million m³ have occurred at least six times at San Vicente volcano over the past 225 yr (Table 1), and at least five times in the past 90 yr. These documented events suggest that the average annual probability of occurrence of small to moderate debris flows at this volcano may range from ~1 in 50 to ~1 in 15. The extents of historical debris flows that reached the communities of San Vicente, Guadalupe, Verapaz, and Tepetitán, as well as the highway between Tecoluca and Zacatecoluca (see Fig. 15), suggest that these flows had volumes on the order of 300,000 m³ to 500,000 m³. Combining again the generalized recurrence frequencies of earthquake-triggered landslides across the country with the historical occurrences of debris flows at the volcano, we suggest that the annual probability of a debris flow ≤500,000 m³ at San Vicente volcano is bounded, to orders of magnitude, within the range from 1 in 100 to 1 in 10 (0.01 < p < 0.1).

San Miguel Volcano

Deposits of large debris flows are neither common nor extensive at San Miguel volcano. Ages and extents of the few documented prehistoric debris-flow deposits at San Miguel volcano (J.J. Major, C.A. Chesner, and C.R. Pullinger, unpublished data) are not known, and modern debris flows have had predominantly small volumes (<100,000 m³; Table 1). On the basis of a lack of such deposits, debris flows on the order of 1 million m³ at San Miguel volcano have an annual probability of occurrence similar to that at San Salvador and San Vicente volcanoes (p < 0.0001). Furthermore, debris flows at San Miguel volcano form predominantly by erosion of scoriaceous tephra on the slopes of the volcano and by small slumps off channel walls, rather than from channel-head landslides as at the other volcanoes. It is unlikely that such an initiation mechanism can readily generate a debris flow on the order of 1 million m³ at this volcano.

Debris flows ≤300,000 m³ have occurred 11 times in the past 60 yr at San Miguel volcano, and seven of these events were <100,000 m³ (Table 1). These historical events suggest that small- to moderate-magnitude debris flows occur more frequently here than at San Salvador and San Vicente volcanoes. On the basis of historical events, the annual probability of occurrence of small to moderate debris flows at San Miguel volcano falls in the range 0.1 < p < 0.2. The modern debris flows did not significantly affect population centers, but they damaged infrastructure including the main road leading from the Pan-American highway to San Jorge (see Figs. 12 and 16).

DEBRIS-FLOW–HAZARD FORECASTS AND WARNINGS

Debris flows associated with volcanic activity cannot be predicted precisely, but they can be anticipated if a volcano shows signs of unrest. Fortunately, scientists normally can recognize and monitor several indicators of impending volcanic eruptions (e.g., Ewert and Swanson, 1992), although they commonly can make only very general statements about the probability, type, and scale of an impending eruption (e.g., Newhall and Hoblitt, 2002). As an example of anticipating conditions favorable for generating eruption-related debris flows, consider San Miguel volcano. When it erupts again, tephra fall will be deposited on its flanks. Subsequent erosion of that tephra (see Fig. 11) may generate debris flows similar to those that have occurred during the past 60 yr at that volcano. In this case, the eruption of the volcano can serve as a warning that conditions are favorable for debris-flow formation, and the distribution of tephra can indicate which flanks are most prone to debris flow. Volcanic unrest, however, need not culminate in an eruption before landslides or debris flows are triggered. Intrusion of magma into a volcano could trigger a landslide before culminating in an eruption. In the rare event of a large debris avalanche triggered by intrusion of magma into a volcano (e.g., McGuire, 1996; Glicken, 1998), deformation of the volcano during intrusion can serve as warning that conditions are hazardous.

It is difficult, if not impossible, to predict the precise occurrence of debris flows triggered by earthquakes or torrential rains, although areas susceptible to such events can be recognized (e.g., Parise and Jibson, 2000; Baeza and Corominas, 2001). Governmental authorities and the public need to realize that potentially lethal events can occur within debris-flow–hazard zones with little or no warning. However, generally hazardous conditions that favor formation of landslides and debris flows can be recognized to some extent. Forecasts for very heavy rainfall, which commonly trigger flood warnings, can serve broadly as indicators that conditions are favorable for landslides and debris flows, especially if prior investigations have defined broad rainfall-intensity thresholds that delimit occurrences of landslides and debris flows (e.g., Keefer et al., 1987). If a volcano watershed experiences a significant wildfire, debris flows should be anticipated at the onset of subsequent rainfall (e.g., Cannon et al., 2001). A primary earthquake shock can serve as a warning that a debris flow may have been triggered, and ground tremor from an approaching flow sometimes can be perceived. Disaster can be averted if those living, working, or recreating within debris-flow–hazard zones seek higher ground without delay if any ground tremor is perceived (Scott, 2000).

PROTECTING COMMUNITIES AND CITIZENS FROM DEBRIS-FLOW HAZARDS

Communities and citizens must plan ahead to mitigate the effects of future debris flows from San Salvador, San Vicente, and San Miguel volcanoes. Long-term mitigation efforts must include using information about debris-flow and other volcano hazards (see, for example, Chesner et al., this volume, Chapter 16; Sofield, this volume, Chapter 11) when making decisions about land use, development, and the siting of critical facilities. Future development should avoid areas that have an unacceptably high risk, or be planned and designed to reduce the level of risk.

Owing to typical debris-flow velocities (3–10 m/s), the hazard zones depicted on the maps herein are areas that can be inundated within a few minutes to a few tens of minutes after the onset of a debris flow. Beyond 10 km from the summits of the volcanoes, escape from debris flows may be possible if people are given sufficient warning; within 10 km debris flows may form and advance too quickly for governmental authorities or others to provide effective warning. Therefore, citizens must learn to recognize for themselves hazardous conditions that favor formation of landslides and debris flows (Major et al., 2003).

Debris flows can occur without warning; therefore, suitable emergency plans for dealing with them should be made in advance. Although it is uncertain when debris flows will occur again at these volcanoes, they occur sufficiently frequently that public officials need to consider issues such as public education, communications, and evacuations as part of a response plan. Emergency plans already developed for floods may apply to some extent, but will need modifications. For inhabitants in low-lying areas, a map showing the shortest route to safe havens would be helpful.

Knowledge and advance planning are the most important items for dealing with debris-flow hazards. Especially important is a plan of action based on the knowledge of relatively safe areas around locations of common gathering, such as homes, schools, and workplaces. Debris flows pose the biggest threat to people living, working, or recreating along river channels that drain the volcanoes. The best strategy for avoiding a debris flow is to move to the highest possible ground. A safe height above river channels depends on many factors including the size of the debris flow, distance from the volcano, and shape of the valley. In general, all but the largest debris flows from these volcanoes will rise less than ~20 m above river level. Volcanic debris flows will recur in El Salvador, and the best way to mitigate their effects is through advance planning.

ACKNOWLEDGMENTS

The tragedy at Casita volcano, Nicaragua, during the passage of Hurricane Mitch motivated the analyses of debris-flow hazards at volcanoes in El Salvador. The U.S. Agency for International Development (USAID) funded the U.S. Geological Survey (USGS) to conduct these analyses. Logistics of working in El Salvador were greatly facilitated by Elizabeth Gonzalez (USAID) and by Randy Updike, Paul Hearn, Verne Schneider, and Willie Rodríguez (all from USGS). Elizabeth Gonzalez, Jorge Rodríguez (Centro de Investigaciones Geotécnicas), and Manuel Diaz (Servicio Nacional de Estudios Territoriales) provided in-country support. Discussions with Manuel Diaz, Luz Barrios, Arturo Quezada, Bill Rose, James Vallance, Tony Crone, Ed Harp, and Craig Chesner stimulated our thinking and aided our efforts in these analyses. Michelle Howell assisted with development of the hazard maps. Reviews of various versions of this chapter by Tony Crone, Dan Miller, Willie Scott, Cynthia Gardner, Lee Siebert, John Ewert, Russ Graymer, Wendy Gerstel, Ed Harp, and Darrell Sofield focused our thoughts and clarified our presentation. We appreciate their thoughtful suggestions for improvement.

REFERENCES CITED

Baeza, C., and Corominas, J., 2001, Assessment of shallow landslide susceptibility by means of multivariate statistical techniques: Earth Surface Processes and Landforms, v. 26, p. 1251–1263.

Baum, R.L., Crone, A.J., Escobar, D., Harp, E.L., Major, J.J., Martinez, M., Pullinger, C.R., and Smith, M.E., 2001, Assessment of landslide hazards resulting from the February 13, 2001, El Salvador earthquake: U.S. Geological Survey Open-File Report 01-119, 22 p.

Black, K.D., 1983, The Zapotitán Valley archeological survey, in Sheets, P.D., ed., Archeology and volcanism in Central America: Austin, University of Texas Press, p. 62–97.

Blanco, F.A., Burgos, E.A., and Mejia, M., 2002, Estudio de amenazas por lahar en El Salvador: Revisión de casos históricos y calibración de herramientas para la evaluación de amenaza [senior thesis]: El Salvador, Universidad Centroamericana "José Simeón Cañas," 158 p.

Bommer, J.J., and Rodríguez, C.E., 2002, Earthquake-induced landslides in Central America: Engineering Geology, v. 63, p. 189–220.

Brauer, J., Smith, J., and Wiles, V., 1995, On your own in El Salvador: Portland, Oregon, On Your Own Publications, 260 p.

Cannon, S.H., Kirkham, R.M., and Parise, M., 2001, Wildfire-related debris-flow initiation processes, Storm King Mountain, Colorado: Geomorphology, v. 39, p. 171–188.

Chesner, C.A., Pullinger, C.R., and Escobar, C.D., 2004, Physical and chemical evolution of San Miguel Volcano, in Rose, W.I., et al., eds., Natural hazards in El Salvador: Boulder, Colorado, Geological Society of America Special Paper 375, p. 213–226 (this volume).

Costa, J.E., and Schuster, R.L., 1988, The formation and failure of natural dams: Geological Society of America Bulletin, v. 100, p. 1054–1068.

Crandell, D.R., 1989, Gigantic debris-avalanche of Pleistocene age from ancestral Mount Shasta volcano, California, and debris avalanche hazard zonation: U.S. Geological Survey Bulletin 1861, 32 p.

Crone, A.J., Baum, R.L., Lidke, D.J., Sather, D.N.D., Bradley, L.A., and Tarr, A.C., 2001, Landslides induced by Hurricane Mitch in El Salvador—An inventory and descriptions of selected features: U.S. Geological Survey Open-File Report 01-444, 24 p.

Day, S.J., 1996, Hydrothermal pore fluid pressure and the stability of porous, permeable volcanoes, in McGuire, et al., eds., Volcano instability on the Earth and other planets: Geological Society [London] Special Publication 110, p. 77–93.

Dull, R.A., Southton, J.R., and Sheets, P., 2001, Volcanism, ecology, and culture: A reassessment of the Volcán Ilopango TBJ eruption in the southern Maya realm: Latin American Antiquity, v. 12, p. 25–44.

Elsworth, D., and Voight, B., 1996, Evaluation of volcano flank instability triggered by dyke intrusion, in McGuire, et al., eds., Volcano instability on the Earth and other planets: Geological Society [London] Special Publication 110, p. 45–53.

Endo, K., Sumita, M., Machida, M., and Furuichi, M., 1989, The 1984 collapse and debris avalanche deposits of Ontake Volcano, Central Japan, in Latter, J.H., ed., Volcanic hazards assessment and monitoring: Berlin, Springer Verlag, p. 210–229.

Escobar, C.D., Mendez, I., and Ramirez, O.R., 1993, Estudio geologico preliminar del Volcán de San Miguel: Peligrosidad eruptiva, estado erosivo, y sus consecuencias: Trabajo de Graduacion, Facultad de Ingenieria y Arquitectura, Universidad Tecnológica, San Salvador, El Salvador, 176 p.

Ewert, J.E., and Swanson, D.A., eds., 1992, Monitoring volcanoes: Techniques and strategies used by the staff of the Cascades Volcano Observatory, 1980–1990: U.S. Geological Survey Bulletin 1966, 223 p.

Feldman, L.H., 1993, Mountains of fire, lands that shake: Culver City, California, Labyrinthos Press, 295 p.

Finn, C.A., Sisson, T.W., and Deszcz-Pan, M., 2001, Aerogeophysical measurements of collapse-prone hydrothermally altered zones at Mount Rainier volcano: Nature, v. 409, p. 600–603.

Finnson, H., Bäcklin, C., and Bodare, A., 1996, Landslide hazard at the San Salvador volcano, El Salvador, *in* Senneset, K., ed., Proceedings of the 7th International Conference on Landslides, Trondheim, Norway, June 17–21: Rotterdam, A.A. Balkema, p. 215–220.

Folk, R.L., 1984, Petrology of sedimentary rocks: Austin, Texas, Hemphill, 184 p.

Francis, P.W., and Self, S., 1987, Collapsing volcanoes: Scientific American, v. 255, p. 90–97.

Glicken, H., 1998, Rockslide–debris avalanche of May 18, 1980, Mount St. Helens volcano, Washington: Bulletin of the Geological Survey of Japan, v. 49, p. 55–106.

Hayashi, J.N., and Self, S., 1992, A comparison of pyroclastic flow and debris avalanche mobility: Journal of Geophysical Research, v. 97, p. 9063–9071.

Iverson, R.M., Reid, M.E., and LaHusen, R.G., 1997, Debris-flow mobilization from landslides: Annual Reviews of Earth and Planetary Sciences, v. 25, p. 85–138.

Iverson, R.M., Schilling, S.P., and Vallance, J.W., 1998, Objective delineation of lahar-hazard zones downstream from volcanoes: Geological Society of America Bulletin, v. 110, p. 972–984.

Iverson, R.M., Reid, M.E., Iverson, N.R., LaHusen, R.G., Logan, M., Mann, J.E., and Brien, D.L., 2000, Acute sensitivity of landslide rates to initial soil porosity: Science, v. 290, p. 513–516.

Janda, R.J., Daag, A.S., Delos Reyes, P.J., Newhall, C.G., Pierson, T.C., Punongbayan, R.S., Rodolfo, K.S., Solidum, R.U., and Umbal, J.V., 1996, Assessment and response to lahar hazard around Mount Pinatubo, 1991–1993, *in* Newhall, C.G., and Punongbayan, R.S., eds., Fire and mud: Eruptions and lahars of Mount Pinatubo, Philippines: Seattle, University of Washington Press, p. 107–140.

Jibson, R.W., and Crone, A.J., 2001, Observations and recommendations regarding landslide hazards related to the January 13, 2001 M-7.6 El Salvador earthquake: U.S. Geological Survey Open-File Report 01-141, 19 p.

Jibson, R.W., Crone, A.J., Harp, E.L., Baum, R.L., Major, J.J., Pullinger, C.R., Escobar, C.D., Martínez, M., and Smith, M.E., 2004, Landslides triggered by the 13 January and 13 February 2001 earthquakes in El Salvador, *in* Rose, W.I., et al., eds., Natural hazards in El Salvador: Boulder, Colorado, Geological Society of America Special Paper 375, p. 69–88 (this volume).

Keefer, D.K., 1999, Earthquake-induced landslides and their effects on alluvial fans: Journal of Sedimentary Research, v. 69, p. 84–104.

Keefer, D.K., Wilson, R.C., Mark, R.K., Brabb, E.E., Brown, W.M., III, Ellen, S.D., Harp, E.L., Wieczorek, G.F., Alger, C.S., and Zatkin, R.S., 1987, Real-time landslide warning during heavy rainfall: Science, v. 238, p. 921–925.

Kerle, N., and van Wyk de Vries, B., 2001, The 1998 debris avalanche at Casita volcano, Nicaragua—Investigation of structural deformation as the cause of slope instability using remote sensing: Journal of Volcanology and Geothermal Research, v. 105, p. 49–63.

Lavigne, F., and Thouret, J.C., 2002, Sediment transportation and deposition by rain-triggered lahars at Merapi Volcano, Central Java, Indonesia: Geomorphology, v. 49, p. 45–69.

Lavigne, F., Thouret, J.C., Voight, B., Suwa, H., and Sumaryono, A., 2000a, Lahars at Merapi volcano, Central Java: An overview: Journal of Volcanology and Geothermal Research, v. 100, p. 423–456.

Lavigne, F., Thouret, J.C., Voight, B., Young, K., LaHusen, R., Marso, J., Suwa, H., Sumaryono, A., Sayudi, D.S., and Dejean, M., 2000b, Instrumental lahar monitoring at Merapi Volcano, Central Java, Indonesia: Journal of Volcanology and Geothermal Research, v. 100, p. 457–478.

López, D.L., and Williams, S.N., 1993, Catastrophic volcanic collapse: relation to hydrothermal processes: Science, v. 260, p. 1794–1796.

Major, J.J., Janda, R.J., and Daag, A., 1996, Watershed disturbance and lahars on the east side of Mount Pinatubo during the mid-June 1991 eruptions, *in* Newhall, C.G., and Punongbayan, R.S., eds., Fire and mud: Eruptions and lahars of Mount Pinatubo, Philippines: Seattle, University of Washington Press, p. 895–919.

Major, J.J., Pierson, T.C., Dinehart, R.L., and Costa, J.E., 2000, Sediment yield following severe volcanic disturbance—A two decade perspective from Mount St. Helens: Geology, v. 28, p. 819–822.

Major, J.J., Schilling, S.P., and Pullinger, C.R., 2003, Volcanic debris flows in developing countries—The extreme need for public education and awareness of debris-flow hazards, *in* Rickenmann, D., and Chen, C.L., eds., Debris-flow hazards mitigation, Proceedings of the 3rd International Conference on Debris-Flow Hazards Mitigation:Mechanics, Prediction, and Assessment, Davos, Switzerland, 10–12 September 2003: Rotterdam, MillPress, p. 1185–1196.

Malin, M.C., and Sheridan, M.F., 1982, Computer-assisted mapping of pyroclastic surges: Science, v. 217, p. 637–640.

Manville, V., 2002, Sedimentary and geomorphic response to ignimbrite emplacement: Readjustment of the Waikato River after the A.D. 181 Taupo eruption, New Zealand: Journal of Geology, v. 110, p. 519–541.

Martinez, J.M., Avila, G., Agudelo, A., Schuster, R.L., Casadevall, T.J., and Scott, K.M., 1995, Landslides and debris flows triggered by the 6 June 1994 Paez earthquake, southwestern Colombia: Landslide News, v. 9, p. 13–15.

McGuire, W.J., 1996, Volcano instability: A review of contemporary themes, in McGuire, et al., eds., Volcano instability on the Earth and other planets: Geological Society [London] Special Publication 110, p. 1–23.

Meyer, D.F., and Janda, R.J., 1986, Sedimentation downstream from the 18 May 1980 North Fork Toutle River debris avalanche deposit, Mount St. Helens, Washington, in Keller, S.A.C., ed., Mount St. Helens: Five years later: Cheney, Eastern Washington University Press, p. 68–86.

Newhall, C.G., and Hoblitt, R.P., 2002, Constructing event trees for volcanic crises: Bulletin of Volcanology, v. 64, p. 3–20.

Parise, M., and Jibson, R.W., 2000, A seismic landslide susceptibility rating of geologic units based on an analysis of characteristics of landslides triggered by the 17 January 1994 Northridge, California earthquake: Engineering Geology, v. 58, p. 251–270.

Pierson, T.C., 1995, Flow characteristics of large eruption-triggered debris flows at snow-clad volcanoes: Constraints for debris-flow models: Journal of Volcanology and Geothermal Research, v. 66, p. 283–294.

Pierson, T.C., Janda, R.J., Thouret, J-C., and Borrero, C.A., 1990, Perturbation and melting of snow and ice by the 13 November 1985 eruption of Nevado del Ruiz, Colombia, and consequent mobilization, flow, and deposition of lahars: Journal of Volcanology and Geothermal Research, v. 41, p. 17–66.

Pierson, T.C., Daag, A.S., Delos Reyes, P.J., Regalado, M.T.M., Solidum, R.U., and Tubianosa, B.S., 1996, Flow and deposition of posteruption hot lahars on the east side of Mount Pinatubo, July–October 1991, *in* Newhall, C.G., and Punongbayan, R.S., eds., Fire and mud: Eruptions and lahars of Mount Pinatubo, Philippines: Seattle, University of Washington Press, p. 921–950.

Reid, M.E., Sisson, T.W., and Brien, D.L., 2001, Volcano collapse promoted by hydrothermal alteration and edifice shape, Mount Rainier, Washington: Geology, v. 29, p. 779–782.

Rodolfo, K.S., 1989, Origin and early evolution of lahar channel at Mabinit, Mayon Volcano, Philippines: Geological Society of America Bulletin, v. 101, p. 414–426.

Rodolfo, K.S., and Arguden, A.T., 1991, Rain-lahar generation and sediment-delivery systems at Mayon Volcano, Philippines, in Fisher, R.V., and Smith, G.A., eds., Sedimentation in volcanic settings: SEPM (Society for Sedimentary Geology) Special Publication 45, p. 71–87.

Romano, L.E., 1997, Catálogo de desastres, accidentes y ecologia (1915–1990): San Salvador, El Salvador, Centro de Protección para Desastres (CEPRODE), 61 p.

Rose, W.I., Conway, F.M., Pullinger, C.R., Deino, A., and McIntosh, W.C., 1999, An improved age framework for late Quaternary silicic eruptions in northern Central America: Bulletin of Volcanology, v. 61, p. 106–120.

Rotolo, S.G., and Castorina, F., 1998, Transition from mildly tholeiitic to calc-alkaline suite: The case of Chichontepec volcanic center, El Salvador, Central America: Journal of Volcanology and Geothermal Research, v. 86, p. 117–136.

Rotolo, S.G., Aiuppa, A., Pullinger, C.R., Parello, F., and Tenorio-Mejica, J., 1998, An introduction to San Vicente (Chichontepec) volcano, El Salvador: Revista Geológica de América Central, v. 21, p. 25–36.

Rymer, M.J., and White, R.A., 1989, Hazards in El Salvador from earthquake-induced landslides, *in* Brabb, E.E., and Harrod, B.L., eds., Landslides: Extent and economic significance: Rotterdam, Balkema, p. 105–109.

Schilling, S.P., 1998, LAHARZ: GIS programs for automated mapping of lahar-inundation hazard zones: U.S. Geological Survey Open-File Report 98-638, 80 p.

Scott, K.M., 1988, Origins, behavior, and sedimentology of lahars and lahar-runout flows in the Toutle-Cowlitz River system: U.S. Geological Survey Professional Paper 1447-A, 76 p.

Scott, K.M., 2000, Precipitation-triggered debris-flow at Casita Volcano, Nicaragua: Implications for mitigation strategies in volcanic and tectonically active steeplands, *in* Wieczorek, G.F., and Naeser, N.D., eds., Debris-flow hazards mitigation, Proceedings of the 2nd International Conference on Debris-Flow Hazards Mitigation, Taipei, Taiwan, 16–18 August 2000: Rotterdam, A.A. Balkema, p. 3–13.

Scott, K.M., Vallance, J.W., and Pringle, P.T., 1995, Sedimentology, behavior, and hazards of debris flows at Mount Rainier, Washington: U.S. Geological Survey Professional Paper 1547, 56 p.

Scott, K.M., Janda, R.J., de la Cruz, E.G., Gabinete, E., Eto, I., Isada, M., Sexon, M., and Hadley, K.C., 1996, Channel and sedimentation responses to large volumes of 1991 volcanic deposits on the east flank of Mount Pinatubo, *in* Newhall, C.G., and Punongbayan, R.S., eds., Fire and mud: Eruptions and lahars of Mount Pinatubo, Philippines: Seattle, University of Washington Press, p. 971–988.

Scott, K.M., Macías, J.L., Naranjo, J.A., Rodríguez, S., and McGeehin, J.P., 2001, Catastrophic debris flows transformed from landslides in volcanic terrains: Mobility, hazard assessment, and mitigation strategies: U.S. Geological Survey Professional Paper 1630, 59 p.

Scott, W.E., Pierson, T.C., Schilling, S.P., Costa, J.E., Gardner, C.A., Vallance, J.W., and Major, J.J., 1997, Volcano hazards in the Mount Hood region, Oregon: U.S. Geological Survey Open-File Report 97-89, 14 p.

Siebert, L., 1984, Large volcanic debris avalanches: Characteristics of source areas, deposits, and associated eruptions: Journal of Volcanology and Geothermal Research, v. 22, p. 163–197.

Siebert, L., Glicken, H.X., and Ui, T., 1987, Volcanic hazards from Bezymianny- and Bandai-type eruptions: Bulletin of Volcanology, v. 49, p. 435–459.

Siebert, L., Kimberly, P., and Pullinger, C.R., 2004, The voluminous Acajutla debris avalanche from Santa Ana volcano, western El Salvador, and comparison with other Central American edifice-failure events, *in* Rose, W.I., et al., eds., Natural hazards in El Salvador: Boulder, Colorado, Geological Society of America Special Paper 375, p. 5–23 (this volume).

Simkin, T., and Siebert, L., 1994, Volcanoes of the world (2nd edition): Tucson, Arizona, Geosciences Press, 349 p.

Smithsonian Institution, 2002, Summary of recent volcanic activity—San Miguel volcano, El Salvador: Global Volcanism Network Bulletin, v. 27, no. 2, p. 5.

Sofield, D.J., 1998, History and hazards of Volcán San Salvador, El Salvador [M.S. thesis]: Houghton, Michigan Technological University, 116 p.

Sofield, D., 2004, Eruptive history and volcanic hazards of Volcán San Salvador, *in* Rose, W.I., et al., eds., Natural hazards in El Salvador: Boulder, Colorado, Geological Society of America Special Paper 375, p. 147–158 (this volume).

Sutherland, D.G., Ball, M.H., Hilton, S.J., and Lisle, T.E., 2002, Evolution of a landslide-induced sediment wave in the Navarro River, California: Geological Society of America Bulletin, v. 114, p. 1036–1048.

Suwa, H., and Yamakoshi, T., 1999, Sediment discharge by storm runoff at volcanic torrents affected by eruption: Zeitschrift für Geomorphologie, Supplement-Band 114, p. 63–88.

Thouret, J.C., Abdurachman, K.E., Bourdier, J.L., and Bronto, S., 1998, Origin, characteristics, and behaviour of lahars following the 1990 eruption of Kelud volcano, eastern Java (Indonesia): Bulletin of Volcanology, v. 59, p. 460–480.

Umbal, J.V., and Rodolfo, K.S., 1996, The 1991 lahars of southwestern Mount Pinatubo and evolution of the lahar-dammed Mapanuepe Lake, *in* Newhall, C.G., and Punongbayan, R.S., eds., Fire and mud: Eruptions and lahars of Mount Pinatubo, Philippines: Seattle, University of Washington Press, p. 951–970.

Vallance, J.W., and Scott, K.M., 1997, The Osceola mudflow from Mount Rainier: Sedimentology and hazards implications of a huge clay-rich debris flow: Geological Society of America Bulletin, v. 109, p. 143–163.

Varnes, D.J., 1978, Slope movement types and processes, *in* Schuster, R.L., and Krizek, R.J., eds., Landslides—Analysis and control: National Academy of Sciences Transportation Research Board Special Report 176, p. 11–33.

Voight, B., 1981, Time scale for the first moments of the May 18 eruption, *in* Lipman, P.W., and Mullineaux, D.R., eds., The 1980 eruptions of Mount St. Helens, Washington: U.S. Geological Survey Professional Paper 1250, p. 69–86.

Voight, B., Glicken, H., Janda, R.J., and Douglass, P.M., 1981, Catastrophic rockslide avalanche of May 18, *in* Lipman, P.W., and Mullineaux, D.R., eds., The 1980 eruptions of Mount St. Helens, Washington: U.S. Geological Survey Professional Paper 1250, p. 347–377.

Voight, B., Janda, R.J., Glicken, H., and Douglass, P.M., 1983, Nature and mechanics of the Mount St. Helens rockslide-avalanche of 18 May 1980: Geotechnique, v. 33, p. 224–273.

Watters, R.J., Zimbelman, D.R., Bowman, S.D., and Crowley, J.K., 2000, Rock mass strength assessment and significance to edifice stability, Mount Rainier and Mount Hood, Cascade Range volcanoes: Pure and Applied Geophysics, v. 157, p. 957–976.

Williams, H., and Meyer-Abich, H., 1955, Volcanism in the southern part of El Salvador: University of California Publications in Geological Sciences, v. 32, 64 p.

Yamagishi, H., 1996, Destructive mass movements associated with Quaternary volcanoes in Hokkaido, Japan, *in* McGuire, et al., eds., Volcano instability on the Earth and other planets: Geological Society [London] Special Publication 110, p. 267–279.

MANUSCRIPT ACCEPTED BY THE SOCIETY JUNE 16, 2003

Geological Society of America
Special Paper 375
2004

Apocalypse then: Social science approaches to volcanism, people, and cultures in the Zapotitán Valley, El Salvador

Payson Sheets

Anthropology, University of Colorado, Boulder Colorado 80309-0233, USA

ABSTRACT

Preindustrial human societies living in ancient Central America apparently varied considerably in their vulnerabilities to the sudden massive stresses caused by explosive volcanic eruptions. The focus is on ancient complex societies and volcanic eruptions in the Zapotitán Valley during the past two millennia in what is now El Salvador, within a contextual and comparative framework of other eruptions and societies from other areas of Central America. Comparing eruptions and their societal effects indicates that there was a threshold of magnitude of these eruptions beyond which the society affected did not recover, and reoccupation was by a different ethnic group. An example is the TBJ (Tierra Blanca Joven) eruption of Ilopango where the original inhabitants never recovered to reoccupy the area. In contrast, the smaller Boquerón eruption devastated the eastern half of the Zapotitán Valley, but recovery was achieved by the indigenous society. The concept of "scalar vulnerabilities" is introduced, where physical factors such as the magnitude of eruptions are compared along with cultural factors such as variation in organizational complexity and institutionalized hostility. Simpler egalitarian societies in this sample were more resilient to sudden massive stresses than were the more complex societies. Simpler societies relied less on intensive agriculture, staple crops, fixed facilities, redistributive economies, and hierarchical institutions. Cultural factors such as chronic political hostilities can greatly increase the vulnerability of societies at any point on the simple-to-complex range. It is argued here that assessing a society's vulnerability must include the magnitude of stress, the complexity of the society (including the related factors of adaptation, demography, and the "built environment"), and the political landscape.

Keywords: scalar vulnerability, societies, risk, stress, recovery.

INTRODUCTION

It is suggested here that scalar vulnerabilities are helpful in exploring the impact and repercussions of explosive volcanic eruptions on pre-Columbian societies. These include phenomena in the natural sciences such as the magnitude of the eruption, the flora and fauna, climate, and soils in the area of impact. These also include phenomena studied by social scientists such as demography, societal complexity, adaptation, and the political landscape (Sheets, 1980). Variation along any of these scales can affect the vulnerability of a society attempting to cope with sudden massive stress.

Numerous volcanic eruptions have affected the Zapotitán Valley of El Salvador during the past two millennia, and the emphasis here is on the interplay among the factors of the volcanic impact, effects on flora, fauna, soils, and human societies, as well as the recoveries or lack of recoveries as they are documented by research. The sample size is small, particularly when examined carefully, as the cases of eruptions, effects, and recoveries are very uneven in how thoroughly each is documented. Therefore, before turning to eruptions within the valley, a wider framework for comparison and interpretation is established by examining two other areas in Central America, northwestern Costa Rica and western Panama.

Sheets, P., 2004, Apocalypse then: Social science approaches to volcanism, people, and cultures in the Zapotitán Valley, El Salvador, *in* Rose, W.I., Bommer, J.J., López, D.L., Carr, M.J., and Major, J.J., eds., Natural hazards in El Salvador: Boulder, Colorado, Geological Society of America Special Paper 375, p. 109–120. For permission to copy, contact editing@geosociety.org. © 2004 Geological Society of America.

The Arenal Cases, Costa Rica

Human societies vary considerably in their vulnerability to disasters, including hazard perception, ability to cope with a sudden massive stress, make adjustments, and recover or attempt to recover. A study of ancient Mexican–Central American cases of explosive volcanism found that "simple" egalitarian sedentary communities were considerably more resilient to explosive volcanism than their more complex chiefdom and state neighbors (Sheets, 2001). The striking resilience of egalitarian ancient Costa Rican societies to some ten explosive eruptions (Melson, 1994) of Arenal volcano (Fig. 1) is attributed to low population densities, availability of refuge areas, utilization of agriculture for a small percentage of the diet, reliance on a wide variety of wild food sources in the rich tropical rainforest for a majority of the diet, lack of hostility among groups, and minimal investment in the "built environment" of facilities constructed for living, storage, workshops, or agriculture. The elaborate post-interment feasting in cemeteries may have involved multiple communities, and if so, could have facilitated exchanges and alliances that could prove useful in times of emergencies when refuge areas

were needed. Complex state-level societies in ancient Mesoamerica evidently were more vulnerable to sudden massive stress because of their greater reliance on the built environment as well as high population densities, scarcity of refuge areas, reliance on intensive agriculture, and domesticated dietary staples. The utilization of a wide range of food sources, from seed crops such as maize and beans, through root crops such as manioc, to tree crops such as palm fruits, may provide resilience to sudden stresses. A significant tephra fall of 20 cm could be very damaging to young seed crops, mildly damaging to some root crops, and minimally deleterious to tree crops. The actual tephra depths in the research area range from only a few cm to a meter or two per eruption. The thickness of each tephra layer thins from the eastern end, closer to the source, toward the western end.

The Barriles Case, Panama

One case highlights the importance of understanding cultural factors in the complex blend of significant physical and social aspects of disasters (Sheets, 2001). Volcán Baru (Fig. 1) in western Panama erupted, probably in the seventh century A.D.,

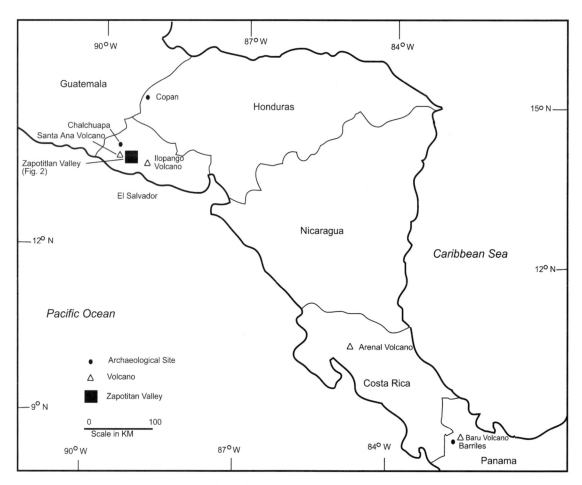

Figure 1. Map of Central America.

and deposited a relatively thin tephra layer, perhaps some 10–15 cm originally, on settlements in the Barriles area (Linares and Ranere, 1980). The social organization of Barriles, and Aguas Buenas culture generally, are controversial (Hoopes, 1996). Hoopes suggests the ranking that is detectable in sites may be only local "big men" rather than being more centralized chiefdoms. Barriles society consisted of a series of culturally similar but politically independent polities packed into the flood plain of the Rio Chiriqui Viejo. The tephra depth is comparable to Arenal tephra depths deposited on egalitarian villages toward the western end of Lake Arenal, where egalitarian communities recovered rapidly, probably within a few decades, with no detectable changes in their architecture, artifacts, adaptation, or any other measure that could be found. So, based solely on volcanological and ecological factors one could predict a reasonably short-term dislocation of Barriles peoples until soil and vegetation recovery had proceeded sufficiently to allow reoccupation. Or, if we focused on vulnerability related to social complexity, based upon Barriles society being mid-range between egalitarian and a state, one might expect relatively mild to negligible long-term effects from a relatively small eruption.

In fact, the stress from the thin tephra blanket from Volcán Baru forced a drastic cultural adjustment. The Barriles chiefdoms in a highland tropical moist seasonal environment in the upper reaches of the river abandoned their territories, never to return, and thus cultural recovery never occurred. They migrated over the continental divide to an environment with much greater precipitation and had to change their architecture and agriculture to adapt to the changed conditions (Linares and Ranere, 1980). Physical factors such as tephra thickness, chemistry, natural environment, flora, or fauna do not explain the dramatic differences in human reactions. Cultural factors are relevant and include the following. The Barriles societies relied on more intensive agriculture than did Arenal societies, based upon a maize staple, and they built more elaborate architecture, especially in the chief's villages. Full-size human sculptures (Haberland, 1973) depicted the social order graphically, with nude males upholding "big men" or chiefs on their shoulders, although other interpretations are conceivable. The "big men" or chiefs wear the peaked caps and necklaces with pendant as symbols of superior status, and often hold ceremonial axes in one hand and a bowl or human head in the other.

So, much higher population density, intensive agriculture, and greater built environment than at Arenal evidently put the Barriles societies at greater risk to sudden stresses, but one other factor should be considered. Although they shared a common culture, the polities were competitive and often at war with each other. The evidence of hostility is recorded in stone sculpture (Linares and Ranere, 1980) with severed heads, the aforementioned sculptures with severed heads, and the special axes made for decapitations. Thus, a hostile political landscape obviated tephra-stressed people taking refuge in nearby areas, in contrast to impacted Arenal societies, and a drastic solution of long-distance migration to a wet lowland nonseasonal tropical rainforest environment ensued. Thus, a tradition of competition and hostil-

ity between adjacent social groups can lay the groundwork for more drastic adjustments in times of emergency.

Archaeology and volcanology can provide examples of multidisciplinary research that can examine disasters and their effects on the environment as well as on human societies of all kinds. Archaeologists, in various parts of the world, have studied how humans have coped with volcanic eruptions, ranging from late Pliocene eruptions in east Africa to late Holocene eruptions throughout the world. The long time scales analyzed by both disciplines allow researchers to study long-range interactions as societies cope with sudden massive stresses with varying degrees of success. This contrasts with most disaster studies that focus on the immediacy of the impact, the deaths, injuries and destruction, and the early stages of people trying to adjust to the disaster. The most important contribution that archaeologists can make in studying disasters is the long-run, over decades or centuries, because this adds a dimension that is largely lacking in disaster research. The following section focuses on some cases of volcanic eruptions and human societies in a small area of Central America.

VOLCANIC ACTIVITY, HUMAN IMPACTS, AND RESPONSES IN THE ZAPOTITAN VALLEY

Although the natural and cultural records of the interplays among volcanic eruptions, societies, and people in El Salvador's Zapotitán Valley (Fig. 2) are far from complete, at least the outlines of these interactions can be traced over the past couple thousand years. There are a few islands of knowledge, but they are surrounded by seas of ignorance that need to be explored. A useful way to examine coping strategies to volcanic eruptions is to utilize the human behavioral adjustments that people commonly use when faced with extreme environmental changes, as developed by Burton et al. (1978). The minimal adjustment is *loss absorption*, but when the first threshold of *awareness* is crossed, the second level of adjustment is *loss acceptance*. However, if the stress is greater and the second threshold, *direct action*, is crossed, then *loss reduction* is the result. If the stress is yet greater, then the third threshold, *intolerance,* is crossed and the result is *radical action*. An example of radical action is migration to a location far from the disaster, such as the uppermost polities near Volcán Baru, and significant changes in the culture and/or adaptation may result.

Coatepeque Eruptions

The Coatepeque eruptions during the late Quaternary, from the Santa Ana volcanic complex, have been studied for about a half a century, beginning with Williams and Meyer-Abich (1955) describing them as a caldera collapse and/or explosion of an earlier edifice. Estimates of the age ranged from 10,000 to 40,000 yr, and the possibility that paleo-Indian sites, which could date to ~8000 to 12,000 yr B.P., might be buried under Coatepeque tephra deposits was raised (Sheets, 1984). However, recent volcanological research has pushed the dating of the Coatepeque eruptions

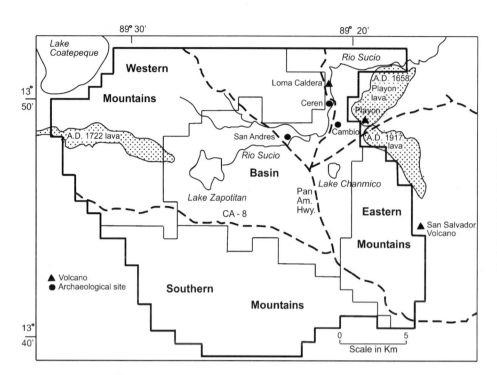

Figure 2. Map of the Zapotitán Valley, El Salvador. The four geographic provinces of Basin, Western Mountains, Eastern Mountains, and Southern Mountains are indicated, along with volcanoes, lava flows, and archaeological sites. The thick kinked line is the limit of the Zapotitán Valley drainage, and the limit of the archaeological survey by Black (1983). Dashed lines are the major paved highways.

farther back in time, thus diminishing the probability that people were affected by any of them. The chances are slight that people were living in what is now El Salvador before 15,000 yr ago.

Pullinger (1998) found and radiometrically dated three eruptions of Coatepeque, including the caldera formation. Rose et al. (1999) and Pullinger (1998) date the Bellavista eruption to 77,000 ± 2000 yr ago. The greater Arce eruption followed shortly thereafter, at 72,000 ± 3000 yr ago, and probably included the caldera formation. The Arce event involved the greatest volume, at ~40 km³ (Pullinger, 1998). The final eruption, called the "Congo event," is dated to 56,900 +2,800 –2100 yr ago. Each of these Coatepeque eruptions could be considered a true natural disaster in the physical sense, as each drastically affected the local flora and fauna, but not people. The eruptions considered below will not be described only as "natural disasters" because they are complex networks of natural and cultural factors. People and their societies were deeply involved, prior to each disaster, during its impact, and in the aftermath and recovery, or lack of recovery. Thus, these disasters are explored in multiple dimensions, including natural, cultural, economic, and political domains, at least insofar as the available documentation will allow. Often only one or two domains are well published in particular cases, thus limiting understanding.

Southern Mesoamerican Society prior to the Ilopango TBJ (Tierra Blanca Joven) Eruption

The pre-eruption southern Mesoamerican society was complex, with advanced chiefdoms or small city-states domi-

nating the broad intermontane valleys in the four centuries before and after Christ (±400 yr). What is now western El Salvador played a major role in the late Preclassic "Miraflores" cultural sphere that extended well into the Guatemalan highlands (Sharer, 1974; Dull et al., 2001). This cultural florescence was characterized by elite-dominated primary sites in the midst of landscapes dotted by secondary sites, towns, and villages, as documented by Black (1983) in the Zapotitán Valley. Elites are the upper class in a state, or the chief's household and retainers in a chiefdom. Sophisticated earthen architecture included adobe brick construction as well as rammed earth and bajareque (wattle-and-daub) construction. The latter is remarkably earthquake-resistant and was favored for commoner housing before and after that eruption (Sheets, 1992). Stelae were carved with calendrical and textual hieroglyphics at Chalchuapa (Sharer, 1974). Large carved stone jaguar faces were placed at the bases of stairways surmounting earthen-fill pyramids. Ceramic innovations were notable, with Usulután resist decoration being made in this cultural sphere, particularly at Chalchuapa (Sharer, 1974) and imitated by ceramicists elsewhere in southern Mesoamerica. Lithic specialists, attached to the elites at primary and secondary centers, were producing prismatic blades, macroblades, tanged macroblades, and scrapers for use in households at all levels of society (Sheets, 1983b).

Ilopango TBJ (Tierra Blanca Joven) Eruption

The Coatepeque and Ilopango eruptions were physically similar massive plinian eruptions of acidic tephra that certainly were devastating to the tropical flora and fauna of central and western El Salvador. They also are similar in that each source had multiple eruptions. Rose et al. (1999) describe four eruptions from Ilopango, from earliest to the most recent, called the TB4, TB3, TB2, and the TBJ. The TB4 probably occurred prior to human habitation of the area, but the TB3 and TB2 eruptions may have affected people. Loci where the TB3 and TB2 tephra units overlie paleosols bearing artifacts should be sought and studied.

Porter (1955) excavated over 3000 potsherds, 47 obsidian artifacts, two metates, and charcoal in a paleosol buried by a thick deposit of white volcanic ash at Barranco Tovar in San Salvador. She noted that ash layer to be 10 to 20 m thick around San Salvador, which sounds like the Ilopango TBJ tephra with the associated pyroclastic flows. She obtained one ^{14}C date, run in the early 1950s when radiocarbon dating was early in its development. That date is 1040 ± 360 B.C. There are at least two possibilities, (1) the site is buried by the TBJ and the date is considerably too early, or (2) the site is earlier and the date is related to the TB2 or TB3 eruption. The site unfortunately is no longer accessible because a large apartment building occupies the location, but other parts of the site may extend beyond the building's foundations. Most volcanological and archaeological research on Ilopango has focused on the latest large explosive eruption, the TBJ, and that is the focus of the rest of this section.

In the early decades of the twentieth century the Salvadoran scholar Jorge Larde (1926) discovered various exposures in the San Salvador region where white volcanic ash buried ceramic and lithic artifacts. He reasoned an ancient eruption had buried archaeological sites. He showed the deposits and artifacts, in what he called the "C horizon," to Sam Lothrop, who described the ash and artifact exposures in considerable detail (Lothrop, 1927). Lothrop focused on the Cerro Zapote site, in southeastern San Salvador. He recognized the early character of the ceramics, dating to what is now called the Preclassic period. Lothrop also identified the artifacts stratigraphically above the volcanic ash layer as Maya in character, in what he called the "A horizon," thus indicating the cultural affiliation of the group that reoccupied the area. His identification of horizons was by arbitrary letters that are not related to how soil horizons are designated today. He speculated that Ilopango could be the source of the volcanic ash. Later research has confirmed Lothrop's suspicion of source, as it is now known that the archaeological sites were buried by tephra from Ilopango, called the Tierra Blanca Joven (i.e., young white earth) tephra (summarized in Sheets, 1983b). Curiosity about the possibility that the numerous other exposures with visually similar tephra and artifacts that were recorded by various scholars in the middle decades of the twentieth century (Sheets, 1983b) could have been affected by the same massive eruption led to regional research by Sheets, Steen-McIntyre, Hart, and others. Later research demonstrated that most tephra deposits mentioned in these studies were from the same massive eruption, the TBJ tephra from Ilopango (Hart and Steen-McIntyre, 1983).

The Ilopango TBJ eruption was dated to A.D. 260 ± 114 by radiocarbon analyses (calibrated 1-sigma composite) conducted in the 1970s (Sheets, 1983a). That date may be too early, based on recent efforts to obtain more precise radiocarbon dates by careful sampling and AMS (accelerator mass spectrometry) dating. The results indicate that eruption may have occurred a century and a half later at A.D. 421 (429) 526 calibrated one sigma, or A.D. 408 (429) 536 calibrated two sigma (Dull et al., 2001).

This eruption was one of the greatest Holocene eruptions in Central America, and its ecological and societal impacts were felt throughout El Salvador and adjoining Guatemala and Honduras. The plinian and phreatomagmatic eruption resulted in the elimination of flora and fauna in tens of thousands of square kilometers. Although population density prior to the eruption has not been established, a conservative estimate suggests there must have been at least 30 people per square kilometer on a general regional basis. The eruption ejected a total volume of ~18 km³ of dense rock equivalent (DRE) into the air, creating a zone of total devastation in an area of some 1000 km² surrounding Ilopango (Dull et al., 2001) where no people, animals, or plants would have survived. The area was devastated by huge pyroclastic flows that uprooted and carbonized trees even at distances of 25 km from the vent. The death toll in this area is estimated at 30,000 people (Dull et al., 2001).

Beyond the area of total devastation was a zone of at least 10,000 km² where the ash blanket was more than 50 cm thick (Dull et al., 2001). This has been mildly called the zone of depopulation by Dull et al. (2001). Based on tephra-induced roof collapses at Ceren (see below), all roofs were collapsed in this zone. An estimated 300,000 people lived in this zone, and some of them could have survived by taking protective measures such as breathing through cotton cloth to filter out the fine tephra, but many must have perished from asphyxiation. Certainly, no survivors could continue living in the area with contaminated fresh water and agricultural fields deeply buried by the sterile white ash. Segerstrom (1950) found that volcanic ash depths greater than 10–25 cm from the recent eruption of Paricutin exceeded the coping abilities of traditional Mesoamerican maize agriculturalists. Survivors would have had to migrate to arable land outside the zone, thus adopting the radical action response mode (Burton et al., 1978).

Archaeological research in El Salvador has shown that the favored loci for settlement in the Preclassic and Classic periods were low-lying valleys, alluvial flood plains, and basins (Black, 1983; Sheets, 1984), because of the access to fresh water, a higher water table for natural vegetation and cultigens, and more arable soils for agriculture. These were the zones most deleteriously affected by pyroclastic flows and lahars during the Ilopango eruption. Thus, in intensively exploiting the prime loci for settlement and adaptation, people in the centuries prior to the TBJ eruption were inadvertently concentrating themselves where the eruption would have the greatest impacts. Generations of people had benefited from living in the "ideal" areas, but their descendents paid the price when Ilopango erupted. The long-run benefits were truncated by the apocalyptic eruption.

People beyond the zone of disruption would have dealt with the thinner ash levels by loss acceptance, as yields diminished under the tephra-induced stress but maize and other cultigens were not killed, and people could still plant through the ash blanket. And, with weathering and incorporation of the tephra, soils could have improved within a few years of the eruption. Still farther from the eruption, in areas like the Southern Maya Lowlands, the thin dusting of TBJ tephra could have improved agricultural productivity in the first year of emplacement by providing a mulch layer and increasing soil porosity, and perhaps suffocating some insect pests.

The Ilopango TBJ eruption was about the same scale as the first century A.D. eruption of Popocatépetl in highland central Mexico, also rated a VEI (volcanic explosivity index) 6 (P. Plunket and G. Uruñuela, 2002, personal commun.). Large areas of the northeastern flanks of Popocatépetl had to be abandoned, and Plunket and Uruñuela think Cholula benefited from the arrival of refugees that may have helped in the massive construction programs. The northwestern flanks also were devastated, and Plunket and Uruñuela suggest a solution to two longstanding archaeological problems. Archaeologists have long wondered why the most fertile and moist area of the Basin of Mexico was abandoned at the same time some 50,000 people arrived at Teotihuacan (P. Plunket and G. Uruñuela, 2002, personal commun.), and the volcanic impact on the slopes of Popocatépetl is suggested as the likely push factor (P. Plunket and G. Uruñuela, 2002, personal commun.). Archaeologists also have long wondered why so many people moved into Teotihuacan almost 2000 yr ago. It appears the huge eruption drastically affected populations on both sides of the volcano, but in quite different ways, and immigrants affected the social dynamics of the two largest settlements in central Mexico.

Human Recovery from the Ilopango TBJ Eruption

Recoveries from tephra-induced stresses occurred earlier at greater distances from Ilopango. As suggested by Lothrop (1927) for central El Salvador, the reoccupation of the Zapotitán Valley emanated from the north, judging from the close similarities in architecture and ceramics of the sixth-century immigrants with those from the Copan area of Honduras. The close cultural and economic relationships of western Salvador and highland Guatemala that existed prior to the eruption were permanently severed by the eruption, and they never were reestablished. Demographic recovery from the Ilopango eruption occurred within about a century, but cultural recovery in the sense of resilience and reestablishment of the same pre-eruption cultural tradition never occurred. The Preclassic, and barely into the Classic, indigenous cultural florescence of the Miraflores cultural tradition in central and western El Salvador was permanently truncated (Dull et al., 2001; Dull, this volume, Chapter 18). While the western component of the Miraflores tradition continued in the Guatemalan highlands, the eastern component was devastated and never recovered in El Salvador.

One of the earliest villages yet found representing the human recovery from the Ilopango eruptions is Ceren (Fig. 3), in the

northern end of the Zapotitán Valley (Sheets, 1992). We estimate about a century elapsed between the Ilopango devastation and the founding of the Ceren village, and it thrived for about another century before it was entombed by the tephra from the Loma Caldera eruption.

The Ceren Site

The Ceren site, called "Joya de Ceren" in El Salvador, was a thriving agricultural village on the left bank of the Rio Sucio (Sheets, 1992). All domestic architecture and even the ridges in the agricultural fields run parallel to or perpendicular to the bank of the river that runs 30 degrees east of north. This regular alignment of public and household architecture as well as agriculture had been a puzzle until its correspondence with the riverbank was noted recently. Only the religious buildings diverge from this orientation, in what appears to be a deliberate exception, presumably for supernatural objectives.

Ceren provides detailed information on the timing and nature of recovery from the Ilopango eruption. If the central dates from the calibrated AMS radiocarbon dates, A.D. 429 and A.D. 656, are used as a way to estimate the elapsed time between the eruptions of Ilopango and Loma Caldera, they appear to have been separated by about two centuries. The Ilopango tephra in the Ceren area was measured at the Cambio site, only 2 km south of Ceren, at 1.2 m (Hart and Steen-McIntyre, 1983). Dan Miller (2002, personal

Figure 3. Stratigraphy and cultural features at the Ceren site, El Salvador. Paleosol at bottom is well developed and contains occasional artifacts dating to the Preclassic period. The white Tierra Blanca Joven tephra overlies the paleosol. A soil developed on top of it (below "A"), and people reoccupied the area and built houses on it. The borrow pit for this Structure 1 at Ceren is visible to the left of the word "Paleosol," where they dug below the TBJ for the construction clay. The thatch roof was preserved, collapsed onto the floor by the tephra overburden. The Loma Caldera tephra is 4 m thick at this locality, extending from "A" up to "B" in this photograph. Phreatomagmatic surge units are light-colored, and direct airfall units are more coarse and dark-colored. Above "B" is a mixed deposit of later tephra units, including the San Andres tuff and tephra from Playon. Photograph by Payson Sheets.

commun.) believes this is excessive, based on his measurements, and it may be slightly less than a meter. Erosion and compaction affected that thickness, so its original thickness was somewhat greater. That would place the Ceren area, and the northern Zapotitán Valley, within the zone of complete devastation from the Ilopango eruption. The eruption would have changed a tropical gallery forest environment interspersed with intense agriculture and relatively dense human populations into a white sterile desert overnight. The sialic nature of the Ilopango TBJ tephra would have slowed soil formation and vegetation recolonization. Floral, faunal, and soils recovery would have preceded maize-based human agricultural recovery. Maize is highly nitrogen demanding, and a very juvenile soil on a sialic tephra would be deficient in nitrogen. Perhaps a century of time between the eruptions was needed for the natural processes of weathering, soil formation, vegetation recovery, and faunal reoccupations. The next century, the sixth century A.D., witnessed the human aspects of recovery from the Ilopango eruption, with the immigrants founding Ceren and other communities in the Zapotitán Valley.

The conservation ethic at Ceren, a World Heritage site (administered by the United Nations), prohibits excavations into structures, so our only evidence of the earliest buildings at Ceren come from examining the bulldozer cut that sliced into the northern end of Structure 1 in Household 1. That cut divulged several stages in remodeling and replastering that could well have taken place over a few generations of people. Another indicator of the duration of occupation is provided by the midden in Operation 2 (McKee, 2002), where the amount of trash accumulation indicates at least many years of occupation, and probably a few generations, but not centuries of household discard activities. It therefore appears that Ceren was occupied for perhaps a century before its burial by tephra from the Loma Caldera volcanic vent.

Because Ceren was one of the earlier communities established in the Zapotitán Valley after the Ilopango eruption, it is pertinent to this paper to note the internal nature of the Ceren community, as a part of exploring the rate and the nature of social recovery from Ilopango. Ceren was a village of commoners, with perhaps 100 to 150 inhabitants and no elite or attached occupational specialists (Sheets, 2002). To date a total of four households has been excavated, and only one of them in its entirety. A civic core of the village has been found, consisting of a formal plaza surrounded at least on its south and west sides by solid earthen-walled public buildings. A religious core of two special structures was found at the easternmost edge of the site, at the highest elevation of the site, on the bank overlooking the Rio Sucio. The two buildings were constructed to be different from household and public buildings in that their orientation was different from the aforementioned 30° east of north. In addition, both buildings contrasted with all other buildings at the site with their white painted walls with red hematite decoration, their unique construction of earthen columns, and the multiple floor levels from outside to the most elevated back room.

The two categories of material culture that are the most sensitive to the ethnicity of the Ceren immigrants and thus their home territory are architecture and artifacts. Each Ceren household built at least three functionally differentiated buildings, including a domicile, a storehouse, and a kitchen (Sheets, 2002). This kind of household structure differentiation is a Maya characteristic that existed in the Copan valley at the time of Ceren (Webster et al., 1997). The Chorti Indians of the same area near Copan continue to construct functionally differentiated household structures (Wisdom, 1940). The artifacts with the highest ethnic sensitivity are the polychrome food and drink serving vessels that make up ~22% of each household's fired clay vessel inventory (Beaudry-Corbett, 2002). The majority of polychrome vessels are of the Copador type, evidently made in one or more localities in the Copan area and imported into the Zapotitán Valley. The Copador polychromes were made available to the commoners by local elites, for exchange by barter, presumably in the marketplace in primary or secondary centers. The locally made utilitarian ceramics, for cooking and storage, are very different from those at Copan and Quirigua. Because no hieroglyphic texts have been found, or are likely to be found, it is not known what language was spoken in Ceren. However, the architecture and artifacts of Ceren are so close to those of the Maya of Copan to state that the Ceren residents were closely Maya related. The vibrancy of the pre-Ilopango "Miraflores cultural sphere" was never reestablished in El Salvador, as cultural recovery never occurred locally. In fact the people moving into the ecologically recovering area were of a different ethnicity. Demographic recovery in the valley and most of central and western El Salvador was complete within three centuries after the eruption.

Loma Caldera Eruption

The Loma Caldera vent, only 600 m north of the thriving Ceren village, erupted early one evening in August, dated by artifact patterns and vegetation maturity (Sheets, 1992). Vegetation, including annuals such as maize, and seasonally sensitive trees such as guayaba, indicate the middle of the rainy season, i.e., August. Artifacts indicate a meal was just served before the eruption, but the dishes were not yet washed, as evidenced by the finger swipes of food left in hemispherical serving bowls. In tropical climates dishes are washed soon after a meal is completed in order to avoid attracting noxious insects, so the eruption occurred very soon after the meal was served. It must have been the evening meal, because dried and stored maize was placed in soaking vessels to soften overnight, agricultural tools were back from the fields, and pots were removed from the three-stone hearths. That would mean the Loma Caldera eruption began ~6–7 p.m. if Cerenians ate dinner about when traditional Salvadorans do now. The dating by year is more challenging, and the radiocarbon dates, when averaged, yielded a composite 2-sigma calibrated range from A.D. 610 to 671 (McKee, 2002).

The extraordinary preservation of the small Ceren village, with thatch roofs, trees, orchards, plants in agricultural fields, and organic artifacts in houses, is of great archaeological importance. The Ceren residents certainly were not thinking of that when they were fleeing for their lives, and whether they were success-ful may be resolved by searching for body cavities in the tephra south of the site with high-frequency ground-penetrating radar antennas. Volcanologically the eruption from the nearby Loma Caldera vent was miniscule in comparison to the Ilopango TBJ eruption, as it only affected a few km^2 (Miller, 1992). People within 10 to 20 km^2 would have had to have abandoned their homes and take up farming elsewhere. The estimated 165–440 people per km^2 in this very fertile basin province (Black, 1983) would indicate some 1500 to 9000 displaced people, with only the people located very close to the vent having been killed by the eruption. When compared to the Ilopango eruption, natural recovery from the Loma Caldera eruption was faster, because the tephra was more mafic and therefore would weather more rap-idly. Also, the source areas for recolonizing flora and fauna were no farther than two kilometers from any point on the tephra blan-ket. Human demographic and cultural recoveries were relatively quick in the area, probably requiring only a few decades.

The Cambio Site: Recovery from the Loma Caldera Eruption

Rarely can regional processes and phenomena be understood from a single site, and Ceren is no exception. The Cambio site (Fig. 4), two km south of Ceren, provides a more complete strati-graphic record of the physical and cultural processes shortly after the Loma Caldera eruption (Chandler, 1983). Cambio received a thin layer of the Loma Caldera tephra, measured by Miller (2002) at 10 cm. Weathering, compaction, and human disturbance likely had thinned it, but how much thicker it was after emplacement is unknown. Human reoccupation of Cambio occurred soon after the Loma Caldera eruption, probably within a few decades. The cultural changes from Ceren to Cambio are almost impercepti-ble; Beaudry-Corbett (2002) notes that a new kind of polychrome pottery called "Arambala" was present at Cambio but was absent at Ceren. Arambala seems to be a local imitation of the imported Copador and Gualpopa polychrome pottery styles. Other than that small addition to the ceramic inventory, all other artifacts and material culture are indistinguishable from Ceren before Loma Caldera erupted. While the effects of the Ilopango TBJ eruption echo through the centuries, the Loma Caldera eruption affected only a small area for a short time.

Boquerón Eruption: The San Andres Talpetate Tuff

During the Late Classic period San Salvador volcano erupted a fine wet tephra through its large crater (Boquerón means "big mouth") that coated the Zapotitán Valley. When it dried it hardened into a tough compact unit. The tephra is 6 m thick atop Boquerón (Hart, 1983), and the nature of tephra deposits convinced Hart that the eruption was phreatomagmatic, through a crater lake. The tephra was formally named the "San Andres Talpetate Tuff" by Hart (1983) after the San Andres site, where it had been noted by various archaeologists for many years, and finally described by Hart at that site and published.

Figure 4. Stratigraphy and cultural features at the Cambio site, El Sal-vador. The Paleosol at the bottom is the same as at Ceren. Note partial vessel in foreground, ~2000 yr old. The Ilopango TBJ tephra overlies that paleosol. The Loma Caldera tephra is thin, with a well-developed paleosol on top. The San Andres tuff is preserved as a solid uninter-rupted layer with a very thin paleosol developed on top of it. "B" marks a building foundation, of river rocks, dating to ~1000 yr ago. The dark layer at the top, overlying the building foundation, is tephra from the A.D. 1658–1659 eruption of Playon volcano. Photograph by Payson Sheets.

The tephra was thick enough on the eastern portion of the valley to eliminate flora, fauna, and people, and it thinned rapidly on the western side. Because the tuff was over 30 cm thick at the San Andres site (Hart, 1983), it probably caused abandonment of that city for a few decades. It was over 20 cm thick at Cambio (Chandler, 1983), and would have had almost as long a deleteri-ous effect there. Hart estimated at least 300 km^2 were covered by the tuff with depths over 7 cm. Because this tephra is so compact and hardened, it would have been more damaging to traditional agriculture than would a loose, dry airfall tephra. Seven to 10 cm of this hard tephra may be about the maximum with which tra-ditional maize-based agriculturalists could cope, as I extrapolate from Segerstrom's (1950) finding that traditional maize agricul-turalists could cope with unconsolidated tephra deposits up to ~25 cm in depth. If this is correct, slightly more than half of the Zapotitán Valley was rendered uncultivatable for a few decades.

Black's (1983) Zapotitán Valley survey estimated the overall regional population densities in the Late Classic to be between 70 and 180 people per square kilometer. These figures can be used along with Hart's 300 km^2 estimate to establish a range of how many people could not have continued living and farming where they were before the eruption. Between 21,000 and 54,000 peo-ple would have had to migrate, a much greater disaster than that caused by the Loma Caldera eruption, but an order of magnitude smaller than the Ilopango TBJ disaster.

Unfortunately, virtually all of the archaeological data recov-ered at San Andres remain unpublished and thus are of little help in dating the eruption, studying its immediate effects, and documenting recovery processes. The best dating is provided by the stratigraphy at the Cambio site, where the tuff is sandwiched between two Late Classic occupation layers that postdate Ceren and the Loma Caldera tephra (Chandler, 1983). The eruption most likely dates to the ninth century A.D., but this is not certain.

In terms of ecological and cultural impacts, this eruption would have been about midway between Ilopango TBJ and Loma Caldera. Cultural and demographic recoveries from the Boquerón eruption occurred relatively rapidly as shown by Cambio stratigraphy (Chandler, 1983). There was no cultural disjuncture that has been detected, in contrast to the Ilopango aftermath. This suggests important lessons about physical scales of eruptions interacting with human scales of vulnerability. These three cases, in the same environment and with similar societal complexities, indicate a significant physical threshold. An eruption the scale of Ilopango created disruptions on such a great regional scale that surviving refugee households were not able to maintain cultural continuity and reoccupy their native territory when sufficient environmental recovery had occurred. Ilopango overwhelmed many political units. It also overwhelmed the regional economic system of subsistence embedded in elite economic interchange of exotic goods. The nodes of still functioning pre-Ilopango society, which could have provided recovery sources maintaining the pre-eruption culture, were nowhere to be found in the region. In contrast, the Boquerón eruption devastated over half of one intermontane valley, causing an estimated tens of thousands of deaths and migrating refugees, but within a few decades full demographic and cultural recovery occurred. The sources for recovery probably were surviving settlements in the western part of the valley and perhaps settlements outside the valley. Probably a key factor is that the cultures surrounding the Zapotitán Valley in the Late Classic were virtually identical to that within the valley.

Playon Eruption A.D. 1658–71

The fissure system that runs northwest from San Salvador volcano has been active often in the past few thousand years, and the Boquerón eruption as well as the seventh-century eruption of Loma Caldera provide examples. A millennium after the Loma Caldera eruption, on 3 November 1658, the area was rocked by major earthquakes that killed many people and destroyed settlements including Quetzaltepeque and San Salvador. The Playon eruption began in earnest the next day as volcanic ash spewed from the vent and an andesitic lava began flowing downhill to the north (Meyer-Abich, 1956). Both lava and tephra caused great destruction to agriculture, the local indigo industry, and livestock, according to colonial documents (Gallardo, 1997). The lava came within 2 km of the town of Quetzaltepeque and largely encircled and partially buried the town of Nexapa, causing it to be relocated to its present location at Nejapa. The lava covered at least 8 km^2 (Hart, 1983). At its northernmost extent the lava blocked the Rio Sucio (formerly known as the Rio Nixapa, or

"river of ash" in Nahuat) and formed a sizeable lake in the center of the Zapotitán Valley. The lacustrine sorting and redepositing of Playon tephra are clearly seen in Gallardo's (1997) photographs of the devastated indigo processing plant at San Andres. The lake lasted for only a few weeks, until the river eroded the lava dam. The soils formed on Playon tephra presently support an exuberant vegetation and intensive agriculture, but the soils formed on the lava are patchy and thin, and the present vegetation is sparse and scrubby, consisting of scattered bushes and seasonal grasses.

Browning (1971) describes in considerable detail the struggles of the displaced Nejapa residents. Nexapa was a Pipil Indian community, functioning at the bottom of the colonial social hierarchy, and their common lands were largely buried under the lava. They moved eastward, to what is now Nejapa, and petitioned the colonial government for land owned by the Crown to be changed for use as their sustaining area. Browning documents resistance by local Spanish hacienda owners and indigo processors, and the frustrated struggles by Nejapa residents to be recognized as land owners rather than simply land occupiers. Those struggles continued for almost 80 yr, and finally Nejapa was rewarded in 1736 by the granting of legal title to the land on which the town stood and a narrow strip of common land extending almost 30 km to the southwest up the slope of San Salvador volcano. Even at this early historic time land was not freely available, and struggles for clear land ownership were particularly difficult for disenfranchised disaster victims.

San Marcelino Eruption A.D. 1722

Pullinger (1998) studied the 1722 lava flow that emerged from the eastern side of the Santa Ana volcanic complex. The flow headed eastward 11 km from its origin from two vents at 1325 m and destroyed the traditional village of San Juan Tecpan. The destruction of some 15 km^2 of prime agricultural land, for subsistence crops as well as cacao, was a major blow to local eighteenth-century residents on the western side of the Zapotitán Valley. Perusal of colonial documents should divulge information on impact, refugees, search for replacement land, and legal struggles, perhaps analogous to Nejapa residents after the Playon eruption. The soils and vegetation on top of this lava flow at present are patchy and sparse, similar to those on top of the Playon lava described above.

Izalco Eruptions A.D. 1770–1966

Izalco is El Salvador's most famous volcano, as its almost continuous eruptions from the eighteenth into the twentieth century were visible well into the Pacific Ocean, earning it the title "Lighthouse of the Pacific" for sailors. Meyer-Abich (1956) conducted the earliest serious geological study of Izalco, and Pullinger (1998) recently surveyed current knowledge. The volcano started building itself from a vent at 1300 m on the southern flank of Santa Ana volcano in 1770, and ceased eruptions in 1966 when the crater rim reached 1952 m. A pyroclastic flow buried the traditional village of Matazano in 1926, killing 56 people. Many lava flows were in the 5–8 km long range, and many ashfalls affected people and vegetation within a few km of the volcano, on and just beyond the southeastern edge of the Zapotitán Valley. The net decline in local agricultural production because of the volcano was, and continues to be, considerable. The large area occupied by the volcano had been a highly productive gentle lower southern slope of the Santa Ana complex, producing coffee, cacao, indigo, and sugar cane. Once again volcanism resulted in a dramatic decline in the life-support capacity of the environment in this crowded nation.

San Salvador Volcano Lava Flow A.D. 1917

The large crater of San Salvador volcano, or "Boquerón," contained a crater lake in the early years of the twentieth century. However, in June of 1917 the lake boiled away in the same week that some earthquakes did considerable damage to cities on all sides of the volcano, and lava began to flow from fissures on the volcano's north side (Meyer-Abich, 1956). The lava flows destroyed highly productive coffee farms. The black basaltic lava covered ~16 km^2 and presently is devoid of vegetation other than some lichens and occasional grasses. The principal use of the flow today is to filter rainwater that is collected and pumped over the volcano to supply San Salvador's needs for water. A grim use of the area developed in the civil war from 1979 through 1992 as paramilitary squads used the lava flow as their favored place to dispose of bodies.

At first glance the record of volcanic eruptions affecting the environment and residents of the Zapotitán Valley during the last 2000 yr may seem impressive. However, the record is not complete. For instance, some sizeable lava flows are not included for lack of dating and documentation, such as that near Cerro Chino, which covers over 10 km^2. Miller (2002) noted a few eruptions in the past two to three thousand years that affected the northern Zapotitán Valley that are not yet understood and thus not a part of the record that can be included here.

CONCLUSIONS AND IMPLICATIONS

An overarching model that could be used to compare eruptions and human responses is here called "scalar vulnerabilities." As the scales or magnitudes of eruptions are compared—along with the scales of societal complexities, dependence on facilities of the built environment, political hostilities, and intensities of agriculture—some patterns begin to emerge. At the extremes of the greatest eruptions, volcanic explosivity index (VEI) 6 or greater, no human societies survive unscathed in the long run. The Ilopango TBJ eruption had massive and long-lasting effects on complex societies in the Zapotitán Valley and adjoining areas in southeastern Mesoamerica, as did the Late Preclassic eruption of Popocatépetl in highland Mexico. However, at a lesser order of magnitude, the large eruptions such as Arenal and Boquerón had significant effects on local societies, necessitating radical action

in terms of outmigration. Both the egalitarian and complex societies recovered thoroughly, as far as can be detected archaeologically. In striking contrast is the modest eruption of Volcán Baru in Panama, which drastically affected the upper Barriles polities, necessitating radical action of outmigration with no recovery and reoccupation. The hostile political landscape among the polities is the apparent explanation of the consequences of the eruption being so much greater than the scale of the physical event or the nature of the societies would predict. Relatively small eruptions, such as Loma Caldera or various lava flows in the Zapotitán Valley, had deleterious effects on the settlements directly affected but they did not have long-lasting societal repercussions. In general, egalitarian societies with low levels of built environments, minimal reliance on intensive agriculture and domesticated staples, and low population densities exhibited the greatest resilience to sudden massive volcanic stresses. Along the scale of societal complexity, the more complex societies were the most vulnerable to volcanically induced stresses. Apart from social complexity, the scale of political hostility can dramatically affect a society's vulnerability, as a small perturbation among societies with chronic warfare can require more extreme coping strategies than would be predicted by the other physical science and societal complexity scales.

The predictability or unpredictability of changes in a society's environment can have different implications. An important cross-cultural study found that unpredictable sudden stresses were much more difficult for nonwestern societies to adjust to than were long-term stresses such as sustained droughts (Ember and Ember, 1992). Unpredictable stresses or disasters led to warfare to obtain resources from others more often than chronic stresses or more predictable disasters.

The various components of volcanic eruptions affected human populations in quite different ways. Lava flows covering a few km^2 eliminated agriculture for many centuries and caused resettlements. Tephra deposits affected a few km^2 to thousands of km^2, but only the really great tephra deposits such as the Ilopango TBJ fundamentally altered the cultural trajectory of ancient societies. The medium- to smaller-size tephra aprons caused short-term difficulties, but weathering, plant succession, and human recovery were relatively rapid, and no long-term cultural effects resulted. A punctuated equilibria model could apply in a general fashion to the cases of the dynamic interactions among volcanic eruptions, people, cultures, and the organizations of societies in the past two millennia in the Zapotitán Valley. Small and middle-scale eruptions temporarily interrupted the equilibria of complex societies, but the nearby sources of flora, fauna, and people allowed for full recovery in the short to medium run of years to decades. The punctuated equilibria model also applies with the great eruptions as analogous to extinction events. The biggest eruptions created stresses so vast, in all domains of ecology, economics, politics, and society, that recoveries of the original social orders simply did not occur. The demographic and cultural reoccupations following the mega-disasters were by different societies.

So, in a general way punctuated equilibria models seem to apply to many of the cases considered here. However, on closer examination equilibrium conditions are an idealization, because some variation in environmental conditions and social environments occurs with all societies through time. Even the most stable society mentioned in this paper, Arenal in Costa Rica, exhibited changes in material culture through the centuries, especially in ceramic form and decoration. Because none of these changes correlate with the documented disasters, no relationships are suggested here.

Tropical botanists are considering the "intermediate disturbance hypothesis," which proposes that intermediate scales of disturbance promote maximum species diversity (Molino and Sabatier, 2001). Their test of the hypothesis in a South American rainforest was sustained. An analog with cultures could be proposed, that small- to medium-scale disasters can have creative aspects in addition to their much-heralded destructive aspects. Some disasters are so immense that they overwhelm all cultures, but medium-scale disasters could promote cultural diversity, redundancy of social units, and societal resilience.

Volcanic/human disasters are not egalitarian, in that they do not affect all components of societies uniformly. People can consider the degree to which they wish to put themselves at risk from the range of hazards that they perceive in their environments. They can then make decisions regarding where they will live. And it is my personal observation that, in complex societies, the degree of freedom of choice is not uniform among social classes. The upper class, the people with greater wealth and power, has greater freedom of choice in where to locate their residences and other facilities than do the poor or politically disenfranchised.

The population of what is now El Salvador has changed dramatically in ancient and historic-to-modern times (Sheets, 1992). The population in that area generally was between 250,000 and 750,000 people in pre-Columbian times, with regional population densities ranging from 10 to 50 per km^2. After the decimation from imported Old World diseases in the sixteenth and seventeenth centuries, the population recovered back to its pre-Columbian peak of 750,000 in 1900. The population of El Salvador has burgeoned in the twentieth century to more than 6,000,000 people, with a density of ~300 per square kilometer. And population is continuing to surge upward as church and state are reluctant to support family planning and thus some kind of population control. Therefore more people are forced to live and work in harm's way, where volcanic and other disasters will take progressively greater tolls of death and destruction. Many of the urban and rural poor are crowded into especially hazardous locations in El Salvador today, such as low-lying barrancas (gullies) that are particularly vulnerable to floods, lahars, and pyroclastic flows.

The papers in this volume are multidisciplinary. On a broader scale, one can inquire how well have social and physical sciences worked together in the way suggested by Kennedy (2002), with the objective of an integrated understanding of disasters? Alexander (1995, 1997) has answered this question

in his two extensive surveys of hazards and disasters, including the full range of physical and social science and engineering disciplines. He tabulated a total of 30 academic disciplines conducting disaster research, which at first glance might indicate a robust thriving field of study. However, he found most studies were intradisciplinary. Further, he noted that the assumptions and understandings of the relationships among people, societies, and nature varied markedly among the various disciplines. He also found a great imbalance in support, with 95% of disaster research funding going to the physical sciences and technology, and only 5% going to social science studies. Based on the above findings, it is little surprise that Alexander (1995) found hazard and disaster theory to be fragmented and poorly developed across the diversity of disciplines.

Ironically, disaster research began with the pioneering social science work of the geographer Gilbert White (1945). Since then many physical and social science disciplines have begun investigating aspects of hazards and disasters, but anthropology has been one of the slowest to join in. Perhaps one reason for that reluctance is that the excesses of environmental determinism in the early decades of anthropology's founding has made scholars leery of exploring the field.

Disaster research is clearly multidisciplinary today, but very few studies or projects are interdisciplinary. Because different disciplines vary considerably in their assumptions and theories about human-environmental interaction, and an individual discipline changes over the decades, comparative studies are often difficult. Not only do different disciplines view hazards and disasters in different ways, so to do different societies. Modern western societies, with extensive education in the physical sciences, view disasters with a heavy emphasis on the geophysical components of their initiations and impacts. However, most human societies have not studied the geophysical mechanics of disaster initiation. Rather, most human societies have explained disasters in religious terms, often explaining extreme events as caused by failures in the sacred covenant between people and deities, and when hazards are perceived they often turn to ritual and sacrifice to try to decrease risks.

ACKNOWLEDGMENTS

I want to express my appreciation to my fine helpful colleagues who critiqued an early version of this paper: E. James Dixon, Cathy Cameron, Linda Cordell, Steve Lekson, Art Joyce, and Frank Eddy. They strengthened the social science aspects of the paper. Dr. Dixon also critiqued a later version, and his assistance is greatly appreciated. Carlos Pullinger was very helpful in volcanological and dating details of the Coatepeque eruptions. Brian McKee gave the manuscript a very careful copyediting, for which I am most grateful. The comments by Dan Miller, Karen Bruhns, and John Hoopes, the three GSA peer reviewers, were particularly helpful. The beginning of the chapter's title is borrowed from the 2002 Chacmool conference theme in Calgary, with appreciation.

120 *P. Sheets*

REFERENCES CITED

Alexander, D., 1995, A survey of the field of natural hazards and disaster studies, *in* Carrara, A., and Guzzetti, F., eds., Geographical information systems in assessing natural hazards: Dordrecht, Kluwer Academic Publishers, p. 1–19.

Alexander, D., 1997, The study of natural disasters, 1977–1997: Some reflections on a changing field of knowledge: Disasters, v. 21, no. 4, p. 284–304.

Beaudry-Corbett, M., 2002, Ceramics and their use at Ceren, *in* Sheets, P., ed., Before the volcano erupted: The ancient Ceren village in Central America: Austin, University of Texas Press, p. 117–138.

Black, K., 1983, The Zapotitán Valley archaeological survey, *in* Sheets, P., ed., Archeology and volcanism in Central America: The Zapotitán Valley of El Salvador: Austin, University of Texas Press, p. 62–97.

Browning, D., 1971, El Salvador: Landscape and society: Oxford, Clarendon Press, 329 p.

Burton, I., Kates, R., and White, G., 1978, The environment as hazard: New York, Oxford University Press, 240 p.

Chandler, S., 1983, Excavations at the Cambio site, *in* Sheets, P., ed., Archeology and volcanism in Central America: The Zapotitán Valley of El Salvador: Austin, University of Texas Press, p. 98–118.

Dull, R., Southon, J., and Sheets, P., 2001, Volcanism, ecology, and culture: A reassessment of the Volcán Ilopango TBJ eruption in the southern Maya realm: Latin American Antiquity, v. 12, no. 1, p. 25–44.

Ember, C., and Ember, M., 1992, Resource unpredictability, mistrust, and war: Journal of Conflict Resolution, v. 36, no. 2, p. 242–262.

Gallardo, R., 1997, El Obraje de Añil de San Andres: Mexico City, Litografica Turmex, 31 p.

Haberland, W., 1973, Stone sculpture from southern Central America, *in* Easby, D., ed., The iconography of Middle American sculpture: New York, Metropolitan Museum of Art, p. 134–152.

Hart, W., 1983, Classic to postclassic tephra layers exposed in archaeological sites, eastern Zapotitán Valley, *in* Sheets, P., ed., Archeology and volcanism in Central America: The Zapotitán Valley of El Salvador: Austin, University of Texas Press, p. 44–51.

Hart, W., and Steen-McIntyre, V., 1983, Tierra Blanca Joven tephra from the A.D. 260 eruption of Ilopango Caldera, *in* Sheets, P., ed., Archeology and volcanism in Central America: The Zapotitán Valley of El Salvador: Austin, University of Texas Press, p. 14–34.

Hoopes, J.W., 1996, Settlement, subsistence, and the origins of social complexity in greater Chiriquí: A reappraisal of the Aguas Buenas tradition, *in* Lange, F.W., ed., Paths to Central American prehistory: Niwot, Colorado, University Press of Colorado, p. 15–48.

Kennedy, D., 2002, Science, terrorism, and natural disasters: Science, v. 295, p. 405.

Larde, J., 1926, Arqueologia Cuzcatleca: San Salvador, Revista de Ethnologia, Arqueologia, y Linguistica, v. 1, p. 3–4.

Linares, O., and Ranere, A., editors, 1980, Adaptive radiations in prehistoric Panama: Cambridge, Massachusetts, Peabody Museum, Harvard University, 529 p.

Lothrop, S., 1927, Pottery types and their sequence in El Salvador: Indian notes and monographs, v. 1, no. 4, p. 165–220.

McKee, B., 2002, Household 2 at Ceren: The remains of an agrarian and craft-oriented Corporate Group, *in* Sheets, P., ed., Before the volcano erupted: The ancient Ceren village in Central America: Austin, University of Texas Press, p. 58–71.

Melson, W., 1994, The eruption of 1968 and tephra stratigraphy of Arenal Volcano, *in* Sheets, P., and McKee, B., eds., Archaeology, volcanism, and remote sensing in the Arenal region, Costa Rica: Austin, University of Texas Press, p. 24–47.

Meyer-Abich, H., 1956, Los Volcanes Activos de Guatemala y El Salvador: San Salvador, Ministerio de Obras Publicas, Anales del Servicio Geologico Nacional de El Salvador, p. 3–129.

Miller, D., 1992, Summary of 1992 geological investigations at Joya de Ceren, *in* Sheets, P., and Kievit, K., eds., 1992 investigations at the Ceren site, El Salvador: A preliminary report: Boulder, Colorado, Department of Anthropology, University of Colorado, p. 5–9.

Miller, D., 2002, Volcanology, stratigraphy, and effects on structures, *in* Sheets, P., ed., Before the volcano erupted: The ancient Ceren village in Central America: Austin, University of Texas Press, p. 11–23.

Molino, J., and Sabatier, D., 2001, Tree diversity in tropical rain forests: A validation of the Intermediate Disturbance Hypothesis: Science, v. 294, p. 1702–1704.

Porter, M., 1955, Material Preclasico de San Salvador: San Salvador, Communicaciones del Instituto Tropical de Investigaciones Cientificas, v. 3/4, p. 105–112.

Pullinger, C., 1998, Evolution of the Santa Ana Volcanic Complex, El Salvador [M.A. thesis]: Houghton, Michigan Technological University, 151 p.

Rose, W.I., Conway, F.M., Pullinger, C.R., Deino, A., and McIntosh, W.C., 1999, A more precise age framework for late Quaternary silicic eruptions in northern Central America: Bulletin of Volcanology, v. 61, p. 106–120.

Segerstrom, K., 1950, Erosion studies at Paricutin: Washington, D.C., U.S. Geological Survey Bulletin 965A, 164 p.

Sharer, R., 1974, The prehistory of the southeastern Maya periphery: Current Anthropology, v. 15, no. 2, p. 165–187.

Sheets, P., 1980, Archaeological studies of disaster: Their range and value: Boulder, Colorado, Working Paper 38, Hazards Center, Institute of Behavioral Science, University of Colorado, 35 p.

Sheets, P., 1983a, Introduction, *in* Sheets, P., ed., Archaeology and volcanism in Central America: The Zapotitán Valley of El Salvador: Austin, University of Texas Press, p. 1–13.

Sheets, P., 1983b, Summary and conclusions, *in* Sheets, P., ed., Archaeology and volcanism in Central America: The Zapotitán Valley of El Salvador: Austin, University of Texas Press, p. 275–293.

Sheets, P., 1984, The prehistory of El Salvador: An interpretive summary, *in* Lange, F., and Stone, D., eds., The archaeology of lower Central America: Albuquerque, University of New Mexico Press, p. 85–112.

Sheets, P., 1992, The Ceren site: A prehistoric village buried by volcanic ash in Central America: Fort Worth, Texas, Harcourt Brace, 150 p.

Sheets, P., 2001, The effects of explosive volcanism on simple to complex societies in ancient Middle America, *in* Markgraf, V., ed., Interhemispheric climate linkages: San Diego, California, Academic Press, p. 73–86.

Sheets, P., editor, 2002, Before the volcano erupted: The ancient Ceren village in Central America: Austin, University of Texas Press, 225 p.

Webster, D., Gonlin, N., and Sheets, P., 1997, Copan and Ceren: Two perspectives on ancient Mesoamerican households: Ancient Mesoamerica, v. 8, p. 43–61.

White, G., 1945, Human adjustment to flood: Chicago, University of Chicago Press, 225 p.

Williams, H., and Meyer-Abich, H., 1955, Volcanism in the southern part of El Salvador: University of California Publications in the Geological Sciences, v. 32, p. 1–64.

Wisdom, C., 1940, The Chorti Indians of Guatemala: Chicago, University of Chicago Press, 490 p.

Manuscript Accepted by the Society June 16, 2003

Geological Society of America
Special Paper 375
2004

The acid volcanic lake of Santa Ana volcano, El Salvador

Alain Bernard*

BRUEGEL, Université Libre de Bruxelles, CP 160/02, 50 Avenue Roosevelt, 1050 Brussels, Belgium

Carlos Demetrio Escobar

*Seccion Vulcanología, Servicio Geológico de El Salvador, c/o Servicio Nacional de Estudios Territoriales,
Alameda Roosevelt y 55 Avenida Norte, Edificio Torre El Salvador, Quinta Planta, San Salvador, El Salvador*

Agnès Mazot

BRUEGEL, Université Libre de Bruxelles, CP 160/02, 50 Avenue Roosevelt, 1050 Brussels, Belgium

Ruben Eduardo Gutiérrez

*Seccion Vulcanología, Servicio Geológico de El Salvador, c/o Servicio Nacional de Estudios Territoriales,
Alameda Roosevelt y 55 Avenida Norte, Edificio Torre El Salvador, Quinta Planta, San Salvador, El Salvador*

ABSTRACT

Physical and chemical parameters were obtained from the crater lake of Santa Ana during a two-year monitoring program between 2000 and 2002. The lake contains cool (20 °C) acid-sulfate-chloride waters with a pH ~1, SO_4 = 11,000 mg/kg, Cl = 7000 mg/kg, and total dissolved solids concentration = 23,000 mg/kg. A bathymetric survey revealed a shallow lake with a maximum depth of 27 m and a volume of 0.47 million m^3. Chemical data obtained from the lake show that the major cations are derived essentially from the congruent dissolution of the basaltic andesite host rock. Thermodynamic modeling shows that the acid waters last equilibrated with the host andesite at low temperature, ~100 °C. Stable isotopic data of the lake waters indicate that D/H and $^{18}O/^{16}O$ isotopic ratios reflect the combination of evaporation effects at the lake surface and the contribution of deep magmatic fluids. $\delta^{34}S_{HSO_4}$ = 16.3‰ suggests that the main source of dissolved bisulfate ions is magmatic SO_2. No $\delta^{18}O$ equilibrium is observed between water and bisulfate ion, suggesting slow kinetics of the isotopic exchange at the low-temperature environment of the lake. Gas emissions from the fumarolic field increased in May 2000; lake temperature increased to 30 °C, and dissolved chloride and sulfate increased as well. Following this change in activity, deuterium and oxygen isotopic ratios shifted toward heavier compositions due to enhanced evaporation at the lake surface.

Keywords: Santa Ana volcano, El Salvador, crater lake, hydrothermal system, acid fluids.

INTRODUCTION

Santa Ana is the highest (2381 masl) and one of the most active stratovolcanoes of El Salvador. Twelve historical eruptions have been recorded since the year 1500 with at least one eruption every century; the last eruption occurred in 1904 (Mooser et al.,

1958). This frequent activity of Santa Ana volcano indicates that an eruption in the near future is likely. The volcano represents a significant threat for one million people (~15% of the total population of El Salvador) living within 25 km from the volcano. Two major cities are lying at ~20 km from the volcano, Santa Ana (pop. 522,000) to the north and Sonsonate (pop. 420,000) to the south (Pullinger, 1998). The activity during the last few thousand years was mostly characterized by phreatic to phreatomagmatic

*abernard@ulb.ac.be

Bernard, A., Escobar, C.D., Mazot, A., and Gutiérrez, R.E., 2004, The acid volcanic lake of Santa Ana volcano, El Salvador, in Rose, W.I., Bommer, J.J., López, D.L., Carr, M.J., and Major, J.J., eds., Natural hazards in El Salvador: Boulder, Colorado, Geological Society of America Special Paper 375, p. 121–133. For permission to copy, contact editing@geosociety.org. © 2004 Geological Society of America.

eruptions at the central summit vent (Pullinger, 1998). This suggests the presence of a large and permanent hydrothermal system at shallow levels within Santa Ana volcano. A lake has been observed in the summit crater since the 1904 eruption. Changes in the activity of the lake were reported periodically, for example in 1920, when a near-boiling of the lake was observed. More recently, in July 1992, an increase in temperature and gas emissions was recorded in the fumarolic area adjacent to the lake. A few geochemical data were collected at that time and showed that the lake was strongly acid and contained elevated concentrations in SO_4 and Cl (Gutierrez and Escobar, 1994).

The presence of a lake in the crater of an active volcano represents a unique situation for long-term monitoring of passive degassing. The lake acts as a calorimeter and chemical condenser and integrates most of the flux of heat and volatiles released by a shallow magma. Marked changes in key chemical and physical parameters are often observed several weeks or months before the onset of a new eruption (Giggenbach and Glover, 1975; Takano, 1987; Badrudin, 1994; Rowe et al., 1992b; Vandemeulebrouck et al., 2000; Varekamp et al., 2001).

This paper describes the principal physical and chemical features of Santa Ana crater lake obtained during a two-year monitoring program. These data will serve as a reference baseline for monitoring future changes in the activity of the lake.

PHYSICAL SETTING OF THE CRATER LAKE

The lake is almost circular with a diameter of ~200 m and occupies the deepest part of Santa Ana crater. The lake has no inlets or outlets and is confined to a deep and narrow crater (Fig. 1). A preliminary bathymetric survey, by means of an echo-sounder, was made in 2000 and completed in 2002 (Figs. 2A, 2B and 3). The maximum depth recorded in 2000 was 27 m with a volume estimated at 4.7×10^5 m^3. In February 2002, the lake level had decreased by ~5–6 m and its volume was reduced to 3.1×10^5 m^3. This significant reduction in lake volume is not correlated with an unusual lack of precipitation nor an increase in the evaporation at the lake surface because lake temperatures were close to ambient in 2002. The occurrence of two major tectonic earthquakes in January (M 7.6) and February 2001 (M 6.6) could have promoted fracturing and enhanced the permeability within the volcanic edifice, leading to an increase in the downward seepage of the lake waters.

GAS FLUXES AND TEMPERATURE

A fumarolic field has been present on the west side of the lake for at least several decades. A temperature of 532 °C was measured from one high-temperature fumarole in January 2000. A sudden increase in degassing rate was observed in May 2000, with the appearance of new fumaroles and an increase in outlet pressure of some fumaroles. Red-glowing areas were observed at night in January 2002. A temperature of 632 °C was measured in February 2002, and 875 °C by Tobias Fischer in June 2002.

SO_2 fluxes measured by COSPEC ranged between 390 and 244 t/d in 2001 and 100 t/d in 2002 (Smithsonian Institution, 2001; W.I. Rose, 2002, personal commun.).

The CO_2 flux emitted at the surface of the lake was measured in 2002 by IR spectrophotometry using a Dräger Polytron. We modified the technique initially developed for soil gas flux monitoring by Chiodini et al. (1996) in order to work on a crater lake by using a floating accumulation chamber. Forty-six measurements were obtained along five transects with a 25 m spacing. The flux of CO_2 ranged from 49 to 2971 g m^{-2} d^{-1} with a mean value of 220 g m^{-2} d^{-1}. The highest fluxes were found in the W zone of the lake where shallow subaqueous fumaroles discharged gases directly into the lake. The total CO_2 output calculated for the lake surface area was estimated at 7 t/d. This value is about one order of magnitude lower than the CO_2 flux recently measured on Kelud crater lake, Indonesia (Bernard and Mazot, Table DR1[1]).

LAKE CHEMISTRY

Chemical data of the lake waters were collected over a two-year period (2000–2002) (Table 1). Most of the lake water samples were collected during the dry season (from December to March in El Salvador) except samples SAN1, SAN 2, and APR02. The pH was measured in the field after standard calibration with pH 1 and 4 reference buffers, and redox potential was measured with a Pt electrode. Cations were analyzed by inductively coupled plasma–optical-emission spectroscopy (ICP-OES) and anions by high-pressure liquid chromatography (HPLC). Ferrous iron was analyzed by spectrophotometry after reaction with orthophenantroline reagent. All samples attain a charge balance of within ±10% except those collected in January 2000. Equilibrium pH calculated with PHREEQC (Parkhurst and Appelo, 1999) show a larger discrepancy with the measured values for these samples (Table 1).

Santa Ana lake is chemically homogeneous; samples collected from different locations at the lake surface or from different depths show only minor variations within analytical errors. Thermal springs were found in 2000 on the SW shoreline of the lake, with temperatures ranging from 41.2 to 80.2 °C and low discharge rates (<1 L/min). Their chemistry (Table 1) is close to the lake waters, suggesting that they represent the main source of the lake waters. The lake waters have a chemical composition typical of acid sulfate-chloride waters with SO_4/Cl ~1.5. The Santa Ana lake is moderately concentrated with a total dissolved solids (TDS) concentration of ~23,000 mg/kg, 5 to 6 times less concentrated than some hyperacidic lakes like Poas or Ijen (Rowe et al., 1992a; Delmelle and Bernard, 1994). Santa Ana lake bears many resemblances to Ruapehu's acid lake in New Zealand (Giggenbach, 1974; Christenson and Wood, 1993;

---

[1]GSA Data Repository Item 2004044, Fluxes of CO_2 gas measured at surface of Kelud crater lake in 2002, is available on request from Documents Secretary, GSA, P.O. Box 9140, Boulder, CO 80301-9140, USA, or editing@geosociety.org, at www.geosociety.org/pubs/ft2004.htm.

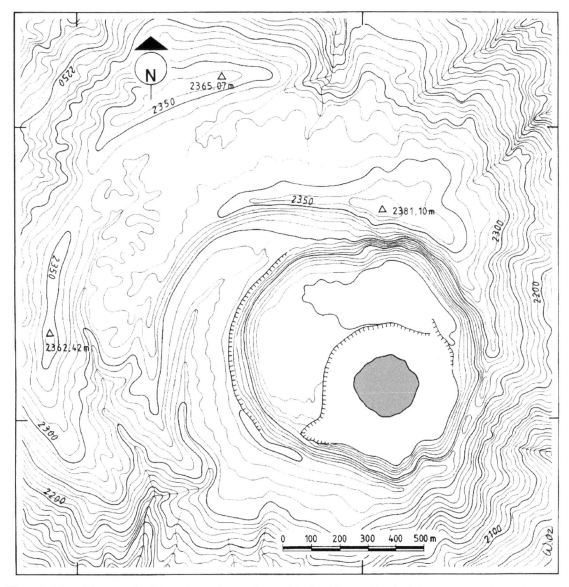

Figure 1. Map of the summit of Santa Ana volcano. Current lake level is noted by shaded area. Contour interval = 10 m.

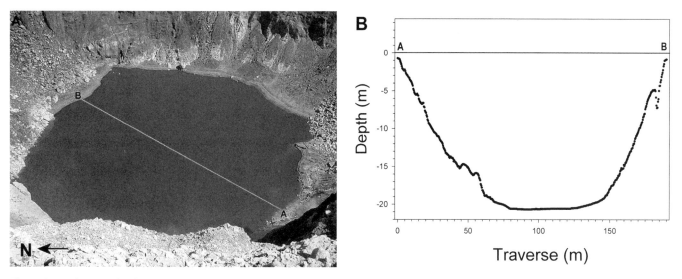

Figure 2. A: Photograph of Santa Ana lake in February 2002. Line A–B is the bathymetric traverse shown on Figure 2B. B: Bathymetric profile along the traverse shown on Figure 2A.

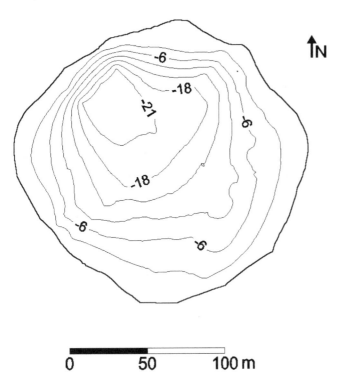

Figure 3. Bathymetric map of Santa Ana crater lake.

correlated with volcanic activity, with high Fe^{2+}/Fe^{3+} corresponding to quiescent period and low lake temperatures (Rowe, 1994). Polythionate concentrations in 2000 were respectively 36.3, 87.6, and 48.2 mg/L for S_4O_6, S_5O_6, and S_6O_6 (Takano, 2002, personal commun.) and are evidence of the discharge of H_2S+SO_2 gases in the lake–hydrothermal system (Takano, 1987; Takano and Watanuki, 1990). The isotopic composition of dissolved HSO_4^- also supports the evidence of the injection of magmatic SO_2 into the lake–hydrothermal system. Observed values of $\delta^{34}S_{HSO_4}$ range between 16.0 and 16.7‰ (Table 3). Several crater lakes have their dissolved sulfates strongly enriched in ^{34}S, and it is commonly accepted that this enrichment is the consequence of the hydrolysis (disproportionation) of SO_2 occurring within the hydrothermal system beneath the lake (Kusakabe et al., 2000, and reference therein). No elemental sulfur (S°) was observed floating on the lake waters.

ORIGIN OF ROCK-FORMING ELEMENTS (RFES) IN THE LAKE WATERS

Different processes contribute to the chemistry of a crater lake: injection of acid volatiles, dissolution of wall rocks, dilution by meteoric waters, evaporation, and reactions of precipitation/dissolution of secondary minerals in lake waters.

In Figure 4, the RFE concentrations in the lake waters are compared to their abundance in volcanic rocks. An average composition of nine samples of olivine basalt and basaltic andesite reported by Pullinger (1998) was used as the reference material for the wall rock (Table 2). The close correspondence of most RFE between lake waters and basaltic andesite suggests that congruent dissolution is the main source of RFE in the lake waters. The concentrations observed correspond to the congruent dissolution of 12 g of andesite per kg of lake waters in 2000 and 18 g in 2002. Silica and Ti, together with Ba and Sr (not shown) are strongly depleted in the lake waters compared to their rock abundances. This is the consequence of the low solubility of these elements in strongly acid waters. Precipitates of these elements are frequently observed in suspension in lake waters like Kawah Ijen, Indonesia (Delmelle and Bernard, 1994).

In many crater lakes, however, significant discrepancies in the relative abundances of RFE between rock compositions and lake chemistry are typically observed (Christenson and Wood, 1993; Delmelle and Bernard, 1994; Ohba et al., 1994; Christenson, 2000; Varekamp et al., 2000). Incongruent dissolution of the host rocks or the nonconservative behavior of some RFE in these acidic fluids are the main causes of these discrepancies and can explain most of the temporal fluctuations in the relative concentrations of RFE observed in crater lakes. Incongruent dissolution is commonly observed when fresh magma intrudes into a hydrothermal system (Giggenbach, 1974; Giggenbach and Glover, 1975). Differences in the rate of leachability of some elements or in the kinetics of dissolution of primary magmatic minerals are responsible for this incongruent dissolution (Rowe et al., 1992a; Christenson and Wood, 1993; Varekamp et al.,

Christenson, 2000) with the notable exception of a much higher Fe content in Santa Ana lake. This partly reflects the difference in wall-rock composition: olivine basalt or basaltic andesite for Santa Ana instead of andesite for Ruapehu. Acid waters are chemically very reactive and will promote extensive water-rock interactions. The main consequence of rock dissolution is the consumption of acidity (H^+) by hydrolysis reactions of silicates. In order to keep the observed low pH values, acid volatiles must be regularly supplied to the lake–hydrothermal system. The significant SO_2 fluxes measured in the plume emitted by the high-temperature fumaroles adjacent to the lake provide clear evidence of a transfer of acid gases to the shallow environment of Santa Ana volcano.

Iron is essentially present in the form of ferrous (Fe^{2+}) iron. A few measurements of Eh with a platinum redox electrode were obtained from the lake waters and ranged between 570 and 610 mV. These measured Eh values are consistent with equilibrium Eh calculated from the Fe^{2+}/Fe^{3+} ratio. This agreement is typically observed in acid waters (pH <2.5) with a high dissolved Fe content where potential redox measurements are quantitatively correlated with the dominant redox couple Fe^{2+}/Fe^{3+} (Nordstrom et al., 1979; Nordstrom and Alpers, 1999). It is unlikely that a true redox equilibrium is reached at low temperature between Fe and sulfur species, but variations in Fe^{2+}/Fe^{3+} could be correlated to changes in the H_2S/SO_2 ratios of gases injected in the lake–hydrothermal system. At Poas volcano, Fe^{2+}/Fe^{3+} is strongly

TABLE 1. CHEMICAL COMPOSITIONS OF LAKE WATERS AND THERMAL SPRINGS (SAS) OF SANTA ANA VOLCANO

Sample name	Date	Depth (m)	T (°C)	pH$_{mes}$	pH$_{calc}$	Na (mg/L)	K (mg/L)	Ca (mg/L)	Mg (mg/L)	Fe$_{tot}$ (mg/L)	Fe^{2+} (mg/L)	Mn (mg/L)	Al (mg/L)	SiO$_2$ (mg/L)	B (mg/L)	Ti (mg/L)	Ba (mg/L)	Sr (mg/L)	F (mg/L)	Cl (mg/L)	SO$_4$ (mg/L)	TDS* (g/L)
SAL1	Jan. 2000	0	18.9	0.9	1.3	378	136	799	294	738	688	16.0	1143	235	6.6	1.38	0.17	5.2	215	5472	8417	17.6
SAL2	Jan. 2000	0	18.9	0.9	1.2	378	137	812	296	756	672	16.2	1147	240	6.7	1.38	0.128	5.2	211	5653	8747	18.1
SAL3	Jan. 2000	0	18.9	0.9	1.2	374	134	802	296	750	680	16.0	1135	237	6.9	1.41	0.104	5.3	235	5655	8744	18.1
SAL4	Jan. 2000	0	18.9	0.9	1.3	376	138	840	308	779	687	16.8	1177	250	7.1	1.46	0.104	5.5	209	5517	8634	18.0
SAL5	Jan. 2000	20	18.9	0.9	1.3	377	137	831	300	771	682	16.9	1165	252	7	1.42	0.128	5.4	223	5609	8704	18.1
SAL6	Jan. 2000	25	18.9	1.0	1.3	377	137	832	302	773	N.A.	17.0	1162	255	7.1	1.43	0.132	5.5	212	5426	8633	17.8
SAN1	July 2000	0	30	0.9	1.0	395	145	862	307	832	700	17.0	1214	282	6	1	0.055	4.1	194	7603	10907	22.5
SAN2	Aug. 2000	0	30	0.9	1.0	367	139	805	283	720	700	16.0	1161	276	5.7	1.3	0.082	3.9	194	7267	10661	21.6
SAP1	Feb. 2001	0	26	0.9	1.1	506	209	1144	392	1065	1055	20.0	1654	359	12.8	2.76	0.074	6.6	379	8369	12464	26.2
SAP2	Feb. 2001	0	26	0.9	1.0	485	196	1108	366	974	N.A.	18.4	1580	357	11.4	2.27	0.079	5.8	386	8758	12899	26.8
ANA21	Feb. 2002	0	20.6	1.1	1.3	487	174	1010	417	1308	N.A.	21.0	1717	303	10.2	3.10	0.099	6.9	341	7615	11318	24.3
ANA22	Feb. 2002	0	20.6	1.1	1.3	481	173	1028	424	1307	1256	20.8	1713	301	10.2	3.20	0.090	7.2	333	7631	10918	23.9
ANA23	Feb. 2002	0	20.6	1.1	1.2	481	171	1031	428	1321	1233	21.1	1712	305	N.A.	N.A.	N.A.	N.A.	314	7700	11157	24.2
ANA24	Feb. 2002	0	20.6	1.1	1.2	491	181	1052	434	1326	N.A.	21.1	1744	309	N.A.	N.A.	N.A.	N.A.	337	7941	11448	24.8
MARC02	Mar. 2002	0	21	1.1	1.1	463	177	1013	423	1314	1308	23.2	1743	302	10.2	3.20	0.053	7.3	372	8401	12598	26.4
APR02	Apr. 2002	0	21	1.1	1.1	457	178	1006	422	1330	1314	23.3	1741	295	9.1	3.10	0.080	7.3	357	8402	12378	26.2
SAS1	Jan.2000	0	52.3	0.9	1.1	573	193	950	384	998	N.A.	23.4	1622	274	8.3	1.51	0.196	6.3	170	8550	11678	25.0
SAS2	Jan.2000	0	80.2	0.9	1.3	731	204	1039	463	1155	N.A.	28.8	1792	278	8.8	1.02	0.148	6.2	170	8277	10534	24.2
SAS3	Jan.2000	0	41.2	0.9	1.1	359	137	817	278	738	N.A.	16.6	1126	252	7.3	0.882	0.112	6.3	253	6088	9528	19.3

*Total dissolved solids.

TABLE 2. AVERAGE CHEMICAL COMPOSITION OF ANDESITE RECALCULATED FROM PULLINGER (1998)

SiO_2 (wt%)	TiO_2 (wt%)	Al_2O_3 (wt%)	FeO (wt%)	MnO (wt%)	MgO (wt%)	CaO (wt%)	Na_2O (wt%)	K_2O (wt%)
52.81	1.08	18.96	9.41	0.18	3.92	8.43	3.86	1.33

TABLE 3. STABLE ISOTOPE DATA FOR SANTA ANA LAKE WATERS
AND THERMAL SPRING (SAS)

Sample name	Date	δD (‰)	$\delta^{18}O_{H2O}$ (‰)	$\delta^{18}O_{HSO4}$ (‰)	$\delta^{34}S_{HSO4}$ (‰)
SAL1	Jan. 2000	−34.9	−2.32	14.4	15.96
SAL2	Jan. 2000	−35.1	−2.35	13.8	N.A.
SAN1	July 2000	N.A.	N.A.	N.A.	16.09
SAN2	Aug. 2000	N.A.	N.A.	N.A.	15.97
SAP1	Feb. 2001	−21	0.2	14.72	15.99
SAP2	Feb. 2001	−23	0.3	14.97	N.A.
ANA21	Feb. 2002	−19	−0.2	14.64	16.65
SAS2	Jan. 2000	−31.1	−1.86	13.6	N.A.

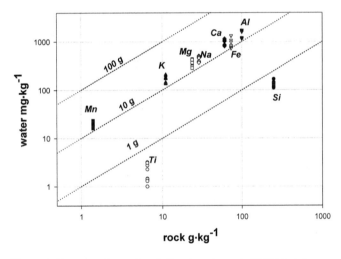

Figure 4. Concentrations of rock-forming elements (RFE) in lake waters compared to Santa Ana basaltic andesite (Table 2).

2001). Temperature or chemical variations of acid fluids recycled in the sublimnic zone beneath the lake can change the degree of saturation of minerals and promote precipitation or dissolution reactions, significantly altering the fluid composition.

SATURATION STATE OF THE LAKE WATERS

An ion-association aqueous speciation model PHREEQC (Parkhurst and Appelo, 1999) with the thermodynamic database WATEQ4f (Ball and Nordstrom, 1991) was used to evaluate the saturation state of the lake waters. All of the lake waters are slightly supersaturated or close to saturation with silica (cristo-

balite or amorphous silica), barite, and gypsum at the observed lake temperatures (19–30 °C).

Only trace amounts of precipitates were found in suspension in the lake waters: amorphous silica, Pb+Sr-rich barite ($BaSO_4$), pyrite, and gypsum. Along the shoreline, where lake waters evaporate, only sporadic precipitates of tamarugite [$NaAl(SO_4)_2 \cdot 6(H_2O)$] were found. PHREEQC was also used to model the saturation state of the lake waters at higher temperatures in the subsurface zone where the convective recycling of the lake waters is assumed, in order to identify which minerals are possibly present at depth. The analytical compositions reported in Table 1 were used, and temperature was increased to simulate conductive heating of lake waters during their seepage beneath the lake. Results of conductive heating up to 250 °C for two representative lake waters are presented in Figure 5.

All the samples give similar results: silica polymorphs, anhydrite, alunite, and diaspore are supersaturated within the range of temperatures studied. With the exception of diaspore, these minerals were frequently reported in the alteration products of Ruapehu volcano (Giggenbach, 1974; Christenson and Wood, 1993; Christenson, 2000). Pyrite is absent from the saturated phases because no H_2S was added to the lake waters. Hydrogen sulfide is likely to be present in the lake waters but its concentration is not known. Using a somewhat arbitrary concentration of 0.1 mg/L of $H_2S(aq)$ in the lake waters, pyrite is supersaturated (log IAP/K > 0) at all temperatures. With this amount of $H_2S(aq)$, elemental sulfur (S°) never reached saturation. Most samples give SI values close to sample ANA21, with anhydrite and alunite reaching saturation in a relatively narrow range of temperatures between 80 and 120 °C. Only samples SAN1 and SAN2 give contrasting SI values for alunite and probably reflect a dilution by acid fluids as discussed below.

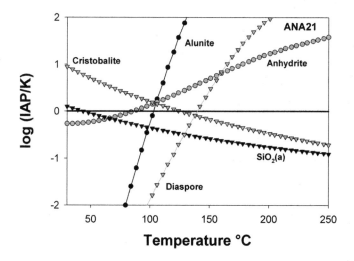

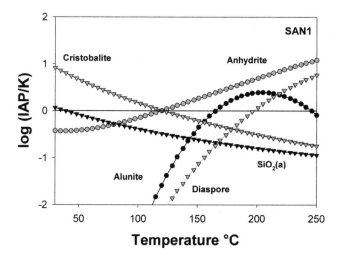

Figure 5. Saturation indexes of mineral phases as a function of temperature for samples ANA21 and SAN1. IAP is the ion activity product and K is the dissolution constant for the mineral phase considered. $SiO_2(a)$ is amorphous silica.

The model was finally used to study water-rock interaction occurring at high temperatures in the sublimnic zone beneath the lake and its possible influence on fluid chemistry. The starting composition was a pure acid fluid based on the analytical concentrations of F, Cl, and SO_4 observed in sample ANA21 and can be considered as a crude approximation of acid fluids formed by the condensation of magmatic vapors that mixed with recycled lake waters. The equilibrium pH calculated for this initial composition is 0.56. To this initial acid fluid, increasing amounts of the average Santa Ana andesite (Table 2) were reacted in successive increments. Because of its conservative behavior in acid fluids, Mg was chosen to constrain the final water/rock ratio. Computation was halted when the Mg concentration calculated by the model reached the concentration observed in the lake. This was obtained after addition of 18 g of rock for 1 kg fluid. Model-

ing of water-rock interaction was carried out at the successive temperatures of 20, 100, 150, and 200 °C. As expected, the pH of the solutions increased during the course of the titration corresponding to the consumption of H$^+$ by hydrolysis reaction. The final pH obtained (1.26 at T = 20 °C) is consistent with the observed pH (1.1). Natroalunite $[Na_{0.5}K_{0.5}Al_3(SO_4)_2(OH)_6]$ instead of alunite $[KAl_3(SO_4)_2(OH)_6]$ was used in the calculations to study the effect of its precipitation on the aqueous concentrations of both Na and K. Natroalunite is more likely to be present in equilibrium with Na > K rich fluids than a pure K end member. Also, the presence of natroalunite with a stoichiometry of $[Na_{0.6}K_{0.4}Al_3(SO_4)_2(OH)_6]$ has been frequently reported in the alteration products from Ruapehu crater lake (Giggenbach, 1974; Christenson and Wood, 1993). Diaspore is not present in the equilibrium composition because this phase is in competition for Al with natroalunite, which is more stable than diaspore in these sulfate-rich fluids.

The calculated aqueous compositions obtained at different temperatures are plotted in Figure 6 together with the compositions observed in lake waters and thermal springs. Significant depletion in Na, K, Al, and Ca compared to lake water compositions are observed for water-rock interactions above 100 °C because of the precipitation of anhydrite and natroalunite removing these elements from the aqueous phase. Similar trends can be expected for most crater lake waters reacting with an andesitic host rock. Using higher SO_4 and Cl concentrations in the starting fluid will not dramatically change these results if sufficient quantities of rocks are allowed to react with an initial fluid of higher acidity. Because of the precipitation of pyrite, Fe is likely to be depleted in the reacting solutions if significant amounts of H_2S are reacting with the waters.

Much geophysical and geochemical evidence suggests that temperatures of 150 °C or above are likely in the sublimnic zone of many crater lakes (Delmelle et al., 2000; Kusakabe et al., 2000; Christenson and Wood, 1993). The results of the model explain why significant discrepancies in the RFE abundances between a lake and its host rock are common in many acid crater lakes, even if quiescent conditions necessary for congruent dissolution are reached (i.e., no intrusion of fresh magma). Large variations in Na, K, Al, and Ca concentrations (and probably Fe if sufficient H_2S is allowed to react) can be expected due to precipitation or dissolution reactions of sulfates and sulfides of these elements. Christenson and Wood (1993) and Christenson (2000) have shown that part of the Na, K, and Al fluctuations observed in Ruapehu crater lake can be explained by precipitation or redissolution of these sulfates within the hydrothermal system.

For Santa Ana, the close correspondence in the RFE abundances between lake waters and basaltic andesite is intriguing and may be fortuitous. These results tend to show that the acid fluids last equilibrated with the host rock at low temperature, ~100 °C or slightly above, and suggest a low rate of convective circulation of lake waters in the subsurface system. Both lake waters and hot springs show no Fe depletion, which tends to suggest that most of

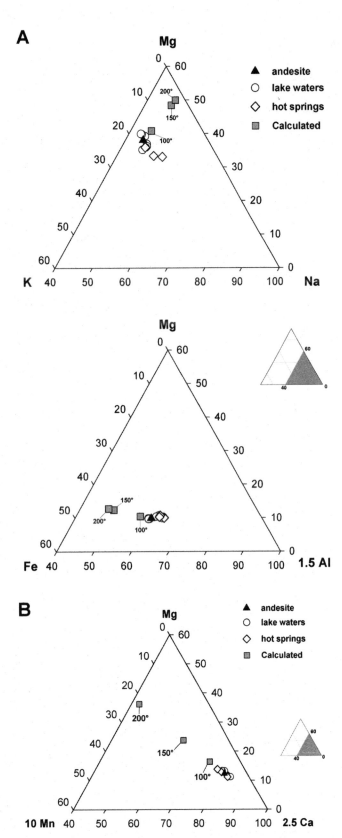

the H_2S is readily oxidized by Fe^{3+} during water-rock interactions by the following reaction, as suggested by Rowe (1994): 16 Fe^{3+} + 8 $H_2S \Rightarrow 16Fe^{2+} + S_8 + 16H^+$.

TEMPORAL EVOLUTION OF LAKE TEMPERATURE AND CHEMISTRY

The only notable change that occurred during the two-year monitoring period was a sudden increase in lake temperatures observed for the first time in May 2000. Temperatures increased from 19 to a maximum of 30 °C in July and August 2000, and bubbling was observed in several areas in the lake as well as an intensification of the fumarolic activity. The heating episode ended between February 2001 and January 2002, when lake temperatures were back to near ambient atmospheric temperature (17 °C). Heating episodes, sometimes cyclic, are relatively frequent in some crater lakes and reflect changes in the flow rate or in the enthalpy of hot fluids entering the lake. These heating episodes always represent an alarming situation because an increasing lake temperature can be a precursory signal for the renewal of magmatic activity. For example, in Kelud volcano (Indonesia) an increase in temperature and bubbling of the lake was observed during the three months preceding the February 1990 eruption (Vandemeulebrouck et al., 2000).

Heat balance calculations based on the model of Stevenson (1992) show that the lake surface dissipated a thermal power equivalent to 16 MW when it was at its peak temperature (30 °C) in July–August 2000. This value is very small in comparison with heat fluxes observed at hot (>40 °C) crater lakes with a large surface area, where values as high as several hundred MW were estimated (Delmelle and Bernard, 2000). The surface area of Santa Ana lake is small with 3.10^4 m², and since heat is lost essentially by evaporation at the lake surface, its capacity to dissipate energy is also limited. For small lakes, a slight increase in the heat flow is sufficient to raise the water temperature several degrees as discussed by Pasternack and Varekamp (1997).

Lake waters collected in July–August 2000 revealed a net increase in Cl and SO_4 (Table 1). The only likely source for chloride is by condensation of magmatic vapors. The variations observed in the lake should thus reflect an increase in the injection of acid fluids into the lake–hydrothermal system. SO_4/Cl remained remarkably constant during all the monitoring period (Fig. 7) as well as $\delta^{34}S$, suggesting that no major changes occurred in the composition of acid gases entering the hydrothermal system beneath the lake. The concentrations of RFE remained unchanged from July–August 2000 until February 2001, when they started to increase. The percentage of residual acidity (PRA) varied also during the heating period and is strongly correlated with temperature (Fig. 8). PRA, introduced by Varekamp et al. (2000), is defined as the percentage of initial acidity left after neutralization by water-rock interaction. PRA reflects the relative rate of the supply of acid gases versus the rate of neutralization by water-rock interaction. PRA has the advantage of being independent of surficial processes (dilution or evaporation) and can give insight on the dynamics of

Figure 6. Relative abundance of rock-forming elements (RFE) in Santa Ana lake waters and thermal springs. Calculated data are equilibrium compositions calculated at various temperatures with PHREEQC.

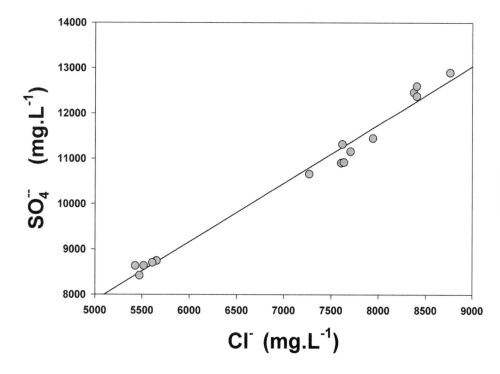

Figure 7. Variations in chlorides and sulfates in Santa Ana lake waters during 2000–2002.

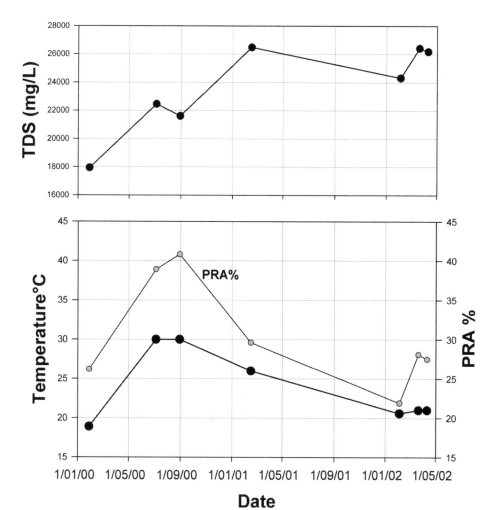

Figure 8. Evolution of total dissolved solids (TDS), percentage of residual acidity (PRA), and temperature for Santa Ana lake during 2000–2002.

these acidic hydrothermal systems. Variations in PRA and compositions show that the lake–hydrothermal system was not able to neutralize the increased input in SO_4 and Cl at least during the first three months. In February 2001, only half of this excess acidity was neutralized by water-rock interaction. It is surprising how the lake–hydrothermal system reacted so sluggishly to this net increase in acidity. The presence of an already leached barren rock protolith in contact with the acid fluids could explain this low buffering capacity. But the close correspondence in RFE abundances between the acid fluids and the basaltic andesite rock (Fig. 4) show that this hypothesis is unlikely. The low buffering power could indicate slow kinetics in water-rock interaction due to relatively low temperatures of the hydrothermal system. Alternatively, the acidity increase could have resulted from the dissolution of fumarolic gases directly injected into the lake from the shore fumaroles and not at deep levels within the hydrothermal system.

ISOTOPE GEOCHEMISTRY

Deuterium and Oxygen Isotopic Compositions

The hydrogen (D/H) and oxygen ($^{18}O/^{16}O$) isotopic ratios for Santa Ana waters (Table 3) are compared (Fig. 9) to the compositions of local meteoric waters and high-temperature gases from volcanic arc magmas (Taran et al., 1989; Giggenbach, 1992). The local meteoric water line (LMWL) was constructed from the GNIP data (IAEA/WMO, 2001) covering the period 1968–1974 (data from Ilopango station, 715 masl). The amount-weighted annual average provided by IAEA (IAEA/WMO, 2001), taking into account the seasonal change in the isotopic composition of

precipitation, is used as the isotopic composition representative of local meteoric waters. The Santa Ana lake waters and thermal springs show enrichments toward heavier isotopic compositions compared to the LMWL composition. In several crater lakes, isotopic compositions can be explained by the contribution of magmatic fluids superimposed on evaporation effects (Rowe, 1994; Delmelle et al., 2000; Sriwana et al., 2000; Ohba et al., 2000; Varekamp and Kreulen, 2000).

The main effect of evaporation in a lake is an enrichment of the water in heavy isotopes with respect to the vapor (Gonfiantini, 1986); this effect is essentially a function of the temperature and atmospheric humidity.

The evaporation effect for Santa Ana lake was evaluated following the procedure of Gonfiantini (1986). The temperature effect on the fractionation factors $\alpha = R_{liquid}/R_{vapor}$ was obtained from the relationship $\ln \alpha = AT^{-2} + BT^{-1} + C$ with A, B, and C coefficients from Majoube (1971). The shift in isotopic composition is defined as the equilibrium enrichment factor: $\Delta\varepsilon_{equi} = (\alpha - 1) \times 1000$. Secondly, the fractionation factors as a function of atmospheric humidity, also called the kinetic enrichment factors, were calculated according to Gonfiantini (1986) and Varekamp and Kreulen (2000) as:

$$\Delta\varepsilon_{kinetic}\ ^{18}O‰ = 14.2\ (1 - h^*)$$

$$\Delta\varepsilon_{kinetic}\ ^{2}H‰ = 12.5\ (1 - h^*)$$

where h^* is a temperature-normalized humidity factor. This factor, h^*, is used instead of the relative atmospheric humidity (h) to take into account an enhanced kinetic fractionation supposed

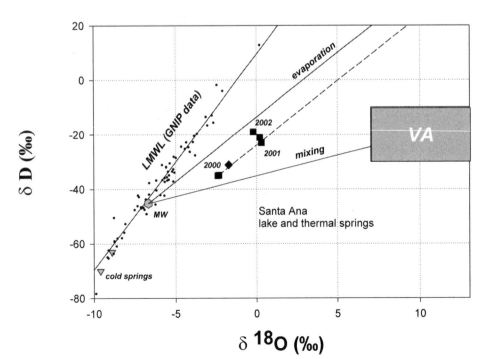

Figure 9. Oxygen and hydrogen isotopic compositions for meteoric waters, thermal springs and Santa Ana lake waters. LMWL is the local meteoric waters reference line and MW is the amount-weighted annual average of local meteoric waters (IAEA/WMO, 2001). Evaporation line is the theoretical evaporation effect calculated for MW according to Gonfiantini (1986) and Varekamp and Kreulen (2000) (see text). Dashed line corresponds to the theoretical evaporation calculated for the 2000 lake water composition. Volcanic arc isotopic composition (VA) is from Taran et al. (1989) and Giggenbach (1992).

to be more adapted to model evaporation processes in hot crater lakes (Varekamp and Kreulen, 2000). h* is defined as:

$$(h*) = P_{sat}^{atm} \times \frac{h}{P_{sat}^{lake}} \; .$$

P_{sat}^{lake} and P_{sat}^{atm} are respectively the saturated vapor pressure at lake and atmospheric temperatures (Varekamp and Kreulen, 2000). Isotopic fractionations due to this enhanced evaporation effect were calculated for the maximum lake temperature observed (30 °C). For T_{atm} = 17 °C (average of measurements with a datalogger during a 24-hour cycle in February 2002) and a relative atmospheric humidity of 65%, h* = 0.30.

The total evaporation effect (Fig. 9) is obtained as:

$$\Delta\varepsilon_{equi} + \Delta\varepsilon_{kinetic}.$$

Santa Ana lake waters and thermal springs are offset from this evaporation line, suggesting that some contribution of isotopically heavy magmatic fluids to the crater lake–hydrothermal system is likely to be present. The shift observed in the compositions following the heating episode during 2000–2001 is best explained by an enhanced evaporation effect due to the higher temperature of the lake surface. The 2001 compositions plot on an evaporation line calculated as above (dashed line on Fig. 9), and no increase in the contribution of magmatic fluids is detected.

Oxygen Isotopes

$\delta^{18}O$ isotopic exchange between water and aqueous HSO_4^- can provide insight on the dynamic of the circulation of lake waters beneath the surface. For several crater lakes, the distribution of $\delta^{18}O$ between water and HSO_4^- is linearly correlated, suggesting an equilibrium oxygen isotopic fractionation at temperatures ~140 °C (Kusakabe et al., 2000). According to these authors, this correlation is good evidence of the recycling and re-equilibration of lake waters within the subsurface hydrothermal system.

$\delta^{18}O_{H_2O}$ and $\delta^{18}O_{HSO_4}$ values for Santa Ana lake (Table 3; Fig. 10) show no such correlation, pointing to the lack of an isotopic equilibrium between water and bisulfate ion. A net increase in the $\delta^{18}O_{H_2O}$ values was observed in 2001, whereas $\delta^{18}O_{HSO_4}$ remained almost constant. This shift in $\delta^{18}O_{H_2O}$, as suggested above, is probably related to an increase in the evaporation effect at the lake surface following the heating episode of 2000. At the low temperature of the lake, slow kinetics of the isotopic exchange prevented the re-equilibration of $\delta^{18}O_{HSO_4}$.

CONCLUSIONS

Most of the physical, chemical, or isotopic data suggest that Santa Ana lake and its associated subsurface hydrothermal system are in a quiescent state. The amount of heat flow supplied to the lake is rather low, and may well explain the poor rate of convective recirculation of lake waters within the subsurface hydrothermal system.

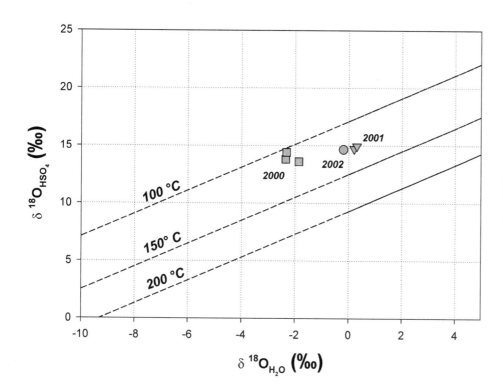

Figure 10. Oxygen isotopic ratios of water and bisulfate in Santa Ana lake waters. Isotherms are based on the equation of Mizutani and Rafter (1983).

These results present a striking contrast to the sustained aerial degassing observed in the fumarolic area adjacent to the lake. High-temperature fumaroles (875 °C) and significant SO_2 fluxes (100–350 t/d) indicate the presence of a high-temperature vapor-dominated zone at some depth below the subsurface hydrothermal system. These gases, released at depth, reach the surface largely unaffected by condensation or interaction with the subsurface hydrothermal system.

The volume of lake waters is low, reducing the threat of formation of dangerous lahars. The small surface area of the lake is confined in a deep and narrow crater, which limits its capacity to dissipate thermal energy and makes this lake a very sensitive calorimeter. If a large amount of heat is supplied to the lake by the underlying system, the lake will quickly boil and evaporate.

Much of the useful information regarding our understanding of these hydrothermal systems was obtained in the past from the interpretation of time-series chemical data collected over a long period. Our understanding of the Santa Ana lake–hydrothermal system would certainly improve if the chemical monitoring of the lake is continued on a regular basis.

ACKNOWLEDGMENTS

We acknowledge Bill Rose who inspired this study on Santa Ana lake and provided financial support for the 2000 and 2001 fieldwork. Carlos Pullinger, Marisa Orantes, Keith Mac Phail, Uwe Grunewald, and la Companera de Bomberos de San Salvador (Section Santa Anita) provided much assistance during fieldwork. This research was funded in part by a grant from the Fonds National de la Recherche Scientifique (FNRS) in Belgium. We are grateful to Minoru Kusakabe and Saskia Gevaert for their help with the isotopic analyses and to Boku Takano for the analyses of polythionates. Comments and corrections by Yuri Taran, Joop Varekamp, and Gregg Bluth greatly helped to improve and clarify this paper.

REFERENCES CITED

Badrudin, M., 1994, Kelut volcano monitoring: Hazards, mitigation and changes in water chemistry prior to the 1990 eruption: Geochemical Journal, v. 28, no. 3, p. 233–241.

Ball, J.W., and Nordstrom, D.K., 1991, WATEQ4F—User's manual with revised thermodynamic database and test cases for calculating speciation of major, trace and redox elements in natural waters: U.S. Geological Survey Open-File Report 90-129, 185 p.

Chiodini, G., Frondini, F., and Raco, B., 1996, Diffuse emission of CO_2 from the Fossa crater, Vulcano Island (Italy): Bulletin of Volcanology, v. 58, p. 41–50.

Christenson, B.W., 2000, Geochemistry of fluids associated with the 1995–1996 eruption of Mt. Ruapehu, New Zealand: Signatures and processes in the magmatic-hydrothermal system: Journal of Volcanology and Geothermal Research, v. 97, p. 1–30.

Christenson, B.W., and Wood, C.P., 1993, Evolution of a vent-hosted hydrothermal system beneath Ruapehu Crater Lake, New Zealand: Bulletin of Volcanology, v. 55, p. 547–565.

Delmelle, P., and Bernard, A., 1994, Geochemistry, mineralogy, and chemical modeling of the acid crater lake of Kawah Ijen Volcano, Indonesia: Geochimica et Cosmochimica Acta, v. 58, p. 2445–2460.

Delmelle, P., and Bernard, A., 2000, Volcanic lakes, in Sigurdsson, H., et al., eds., Encyclopedia of volcanoes: San Diego, California, Academic Press, p. 877–895.

Delmelle, P., Bernard, A., Kusakabe, M., Fischer, T.P., and Takano, B., 2000, Geochemistry of the magmatic hydrothermal system of Kawah Ijen volcano, East Java, Indonesia: Journal of Volcanology and Geothermal Research, v. 97, p. 31–53.

Giggenbach, W.F., 1974, The chemistry of Crater Lake, Mt. Ruapehu (New Zealand) during and after the 1971 active period: New Zealand Journal of Science, v. 17, p. 33–45.

Giggenbach, W.F., 1992, Isotopic shifts in waters from geothermal and volcanic systems along convergent plate boundaries and their origin: Earth Planetary Science Letters, v. 113, p. 495–510.

Giggenbach, W.F., and Glover, R.B., 1975, The use of chemical indicators in the surveillance of volcanic activity affecting the crater lake on Mt. Ruapehu, New Zealand: Bulletin of Volcanology, v. 39, p. 132–145.

Gonfiantini, R., 1986, Environmental isotopes in lake studies, in Frits, P., and Fontes, J.C.H., eds., Handbook of environmental isotope geochemistry: Amsterdam, Elsevier, p. 113–168.

Gutierrez, R.E., and Escobar, C.D., 1994, Crisis en la actividad del volcán de Santa Ana (Ilamatepec), del 22 de Julio al 21 de Agosto 1992, El Salvador, America Central: Centro de Investigaciones Geotecnicas, unpublished report, 14 p.

IAEA/WMO (International Atomic Energy Agency/World Meteorological Organization), 2001, Global network of isotopes in precipitation: The GNIP database: http//isohis.iaea.org (June 2003).

Kusakabe, M., Komoda, Y., Takano, B., and Abiko, T., 2000, Sulfur isotopic effects in the disproportionation reaction of sulfur dioxide in hydrothermal fluids: Implications for the 34S variations of dissolved bisulfate and elemental sulfur from active crater lakes: Journal of Volcanology and Geothermal Research, v. 97, p. 287–307.

Majoube, M., 1971, Fractionnement en oxygène-18 et deutérium entre l'eau et sa vapeur: Journal de Chimie Physique, v. 197, p. 1423–1436.

Mizutani, Y., and Rafter, T.A., 1983, Oxygen isotopic composition of sulphates. Part 3. Oxygen isotopic fractionation in the bisulphate ion-water system: New Zealand Journal of Science 12, p. 54–59.

Mooser, F., Meyer-Abich, H., and McBirney, A.R., 1958, Catalogue of the active volcanoes of the world and solfatara fields of Central America, part VI: Napoli, Italy, International Volcanological Association, 146 p.

Nordstrom, D.K., and Alpers, C.N., 1999, Geochemistry of acid mine waters, in Plumlee, G.S., and Logsdon, M.J., eds., The environmental geochemistry of mineral deposits: Processes, techniques, and health issues: Society of Economic Geologists Reviews in Economic Geology, v. 6A, p. 133–160.

Nordstrom, D.K., Jenne, E.A., and Ball, J.W., 1979, Redox equilibria of iron in acid mine waters, in Jenne, E.A., ed., Chemical modelling in aqueous systems: Washington, D.C., American Chemical Society Symposium Series 93, p. 51–80.

Ohba, T., Hirabayashi, J., and Nogami, K., 1994, Water, heat, and chloride budgets of the crater lake, Yugama at Kusatsu-Shirane volcano, Japan: Geochemical Journal, 28, p. 217–231.

Ohba, T., Hirabayashi, J., and Nogami, K., 2000, D/H and $^{18}O/^{16}O$ ratios of water in the crater lake at Kusatsu-Shirane volcano, Japan: Journal of Volcanology and Geothermal Research, v. 97, p. 329–346.

Parkhurst, D.L., and Appelo, C.A.J., 1999, User's guide to PHREEQC—A computer program for speciation, batch-reaction, one-dimensional transport, and inverse geochemical calculations: U.S. Geological Survey Water-Resources Investigations Report 99-4259, 312 p.

Pasternack, G.B., and Varekamp, J.C., 1997, Volcanic lake systematics. I. Physical constraints: Bulletin of Volcanology, v. 58, p. 528–538.

Pullinger, C., 1998, Evolution of the Santa Ana volcanic complex, El Salvador [M.S. thesis]: Houghton, Michigan Technological University, 151 p.

Rowe, G.L., 1994, Oxygen, hydrogen, and sulfur isotope systematics of the crater lake system of Poas volcano, Costa Rica: Geochemical Journal, v. 28, p. 263–287.

Rowe, G.L., Brantley, S.L., Fernandez, M., Fernandez, J.F., Borgia, A., and Barquero, J.H., 1992a, Fluid-volcano interaction in an active stratovolcano: The crater lake system of Poas Volcano, Costa Rica: Journal of Volcanology and Geothermal Research, v. 49, p. 23–51.

Rowe, G.L., Ohsawa, S., Takano, B., Brantley, S.L., Fernandez, J.F., and Barquero, J.H., 1992b, Using crater lake chemistry to predict volcanic activity at Poas Volcano, Costa Rica: Bulletin of Volcanology, v. 54, p. 494–503.

Smithsonian Institution, Bulletin of the Global Volcanism Network, 2001, v. 26, no. 4, p. 9–11.

Sriwana, T., van Bergen, M.J., Varekamp, J.C., Sumarti, S., Takano, B., van Os, B.J.H., and Leng, M.J., 2000, Geochemistry of the acid Kawah Putih lake, Patuha Volcano, West Java, Indonesia: Journal of Volcanology and Geothermal Research, v. 97, p. 77–104.

Stevenson, D.S., 1992, Heat transfer in active volcanoes: Models of crater lake systems [Ph.D. thesis]: Cambridge, UK, Department of Earth Sciences, The Open University, 234 p.

Takano, B., 1987, Correlation of volcanic activity with sulfur oxyanion speciation in a crater lake: Science, v. 235, p. 1633–1635.

Takano, B., and Watanuki, K., 1990, Monitoring of volcanic eruptions at Yugama crater lake by aqueous sulfur oxyanions: Journal of Volcanology and Geothermal Research, v. 40, p. 71–87.

Taran, Y.A., Pokrovsky, B.G., and Dubik, Y.M., 1989, Isotopic composition and origin of water from andesitic magmas: Doklady (Translations) Academy of Science USSR, v. 304, p. 440–443.

Vandemeulebrouck, J., Sabroux, J.-C., Halbwachs, M., Surono, N., Poussielgue, J., Grangeon, J., and Tabbagh, J., 2000, Hydroacoustic noise precursors of the 1990 eruption of Kelut Volcano, Indonesia: Journal of Volcanology and Geothermal Research, v. 97, p. 443–456.

Varekamp, J.C., and Kreulen, R., 2000, The stable isotope geochemistry of volcanic lakes, with examples from Indonesia: Journal of Volcanology and Geothermal Research, v. 97, p. 309–327.

Varekamp, J.C., Pasternack, G.B., and Rowe, G.L., Jr., 2000, Volcanic lake systematics II. Chemical constraints: Journal of Volcanology and Geothermal Research, v. 97, p. 161–179.

Varekamp, J.C., Ouimette, A.P., Herman, S.W., Bermudez, A., and Delpino, D., 2001, Hydrothermal element fluxes from Copahue, Argentina: A "beehive" volcano in turmoil: Geology, v. 29, p. 1059–1062.

MANUSCRIPT ACCEPTED BY THE SOCIETY JUNE 16, 2003

Geological Society of America
Special Paper 375
2004

Spatial and temporal variations of diffuse CO_2 degassing at the Santa Ana–Izalco–Coatepeque volcanic complex, El Salvador, Central America

José M.L. Salazar
Pedro A. Hernández
Nemesio M. Pérez*
*Environmental Research Division, Instituto Tecnológico y de Energías Renovables (ITER),
38611 Granadilla, S/C de Tenerife, Spain*

Rodolfo Olmos
Francisco Barahona
Rafael Cartagena
Tomás Soriano
Instituto de Ciencias de la Tierra, Universidad de El Salvador, El Salvador, Central America

Dina L. López
Department of Geological Sciences, 316 Clippinger Laboratories, Ohio University, Athens, Ohio 45701, USA

Hirochika Sumino
Kenji Notsu
*Laboratory for Earthquake Chemistry, Graduate School of Science, University of Tokyo,
Hongo, Bunkyo-Ku 113-0033, Tokyo, Japan*

ABSTRACT

We report the first detailed study of spatial and temporal variations on the diffuse emission of carbon dioxide from the Santa Ana–Izalco–Coatepeque volcanic complex. Soil CO_2 efflux measurements were performed at 447 sampling sites and reached values up to 293 g m^{-2} d^{-1} for the March 2001 survey. Most of the diffuse CO_2 degassing is occurring at the center of this volcanic complex, near the Cerro Pacho dome where the intersection of one of the regional NW-SE fault/fracture systems and the south Coatepeque caldera rim occurs. Soil CO_2 efflux measurements were not performed inside Santa Ana's summit crater due to accessibility problems at the time of this survey. The total diffuse CO_2 emission for this volcanic complex was estimated at ~600 t/d. Low CO_2 efflux values were identified on the flanks and summit regions of the Izalco and Santa Ana volcanoes. Isotopic data of gases collected from low-temperature fumaroles in the study area showed $^3He/^4He$ ratios close to the atmospheric composition, except for the fumarolic discharges inside Santa Ana summit crater (7.54). Carbon isotopic signature of these gases suggested that 74% of CO_2 is limestone-derived from thermodecarbonation processes, while the fumarole sample

*nperez@iter.rcanaria.es

Salazar, J.M.L., Hernández, P.A., Pérez, N.M., Olmos, R., Barahona, F., Cartagena, R., Soriano, T., López, D.L., Sumino, H., and Notsu, K., 2004, Spatial and temporal variations of diffuse CO_2 degassing at the Santa Ana–Izalco–Coatepeque volcanic complex, El Salvador, Central America, *in* Rose, W.I., Bommer, J.J., López, D.L., Carr, M.J., and Major, J.J., eds., Natural hazards in El Salvador: Boulder, Colorado, Geological Society of America Special Paper 375, p. 135–146.

from Santa Ana crater showed ~48% of limestone-derived CO_2. Temporal variations of the diffuse CO_2 degassing observed at the Cerro Pacho dome ranged from 4.3 to 327 g m^{-2} d^{-1}, with a median value of 98 g m^{-2} d^{-1}. Time domain analysis of the soil CO_2 efflux showed a strong autoregressive behavior, whose covariance slowly dies through time. Spectral analysis showed the existence of soil CO_2 efflux variations at diurnal and semidiurnal frequencies. This finding suggests that soil CO_2 efflux variations are coupled with those of meteorological variables (i.e., barometric pressure, wind speed, etc.), accounting for much of the behavior of the soil CO_2 efflux time series at the short-term scale.

Keywords: carbon dioxide, diffuse degassing, spatial distribution, time series analysis, Santa Ana–Izalco–Coatepeque, El Salvador.

INTRODUCTION

On January 13, 2001, a 7.6 magnitude earthquake (12.8° N, 88.8° W) occurred off the El Salvador coastline within the subduction zone of the Cocos plate (60 km depth), causing extensive damage throughout the entire country (Bulletin of the Global Volcanism Network, 2001). By 19 January 2001, approximately 660 aftershocks had been registered. On February 13, 2001, another 6.6 magnitude earthquake with an inland epicenter ~20 km west of the Chichontepeque (San Vicente) volcano destroyed numerous towns and villages in the north area of this volcano, causing many fatalities and leaving thousands of people homeless.

Soon after the devastating earthquakes, the University of El Salvador requested the urgent collaboration of the Spanish Aid Agency (AECI) in El Salvador. The main goal of this international collaboration was to provide a multidisciplinary approach for volcano monitoring at El Salvador. The Santa Ana–Izalco–Coatepeque volcanic complex was one of volcanoes investigated because about one million inhabitants reside within a 30 km radius of it. Santa Ana volcano started a new period of vigorous plume degassing by May 2000, though no abnormal seismicity was recorded beneath the volcano during this period. These changes were associated with an increased venting of a well-developed hydrothermal system through the crater lake, hot springs, and fumaroles, reflecting the fracturing or leaking of the hydrothermal cap, and not due to the injection of new magma beneath the crater. Puffing of gases from the crater was observed during aftershocks (Bulletin of the Global Volcanism Network, 26:04, 2001). Santa Ana has been one of the most active volcanoes of El Salvador with several historical eruptions. In addition, a growing economy is being built up in the region due to the activity of the nearby port of Acajutla and the increasing production of large coffee plantations surrounding the volcano (Pullinger, 1998).

Three hazard-mitigation actions consisted of a soil gas and CO_2 efflux survey at the Santa Ana–Izalco–Coatepeque volcanic complex, the installation of an automatic geochemical station for continuous monitoring of the diffuse CO_2 degassing at the study area (the first of a total of six geochemical stations installed in El Salvador), and the training of Salvadorans to operate the geochemical monitoring network. The survey carried out at the Santa Ana–Izalco–Coatepeque volcanic complex has helped to identify areas where anomalous degassing occurs. The observed spatial distribution of soil CO_2 efflux will be tremendously beneficial for the Santa Ana–Izalco–Coatepeque volcano monitoring since spatial variations of CO_2 efflux anomalies can be related to magma movement or seismic activity changes. In addition, this survey supported the site selection for the installation of instruments. The first results of the continuous monitoring of diffuse CO_2 efflux are also providing a useful database for the medium-term eruption forecasting because an increase of CO_2 efflux can be related to a pre-eruptive stage of the Santa Ana–Izalco–Coatepeque volcanic complex. Definitively, these three actions are fundamental for future spatial and temporal studies related to diffuse CO_2 degassing phenomena and contributed to the integration of new approaches to achieve optimum volcano monitoring in the area.

Carbon dioxide has been used as a tracer of subsurface magma degassing due to the fact that it is the major gas species after water vapor in both volcanic hydrothermal fluids and magmatic fluids (Gerlach and Graeber, 1986). As a consequence, gas anomalies appear at the surface though some of them are not visible to the human eye (Allard, 1992). Mapping and monitoring of these diffuse emanations can help to identify the active structural features of volcanic edifices and, possibly, to infer the location of potential eruptions (Hernández et al., 2001a). Extensive work has been performed at volcanic and geothermal areas over the last 13 yr, suggesting that even during periods of inactivity, volcanoes release large amounts of carbon dioxide through their flanks (Allard et al., 1987, 1992; Baubron et al., 1990; Farrar et al., 1995; Chiodini et al., 1996, 1998; Pérez et al., 1996; Hernández et al., 1998, 2001b; Gerlach et al., 1998; Giammanco et al., 1998; Salazar et al., 2001a). However, few works related to continuous monitoring of the diffuse degassing of carbon dioxide have been published (Salazar et al., 1999, 2000, 2001b, 2002; Notsu et al., 2000; Padrón et al., 2001; Olmos et al., 2001; Rogie et. al., 2001; Mori et al., 2002).

Here we present the first study of the spatial and temporal features of the diffuse CO_2 degassing at the Santa Ana–Izalco–Coatepeque volcanic complex. Section 2 will briefly present the geological setting of the study area while in section 3 procedures and methods are detailed. In section 4 the results are presented and discussed. Two objective-based methods are used to identify potential geochemical populations within the empirical frequency distribution of soil CO$_2$ efflux at the study area. The Gap Statistic method (Miesch, 1981), a method not commonly cited in the specialized scientific literature, is illustrated and compared with the results obtained with the Sinclair method. Section 4 will deal with the isolation of a threshold value to separate background from anomalous geochemical populations and to define a background mean. A discussion on the spatial distribution of the diffuse CO$_2$ degassing will be presented together with the results of soil gas analysis. Fumarole gas compositions will also be discussed. Section 4 ends with the time series analysis of the dynamics of diffuse CO$_2$ degassing at Cerro Pacho dome. Finally, the conclusions are presented in section 5.

GEOLOGICAL SETTING

The Santa Ana–Izalco–Coatepeque volcanic complex (Fig. 1) is located at the intersection of a NW-SE system of regional faults (Williams and Meyer-Abich, 1955) and the southern boundary of the so-called "Central American Graben" or median trough, which is an extensional structure parallel to the Pacific coastline of El Salvador (Rotolo and Castorina, 1998). The NW-SE system of faults is considered a major structural feature of El Salvador's geology (Wiesemann, 1975), found over the entire country and evidenced in the Santa Ana region by fissures and alignments of volcanic edifices and eruptive centers.

The Santa Ana–Izalco–Coatepeque volcanic complex consists of the Coatepeque collapse caldera (a 6.5 × 10.5 km elliptical depression) and the Santa Ana and Izalco volcanoes, as well as numerous cinder cones and explosion craters. The Santa Ana volcano (0.22 Ma), located 40 km west of San Salvador, is a massive stratovolcano rising 2365 m above sea level. The Coatepeque depression is the result of the Coatepeque volcano collapse ca. 50–70 ka (Pullinger, 1998). Volcanic activity then resumed, shifting 7 km SW to the present location of the Santa Ana volcanic edifice. A new edifice collapse occurred after the Coatepeque collapse, resulting in a large debris avalanche that covered an area of ~300 km² and formed the present Acajutla peninsula. Today, the Santa Ana volcano is active and fills the old Santa Ana collapse caldera. Historical activity of the Santa Ana volcano is well documented going back to the sixteenth century. The San Marcelino eruption in 1722 produced 13 km long lava flows. The summit of the Santa Ana volcano contains an acid lake within a 0.5 km diameter explosion crater that was formed during the most recent eruption in 1904. Hot springs, gas bubbling, and intense fumarolic emissions are observed along the shoreline of this crater lake. A volcanic plume, usually driven by the NE trade wind, may be seen rising up to 500 m from the summit crater of the Santa Ana volcano (Pullinger, 1998, and references therein). Acid rainfall associated with its volcanic plume produced extensive damage to surrounding vegetation and plantations on the SW flanks of the volcano during 2001 and 2002.

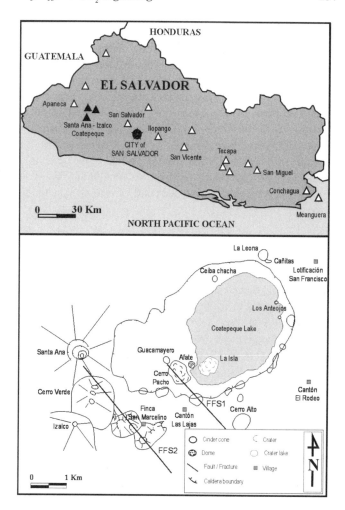

Figure 1. Map of the Central American Volcanic Belt in El Salvador showing the location of the Santa Ana–Izalco–Coatepeque volcanic complex. At the bottom, simple geological map with the major volcano-structural features of the study area.

Weak fumaroles (61.7 °C) are observed along the NW-SE system of regional fractures between the Santa Ana and Cerro Verde edifices and probably are also associated with the local radial fracture system of the Santa Ana volcano. The crater and upper flanks of the Santa Ana volcano show thin phreatomagmatic tephra deposits that act as impermeable barriers to vertical gas migration.

Izalco volcano is a young stratovolcano located south of the Santa Ana volcano. The formation of Izalco volcano started in 1779 and proceeded almost continuously until 1966. Since then, the Izalco volcano has not erupted. Low-temperature fumaroles (73.8 °C, March 2001) were observed at the crater walls of the Izalco volcano. Although the Izalco volcano might not be considered as part of the Santa Ana volcanic system (Pullinger, 1998), it was included within the study area because of its intense and recently frequent volcanic activity.

Post-caldera volcanic activity at Coatepeque is dominated by minor effusive eruptions forming cinder cones inside and outside the Coatepeque caldera. Intracaldera dome growth and hydrothermal activity still continue on its southwestern area. Cerro Pacho (Fig. 1) is probably the most recently emplaced volcanic dome at the Coatepeque caldera, where hydrothermal activity is visible through diffuse steaming and warm ground ("Agua Caliente" spot, 95.2 °C). In 1992, CEL (Comisión Ejecutiva Hidroeléctrica del Río Lempa) investigated the potential geothermal resources in the Coatepeque caldera, finding a significant potential for electricity generation at the Cerro Pacho and La Isla volcanic domes (Fig. 1).

PROCEDURES AND METHODS

In March 2001, a soil gas and CO_2 efflux survey was performed to evaluate the spatial distribution of the diffuse CO_2 degassing rates at Santa Ana volcanic complex. A total of 447 soil gas and diffuse CO_2 efflux measurements were carried out following geological, structural, and accessibility criteria, and covered an area of 210 km^2. This survey was performed during the dry season in El Salvador, under sunny skies and low wind. These conditions provided minimal soil gas efflux variations from atmospheric changes.

Soil CO_2 efflux measurements were performed in accordance with the accumulation chamber technique (Chiodini et al., 1998), by means of RIKEN 411A and Dräger Polytron NDIR CO_2 portable sensors. Soil CO_2 efflux values (g m^{-2} d^{-1}) were determined from the rate of CO_2 concentration increase in the chamber at each site, accounting for local values of atmospheric pressure and temperature to convert volumetric concentrations to mass concentrations. The reproducibility of measurements for the range 2–35,000 g m^{-2} d^{-1} was estimated at about ±10% from the replicate measurements carried out on known CO_2 efflux rates in the laboratory. Soil gases were sampled at a depth of 30–40 cm using a metallic soil probe and 10 cc hypodermic syringes following the method of acidified distilled water displacement on 10 cc vacutainers. Soil gas concentrations were measured using an in situ laboratory with a VARIAN microGC 2002P. Isotope analysis of fumarolic gases was determined at the Laboratory for Earthquake Chemistry, University of Tokyo, Japan.

The CO_2 efflux values were used to construct a contour map of the study area using kriging as the interpolation technique. Due to the chosen sampling strategy and the lack of information at short sampling distances, the selection of a variogram model is not straightforward. Because the majority of usual variogram models behave in the same manner close to the origin, an isotropic linear variogram with no nugget effect was fitted to the CO_2 efflux data (Journel, 1988; Cressie, 1990).

Temporal variations of the soil CO_2 efflux, together with barometric pressure, air temperature and relative humidity, wind speed and direction, and soil temperature and water content, were monitored on an hourly basis using an automatic self-powered geochemical station (West Systems) set up at the Cerro Pacho

volcanic dome. To power the geochemical station, a photovoltaic panel, a charge regulator, and a battery were used. The battery voltage was also monitored to provide information on the solar radiation relative time variations. The geochemical station is designed to temporarily store all the data and radio telemeter it to a proximal station located 5 km away on the NE shore of the Coatepeque Lake. From this remote control station, data retrieval is feasible through a modem connection. The observation started on May 2001, but stopped on August 2001 due to instrumental problems. In November 2001, the observation resumed and still continues, though with some difficulties associated with very frequent electrical power shutdowns in the region of Santa Ana.

RESULTS AND DISCUSSION

Statistical Analysis of Diffuse CO_2 Efflux Survey Data

The CO_2 efflux ranged from undetectable values up to 293 g m^{-2} d^{-1}, with a median of 8.9 g m^{-2} d^{-1} and a quartile range of 4.5 g m^{-2} d^{-1}. The sampling frequency distribution of the diffuse CO_2 efflux values showed a polymodal shape (Fig. 2A) with pronounced high-order statistics: skewness (7.13) and kurtosis (61.7). The lack of symmetry (tail to higher fluxes) and high observed kurtosis indicate the existence of potential outliers within the data set. Before any data analysis, the CO_2 efflux values were log-transformed to make the previous skewed distribution more symmetrical. The shape of the sampling frequency distribution suggests the presence of at least two normal populations (Normal I and Normal II), with an intermediate population separating them (Mixing population). A total of 246 measurements (55%) of the original data were below the lower quantification limit of the instruments, IQL, estimated to be 2 g m^{-2} d^{-1} from laboratory calibrations. The log-transformed data set also showed very irregular behavior when it was graphed on a normal probability plot (Fig. 2B). Again, the lower instrumental quantification limit is depicted at the 2 g m^{-2} d^{-1} level. CO_2 efflux values higher than this limit plotted almost linearly with some inflection zones, suggesting the existence of different lognormal populations within the data set of CO_2 efflux values higher than IQL. If the first 246 CO_2 efflux values are temporally discarded, the sampling frequency distribution shows a multinormal-like shape (Fig. 3A). At least three distinct modes are found: Normal I, Normal II, and the mode between values referred to as GAP-1 and GAP-2.

Two different statistical approaches were applied to the CO_2 efflux values higher than the IQL (Fig. 3B) in the search of different geochemical populations within the diffuse CO_2 degassing phenomena: Sinclair's method (Sinclair, 1974) and Miesch's Gap Statistic method (Miesch, 1981). Both methods, the so-called Model-Based-Objective methods (Stanley and Sinclair, 1989), pursue the objective threshold selection. Once this target is accomplished, anomaly and background values may be defined and separated.

In accordance with the Sinclair method, a visual inspection of the frequency distribution showed an inflection region around

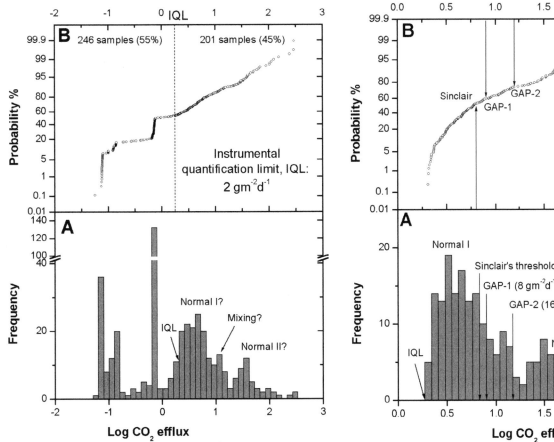

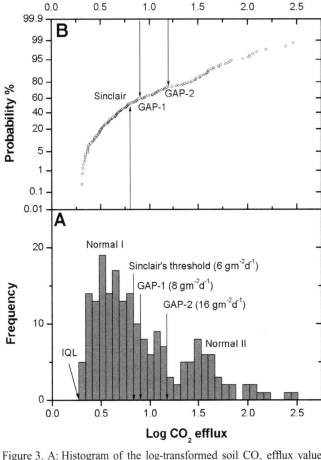

Figure 2. A: Histogram of the log-transformed soil CO_2 efflux values showing the existence of three modes within the sampling frequency distribution. B: Probability plot of the log CO_2 efflux values observed in 447 sampling sites at the Santa Ana–Izalco–Coatepeque volcanic complex. 246 samples were below the instrumental quantification limit (IQL).

Figure 3. A: Histogram of the log-transformed soil CO_2 efflux values showing that at least two geochemical populations can be separated: background (CO_2 efflux < 6 g m^{-2} d^{-1}) and anomalous (CO_2 efflux > 16 g m^{-2} d^{-1}) populations. B: Probability plot of the log CO_2 efflux values observed in 201 sampling sites at the Santa Ana–Izalco–Coatepeque volcanic complex. Three similar thresholds are identified as Sinclair, GAP-1 and GAP-2

the 51 percentile, suggesting a threshold value, *T*, separating background from anomalous values, of 6 g m^{-2} d^{-1}.

The Gap Statistic method involves a comparison of the normal distribution with the observed geochemical data distribution, which has been monotonically converted to a zero-skewness standardized (z-score) distribution by a three-parameter logarithmic transform (average, standard deviation, and symmetry parameter, alpha; Stanley and Sinclair, 1989). The gap statistic is the maximum computed gap between successive ordered and standardized observations after adjusting each individual gap for the expected frequency. This expected frequency is taken from a fitted normal curve with unit area (for details see Miesch, 1981). The significance level of the largest adjusted gap between successive values, relative to what might be expected from a sampled normal distribution of similar size, is then compared with critical-value tables to determine its level of significance as a threshold identification of anomalous samples follows directly (Stanley and Sinclair, 1989).

To get the adjusted gaps (difference between the observed and expected normal gaps), a modification of the Gap Statistic method was used. Due to the instrumental limitations and degree of degassing at the study area, we do not know accurately how the 55% of data plot, but we may account for their statistical weight in the Gap Statistic analysis considering that the CO_2 efflux values over the IQL correspond to relative frequencies from 55 to 100% of the total data set (201 sampling sites from a total of 447). After transforming the reduced data set to improve the symmetry (log[efflux-α], with $\alpha = 1.9153$), two GAP values (GAP-1 = 8.0 g m^{-2} d^{-1}, at the 82% absolute percentile; and GAP-2 = 16 g m^{-2} d^{-1}, at the 89% absolute percentile) were found above Sinclair's threshold, although both were included within the inflection region of the sampling frequency distribution.

To test the statistical significance of the detected gaps, we compared the midpoint of the adjusted gaps with their corresponding critical values. These critical values are derived from interpolation on the tables presented by Miesch (1981) for a total of 447 samples at the 20% (critical gap = 0.0171) and 10% (critical gap = 0.0187) statistical significance. GAP-1 (gap statistic = 0.0187) has statistical significance at the 20% level, while GAP-2 (gap statistic = 0.0216) has a statistical significance at the 5% level.

In spite of the low observed CO_2 efflux range (293 g m^{-2} d^{-1}) at the Santa Ana–Izalco–Coatepeque volcanic complex, the previous results support the existence of at least two normal populations clearly identified within the CO_2 efflux data set, namely, a background geochemical population (lower population) and an anomalous geochemical population (upper population). Both geochemical populations mix somewhere between Sinclair's threshold and the GAP-2 (Fig. 3B) at the inflection region of the sampling frequency distribution. To get a representative index of the background population, we read the geometric mean of the lower population at the 50 percentile assuming a threshold equal to the geometric mean of the CO_2 efflux data interval 3–16 g m^{-2} d^{-1} (threshold, T = 9.2 g m^{-2} d^{-1}). The background population may be then defined as having a geometric mean of ~2 g m^{-2} d^{-1}, close to the IQL value, and includes 83% of the total data.

We consider that the existence of distinct geochemical populations reflect the influence of different physical-chemical processes (external and internal) acting at different spatial scales on the diffuse degassing of CO_2. The working hypothesis underlying this study is that the surface diffuse degassing phenomena at the Santa Ana–Izalco–Coatepeque volcanic complex is the surface evidence of a perturbation of deep origin at the roots of this volcanic complex, responsible for the observed spatial distribution of soil CO_2 efflux rates.

Spatial Distribution of the Diffuse CO_2 Degassing

To evaluate the spatial distribution of the observed CO_2 efflux rates, a contour map based on multiples of the threshold value, T, was elaborated (Fig. 4). This contour map reveals the existence of areas with CO_2 efflux values significantly higher than the estimated threshold. Soil CO_2 efflux values higher than 73.6 g m^{-2} d^{-1} (> 8 × T) are associated with the fault/fracture systems of NW-SE orientation (labeled as FFS1 and FFS2 on Fig. 4). The highest CO_2 efflux values (> 16 × T) were observed at the intersection of one of the NW-SE faults (FFS1) at the southernmost part of the Coatepeque caldera rim and extended to the east flank of the Cerro Verde cinder cone, which is traversed by the fault/fracture system FFS2. Relatively high CO_2 efflux values were also found at the Cerro Pacho dome inside the Coatepeque caldera. Areas

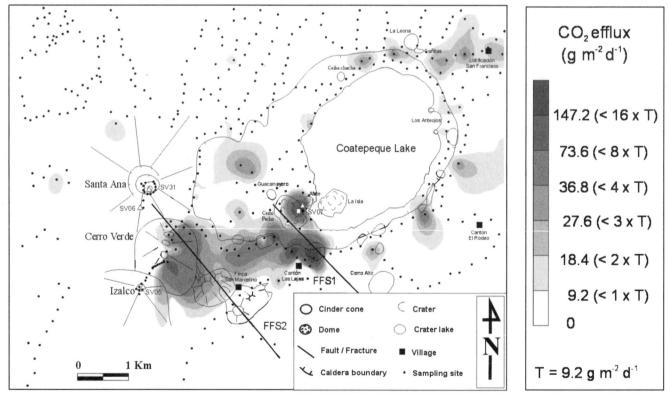

Figure 4. Contour map of the soil CO_2 efflux values (g m^{-2} d^{-1}) measured at the Santa Ana–Izalco–Coatepeque volcanic complex. Two major NW-SE fault/fracture systems are pictured as FFS1 and FFS2. Fumarole gas sample sites are also shown (SV06—Santa Ana south flank; SV31—Santa Ana summit crater; SV07—Cerro Pacho dome; and SV05—Izalco crater) as open triangles. An open square indicates the position of the automatic station for the continuous monitoring of the soil CO_2 efflux.

with CO_2 efflux values higher than 27.6 g m^{-2} d^{-1} (> 3 × T) were found dispersed at the NE sectors of the Coatepeque caldera, most of them outside the caldera rim, and may be associated with biological production of carbon dioxide within cultivated areas. However, these anomalies align on NE-SW direction, which might suggest the existence of some fractures/faults intersecting the ring structure of Coatepeque caldera. Similar anomalous CO_2 efflux values were observed at the SE and NW sector of the study area, though they are were defined by single measurements.

The January 13, 2001, earthquake damaged the steep paths that lead up to the crater's internal flanks, so that it was not possible to reach there and to measure the diffuse CO_2 degassing close to the high-temperature fumarolic field of the Santa Ana volcano crater. The summit of the Santa Ana volcano showed negligible CO_2 efflux values where impermeable phreatomagmatic deposits were observed. Neither the summit nor flanks of the Izalco volcano showed measurable soil CO_2 efflux rates. The rest of the study area showed background values (<T) typical of CO_2 production by decomposition of organic matter and microbiological activity in the forested soils in El Salvador.

The contour map of the diffuse CO_2 degassing at the Santa Ana–Izalco–Coatepeque volcanic complex reveals an interesting spatial correlation with some of its prominent geological and structural features. Negligible diffuse CO_2 degassing was observed at the Izalco volcano, in good agreement with the absence of geophysical and geochemical features of activity. It is interesting to note that even though the Santa Ana edifice exhibited a volcanic plume during the survey, there was no evidence of diffuse emanations of carbon dioxide at its summit and south flank. In this case the diffuse degassing may be limited by the low vertical and lateral permeability of the volcanic deposits on its flanks. The degassing phenomenon of carbon dioxide is restricted to areas such as the recent Cerro Pacho dome and the fault/fracture system FFS1 at the intersection with the Coatepeque caldera rim. These findings suggest that the degassing is structurally controlled by the enhanced permeability of very localized fractured terrains.

In order to provide a baseline for future works on the area, a rough estimation of the total output of diffuse CO_2 from the Santa Ana–Izalco–Coatepeque volcanic complex was computed. Considering the volume bounded by the three-dimensional surface built up by the contouring on the study area will produce an overestimation of the total diffuse CO_2 output (1400 t/d) because the sampling grid was very irregular. In addition, the behavior of soil CO_2 efflux at short distances (nugget effect) could not be conveniently estimated by means of variogram analysis. To avoid this bias, it is more convenient and simple to calculate the sum of every single-measurement contribution to the total output by multiplying each CO_2 efflux value by an average area (i.e., the median of half-distances between sampling sites: 125 m). This procedure shows that the 447 measurements covered an effective area of 7 km² (instead of 210 km²). The diffuse CO_2 outputs occurring at background and anomalous areas were estimated at ~34 and ~162 t/d, respectively. If we assume that the remaining area (203 km²) releases background soil CO_2 efflux values (~2 g m^{-2} d^{-1}), the total contribution of the background population can be estimated at ~440 t/d.

This conservative estimation shows that ~600 t/d of CO_2 are released by the Santa Ana–Izalco–Coatepeque volcanic complex in a diffuse form. If this approach becomes true, ~162 t/d CO_2 (27% of the total output) may be derived from deep sources beneath the volcanic complex.

Most of the soil gas analysis (Table 1) showed a median H_2 and He composition close to that of atmospheric air with a partial CO_2 enrichment (up to 5.6 vol%). Ternary diagrams (Figs. 5A and 5B) showed that soil gas composition plot along a mixing line between atmospheric air and a CO_2-rich end member. This finding suggests a wide range of air contamination. In fact, the majority of gas samples showed a N_2/O_2 ratio of ~3.7.

Fumarolic and Diffuse Gas Emission Compositions

Fumaroles on the west side of the Santa Ana crater lake had a maximum temperature, recorded in January 2000, of 523 °C (Bulletin of the Global Volcanism Network, 26:04, 2001). This fumarolic activity is responsible for Santa Ana's volcanic plume. COSPEC measurements of the plume were made using the Guatemalan COSPEC on 8 and 9 February 2001. Tripod-based surveys were made from Cerro Verde (elevation ~2000 m), 2 km south of the crater. Automobile-based traverses conducted along the Santa Ana–Sonsonate highway (5 km west of Santa Ana) were made on 9 February. These surveys resulted in an average SO_2 flux of 393 t/d on 8 February and 244 t/d on 9 February (Bulletin of the Global Volcanism Network, 26:04, 2001).

Fumarole gases collected at the Cerro Pacho dome on March 23, 2001 (Table 2), showed a CO_2 concentration of 13.53 vol%, a $\delta^{13}C(CO_2)$ of –5.65‰, a corrected $^3He/^4He$ ratio of 1.34 ± 0.057 R_A (where R_A is the atmospheric $^3He/^4He$ ratio: 1.4 × 10^{-6}), and a $CO_2/^3He$ ratio of 1.47 × 10^{10}, showing a major contribution of atmospheric air to the helium composition of this gas sample.

TABLE 1. ANALYTICAL RESULTS OF THE SOIL GASES FROM
THE SANTA ANA–IZALCO–COATEPEQUE VOLCANIC COMPLEX

Location	H_2 (ppmv)	He (ppmv)	CO_2 (ppmv)	N_2/O_2	He/CO_2	H_2/CO_2
Minimum	<0.1	4.43	34	0.35	9.6 × 10^{-5}	1.6 × 10^{-6}
Maximum	2.4	9.65	55.74	5.23	5.9 × 10^{-2}	2.3 × 10^{-3}
Median	0.18	5.52	2.40	3.67	2.3 × 10^{-3}	7.3 × 10^{-5}

A

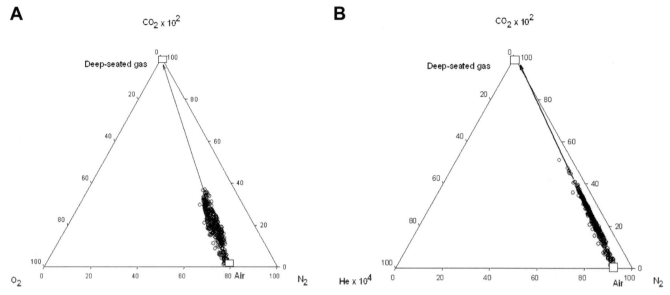

Figure 5. CO_2–O_2–N_2 (A) and CO–N_2–He (B) ternary diagrams of soil gases from the study area.

TABLE 2. ISOTOPIC COMPOSITION OF FUMAROLES SAMPLED AT THE STUDIED AREA

Sample	Date (dd/mm/yy)	Location	Temp. (°C)	He (ppm)	$(^3He/^4He)_o$ (Ra)	$^4He/^{20}Ne$	$(^3He/^4He)_c$ (Ra)	CO_2 (%)	$\delta^{13}C$-CO_2 (vs. PDB)
SV05	24/03/01	Izalco summit	73.8	3.6	1.03 ± 0.04	0.241	0.89	0.79	−22.9
SV06	27/03/01	Santa Ana South flank	61.7	5.2	1.43 ± 0.04	0.464	2.03	6.21	−1.94
SV07	23/03/01	Cerro Pacho	95.2	5.0	1.34 ± 0.06	0.304	4.58	13.53	−5.65
SV31	15/08/02	Santa Ana crater	123.5	4.4	7.54 ± 0.18	225	7.54	13.03	−2.04

Mass balance of the CO_2 in the fumarolic gas using the method of Sano and Marty (1995) yielded 73.19% of limestone-derived CO_2, while 16.59% and 10.22% contributions are associated with sediments and mantle-derived CO_2, respectively. These results suggest that the composition of the Cerro Pacho fumaroles comes up from the mixing of the atmospheric and limestone-derived CO_2, with a minor contribution of the magmatic end member. CEL (1992) reported a CO_2 concentration of 3.7–3.9 vol%, a H_2S concentration of 200 ppm, and 123 mg/kg and 0.041 mg/kg of condensed vapor of NH_3 and B, respectively, suggesting the contribution of a geothermal reservoir to the present fumarole composition. However, CEL reported N_2/O_2 ratios of 3.73, suggesting that atmospheric contamination is implicitly present at Cerro Pacho fumaroles (CEL, 1992). Fumarolic gas samples were also collected at the south flank of the Santa Ana volcano (March 27, 2001) and the summit crater of the Izalco volcano (March 24, 2001). $^3He/^4He$ ratios were close to the atmospheric composition: 1.43 ± 0.040 R_A and 1.03 ± 0.043 R_A, respectively. On August 15, 2002, a gas sample from the low-temperature fumaroles inside

the Santa Ana summit crater was collected (Table 1). The carbon isotope analysis yielded similar results for the Santa Ana flank and summit crater gas samples, −1.94‰ and −2.04‰, respectively. Despite the low CO_2 efflux values observed at the flanks of Santa Ana volcano, the carbon isotope signatures suggest that a fraction of the carbon dioxide released through the flanks of the volcano might be derived from thermodecarbonation of marine carbonates proximal to or incorporated within the magmatic plumbing system.

Dynamics of the Diffuse CO_2 Degassing

Based on the spatial distribution of the soil CO_2 diffuse emissions and chemical and isotopic composition of the fumarolic gas samples, we decided to set up the automatic geochemical station at the Cerro Pacho dome inside the Coatepeque caldera. Temporal variations of the soil CO_2 efflux (Fig. 6) measured at the Cerro Pacho dome ranged from 4.3 to 327 g m^{-2} d^{-1}, with a median value of 98 and a quartile range of 26 g m^{-2} d^{-1}. The baro-

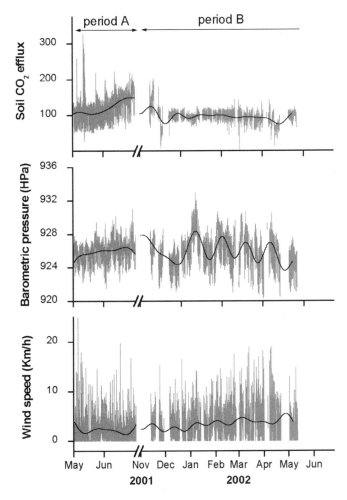

Figure 6. Time series plot of (A) the diffuse CO_2 efflux, (B) barometric pressure, and (C) wind speed from May 2001 to May 2002 at the Santa Ana–Izalco–Coatepeque volcanic complex. The black line depicts a 24-hour-FFT low band-pass filter.

volcano from May 2000 and at Cerro Pacho dome during period A. Sites well connected at depth such as those described anomalous degassing areas in the March 2001 soil CO_2 efflux survey are expected to respond to fluid pressure changes faster than background and less fractured areas. If the CO_2 efflux time series observed at Cerro Pacho dome is extrapolated to the whole study area, the estimation of total diffuse CO_2 output will be quite different. If a twofold increase is assumed in each single measuring site, the total output could be estimated at 1200 t/d and so on. Taking into account that only one site was being monitored in a continuous base after carrying out the March 2001 CO_2 efflux survey, it is not possible to infer how the degassing of Santa Ana volcano might have affected the entire study area. Therefore, the constraints of the adopted procedure to estimate the total diffuse CO_2 output seem to be more appropriate.

Moderate barometric pressure high and low fronts affecting the region of Santa Ana have been clearly identified during the period B, with characteristic frequencies of ~2–4 weeks, with a slight decreasing trend from 928 HPa in January 2002 to 924 HPa in May 2002. A succession of wind speed cycles can also be observed during the period B, though at higher frequencies than those observed in the barometric pressure. The weather conditions at the observation site in the Cerro Pacho dome are buffered and controlled by the Coatepeque lake, which guarantees low wind speed amplitudes and quite variable air humidity levels.

To test for the existence of periodicity or seasonality within the observed soil CO_2 efflux data, both time and frequency domain classical statistical techniques were applied. A set of 2229 valid CO_2 efflux observations (i.e., data for all the variables were available for the same measurement) were selected in period B to avoid the frequent interruptions within the time series. By means of the correlogram, which graphically displays the autocorrelation function (ACF) of the observed time series, it is found that the soil CO_2 efflux temporal variations exhibit a highly significant autoregressive behavior (Fig. 7A). The estimate of the ACF provides a picture of serial dependencies (serial correlation coefficients) for consecutive lags (Box and Jenkins, 1976). This ACF shows a main seasonality reproduced every 24 hours (lags 1, 2, 3, ...) and a low-amplitude secondary seasonality reproduced every 12 hours identified as single peaks up to lag 4. These serial correlations extend through the correlogram, becoming equivalent to those of a white noise process after ~300 observations (14–15 days).

To define the seasonal fluctuations of different lengths suggested by the ACF analysis, the spectral density of the soil CO_2 efflux time series was computed (Fig. 7B). Spectral density estimates were computed by smoothing the periodogram values with a Hamming 11-weighted moving average transformation. To avoid frequency "leakage," the time series length was truncated to a power of 2 (2048 data), mean subtracted and detrended, and finally tapered 15%. Therefore, we searched for the frequency regions with the greatest spectral densities that explain most of the periodic behavior of the soil CO_2 efflux time series on the short-term scale (from a few hours to days). Two frequency regions cor-

metric pressure and wind speed time series are also presented. To avoid the noisy appearance of these time series on the short-term scale, a low-pass FFT filter with a cutoff frequency of 0.04167 Hz (equivalent to one cycle per day) was applied (solid line in Fig. 6). An increasing trend was observed in the diffuse CO_2 degassing during the period May–July 2001 (period A). This finding may be related to (1) the period of anomalous plume activity at the Santa Ana volcano during 2001 and soon after the earthquakes, and/or (2) the beginning of the rainy season in El Salvador. This last cannot be tested due to the lack of rainfall data within the study area and the lack of CO_2 efflux data covering both dry and rainy seasons. However, a second-order or weak stationarity (Chatfield, 1996) is observed in the soil CO_2 efflux time series throughout the observation period covering November 2001 to May 2002 (period B). These observations raise the question of whether the entire study area was affected by the intense degassing observed at the summit of Santa Ana

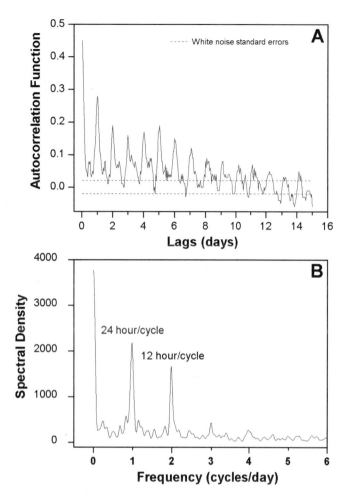

Figure 7. A: Autocorrelation function of the soil CO_2 efflux time series for the period B of Figure 6. B: Spectral density of the soil CO_2 efflux time series for the period B.

responding to the highest spectral densities were found centered at 24 hours (diurnal) and 12 hours (semidiurnal), suggesting a close relationship between the temporal variations of the soil CO_2 efflux and those of the meteorological variables (external variables), which show diurnal and semidiurnal periodicity (i.e., barometric pressure, wind speed, solar radiation, air temperature, etc.).

It has been widely recognized that changing meteorological and soil physical variables play a central role in the short-term variability of soil gas concentrations (Hinkle, 1991, 1994; Massmann and Farrier, 1992; Woodcock, 1987). Once the gas reaches the unsaturated zone, combined pressure and concentration gradients are developed. Atmospheric disturbances acting at the soil horizon introduce additional gas concentration and pressure gradients, resulting in a complex coupling of internal and external forces responsible for the observed concentration profiles and emission rates. Gas concentration gradients within the soil horizon produce diffusive fluxes, whereas pressure gradients account for advective or viscous flows (Thorstenson and Pollock,

1989). Among such atmospheric disturbances, those of barometric pressure and wind velocity are of great importance. Diffuse CO_2 degassing is also influenced by the particular topographic situation of the observation site as well as the forced soil ventilation driven by barometric pressure changes and wind conditions (Rogie et al., 2001; Salazar et al., 2001b, 2002). Rainfall events must be also considered because the percolating water forms an impermeable barrier that increases the soil gas concentrations and effluxes, trapping gases from below (Hinkle, 1994).

In spite of this complex scenario, the spectral coherence of the observed time series may provide additional information so as to understand which external variables, and to what extent, drive the temporal variations of the soil CO_2 efflux. The spectral coherency computes the squared correlation between the cyclical components in two time series at different frequencies. Figure 8 shows the spectral coherence between barometric pressure and soil CO_2 efflux (Fig. 8A), wind speed and soil CO_2 efflux (Fig. 8B), and wind speed and barometric pressure (Fig. 8C). In this case, a higher tapering level was applied (25%) to get smoother squared coherence plots. Barometric pressure displays a high spectral coherence with the soil CO_2 efflux time series at 1 (0.47) and 2 (0.66) cycles per day. The analysis of the cross correlation between soil CO_2 efflux and the barometric pressure yields asymmetrical serial correlation values at lags of 6 and 18 hours, whereas positive correlation is observed at 0 lag. These results suggest that the observed diurnal and semidiurnal barometric pressure changes lead to the observed temporal variations of the soil CO_2 efflux with an average delay of 6 hours. Spectral coherence of wind speed versus soil CO_2 efflux does not show single frequency peaks, despite the adopted means to avoid frequency leakage. Maximum values of spectral coherence (0.43) are found enclosed in a frequency region between 1 and 2 cycles per day. Nonnegligible spectral coherence is also distributed at harmonic frequencies (4, 6, 8 and 3, 5 cycles per day). This finding suggests that the wind speed behavior at the observation site at Cerro Pacho follows a nontrivial pattern, where the common diurnal and semidiurnal variations are present, superimposed on lesser-known high-frequency variations. Wind speed shows a strong spectral coherence peak (0.84) with barometric pressure (Fig. 8C) at diurnal frequencies. Wind speed and barometric pressure are positively correlated up to lag 6, pointing out that changes in the barometric pressure within a time window of 6 hours may drive the wind conditions.

CONCLUSIONS

This paper presents the first detailed study of the spatial and temporal features of diffuse CO_2 degassing at the Santa Ana–Izalco–Coatepeque volcanic complex in El Salvador. Most of the diffuse CO_2 degassing anomaly is limited to the central area of the volcanic complex, between the San Marcelino and Cerro Verde cinder cones and the intersection of a regional fracture/fracture system (FFS1) with the Coatepeque caldera, including the Cerro Pacho volcanic dome. This observation highlights the

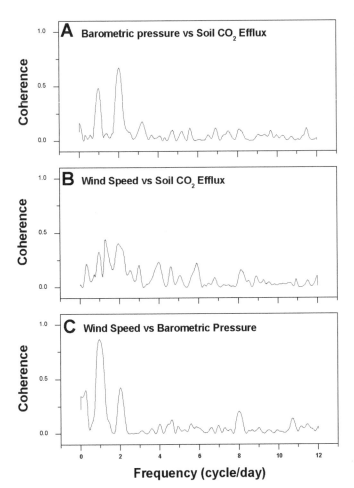

Figure 8. Spectral coherence of (A) the barometric pressure and (B) wind speed versus the soil CO_2 efflux time series for the period B of Figure 6. C: Spectral coherence plot between the barometric pressure and wind speed time series measured at the Cerro Pacho dome.

structural control on CO_2 degassing that is channelized from depth toward the surface through terrains with enhanced permeability due to fracturing. Two different methods, Sinclair and Gap Statistic, have been used to identify different geochemical populations in the spatial distribution of soil CO_2 efflux at the study area. After identifying potential threshold values in the analysis, a threshold of 9.2 g m^{-2} d^{-1} was selected to separate background and anomalous geochemical populations. The background mean was then estimated at ~2 g m^{-2} d^{-1}. The estimated diffuse emission rate of CO_2 for the Santa Ana–Izalco–Coatepeque volcanic complex is ~600 t/d.

The observed temporal variations of diffuse CO_2 degassing at the Cerro Pacho dome reveal a strong autoregressive behavior, whose covariance slowly dies out through time. Changes in the meteorological variables (i.e., barometric pressure, wind speed, etc.) account for much of the behavior of the soil CO_2 efflux time series on a short-term scale. However, a larger observation time span is needed to understand the soil CO_2 efflux medium- and long-term responses to the influence of the rainy season and the meteorological variable changes.

These results constitute the first baseline data for future work on the diffuse degassing phenomena at the Santa Ana–Izalco–Coatepeque volcanic complex.

ACKNOWLEDGMENTS

We are grateful to the Office of the Spanish Agency for International Cooperation (AECI) in El Salvador, the Spanish Embassy in El Salvador, the Ministry of Environment and Natural Resources of the Government of El Salvador, and Geotérmica Salvadoreña (GESAL) for their assistance during the fieldwork. We are indebted to the Grupo de Reacción Policial (GRP) of El Salvador's Civil National Police for providing security during our stay in El Salvador and John M. Murray for help with English. Isotopic analyses were done by H. Sumino, A. Shimizu, and professor K. Nagao at the Laboratory for Earthquake Chemistry, University of Tokyo, Japan. This research was mainly supported by the Spanish Aid Agency (Agencia Española de Cooperación Internacional—AECI), but additional financial aid was provided by the Cabildo Insular de Tenerife, Caja Canarias (Canary Islands, Spain), the European Union, and the University of El Salvador.

REFERENCES CITED

Allard, P., 1992, Diffuse degassing of carbon dioxide through volcanic systems: Observed facts and implications: Report of the Geological Survey of Japan, no. 279, p. 7–11.

Allard, P., Le Bronce, J., Morel, P., Vavasseur, C., Faivre-Pierret, R., Robe, M.C., Roussel, C., and Zettwoog, P., 1987, Geochemistry of soil gas emanations from Mt. Etna, Sicily: Terra Cognita, v. 7(G), p. 17–52.

Baubron, J.C., Allard, P., Sabroux, J.C., Tedesco, D., and Toutain, J.P., 1990, Diffuse volcanic emissions of carbon dioxide from Vulcano Island, Italy: Nature, v. 344, p. 51–53.

Bulletin of the Global Volcanism Network, 2001, v. 26, no. 4, p. 9–11.

CEL (Comisión Ejecutiva Hidroeléctrica del Río Lempa), 1992, Desarrollo de los recursos geotérmicos del area Centro-Occidental de El Salvador. Prefactibilidad geotérmica del área de Coatepeque. Reconocimiento geotérmico. Informe final. Geotérmica Italiana, Unpublished internal report.

Chatfield, C., 1996, The analysis of time series, an introduction (5th edition): London, UK, Chapman & Hall/CRC, p. 279.

Chiodini, G., Frondini, F., and Raco, B., 1996, Diffuse emission of CO_2 from the Fossa crater, Vulcano Island (Italy): Bulletin of Volcanology, v. 58, p. 41–50.

Chiodini, G., Cioni, R., Guidi, M., Raco, B., and Marini, L., 1998, Soil CO_2 flux measurements in volcanic and geothermal areas: Applied Geochemistry, v. 13, p. 543–552.

Cressie, N.A.C., 1990, The origins of kriging: Mathematical Geology, v. 22, p. 239.

Farrar, C.D., Sorey, M.L., Evans, W.C., Howle, J.F., Kerr, B.D., Kennedy, B.M., King, Y., and Southon, J.R., 1995, Forest-killing diffuse CO_2 emission at Mammoth Mountain as a sign of magmatic unrest: Nature, v. 376, p. 675–678.

Gerlach, T.M., and Graeber, E.J., 1986, Volatile budget of Kilauea volcano: Nature, v. 313, p. 273–277.

Gerlach, T., Doukas, M., McGee, K., and Kessler, R., 1998, Three-year decline of magmatic CO_2 emission from soils of a Mammoth Mountain tree kill: Horseshoe Lake, California, 1995–1997: Geophysical Research Letters, v. 25, p. 1947–1950.

Giammanco, S., Gurrieri, S., and Valenza, M., 1998, Anomalous soil CO_2 degassing in relation to faults and eruptive fissures on Mount Etna (Sicily): Bulletin of Volcanology, v. 69, p. 252–259.

Hernández, P.A., Pérez, N.M., Salazar, J.M., Notsu, K., and Wakita, H., 1998, Diffuse emission of carbon dioxide, methane and helium-3 from Teide volcano, Tenerife, Canary Islands: Geophysical Research Letters, v. 25, p. 3311–3314.

Hernández, P.A., Salazar, J.M., Shimoike, Y., Mori, T., Notsu, K., and Pérez, N., 2001a, Diffuse emission of CO_2 from Mijake-jima volcano, Japan: Chemical Geology, v. 177, p. 175–185.

Hernández, P.A., Notsu, K., Salazar, J.M.L., Mori, T., Natale, G., Okada, H., Virgili, G., Shimoike, Y., Sato, M., and Pérez, N.M., 2001b, Carbon dioxide degassing by advective flow from Usu volcano, Japan: Science, v. 292, p. 83–86.

Hinkle, M.E., 1991, Seasonal and geothermal production variations in concentrations of He and CO_2 in soil gases, Roosevelt Hot Springs Known Geothermal Resource Area, Utah, USA: Applied Geochemistry, v. 6, p. 35–47.

Hinkle, M.E., 1994, Environmental conditions affecting concentrations of He, CO_2, O_2 and N_2 in soil gases: Applied Geochemistry, v. 9, p. 53–63.

Journel, A., 1988, Principles of environment sampling: Washington, D.C., American Chemical Society, p. 45–72.

Massmann, J., and Farrier, D.F., 1992, Effects of atmospheric pressure on gas transport in the vadose zone: Water Resources Research, v. 28, no. 3, p. 777–791.

Miesch, A.T., 1981, Estimation of the geochemical threshold and its statistical significance: Journal of Geochemical Exploration, v. 16, p. 49–76.

Mori, T., Hernández, P.A., Salazar, J.M.L., Pérez, N.M., Virgili, G., and Okada, H., 2002, Continuous monitoring of soil CO_2 efflux from the summit region of Usu volcano, Japan: Bulletin of the Volcanological Society of Japan, v. 47, p. 339–345 (in Japanese).

Notsu, K., Mori, T., Hernández, P.A., Virgili, G., Salazar, J.M.L., Pérez, N.M., and Okada, H., 2000, Continuous monitoring of soil CO_2 diffuse emission at Usu volcano, Hokkaido, Japan: Eos (Transactions, American Geophysical Union), v. 81, p. 1320.

Olmos, R., Barahona, F., Cartagena, R., Soriano, T., Salazar, J.M.L., Hernández, P.A., Pérez, N.M., Notsu, K., and López, D., 2001, Diffuse carbon dioxide degassing monitoring at Santa Ana–Izalco–Coatepeque volcanic system, El Salvador, Central America: Eos (Transactions, American Geophysical Union), v. 82, p. 1132.

Padrón, E., Salazar, J.M.L., Hernández, P.A., and Pérez, N.M., 2001, Continuous monitoring of diffuse CO_2 degassing from Cumbre Vieja volcano, La Palma, Canary Islands: Eos (Transactions, American Geophysical Union), v. 82, p. 1132.

Pérez, N.M., Wakita, H., Lolok, D., Patia, H., Talai, B., and McKee, C.O., 1996, Anomalous soil gas CO_2 concentrations and relation to seismic activity at Rabaul caldera, Papua New Guinea: Geogaceta, v. 20, no. 4, p. 1000–1003.

Pullinger, C., 1998, Evolution of the Santa Ana volcanic complex, El Salvador [M.S. thesis]: Houghton, Michigan Technological University, 152 p.

Rogie, J.D., Kerrick, D.M., Sorey, M.L., Chiodini, G., and Galloway, D.L., 2001, Dynamics of carbon dioxide emission at Mammoth Mountain, California: Earth and Planetary Science Letters, v. 188, p. 535–541.

Rotolo, S.G., and Castorina, F., 1998, Transition from mildy-tholeiitic to calc-alkaline suite: The case of Chintepec volcanic centre, El Salvador, Central America: Journal of Volcanology and Geothermal Research, v. 86, p. 117–136.

Salazar, J.M.L., Pérez, N.M., and Hernández, P.A., 1999, Continuous monitoring of soil CO_2 flux levels at the summit of Teide volcano, Tenerife, Canary Islands: Eos (Transactions, American Geophysical Union), v. 80, p. 1149.

Salazar, J.M.L., Pérez, N.M., and Hernández, P.A., 2000, Secular variations of soil CO_2 flux levels at the summit cone of Teide volcano, Tenerife, Canary Islands: Eos (Transactions, American Geophysical Union), v. 81, p. 1317.

Salazar, J.M.L., Hernández, P.A., Pérez, N.M., Melián, G., Álvarez, J., Segura, F., and Notsu, K., 2001a, Diffuse emission of carbon dioxide from Cerro Negro volcano, Nicaragua: Geophysical Research Letters, v. 28, p. 4275–4278.

Salazar, J.M.L., Hernández, P.A., Pérez, N.M., Barahona, F., Olmos, R., Cartagena, R., Soriano, T., Notsu, K., and López, D., 2001b, Anomalous diffuse CO_2 emission changes at San Vicente volcano related to earthquakes in El Salvador, Central America: Eos (Transactions, American Geophysical Union), v. 82, p. 1371.

Salazar, J.M.L., Pérez, N.M., Hernández, P.A., Soriano, T., Barahona, F., Olmos, R., Cartagena, R., López, D.L., Lima, R.N., Melián, G., Galindo, I., Padrón, E., Sumino, H., and Notsu, K., 2002, Precursory diffuse carbon dioxide degassing signature related to a 5.1 magnitude earthquake in El Salvador, Central America: Earth and Planetary Science Letters, v. 205, p. 81–89.

Sano, Y., and Marty, B., 1995, Origin of carbon in fumarolic gas from island arcs: Chemical Geology, v. 199, p. 265–274.

Sinclair, A.J., 1974, Selection of thresholds in geochemical data using probability graphs: Journal of Geochemical Exploration, v. 3, p. 129–149.

Stanley, C.R., and Sinclair, A.J., 1989, Comparison of probability plots and the gap statistic in the selection of thresholds for exploration geochemistry data: Journal of Geochemical Exploration, v. 32, p. 355–357.

Thorstenson, D.C., and Pollock, D.W., 1989, Gas transport in unsaturated porous media: The adequacy of Fick's law: Reviews of Geophysics, v. 27, no. 1, p. 61–78.

Wiesemann, G., 1975, Remarks on the geologic structure of the Republic of El Salvador, Central America: Mitt. Geologisch-Paläontologisches Institut, University of Hamburg, v. 44, p. 557–574.

Williams, H., and Meyer-Abich, H., 1955, Volcanism in the southern part of El Salvador, with particular reference to the collapse basins of lakes Coatepeque and Ilopango: University of California Publications in Geological Sciences, v. 32(1), p. 1–64.

Woodcock, A.H., 1987, Mountain breathing revisited—The hyperventilation of a volcano cinder cone: Bulletin of the American Meteorological Society, v. 68, p. 125–130.

MANUSCRIPT ACCEPTED BY THE SOCIETY JUNE 16, 2003

Geological Society of America
Special Paper 375
2004

Eruptive history and volcanic hazards of Volcán San Salvador

Darrell Sofield*

GeoEngineers, Inc., 1101 S. Fawcett Avenue, Tacoma, Washington, 98402, USA

ABSTRACT

Volcán San Salvador (110 km³) looms over the capital of El Salvador, with the same given name. The volcano has been dormant since its last eruption in 1917. Meanwhile, a metropolis with more than two million people has developed around and encroached upon the volcano flanks. This paper details the volcano's eruptive history and discusses the particular hazards associated with Volcán San Salvador. Additional thickness measurements (new data) are used to recalculate and improve aerial extent and volume estimates of widespread tephra. Five eruption scenarios are outlined that describe the major types of eruptions observed in the geologic record. They include monogenetic (magmatic and hydromagmatic) flank eruptions and three increasingly explosive eruption scenarios originating from a central vent. A map delineates hazard zones based on the types of volcanic hazards from the central vent. The relative risk of a monogenetic flank vent eruption is identified. This paper summarizes the most accurate data available to provide better understanding of the volcanic hazards so that the risks associated with the next volcanic eruption can be minimized.

Keywords: San Salvador, volcanic hazards, volcano, Boquerón.

INTRODUCTION

Volcán San Salvador is one of five historically active volcanic centers in El Salvador (Fig. 1). It is located within the most populated region of the country, towering above San Salvador, the country's capital and economic heart. Other major industrial centers such as Nueva San Salvador (Santa Tecla), Lourdes, Antiguo Cuscatlán, and Apopa are situated on the lower flanks of the volcano (Fig. 2). Volcán San Salvador has a long history of repeated, and sometimes violent, eruptions, but it has not erupted since 1917. In subsequent years, the population increased tenfold and the urban areas have encroached upon the flanks of the volcano. A population unacquainted with volcanic phenomena and urban encroachment into hazardous locations increase the risk that even small volcanic events may have serious societal consequences.

Volcán San Salvador is composed of a complex accumulation of central vent deposits and monogenetic flank vent deposits. In the center is an edifice, known as El Boquerón, with a well-defined cone. Boquerón contains a circular crater ~1600 m in

diameter and 500 m deep. El Picacho (1959 m) is a prominent peak northeast of Boquerón. El Jabalí (1400 m) is a peak located northwest of Boquerón. Both peaks are remnants of a crater wall (4 km by 6 km wide) from an older edifice referred to as San Salvador (Fig. 2).

To a casual observer, Volcán San Salvador appears benign. Currently, coffee plantations and small farms extend up the flanks to nearly to the rim of Boquerón. Boquerón's crater walls are lush with vegetation, which obscure intercalated lavas and ash. The only signs of recent volcanic activity are a small (30 m high) cinder cone in the crater, Boqueróncito, and relatively young a'a lava flows on the northwest flank.

In the past 70,000 yr, Volcán San Salvador erupted repetitively from a central vent as well as from smaller vents and fissures on its flanks. Since ~1580 yr B.P., eruptions have occurred most frequently on the northwest flank of the volcano. These small eruptions have formed monogenetic cinder cones and explosion craters and in some cases andesitic lava flows. However, Volcán San Salvador also experienced violent explosive eruptions, the most recent of which occurred 800 yr B.P. History shows it is only a matter of time before the volcano will erupt again in some fashion (Sofield, 1998).

*dsofield@geoengineers.com

Sofield, D., 2004, Eruptive history and volcanic hazards of Volcán San Salvador, *in* Rose, W.I., Bommer, J.J., López, D.L., Carr, M.J., and Major, J.J., eds., Natural hazards in El Salvador: Boulder, Colorado, Geological Society of America Special Paper 375, p. 147–158. For permission to copy, contact editing@geosociety.org.

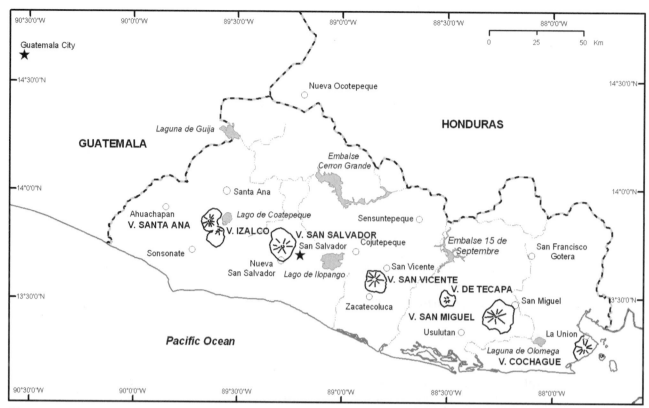

Figure 1. Location of significant Quaternary volcanoes in El Salvador. Open circles indicate major cities. Stars represent the country's capital and largest city. Radiating splays enclosed with a circular polygon approximates the shape of the composite cone and the footprint of the volcano. Two of the major lakes (gray polygons), Lago de Ilopango and Lago de Coatepeque are Quaternary silicic calderas.

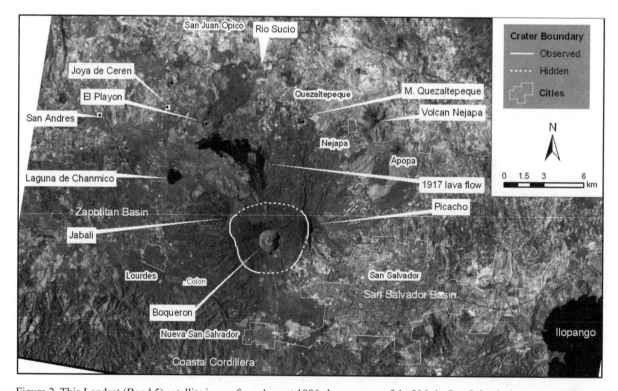

Figure 2. This Landsat (Band 5) satellite image from August 1986 shows many of the Volcán San Salvador's structural features and recent volcanic deposits. The view details how the urban areas (gray stippled regions) are encroaching onto the volcano's flanks. El Boquerón is located in the center of the image. The 1917 eruption deposit, El Boqueróncito (the dark spot in the middle of the El Boquerón's crater) and lava flow (the dark form on the north flank of El Boquerón) are clearly visible. Multiple circular pits and lakes (e.g., Laguna de Chanmico) on the volcano's flanks are explosion craters. The peaks, Picacho and Jabali, are prominent features of an older larger edifice. It is likely that a large eruption destroyed the edifice ~40,000 yr ago, forming a large crater (dashed lines). The portion of the crater wall that is not covered by younger lavas is visible from space.

This paper describes the geological history of the San Salvador volcano based upon physical volcanology and historical accounts. It compiles the following: data first published in Sofield (1998), unpublished tephra thickness measurements collected in November 1998, UTM-projected base maps digitized by U.S. Agency for International Development and the U.S. Geological Survey in 2001, and isopach maps recalculated in 2002. The paper also includes five eruption scenarios describing the major types of eruptions deduced from the geologic record. A hazard zonation map for the central vent and relative risk map for a monogenetic flank eruption provides tools for policymakers to assess the risk of a future eruption.

REGIONAL SETTING

Volcán San Salvador is one of 33 volcanic centers that make up the Central American Volcanic Front. It is a narrow zone (20 km wide and 1100 km long) of closely spaced (~30 km) volcanic centers that runs parallel to the Pacific Coast and the Middle American Trench (Stoiber and Carr, 1973). In El Salvador, all of the volcanic centers have developed within an elongate structural depression known as the Median Trough (Carr and Stoiber, 1981). The Median Trough (Fig. 1) is bounded by a series of normal faults that offset Miocene to late Pliocene lavas of the Balsámo and Cuscatlán Formations (Reynolds, 1980) and exposed the heavily eroded mountain chain, known as the Coastal Cordillera, located south of the volcano. Volcán San Salvador and Pleistocene volcanic centers (Volcán Nejapa and Gazapa) built up composite cones on the floor of the Median Trough. The Median Trough is composed of a thick accumulation of Tertiary volcanics (Cuscatlán Formation). Two relatively flat basins, Zapotitán and San Salvador, are located east and west respectively of the volcano (Fig. 2). Lavas from Volcán San Salvador rest against down-dropped blocks of the Balsámo and Cuscatlán Formation, on the south side of the regional Median Trough.

ERUPTION HISTORY OF VOLCÁN SAN SALVADOR

Introduction

Volcán San Salvador's evolution includes two periods of central vent edifice construction (San Salvador and Boquerón), each ended with a period of explosive eruption(s) and crater formation (G-1 and San Andrés). For at least the last 1580 yr, the majority of the eruptive events occurred on the volcano's flanks. Flank deposits include lava, cinder cones, pyroclastic base surge deposits, and tephra-fall tephra. The last eruption in 1917 produced a cinder cone in the base of the Boquerón crater with spatter cones and a large lava flow on the northwest flank of the volcano. The sections below describe the eruption history in detail.

San Salvador Edifice (>70,000 yr B.P.)

The oldest exposed rocks of the volcano are altered blocky basaltic-andesitic lavas and tephra deposits exposed at the base of the Picacho and Jabalí peaks (Fig. 3). These peaks consist of intercalated lava flows, breccias, and tephra-fall deposits that dip from 15° to 25° away from the center of the volcano.

Williams and Meyer-Abich (1955) used geomorphology and the attitude of the layered deposits to infer that Picacho and Jabalí are remnants of a large edifice that was 10 to 14 km wide and ~3000 m above sea level. Stratigraphically, the youngest of these deposits underlie ash from the 72 ka eruption of Coatepeque, a caldera 33 km west of Volcán San Salvador (Rose et al., 1999) (Fig. 1). Geochemical whole rock analysis on a stratigraphic sequence of lava flows from the edifice indicated there was an increase in SiO_2 and K_2O and decrease in CaO and MgO as the edifice grew (Sofield, 1998). Unconformities or paleosol development in these deposits suggest that they developed from many small explosive eruptions.

G-1 "Plinian" Explosive Episode (~40,000 yr B.P.)

The lava and ash fall deposits from the ancestral San Salvador edifice and Coatepeque ash are overlain by a series of pyroclastic surge, pyroclastic flow, and air fall deposits. This episode is distinguishable in outcrops by the presence of a dacitic pumice-fall deposit, known as G-1. G-1 is identified by its gray hue and a distinctive coarse basal layer that is approximately one-tenth of the total thickness of the deposit. Most of the regional tephras have a lighter off-white hue. The pumice and lithic fragments within the basal layer are roughly twice the diameter as the rest of the deposit. Outcrops within 10 km of the volcano's summit are more than a meter thick and contain up to 20% altered, olivine-bearing basaltic lithics similar in mineralogy to the lavas exposed in the San Salvador edifice. The distinctive gray pumice-fall deposit is found in outcrops up to 10 km away. It is estimated that 10 cm of tephra fall was deposited over ~905 km² (Fig. 4). The G-1 pumice fall has a southwesterly dispersal axis and an estimated volume of 1.1–3.0 km³.

The G-1 pumice fall is thought to be part of an episode of violent Plinian eruptions and crater formation. Picacho and Jabalí are remnants of the crater wall that is 4.5 km wide and 6.0 km long. Pyroclastic surge and pyroclastic flow deposits (up to 9 m thick) are observed within the valleys to the north, south, and east of the volcano (Fig. 4). Isopach data suggest the total volume of the G-1 volcanic deposits from the episode to be ~2 to 8 km³.

The timing of the G-1 eruption is constrained by stratigraphic relations of tephra-fall deposits from eruptions of Ilopango Caldera, located 28 km east of the Volcán San Salvador (Fig. 1). The G-1 pumice fall lies between Ilopango caldera tephra-fall deposits Tierra Blanca 3 and Tierra Blanca 4. On the basis of paleosols formed on these tephra-fall deposits, G-1 is estimated to have erupted ~40,000 to 50,000 yr B.P. (Rose et al., 1999).

Boquerón Edifice Construction (~40,000 to ~1580 yr B.P.)

Central vent eruptions filled the G-1 eruption crater with lava flows and tephra-fall deposits, creating the Boquerón

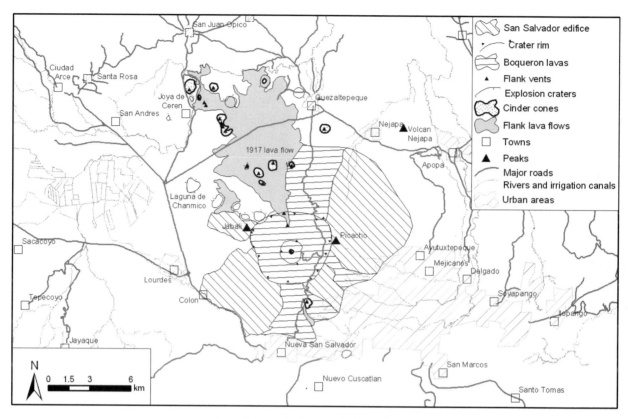

Figure 3. Surficial geologic map of Volcán San Salvador.

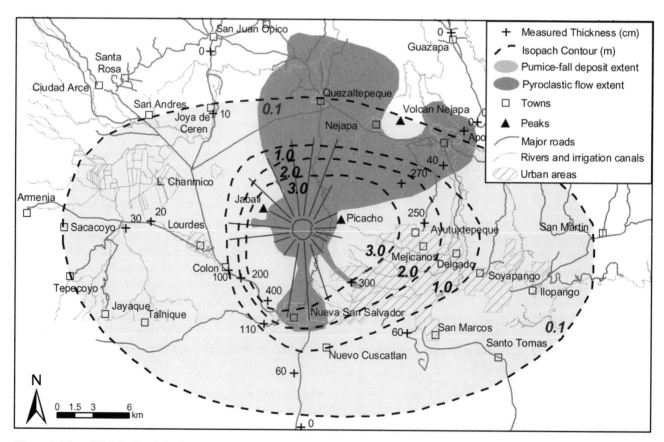

Figure 4. Map of Volcán San Salvador's G-1 pumice-fall and pyroclastic flow deposits (40,000 yr B.P.). The map shows the stratigraphic locations of the pumice-fall deposit and estimated isopach boundaries for 0.1, 1, 2, and 3 m. The estimated extent of the pyroclastic flow deposits is based on stratigraphic sample locations and topography.

edifice. The best exposures of Boquerón's deposits are located in the inner walls of the summit crater. They consist of intercalated blocky lava flows and tephra-fall deposits. Fairbrothers et al. (1978) studied a sequence of 14 lavas from Boquerón's southern crater wall. Results of this study indicated that these basaltic-andesites and andesites are richer in alkali elements and FeO than lavas from the San Salvador edifice (Sofield, 1998), making them chemically distinct.

The most recent lava flows from Boquerón spilled over the north, east, and south crater rim of the ancestral San Salvador edifice. Figure 3 shows the surficial geology including the extent of San Salvador and Boquerón edifice deposits. The construction of Boquerón edifice ended prior to 1580 yr B.P. Tierra Blanca Joven, the youngest tephra fall associated with the Ilopango caldera (dated at A.D. 429, Dull et al., 2001) covers all the exposed lavas on the Boquerón edifice.

San Andrés "sub-Plinian" Explosive Episode (800 yr B.P.)

Eight hundred yr B.P., the Boquerón edifice exploded violently, creating the present 1 km wide crater and depositing ash and mud west of the volcano. Hart (1983) correlated a thick tephra-fall deposit on the southwest rim of Boquerón with an indurated volcanic ash in the Zapotitán Valley west of the volcano. The deposit was named after an archaeological site, San Andrés, from which the age of the deposit was inferred. Using archaeometric dating of pottery, Sheets (1994) determined that the San Andrés tephra fall erupted ~800 yr B.P. Further study determined that a small pyroclastic flow preceded the tephra-fall deposition, and at least three debris avalanches were triggered on the north and west slopes of the volcano after the tephra fall was deposited. An isopach study determined that a large area (~293 km²) lying west and south of the volcano was covered with 10 cm or more of the tephra fall (Fig. 5). This tephra-fall deposit had a westerly dispersal axis. Armored lapilli, convolute laminations, and the deposits' indurated habit suggest that the eruption may have been phreatomagmatic. The deposits have a total volume of ~0.3 to 0.5 km³, which includes the 0.18 to 0.30 km³ tephra-fall deposit and the 0.15 km³ pyroclastic flow deposit.

Flank Vent Activity (>1580 yr B.P. to present)

Sometime prior to the eruption of Tierra Blanca Joven, volcanic eruptions from the central vent stopped and magma

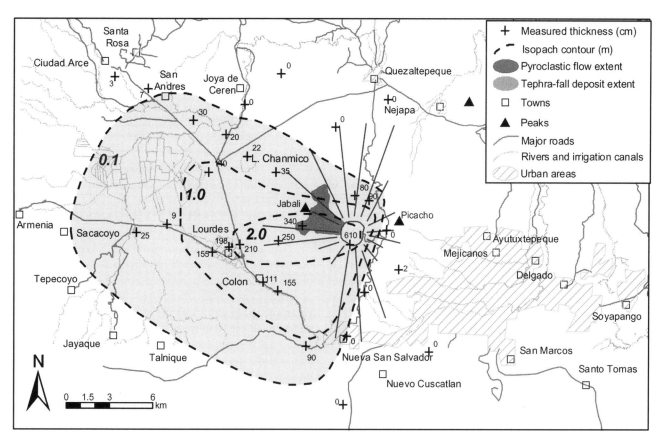

Figure 5. Map of Volcán San Salvador's San Andrés tephra-fall and pyroclastic flow deposits (800 yr B.P.). The map shows the stratigraphic locations of the tephra-fall deposit and estimated isopach boundaries for 0.1, 1, and 2 m. The estimated extent of the pyroclastic flow deposits is based on stratigraphic sample locations and topography.

surfaced from monogenetic vents on the flanks of the volcano. In all, 24 flank vent deposits are stratigraphically distinguishable from one another. Whole rock analyses of the majority of the different flank deposits suggested they are geochemically distinct from one another but were derived from a similar magma source (Sofield, 1998). The majority of the flank deposits are monogenetic eruptions from vents or fissures along two fault zones, trending N40W and N65E. Some deposits observed indicate that the interaction with near-surface waters played a large part in the type of volcanic activity observed. Roughly 30% of the flank eruption vents are explosion craters. These craters commonly include layered volcaniclastic sediments with sandwave, planar, and massive bedforms indicative of a pyroclastic base surge eruption. The rest of the deposits formed either cinder or spatter cones and associated lava flows.

The three historical eruptions (1917, 1658, 1575) formed cinder or spatter cones, tephra-fall deposits, and lava flows. Cinder cones located on the flanks of Volcán San Salvador are conical in shape and are commonly composed of unconsolidated blocks and scoria. The block fragments are highly vesicular. Montaña Quezaltepeque, northwest of San Salvador, is a classic example of a cinder cone. It is excavated for rock aggregate, which exposed the temporal tephra layering. In addition to a cinder cone, the eruptions produce significant tephra-fall deposits downwind of the vent. Flank eruptions may include an effusive stage producing lava flows 8 to 30 m thick. The largest flow, El Playón flow, covers ~10 km^2. The flows have both a'a and blocky surfaces. The surfaces of blocky flows comprise smooth-sided blocks as large as four meters in size. The a'a flows are composed of small irregularly shaped vesicular blocks, commonly with sharp terminations.

Most Recent San Salvador Activity (1917–present)

The most recent volcanic activity began shortly after an earthquake (magnitude estimated at 5.6 on the Richter scale) in June 1917. Several fissures opened along the northwestern flank of the volcano. Eighteen days after the eruption began, steam billowed from El Boquerón crater (or the central vent). A crater lake, ~80 m deep, boiled off within a month of the activity. A local elderly herdsman recalled how explosions shot rocks and debris over the crater rim (Sofield, 1998). When the eruption stopped, a small (30 m high) cinder cone, Boqueróncito, formed on the floor of the crater, and the lake has never reappeared. The majority of the magma's total volume (97%) extruded from three vents on the northwestern flank of the volcano. Fairbrothers et al. (1978) compared the composition of lavas from the flank and Boqueróncito. Although the composition varied slightly, they concluded the two deposits did derive from the same magma source. Since the 1917 eruption, there have been sporadic reports of fumarolic activity within the central crater and from two fissures located on the N40W-trending fault. No fumarolic activity was observed in 1996.

THE FREQUENCY OF ERUPTIONS

Stratigraphy and chemical analyses indicate the occurrence of at least 42 volcanic events at Volcán San Salvador in the past 40,000 yr, including 14 chemically distinct lavas in the Boquerón crater, two explosive episodes, and 26 separate flank eruption vent deposits postdate the eruption of G-1 (Fairbrothers et al., 1978; Sofield, 1998). Based on those data, the estimated recurrence interval for an eruption is ~1 per 950 yr. The volcano has undoubtedly erupted more frequently, because some eruptions do not leave conspicuous deposits in the geologic record. If eruptive events are separated into discrete, datable time periods, then the following estimated recurrence interval for eruptions is established: Between 40,000 yr B.P. and 1580 yr B.P. ~25 identifiable eruptive events (14 chemically distinct lavas in the Boquerón crater and 11 separate flank eruption vent deposits that postdate the eruption of G-1 and predate the eruption of Tierra Blanca Joven) indicate an eruption recurrence interval of ~1 event per 1500 yr (Fairbrothers et al., 1978; Sofield, 1998). Between 1580 yr B.P. and 800 yr B.P. (the eruption of the Tierra Blanca Joven and the San Andrés tephra fall), eight eruptive events were identified, indicating an eruption recurrence interval of ~1 event per 113 yr. Between 800 yr B.P. (the eruption of the San Andrés tephra fall) and present, there were six identifiable eruptive events indicating an eruption recurrence interval of ~1 event per 133 yr (Sofield, 1998). Please note that these numbers are different than reported in Major et al. (2001) where some events were mistakenly left off and others were counted twice. These eruption frequencies are only statistically averaged values. It is possible that multiple events are associated with a single eruptive episode, such as extrusion of lava flows from two separate vents. Nevertheless, even accounting for the imperfections of the geologic record, evidence clearly indicates that Volcán San Salvador erupts frequently, with an estimated recurrence interval of somewhere between every 113 to 950 yr. The most recent eruptive history suggests an estimated recurrence interval of one flank eruption every 133 yr.

VOLCANIC HAZARDS OF VOLCÁN SAN SALVADOR

The geologic record of Volcán San Salvador indicates there are many volcanic hazards associated with the volcano. Most of these hazards are associated with the eruption of magma. However, debris avalanches, lahars, and the buildup of corrosive or asphyxiating gases can occur without an eruption. Historic deposits indicate that magma erupted from two types of vents, the central vent (or Boquerón) and monogenetic vents on the volcano's flank. The nature and scale of an eruption depends in part on the type of vent, the amount and composition of the magma, and the interaction of the magma with water.

Lava flows and tephra are the most frequent type of volcanic hazard observed in the eruption record. However, pyroclastic base surges (associated with hydromagmatic eruptions), and pyroclastic flow deposits are clearly visible in outcrops. Historically (last 425 yr), Volcán San Salvador's activity included Vulcanian, Strombolian, and effusive eruptions (Larde, 1956; Larde and Lardin, 1948). Reportedly, these eruptions caused no deaths, but they destroyed productive agricultural land, a section of a railroad, and in 1658 forced the relocation of a town.

Eruptions of Volcán San Salvador have been more diverse and destructive over the last 1800 yr. Types of eruption include additional monogenetic magmatic flank eruption, violent hydro-magmatic explosions, and a sub-Plinian eruption (San Andrés event) consisting of a pyroclastic flow, tephra fall, and debris avalanches. Today an estimated two million people live in the capital, San Salvador, which is the economic and political center of the entire country. The rim of Boquerón is only 7 km west of the capital's main plaza. If an explosive eruption like the San Andrés event were to occur today without warning, it could cause the death of thousands of people and would deal a severe blow to the entire country of El Salvador. Potentially much of El Salvador's people, industry, agriculture, and infrastructure are at risk from a future eruption of Volcán San Salvador.

LIKELY FUTURE ERUPTION ACTIVITY

The following five eruption scenarios (two from a flank vent and three from the central vent) are based on the type of volcanic event and the frequency of recurrence in the stratigraphic record. Physical characteristics of the deposits, chemical analyses, historic meteorological data, archaeological data, and historical accounts were used to develop likely scenarios for future eruptions of Volcán San Salvador. Volcanic explosivity index (VEI) for each scenario is based on estimates outlined by Simkin and Siebert (1994). A VEI value of 1 is the least explosive and 8 is the most explosive based on the volume of tephra deposited. These scenarios include

1. A hydromagmatic flank eruption with a VEI value between 1 and 3. A magmatic flank eruption with a VEI value between 1 and 3. This is the most likely future type of eruption.

2. A small-scale eruption within the Boquerón crater with a VEI value between 1 and 3.

3. A sub-Plinian eruption from the central vent, equivalent to the San Andrés event with an estimated VEI value of 4 or 5.

4. A Plinian eruption from the central vent, similar to the G-1 event, with an estimated VEI value of 6. The least-likely future type of eruption or "worst-case" scenario.

Flank Eruption Scenarios

Field observations indicate that the majority of flank eruptions form new vents producing either a magmatic (effusive, Strombo-lian, or vulcanian) or hydromagmatic eruption with an estimated VEI value of 3 or less. In the last 1700 yr flank vent eruptions were more common than an eruption from the central vent, suggesting it is the most likely eruption type to occur in the future.

Monogenetic Magmatic Eruption (VEI 1 to 3)

Tephrochronology discussed in Sofield (1998) suggests the most frequent type of flank eruption is monogenetic mag-

matic eruption along the northwestern flank of the volcano. The majority of the flank vent eruptions (12 of 13) in the last 1800 yr occurred in the northwestern flank of the volcano along the N40W-trending fault zone. The hazards associated with this type of magmatic volcanic eruption are well documented in literature. Luhr and Simkin's (1993) collection of work regarding the Pari-cutin Volcano eruption in Michoacán, Mexico, reveals a valuable record for reference. Although the exact circumstances may vary, Paricutin's eruption bears a striking similarity to the erup-tion of El Playón, a flank vent of San Salvador, in 1658 (Larde and Lardin, 1948). In both cases, the eruption began explosively with the development of cinder cones and tephra deposits down-wind of the vent. The explosive nature of the eruption dwindled and an effusive stage began. Crops were ruined by tephra fall, a town and valuable cropland were buried by lava flows, and people were relocated. Luhr and Simkin (1993) documented the personal trauma and quality-of-life changes that people near Paricutin experienced.

On Volcán San Salvador, at least four flank eruptions have produced thick tephra-fall deposits. Limited meteorological and isopach data suggest future tephra-fall deposits from San Salvador's northwestern flank vent eruptions will most likely accumulate west of the vent. Hill et al. (1998) determined that wind direction and speed are the major factors in the distribu-tion of tephra-fall deposits of small eruptions. The surficial wind conditions are used rather than high-altitude measurements commonly used in Plinian tephra-fall deposition models. Long-term or recent surficial meteorological data are hard to find in El Salvador. The most detailed data were collected from 1960 to 1965 and show that the surficial wind direction is different in San Salvador and Zapotitán Valley, where the recent eruptions took place. The data from the San Andrés airport determined that the average monthly surface wind direction "throughout the year" is to the west (Servicio Meteorologico Nacional, 1966). Tephra fall from the El Playón eruption is 65 cm thick 2 km west of the vent. However, no tephra is present 1 km east of the vent, suggesting that the wind was blowing westward during the 1658 eruption as well (Sofield, 1998).

Volcán San Salvador's 1917 lava flow erupted from mul-tiple fissures along the N40°W fault zone. The flow traveled a distance of ~4 km downslope and covered a section of railroad tracks. Flank vent lava flows have also dammed the Río Sucio (Fig. 2) twice (Sofield, 1998). In 1575 and 1658, the lava dam caused flooding in two upstream locations and may have caused flooding downstream if the dam failed catastrophically.

Hydromagmatic Eruptions (VEI 1 to 3)

Hydromagmatic eruptions commonly produce violent pyro-clastic base surges (turbulent, high-velocity, low-density flows that blast out from the base of the eruption column) in addition to tephra fall and ballistic volcanic bombs (Fisher and Schmincke, 1984). Hydromagmatic eruptions are the second most frequent type of flank eruption to occur at Volcán San Salvador. Field relationships and tephrochronology indicate that three separate hydromagmatic eruptions occurred northwest of the volcano between 1580 and 800 yr B.P.

Joya de Cerén, an archaeological site northwest of the volcano, provides some of the best information about a hydromagmatic eruption. At the site, years of research unearthed a modest village covered in 4 m of volcanic debris. Multiple pulses of pyroclastic base surges, tephra-fall deposits, and ballistic bombs from the Loma Caldera buried the village. Loma Caldera is the name of an explosion crater 650 m northeast of the village (Miller and Conyers, 1994). Sheets (1994) determined that the pyroclastic base surges knocked down reinforced adobe walls, and volcanic bombs ignited thatch roofs. Although deposits buried domesticated animals, storage pots, and personal possessions, no human remains were found. This suggests that the residents were warned by an eruption precursor (possibly a tremor or steam explosion) and had time to flee, but not with enough time to take their belongings with them (Conyers, 1996).

Tephra fall from the Loma Caldera had a westerly dispersal axis. Tephra fall from a hydromagmatic eruption has the same dispersal mechanism as a magmatic flank eruption, hence it deposited downwind from the vent. The full extent of the of Loma Caldera's pyroclastic base surge deposits are unknown. However, at Laguna de Chanmico, another hydromagmatic explosion crater, pyroclastic base surge bedforms were observed up to 1.3 km from the center of the crater.

Central Vent Eruption Scenarios

The 1917 eruption was the only observed central vent eruption since Boquerón crater formed during the explosive eruption 800 yr B.P. Given that only one central vent eruption has occurred since the formation of Boquerón crater, a recurrence interval was not calculated. However, it is generally understood that small-scale eruptions occur more frequently than large cataclysmic episodes. The geologic record supports that hypothesis. Therefore, the following scenarios for central vent eruptions are discussed.

Small Eruption within the Boquerón Crater (VEI 1 to 3)

Accounts about the 1917 eruption provide an understanding of volcanic hazards from small-scale eruptions from the central vent, which include ballistic bombs, tephra fall, volcanic gases, and possible hydromagmatic surges. Ash, tephra, and ballistic bombs reportedly cleared the crater rim during the 1917 eruption (Larde, 1956; Sofield, 1998). Boquerón's crater walls presently protect the cities surrounding the volcano from future lava flows and hydromagmatic base surges. However, future eruptions could fill the Boquerón crater (500 m deep) with volcanic deposits, thus increasing the risk to the towns and plantations on the volcano's flanks from volcanic hazards and debris avalanches much like Costa Rica's Volcán Arenal experienced in 1998 (Wunderman and Venzke, 1998). Tephra fall, ballistic volcanic bombs, and to some degree volcanic gases represent a hazard to the slopes of Volcán San Salvador, whether the crater fills in or not.

Sub-Plinian Eruption from the Central Vent (VEI 4 to 5)

Sub-Plinian eruptions produce pyroclastic flow and surges in addition to tephra. The geologic record of Volcán San Salvador indicates two eruptions of this size or larger have occurred in the last 72,000 year B.P. If a sub-Plinian eruption (the size of the San Andrés event) occurred today, the area affected by tephra fall would be largely determined by the prevailing high-altitude wind directions. Airlines and airports collect radiosonde data on regional atmosphere wind speed and direction. The closest data collection station is in Guatemala City, 140 km west-northwest of the volcano. Rose and Mercado (1986) gathered two years of radiosonde wind direction data for altitudes between 10,000 and 50,000 ft above Guatemala City. If one assumes the wind direction above Volcán San Salvador to be the same as that above Guatemala City, the data indicate that the prevailing wind direction from January through March (the rainy season) is toward the east. From June to October (the dry season) it prevails toward the west. In April, May, November, and December the weather patterns are transitional and wind is more likely to fluctuate from westerly to northerly and easterly.

The area potentially affected by pyroclastic flows from an eruption the size of the San Andrés event is more difficult to determine. The Jabalí and Picacho peaks are topographic barriers that could potentially shelter the region northeast and west of Boquerón from small pyroclastic flows. During and after the eruption, thick pyroclastic deposits are prone to remobilize on steep or rain-washed slopes resulting in debris avalanches or a lahar. Debris avalanches and lahars from a large eruption of Volcán San Salvador represent a significant hazard to communities within topographic channels and on the valley floor (i.e., Lourdes and Apopa).

Plinian Eruption from the Central Vent (VEI 6)

The largest explosive episode in Volcán San Salvador stratigraphic history most likely occurred ~40,000 yr B.P. This eruptive episode was cataclysmic, producing a widespread tephra-fall deposit (1.7 to 3.0 km³) and multiple pyroclastic flows. With a Plinian eruption the topographic barriers Jabalí and Picacho peaks would provide little protection and pyroclastic flows would likely reach the valley floors. An event similar in magnitude to the G-1 eruption would be likely to blanket the capital and towns surrounding the volcano in a meter of ash, irrespective of the prevailing wind direction. The same diurnal wind pattern discussed above can carry ash far from the volcano. Blong (1984) suggests that a deposition of 10 cm of ash (approximately the amount of ash needed to collapse a roof) constitutes a severe hazard. Ten cm of ash from the G-1 eruption was found in Joya de Cerén 12 km northwest of the volcano. The isopach study estimated that 4.3% of El Salvador's total area was covered in at least 10 cm of ash.

An eruption of this magnitude has never occurred (anywhere in the world) so close to a population center of San Salvador's current size. Again, Boquerón crater is a mere 7 km from the city's center. With estimates of 60% of the nation's gross domestic product being created within greater San Salvador, a Plinian eruption would likely cripple the entire country for a year or more. Secondary hazards, such as the remobilization of ash and debris would likely cause lahars and dump high-sediment loads into the rivers for many years after the eruption. For comparison, the International Trade Commission estimates the clean-up and property damage associated with the Mount St. Helens eruption to be 1.1 billion dollars, and no one lives within 10 km of the volcano (Tilling et al., 1990). Thankfully an eruption of this magnitude is the least-likely possible scenario and would not occur without some warning.

HAZARD ZONATION MAPS FOR VOLCÁN SAN SALVADOR

Two volcano hazard maps have been produced based upon the eruption history of Volcán San Salvador and the characteristics of its deposits. The hazard zones associated with an eruption from the Boquerón crater are delineated in Figure 6. Figure 7 delineates the areas of relative risk from a flank eruption. The

hazards and risk maps are based on 1:50,000-scale topographic maps that the U.S. Geological Survey (USGS) digitally reproduced (Major et al., 2001). Although hazard zone boundaries and risk regions are shown as lines, these boundaries are only approximate and the area immediately beyond the hazard zones should not be regarded as hazard-free.

Hazard Zones Associated with an Eruption from the Boquerón Crater

The primary tephra-fall hazard zone (Fig. 6) identifies areas around Volcán San Salvador where tephra-fall accumulations of the San Andrés event are more than 10 cm thick (Hart, 1983; Sofield, 1998). Since wind direction is the greatest factor in determining tephra-fall distribution for sub-Plinian or Plinian central vent eruptions, two regions of high risk were designated on the basis of predominant wind directions during the rainy and dry seasons.

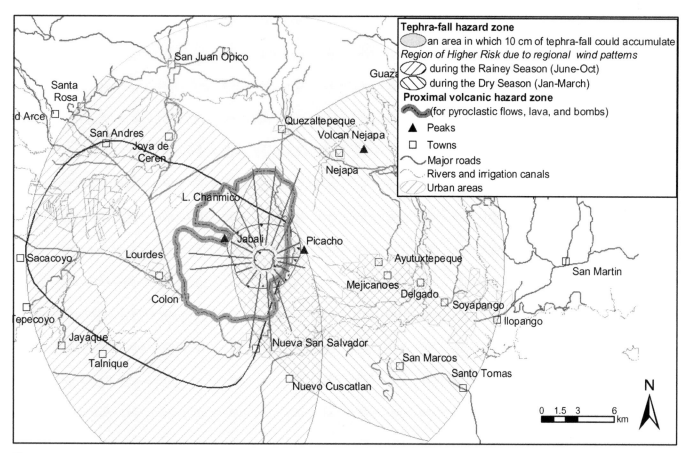

Figure 6. Hazard zonation map for Volcán San Salvador: Boquerón's proximal hazard zone boundary (for pyroclastic flows, lava flow, and ballistic volcanic bombs) and tephra-fall hazard zone. Tephra-fall hazard zone defined as an area in which 10 cm of tephra fall could accumulate based on an eruption the size of the San Andrés eruption (800 yr B.P.). Regions of higher risk are due to dominant regional wind pattern that occurs during the dry (January through March) and rainy (June through October) seasons. During these months the likelihood of the tephra accumulating in these zones is higher than the rest of the tephra-fall hazard zone. The proximal hazard zone boundary estimates the distance pyroclastic flows, lava, and ballistic bombs are likely to travel from the Boquerón crater rim. It is based on the L/H ratio a pyroclastic flows from San Andrés eruption (800 yr B.P.). Volcán San Salvador proximal volcanic hazard zone defined by either 1 km from rim of Boquerón or L/H ratio greater than 0.2.

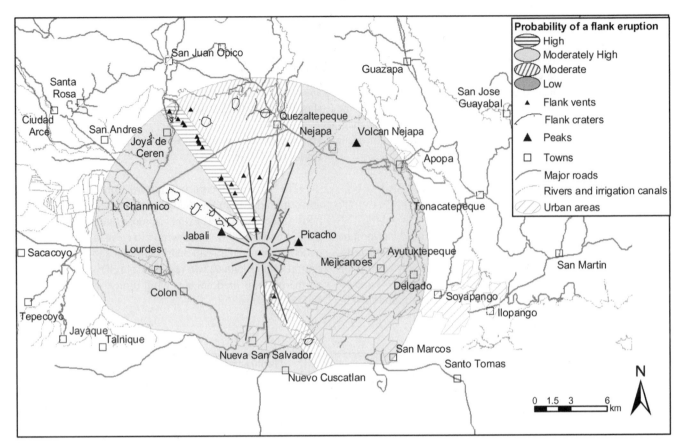

Figure 7. Map of areas of relative probability for a volcanic eruption on the flanks of Volcán San Salvador. The region of "highest" risk has an eruption recurrence rate of one eruption in the area every 142 yr during the last 1580 yr. The region of "moderately high" risk has an eruption recurrence rate of one eruption every 600 yr. The region of "moderate" risk is an area with flank vents that have not erupted within the past 1580 yr. The region of "low" risk encompasses the rest of the volcano's flanks, in which no vents have been observed.

The proximal hazard zone boundary (Fig. 6) estimates the distance pyroclastic flows, lava, and ballistic bombs are likely to travel from the Boquerón crater rim, based on the distance pyroclastic flows have traveled during the past central vent eruptions of Volcán San Salvador. This distance is estimated by determining the difference in elevation from the eruptive vent (in this case the rim of Boquerón) to the farthest point that any pyroclastic flow (in this case San Andrés tuff deposits) had reached in the past (H). H is divided by the horizontal distance between the points (L). The H/L value (0.23) is similar to the 0.2 ratio used by the USGS for pyroclastic flows in the Cascade volcanoes (Major et al., 2001). This ratio was then applied to the flanks of Volcán San Salvador to produce in an irregularly shaped zone around the Boquerón edifice. The proximal hazard zone also includes all land within 1 km of the crater rim to account for ballistic projectiles. The irregular shape of this zone is a result of differences in the volcano's topography. Pyroclastic flows or lavas may only affect a portion of the proximal hazard zone during an eruption. Jabalí and Picacho peaks are not included in this hazard zone because they are significant topographic barriers extending past the 0.2 H/L ratio and would deflect small pyroclastic flows and any lava flow.

Hazard Zones Associated with a Flank Eruption

Volcanic deposits from vents are peppered all over the north, west, and southeastern flanks of Volcán San Salvador (Fig. 6). In order to assess the hazards associated with flank eruptions, it is important to identify the location where future flank vent eruption will occur. Most eruptions in the past 2000 yr have occurred on the flanks along fault zones, but there are no consistent temporal trends to suggest the location of the next vent (Sofield, 1998). Therefore, we consider the risk to be the same along the entire fault trend. Four regions of relative risk are identified based on estimated recurrence intervals and how recently the last eruption in the region occurred (Fig. 7).

The region of "highest" risk is a narrow band along the N40W fault trend on the northwestern flank of the volcano. This region has an eruption recurrence rate of one eruption every 142 yr during the last 1580 yr. The region of "moderately high" risk is a narrow band along the N65W fault trend. This region has an eruption recurrence rate of one eruption every 600 yr over the last 1580 yr. There are two areas of "moderate" risk; one is located on the N40W fault trend southeast flank of the volcano and the other is a section of the volcano's north flank. This region includes flank vents that have not erupted within the past 1580 yr. The region of "low" risk encompasses the rest of the volcano's flanks in which no vents have been observed. This low-risk region is delineated by the Median Trough and a circle around the volcano determined by the distance of the farthest known vent (Laguna Caldera) from Boquerón.

The area around a vent that will be buried or affected by an eruption depends upon the type of eruption, the near-surface wind direction, and local topography. The nature of a flank eruption is dependent upon the magma chemistry, dissolved volatiles, proximity of surface water or shallow aquifers, and eruptive rate of the magma. Based upon deposits from other flank vents, the following hazard zones should be established for a new vent depending on the nature of the eruption: lava flows—4 km downslope of the vent; tephra fall—up to 5 km downwind of the vent (depending on the wind speed); a pyroclastic base surge—1.5 km radius around the vent.

The lava flow hazard zone is based on the premise that the flows will follow topography and travel less than 4 km from their source at Volcán San Salvador. Isopach studies of tephra-fall deposits from the Loma Caldera (Joya de Cerén) and El Playón vent (Hart, 1983; Sofield, 1998) indicate that 10 cm of tephra fall was deposited up to 5 km downwind of these vents. With the data collected on base surge deposits in 1996 through 1998, there is not an accurate way to determine the distance a pyroclastic base surge will travel from an explosion crater. Therefore, 0.2 km was added to the maximum radial distance from which a surge deposit was observed from Laguna de Chanmico. Thus, 1.5 km is suggested as the maximum distance that pyroclastic base surges will travel from their source.

OTHER VOLCANO-RELATED HAZARDS

The release of corrosive or asphyxiating volcanic gases and debris avalanches represent two other volcanic hazards that will not be addressed in detail, but which both pose significant risk at Volcán San Salvador. Volcanic gases released into the atmosphere include H_2O vapor, SO_2, CO_2, minor amounts of CO, HS, Cl-, and F- (Williams and McBirney, 1979). Large amounts of gases are usually emitted during an eruption, but dangerous amounts of gases can vent to the surface through fumaroles between eruptions. Boquerón's deep crater walls create a confined basin that could allow dangerous gases to build up, creating a high risk to anyone venturing inside. Thus, any increase in magmatic activity within the edifice could result in a volcanic gas hazard within the craters.

Steep slopes on Volcán San Salvador are particularly susceptible to debris avalanches because they are composed of sloping layers of loosely consolidated rock. Hydrothermal alteration and hydrologic undercutting increase the susceptibility of the material to slide by weakening stable rock masses. Debris avalanches may be set in motion by eruptions, but also are triggered by earthquakes and water saturation. A debris avalanche may entrain water and mobilize into a lahar. Major et al. (2001) studied some of these his-

toric debris avalanches and created hazard zones for these events at Volcán San Salvador (Major et al., this volume, Chapter 7).

CONCLUSIONS

The geologic history described above provides a summary of volcanic activity on Volcán San Salvador in the past. The volcano has erupted intermittently for more than 70,000 yr. Approximately 40,000 yr ago a Plinian eruption created a large crater within a broad basalt and basaltic-andesite edifice and extensive pumice-fall and pyroclastic flow deposits. By 1580 yr B.P., a second edifice composed of basaltic-andesitic lavas and ash had formed within the crater. Eight hundred yr B.P. a second smaller-scale explosive event occurred, forming the current crater Boquerón and depositing tephra and a small pyroclastic flow to the west of the volcano. Some of the youngest volcanic features and deposits are on the north and northwestern flanks. These features consist of monogenetic cinder cones, lava flows, and explosion craters that are generally concentrated along two fault zones. The most recent activity in 1917 included the formation of cinder cones and lava flows from vents with a central crater and along a N40°W-trending fault zone.

Both flank and central vent eruptions are possible in the future. The average recurrence interval of an eruption is from one every 113 to 950 yr. Monogenetic magmatic flank eruptions have the shortest average recurrence interval for any type of eruption (over the last 800 yr it was one every 133 yr). These statistics suggest that some type of eruption will likely occur within this century.

Due to the significant population density on and near the volcano, any future volcanic activity will affect the surrounding communities. The ultimate goal of this work is to identify the hazards to the surrounding communities. It is the responsibility of policymakers to understand these hazards, determine the risks associated with a future eruption of Volcán San Salvador, and reduce the loss of life and property should an eruption occur.

Little can be done to stop a volcanic eruption. But, successful measures for decreasing the risks associated with a volcanic eruption have been developed in volcanically hazardous regions around the world. Recently, El Salvador has taken positive measures toward reducing the risks of Volcán San Salvador, including rezoning and public education. A partial list of the measures that decrease the risks associated with a volcanic eruption includes the following.

SUGGESTIONS FOR FURTHER ACTION

1. Avoiding further growth in hazardous areas, or limiting the type of growth allowed.
2. Offering incentives to deter development in areas prone to hazards. Land use zoning restrictions, building code laws, or tax and utilities services rates are common disincentives used to discourage development in hazard-prone areas.
3. Creating and practicing warning and evacuation plans for people in the hazard zones. A collaboration of the government,

churches, and social groups appears to work the best in reaching the entire population.

4. Investing in instruments and warning systems (and the people to run them) to enable eruption forecasts and rapidly disseminate the information.

5. Educating people who live, work, and visit the areas of volcanic hazard about the hazards they face, and what to do in the event of an emergency.

All of these changes imply that the hazards of Volcán San Salvador and other regional natural phenomena are understood. Hopefully, this work will assist the Salvadorians in understanding their vulnerability to volcanic hazards.

ACKNOWLEDGMENTS

This work could not have been completed without the kindness and constructive criticism of many people. I would like to thank my colleagues in El Salvador, Michigan Technological University, the U.S. Geological Survey, and GeoEngineers for thoughtful assistance. Special thanks to the Pullinger family for their hospitality, Comisión Ejecutiva del Rio Lempa and Servicio Nacional de Estudios Territoriales for the logistical support. Demetrio Escobar and Carlos Pullinger deserve a special commendation for their companionship. Katie Hinman, Mary Ann Reinhart, and the GSA editors carefully reviewed the draft documents. Thank you to all.

REFERENCES CITED

Blong, R., 1984, Volcanic hazards: A sourcebook on the effects of eruptions: London, Academic Press, p. 201–204.

Carr, M., and Stoiber, R., 1981, Lava and faults characterize El Salvador: Geotimes, v. 26, no. 7, p. 20–21.

Conyers, L., 1996, Clues from a village; dating a volcanic eruption: Geotimes, v. 41, no. 11, p. 20–23.

Dull, R., Southon, J., and Sheets, P., 2001, Volcanism, Ecology, and Culture: A reassessment of the Volcan Ilopango TBJ eruption in the southern Maya realm: Latin American Antiquity, v. 12, no. 1, p. 24–44.

Fairbrothers, G., Carr, M., and Mayfield, D., 1978, Temporal magmatic evolution at Boquerón Volcano, El Salvador: Contributions to Mineralogy and Petrology, v. 67, p. 1–9.

Fisher, R., and Schmincke, H.-U., 1984, Pyroclastic rocks: Berlin, Springer Verlag, 472 p.

Hart, W., 1983, Classic to postclassic tephra layers exposed in archeological sites, eastern Zapotitán Valley, *in* Sheets, P.D., ed., Archeology and volcanism in Central America: The Zapotitan Valley of El Salvador: Austin, University of Texas Press, p. 44–51.

Hill, B., Conner, C., Jarzemba, M., La Femina, P., Navarro, M., and Strauch, W., 1998, 1995 eruption of Cerro Negro Volcano, Nicaragua and risk assessment for future eruptions: Geological Society of America Bulletin, v. 110, p. 1231–1241.

Larde, J., 1956, El Quezaltepeque: su eruption y terremoto de junio de 1917: San Salvador, Tribuna Libre, April 4. Copies of the newspaper strip are available at Michigan Technological University Geology Department El Salvador Data Archive.

Larde, J., and Lardin, J., 1948, Genesis del volcán del Playon: Volcanologica Salvadoreña, Revista del Ministerio de Cultura, v. 7-24, p. 101–111.

Luhr, J., and Simkin, T., editors, 1993, Paricutin: The volcano born in a Mexican cornfield: Phoenix, Geoscience Press, 427 p.

Major, J., Schilling, S., Sofield, D., Escobar, C., and Pullinger, C., 2001, Volcano hazards in the San Salvador region, El Salvador: U.S. Geological Survey Open-File Report 01-366, 17 p.

Major, J.J., Schilling, S.P., Pullinger, C.R., and Escobar, C.D., 2004, Debris-flow hazards at San Salvador, San Vicente, and San Miguel Volcanoes, El Salvador, *in* Rose, W.I., et al., eds., Natural hazards in El Salvador: Boulder, Colorado, Geological Society of America Special Paper 375, p. 89–108 (this volume).

Miller, C.D., and Conyers, L., 1994, Character of hydromagmatic deposits which buried the Mayan village of Cerén, El Salvador: Geological Society of America Abstracts with Programs, v. 26, no. 7, p. 157.

Reynolds, J., 1980, Late Tertiary volcanic stratigraphy of northern Central America: Bulletin of Volcanology, v. 41, no. 3, p. 601–607.

Rose, W., and Mercado, R., 1986, Reporte a UNDRO/OFDA mision al Volcán Tacaná, Instituto Nacional de Sismología Volcanalogía, Meteorología e Hidrología (INSIVUMEH, Guatemala), p. 40.

Rose, W., Conway, F., Pullinger, C., Deino, A., and McIntosh, W., 1999, An improved age framework for late Quaternary silicic eruptions in northern Central America: Bulletin of Volcanology, v. 61, p. 106–120.

Servicio Meteorologico Nacional, 1966, Almanaque Salvadoreño: San Salvador, Ministerio De Obras Publicas, Republica de El Salvador Centro America, 46 p.

Sheets, P., 1994, Tropical time capsule: Archaeology, v. 47, no. 4, p. 20–26.

Simkin, T., and Siebert, L., 1994, Volcanoes of the world (2nd edition): Tucson, Geoscience Press, 368 p.

Sofield, D., 1998, History and hazards of Volcán San Salvador, El Salvador [M.S. thesis]: Houghton, Michigan Technological University, 116 p.

Stoiber, R., and Carr, M., 1973, Quaternary volcanic and tectonic segmentation of Central America: Bulletin of Volcanology, v. 37, no. 3, p. 304–325.

Tilling, R., Topinka, L., and Swanson, D., 1990, Eruptions of Mount St. Helens: past, present, and future: U.S. Geological Survey—Special Interest Publication.

Williams, H., and McBirney, A., 1979, Volcanology: San Francisco, Freeman, Cooper, 397 p.

Williams, H., and Meyer-Abich, H., 1955, Volcanism in the southern part of El Salvador: University of California Publication in Geological Sciences, v. 32, p. 1–64.

Wunderman, R., and Venzke, E., 1998, Arenal, Costa Rica: Bulletin of the Global Volcanism Network, v. 23, no. 3, p. 14.

MANUSCRIPT ACCEPTED BY THE SOCIETY JUNE 16, 2003

Geological Society of America
Special Paper 375
2004

Subaqueous intracaldera volcanism, Ilopango Caldera, El Salvador, Central America

Crystal P. Mann*
John Stix
Department of Earth and Planetary Sciences, McGill University, 3450 University Street, Montréal, Québec H3A 2A7, Canada

James W. Vallance*
Department of Civil Engineering, McGill University, 817 Sherbrooke St. West, Montréal, Québec, H3A 2K6, Canada

Mathieu Richer
Department of Earth and Planetary Sciences, McGill University, 3450 University Street, Montréal, Québec H3A 2A7, Canada

ABSTRACT

The Ilopango Caldera, located 10 km east of San Salvador, has erupted voluminous silicic pyroclastics four times in the last 57,000 years. The present caldera has a quasi-rectangular shape and is filled by Lake Ilopango. This paper provides a detailed description of a segment of the intracaldera stratigraphy at Ilopango caldera, with emphasis on the San Agustín Block Unit. Physical volcanology, petrology, and geochemistry establish the depositional environment and eruptive conditions of the intracaldera sequence and help to model the emplacement of the San Agustín Block Unit. The intracaldera stratigraphy comprises a sequence of pyroclastic density currents, unconformably overlain by lacustrine sediments and conformably overlain by the San Agustín Block Unit. A new radiocarbon age on wood near the top of the Lacustrine Unit indicates that a lake was present ≥43,670 years ago. The intracaldera sequence displays abundant evidence of emplacement in a subaqueous environment.

The San Agustín Block Unit comprises a basal Fine Ash facies and an overlying Pumice Breccia facies. The basal Fine Ash facies is a hydromagmatic layer containing pumiceous and blocky angular glass shards, aggregates of fine ash and phenocryst fragments, and phenocrysts with a fine ash coating. The overlying Pumice Breccia facies is composed of pumice clasts up to three meters in length. The pumice clasts display a series of jointing textures indicative of hot emplacement and rapid cooling. These two facies suggest an initial subaqueous explosive eruption in which a vesiculated silicic melt fragmented upon contact with the water. When the magma had degassed sufficiently, the eruption style evolved to subaqueous dome growth that spalled quenched pumice clasts from a moderately vesiculated carapace.

RESUMEN

La Caldera de Ilopango se encuentra a 10 km al este de San Salvador. Ilopango ha tenido cuatro grandes erupciones de piroclastos ricos en sílice durante los últimos

*E-mail, Mann: cpmann@eps.mcgill.ca. Present address, Vallance: U.S. Geological Survey, 1300 SE Cardinal Court, Bldg. 10, Suite 100, Vancouver, WA, 98683-9589, USA

Mann, C.P., Stix, J., Vallance, J.W., and Richer, M., 2004, Subaqueous intracaldera volcanism, Ilopango Caldera, El Salvador, Central America, *in* Rose, W.I., Bommer, J.J., López, D.L., Carr, M.J., and Major, J.J., eds., Natural hazards in El Salvador: Boulder, Colorado, Geological Society of America Special Paper 375, p. 159–174. For permission to copy, contact editing@geosociety.org. © 2003 Geological Society of America.

57,000 años. La caldera actualmente tiene forma rectangular y el Lago de Ilopango se encuentra dentro de ella. Este artículo proporciona una descripción detallada de un segmento de la estratigrafía en el interior de la caldera de Ilopango con énfasis en una unidad que se llama "Unidad de Bloque San Agustín." La volcanología física, petrología y geoquímica describen el medio ambiente deposicional y las condiciones eruptivas de la secuencia en el interior de la caldera y ayudan a modelar el emplazamiento de la "Unidad de Bloque San Agustín." La secuencia en el interior de la caldera es constituida de flujos piroclásticos discordantes sobre ella yacen sedimentos lacustres y sobre estos últimos yace de manera concordante la "Unidad de Bloque San Agustín." Una datación reciente con el método de radiocarbón de madera encontrada cerca del techo de la secuencia lacustre indica que un lago ya existia hace ≥43,670 años. La secuencia en el interior de la caldera muestra evidencia de un emplazamiento en un medio ambiente subacuático.

La "Unidad de Bloque San Agustín" constituye una facie basal de ceniza fina y una facie superpuesta de pómez brecha. La facie basal de ceniza fina es un estrato hidromagmático conteniendo pedazos de vidrio, algunos en forma de bloque y otros vesiculares con textura de pómez. Contiene además un agregado de ceniza fina con fragmentos fenocristales y fenocristales con una recubierta de ceniza fina. La pómez brecha está compuesta de pómez con una elongación a trece metros. Los clastos pómez muestran diaclasas radiales, diaclasas concéntricas y diaclasas perpendiculares a la superficie que indican una deposición caliente y un enfriamiento rápido. Las dos facies sugieren una erupcíon inicialmente explosiva donde un magma sílice vesiculado es fragmentado cuando se pone en contacto con el agua. Cuando el magma está lo suficientemente desgasificado, la erupcíon desarrolla un domo subacuático desboronando clastos templados de pómez proveniendo de una caparazón vesiculada.

Keywords: Ilopango caldera, El Salvador, subaqueous intracaldera volcanism, hydromagmatic fragmentation, subaqueous lava dome.

INTRODUCTION

Caldera-forming eruptions are among the most devastating volcanic eruptions on Earth. Well-known examples include Krakatau, Indonesia (1883), Taupo, New Zealand (A.D. 186), Toba, Sumatra (75 ka), and Long Valley, USA (ca. 760 ka). Once a caldera forms, the depression may fill with water, increasing the probability of explosive eruptions. Ilopango Caldera in El Salvador contains Lago de Ilopango, an 8 × 11 km lake located ~10 km east of San Salvador, the capital city (Fig. 1). At least four large silicic explosive eruptions have occurred in the past 57,000 years (Rose et al., 1999). The last violent eruption in A.D. 429 contributed to the devastation of Early Classic Mayan civilization in El Salvador and eastern Guatemala (Lothrop, 1927; Sheets, 1979; Hart, 1981; Dull et al., 2001). In A.D. 1879 and 1880 dome growth formed the Islas Quemadas (Goodyear, 1880; Richer et al., this volume). This recent activity strongly indicates that a body of magma still exists below the Ilopango Caldera. These eruptions demonstrate that Ilopango is still active and clearly capable of explosive and effusive eruptions in the future.

In order to fully understand the effects of water during subaqueous eruptions in a caldera lake and the subsequent depositional processes, a comparison of subaerial and subaqueous deposits is necessary. In this regard, Ilopango provides an unsurpassed oppor-

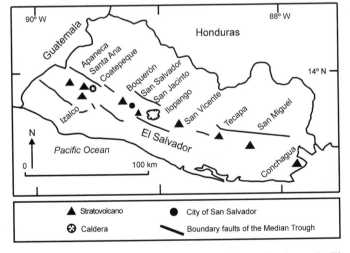

Figure 1. Location map of the Central American volcanic arc in El Salvador. Structural failure at the end of the Pliocene or early Pleistocene resulted in a tectonic depression called the Median Trough, which includes the volcanic chain (Williams and Meyer-Abich, 1955). Ilopango is bordered by three large volcanic structures. Boquerón, a large composite volcano, borders Ilopango to the west-northwest. San Jacinto, a cluster of andesitic domes, borders Ilopango to the west. San Vicente, a large composite volcano, borders Ilopango to the southeast. Modified from Carr et al. (1981).

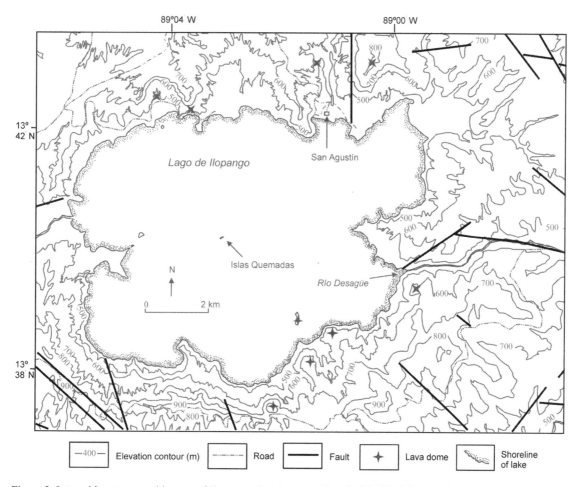

Figure 2. Intracaldera topographic map of Ilopango showing area of study. Modified from Instituto Geográfico Nacional (1981, 1984) and Weber and Wiesemann (1978).

tunity to compare intracaldera and extracaldera deposits, since the pyroclastic facies found outside the caldera are all emplaced subaerially, whereas many deposits within the caldera appear to be emplaced subaqueously. Until now, studies of Ilopango have focused primarily on the extracaldera deposits. In reconnaissance studies, Williams and Meyer-Abich (1955) divided the intracaldera deposits into (1) older fluvial and lacustrine deposits with interbedded "pyroclastic avalanche" deposits on the southern and northeastern walls of the caldera and (2) younger dacite pumice and ash deposits emplaced in water, and "pyroclastic avalanches, pyroclastic falls and lahars" on the northern, western, and upper parts of the eastern caldera walls. Weber and Wiesemann (1978) mapped the intracaldera deposits in the north to northwest and north-northeast part of the caldera as epiclastic rocks, and deposits in the south to southeast and south-southwest part as ignimbrite. Hart (1981) described the intracaldera deposits on the northern caldera wall as "undifferentiated Ilopango-derived pyroclastic and epiclastic deposits with the occasional lahar deposit."

In this paper we provide a detailed description of a segment of the intracaldera stratigraphy at Ilopango caldera and focus on a

sequence that we have informally named the San Agustín Block Unit, which is well exposed to the east and west of the northern lakeshore village of San Agustín (Fig. 2). Found only inside the caldera, the San Agustín Block Unit comprises a thin, basal fine-grained ash and a thick, overlying pumice breccia. The pumice breccia is a spectacular facies with pumice clasts up to 3.0 m across, bed thicknesses up to 15 m, and abundant cooling-induced textures. These features make the Pumice Breccia facies a key intracaldera marker horizon. In this report, we use physical volcanology, petrology, and geochemistry to establish the depositional environment, eruptive conditions, and mode of emplacement of the intracaldera sequence. We first describe the associated intracaldera units and present evidence that a lake occupied Ilopango at the time these units were emplaced. We then demonstrate that the San Agustín Block Unit was both subaqueously erupted and subaqueously emplaced and discuss a plausible model of formation. The various units described below are named informally, for convenience, Pyroclastic Density Current Sequence (PDCS), Lacustrine Unit (LU), San Agustín Block Unit (SABU), Fine Ash facies (FAf), and Pumice Breccia facies (PBf).

GEOLOGIC SETTING

Regional Geology

El Salvador is part of the Central American volcanic arc that extends 1100 km from the Guatemala-México border to northern Panama (Carr and Stoiber, 1990). Tectonism and volcanism largely shape the regional geology of El Salvador. The volcanic chain occurs in a tectonic depression bounded to the south by fault blocks consisting of, from west to east, the Tacuba, Balsam, and Jucuarán mountain ranges (Williams and Meyer-Abich, 1955; Wiesemann, 1975; Carr and Stoiber, 1981). The volcanic chain includes composite volcanoes, dacitic domes, and two calderas, Coatepeque and Ilopango (Fig. 1).

Ilopango Caldera and the Intracaldera Setting

The Ilopango Caldera is located at the junction of the east-west, north-south, and northeast-southwest fault zones (Wiesemann, 1975). The present caldera has a quasi-rectangular shape measuring 16 km east-west and 13 km north-south, and its morphology seems to be controlled by tectonics and mass wasting processes. The southern wall of the caldera is defined by the upthrown Balsam mountain range. The youngest dome, Islas Quemadas, is located in the approximate center of the caldera and may lie along east-west fault structures (Williams and Meyer-Abich, 1955). The Río Desagüe drains the lake to the east along an east-west fault (Fig. 2). Many domes ranging in composition from andesite to dacite occur within the caldera (Weber and Wiesemann, 1978). These domes also may lie along established faults. The upthrown southern wall of the caldera (elevations of 500 to 1000 m) is 100 to 200 m higher above lake level than the northern caldera wall (elevations of 500 to 800 m).

Eruptive History of Ilopango

Since the initial structural failure at the end of the Pliocene or early Pleistocene that formed the Ilopango depression, a series of large explosive eruptions from the caldera have formed voluminous deposits of pyroclastic material. The four pyroclastic events from oldest to youngest are informally named Tierra Blanca 4 (TB4), Tierra Blanca 3 (TB3), Tierra Blanca 2 (TB2), and Tierra Blanca Joven (TBJ) (CEL, 1992). These eruptions have been temporally constrained to the last 56.9 +2.8 −2.1 ka (Rose et al., 1999). TB4, TB3, and TB2 are interpreted to be due to eruptions from Ilopango, but detailed studies have not yet been done. The youngest pyroclastic eruption, TBJ, is well established at A.D. 429 ± 20 years (Dull et al., 2001).

OBSERVATIONS AND RESULTS

Terminology

Pyroclastic terms used here are based on the size classification of volcaniclastic fragments by Fisher (1961) in conjunction with a constituent compositional classification established by Cook (1965). Nomenclature and stratigraphic thicknesses are based on the recommendations of Ingram (1954).

Field Description, Petrology, and Chemistry of Intracaldera Units

The stratigraphic sequence was traceable around the lake but laterally discontinuous (Fig. 3). Not all units are observed at each locality, and in some instances the lowermost unit extends beneath the lake. For this reason, absolute thicknesses could not be established, and the thicknesses reported here are considered minimum values.

Pyroclastic Density Current Sequence (PDCS)

With an overall thickness of 1 to 30 m, this sequence is the lowermost unit of the intracaldera stratigraphy. The Pyroclastic Density Current Sequence comprises two lithified, moderately sorted subunits made up of lithic fragments, pumice, and fine ash (Fig. 3). There is an absence of paleosols and lacustrine layers throughout the sequence. No welding textures, cooling joints, elutriation pipes, or charcoal were observed.

The lower unit and the upper unit display distinct textures. The lower unit is massive at its base and middle and thickly laminated (3 to 5 mm) toward the top; pumice and lithic material are normally graded with a decrease in grain size upward. The contact between the lower unit and upper unit is sharp, with the upper unit directly conformable above the lower unit. The upper unit is laminated on a cm scale at its base and becomes more massive toward the middle and top, with no apparent grading. At the top of the upper unit, accretionary lapilli up to 1 cm in diameter and rounded to subrounded clots of friable, brown silt are observed.

In both units ~85 to 90% of the volume is matrix, which is made up of fine ash and phenocrysts of hornblende, feldspar, and Fe-Ti oxides. Pumice clasts compose 5 to 10 vol% of each unit and are subangular to subrounded, ranging in size from 0.5 to 4 cm with phenocrysts of 1 to 2 vol% hornblende, <1 vol% feldspar, and traces of Fe-Ti oxides. Three types of lithic fragments are present: (1) black obsidian, (2) altered reddish-brown, porphyritic fragments of lava with feldspar needles, and (3) gray fragments of porphyritic lava with hornblende needles and a glassy groundmass. Lithic fragments compose <7 vol% of each unit; they are subangular to subrounded in shape, ranging in size from 0.2 to 3 cm. Electron microprobe analysis of glass shards from the pumice clasts indicates a rhyolitic composition with 75.5 to 78.7 wt% SiO_2 and 6.4 to 7.4 wt% $Na_2O + K_2O$ (Fig. 4; Table 1).

The Pyroclastic Density Current Sequence is unconformably overlain by the Lacustrine Unit, as indicated by an erosional contact between the top of the upper unit of the Pyroclastic Density Current Sequence and the base of the Lacustrine Unit.

Lacustrine Unit (LU)

This unit ranges from 1 to 6 m in thickness and is characterized by an upward transition of finely laminated (0.5 to 3 mm) to

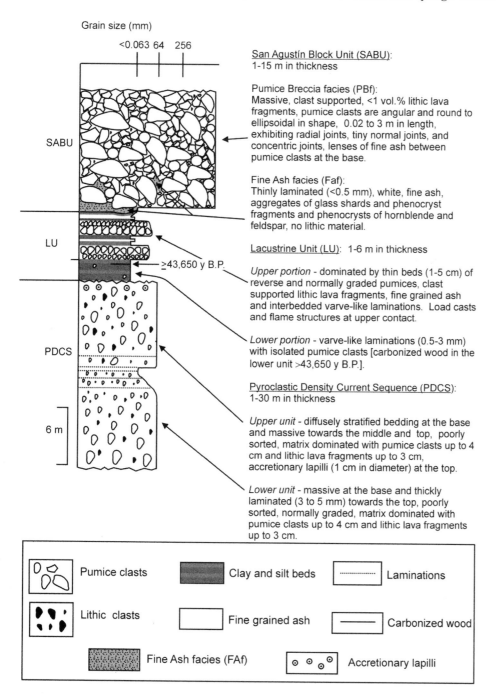

Grain size (mm)

San Agustín Block Unit (SABU):
1-15 m in thickness

Pumice Breccia facies (PBf):
Massive, clast supported, <1 vol.% lithic lava fragments, pumice clasts are angular and round to ellipsoidal in shape, 0.02 to 3 m in length, exhibiting radial joints, tiny normal joints, and concentric joints, lenses of fine ash between pumice clasts at the base.

Fine Ash facies (Faf):
Thinly laminated (<0.5 mm), white, fine ash, aggregates of glass shards and phenocryst fragments and phenocrysts of hornblende and feldspar, no lithic material.

Lacustrine Unit (LU): 1-6 m in thickness

Upper portion - dominated by thin beds (1-5 cm) of reverse and normally graded pumices, clast supported lithic lava fragments, fine grained ash and interbedded varve-like laminations. Load casts and flame structures at upper contact.

≥43,650 y B.P.

Lower portion - varve-like laminations (0.5-3 mm) with isolated pumice clasts [carbonized wood in the lower unit >43,650 y B.P.].

Pyroclastic Density Current Sequence (PDCS):
1-30 m in thickness

Upper unit - diffusely stratified bedding at the base and massive towards the middle and top, poorly sorted, matrix dominated with pumice clasts up to 4 cm and lithic lava fragments up to 3 cm, accretionary lapilli (1 cm in diameter) at the top.

Lower unit - massive at the base and thickly laminated (3 to 5 mm) towards the top, poorly sorted, normally graded, matrix dominated with pumice clasts up to 4 cm and lithic lava fragments up to 3 cm.

Figure 3. Schematic stratigraphic section of Ilopango intracaldera units. See text for details.

Pumice clasts	Clay and silt beds	Laminations
Lithic clasts	Fine grained ash	Carbonized wood
Fine Ash facies (FAf)	Accretionary lapilli	

thin beds (1 to 5 cm) (Fig. 3). There are no paleosols, fossils, or evidence of bioturbation in the unit.

The lower portion of the sequence displays abundant fine horizontal laminations (0.5 to 3 mm) of fine ash, silt, and clay beds. The laminated beds are usually flat and horizontal, but occasionally are deformed and exhibit irregular patterns. Isolated round to subrounded sand to pebble-sized pumice clasts are found throughout the lower portion (Fig. 5A).

A single sample of charred wood collected near the top of the lower portion of the Lacustrine Unit was found in a canyon 1.5 km northeast of the village of San Agustín. The sample (Beta 166558) was collected with tweezers, wrapped in aluminum foil and submitted to Beta Analytic for ^{14}C dating. The radiocarbon contained in this sample was statistically indistinguishable from radiocarbon background standards, yielding a result of ≥43,670 yr B.P.

The upper portion of the sequence is very thin to thinly bedded (1 to 5 cm), alternating between rounded to subrounded, clast-supported pumice beds that are both reversely and normally graded with pumice up to 2 cm in diameter, and well-sorted, subrounded

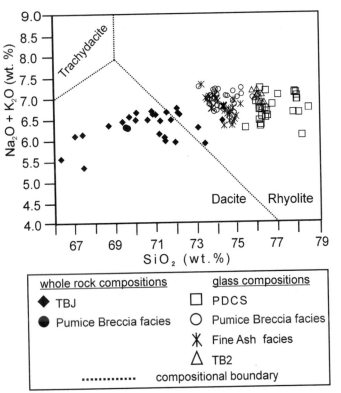

Figure 4. Whole rock and glass compositions from selected units erupted from Ilopango caldera on a SiO$_2$ versus Na$_2$O + K$_2$O classification diagram. All values are normalized to 100 wt% anhydrous. The compositions of TBJ are from Carr and Rose (1987). In general, glass compositions of the Pumice Breccia facies are rhyolitic in composition, whereas whole rock compositions are dacitic. Modified from LeBas et al. (1986).

to subangular clast-supported beds, rich in lithic fragments, with clasts up to 1 cm in diameter. These beds are typically separated by fine ash beds and units consisting of moderately sorted pumice and/or lithic fragments in a fine-grained matrix. Lenticular clots of fine white ash ranging in length from 3 to 6 cm are observed close to the top of the unit, which are associated with thin beds containing both lithic fragments and pumice. The contact of the Lacustrine Unit and the overlying San Agustín Block Unit is sharp, and soft-sediment deformation is indicated by the presence of flame structures and load casts compacting sediment up to 20 cm in depth (Fig. 5B). No paleosols are observed at the contact.

San Agustín Block Unit (SABU)

The San Agustín Block Unit is found only inside the caldera and is traceable discontinuously around the lake. This unit drapes preexisting topography with thicknesses ranging from 1 to 15 m. Two distinct facies define the San Agustín Block Unit: a basal layer of fine ash and an overlying pumice breccia (Figs. 3 and 6).

Fine Ash Facies (FAf)

The fine-grained ash layer at the base and in lenses within the basal portion of the overlying pumice breccia is a white, friable, moderately sorted, vitric-crystal fine ash that lacks lithic fragments (Fig. 6). No diatoms are present. This layer is laterally discontinuous and has a maximum thickness of 30 cm. It is characterized by thin (<0.5 mm) yellowish-orange lamellae composed of juvenile material such as coarse pumice shards (vesicular glass), fine ash, and phenocryst fragments.

The coarse pumice shards (0.063 to 1 mm) display two dominant morphologies. The first type is thin, fibrous grains

TABLE 1. ELECTRON MICROPROBE DATA ON GLASS COMPOSITIONS

Unit: Clast type: Analysis*:	PDCS Pumice 30	SABU-FAf Fine ash 32	SABU-PBf Pumice 29	TB2 Pumice 11
wt% SiO$_2$	73.38 ± 0.62	70.50 ± 0.89	71.40 ± 0.69	72.41 ± 0.29
TiO$_2$	0.18 ± 0.02	0.27 ± 0.03	0.26 ± 0.03	0.23 ± 0.02
Al$_2$O$_3$	12.13 ± 0.18	13.36 ± 0.25	13.45 ± 0.29	12.51 ± 0.08
FeO	1.25 ± 0.08	1.69 ± 0.12	1.69 ± 0.14	1.38 ± 0.05
MnO	0.07 ± 0.02	0.11 ± 0.02	0.10 ± 0.03	0.07 ± 0.03
MgO	0.24 ± 0.03	0.46 ± 0.11	0.43 ± 0.05	0.27 ± 0.01
CaO	1.38 ± 0.06	2.03 ± 0.12	2.05 ± 0.18	1.51 ± 0.03
Na$_2$O	3.96 ± 0.28	4.22 ± 0.23	4.51 ± 0.17	4.13 ± 0.14
K$_2$O	2.48 ± 0.09	2.20 ± 0.06	2.21 ± 0.07	2.54 ± 0.05
P$_2$O$_5$	0.03 ± 0.01	0.06 ± 0.01	0.06 ± 0.01	0.03 ± 0.01
Total	95.10 ± 0.94	94.89 ± 0.91	96.15 ± 0.54	95.08 ± 0.31
Cl	0.18 ± 0.02	0.14 ± 0.04	0.14 ± 0.01	0.16 ± 0.01
O	−0.04 ± 0.00	−0.03 ± 0.01	−0.02 ± 0.00	−0.04 ± 0.00

Note: PCDS—Pyroclastic Density Current Sequence; SABU—San Agustín Block Unit; FAf—Fine Ash facies; PBf—Pumice Breccia facies; TB2—Tierra Blanca 2. Electron microprobe analyses of volcanic glass were performed with a JEOL JXA-8900R at McGill University, Canada. Analyses were performed with an accelerating voltage of 15 kV, a beam current of 20 nA, and a beam diameter of 15 μm. These parameters provided the best conditions to minimize Na loss.

*Total number of analyses of representative unit. Compositions are displayed as mean ± 1σ.

with elongate parallel vesicles displaying both ragged and blunt edges. Some of these grains are delicately twisted with occasional ruptured circular vesicles (Fig. 7A). The second type is equant grains, with ovoid to circular vesicles enclosed by glass walls (Fig. 7B). Both types of shards are colorless with no visible microlites. No hydration structures are observed (Wohletz, 1987). Electron microprobe analysis of the glass shards indicates a rhyolitic composition with 73.2 to 75.3 wt% SiO_2 and 6.4 to 7.4 wt% $Na_2O + K_2O$ (Fig. 4; Table 1).

The fine ash is present as aggregates, shards of glass, and adhering dust and composes 60 vol% of the bed. The aggregates are rounded to subrounded clusters of tiny fragmented phenocrysts and angular glass shards (Fig. 7C). Some of the aggregates are slightly altered to a dirty orange color. The glass shards composing the fine fraction (<0.063 mm) display two principal morphologies: (1) angular and blocky with planar to curviplanar fracture surfaces and a scarcity of ovoid vesicles (Fig. 7D), and (2) Y-shaped forms representing bubble wall junctions.

Phenocrysts and phenocryst fragments of hornblende and plagioclase also are observed. The phenocrysts are subhedral and range in size from 0.063 to 1 mm (Fig. 7E). Phenocryst fragments are <0.063 mm. Fine ash adheres to many phenocrysts as observed on the hornblende in Figures 7E–F.

Pumice Breccia Facies (PBf)

This facies is a massive, well-sorted, clast-supported bed with thicknesses ranging from 1 to 15 m (Fig. 8A). The pumice clasts display angular to spherical to ellipsoidal shapes, with aspect ratios (vertical axis/horizontal axis) of 0.2 to 1, ranging in length from 0.02 to 3 m (Figs. 8B–C). Pumice clasts up to 60 cm in length tend to be quite well rounded and have aspect ratios of 1. Pumice clasts >60 cm are subangular to subrounded in shape with aspect ratios of 0.2 to 0.9. The larger pumice clasts (>20 cm in length) are heavily fractured and break into smaller fragments when disturbed. The pumice clasts 2 to 20 cm in length are prismatic in shape. The bulk of the pumice clasts are between 60 and 90 cm in length. This facies loads the underlying Lacustrine Unit, forcing sediment and the basal fine ash layer to be pushed up between the pumice clasts in the form of flame structures. Lenses of the fine ash are observed toward the base of this facies as isolated pockets between pumice clasts.

Three types of jointing structures dominate the pumice clasts. (1) Radial columnar joints up to 3 to 4 cm in width and 8 to 10 cm in length form fan-shaped structures perpendicular to the curving clast surface (Fig. 8D) and also to concentric joints defined below. (2) Tiny normal joints 1 cm in width and up to 2 cm in length are observed at the edges of the pumice clasts. (3) Concentric joints are curved and concentric to the pumice core (Fig. 8E).

The vesicularity of pumice clasts ranges from 40 to 80 vol% with phenocrysts of hornblende, plagioclase, and ilmenite. Two main variants of the vesicular texture dominate: (1) irregular shapes range from round to more elongate up to 3.5 mm in length and (2) smaller round vesicles as small as 0.020 mm. Phenocrysts range in size from 0.8 mm to 1.2 cm. The larger

Figure 5. Examples of sedimentary structures in the Lacustrine Unit. A: Laminated beds of the lower portion of the Lacustrine Unit. Machete handle for scale is 6 cm long. B: Lacustrine Unit (LU) overlain by the Pumice Breccia facies (PBf). Flame structures indicated by the white arrow. Hammer for scale is 20 cm long.

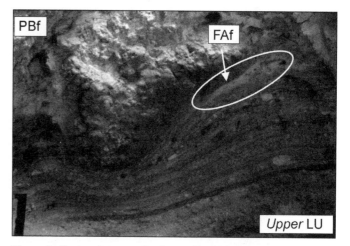

Figure 6. Contact between the Lacustrine Unit and the San Agustín Block Unit. PBf—Pumice Breccia facies, FAf—Fine Ash facies, Upper LU—upper portion of the Lacustrine Unit. A discontinuous lens of white volcanic ash, which is the Fine Ash facies, is observed at the base of the Pumice Breccia facies.

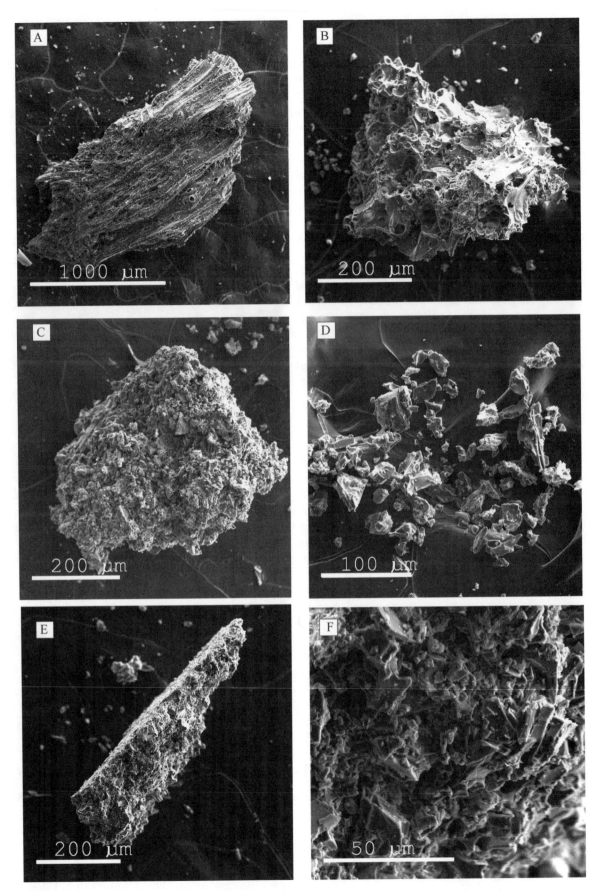

Figure 7. Secondary electron imaging (SEI) of the Fine Ash facies grain morphology. A: Fibrous pumice shards with parallel elongate vesicles. B: Equant pumice shards with ovoid to spherical vesicles. C: Aggregate of fine ash, phenocryst fragments, and larger glass shards. D: Angular, blocky glass shards with planar and curviplanar fractures across bubble junctions. E: Hornblende with adhering dust. F: Dust adhering to hornblende in image E. Note that scales are different in each photo. SEI images were taken using a JEOL JXA-8900L electron microprobe.

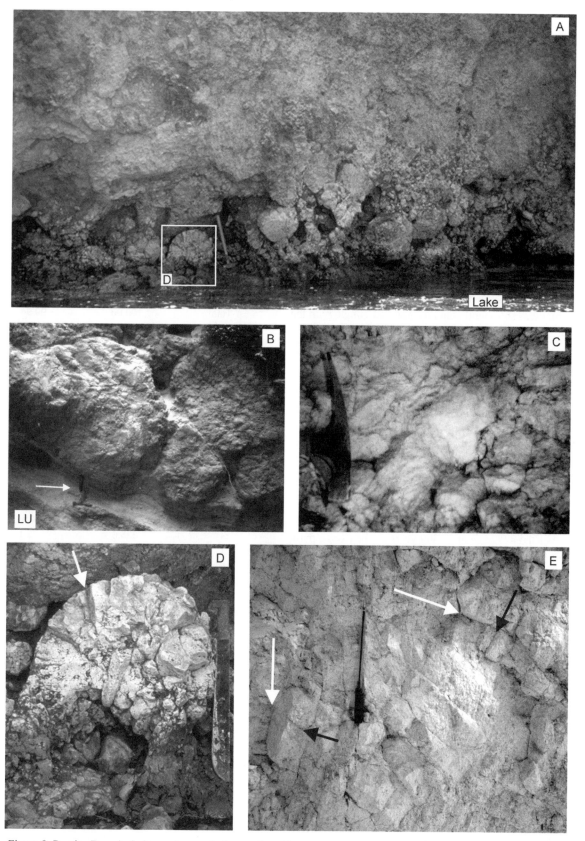

Figure 8. Pumice Breccia facies grain morphology and cooling textures. A: Massive, clast-supported pumice clasts; note large round pumices at the base. White box is magnified in D below. Hammer for scale is 80 cm long. B: Larger pumice clasts displaying subangular to subrounded clast morphology. White arrow points to a hammer for scale, which is 20 cm high. C: Smaller (<20 cm) angular pumice clasts. These clasts represent the fractured surface of a larger pumice. Hammer head for scale is 13 cm long. D: Radially jointed pumice (indicated with white arrow) with fragmented core, machete for scale is 45 cm long. E: Concentric joints (indicated with the white arrows) with radial joints (indicated with black arrows) normal to concentric joints, pen for scale is 14 cm long.

phenocrysts mostly appear as glomeroporphyritic clusters of plagioclase, hornblende, and ilmenite and clusters of plagioclase with hornblende. Bubbles radiate from many of the large phenocrysts and glomeroporphyritic clusters.

Lithic fragments compose <1 vol% of this facies, consisting of gray fragments of porphyritic lava with hornblende and feldspar needles in a glassy matrix. The fragments are 1 to 2 cm in size and subangular to subrounded in shape.

Electron microprobe analysis of the pumice glass shards indicates a rhyolitic composition of 73 to 75.9 wt% SiO_2 and 6.6 to 7.4 wt% $Na_2O + K_2O$ (Fig. 4; Table1), while whole rock powders analyzed by XRF have a dacitic compositions of 69.5 wt% SiO_2 and 6.3 wt% $Na_2O + K_2O$ (Fig. 4; Table 2).

Mingling textures occur as rounded clots of more mafic compositions surrounded by the dacite and as planar dark bands alternating with the dacitic host. These textures indicate that the mafic material was sufficiently fluid to flow and mingle with the dacitic host.

DISCUSSION

Intracaldera Framework

Constrained by the age of the lower unit of the overlying Lacustrine Unit, the Pyroclastic Density Current Sequence may be the intracaldera expression of one of the older Tierra Blanca eruptions. A minimum limiting ^{14}C age of ≥43,670 years has been established on the lower portion of the Lacustrine Unit. The presence of lacustrine sediments suggests that a lake, possibly a water-filled caldera, occupied Ilopango at this time. Rose et al. (1999) established a ^{14}C age of 57,000 years on the Congo pyroclastic flow from Coatepeque Caldera. The Congo pyroclastic flow underlies Tierra Blanca 4 (TB4), a fallout deposit from the oldest of the documented Ilopango eruptions. Therefore the TB4 eruption is younger than 57,000 years.

Depositional Setting

The sharp contact and the absence of paleosols and lacustrine layers between the two Pyroclastic Density Current subunits indicate that these two units were deposited in rapid succession. In the lower unit, the normally graded pumice and lithic material and the gradation from a massive base to a finely laminated top suggest a single depositional unit with a change in depositional energy. The more massive lower portion suggests a dense concentrated flow, whereas the laminations at the top suggest gradual sorting of dilute suspended ash as observed in the subaqueous environment (Fiske and Matsuda, 1964).

The sharp contact between the lower and upper units indicates a change in depositional environment. The upper unit is characterized by a massive, moderately sorted, nongraded texture, whereas the laminations at its base are moderately sorted. The laminated layers may represent initial pulses as the pyroclastic density current gradually became more sustained.

TABLE 2. WHOLE ROCK GEOCHEMISTRY

Unit: Clast type: Sample:	SABU-PBf Pumice CM261-01	SABU-PBf Pumice CM252-01
wt%		
SiO_2	66.49	66.31
TiO_2	0.38	0.39
Al_2O_3	15.22	15.23
Fe_2O_3	3.47	3.33
MnO	0.13	0.13
MgO	1.15	1.19
CaO	3.66	3.55
Na_2O	4.03	3.98
K_2O	1.95	1.97
P_2O_5	0.14	0.13
LOI	3.44	3.73
Total	100.06	99.94
ppm		
BaO	1059	1053
Ce	0	26
Co	0	11
Cr	0	0
La	13	0
Ni	0	0
Sc	0	0
V	46	39
Ga	14.4	14.9
Nb	5.1	5.8
Pb	2.9	2.1
Rb	43.4	44.9
Sr	325.6	325.4
Th	3.9	3.5
U	1.8	1.9
Y	19.7	20.0
Zr	159.1	158.1

Note: SABU—San Agustín Block Unit; PBf— Pumice Breccia facies. X-ray fluorescence whole rock analyses were performed with a Philips PW2440 4 kW automated XRF spectrometer system at the Geochemical Laboratories, McGill University, Canada, by Tariq Ahmedali and Glenna Keating. Major elements (Si, Ti, Al, Fe, Mn, Ca, Na, K, and P) were determined using 32 mm diameter fused beads composed of 5 parts lithium tetraborate and 1 part sample, and trace elements (Ba, Ce, Co, Cr, La, Ni, Sc, V, Ga, Nb, Pb, Rb, Sr, Th, U, Y, Zr) were determined using 40 mm diameter pressed pellets.

It is difficult to determine the temperature of the Pyroclastic Density Current Sequence at the time of emplacement. Key indicators for hot emplacement in the Pyroclastic Density Current Sequence are absent, which may suggest that this sequence was emplaced at lower temperatures.

The presence of accretionary lapilli at the top of the upper unit suggests a water vapor–rich environment (Fisher and Schmincke, 1984). Accretionary lapilli will form in the subaerial environment during magmatic eruptions through clouds (Alvarado and Soto, 2002) and during phreatomagmatic eruptions when lake water is vaporized, such as during the 1965 eruption of Taal volcano (Moore et al., 1966).

On the basis of the available evidence, we conclude that water was present during eruption. The lower unit and the upper unit are clearly temporally related. The normally graded beds and sedimentary structures of the lower unit suggest deposition in a subaqueous environment. Although the temperature of the unit is undetermined, the absence of indicators of hot emplacement is consistent with deposition in a subaqueous environment. The textures and sedimentary structures of the upper unit could be indicative of a subaerial or subaqueous environment, but the presence of accretionary lapilli in the upper unit could indicate that the eruption column was partly emergent in a humid subaerial environment.

The unit overlying the Pyroclastic Density Current Sequence is lacustrine in origin, suggesting a lake was present after emplacement of the Pyroclastic Density Current Sequence. The lower part of the Lacustrine Unit is marked by finely laminated beds of volcanic ash, silt, and clay scattered with sand-sized pumice and intermittent deformed beds. The varve-like laminations are structures indicative of a quiet environment below wave base, such as a lake (Pettijohn and Potter, 1964). Similar sedimentary structures well documented at Laguna de Ayarza in Guatemala have been interpreted to result from the sedimentary infilling of a caldera lake (Poppe et al., 1985). Sedimentation of the fine-grained material was the result of suspension settling, with the isolated, larger pumice clasts representing intermittent waterlogging and sinking of floating pumice (Manville et al., 1998). The subtle deformation of the laminations could be due to tectonic activity such as earthquakes causing the disturbance, but the calm environment continued, as shown by the fine horizontal laminations above.

Clast-supported, reversely and normally graded pumice beds with load casts and accompanying flame structures characterize the uppermost portion of the unit. Pumice that is erupted subaerially and then emplaced subaqueously becomes water-saturated as a function of its surface area (Whitham and Sparks, 1986); therefore, the larger the pumice, the longer it takes to sink (e.g., the A.D. 181 Taupo eruption: White et al., 2001). Evidence of such a process is the reversely-graded accumulation of pumice clasts observed in the field. In the upper portion of the unit, the clast-supported layers of pumice alternate between reversely graded and normally graded. Once pumice becomes saturated with water, the weight of the larger pumice is greater. If remobilized, the redeposited bed would then be normally graded.

The load casts and flame structures observed at the uppermost part of the unit indicate soft-sediment deformation. For plastic deformation to occur in fine-grained sediments instead of brittle faulting requires the sediment to be saturated with at least 15–20 wt% water (Heiken, 1971). In the subaerial environment the deposition of such finely laminated beds is difficult because the environment is not quiet for a sufficiently long period.

Subaqueous Eruption and Emplacement of the San Agustín Block Unit

Relationship between the Fine Ash Facies and the Pumice Breccia Facies

On the basis of the stratigraphic relationships and chemical similarities, we conclude that the basal Fine Ash facies and the overlying Pumice Breccia facies are coeval. The Fine Ash facies is found at the base of the unit and is directly overlain by the Pumice Breccia facies, with an absence of paleosols and lacustrine beds between them. Electron microprobe analysis shows that the glass shards from the Fine Ash facies and glass from the Pumice Breccia facies are rhyolitic in composition with values of 73 to 75.5 wt% SiO_2 and 6.3 to 7.5 wt% $Na_2O + K_2O$ (Fig. 4; Table 1). Assuming that the glass is a quenched record of the evolved magmatic melt, the geochemical data of the glasses show that the Fine Ash facies and Pumice Breccia facies are chemically similar (Figs. 9A–C).

Fragmentation of the Fine Ash

To determine the eruptive environment of the basal Fine Ash facies, it is important to understand the fragmentation process or processes taking place by evaluating the grain morphology of the juvenile material and grain size distribution of the bed.

During magmatic fragmentation, bubbles rupture in response to large differences between internal pressure and ambient pressure (Sparks, 1978). Controlled by bubble morphology, tephra produced by magmatic fragmentation records the vesicularity of the melt, and the glass shards reflect the thinned bubble walls or bubble junctions at the time of eruption (Heiken and Wohletz, 1991).

Hydromagmatic fragmentation takes place when magma interacts with external water, causing the quenched melt to shatter (Wohletz, 1983). Therefore, the juvenile products of a phreatomagmatic eruption are not primarily controlled by bubble morphology but instead display angular, blocky grains with low vesicularity, whereas thicker plate-like fragments are representative of the bubble walls (Wohletz, 1983).

In the Fine Ash facies, the coarse pumice shard morphology indicates magmatic fragmentation, whereas the finer-grained shards indicate hydromagmatic fragmentation. The coarser-grained (>0.063 mm) pumice shards are elongate with pipe vesicles and equant with ovoid vesicles, the shard margins being clearly controlled by vesicle shape (Figs. 7A–B). However, the finer-grained glass shards (<0.063 mm) are dominated by fracture-bounded grains, giving rise to blocky, angular shards with curviplanar and planar surfaces (Fig. 7C).

Ash Aggregates in the Fine Ash Facies

The presence of angular, blocky fragments of glass argues for the involvement of water during fragmentation, whereas aggregates of ash and abundance of adhering dust are evidence

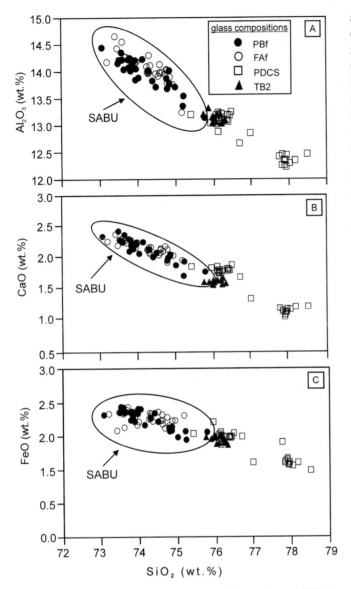

Figure 9. Glass compositions for two intracaldera units and TB2. A: Al_2O_3 vs. SiO_2. B: CaO vs. SiO_2. C: FeO vs. SiO_2. SABU is the San Agustín Block Unit; ellipse encompasses the compositional field of the San Agustín Block Unit. PBf—Pumice Breccia facies, FAf—Fine Ash facies, PDCS—Pyroclastic Density Current Sequence, TB2—Tierra Blanca 2. All values are normalized to 100 wt% anhydrous, and total Fe is calculated as FeO.

atomagmatic, and vulcanian eruptions of varying compositions (Clanton et al., 1983; Heiken and Wohletz, 1985). The mechanism of aggregate formation is poorly understood, but Sorem (1982) reports ash clusters in the distal fall deposit of the 1980 Mount St. Helens eruption and interprets the morphology as a result of both mechanical interlocking of the grains and electrostatic attraction. Houghton et al. (2000) expand on this and attribute aggregate formation to many variables, including the amount of liquid (water and dissolved gas species) present.

We have previously presented evidence for the presence of a lake at the time of eruption. Thus, we can make three important inferences. (1) The lake was the source of the water that contributed to fragmentation of the magma. (2) The lake was the source of the water that caused ash aggregates to form. (3) The eruption column was partly subaqueous and partly subaerial.

The occurrence of both magmatic and hydromagmatic shards illustrates magma interaction with external water, whereas the absence of lithic material indicates little erosion of the surrounding country rock. The fragmentation may have occurred near the top of the conduit in shallow water depth. Fragmentation in the upper portions of the conduit would reduce the amount of lithic material entrained, whereas shallow depths would promote both substantial vesiculation of the magma and hydromagmatic fragmentation.

Eruption and Emplacement of the Pumice Breccia Facies

The dacitic pumice clasts are brecciated and display textures indicative of hot emplacement and rapid cooling. The original pumice clasts have been fractured and many have brecciated into smaller pumice fragments ranging in size from 2 to 20 cm. Such brecciation is associated with the rapid cooling of lavas and is common in both the subaerial and subaqueous environment (Pichler, 1965; Fink, 1983).

Abundant cooling textures occur in the large pumice breccia clasts (>20 cm) as radial joints, columnar joints, tiny normal joints, and concentric joints. Although cooling joints are observed in both the subaqueous and subaerial environment, the diversity of joint types and patterns at Ilopango are associated with subaqueous eruptions elsewhere (Furnes et al., 1980; Kano et al., 1991; McPhie et al., 1993; Sakamoto and Tanimoto, 1996; Allen and McPhie, 2000). For example, radial joints are observed in subaqueous lobes of rhyolitic lavas perpendicular to the lobe surface (Furnes et al., 1980; Kano et al., 1991), whereas tiny normal joints are observed perpendicular to quenched margins in the subaqueously erupted and emplaced cobble-boulder facies of the rhyolitic Yali pumice breccia (Allen and McPhie, 2000). There is also an absence of microlites in the glass of the pumice clasts, suggesting rapid quenching of the clasts.

The efficiency of brecciation, the abundance and morphology of the different joints, and the absence of microlites throughout the clasts, together argue for hot clasts that were rapidly quenched. We have shown that a lake was present at the time of emplacement of the San Agustín Block Unit; based on the textures attributed to cooling, we infer that the Pumice Breccia facies was rapidly quenched in the lake.

for humid atmospheric conditions or water vaporized from the lake at the time of eruption. Secondary electron images (SEI) of the fine ash reveal abundant dust adhering to phenocrysts and glass shards, as well as rounded to subrounded aggregates of juvenile material consisting of fragmented phenocrysts, blocky glass shards, and fine ash. These textures are not a product of fragmentation, but instead are indicators of conditions in and surrounding the eruption column. Similar aggregates have been observed in deposits from subaerial plinian, phreatic, phre-

Extremely large clasts of pumice are documented in lake-filled intracaldera environments and in lake environments proximal to volcanic activity. In both environments, lava domes are present within the lake and at the lakeshore. Pumice clasts up to 12 m are observed at Sierra La Primavera Caldera, México (Mahood, 1980), and pumice clasts up to 17 m are documented from the A.D. 186 Taupo eruption, New Zealand (Wilson and Walker, 1985; von Lichtan et al., 2002). At Mono Lake, eastern California, USA, Stine (1984) records pumice clasts at the lakeshore ≤2 m. Floating rafts of pumice have been documented during the 1953–1957 marine eruption of the Tuluman volcano, Papua New Guinea (Reynolds and Best, 1976). On the basis of the similar geochemistry of pumice clasts and dome material, the large clasts at Sierra La Primavera and Taupo were interpreted as pumice clasts spalled from the carapace of a growing subaqueous lava dome (Mahood, 1980; Wilson and Walker, 1985). At Mono Lake, geochemical analyses have not been performed on the pumice blocks to test the link with the dome material, but Stine (1984) proposed that they formed during a sublacustrine eruption and rose to the surface (Stine, 1984). The Tuluman volcano eruption ended with formation of an underwater dome located just below sea level (Reynolds and Best, 1976).

The cooling-induced textures found at Sierra La Primavera and Taupo resemble those at Ilopango, although some of the depositional characteristics are quite different. At Sierra La Primavera laminated layers intermittently drape the large clasts of pumice, whereas at Ilopango the pumice breccia bed is massive and clast-supported without intermittent laminated layers. At Sierra La Primavera, the large clasts of pumice became waterlogged at different rates and thus sank at different times (Mahood, 1980). At Taupo, the large clasts of pumice occur as isolated blocks around the lake and appear to be the youngest products of the A.D. 186 eruption (Wilson and Walker, 1985). At Mono Lake, the pumice blocks also are isolated and concentrated mostly on the western and northern shores of the lake (Stine, 1984). At Ilopango, the unit is massive and discontinuous owing to erosion, although it is also traceable around the lake. From this we infer that (1) the pumice clasts at Ilopango sank rapidly, not allowing time for the deposition of lacustrine layers, and (2) the pumice clasts at Ilopango also may record lake level at the time they were deposited.

Considering the evidence for rapid cooling and rapid deposition, we suggest that the pumice clasts became quickly waterlogged after eruption. The massive, clast-supported nature of the Pumice Breccia facies indicates rapid deposition. When immersed in water, hot pumice will rapidly ingest the surrounding water, increase in density, and sink (Whitham and Sparks, 1986; Cashman and Fiske, 1991; Kano, 1996). The presence of lithic fragments 1 to 2 cm in size that coexist with pumices ranging up to 3 m in length indicates that both materials were deposited at the same time. Once sufficiently waterlogged, the pumice clasts will settle out of the water column at the same rate as a smaller clast with a much higher initial density (Cashman and Fiske, 1991), suggesting they are hydraulically equivalent.

The range of vesicularity of the pumice also suggests that some of these clasts were comparatively dense and sank immediately.

Model of Eruption and Emplacement of the San Agustín Block Unit

We propose a sequence of magmatic and phreatomagmatic eruptions that formed the San Agustín Block Unit (Fig. 10). Inside the caldera a shallow subaqueous eruption began, and vesiculating rhyolitic melt fragmented explosively. The eruption column progressed from subaqueous to subaerial. In the subaerial environment, the moist particles in the eruption plume aggregated, and fine ash adhered to the phenocrysts and pumice shards. The larger and denser material fell out of the eruption plume and settled through lake water to be deposited on the bottom. Some ash may have been carried away by wind. This explosive phase was followed by construction of a subaqueous dome. As the hot dacitic magma was extruded from the vent, it cooled rapidly and brecciated upon contact with the lake water. Quenched clasts of pumice spalled from the vesiculated carapace of the growing dome. The denser clasts sank immediately, whereas others ingested water and sank when sufficiently waterlogged. Some clasts sank close to the source, while other clasts floated to the edge of the lake. The first pumice clasts to sink made contact with the layer of fine ash, disrupting it and causing the fine sediment to become caught up between depositing pumice clasts.

CONCLUSIONS

1. We have investigated a portion of the intracaldera stratigraphy at Ilopango caldera consisting of a sequence of pyroclastic density currents followed by a lacustrine deposit, in turn followed by the San Agustín Block Unit. The lacustrine sequence, containing driftwood, indicates that a lake occupied Ilopango caldera at ≥43,670 yr B.P.

2. The San Agustín Block Unit comprises a basal Fine Ash facies and an overlying Pumice Breccia facies found only within the caldera. The fine ash consists of coarse pumiceous glass shards and fine blocky glass shards, aggregates of fine ash with phenocryst fragments, and phenocrysts with abundant adhering dust. The blocky shards are hydromagmatic in origin, whereas the aggregates and phenocrysts with adhering dust support a water-rich environment. The Pumice Breccia facies contains pumice clasts whose shapes are angular to ellipsoidal, with aspect ratios of 0.2 to 1 and sizes ranging up to 3 m in length. The pumice clasts display radial joints, tiny normal joints, and concentric jointing, all indicative of hot emplacement and rapid quenching. The distinguishable characteristics of the Pumice Breccia facies, such as pumice diameter, thickness, and jointing textures, indicate that this unit is a key stratigraphic marker horizon with which to conduct further intracaldera studies.

3. Through observation of the individual intracaldera stratigraphic units, we infer that a lake occupied the caldera, at least

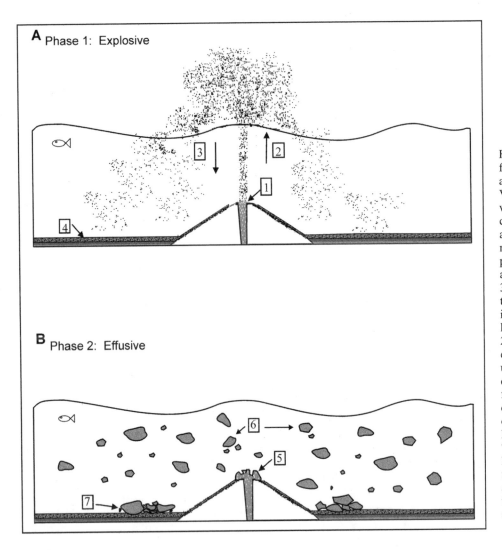

Figure 10. Schematic diagram of model for the San Agustín Block Unit eruption and emplacement. A: Phase 1. 1—Vesiculating rhyolitic melt erupts into water, fragmenting explosively upon contact. 2—Plume begins subaqueously and progresses to the subaerial environment. Once subaerial, the fine ash and phenocryst fragments cluster together, and fine ash adheres to phenocrysts. 3—The denser particles fall back into the lake. 4—The basal Fine Ash facies is deposited onto the lake bottom overlying lacustrine sediments. B: Phase 2. 5—A more degassed dacitic melt extrudes from the vent, cooling rapidly upon contact with lake water. 6—The quenched pumice clasts are spalled from the carapace of the dome. The dense pumice clasts sink immediately; others sink when sufficiently waterlogged, and some float to the lake edge. 7—Accumulating pumice clasts on the lake bottom form the Pumice Breccia facies, which directly overlies the fine ash layer, disturbing it and causing the fine sediment to become caught up among the depositing pumice clasts.

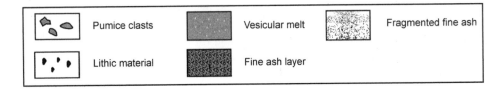

periodically, for an extended length of time and has played an active role in explosive volcanism at Ilopango.

ACKNOWLEDGMENTS

We extend our gratitude to the Fundación de Amigos del Lago de Ilopango, especially Ernesto Freund, Jeannette Monterrosa, and Fred Thill, for their help with logistics and incredible support while in El Salvador. We thank Carlos Pullinger of the Servicio Geológico division of the Servicio Nacional de Estudios Territoriales for taking the time to familiarize us with the outer caldera deposits and for his ongoing enthusiasm to understand the San Agustín Block Unit. We are forever indebted to our new friends at San Agustín, especially Natividad Rauda and his family, for their patience while in the field and acceptance of C.P.M. and M.R. into their village and homes. We thank Chris Harpel of the Cascades Volcano Observatory for his help tracking down elusive references and Sergio Espinosa for his help with translation of the abstract to Spanish. We also thank Guillermo Alvarado and Barry Cameron for their reviews, which significantly improved this manuscript. We are most grateful to Mike Carr for his input and suggestions to the paper. This project was funded by a bursary awarded through the Canadian Bureau of International Education by the Canadian International Development Agency to C.P.M., by a research grant from the Natural Sciences and Engineering Research Council of Canada to J.S., and by National Science Foundation Grant EAR-9903291 to J.V.

REFERENCES CITED

Allen, S.R., and McPhie, J., 2000, Water-settling and resedimentation of submarine rhyolitic pumice at Yali, eastern Aegean, Greece: Journal of Volcanology and Geothermal Research, v. 95, p. 285–307.

Alvarado, G.E., and Soto, G.J., 2002, Pyroclastic flow generated by crater-wall collapse and outpouring of the lava pool of Arenal Volcano, Costa Rica: Bulletin of Volcanology, v. 63, p. 557–568.

Carr, M.J., and Rose, W.I., 1987, CENTAM-a database of analyses of Central American volcanic rocks: Journal of Volcanology and Geothermal Research, v. 33, p. 231–252.

Carr, M.J., and Stoiber, R.E., 1981, Lavas and faults characterize El Salvador: Geotimes, v. 26, no. 7, p. 20–21.

Carr, M.J., and Stoiber, R.E., 1990, Volcanism, *in* Dengo, G., and Case, J.E., eds., The Caribbean region: Boulder, Colorado, Geological Society of America, Geology of North America, v. H, p. 375–391.

Carr, M.J., Mayfield, D.G., and Walker, J.A., 1981, Relation of lava compositions to volcano size and structure in El Salvador: Journal of Volcanology and Geothermal Research, v. 10, p. 35–48.

Cashman, K.V., and Fiske, R.S., 1991, Fallout of pyroclastic debris from submarine volcanic eruptions: Science, v. 253, p. 275–280.

CEL (Comisión Ejecutiva Hidroelectrica del Rio Lempa), 1992, Desarrollo de los recursos geotérmicos del area centro-occidental de El Salvador. Prefactibilidad geotérmica del area de Coatepeque: Comisión Ejecutiva Hidroelectrica del Rio Lempa Reconocimiento Geotérmico, Internal Report, p. 1–104.

Clanton, U.S., Gooding, J.L., and Blanchard, D.P., 1983, Volcanic ash "clusters" in the stratosphere after the El Chichon, Mexico, eruption [abs.]: Eos (Transactions, American Geophysical Union), v. 64, p. 1139.

Cook, E.F., 1965, Stratigraphy of Tertiary volcanic rocks in eastern Nevada: Nevada Bureau of Mines Report 11, 66 p.

Dull, R., Southon, J., and Sheets, P., 2001, Volcanism, ecology and culture: A reassessment of the Volcán Ilopango TBJ eruption in the southern Maya realm: Latin American Antiquity, v. 12, p. 25–44.

Fink, J.H., 1983, Structure and emplacement of a rhyolitic obsidian flow: Little Glass Mountain, Medicine Lake Highland, northern California: Geological Society of America Bulletin, v. 94, p. 362–380.

Fisher, R.V., 1961, Proposed classification of volcaniclastic sediments and rocks: Geological Society of America Bulletin, v. 72, p. 1409–1414.

Fisher, R.V., and Schmincke, H.-U., 1984, Pyroclastic rocks: Berlin, Springer Verlag, 472 p.

Fiske, R.S., and Matsuda, T., 1964, Submarine equivalents of ash flows in the Tokiwa Formation, Japan: American Journal of Science, v. 262, p. 76–106.

Furnes, H., Fridleifsson, I.B., and Atkins, F.B., 1980, Subglacial volcanics—On the formation of acid hyaloclastites: Journal of Volcanology and Geothermal Research, v. 8, p. 95–110.

Goodyear, W.A., 1880, Earthquake and volcanic phenomena. December 1879 and January 1880 in the Republic of Salvador, Central America: Panama, Star and Herald, p. 56.

Hart, W.J.E., 1981, The Panchimalco tephra, El Salvador, Central America [M.S. thesis]: New Brunswick, New Jersey, Rutgers University, 101 p.

Heiken, G.H., 1971, Tuff rings: Examples from the Fort Rock–Christmas Lake Valley Basin, south-central Oregon: Journal of Geophysical Research, v. 76, p. 5615–5626.

Heiken, G., and Wohletz, K., 1985, Volcanic ash: Berkeley, University of California Press, 246 p.

Heiken, G., and Wohletz, K., 1991, Fragmentation processes in explosive volcanic eruptions: Sedimentation in volcanic settings: SEPM (Society for Sedimentary Geology) Special Publication 45, p. 19–26.

Houghton, B.F., Wilson, C.J.N., Smith, R.T., and Gilbert, J.S., 2000, Phreatoplinian eruptions, *in* Sigurdsson, H., et al., eds., Encyclopedia of volcanoes: San Diego, California, Academic Press, p. 513–525.

Ingram, R.L., 1954, Terminology for the thickness of stratification and parting units in sedimentary rocks: Geological Society of America Bulletin, v. 65, p. 937–938.

Instituto Geográfico Nacional "Ingeniero Pablo Arnoldo Guzmán", 1981, América Central República de El Salvador Lago de Ilopango: map SAL-LL1, scale 1:20,000, 1 sheet.

Instituto Geográfico Nacional "Ingeniero Pablo Arnoldo Guzmán", 1984, República de El Salvador: maps 2356 I, 2357 II, 2457 III, 2456 IV, scale 1:50,000, 4 sheets.

Kano, K., 1996, A Miocene coarse volcaniclastic mass-flow deposit in the Shimane Peninsula, SW Japan: Product of a deep submarine eruption?: Bulletin of Volcanology, v. 58, p. 131–143.

Kano, K., Takeuchi, K., Yamamoto, T., and Hoshizumi, H., 1991, Subaqueous rhyolite block lavas in the Miocene Ushikiri Formation, Shimane Peninsula, SW Japan: Journal of Volcanology and Geothermal Research, v. 46, p. 214–253.

LeBas, M.J., LeMaitre, R.W., Streckeisen, A.L., and Zanettin, B., 1986, A chemical classification of volcanic rocks based on the total alkali-silica diagram: Journal of Petrology, v. 27, p. 247–750.

Lothrop, S.K., 1927, Pottery types and their sequence in El Salvador: Indian Notes and Monographs, v. 1, p. 165–220.

Mahood, G.A., 1980, Geological evolution of a Pleistocene rhyolitic center—Sierra La Primavera, Jalisco, Mexico: Journal of Volcanology and Geothermal Research, v. 8, p. 199–230.

Manville, V., White, J.D.L., Houghton, B.F., and Wilson, C.J.N., 1998, The saturation behavior of pumice and some sedimentological implications: Sedimentary Geology, v. 119, p. 5–16.

McPhie, J., Doyle, M., and Allen, R., 1993, Volcanic textures: A guide to the interpretation of textures in volcanic rocks: Tasmania, CODES Key Centre, 198 p.

Moore, J.G., Nakamura, K., and Alcaraz, A., 1966, The 1965 eruption of Taal volcano: Science, v. 151, p. 955–960.

Pettijohn, F.J., and Potter, P.E., 1964, Atlas and glossary of primary sedimentary structures: New York, Springer Verlag, 370 p.

Pichler, H., 1965, Acid hyaloclastites: Bulletin of Volcanologique, v. 28, p. 293–310.

Poppe, L.J., Paull, C.K., Newhall, C.G., Bradbury, J.P., and Ziagos, J., 1985, A geophysical and geological study of Laguna de Ayarza, a Guatemalan caldera lake: Journal of Volcanology and Geothermal Research, v. 25, p. 125–144.

Reynolds, M.A., and Best, J.G., 1976, Summary of the 1953–1957 eruption of Tuluman Volcano, Papua New Guinea, Volcanism in Australasia: New York, Elsevier Scientific Publishing, p. 287–296.

Richer, M., Mann, C.P., and Stix, J., 2004, Mafic magma injection triggers eruption at Ilopango Caldera, El Salvador, Central America, *in* Rose, W.I., et al., eds., Natural hazards in El Salvador: Boulder, Colorado, Geological Society of America Special Paper 375, p. 175–189 (this volume).

Rose, W.I., Conway, F.M., Pullinger, C.R., Deino, A., and McIntosh, W.C., 1999, An improved age framework for late Quaternary silicic eruptions in northern Central America: Bulletin of Volcanology, v. 61, p. 106–120.

Sakamoto, I., and Tanimoto, H., 1996, Rhyolitic pumice dredged from the bank located between Hachijyoshima and Aogashima Is: Journal of the Faculty of Marine Science and Technology, Tokai University, v. 41, p. 139–156.

Sheets, P.D., 1979, Environmental and cultural effects of the Ilopango eruption in Central America, *in* Sheets, P.D., and Grayson, D.K., eds., Volcanic activity and human ecology: New York, Academic Press, p. 525–564.

Sorem, R., 1982, Volcanic ash clusters: Tephra rafts and scavengers: Journal of Volcanology and Geothermal Research, v. 13, p. 63–71.

Sparks, R.S.J., 1978, The dynamics of bubble formation and growth in magmas: A review and analysis: Journal of Volcanology and Geothermal Research, v. 3, p. 1–37.

Stine, S., 1984, Late Holocene lake level fluctuations and island volcanism at Mono Lake, California, *in* Stine, S., Wood, S., Sieh, K., and Miller, C.D., eds., Holocene paleoclimateology and tephrochronology east and west of the Central Sierra Crest: Lee Vining, California, October 12–14, Friends of the Pleistocene, Guidebook, p. 21–49.

von Lichtan, I.J., White, J.D.L., and Manville, V., 2002, Giant rafted pumice around Lake Taupo from the 1.8 ka eruption [abs.]: Eos (Transactions, American Geophysical Union), Western Pacific Geophysical Meeting, v. 83, p. WP110.

Weber, H.S., and Wiesemann, G., 1978, Mapa geológico general de El Salvador: Instituto Geográfico Nacional de El Salvador, map San Salvador, scale 1:100,000, 1 sheet.

White, J.D.L., Manville, V., Wilson, C.J.N., Houghton, B.F., Riggs, N.R., and Ort, M., 2001, Settling and deposition of A.D. 181 Taupo pumice in lacustrine and associated environments, *in* White, J.D.L., and Riggs, N.R., eds.,

Volcaniclastic sedimentation in lacustrine settings: International Association of Sedimentologists Special Paper, v. 30, p. 141–150.

Whitham, A., and Sparks, R., 1986, Pumice: Bulletin of Volcanology, v. 48, p. 209–223.

Wiesemann, G., 1975, Remarks on the geologic structure of the Republic of El Salvador, Central America: Mitteilungen Aus Den Geologische—Palaeontologische Institute Der Universitäet Hamburg, v. 44, p. 557–574.

Williams, H., and Meyer-Abich, H., 1955, Volcanism in the southern part of El Salvador with particular reference to the collapse basins of Lakes Coatepeque and Ilopango: University of California Publications in the Geological Sciences, v. 32, p. 1–64.

Wilson, C.J.N., and Walker, G.P.L., 1985, The Taupo eruption, New Zealand. 1. General aspects: Royal Society of London Philosophical Transactions, v. 314, p. 199–226.

Wohletz, K.H., 1987, Chemical and textural surface features of pyroclastics from hydrovolcanic eruption sequences, *in* Marshall, J.R., ed., Clastic particles: Scanning electron microscopy and shape analyses of sedimentary and volcanic clasts: New York, Van Nostrand Reinhold Co., p. 79–97.

Wohletz, K.H., 1983, Mechanisms of hydrovolcanic pyroclast formation: Grain-size, scanning electron microscopy, and experimental studies: Journal of Volcanology and Geothermal Research, v. 17, p. 31–63.

MANUSCRIPT ACCEPTED BY THE SOCIETY JUNE 16, 2003

Geological Society of America
Special Paper 375
2004

Mafic magma injection triggers eruption at Ilopango Caldera, El Salvador, Central America

Mathieu Richer
Crystal P. Mann*
John Stix
Department of Earth and Planetary Sciences, McGill University, 3450 University Street, Montreal, Québec H3A 2A7, Canada

ABSTRACT

The most recent activity at Ilopango Caldera occurred in 1880 when a dacitic lava dome (68.4 wt% SiO_2) emerged from the center of the caldera lake. Dark inclusions of basaltic andesite (53.9 wt% SiO_2) represent 2 to 3 vol% of the dacite dome.

Petrographic textures and geochemistry indicate that the mafic inclusions were solidified after injection into partially crystallized dacite magma. During and before solidification, plagioclase crystals observed in the dacite actively transferred into the mafic magma, entraining dacitic melt and producing a reaction rim on crystals. Plagioclase crystals with sodic cores (An_{40-50}) and more calcic overgrowth rims (~An_{70}) derived from the dacitic melt are common inside mafic inclusions. Intimately mingled dacitic glass exhibits small but significant differences in chemistry relative to the host dacite, most notably a potassium enrichment, which we attribute to diffusion processes between the two magmas.

We propose that an injection of mafic magma at the base of partially crystallized dacite magma triggered the 1880 eruption. The overpressure necessary for eruption was generated by the addition of mafic magma, which triggered vesiculation of the rising dacite. The presence of mafic magma in both the A.D. 429 and 1880 eruptions may indicate that this process is of general importance in the eruptive history of Ilopango.

RESUMEN

La última erupción de la caldera Ilopango fue en 1880, cuando emergió un domo dacítico (68.4% SiO_2) del centro del lago. Inclusiones oscuras de andesitas basálticas (53.9% SiO_2) constituyen de 2 a 3% del volumen del domo dacítico.

Las texturas petrográficas y los análisis geoquímicos sugieren que las inclusiones máficas se solidificaron al ser inyectadas en un magma dacítico parcialmente cristalizado. Antes y durante la solidificación, cristales de la dacita fueron transferidos al magma máfico, entrando liquido dacítico y produciendo coronas de reacción sobre los cristales. Los cristales de plagioclasa originalmente derivados de la dacita presentan núcleos sódicos (An_{40-50}) y contornos cálcicos (~An_{70}), encontrándose comunmente dentro de las inclusiones máficas. También, se registró la presencia de vidrio dacítico dentro de las inclusiones, este vidrio presenta pequeñas, pero importantes diferencias en su composición química con respecto a la dacita. Particularmente, se observó un enriquecimiento en potasio, el cual atribuimos a procesos de difusión entre ambos magmas.

*Corresponding author: cpmann@eps.mcgill.ca

Richer, M., Mann, C.P., and Stix, J., 2004, Mafic magma injection triggers eruption at Ilopango Caldera, El Salvador, Central America, *in* Rose, W.I., Bommer, J.J., López, D.L., Carr, M.J., and Major, J.J., eds., Natural hazards in El Salvador: Boulder, Colorado, Geological Society of America Special Paper 375, p. 175–189.

Por lo tanto, proponemos que una inyección de magma máfico en la base de un magma dacítico parcialmente cristalizado originó la erupción de 1880. La presión necesaria para la erupción fue producida por la adición de un volumen de magma máfico, por el recalentamiento y degasificación de la dacita. La presencia de magma máfico tanto en la erupción de 429 DC., como en la erupción de 1880, indica que este proceso es de importancia general en la historia eruptiva de la caldera Ilopango.

Keywords: El Salvador, Ilopango Caldera, silicic dome, magma mixing, dusty plagioclase

INTRODUCTION

Mingling and mixing of magmas of different composition have been documented at many active volcanoes (e.g., Murphy et al., 1998; Eichelberger et al., 2000; Aguirre-Díaz, 2001). In the volcanic environment, evidence of magma interaction is indirect since the roots of these systems lie several kilometers below the surface. Mingling and mixing have been recognized on the basis of disequilibrium phenocryst assemblages in erupted magmas (Tepley et al., 1999), quenched inclusions of magmatic origin found in many silicic lava flows and domes (Bacon, 1986), banded pumices and compositional gradients in ignimbrite deposits (Bacon, 1983), and isotopic variations recorded by phenocrysts (Tepley et al., 2000). These observations indicate that many shallow silicic magma chambers are periodically reinjected by hotter, more mafic magma from a deeper source. Mafic magma replenishment is important because it extends the life of large silicic magmatic systems and provides a mechanism to trigger silicic eruptions (Sparks and Sigurdsson, 1977).

The mixing and mingling of mafic and silicic magmas has been a subject of interest for some time. It was suggested that a variety of petrological suites of intermediate composition could be the result of magma mixing (Eichelberger, 1978). In further studies, it was recognized that magmas with large compositional contrast do not mix readily because of thermal and mechanical constraints (Sparks and Marshall, 1986). Complete mixing of two magmas will occur only if the two magmas are still liquid at thermal equilibrium. When in contact with the comparatively cool silicic liquid, however, the mafic end member commonly crystallizes rapidly and becomes highly viscous before reaching thermal equilibrium. By contrast, the silicic magma can remain liquid at low temperatures, since shallow silicic magma bodies are commonly saturated with volatiles. A large temperature contrast between the two magmas accentuates the degree of undercooling of the mafic magma, promoting its solidification. In this case, the two magmas mingle, but complete mixing cannot occur. Hybridization between magmas of large compositional contrast will occur only when the proportion of mafic magma exceeds the proportion of silicic magma (Sparks and Marshall, 1986). In this regard, the plutonic record allows us to observe in situ the processes occurring in a silicic magma chamber during and after an injection of mafic magma (e.g., Wiebe, 1994).

The Cadillac Mountain Intrusive Complex, Mount Desert Island, Maine, offers exceptional exposures of a shallow gra-

nitic magma chamber that was periodically injected by basaltic magma from deeper sources (Wiebe, 1994). After injection, basaltic magmas ponded at the base of the magma chamber due to their high density relative to the host silicic magma. The large temperature contrast between the two magmas caused rapid crystallization of the mafic magma, resulting in a progressive viscosity increase with the degree of crystallization (Fig. 1).

If the mafic magma remains plastic for a period of time while cooling, the interface between mafic and silicic magma can be a very dynamic environment, both physically and chemically. Many processes may occur after mafic injection, depending on the thermal state and rheology of both magmas: turbulent convective stirring at the interface (Huppert et al., 1984), selective chemical exchange (Watson and Jurewizc, 1984), crystallization and fractionation of mafic magma (Wiebe, 1996), and hybridization (Wiebe, 1994). The viscous drag imparted by the

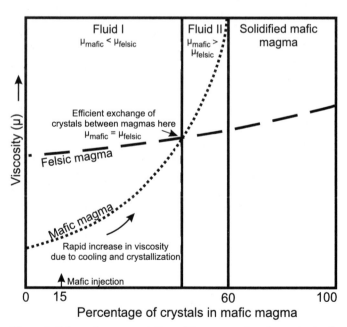

Figure 1. Schematic representation of the progressive change in rheology of mafic magma due to crystallization after injection into cooler dacite. The mafic magma was effectively solid when reaching thermal equilibrium with the dacite, preventing mixing between the two. A small volume of mafic magma and the presence of dissolved volatiles in the dacite promoted solidification. Modified from Sparks and Marshall (1986).

silicic magma during turbulent convection disrupts the interface, acting as an erosive process and generating small mafic enclaves (Wiebe, 1994). These physical processes occur when the viscosity of the mafic magma is still relatively low, possibly in the fluid I region defined by Sparks and Marshall (1986) (Fig. 1).

The higher structural levels of the Cadillac Mountain Intrusive Complex contain ubiquitous hybrid inclusions (Fig. 2A), which range in composition from 55 to >70 wt% SiO_2 (Wiebe et al., 2000). The origin of these inclusions is undoubtedly related to the mafic injection observed at deeper structural levels. These

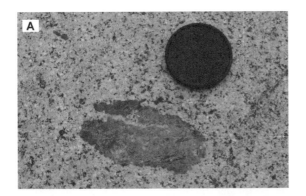

Figure 2. A: Blob of undercooled mafic magma in a host granite from the Cadillac Mountain Intrusive Complex, Mount Desert, Maine. The blob was formed at a deeper structural level at the interface between the injected mafic magma and the host granite and then dispersed throughout the granitic magma chamber by convection (Wiebe, 1994). Lens cap is 5 cm. Photo courtesy of Ben Kennedy. B: Mafic inclusion in the host dacite of the Islas Quemadas. The inclusion is angular, and the sharp contact with the host dacite is slightly irregular. Plagioclase is the main mineral phase in both the host dacite and the basaltic andesite inclusion. The host dacite is vesicular, especially at the contact with the inclusion. The inclusion also contains small vesicles. Scale in cm.

hybrid inclusions represent blobs of mafic magma formed at the mafic/felsic interface, which have fractionated to variable extents during crystallization of the mafic magma and have been entrained to higher levels by convection within the silicic host magma (Wiebe, 1994). If an eruption occurs, these inclusions may be ejected along with the silicic magma and will be found as part of the eruptive products. Therefore, understanding the origin of mafic inclusions in both plutonic and volcanic environments has general importance for the phenomenon of magma mixing and mingling in silicic systems.

Ilopango caldera in El Salvador is a silicic magmatic center that displays evidence for mingling of magmas. Studying Ilopango can help us understand processes related to magma interactions in shallow intrusive and volcanic environments, just as the Cadillac Mountain Intrusive Complex reveals similar relationships in deeper plutonic environments. Thus, the two silicic centers provide complementary information regarding the phenomenon of magma mingling.

Ilopango caldera is part of the Central American volcanic arc. It is bounded to the west-northwest by San Salvador volcano, and to the east-southeast by San Vicente volcano, both of which are andesitic in composition (Fig. 3). The caldera is filled by a lake that is rectangular in shape, with an irregular shoreline that is strongly embayed in many places. These features are the result of faulting in the region and the mass wasting of the caldera walls during and after collapse of the caldera.

The rocks related to the caldera are andesitic to dacitic in composition. Numerous lava domes are found in and around the lake, and voluminous pyroclastic deposits also are exposed inside and outside the caldera. Many of the intracaldera pyroclastic rocks display evidence for being erupted and deposited in a subaqueous environment (Mann et al., this volume). The pyroclastic sequence outside the caldera consists of pyroclastic falls and pyroclastic flows, commonly containing accretionary lapilli and other features indicative of magma-water interaction. The

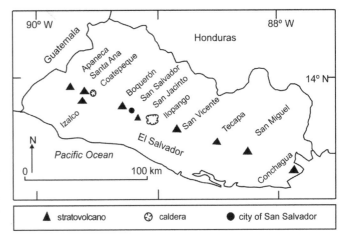

Figure 3. Location map of Ilopango caldera, El Salvador. Ilopango forms part of the active volcanic arc.

pyroclastic rocks comprise four dacitic units, which are informally named, from oldest to youngest, Tierra Blanca 4, Tierra Blanca 3, Tierra Blanca 2, and Tierra Blanca Joven. These units are less than 57,000 yr old (Rose et al., 1999), and Tierra Blanca Joven has been dated at A.D. 429 ± 20 (Dull et al., 2001). The most recent eruption of Ilopango occurred in 1880 when a dacitic lava dome called Islas Quemadas grew in the middle of the lake.

This paper discusses the origin of two types of inclusions found in the Islas Quemadas lava dome. One type of inclusion may have been formed in a similar manner to mafic inclusions found in the Cadillac Mountain Intrusive Complex, after mafic magma was injected at the base of a dacitic magma body (Fig. 2B). The other type could represent inclusions of a crystal "mush" zone of the dacitic magma chamber. The 1880 eruption was likely triggered by the injection of hot mafic magma. The interaction of silicic and mafic magmas, as documented in this paper, has implications for the reactivation of Ilopango caldera in the future.

CHRONOLOGY OF THE 1880 ERUPTION

Montessus De Ballore (1884) summarized earthquake activity and volcanic phenomena associated with the dome formation, which occurred from December 20, 1879, through March 19, 1880. The current summary of the eruption is largely based on short reviews of the eruption by Golombek and Carr (1978) and Newhall and Dzurisin (1988).

The eruption of the Islas Quemadas occurred in three distinct phases. The first phase comprised a period of anomalous seismic activity in the vicinity of Lake Ilopango. The first two earthquakes were felt on December 20 and 21, 1879. Seismicity increased in frequency on December 24–26, and a main shock was felt at midday on December 27. A strong earthquake on December 31, associated with plate convergence, was felt throughout El Salvador, and also in Nicaragua and Guatemala.

The second phase, the extrusion of lava from the center of the lake, is thought to coincide with a sudden rise in the lake level on January 6, 1880. By January 11, the lake had risen 1.2 m above its previous level, an addition of 0.13 km³ of water, approximately corresponding to the subaqueous volume of the dome. The only outlet of the lake, the Río Desagüe, was rapidly downcut by the outflowing water, the valley down the Río Jiboa was entirely flooded, and the town of Atuscatla destroyed. A total of ~1.2 km³ of water was lost from the lake due to the downcutting of the outlet (Newhall and Dzurisin, 1988). The water level stopped rising on January 11, and sulfur gas could be smelled in the vicinity of the lake. Incandescent rocks appeared at the surface of the lake on January 20, and after a strong explosion, a large column of black ash rose from the center. On January 23, an ash column could be seen from a great distance (Anonymous, 1880). The eruption waned until early February 1880. The growth of the dome was largely complete by that time, and small islands with a maximum diameter of 150 m and a height of 40–50 m rose above the surface of the lake (Mooser et al., 1958).

The third phase of the eruption occurred from the end of February until March 19, 1880. This period was associated with violent explosions of steam and ash, and many subsurface rumblings. During the eruption, Golombek and Carr (1978) noticed a correlation between cycles of dome growth, seismicity, and semidiurnal maxima and minima in earth tides. They concluded that the seismic and volcanic phenomena had been triggered by earth tides.

At the time of this writing, only two small islands are preserved. Both are less than 10 m in diameter, and only one rises ~10 m above the lake level. After the strong tectonic earthquake of February 13, 2001, major sections of the Islas Quemadas collapsed under water. Fresh rock is now exposed at the surface.

PETROGRAPHY

Host Dacite

The Islas Quemadas are mainly composed of a light gray dacite (68.4 wt% SiO_2) containing ~20 vol% phenocrysts of plagioclase (10–12%, up to 5 mm in length), hornblende (3–4%, up to 3–4 mm long), pyroxene (<1%, up to 1 mm long), Fe-Ti oxides (<2%, <1 mm), and accessory apatite. The groundmass is composed of microphenocrysts of plagioclase, hornblende, Fe-Ti oxides, abundant microlites, and vesicular glass (Table 1). The glass contains ~20 vol% vesicles and exhibits flow textures.

Most phenocrysts are clear and euhedral in shape, but ~10% of the plagioclase and hornblende phenocrysts have reaction textures. A few plagioclase crystals have a dusty texture consisting of a diffuse zone that mantles the interior of the crystal (Fig. 4A). The dusty zone comprises an interconnected network of fine melt needles, calcic and sodic plagioclase, and void space (Fig. 4B). It varies in thickness, entering into the preexisting crystal from the margin inward and sometimes extending throughout the core of the crystal. The dusty zone is invariably mantled by an overgrowth of clear plagioclase.

Hornblende crystals are usually brown, pleochroic, and euhedral. Most crystals lack any sign of reaction, but some show a reaction texture that consists of a rim of plagioclase, clinopyroxene, orthopyroxene, magnetite, ilmenite ± liquid (Figs. 4C, D). The thickness of the reaction rim varies from tens of microns to ~500 μm and is usually overgrown by euhedral hornblende (Fig. 4C). In a few cases, the overgrowth on the hornblende is absent (Fig. 4D).

The dacite contains two types of inclusions; one type is a crystal-rich inclusion referred to as salt and pepper inclusions, and the other is a fine-grained, porphyritic mafic inclusion. The petrographic textures of these inclusions reveal useful information concerning the processes occurring in the magmatic system below Ilopango caldera.

Salt and Pepper Inclusions

Crystal-rich inclusions, consisting of >80 vol% interlocking crystals, represent 1–2 vol% of the dome. They are typically fine

TABLE 1. ELECTRON MICROPROBE ANALYSES OF GLASSES

Glass host:	Dacite	Mafic inclusion	Mafic inclusion	S and P inclusion	S and P inclusion	S and P inclusion
Analyses*:	5	19	3	20	3	3
Figure†:			7B		5B	5A
wt%						
SiO_2	75.19 ± 0.40	74.05 ± 0.65	74.07 ± 0.53	75.42 ± 0.54	75.98 ± 0.51	74.76 ± 0.46
TiO_2	0.23 ± 0.01	0.26 ± 0.04	0.22 ± 0.05	0.22 ± 0.03	0.21 ± 0.00	0.22 ± 0.03
Al_2O_3	11.97 ± 0.10	12.23 ± 0.35	12.16 ± 0.30	12.17 ± 0.26	12.15 ± 0.26	12.23 ± 0.48
FeO	0.89 ± 0.06	1.18 ± 0.23	1.26 ± 0.40	1.16 ± 0.15	1.08 ± 0.13	1.32 ± 0.34
MnO	0.05 ± 0.03	0.04 ± 0.02	0.04 ± 0.02	0.05 ± 0.02	0.06 ± 0.02	0.06 ± 0.03
MgO	0.04 ± 0.01	0.12 ± 0.07	0.19 ± 0.15	0.16 ± 0.04	0.16 ± 0.06	0.17 ± 0.03
CaO	0.91 ± 0.07	0.75 ± 0.22	0.72 ± 0.17	1.00 ± 0.15	1.00 ± 0.05	0.95 ± 0.16
Na_2O	4.13 ± 0.07	3.89 ± 0.19	3.78 ± 0.30	4.08 ± 0.13	4.20 ± 0.16	4.02 ± 0.11
K_2O	3.48 ± 0.13	4.78 ± 0.19	4.83 ± 0.10	3.45 ± 0.09	3.31 ± 0.09	3.44 ± 0.03
P_2O_5	0.03 ± 0.01	0.05 ± 0.03	0.04 ± 0.02	0.02 ± 0.01	0.03 ± 0.01	0.02 ± 0.01
Total	96.91 ± 0.45	97.44 ± 0.59	97.29 ± 0.98	97.74 ± 0.56	98.17 ± 0.20	97.19 ± 0.63
Cl	0.12 ± 0.02	0.19 ± 0.05	0.20 ± 0.06	0.14 ± 0.02	0.13 ± 0.03	0.15 ± 0.03
BaO	0.17 ± 0.02	0.18 ± 0.03	0.18 ± 0.04	0.17 ± 0.03	0.16 ± 0.06	0.17 ± 0.03

Note: S and P inclusion—salt and pepper inclusion. Electron microprobe analyses of glasses were performed with a JEOL JXA-8900R at McGill University, Canada. Analyses were performed with an accelerating voltage of 15 kV, beam current of 30 nA and a beam diameter of 10 μm.

*Number of analyses. Analyses are reported as mean ±1σ.
†Corresponds to figure number in paper.

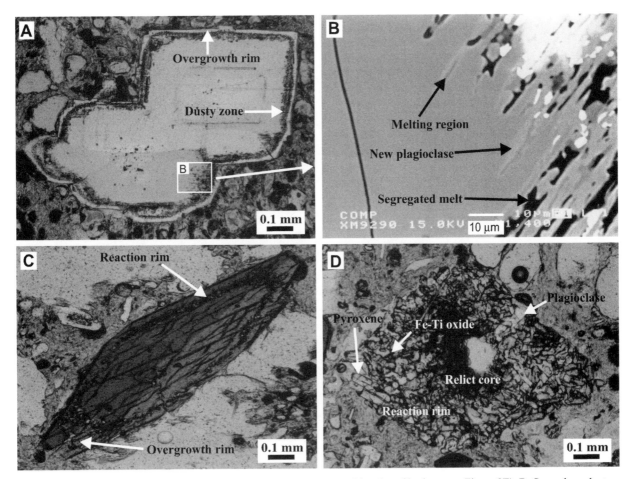

Figure 4. Host dacite phenocryst textures. A: Dusty plagioclase (compositional profile shown on Figure 8E). B: Secondary electron image showing detail of the dusty zone consisting of small melt fronts that penetrate the crystal from the margin inward. Black areas represent void space, probably due to melt segregation. C: Hornblende crystal with a thin reaction rim and a euhedral overgrowth rim. D: Relict core of hornblende with a thick reaction rim. No overgrowth of hornblende is observed.

grained (<0.1 to 1 mm) and usually have sharp contacts with the host dacite. Individual inclusions vary in size between a few millimeters to a few centimeters in diameter. There is no relationship between the size of the inclusions and their grain size.

The microphenocryst assemblage consists of plagioclase (70–75 vol%), hornblende (<2 vol% to 10 vol%), pyroxene (<1 vol% to 10 vol%) and Fe-Ti oxides (1–3 vol%). The inclusions contain 2–5 vol% interstitial glass and 10–12 vol% subrounded vesicles. The microphenocrysts are euhedral and do not show any reaction textures. Plagioclase microphenocrysts have a swallowtail texture. A range of texturally different inclusions can be found in the samples collected. Most inclusions have hornblende as the dominant mafic phase, with pyroxene as a minor component (Fig. 5A). The grain size of these inclusions varies on average from ~0.1 mm to ~0.7 mm. The interlocking textures reveal an order of crystallization that consists of plagioclase + Fe-Ti oxides → plagioclase + hornblende → Fe-Ti oxides. One inclusion contains pyroxene as the dominant mafic phase, and hornblende is minor (Fig. 5B). The crystallization sequence of this inclusion consists of Fe-Ti oxides + apatite → plagioclase → plagioclase + pyroxene → Fe-Ti oxides + hornblende.

A number of texturally different crystals are present inside some inclusions. These crystals are larger in size (>1 mm) and have reaction textures. Dusty plagioclase crystals and relict hornblendes with rims of plagioclase, clinopyroxene, orthopyroxene, and Fe-Ti oxides are found in the microphenocryst groundmass (Figs. 5C, D). Olivine crystals with orthopyroxene rims also were found in one inclusion. The origin and implications of these crystals are discussed below.

Mafic Inclusions

Dark gray, porphyritic, mafic inclusions form a minor (2–3 vol%) but ubiquitous component of the Islas Quemadas dacite. Inclusions range in size from <0.1 to 10 cm and are typically ellipsoidal to subangular with crenulated margins that are convex toward the host dacite (Fig. 2B). The grain size of the groundmass varies according to the size of the inclusions, with

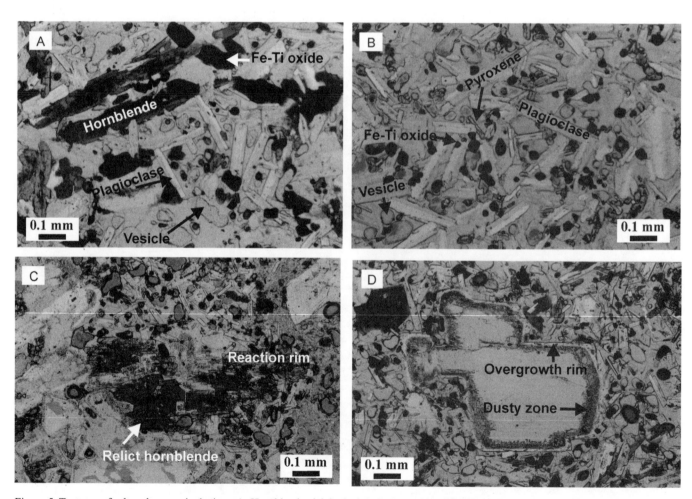

Figure 5. Textures of salt and pepper inclusions. A: Hornblende-rich inclusion. B: Pyroxene-rich inclusion. A and B glass compositions shown in Table 1. C: Hornblende xenocryst with reaction rim comprising pyroxene, plagioclase, and Fe-Ti oxides. D: Plagioclase xenocryst with a dusty texture. The core is more sodic ($An_{40–46}$) while the rim is more calcic ($An_{60–70}$). Compositional profile is shown on Figure 8F.

small inclusions having a very fine-grained groundmass and larger inclusions coarser groundmasses. Chilled margins around inclusions were not observed. The inclusions contain 2–3 vol% irregularly shaped vesicles up to ~1 mm in size.

The phenocryst assemblage of the mafic inclusions consists of plagioclase (7–10 vol%, up to 6 mm long), olivine (2–3 vol%, up to 3 mm long), pyroxene (1–2 vol%, up to 1 mm long), and Fe-Ti oxides (<1 vol%, <0.5 mm). The groundmass is completely crystallized and consists of plagioclase (60 vol%), pyroxene (30 vol%), and Fe-Ti oxides (10 vol%).

In the field, we were able to collect sufficient material from one of the inclusions to analyze its bulk rock chemistry. The analysis by X-ray fluorescence is shown in Table 2 and reveals a basaltic andesite composition for this inclusion. The silica content is 53.87 wt% on an anhydrous basis, and the MgO content is 5.44 wt%.

While some plagioclase crystals are calcic and euhedral (Fig. 6A), a number of plagioclase and hornblende crystals have disequilibrium textures inside the mafic inclusions. Dusty plagioclases and hornblendes with reaction rims consisting of plagioclase, clinopyroxene, orthopyroxene, and Fe-Ti oxides are commonly found inside inclusions (Figs. 6B, C, D). In many cases, the reaction extends into the core of the crystal. Dusty plagioclases are invariably overgrown by clear plagioclase. In some cases only a pseudomorph of hornblende is preserved in the mafic inclusions. In other cases where hornblende is partially preserved, no hornblende overgrowth is observed, in contrast to hornblende in the host dacite.

Phenocrysts of plagioclase and hornblende from the host dacite frequently penetrate mafic inclusions, leaving trails of clear, vesicular glass behind (Figs. 6B, 7A). Figure 7A shows a plagioclase crystal breaking through the inclusion interface. A dusty reaction rim has developed on the crystal's surface, the thickest part pointing toward the inside of the inclusion. The rim is absent at the back of the crystal, and a small trail of dacite melt follows the crystal inside the inclusion. Within mafic inclusions, similar trails of clear, vesicular glass are observed. The trails of melt are sometimes interrupted by mafic groundmass, isolating pockets of dacitic glass and crystals (Fig. 7B). Small acicular crystals of hornblende and plagioclase are sometimes observed inside these glass pockets (Fig. 7C). Unlike the larger crystals, these acicular crystals have no reaction rims. Figure 7C shows a pocket of melt from which acicular hornblende and plagioclase crystals have been concentrated at one end of the pocket by a small-scale gas filter pressing mechanism (Sisson and Bacon, 1999). In summary, the mafic inclusions preserve abundant evidence showing interaction and exchange of material between the host dacite magma and the basaltic andesite magma.

Plagioclase Chemistry

Plagioclase crystals from the different components of the Islas Quemadas dacite have been analyzed by electron micro-probe, and their compositions are summarized on a plot of calcium versus sodium in Figure 8A. The large, unzoned plagioclase crystals mostly found inside mafic inclusions plot in a cluster at the highest calcium contents (An_{90-92}). A representative compositional profile of such a crystal, which is unzoned, is shown by the open diamonds in Figure 8B. The solid circles represent the composition of a second similar crystal that was

TABLE 2. WHOLE ROCK GEOCHEMISTRY

Unit sample	Host dacite CM196-01	Mafic inclusion CM196-01i
wt%		
SiO_2	67.74	53.39
TiO_2	0.42	0.75
Al_2O_3	15.47	18.15
Fe_2O_3	4.02	9.18
MnO	0.12	0.17
MgO	1.41	5.39
CaO	3.83	8.84
Na_2O	4.26	3.05
K_2O	2.08	0.91
P_2O_5	0.14	0.15
LOI	0.33	−0.01
Total	99.82	99.97
ppm		
BaO	1095	526
Ce	27	0
Co	0	30
Cr	0	0
La	0	11
Ni	0	14
Sc	0	22
V	62	208
Ga	15	15.4
Nb	5.4	3.5
Pb	2.8	0
Rb	46.1	20.9
Sr	314.4	417.4
Th	3.2	1.3
U	1.8	1.2
Y	20.14	21.0
Zr	167.3	89.6

Note: X-ray fluorescence whole rock analyses were performed with a Philips PW2440 4 kW automated XRF spectrometer system at the Geochemical Laboratories, McGill University, Canada, by Tariq Ahmedali and Glenna Keating. Major elements (Si, Ti, Al, Fe, Mn, Ca, Na, K, and P) and trace elements (Ba, Ce, Co, Cr, La, Ni, Sc, V, Ga, Nb, Pb, Rb, Sr, Th, U, Y, Zr) were determined using 32 mm diameter fused beads composed of 5 parts lithium tetraborate and 1 part sample.

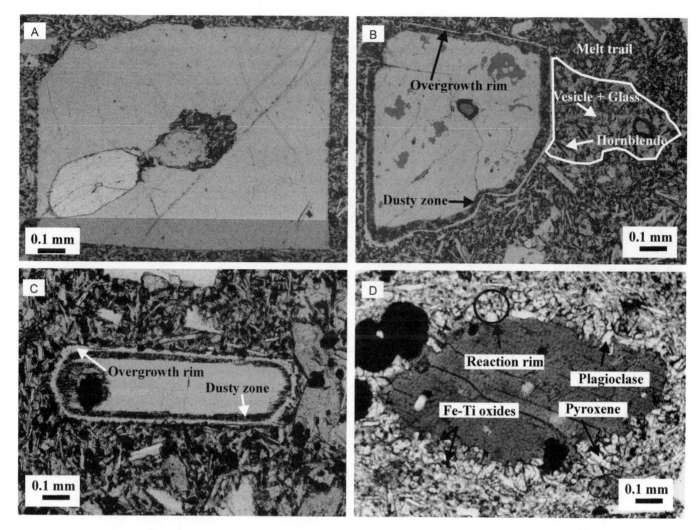

Figure 6. Textures of phenocrysts in mafic inclusions. A: Euhedral calcic plagioclase crystal (compositional profile shown on Figure 8B—open diamonds). B: Dusty plagioclase with sodic core (An_{36-44}) and calcic rim (An_{66-73}). The outlined melt trail of dacitic glass contains acicular hornblende, vesicles, and glass, and was incorporated during crystal transfer from the host dacite to the mafic inclusion. C: Plagioclase from dacitic host with a well-developed dusty texture. The dusty zone is irregular and diffuse inward at the ends of the crystal, while constant and sharp along the width. A compositional profile along the length of the crystal is shown on Figure 8C. D: Hornblende crystal with a coarse reaction rim comprising crystals of plagioclase, pyroxene, and Fe-Ti oxides (the thickness of the rim is outlined by the circle).

found in the host dacite. Two analyses of the core of this crystal give the same calcium content as the first crystal inside the mafic inclusion (An_{90}). However, two analyses of the rim of the crystal, which was in contact with the dacite, give slightly lower calcium contents (An_{78-82}, also shown on Fig. 8B). The horizontal bars on Figure 8A represent the composition of dusty plagioclase crystals found in mafic inclusions. The cores of these crystals are quite sodic (An_{40-50}) while their overgrowth rims are significantly more calcic (An_{70}). A representative compositional profile of this type of crystal is shown in Figure 8C.

Euhedral host dacite plagioclase crystals are represented by open circles and plot at the sodic end on Figure 8A. A representative compositional profile of such crystals is shown in Figure 8D, oscillating in composition between An_{40} and An_{60}. Figure 8E

shows the compositional profile of a dusty plagioclase crystal found in the host dacite. The rim overgrowing the dusty zone has similar composition to the core of the crystal.

The open squares in Figure 8A represent the composition of plagioclase microphenocrysts found in salt and pepper inclusions. The cores of these plagioclase microphenocrysts have intermediate compositions (An_{60-75}) while their rims are more sodic (~An_{46}). A compositional traverse of these microphenocrysts was not performed. The composition of a dusty plagioclase crystal found inside a salt and pepper inclusion is represented by the solid triangles in Figure 8A. The core of the crystal is sodic (An_{40-50}), whereas the rim overgrowing the dusty zone is substantially more calcic (An_{70}) (Fig. 8F), similar to the dusty plagioclase found inside mafic inclusions (Fig. 8C).

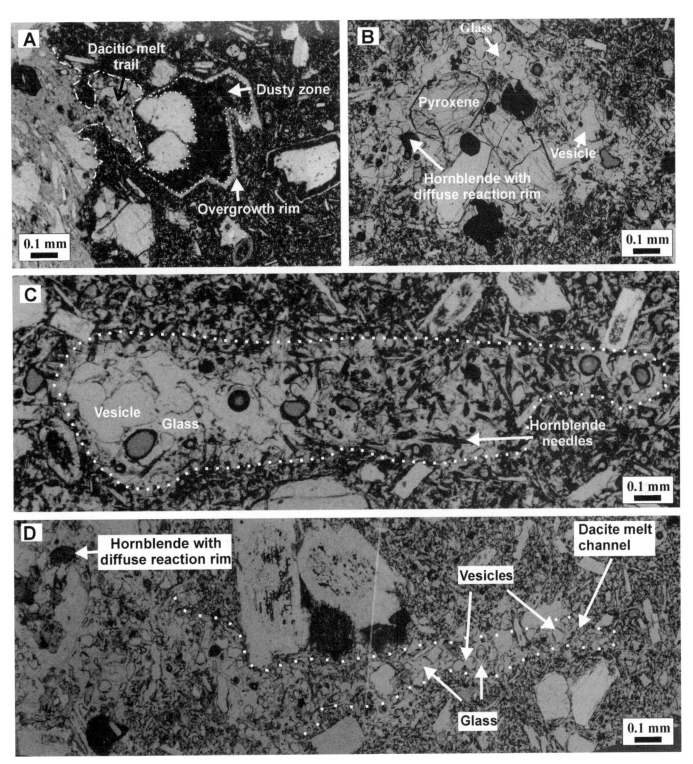

Figure 7. Mingling textures in mafic inclusions. A: Dacitic plagioclase being transferred into a mafic inclusion. The dotted line represents the dusty reaction rim that has developed on the crystal surface. The variation in thickness of the dusty rim corresponds to the amount of time spent in contact with the mafic inclusion. Note the trail of dacitic melt and the absence of the dusty rim at the back of the crystal. B: Isolated pocket of dacitic glass inside a mafic inclusion. The glass composition is listed in Table 1. C: Dacitic glass pocket from which acicular hornblende and plagioclase have crystallized. The residual melt and gas have been squeezed out of the crystal-rich region by a small-scale gas filter-pressing mechanism (Sisson and Bacon, 1999). D: Thin, sinuous channel of dacitic glass inside a mafic inclusion.

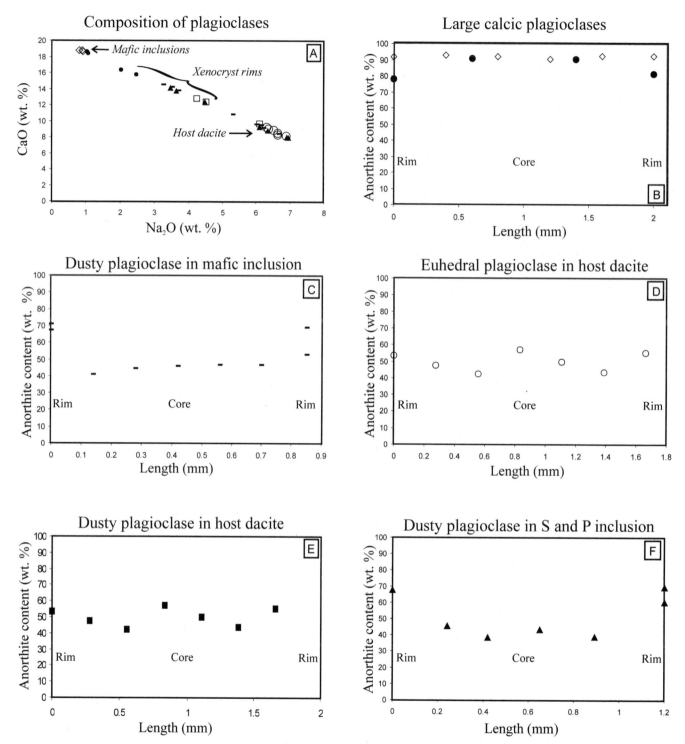

Figure 8. Plagioclase chemistry. A: Composition of different plagioclases on a CaO vs. Na₂O wt% plot. Open diamonds represent a large calcic plagioclase inside a mafic inclusion (profile on B), solid circles represent a large calcic plagioclase inside the host dacite (profile on B), horizontal bars represent a dusty plagioclase inside a mafic inclusion (profile on C), open circles represent a host dacite plagioclase (profile on D), open squares represent plagioclase microphenocrysts inside salt and pepper inclusions, profile on E as closed squares, and solid triangles represent a dusty plagioclase crystal inside a salt and pepper inclusion (profile on F). B: Compositional profiles of two large, euhedral, calcic plagioclase crystals. The open diamonds represent a plagioclase inside a mafic inclusion, and the solid circles represent the same type of plagioclase inside the host dacite. C: Compositional profile of a dusty plagioclase inside a mafic inclusion. D: Compositional profile of a euhedral plagioclase in host dacite. E: Compositional profile of a dusty plagioclase crystal in the host dacite. F: Compositional profile of a dusty plagioclase inside a salt and pepper inclusion.

Glass Chemistry

Glass chemistry was determined by the electron microprobe (Table 1; Fig. 9). Several observations can be made when comparing the glass chemistry of the three lithologies (i.e., host dacite, mingled glass inside mafic inclusions, and salt and pepper inclusions). First, the chemistry of all analyzed glasses is very similar, and only small differences are observed (Fig. 9). All glasses are rhyolitic in composition, with silica contents between 75 and 77 wt% for glass in mafic inclusions, between 76 and 78 wt% for salt and pepper inclusions, and between 77 and 78 wt% for the host dacite. There is considerable scatter for the sodium and calcium contents of the glass, and no distinction between the different lithologies can be made using these elements. Chlorine contents tend to be higher for glass inside mafic inclusions relative to the host dacite and salt and pepper inclusions, but again considerable scatter is observed, and chlorine values as low as in the host dacite are also found inside mafic inclusions. Magnesium and iron show a positive correlation, with the highest values found in the mingled glass inside the mafic inclusions.

Simple mixing between two end members produces straight lines on element-element plots (Langmuir et al., 1978). Mixing calculations between the basaltic andesite end member and the host dacite residual melt were performed using the following equation:

$$(x) \text{ (oxide wt\% in basaltic andesite)} + (1 - x) \text{ (oxide wt\% in host dacite residual melt)} = \text{oxide wt\% in mingled glass inside mafic inclusion,}$$

where x is the proportion of basaltic andesite that has mixed. The equation was applied to silica, aluminum, magnesium, and iron contents of mingled glass inside a mafic inclusion, and yielded 12%, 15%, 4%, and 1.6% of the basaltic andesite component, respectively.

The only characteristic that clearly distinguishes the mingled glass inside mafic inclusions from the glass of host dacite and salt and pepper inclusions is the higher potassium content (Fig. 9A). An inverse relationship exists between the potassium content of the mingled glass inside mafic inclusions and the distance from the analyzed point to the mafic groundmass (Fig. 9B).

DISCUSSION

The Origin of Dusty Plagioclase

Plagioclase forms a solid solution between its sodic and calcic end members. During dissolution under equilibrium conditions, the liquid being formed from a crystal is more sodic than the residual crystal, with their respective compositions being controlled by temperature (Bowen, 1913). The dissolution kinetics of plagioclase is best explained by a solution/re-precipitation mechanism, in which the residual plagioclase constantly readjusts its composition in order to accommodate an increase in temperature (Johannes et al., 1994). The liquid forms at the crystal interface or along crystal defects, from which re-precipitates a more calcic plagioclase that is closer in composition to equilibrium conditions. This process repeats itself until the reaction stops. Melting usually propagates inward in a branch-like manner, following weakness planes such as cleavages (Tsuchiyama and Takahashi, 1983). During melting, melt domains may segregate, leaving needle-like voids. The dissolution of the crystal from the surface inward results in the dusty texture. This texture is commonly encountered in volcanic arc systems because crystals of different composition coexist in a magmatic environment where conditions fluctuate (e.g., Kayamoto, 1992). The dusty texture is associated with the phenomenon of magma mixing because preexisting sodic plagioclases are suddenly exposed to

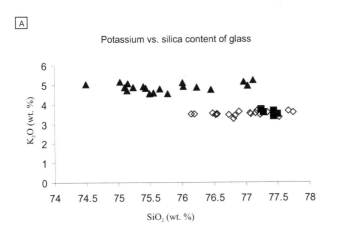

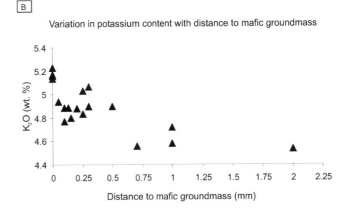

Figure 9. Glass chemistry. A: Potassium vs. silica content of the various glasses. Solid triangles represent mingled dacite glass inside mafic inclusions, open diamonds represent glass inside salt and pepper inclusions, and solid squares represent the host dacite glass. B: Variation in potassium content of the mingled dacite glass inside mafic inclusions with the shortest distance between the location of the analysis and the mafic groundmass. This trend is consistent with diffusion of potassium from the basaltic andesite melt to the dacite melt.

hotter, more mafic conditions, thereby reacting and dissolving to produce dusty rims (Hibbard, 1981; Tsuchiyama, 1985).

Islas Quemadas Host Dacite

The occurrence of dusty plagioclase in the host dacite indicates to us that a small fraction of dacite phenocrysts have undergone partial dissolution during a heating event. Hornblende phenocrysts exhibit textures that could also be explained by a heating event. The dehydration reaction of hornblende, producing an anhydrous assemblage of plagioclase, pyroxenes, and Fe-Ti oxides, may occur in response to depressurization and/or temperature increase (Rutherford and Hill, 1993). It was shown by Rutherford and Hill (1993) that the reaction rims on hornblende phenocrysts of Mount St. Helens dacite were produced during depressurization of the magma; these rims were not overgrown by additional hornblende. In the present study, by contrast, reaction rims are frequently overgrown by more hornblende (Fig. 4C). This implies that conditions must have re-equilibrated in the stability field of hornblende for a period of time after the crystals were heated. Since only ~10% of plagioclase and hornblende phenocrysts show evidence for reaction, heating was likely nonuniform. Hornblende is unstable above ~880 °C (Barclay et al., 1998), thus the temperature must have been in excess of ~880 °C close to the heat source, with subsequent re-equilibration below that temperature. Our analysis suggests that the source of heat is linked to the origin of mafic inclusions.

Mafic Inclusions

Within the mafic inclusions of Islas Quemadas, the grain size of the groundmass varies according to the size of the inclusion. The contact of the inclusions with the host dacite is crenulated and convex toward the host. These features indicate that the mafic inclusions are quenched magmatic products. The mingling textures that we have documented clearly demonstrate that mafic inclusions were once molten in the dacite magma and have solidified progressively in response to undercooling.

The progressive increase in viscosity of mafic inclusions during cooling is best demonstrated by the texture shown on Figure 7A, where a plagioclase crystal has been arrested in the process of being transferred from the dacite to a mafic inclusion. The dusty rim texture, indicative of partial dissolution, is instructive in this case. Tsuchiyama (1985) demonstrated that the thickness of the reaction rim is proportional to the duration of heating. The head of the crystal, which was first in contact with the hot mafic inclusion, has the thickest reaction rim. The rim gradually decreases in thickness on the sides of the crystal and is absent at its tail. These observations indicate that as the blob of mafic liquid became progressively more viscous, the crystal transfer process gradually slowed and eventually stopped, never completely crossing the inclusion interface. Because the crystal was not fully incorporated into the inclusion, its tail was not in direct contact with the mafic magma and did not have an opportunity to react and produce a dusty rim.

We also observe that the crystal transfer created a small trail of dacitic melt inside the inclusion before its complete solidification. The origin of channels and pockets of dacitic melt inside mafic inclusions can be explained by the crystal transfer process. Soon after it was incorporated in the dacite, the mafic magma had a relatively low viscosity, possibly in the fluid I region defined by Sparks and Marshall (1986) (Fig. 1). The low viscosity allowed crystals from the dacite to penetrate more easily into mafic magma, entraining dacitic melt at the same time. This process seems to be dominant and is the most efficient mechanism by which dacite and basaltic andesite have mingled together.

Upon transfer into a hotter, more mafic environment, the dacite plagioclase partially dissolved, producing dusty rims. Some crystals melted more than others, having rounded edges as well as a dusty rim. The sodic composition of the plagioclase cores is evidence that the dusty plagioclase originally grew in the dacite magma. The overgrowth rim is more calcic and at equilibrium with a hotter and more mafic magma, consistent with further crystallization of the plagioclase inside the mafic inclusions.

Hornblende crystals from the host dacite are sometimes found inside mafic inclusions. These hornblendes have reacted to produce rims of plagioclase, pyroxene, and Fe-Ti oxides. This dehydration reaction resulted from heating. By contrast with hornblende in the host dacite, hornblende within the mafic inclusions never overgrows the reaction rim of these crystals. We thus conclude that the composition of the mafic magma was not appropriate for hornblende to crystallize from this melt. The origin of the small acicular hornblende crystals that are associated with dacitic melt inside mafic inclusions results from the cooling of mafic inclusions. In this case, conditions re-equilibrated in the hornblende stability field, and small hornblendes crystallized from the dacitic melt.

Our petrographic and chemical analyses show that the clear vesicular glass found in association with dacite xenocrysts inside mafic inclusions is indeed dacitic melt that has mingled with mafic liquid. Mixing calculations between the basaltic andesite and host dacite residual melt end members cannot successfully explain the small variations in composition of the mingled glass. For example, the proportion of mafic magma needed to produce the observed changes in composition varies from 1.6 to 15% for an analysis of the mingled glass. Clearly, simple mixing is not the only process that has modified the glass composition. Figure 9B shows an inverse correlation between the potassium content of dacitic glass in mafic inclusions and the shortest distance between the location of the analysis and the mafic groundmass. In this case, a process of diffusion better explains the compositional variation. Uphill diffusion of potassium (i.e., from mafic to felsic magma) may be a relatively rapid process in magmas (Watson and Jurewizc, 1984). By contrast, the diffusivity of silica and aluminum is too slow in viscous magmas, and diffusion alone cannot explain their slight variations. Other processes that could cause some of the chemical variation are (a) partial melting of dacite xenocrysts inside mafic inclusions, and (b) subsequent crystallization of hornblende and plagioclase, which varies on a local scale. We conclude that the two magmas have mostly mingled together, but mixing and hybridization were limited by the solidification of the mafic magma. The changes in chemistry are small and could be produced by several processes on a local scale.

Salt and Pepper Inclusions

The fine-grained crystal-rich inclusions are similar in mineralogy to the host dacite, consisting mainly of plagioclase and hornblende. The interstitial glass in these inclusions is also similar in composition to that of the host dacite (Fig. 9A). Unlike the mafic inclusions, there is no correlation between the size of inclusions and the grain size of crystals. The contact between inclusions and host dacite is relatively sharp. These observations suggest that the inclusions crystallized prior to their incorporation in the host dacite and behaved as xenoliths during the eruption. These are interpreted to be cognate plutonic inclusions. The exact origin of these inclusions is speculative since they could have crystallized along the wall or roof of a more liquid body, or they could have been isolated from the main dacite body and incorporated during the ascent of the dacite along the conduit. The differences in mineralogy of these inclusions (i.e., hornblende-rich vs. pyroxene-rich) may reflect crystallization at different depth, temperature, and/or water pressure. In any case, they appear to represent portions of the dacite magma that were partially crystallized when the eruption occurred. The presence of glass in the inclusions indicates that they were transported rapidly from their place of crystallization and quenched by the eruptive process, with vesiculation of the glass resulting from decompression during the eruption. The fine-grained microphenocrysts representing the bulk of the inclusions have no reaction textures, implying that the inclusions themselves did not experience reheating by the mafic magma during the eruption. However, a small fraction of texturally different crystals also are found inside the inclusions. Dusty plagioclase with sodic cores and more calcic rims, hornblende rimmed by plagioclase, pyroxenes and Fe-Ti oxides, and olivine rimmed by orthopyroxene are observed inside these inclusions. As discussed previously, reheating of such crystals produces the reaction textures of plagioclase and hornblende. Since the inclusions themselves were not reheated during the 1880 eruption, these preexisting crystals must have been reheated during a previous mafic magma injection event. The last explosive eruption, the Tierra Blanca Joven dated at A.D. 429 ± 20 (Dull et al., 2001), contains banded pumices (Hart, 1981). This observation confirms that mafic and silicic magmas were concurrently active in the Ilopango magmatic system in previous eruptions.

Eruption Model

Our analysis indicates that mafic inclusions were molten when incorporated into the dacite magma and subsequently cooled rapidly. Based on the petrographic textures of the Islas Quemadas samples and on the magmatic textures of the Cadillac Mountain Intrusive Complex (Wiebe et al., 2000), the following sequence of events leading to the 1880 eruption is proposed (Fig. 10):

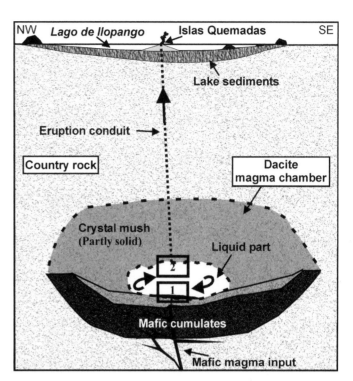

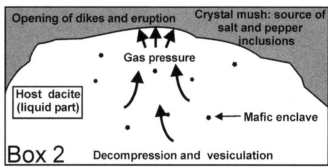

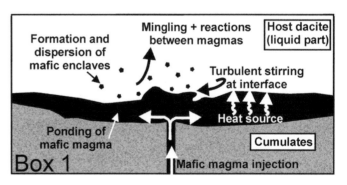

Figure 10. Conceptual model of mafic magma injection and eruption at Ilopango caldera. The upper diagram illustrates the entire magmatic system, while the two lower panels (Boxes 1, 2) display details of magmatic processes. Box 1: Injection of mafic magma and formation of mafic enclaves at the interface between mafic magma and host dacite (Wiebe, 1994). Box 2: Generation of overpressure and eruption following mafic magma injection. The size of the magma chamber is not known.

1. Mafic magma was injected into the base of a partly crystallized dacite body underlying Ilopango caldera. The mafic magma ponded and began to crystallize because of the large density and temperature contrast between the two magmas. The heat supplied by the mafic magma induced turbulent convection in the dacite. Host dacite phenocrysts that came into close contact with the hot mafic magma reacted strongly and were wholly or partly dissolved.

2. Disruption of the mafic/felsic interface occurred due to turbulent convection of the dacite. Individual blobs of mafic magma were formed during this process and were entrained into the dacite by convection. Mingling between dacite and basaltic andesite was controlled mainly by crystal transfer and was most efficient during this early stage, since the mafic magma was still fluid and hot. The dacite phenocrysts that were transferred mechanically into mafic magma produced the observed reaction textures.

3. The mafic injection could have produced the overpressure necessary for the eruption to occur because (a) the injection increased the volume of the magma body, (b) decompression of the dacite during convection induced rapid vesiculation and gas pressure, and/or (c) the crystallization of the mafic magma may itself have liberated volatiles.

4. The overpressure was sufficient to create dikes, allowing magma to move upward and eventually erupt into the lake. The upward flow of the magma was vigorous enough to entrain mafic material and small pieces of the partly crystallized portions of the dacite magma chamber.

The Islas Quemadas were formed in a three-phase eruption that waned after three months of activity. This implies that a single injection of mafic magma, or multiple injections at short time intervals followed by an undetermined period of stewing, preceded the eruption. A discrete injection of a small volume of mafic magma limited the eruption. By contrast, when fresh magma repeatedly replenishes a magma chamber during eruption, magma production rates increase with time, as seen by the Soufrière Hills eruption, Montserrat (Sparks et al., 1998).

CONCLUSIONS

The dacitic lava dome that was formed in 1880 at Ilopango caldera, El Salvador, is likely the result of mafic magma replenishment of the dacitic magmatic system. The Islas Quemadas dacite carries two types of inclusions: basaltic andesite inclusions (quenched mafic inclusions) and crystal-rich dacitic inclusions (cognate plutonic inclusions). The quenched mafic inclusions represent undercooled blobs of magma that were formed after an injection of mafic magma at the base of a dacite magma body. The cognate plutonic inclusions represent partly crystallized dacite. The eruption was likely triggered by injection of hot mafic magma.

Ilopango caldera overlies a crystallizing dacitic magma chamber that experiences periodic replenishment by mafic magma from deeper sources. The interaction of mafic magma and dacite has occurred in the past, as revealed by the crystal-rich inclusions in the Islas Quemadas dacite. The banded

pumices found in the deposits of the A.D. 429 eruption further suggest that mafic magma injection has occurred in the past. Ilopango caldera overlies a complex system of northeast and north-northeast–trending strike-slip faults (Weber et al., 1978). The highly fractured basement and the caldera faults themselves may provide conduits for mafic magmas. Magmas of different composition are able to pond, differentiate, and interact with one another in this setting, sometimes triggering eruptions.

Four large explosive eruptive events have occurred at Ilopango in the last 57,000 yr. Between cycles of explosive eruptions, lava domes are commonly emplaced. The last explosive eruption occurred ~1600 yr ago, and the caldera currently appears to be in a period of more effusive activity. Our data strongly indicate that mafic magma periodically replenishes the magma system, which may trigger silicic eruptions at Ilopango. Such replenishment in the future is likely, given the life span of the caldera, the high frequency of eruptions, and the probability of an active magma body under Ilopango. The next eruption at Ilopango caldera is likely to be similar in nature to the one that formed the Islas Quemadas. Such an eruption could be quite hazardous, since the presence of both mafic magma and the lake serve to enhance explosivity.

ACKNOWLEDGMENTS

We thank Fred Thill, Jeannette Monterrosa, and Ernesto Freund of the Fundación Amigos del Lago de Ilopango for their continued assistance during the course of this project. Carlos Pullinger of the Servicio Geológico, Servicio Nacional de Estudios Territoriales, El Salvador, provided much help, advice, and insight into the geology of Ilopango and El Salvador. We are indebted to Natividad Rauda and the people of the village of San Agustín for their generosity and hospitality. Two anonymous reviewers provided comments that improved the paper, and we are grateful to them. This research was supported financially by grants to M.R. from GEOTOP (UQAM-McGill), the Youth Activities Fund of the Explorers Club, and the Office Québec-Amérique pour la Jeunesse; to C.M. from the Canadian International Development Agency; and to J.S. from the Fonds pour la Formation des Chercheurs et l'aide à la Recherché of Québec, and the Natural Sciences and Engineering Research Council of Canada.

REFERENCES CITED

Aguirre-Díaz, G.J., 2001, Recurrent magma mingling in successive ignimbrites from Amealco caldera, central Mexico: Bulletin of Volcanology, v. 63, p. 238–251.
Anonymous, 1880, A lacustrine volcano: Nature, v. 22, p. 129.
Bacon, C.R., 1983, Eruptive history of Mount Mazama and Crater Lake caldera, Cascade Range, USA: Journal of Volcanology and Geothermal Research, v. 18, p. 57–115.
Bacon, C.R., 1986, Magmatic inclusions in silicic and intermediate volcanic rocks: Journal of Geophysical Research, v. 91, p. 6091–6112.
Barclay, J., Rutherford, M.J., Carroll, M.R., Murphy, M.D., Devine, J.D., Gardner, J., and Sparks, R.S.J., 1998, Experimental phase equilibria constraints on pre-eruptive storage conditions of the Soufriere Hills magma: Geophysical Research Letters, v. 25, p. 3437–3440.
Bowen, N.L., 1913, The melting phenomena of the plagioclase feldspars: American Journal of Science, v. 35, p. 577–599.

Dull, R.A., Southon, J.R., and Sheets, P., 2001, Volcanism, ecology and culture: A reassessment of the volcán Ilopango TBJ eruption in the southern Maya realm: Journal of Latin American Antiquity, v. 12, p. 25–44.

Eichelberger, J.C., 1978, Andesitic volcanism and crustal evolution: Nature, v. 275, p. 21–27.

Eichelberger, J.C., Chertkoff, D.G., Dreher, S.T., and Nye, C.J., 2000, Magmas in collision: Rethinking chemical zonation in silicic magmas: Geology, v. 28, p. 603–606.

Golombek, M.P., and Carr, M.J., 1978, Tidal triggering of seismic and volcanic phenomena during the 1879–1880 eruption of Islas Quemadas volcano in El Salvador, Central America: Journal of Volcanology and Geothermal Research, v. 3, p. 299–307.

Hart, W.J.E., 1981, The Panchimalco tephra, El Salvador, Central America [M.S. thesis]: New Brunswick, New Jersey, Rutgers University, 101 p.

Hibbard, M.J., 1981, The magma mixing origin of mantled feldspars: Contributions to Mineralogy and Petrology, v. 76, p. 158–170.

Huppert, H.E., Sparks, R.S.J., and Turner, J.S., 1984, Some effects of viscosity on the dynamics of replenished magma chambers: Journal of Geophysical Research, v. 89, p. 6857–6877.

Johannes, W., Koepcke, J., and Behrens, H., 1994, Partial melting reactions of plagioclases and plagioclase bearing systems, *in* Parsons, I., ed., Feldspars and their reactions: Dordrecht, Netherlands, Kluwer, 650 p.

Kayamoto, T., 1992, Dusty and honeycomb plagioclase: Indicators of processes in the Uchino stratified magma chamber, Izu Peninsula, Japan: Journal of Volcanology and Geothermal Research, v. 49, p. 191–208.

Langmuir, C.H., Vocke, R.D., Jr., Hanson, G.N., and Hart, S.R., 1978, A general mixing equation with applications to Icelandic basalts: Earth and Planetary Science Letters, v. 37, p. 380–392.

Mann, C.P., Stix, J., Vallance, J.W., and Richer, M., 2004, Subaqueous intracaldera volcanism, Ilopango Caldera, El Salvador, Central America, *in* Rose, W.I., et al., eds., Natural hazards in El Salvador: Boulder, Colorado, Geological Society of America Special Paper 375, p. 159–174 (this volume).

Montessus De Ballore, F., 1884, Temblores y erupciónes volcánicas en Centro-América con un apendice meteorológico: Imprenta Del Doctor Francisco Sagrini, San Salvador, 246 p.

Mooser, F., Meyer-Abich, H., and McBirney, A.R., 1958, Catalogue of the active volcanoes of the world including solfatara fields of Central America, Part VI: Napoli, Italy, International Volcanological Association, 146 p.

Murphy, M.D., Sparks, R.S.J., Barclay, J., Carroll, M.R., Lejeune, A.-M., Brewer, T.S., Macdonald, R., Black, S., and Young, S., 1998, The role of magma mixing in triggering the current eruption at the Soufriere Hills volcano, Montserrat, West Indies: Geophysical Research Letters, v. 25, p. 3433–3436.

Newhall, C.G., and Dzurisin, D., 1988, Historical unrest at large calderas of the world: U.S. Geological Survey Bulletin 1855, Washington, D.C., 1108 p.

Rose, W.I., Conway, F.M., Pullinger, C.R., Deino, A., and McIntosh, W.C., 1999, An improved age framework for late Quaternary silicic eruptions in northern Central America: Bulletin of Volcanology, v. 61, p. 106–120.

Rutherford, M.J., and Hill, P.M., 1993, Magma ascent rates from amphibole breakdown: An experimental study applied to the 1980–1986 Mount St. Helens eruptions: Journal of Geophysical Research, v. 98, p. 19,667–19,685.

Sisson, T.W., and Bacon, C.R., 1999, Gas-driven filter pressing in magmas: Geology, v. 27, p. 613–616.

Sparks, R.S.J., and Marshall, L.A., 1986, Thermal and mechanical constraints on mixing between mafic and silicic magmas: Journal of Volcanology and Geothermal Research, v. 29, p. 99–125.

Sparks, R.S.J., and Sigurdsson, H., 1977, Magma mixing: A mechanism for triggering acid explosive eruptions: Nature, v. 267, p. 315–318.

Sparks, R.S.J., Young, S.R., Barclay, J., Calder, E.S., Cole, P.D., Darroux, B., Druitt, T.H., Harford, C., Herd, R.A., Hoblitt, R., James, M.R., Lejeune, A.-M., Norton, G., Skerrit, G., Stasiuk, M.V., Stevens, N.S., Toothill, J., Wadge, G., and Watts, R., 1998, Magma production and growth of the lava dome of the Soufriere Hills Volcano, Montserrat, West Indies: November 1995 to December 1997: Geophysical Research Letters, v. 25, p. 3421–3424.

Tepley F.J., III, Davidson, J.P., and Clynne, M.A., 1999, Magmatic interactions as recorded in plagioclase phenocrysts of Chaos Crags, Lassen Volcanic Center, California: Journal of Petrology, v. 40, p. 787–806.

Tepley F.J., III, Davidson, J.P., Tilling, R.I., and Arth, J.G., 2000, Magma mixing, recharge and eruption histories recorded in plagioclase phenocrysts from El Chichón Volcano, Mexico: Journal of Petrology, v. 41, p. 1397–1411.

Tsuchiyama, A., 1985, Dissolution kinetics of plagioclase in the melt of the system diopside-albite-anorthite, and origin of dusty plagioclase in andesites: Contributions to Mineralogy and Petrology, v. 89, p. 1–16.

Tsuchiyama, A., and Takahashi, E., 1983, Melting kinetics of a plagioclase feldspar: Contributions to Mineralogy and Petrology, v. 84, p. 345–354.

Watson, E.B., and Jurewizc, S.R., 1984, Behavior of alkalies during diffusive interaction of granitic xenoliths with basaltic magma: Journal of Geology, v. 92, p. 121–131.

Weber, H.S., Wiesemann, G., Lorenz, W., and Schmidt-Thomé, M., 1978, Mapa Geológico de la República de El Salvador/América Central: Bundesanstalt für Geowissenschaften und Rohstoffe, Hannover, scale 1: 100,000, 6 sheets.

Wiebe, R.A., 1994, Silicic magma chambers as traps for basaltic magmas: The Cadillac Mountain Intrusive Complex, Mount Desert Island, Maine: Journal of Geology, v. 102, p. 423–437.

Wiebe, R.A., 1996, Mafic-silicic layered intrusions: The role of basaltic injections on magmatic processes and the evolution of silicic magma chambers: Transactions of the Royal Society of Edinburgh: Earth Sciences, v. 87, p. 233–242.

Wiebe, R.A., Snyder, D., and Hawkins, D., 2000, Correlating volcanic and plutonic perceptions of silicic magma chamber processes: Evidence from coastal Maine plutons, Field Forum, September 14–18, 2000, Ellsworth, Maine: Boulder, Colorado, Geological Society of America Field Guide, 61 p.

Manuscript Accepted by the Society June 16, 2003

Geological Society of America
Special Paper 375
2004

Dynamics of diffuse degassing at Ilopango Caldera, El Salvador

Dina L. López*
Loretta Ransom*
Department of Geological Sciences, Ohio University, 316 Clippinger Laboratories, Athens, Ohio 45701, USA

Nemesio M. Pérez
Pedro A. Hernández
Environmental Research Division, Instituto Tecnológico y de Energías Renovables, ITER, 38611 Granadilla de Abona, Santa Cruz de Tenerife, Spain

Jeannette Monterrosa
Fundación de Amigos del Lago de Ilopango, San Salvador, El Salvador, Central America

ABSTRACT

Ilopango Caldera in central El Salvador is filled by Ilopango Lake. In November and December of 1999, radon, thoron, carbon dioxide, and mercury soil gas concentrations were obtained at 106 points within the caldera, as well as carbon dioxide efflux. The spatial distribution of the concentrations of these gases and carbon dioxide efflux show that the values of Ilopango Caldera are within background levels of other active volcanoes of the world and El Salvador. However, several areas where anomalies of high radon, thoron, carbon dioxide, and carbon dioxide efflux coincide were identified in regions of dense faulting. Heavier carbon isotope values between −13‰ and −20‰ (overall range: −13.1‰ to −29.8‰) in the anomalous regions suggest mixing of minor amounts of volcanic gases with biogenic gases or the presence of C_4 plants. Degassing of carbon dioxide from the lake was calculated using a double boundary layer model. A total carbon dioxide efflux of 644 to 1111 t d^{-1} was calculated for the lake and the soils of the caldera. Considering the low soil degassing of carbon dioxide (0.3–3.9 g m^{-2} d^{-1}), radon (1.2–108.5 pCi L^{-1}), and mercury (0–0.016 mg m^{-3}), Ilopango can be considered a quiescent caldera at the time of this survey.

Keywords: Ilopango Caldera, El Salvador, carbon dioxide, diffuse soil gases, radon, mercury.

INTRODUCTION

Ilopango Caldera in El Salvador, Central America (Fig. 1), has played an important role in the history of Central American people. The volcanic eruption of Ilopango Caldera in A.D. 429 (Dull, 2001) is responsible for the migration of the Mayan people from El Salvador and the highlands of Guatemala to Tikal (Sheets, 1979). The caldera is still considered a threat as it erupted as recently as 1880.

The volcanic front of Central America is the result of the subduction of the Cocos plate under the Caribbean plate, along the Middle American Trench (Molnar and Sykes, 1969). The Cocos plate is currently being subducted at a rate of 5–7 cm a year (White and Harlow, 1993). The chain of volcanoes in El Salvador is bounded by normal faults running northwest to southeast throughout El Salvador. Ilopango Caldera is located at the southern margin of this graben.

Ilopango Caldera is a depression measuring 16 km east-west and 13 km north-south (Mann et al., this volume, Chapter 12). Ilopango Lake measures 11 km east to west and 8 km north to south (Fig. 2). The north rim of the caldera is 200–300 m above

*E-mails: lopezd@ohio.edu, loreta@hotmail.com

López, D.L., Ransom, L., Pérez, N.M., Hernández, P.A., and Monterrosa, J., 2004, Dynamics of diffuse degassing at Ilopango Caldera, El Salvador, *in* Rose, W.I., Bommer, J.J., López, D.L., Carr, M.J., and Major, J.J., eds., Natural hazards in El Salvador: Boulder, Colorado, Geological Society of America Special Paper 375, p. 191–202. For permission to copy, contact editing@geosociety.org. © 2003 Geological Society of America.

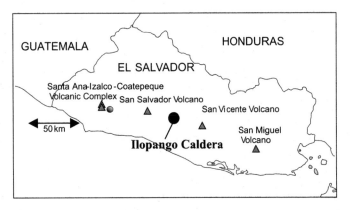

Figure 1. Map of El Salvador's volcanic centers. Ilopango Caldera is located in central El Salvador, 10 km from San Salvador city.

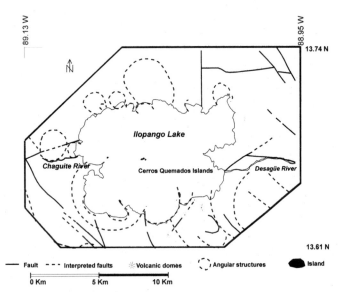

Figure 2. Simplified structural map of Ilopango Caldera (after Geographical Institute of El Salvador, 1970).

lake level and the south rim is 400–500 m above lake level. Lake level is ~400 m above sea level (Meyer-Abich, 1956). The caldera has an area of 185 km² and the lake has a maximum depth of ~240 m. The Chaguite River flows into the lake in the west and the Desagüe River drains the lake to the east.

Williams and Meyer-Abich (1955) state that the caldera formed via three distinct collapse episodes that happened with violent volcanic eruptions. The first collapse occurred during the Plio-Pleistocene formation of the Central American graben. The last caldera collapse generated the Tierra Blanca Joven deposit and occurred ~1600 yr ago, in A.D. 429 (Dull, 2001). This eruption produced a volume of ~18 km³ of rock (Hart, 1981). Four tephra deposits have been identified associated with eruptions from Ilopango (Tierra Blanca 4 to Tierra Blanca

Joven). The main faults and domes found at Ilopango Caldera are shown in Figure 2.

The volcanic rocks of Ilopango are Ca-rich, falling within the calc-alkaline series (Hart, 1981). Williams and Meyer-Abich (1955) identified the majority of the Ilopango Caldera exposed rocks as dacite and rhyodacite pumice deposits with some undifferentiated Pliocene volcanic rocks exposed in the southeast portion of the caldera, whereas Weyl (1980) describes the rocks as being mainly dacitic in composition. The majority of the area is covered with pyroclastic deposits (from the Tierra Blanca Joven eruption).

At the time of this investigation, before the 2001 earthquakes of El Salvador, Ilopango Caldera did not show evidence of fumarolic activity. However, in 2002 hot water had been detected to the south of the caldera, close to the Los Patos Islands (Fig. 2) (C. Pullinger, 2002, personal commun.). Several domes associated with various eruptive episodes are found outside the caldera boundaries, within the caldera walls, and within the lake (Fig. 2). The latest activity, a dome extrusion (Cerros Quemados Islands), occurred December 1879–January 1880 (Meyer-Abich, 1956). Studies of the intracaldera stratigraphic units (Mann et al., this volume, Chapter 12) suggest that a lake has occupied Ilopango Caldera for extended periods of time during the last 44,000 yr and that subaqueous eruptions are common.

Ilopango Caldera experiences frequent seismicity. Recorded seismic events between May 1984 and April 2002 are presented in Figure 3. Note the density of events toward the center, south, and east of the caldera, and the rough alignment in the EW and NNW directions. Recent seismic activity in El Salvador (January and February, 2001) was concentrated to the south and southeast of the Ilopango Caldera. However, the spatial distribution of events from 1984 to 2000 is not different from the distribution observed in Figure 3 during 1984–2002.

Volcanoes can discharge gases to the atmosphere either through the volcanic pipes at craters or as diffuse soil gases. Magmatic gases can be incorporated to the hydrothermal envelope surrounding the magmatic chamber and later discharged to the atmosphere at the soil-air interface together with other gases of nonmagmatic origin. Studies of soil gases at the soil-air interface can give important clues about the magmatic activity of a volcanic system (e.g., Capaldi et al., 1992; Hernández et al., 2001). At Long Valley caldera, Hg spatial distribution patterns in soils have been used as volcanic activity tracers (Varekamp and Buseck, 1983). In this paper, we report our results on diffuse soil degassing studies carried out at Ilopango Caldera during November and December 1999, prior to the high seismicity of 2001.

The purposes of this research are (1) to determine the spatial distribution of CO_2, mercury, and radon concentration in the soils of the caldera and its correlation with the geological structures in the region, (2) to determine the efflux of CO_2 released by the soils and water of Ilopango Caldera and to compare the total efflux with other volcanic systems of the region and the world, and (3) to use all these parameters to monitor the volcanic activity of the caldera.

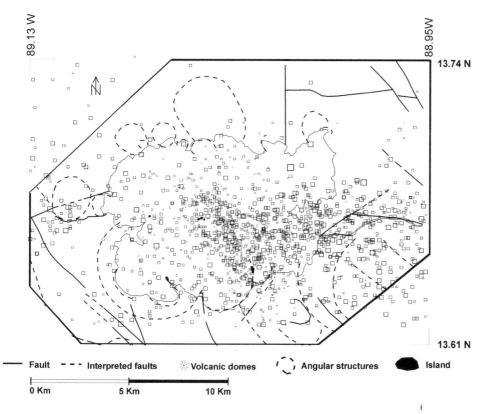

Figure 3. Seismicity of Ilopango Caldera area between May 1, 1984, and April 4, 2002. Unpublished data provided by Servicio Nacional de Estudios Territoriales de El Salvador (SNET). Only events shallower than 40 km are considered. Size of symbol is proportional to magnitude. Highest magnitude is 4.3.

Measurements taken by the authors in January and March 2001 (after January 13 and February 13 earthquakes) showed that emissions of soil CO_2 gas, water chemical composition, and lake temperatures did not change during the 2001 seismic events. Immediate magmatic reactivation as a consequence of the earthquakes does not seem likely to have occurred. However, monitoring of diffuse emissions is needed to determine if these emissions have increased at a later time.

DIFFUSE DEGASSING

Recent studies in other volcanoes around the world (e.g., Etiope et al., 1999; Gerlach et al., 1998; Baubron et al., 1991) have shown that the release and movement of volcanic gases from the magmatic chamber to Earth's surface occurs in two ways: through the volcanic pipe if the volcano is actively degassing with a gas plume, or transferred from the magmatic system through the surrounding hydrothermal envelope. The fraction of gas that has not reacted with the aqueous solution and rocks along the path is finally released at the soil-air interface. Gases can be transported throughout the rocks and soils by two main mechanisms: (1) diffusion, due to a gradient in concentration, and (2) advection (and dispersion), where the gas is transported by a moving phase. However, researchers working in volcanic soil degassing call "diffuse" (e.g., Hernandez et al., 2001; Chiodini et al., 1996) the combination of diffusive and advective gas transfer throughout the volcanic edifice (excluding the volcanic pipe).

Correlations between seismic activity and soil degassing of gases such as radon and carbon dioxide provide evidence that soil gas studies may be useful in forecasting volcanic and seismic events (Reimer, 1981, 1980; King, 1980; Salazar et al., 2002). These studies should be combined with other methods of volcano surveillance such as seismicity and deformation (Heiligmann et al., 1997). If diffuse volcanic gas emanations and central fumarole degassing are correlated, continuous geochemical monitoring at a safe distance from active craters is possible (Baubron et al., 1990).

Previous studies at other volcanoes show that effluxes of diffuse soil emission of carbon dioxide from volcanoes can be similar in magnitude to the efflux at the crater plume (Allard et al., 1991). At some locations, dissolved CO_2 in groundwater can be a significant component of the total volcanic CO_2 efflux (Rose and Davisson, 1996; Sorey et al., 1998). Faults may act as conduits for gas and water flow from depth. Carbon dioxide is also emitted from oxidized organic matter in soils, usually with a microbial mediated path (Drever, 1997).

Gases can also be discharged at the air-water interface if lakes or other surficial water bodies are present in the volcanic region. The exchange of gases between atmospheric air and the underlying water column can be described using the double diffusive layer model. This model considers mass transfer through two thin layers, one in the liquid phase and another in the gas phase, and a Fickian efflux F between both phases (Stumm and Morgan, 1996). The alkalinity of the water can be used to determine the concentration of CO_2 in the water phase and the efflux of CO_2 discharged at the interface, as it is described in the methodology.

Radon (^{222}Rn) monitoring in groundwaters and soil gases at active volcanoes reveals that some temporal changes of radon are associated with pre-eruptive volcanic activity, and laboratory work supports these findings (Capaldi et al., 1992). A significant amount of ^{222}Rn is released from rocks that are subjected to processes related to the ascent of magma, such as heating, deformation, and fracturing. Radon anomalies associated with earthquake activity are probably produced by variations in the local stress field. Variations in soil gas concentrations are more important than absolute values when monitoring volcanic and seismic activity (e.g., Notsu et al., 1991).

Sources of ^{222}Rn gas in volcanic systems include magma, wall-rock, and hydrothermal fluids. All of these contain ^{226}Ra, the parent isotope of ^{222}Rn (radon). In areas where ^{222}Rn is transported by diffusion, the mean depth from which ^{222}Rn escapes from soils is ~2 m (Connor et al., 1996), because the diffusivity of ^{222}Rn in water and soils is rather slow. As a consequence, anomalous concentrations of radon in volcanic regions can occur only if rapidly moving hydrothermal fluids are transporting the gas advectively or convectively (Connor et al., 1996).

Changes in the stress field of the crust modify the porosity and volume of cracks and fissures, affecting the concentration of soil radon gas (Heiligmann et al., 1997). These processes include (1) annual and diurnal variations of wind, rain, temperature, water content in soil, and atmospheric pressure, (2) deformation of the volcanic edifice (increase in rock surface area), (3) magmatic degassing of ^{222}Rn, and (4) increases in temperature of the hydrothermal system (Connor et al., 1996).

Thoron gas (^{220}Rn) is a short-lived isotope of radon. It is part of the thorium (^{232}Th) decay series having a half-life of 55 seconds. Due to its short half-life, the transport distance of thoron is limited to a few centimeters (either by diffusion or advection), compared to several meters for radon (Hutter, 1993). Changes in thoron/radon isotopic ratios suggest an advective component on the transport of soil gases (Hutter, 1993). At volcanic systems, these changes may indicate advective movement of gases, sometimes generated by magmatic input.

Volcanic gases are rich in mercury, which is exsolved early from magmas (Varekamp and Buseck, 1981). In hydrothermal systems, mercury occurs around hot springs reaching concentrations up to several tens of ppm in altered fumarolic areas (Varekamp and Buseck, 1983). The solubility of elemental mercury is favored by temperatures above 200 °C, high pH, low total S, ionic strength, and reducing conditions (Varekamp and Buseck, 1984). Mercury partitions strongly into the vapor phase of hydrothermal systems as Hg^{o}_{vap} and it is transported in two-phase convective transport to the soil-air interface (Varekamp and Buseck, 1984). Williams (1985) notes that differences in the spatial distribution of radon and mercury concentrations can provide information on active or fossil zones of vertical convective heat and gas flow. Mercury is considered an important geochemical indicator for fault activity (Lai, 1999). Unlike radon, which is inert, mercury is reactive and can be sorbed on the clay-sized fraction of soils and organic matter.

METHODOLOGY

Soil Degassing

Soil and soil gas samples were collected at 106 stations during November 22 to December 8, 1999. Figure 4A shows these sample locations. Sampling sites were located around the lakeshore and along the caldera rim to cover as much of the caldera system as possible. The area was covered mainly by coffee plantations and in some cases, corn fields.

Radon concentration was measured in the field by emanometry using a Pylon Portable Radiation Monitor Model AB-5 and a Pylon Model 300A Lucas Cell. Gas was pumped through a metallic probe inserted 40 cm deep in the soil, and collected in the Lucas Cell. The cell was purged three times to avoid atmospheric contamination. Alpha particles produced during the decay process hit the walls of the scintillation cell and produced light pulses. The light pulses caused by the absorption of alpha particles were amplified by a photomultiplier tube and converted into electronic pulses. The number of pulses per minute were converted to radon and thoron concentrations using equations that consider the decay of ^{220}Rn and ^{222}Rn and their daughter products (Pylon Electronics, 1993). This technique gives only an approximate value of the ^{220}Rn and ^{222}Rn concentrations. In order to get accurate measurements of the radon concentrations, 3.5 hours per measurement are needed (Pylon Electronics, 1993). This long period of time is too long for surveys with many points. However, repeated measurements of radon concentration at some points gave a standard error of 15% or better.

Soil gas samples for analysis of CO_2 and mercury concentrations were taken using a sonde (a tube with a gas-permeable section that allows the penetration of soil gas) and extracting the soil gas at ~80 cm depth with a hypodermic syringe through a septum at the top of the sonde. The extracted soil gas was injected either in the LI-COR Gas Hound CO_2 Analyzer or in the Jerome 431-X Mercury Vapor Analyzer. Soil CO_2 efflux and CO_2 concentrations were measured in the field using the LI-COR Gas Hound CO_2 Analyzer Model LI-800. CO_2 efflux was measured using the "accumulation chamber method" (Baubron et al., 1990). Soil samples were collected at each sampling site at a depth of 30–40 cm to avoid the organic-rich layer at the surface. Samples from the organic-rich surface layer do not allow for a good discrimination between soils of different organic matter content. This distinction is important to assign CO_2 anomalies to organic or volcanic sources. Soil pH and the organic matter content were analyzed in the laboratory. Soil pH was determined in a solution formed by 25 grams of air-dried soil with 25 ml of de-ionized water (McLean, 1982). Organic matter content was determined using the dry-ash method (Ben-Dor and Banin, 1989).

Gas samples for carbon isotopes were taken using the same sonde and hypodermic syringe and storing the gas in evacuated

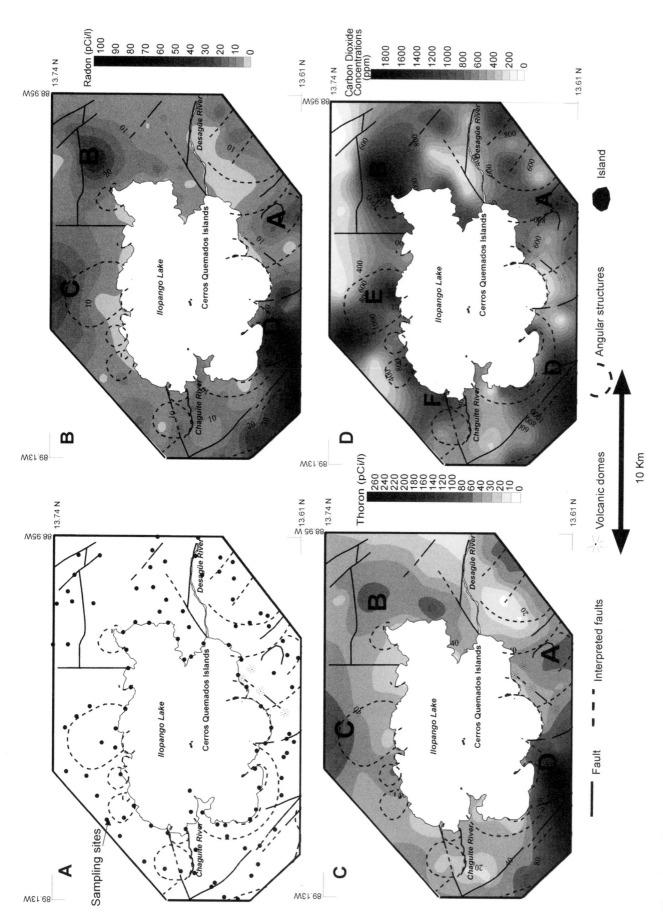

Figure 4. Sampling sites for soil gas samples and CO_2 efflux measurements (A). Concentration of radon (B), concentration of thoron (C), and concentration of CO_2 (D) in soil gases at Ilopango Caldera. Anomalies A, B, C, D, E, and F are shown.

15 cc vacutainers. Carbon isotopes were determined with a Finnigan MAT Delta S mass spectrometer. The $^{13}C/^{12}C$ ratios are reported as $\delta^{13}C$ values (±1‰).

Mercury was analyzed in the field using the Jerome 431-X Mercury Vapor Analyzer. This analyzer used two gold films, one a reference film and the other a sensor film. Mercury is adsorbed to the gold film surface, changing the electrical resistance, which is measured by a Wheatstone bridge circuit and is converted into mg Hg m^{-3} (Arizona Instruments, 1996).

CO$_2$ Degassing from Lake Water

The determination of the CO$_2$ efflux discharged from the water to the atmosphere was based on measurements of alkalinity taken at Ilopango Lake. Water samples were taken at Ilopango Lake at 5 and 50 m depth in January and July 1998. These water samples were filtered using a syringe and a 0.45 µm filter to separate any suspended particles from the dissolved species. Only dissolved carbon is assumed to participate in the diffusive degassing of CO$_2$ from the lake water to the atmosphere. Points were sampled at Corinto station (Fig. 2), close to Cerros Quemados Islands, Los Patos Islands, close to Chaguite River, and close to Desagüe River. The filtered water was titrated with sulfuric acid to determine the alkalinity. A HACH digital titrator was used for that purpose.

The exchange of gases between atmospheric air and the underlying water column at Ilopango Lake can be described using the double diffusive layer model (Thibodeaux, 1996). This model considers that gases are transferred from one phase to the other through thin layers (a few µm thick), one in the liquid phase and another in the gas phase. The dominant mass transfer mechanism within these layers is diffusive (even if the transfer is dominantly advective within the phases). If the gas does not undergo chemical reaction during the diffusive transfer, the following equation describes the Fickian efflux F between both phases (Stumm and Morgan, 1996):

$$F = -\frac{D_a}{z_a}\left(c_a - c_{\frac{a}{w}}\right) = -\frac{D_w}{z_w}\left(c_{\frac{w}{a}} - c_w\right) \qquad (1)$$

where D is the diffusivity in cm^2 s^{-1}, z is the film thickness, and c is concentration. The suffixes a and w refer to the air and water film, respectively, and the a/w suffix, to the air/water interface.

If the gas reacts fast compared to the travel time to cross the film, equation 1 is no longer valid. If the reactions are fast enough to assume chemical equilibrium between reactants and products, a chemical enhancement factor α should be considered (Stumm and Morgan, 1996), and the equation that describes the gas transfer from the water to the air is:

$$F = -\alpha\frac{D_w}{z_w}\left(c_{\frac{w}{a}} - c_w\right) \qquad (2)$$

For the transfer of carbon dioxide, the chemical enhancement factor α can be calculated considering the transfer of total carbon across the distance z_w. α is the ratio between the efflux of total carbon (sum of all carbon species: $H_2CO_3^*$, HCO_3^-, and CO_3^{2-}) and the efflux of only $H_2CO_3^*$ (Stumm and Morgan, 1996). The different carbonate species within the water and at the water-air interface were found using values of alkalinity taken during January and July, 1998, and application of Henry's law for dissolution of gases in liquids. The thickness of the water film can be found by using the empirical equation derived by Kling et al. (1992) for lakes around the world, which relates the boundary layer thickness to the wind velocity, or by using an estimate of z_w based on previous studies at other lakes (e.g., Kling et al., 1992; Wanninkhof et al., 1985, 1990).

DISCUSSION OF RESULTS

Diffuse Soil Degassing

Radon

Radon (^{222}Rn) concentrations ranged from 1.2 to 108.5 pCi L^{-1} with an average of 13.5 pCi L^{-1}. Statistical analyses of the radon concentrations were conducted by plotting concentration versus cumulative frequency percent. Inflection points in this graph indicate the boundary between different populations (Tennant and White, 1959). This analysis allows the identification of three populations for radon: base or background (range 1.2–16.6, average 7.4 pCi L^{-1}), intermediate (range 18.4–24.9, average 20.3 pCi L^{-1}) and peak (range 26.2–108.5, average 46.6 pCi L^{-1}). Figure 4B displays the ^{222}Rn contour map constructed using kriging as an interpolation technique (Swan and Sandilands, 1995), identifying several anomalous regions within the study area. Four high radon anomalies can be observed. They seem to occur roughly around the rim of the caldera. These anomalies have been labeled A to D and have been used for further comparison with other soil gas results.

Thoron (^{220}Rn) concentrations ranged from 7.2 to 271.6 pCi L^{-1} with an average of 50.2 pCi L^{-1}. Based on the probability plot of thoron data, three overlapping populations were identified. Background (range 7.2–26.7, average 19.2 pCi L^{-1}), intermediate (range 28.3–40.6, average 33.6 pCi L^{-1}), and peak (range 42.0–271.6, average 79.4 pCi L^{-1}). An inspection of the ^{220}Rn contour map (Fig. 4C) shows that the anomalies have a similar spatial distribution to those of radon with anomalies A and D occurring close to the southern caldera rim. Thoron/radon isotopic ratios ranged from 1.4 to 26.6 with an average of 5.8. Three overlapping geochemical populations were also identified, background (range 1.4–4.2, average 3.0), intermediate (range 4.4–7.8, average 5.6), and peak (range 7.9–26.6, average 12.9). The spatial distribution of thoron/radon ratios differs from that of radon and thoron, with the highest values located along the inner walls of the caldera, where high fracturing occurs, and faster gas transfer is probably occurring allowing the output of short-lived thoron.

At Ilopango Caldera there is increased faulting and fracturing from the northern to the southern portion of the study area. The southwest portion of the study area showed anomalous radon and thoron concentrations, which are possibly the result of the increased fracturing in the area. Ilopango Caldera radon values are low compared to other active volcanic systems of the world, which can produce values in the tens of thousands pCi L^{-1}; e.g., plume activity at Etna volcano, Sicily (Lambert et al., 1976), and eruption of Cerro Negro cinder cone, Nicaragua (Connor et al., 1996). However, the values at Ilopango Caldera are comparable to pre-eruptive values for Cerro Negro. The radon levels at Ilopango are comparable to those found in San Salvador volcano, which range from 1 to 284 pCi L^{-1} (Pérez et al., this volume, Chapter 17).

Carbon Dioxide Concentrations and Effluxes

Carbon dioxide concentrations ranged from 125 to 1965 ppm with an average of 734.6. Statistical analyses of the carbon dioxide concentrations allow the identification of three populations: base (range 125–156.3, average 143.4 ppm, only 7.5% of the total population), intermediate (range 176.3–548.8, average 357.1 ppm, 30.2% of total population), and peak (range 575–1965, average 989.2 ppm, 62.3% of total population). The base population and part of the intermediate population have values that are below the average value for the atmospheric CO_2 concentration of 365 ppm. To explain these low values we need to consider several facts. Earth's atmosphere does not have a constant value for the concentration of CO_2. The concentration of CO_2 in the atmosphere as well as in the interstitial soil gas has seasonal and diurnal fluctuations and depends on the ambient temperature (e.g., Maljanen et al., 2002; Ball et al., 1999; Granier et al., 2000; Moren and Lindroth, 2000). Seasonal and diurnal changes in temperature strongly affect the production rates of CO_2 in soils. Values as low as 120 ppm for soil CO_2 have been reported in the literature (Amundson et al., 1990). In El Salvador, monitoring of diffuse CO_2 effluxes from San Vicente volcano shows an efflux with periodic diurnal variations (Salazar et al., 2002). It is possible that the low CO_2 concentration values observed in some points at Ilopango Caldera fall within the normal variations produced by diurnal fluctuations in concentration and efflux. Another possibility is that higher water content in the soils of those points could generate low CO_2 values. Many points were taken close to the Ilopango Lake shore (Fig. 4A). This investigation was carried out during the dry season; however, water content in the soils close to the lake could be higher due to the nearer water table and capillary effects. An inverse relationship between soil water tension and CO_2 concentrations in soil gas has been identified (Buyanovski and Wagner, 1983).

Figure 4D shows the spatial distribution of soil CO_2 concentration at the study area. The contour map was elaborated using also the kriging method. Three anomaly areas, labeled A, B, and D and located in the southeast, northeast, and southwest area, respectively, correspond to anomalies A, B, and D for radon and thoron in Figures 4B and 4C. Other distinct anomalies observed in Figure 4D are in the area just north and on the western edge of Ilopango Lake and have been labeled E and F.

Carbon dioxide efflux ranges from 0.3 to 3.9 g m^{-2} d^{-1} with an average of 1.6 g m^{-2} d^{-1}. We also applied a probability plot to distinguish different geochemical populations, but only one normal distribution was identified. The CO_2 efflux values are lower than values observed at other active volcanoes of the world (e.g., maximum efflux of more than 4800 g m^{-2} d^{-1} for Teide volcano, Hernandez et al., 1998; maximum value of 760 for San Salvador volcano, Pérez et al., this volume, Chapter 17). However, they compare to the background values measured in the Santa Ana–Izalco–Coatepeque volcanic complex (from undetectable to 293 g m^{-2} d^{-1}, with background population up to 6 g m^{-2} d^{-1}, Salazar et al., this volume, Chapter 10), and San Salvador volcano (from undetectable up to 760 g m^{-2} d^{-1}, with 97% of the sampling points with values lower than 2 g m^{-2} d^{-1}, Pérez et al., this volume, Chapter 17), and the total range measured at San Miguel volcano (undetectable to 5 g m^{-2} d^{-1}, Cartagena et al., this volume, Chapter 15). Relatively high values are observed to the southeast (A) and northeast (B) (Fig. 5A) in close proximity to the carbon dioxide concentration, radon, and thoron anomalies in those regions. E and F, north and west of the lake, coincide with anomalies observed in the carbon dioxide concentration contour map. G, a high anomaly to the south of the lake, is observed close to anomaly D for CO_2 concentrations.

The CO_2 at dormant or active volcanoes can be derived from the mantle, thermal metamorphism of carbonate rocks, organic process, or mixing of gas from these different sources (Heiligmann et al., 1997). In active volcanoes diffuse efflux of carbon dioxide can reach thousands of grams per square meter per day. For example, soil CO_2 efflux at Mammoth Mountain, California, showed flow rates up to tens of thousands g m^{-2} d^{-1} (Gerlach et al., 1998). In comparison, CO_2 efflux of temperate forest soils can exceed 25 g m^{-2} d^{-1} (Gerlach et al., 1998). The low CO_2 efflux values measured at Ilopango Caldera suggest that most of the CO_2 is produced by biological activity within the soils. Although the values at the Ilopango Caldera are likely to be predominantly biogenic, the spatial correlation with soil radon and CO_2 concentration anomalies (A to D, in Figures 4D and 5A) suggest that a small fraction of these gases could have a different origin. The spatial distribution of CO_2 efflux and the area covered by this survey allow us to compute the total CO_2 output from the soils of Ilopango Caldera as 594 t d^{-1}.

Carbon Isotopes

Carbon isotope values were determined for 31 of the 106 sampling sites. The data ranged from −29.8‰ to −13.1‰, with an average of −22.1‰. Four areas with relatively high isotopic values are observed in Figure 5C. The $\delta^{13}C$ value has often been used to identify the origin of carbon in natural gas samples (Schwarcz, 1969; Hoefs, 1980). Mid-oceanic ridge basalt (MORB) glasses have $\delta^{13}C$ values between −4‰ and −9‰ with an average of −6.5‰, which is considered to represent the upper-mantle C (Faure, 1986; Javoy et al., 1986; Marty and Jambon, 1987). In contrast, $\delta^{13}C$ values of crustal carbon can vary significantly, with two main sources considered: (1) marine limestone including

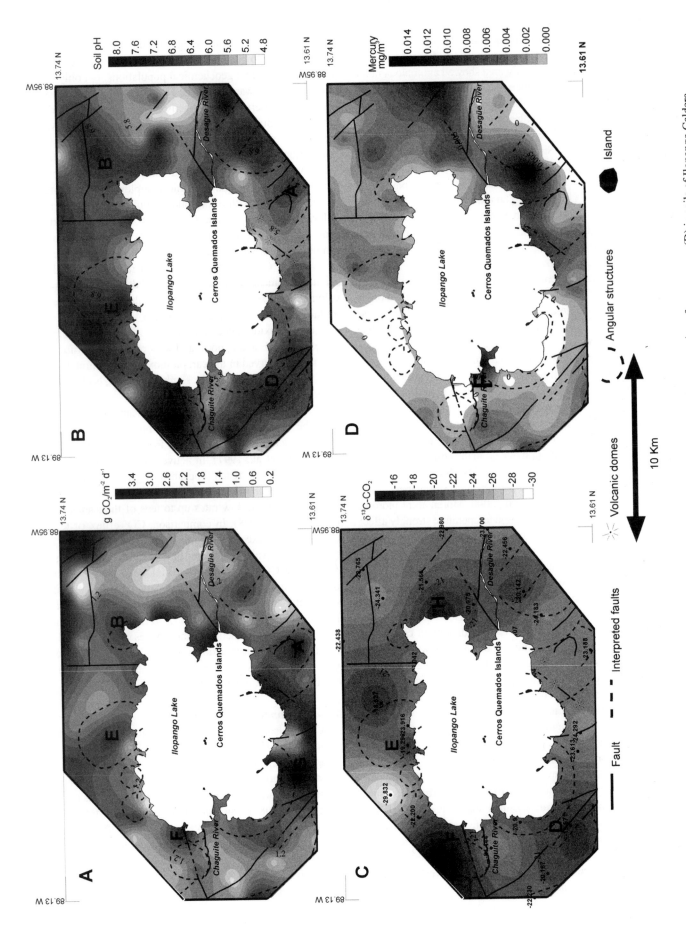

Figure 5. CO_2 efflux in soil gases (A), soil pH (B), isotopic composition of soil gases (C), and concentration of mercury gas (D) in soils of Ilopango Caldera.

slab carbonate, which have an average $\delta^{13}C$ value near 0‰, and (2) organic carbon from sedimentary rocks with $\delta^{13}C$ values lighter than −20‰. The geological map of El Salvador (Geographical Institute of El Salvador, 1970) does not show evidence of marine limestones directly underlying Ilopango Caldera. Any limestone component should come from the subducting slab. Farrar et al. (1995) stated that typical forest soil has carbon isotope values between −19‰ and −25‰. Plants are the main source of carbon for most soils. The isotopic composition of soil gases reflects the value of the flora (Amundson et al., 1998). Most plants throughout the history of Earth have evolved using a photosynthetic pathway that makes them depleted in ^{13}C with respect to the atmospheric CO_2 (Amundson et al., 1998). These are the plants known as C_3 plants that have an average $\delta^{13}C$ of −27‰ (Deines, 1980). However, during the Cenozoic, plants with a new photosynthetic pathway appeared (Amundson et al., 1998). These are the C_4 plants that do not discriminate against $^{13}CO_2$ during photosynthesis and have an average $\delta^{13}C$ value of ~−12 (Deines, 1980). C_4 plants are mainly tropical plants such as corn and sugar cane. At Ilopango Caldera corn plantations are common.

Some of the data points for Ilopango Caldera fall between the values given by Faure (1986) for the upper-mantle C and for C_4 plants (Deines, 1980) and below that stated by Farrar et al. (1995) for forest soils and for C_3 plants (Deines, 1980). This suggests two possible explanations for the high ^{13}C anomalies. One is the mixing of biogenic gases with a minor contribution of volcanic gas, and the other is the presence of C_4 plants in the anomalous area. To understand which hypothesis explains better the isotopic anomalies observed in Figure 5C, we may compare these anomalies with the other gas anomalies in the area. In Figure 5C, locations E and F correspond to anomalies observed in CO_2 efflux and CO_2 concentrations. Anomaly D corresponds to anomalies observed in carbon dioxide, radon, and thoron concentrations. The high value at area H does not appear to correspond to anomalies for the other gases. However, H is close to the area where soil gas radon, thoron, and carbon dioxide concentrations together with CO_2 efflux anomalies were observed (anomaly B). Unfortunately the area of anomaly B was not sampled for isotopic analysis. The fact that some high ^{13}C anomalies coincide with anomalies for CO_2 efflux and concentration, and for radon gases, seems to suggest that weak degassing of the volcanic system could produce gases that mix with organic soil gases and discharge at the soil-air interface. Nevertheless, a scatter diagram of $\delta^{13}C$ versus CO_2 concentrations does not produce a statistically significant mixing line between organic soil gases and mantle gases because the data points are very disperse. For that reason, some contribution of C_4 plants to these high values cannot be rejected.

Soil pH and Organic Matter

Soil pH measured in all soil samples ranged from 4.74 to 8.10, with an average pH of 6.2. Anomalies occur in various locations throughout the study area and can be observed in Figure 5B. High pH correlates with anomalies of CO_2 efflux and CO_2, radon, and thoron concentrations. Areas labeled A and B are associated with

anomalies found in CO_2 efflux and CO_2, radon and thoron concentration. Anomaly D is found in CO_2 concentrations, and radon and thoron concentrations. E and F anomalies are both found in CO_2 efflux and concentrations. For CO_2 derived from the decomposition of organic matter, a correlation between high CO_2 concentration and low pH is expected (Drever, 1997). However, we get an opposite correlation in Ilopango soils. Given the calcium-rich composition of Ilopango volcanic products (Hart, 1981), alteration of the silicates can produce carbonates. Soil mineralogical composition at Ilopango has not been investigated. However, sediments at Ilopango Lake are rich in carbonates and calcareous organisms (Ransom, 2002). If the sediments are derived from the soils and rocks in the Ilopango watershed, soils rich in calcium carbonate may be present. The correlation between high pH values (>7) and high CO_2 concentrations suggests that the pH of the soil is buffered by carbonates in the soils. If carbonate-rich soils are present, the equilibrium of the dissolution of calcium carbonate should buffer the maximum value of the pH (Drever, 1997).

Organic matter content ranged from 1.66 to 12.54% with an average value of 4.44%. The highest values measured occur in the south to southeast portion of the study area and may be associated with the A anomaly located in all soil gas contour maps at the southeast. In general, high values of organic matter content do not correspond with high values of CO_2 efflux and concentration, suggesting that relatively high CO_2 effluxes in the caldera are not associated with high organic matter content.

Mercury

Mercury concentrations at Ilopango Caldera ranged from 0 to 0.016 mg m^{-3} (Fig. 5D). The average value is 0.002 mg m^{-3}. Anomalies occur to the east, southeast and west of Ilopango Lake. The two anomalies located in the east do not correspond to other soil gas anomalies. However, the anomaly F at the west end of Ilopango Lake is close to anomaly F observed in CO_2 efflux and CO_2 concentration. The anomalies to the east and southeast are close to the area between San Vicente volcano and Ilopango Caldera that had high seismicity during 2001.

The results for the concentration of radon, thoron, CO_2, mercury, isotopic carbon composition, soil pH, and organic matter content suggest that anomaly A for carbon dioxide may be associated with the amount of organic matter in that area, but it does not explain the anomaly for radon and thoron. Anomalies A, B, C, E, and F occur in regions of faults and semicircular structural depressions of the caldera and may be produced by high permeability and enhanced release of hydrothermal and volcanic fluids. The low permeability of the Tierra Blanca Joven deposit is probably precluding the emission of gases, except at permeable faults and circular fractures that allow fast fluid movement.

CO_2 Degassing from Lake Water

Water samples taken at Ilopango Lake at 5 and 50 m depth in January and July 1998 indicate that alkalinity at 5 m depth is 268 ± 15 mg $CaCO_3$ L^{-1}, and at 50 m depth is 265 ± 13 mg

CaCO$_3$ L^{-1}. The *t* test for two samples with equal means (Swans and Sandilands, 1995) shows that the two values are not significantly different at the 95% confidence level, suggesting that the alkalinity of the lake does not show big differences with depth. In comparison, pH values decreased with depth significantly, with values as high as 8.8 at the surface and as low as 7.4 at depth in July 1998. At 5 m depth the pH values are 8.63 ± 0.03. The CO$_2$ efflux released by Ilopango Lake can be estimated using equations 1 and 2 in the methodology section of this paper. Possible values for the thickness of the boundary layer can be calculated using the Kling et al. (1992) equation and the average value for the wind velocity. For an average annual value of the wind velocity of 2.45 km/h (0.68 m/s), obtained from the wind velocity registered at Ilopango Airport Meteorological Station, the thickness of the boundary layer is 295 μm. According to Kling et al. (1992) the boundary layer thickness found using this equation could be overestimated. For that reason, calculations using a value of 40 μm were also done. This value is similar to boundary layer thickness reported in other lakes (e.g., Wanninkhof et al., 1985, 1990; Kling et al., 1992).

The CO$_2$ efflux for the nonreactive scenario was calculated first. The concentration at the water/air interface, $c_{w/a}$, was calculated using Henry's law and a partial pressure for CO$_2$ in the air of $10^{-3.5}$ atm. The concentration of dissolved CO$_2$ (or H$_2$CO$_3$*) at the bottom of the diffusive layer, c_w, was calculated assuming that alkalinity is equal to the concentration of bicarbonate and using the pH values and the equation for dissociation of dissolved CO$_2$ into bicarbonate (Drever, 1997). The diffusivity value of CO$_2$ (D_w) in water was 2×10^{-5} cm^2 s^{-1} (Stumm and Morgan, 1996). The efflux of CO$_2$ (equation 1) for the 40 μm diffusive layer was 3.6 g m^{-2} d^{-1} and for the 295 μm layer, 0.5 g m^{-2} d^{-1}.

For the reactive case, the carbon species were assumed to be reacting as they cross the boundary layer. The chemical enhancement factor α was found using the transfer of total inorganic carbon through the diffusive layer ($C_{Tw/a} - C_{Tw}$) and the transfer of only dissolved CO$_2$. At the water bulk side (bottom of the boundary layer), C_{Tw} was found using the dissociation equation for the different carbonate species (Drever, 1997), the pH, and the alkalinity. C_{Tw} was the sum of H$_2$CO$_3$*, HCO$_3^-$, and CO$_3^{2-}$. At the air-water interface, the concentration of H$_2$CO$_3$* was given by Henry's law. Alkalinity and the concentration of H$_2$CO$_3$* allowed the calculation of interface pH (equal to 9.1), HCO$_3^-$, and CO$_3^{2-}$. The chemical enhancement factor

$$\alpha = \frac{\left(\dfrac{c_{Tw} - c_{Tw}}{a} \right)}{\left(c_{H_2CO_{3w}^*} - \dfrac{c_{H_2CO_{3w}^*}}{a} \right)}$$

was 11.3. Substitution of this value in equation 2 gave a CO$_2$ efflux of 38.0 g m^{-2} d^{-1} for the 40 μm boundary layer and 5.2 g m^{-2} d^{-1} for the 295 μm layer. Considering that the range of CO$_2$ efflux from the soils of the caldera is 0.3 to 3.9 g m^{-2} d^{-1}, the 38 g m^{-2} d^{-1}

is too high for the emissions from the water. These values suggest that either the boundary layer is as thick as 295 μm or the chemical reactions as the gas crosses the thin boundary layer can be neglected. The CO$_2$ efflux from Ilopango Lake waters is likely to be somewhere in between 0.5 and 5.2 g m^{-2} d^{-1}.

Total CO$_2$ Emission from Ilopango Caldera

Considering the spatial distribution of CO$_2$ efflux and the area covered, the total CO$_2$ output from the soils of Ilopango Caldera was found to be 594 t d^{-1}. The total emissions from the water for the 0.5 and 5.2 g m^{-2} d^{-1} emission range and the 99.5 km^2 area of the lake, gives a range of 50 to 517 t d^{-1}. Adding the soil and the water gas flows gives a range of 644 to 1111 t d^{-1} of CO$_2$ emitted by Ilopango Caldera. These values are much lower than the CO$_2$ diffuse emissions from nearby San Salvador volcano (4000 t d^{-1}, Pérez et al., this volume, Chapter 17), and Santa Ana–Izalco–Coatepeque volcanic complex (1400 t d^{-1}, Salazar et al., this volume, Chapter 10), but comparable with other volcanoes that are in a quiescent period (e.g., 380 t d^{-1} for Teide volcano, Canary Islands, Spain, Hernández et al., 1998; 111 t d^{-1} for Poás volcano, Melián et al., 2001), suggesting that Ilopango Caldera is in a period of relative quiescence. This is supported by the apparent absence of fumarolic activity in 1999, lake water temperature, and chemical composition of the lake waters (Ransom, 2002).

CONCLUSIONS

The high alkalinity of Ilopango Lake waters (268 ± 15 mg CaCO$_3$ L^{-1}) can be explained by the composition of the rocks and sediments underlying and surrounding the lake. According to Hart (1981) the rocks belong to a calcic volcanic suite. The Cerros Quemados rocks have 4–8% calcium oxide. Minerals like feldspar react with CO$_2$ to produce bicarbonate (Appelo and Postma, 1994) and the high alkalinity of the water. If concentrations are high enough, precipitation of calcite can also occur, as the soil pH seems to indicate in regions of high CO$_2$ emissions. The results presented in this paper, as well as many other studies on diffuse degassing of volcanic systems, should be considered with caution. This study and previous studies of spatial distributions of diffuse degassing do not consider the diurnal and seasonal variations in CO$_2$ efflux and atmospheric and soil gas concentrations. Ideally, measured values should be corrected to reflect the average effluxes and concentrations. However, the location of high anomalies is probably not different for the corrected and uncorrected values. Future studies should be directed to create a methodology to correct the measured values.

Soil diffuse gas concentrations of Ilopango Caldera are within background concentrations of other volcanoes worldwide, including San Salvador volcano, suggesting that the caldera is emitting moderate effluxes of gases, mostly biogenic in the case of CO$_2$. This behavior could be due to a low degassing of the magmatic system, or low-permeability soils covering the caldera, or both. The Tierra Blanca Joven tephra shows low permeability and could explain this low emission of gases.

Various anomalies for radon, thoron, mercury, and carbon dioxide concentrations and efflux do coincide and may be the result of increased fracturing and permeability in those areas. Carbon isotopes in CO_2 ($\delta^{13}C$) show high values at four regions suggesting either mixing of biogenic gases with volcanic gases or the presence of C_4 flora in the region. Anomalies A and D for radon, thoron, and carbon dioxide could be the result of high rock fracturing. Sites D, E, F, and H (Figs. 4 and 5) may have a minor input of volcanic gases or reflect the channeling of shallower gases throughout permeable faults and contacts. Note that anomaly A falls on a fault that follows the NNW trend observed in the seismicity of Figure 3. The high organic matter content at A could explain the high concentration of CO_2 but not radon. The diffuse radon, carbon dioxide, and mercury gases emitted from Ilopango Caldera soils suggest that, in 1999, the caldera was in a quiescent period. Ransom (2002) found that, according to the classification of Varekamp et al. (2000), which is based in the chemical composition of water in volcanic lakes, Ilopango Lake can be classified as a quiescent geothermal lake. However, the identification of warm seepage water within the lake after the 2001 earthquakes suggests that this caldera should be monitored physically and chemically to identify any premonitory signal of magmatic reactivation.

ACKNOWLEDGMENTS

We are deeply grateful to Don Ernesto Freund and the Fundacion de Amigos del Lago de Ilopango for their logistic support and field assistance. We thank K. Notsu for his help in this research. Geotermica Salvadoreña (GESAL) facilitated frequent transportation. The field and technical assistance of Tomas Soriano, Rafael Cartagena, Rodolfo Olmos, and Francisco Barahona from Universidad de El Salvador was very important for the success of this research. We greatly appreciate the work of reviewers J. Varekamp and L.J. Wardell.

REFERENCES CITED

Allard, P., Carbonnelle, J., Dajlevic, D., LeBronec, J., Morel, P., Robe, M.C., Maurenas, J.M., Faivre-Pierret, R., Martin, D., Sabroux, J.C., and Zettwoog, P., 1991, Eruptive and diffuse emissions of CO_2 from Mount Etna: Nature, v. 351, p. 387–391.

Amundson, R., Stern, L., Baisden, T., Wang, Y., 1998, The isotopic composition of soil and soil-respired CO_2: Geoderma, v. 82, p. 83–114.

Appelo, C.A.J., and Postma, D., 1994, Geochemistry, groundwater and pollution: Rotterdam, A.A. Balkema, p. 536.

Arizona Instrument, 1996, Jerome 431-X mercury vapor analyzer operation manual. Part Number SS-086, Doc # 6J21–0001, Rev C., Phoenix, Arizona, p. 60.

Ball, B.C., Scott, A., and Parker, J.P., 1999, Field N_2O, CO_2 and CH_4 fluxes in relation to tillage, compaction and soil quality in Scotland: Soil and Tillage Research, v. 53, p. 29–39.

Baubron, J.C., Allard, P., and Toutain, J.P., 1990, Diffusive volcanic emissions of carbon dioxide from Vulano Island, Italy: Nature (Letters to Nature), v. 344, p. 51–53.

Baubron, J.C., Allard, P., Sabroux, J., Tedesco, P., and Toutain, J.P., 1991, Soil gas emanations as precursory indicators of volcanic eruptions: Geological Society [London] Journal, v. 148, p. 571–576.

Ben-Dor, E., and Banin, A., 1989, Determination of organic matter content in arid-zone soils using a simple "loss-on-ignition" method: Commununications in Soil Science and Plant Analysis, v. 20, no. 15–16, p. 1675–1695.

Buyanovski, G.A., and Wagner, G.H., 1983, Annual cycles of carbon dioxide level in soil air: Soil Science Society of America Journal, v. 47, p. 1139–1145.

Capaldi, G., Pece, R., and Veltri, C., 1992, Radon variation in groundwaters in the Campi Flegrei Caldera (Southern Italy) during and after the 1982–1984 Bradyseismic Crisis: PAGEOPH, v. 138, no. 1, p. 77–93.

Cartagena, R., Olmos, R., López, D.L., Soriano, T., Barahona, F., and Hernández, P., 2004, Diffuse soil degassing of carbon dioxide, radon, and mercury at San Miguel volcano, El Salvador, in Rose, W.I., et al., eds., Natural hazards in El Salvador: Boulder, Colorado, Geological Society of America Special Paper 375, p. 203–212 (this volume).

Chiodini, G., Frondini, F., and Raco, B., 1996, Diffuse emission of CO_2 from the Fossa crater, Vulcano Island (Italy): Bulletin of Volcanology, v. 58, p. 41–50.

Connor, C., Hill, B., LaFemina, P., Navarro, M., and Conway, M., 1996, Soil ^{222}Rn pulse during the initial phase of June–August 1995 eruption of Cerro Negro, Nicaragua: Journal of Volcanology and Geothermal Research, v. 73, p. 119–127.

Deines, P., 1980, The isotopic composition of reduced organic carbon, in Fritz, P., and Fontes, J.Ch., eds., Handbook of environmental isotope geochemistry, 1: The terrestrial environment: Amsterdam, Elsevier, p. 329–406.

Drever, J.I., 1997, The geochemistry of natural waters (3rd edition): New Jersey, Prentice Hall, 436 p.

Dull, R.A., 2001, El bosque perdido: A cultural-ecological history of Holocene environmental change in western El Salvador [Ph.D. thesis]: Berkeley, University of California, 137 p.

Etiope, G., Beneduce, P., Calcara, M., Favali, P., Frugoni, F., Schiattarella, M., and Smriglio, G., 1999, Structural pattern and CO_2-CH_4 degassing of Ustica Island, Southern Tyrrhenian basin: Journal of Volcanology and Geothermal Research, v. 88, p. 291–304.

Farrar, C.D., Sorey, M.L., Evans, W.C., Howie, F.F., Kerr, B.D., Kennedy, B.M., King, C.Y., and Southon, J.R., 1995, Forest-killing diffuse CO_2 emission at Mammoth Mountain as a sign of magmatic unrest: Nature, v. 37, p. 675–678.

Faure, G., 1986, Principles of isotope geology: New York, John Wiley & Sons, 589 p.

Geographical Institute of El Salvador, 1970, Geological map of El Salvador (1964–1970), scale 1:100,000, 6 sheets.

Gerlach, T.M., Doukas, M.P., McGee, K.A., and Kessler, R., 1998, Three-year decline of magmatic CO_2 emissions from soils of a Mammoth Mountain tree kill: Horseshoe Lake, California, 1995–1997: Geophysical Research Letters, v. 25, no. 11, p. 1947–1950.

Granier, A., Ceschia, E., Damesin, C., Dufrene, E., Epron, D., Gross, P., Lebaube, S., Le Dantec, V., Le Goff, N., Lemoine, D., Lucot, E., Ottorini, J.M., Pontailler, J.Y., and Saugier, B., 2000, The carbon balance of a young Beech forest: Functional Ecology, v. 14, p. 312–325.

Hart, W.J.E., 1981, The panchimalco tephra, El Salvador, Central America [M.S. thesis]: New Brunswick, New Jersey, Rutgers University, 101 p.

Heiligmann, M., Stix, J., Williams-Jones, G., Lollar, B.S., and Garzon, G., 1997, Distal degassing of radon and carbon dioxide on Galeras volcano, Colombia: Journal of Volcanology and Geothermal Research, v. 77, p. 267–283.

Hernández, P.A., Perez, N.M., Salazar, J.M., Nakai, S., Notsu, K., and Wakita, H., 1998, Diffuse emission of carbon dioxide, methane, and helium-3 from Teide volcano, Tenerife, Canary Islands: Geophysical Research Letters, v. 25. no. 17, p. 3311–3314.

Hernández, P.A., Mori, T., Natale, G., Notsu, K., Okada, H., Pérez, N.M., Salazar, J.M., Sato, M., Shimoike, Y., and Virgili, G., 2001, Carbon dioxide degassing by active flow from Usu volcano, Japan: Science, v. 292, no. 5514, p. 83–86.

Hoefs, J., 1980, Stable isotope geochemistry: Berlin, Springer Verlag, 208 p.

Hutter, A.R., 1993, Thoron/radon ($^{220}Rn/^{222}Rn$) ratios as indicators of soil gas transport: Geological Society of America Abstracts with Programs, v. 25, no. 6, p. A-195.

Javoy, M., Pineau, F., and Delorme, H., 1986, Carbon and nitrogen isotopes in the mantle: Chemical Geology, v. 57, p. 41–62.

King, C.Y., 1980, Episodic radon changes in Subsurface soil gas along active faults and possible relation to earthquakes: Journal of Geophysical Research, v. 85, p. 3065–3078.

Kling, G.W., Kipphut, G.W., and Miller, M.C., 1992, The flux of CO_2 and CH_4 from lakes and rivers in arctic Alaska: Hydrobiologia, v. 240, p. 23–36.

Lai, L.T., 1999, Geochemical characteristics of radon and mercury distribution in relation to active tectonics in Son La hydropower dam site: Journal of Geology, Series B, v. 13–14, p. 42–46.

Lambert, G., Bristeau, P., and Polian, G., 1976, Emission and enrichments of radon daughters from Etna Volcano magma: Geophysical Research Letters, v. 3, no. 12, p. 724–726.

Maljanen, M., Martikainen, P.J., Aaltonen, H., and Silvola, J., 2002, Short-term variation in fluxes of carbon dioxide, nitrous oxide and methane in cultivated and forested organic boreal soils: Soil Biology & Biochemistry, v. 34, p. 577–584.

Mann, C.P., Stix, J., Vallance, J.W., and Richer, M., 2004, Subaqueous intracaldera volcanism, Ilopango Caldera, El Salvador, Central America, *in* Rose, W.I., et al., eds., Natural hazards in El Salvador: Boulder, Colorado, Geological Society of America Special Paper 375, p. 159–174 (this volume).

Marty, B., and Jambon, A., 1987, C/^{3}He in volatile fluxes from the solid Earth: Implications for carbon geodynamics: Earth and Planetary Science Letters, v. 83, p. 16–26.

McLean, E.O., 1982, Soil pH and lime requirement, *in* Page, A.L., et al., eds., Methods of soil analysis, Part 2. Chemical and microbiological properties—Agronomy Monograph no. 9 (2nd edition): Madison, Wisconsin, American Society of Agronomy and Soil Science Society of America, p. 199–225.

Melián, G., Galindo, I., Salazar, J.M.L., Hernández, P.A., Pérez, N.M., Ramírez, C., Fernández, M., and Notsu, K., 2001, Spatial and secular variations of diffuse CO_2 degassing from Poás volcano, Costa Rica, Central America: Eos (Transactions, American Geophysical Union), v. 82, p. F1332.

Meyer-Abich, H., 1956, Los volcanes activos de Guatemala y El Salvador (America Central): Anales del Servicio Geológico Nacional de El Salvador, 102 p.

Molnar, P., and Sykes, L., 1969, Tectonics of the Caribbean and Middle America regions from focal mechanisms and seismicity: Geological Society of America Bulletin, v. 80, p. 16–1684.

Moren, A.S., and Lindroth, A., 2000, CO_2 exchange at the floor of a boreal forest: Agricultural and Forest Meteorology, v. 101, p. 1–14.

Notsu, K., Wakita, H., and Igarashi, G., 1991, Precursory changes in fumarolic gas temperature associated with a recent submarine eruption near Izu-Oshima volcano, Japan: Geophysical Research Letters, v. 18, p. 191–193.

Pérez, N.M., Salazar, J.M.L., Hernández, P.A., Soriano, T., López, D.L., and Notsu, K., 2004, Diffuse CO_2 and ^{222}Rn degassing from San Salvador volcano, El Salvador, Central America, *in* Rose, W.I., et al., eds., Natural hazards in El Salvador: Boulder, Colorado, Geological Society of America Special Paper 375, p. 227–236 (this volume).

Pylon Electronics, 1993, Pylon Model AB-5 Portable Radiation Monitor Instruction Manual, June 5, Manual Number A900024: Ottawa, Canada, Pylon Electronics, 69 p.

Ransom, L., 2002, Volcanic diffuse soil degassing and lake chemistry of the Ilopango Caldera system, El Salvador, Central America [M.S. thesis]: Athens, Ohio University, 177 p.

Reimer, G.M., 1980, Use of soil-gas helium concentrations for earthquake prediction: Limitations imposed by diurnal variations: Jouranl of Geophysical Research, v. 85, p. 3107–3144.

Reimer, G.M., 1981, Helium soil-gas variations associated with recent Central California earthquakes: Precursor or coincidence?: Geophysical Reseaarch Letters, v. 8, p. 433–435.

Rose, T.P., and Davisson, M.L., 1996, Radiocarbon in hydrologic systems containing dissolved magmatic carbon dioxide: Science, v. 273, p. 1367–1370.

Salazar, J.M.I., Pérez, N.M., Hernández, P.A., Soriano, T., Barahona, F., Olmos, R., Cartagena, R., López, D.L., Lima, N., Melián, G., Castro, L.,

Galindo, I., and Notsu, K., 2002, Precursory diffuse carbon dioxide soil degassing signatures of recent earthquakes in El Salvador, Central America: Earth and Planetary Sciences Letters, v. 205, p. 81–89.

Salazar, J.M.L., Hernández, P.A., Pérez, N.M., Olmos, R., Barahona, F., Cartagena, R., Soriano, T., López, D.L., Sumino, H., and Notsu, K., 2004, Spatial and temporal variations of diffuse CO_2 degassing at Santa Ana–Izalco-Coatepeque volcanic complex, El Salvador, Central America, *in* Rose, W.I., et al., eds., Natural hazards in El Salvador: Boulder, Colorado, Geological Society of America Special Paper 375, p. 135–146 (this volume).

Schwarcz, H.P., 1969, The stable isotopes of carbon, *in* Wedepohl, K.H., ed., Handbook of geochemistry: Berlin, Springer Verlag, 107 p.

Sheets, P.D., 1979, Volcanic disasters and the archaeological record, *in* Grayson, D.K., ed., Volcanic activity and human ecology: New York, Academic Press, 644 p.

Sorey, M.L., Evans, W.C., Kennedy, B.M., Farrar, L.J., Hainsworth, L.J., and Hausback, B., 1998, Carbon dioxide and helium emissions from a reservoir of magmatic gas beneath Mammoth Mountain, California: Journal of Geophysical Research, v. 103, no. B7, p. 15,303–15,323.

Stumm, W., and Morgan, J.J., 1996, Aquatic chemistry (3rd edition): New York, John Wiley & Sons, 1022 p.

Swan, A.R.H., and Sandilands, M., 1995, Introduction to geological data analysis: Cambridge, Massachusetts, Blackwell Science, 446 p.

Tennant, C.B., and White, M.L., 1959, Study of the distribution of some geochemical data: Economic Geology, v. 54, p. 1281–1290.

Thibodeaux, L.J., 1996, Environmental chemodynamics (2nd edition): New York, John Wiley and Sons, 593 p.

Varekamp, J.C., and Buseck, P.R., 1981, Mercury emissions from Mount St. Helens during September 1980: Nature, v. 293, p. 555–556.

Varekamp, J.C., and Buseck, P.R., 1983, Hg anomalies in soils: A geochemical exploration method for geothermal areas: Geothermics, v. 12, p. 29–47.

Varekamp, J.C., and Buseck, P.R., 1984, The speciation of mercury in hydrothermal systems, with application to ore deposition: Geochimica et Cosmochimica Acta, v. 48, p. 177–185.

Varekamp, J.C., Pasternack, G.B., and Rowe, G.L., Jr., 2000, Volcanic lake systematics II. Chemical constraints: Journal of Volcanology and Geothermal Research, v. 97, p. 161–179.

Wanninkhof, R., Ledwell, J.R., and Broecker, W.S., 1985, Gas exchange–wind speed relation measured with sulfur hexafluoride on a lake: Science, v. 227, p. 1224–1226.

Wanninkhof, R., Mulholland, P.J., and Elwood, J.W., 1990, Gas exchange rates for a first-order stream determined with deliberate and natural tracers: Water Resource Research, v. 26, p. 1621–1630.

Weyl, R., 1980, Geology of Central America: Berlin, Gebrüder Borntraeger, 371 p.

White, R., and Harlow, D., 1993, Destructive upper-crustal earthquakes of Central America since 1900: Bulletin of the Seismological Society of America, v. 83, no. 4, p. 1114–1115.

Williams, H., and Meyer-Abich, H., 1955, Volcanism in the southern part of El Salvador, with particular reference to the collapse basins of lakes Coatepeque and Ilopango: University of California Publications in Geological Sciences, v. 32, p. 1–64.

Williams, S.N., 1985, Soil radon and elemental mercury distribution and relation to magmatic resurgence at Long Valley Caldera: Science, v. 229, p. 551–553.

MANUSCRIPT ACCEPTED BY THE SOCIETY JUNE 16, 2003

Geological Society of America
Special Paper 375
2004

Diffuse soil degassing of carbon dioxide, radon, and mercury at San Miguel volcano, El Salvador

Rafael Cartagena
Rodolfo Olmos
Universidad de El Salvador, San Salvador, El Salvador, Central America

Dina L. López*
Department of Geological Sciences, Ohio University, 316 Clippinger Laboratories, Athens, Ohio 45701, USA

Tomás Soriano
Francisco Barahona
Universidad de El Salvador, San Salvador, El Salvador, Central America

Pedro A. Hernández
Nemesio M. Pérez
*Environmental Research Division, Instituto Tecnológico y de Energías Renovables,
38611 Granadilla de Abona, Santa Cruz de Tenerife, Spain*

ABSTRACT

San Miguel volcano in eastern El Salvador is one of the most active volcanoes in Central America. During the last 250 yr, it has erupted at least 28 times. The city of San Miguel (more than 300,000 inhabitants) is located 10 km from the summit. An investigation of the concentration of diffuse soil gases—radon, carbon dioxide, and mercury—and soil fluxes of carbon dioxide was done in December 1999–January 2000. Radon (^{222}Rn) concentrations ranged from 2 to 833 pCi/L (picocuries per liter) with an average of 110 pCi/L. Thoron (^{220}Rn) concentrations ranged from 20 to 2178 pCi/L with an average of 356 pCi/L. These are similar to concentrations measured at other erupting volcanoes of the world.

Carbon dioxide concentrations and fluxes at San Miguel are low, with fluxes ranging from less than 0.1 to 5.0 g m^{-2} d^{-1}, with an average of 1.5 g m^{-2} d^{-1}. These fluxes are within the background levels of San Salvador volcano and the Santa Ana–Izalco–Coatepeque volcanic complex, and also compare with values found at Ilopango Caldera in El Salvador. Mercury soil gas concentrations were also low, with values ranging from zero to 56 µg m^{-3}, with an average of 2 µg m^{-3}. Carbon isotopic compositions indicate that the soil carbon dioxide is predominantly biogenic. The concentrations of the investigated gases as well as the flux of carbon dioxide are generally lower at the higher elevations of the volcanic edifice and higher at the lower elevations, close to NNW-trending faults, and to contacts between the different rock units. The low fluxes of carbon dioxide throughout the soils of San Miguel volcano are possibly due to low permeability of the volcanic cover, which is thicker at the higher elevations, and to ready degassing through the open volcanic pipe. This low flux is also consistent with the reported small but frequent historical eruptions of this volcano and its low SO$_2$ fluxes.

Keywords: diffuse degassing, magmatic hydrothermal system, carbon dioxide, radon, mercury.

---

*dlopez@ohio.edu

Cartagena, R., Olmos, R., López, D.L., Soriano, T., Barahona, F., Hernández, P.A., and Pérez, N.M., 2004, Diffuse soil degassing of carbon dioxide, radon, and mercury at San Miguel volcano, El Salvador, *in* Rose, W.I., Bommer, J.J., López, D.L., Carr, M.J., and Major, J.J., eds., Natural hazards in El Salvador: Boulder, Colorado, Geological Society of America Special Paper 375, p. 203–212. For permission to copy, contact editing@geosociety.org. © 2004 Geological Society of America

INTRODUCTION

San Miguel volcano is a stratovolcano located in eastern El Salvador (Fig. 1) at 13.43° N, 88.27° W, and elevation 2130 m. It is one of the most active volcanoes in Central America, and the second in number of historical eruptions in El Salvador after Izalco volcano. According to Simkin and Siebert (2000), San Miguel has erupted in 1510, 1699, 1762, 1769, 1787, 1819, 1844–1848, 1855, 1857, 1862, 1867–1868, 1882, 1884, 1890–1891, 1919–1920, 1920–1925, 1929, 1930, 1931, 1939, 1954, 1964, 1966, 1967, 1970, 1976–1977, 1985–1986, and 1995, and probably (uncertain) in 1798, 1811, 1854, and 1936. If we consider that before the eighteenth century, observations of this volcano were probably not well documented, the number of eruptions indicates that the probability of an eruption in a given year is 16%. During the last thirteen years, San Miguel has been degassing with a small sulfur dioxide plume and occasional small ash emissions.

San Miguel volcano is located in one of the most populated and economically active regions of El Salvador. The city of San Miguel is located only 10 km east of the summit. It is the third in size in El Salvador, with more than 300,000 inhabitants, and it is the center of economic activity for eastern El Salvador. Several towns and small villages surround the volcano, and its flanks are covered with coffee plantations. Considering its frequent eruptive activity and its location, San Miguel is a volcano that should be investigated and monitored to understand its behavior and to alert the population in case of an incoming eruption. However, as for most of El Salvador's volcanoes, detailed investigation of the volcanic products and gas emissions has been done only recently (Chesner et al., this volume, Chapter 16; Johnson et al., 1999; Major et al., 2001). During more than fifteen years, geologic studies in this country were not possible due to the internal war that ended in 1992 and the lack of Salvadorian geologists.

In this paper, we present the results of the first investigations of diffuse soil degassing at San Miguel volcano. Diffuse degassing has been recognized as an important gas release mechanism of magmatic systems (e.g., Hernández et al., 2001; Giammanco et al., 1995), as well as a good indicator for seismic activity and fault permeability (e.g., Salazar et al., 2002; Etiope et al., 1999). The purpose of this research was to determine the concentrations of radon, thoron, carbon dioxide (CO_2), and mercury in soil gases of San Miguel volcano, as well as the thoron/radon ratio and the flux of CO_2 from the soils to the atmosphere, in order to identify regions of higher concentrations and flux. The spatial distribution of radon, CO_2, and mercury concentrations and CO_2 flux, correlated with the geological structures of the volcano, allows a better understanding of the flow of gases within the volcanic edifice and the state of volcanic activity.

Geological Setting

The volcanic chain in El Salvador runs parallel to the Pacific Coast, from northwest to southeast, and is related to the subduction of the Cocos plate below the Caribbean plate, defined by the

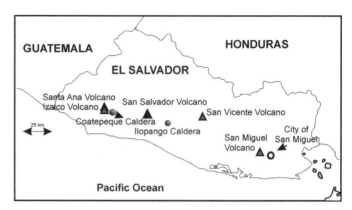

Figure 1. San Miguel volcano is located in eastern El Salvador. San Miguel City is located 10 km from the volcano's summit.

Middle American Trench (Molnar and Sykes, 1969). A normal graben (Median Trough) formed during the Pliocene-Pleistocene crosses El Salvador and continues along Nicaragua parallel to the Middle American Trench (Meyer-Abich, 1956). The volcanic chain in El Salvador (and San Miguel volcano) is aligned with the southern margin of this trough. San Miguel is a stratovolcano with a relief of ~2000 m. The summit crater is ~1 km in diameter.

Meyer-Abich (1956) stated that the older lava flows of San Miguel are olivine-pyroxene basalts. Figure 2 shows a geologic map of the San Miguel volcano area and sampling sites for this investigation. The volcano lies on top of volcanic rocks from the Pliocene-Pleistocene Cuscatlán Formation. The more recent lava flows and pyroclastic deposits of the San Salvador Formation have basaltic to intermediate composition (Geographical Institute of El Salvador, 1978). A stratigraphic sequence of 22 consecutive lava flows has been identified within the crater walls (Chesner et al., this volume, Chapter 16), with 12 lava flows erupted since 1699, 9 of them at flank vents (4 from the SE flank, 3 from the N flank, 1 from the NE flank, and 1 from the WSW flank) and 3 from the summit crater. An upward migration of the vents has been proposed (Chesner et al., this volume) because the older historic lava flows (years 1699, 1762, 1787, and 1819) were erupted at a lower elevation (400–500 m), the next lava flows (years 1844, 1848, 1855, and 1867) were erupted at 800–1200 m, and the more recent ones were erupted at the summit crater (years 1884, 1931, and 1976). The geological record indicates that large ashfalls, scoria falls, phreatomagmatic eruptions, and pyroclastic flows are not frequent. During periods of heavy rainfall, minor debris flows have occurred at the NW flank (Major et al., 2001; Chesner et al., this volume). The composition of San Miguel's products (Chesner et al., this volume) is calc-alkaline, medium-K basalts and basaltic andesites, with SiO_2 ranging between 51 and 53%, and relatively high Cu content, typically between 170 and 210 ppm. The minerals include phenocrysts of plagioclase, olivine, orthopyroxene, and magnetite, with occasional clinopyroxene.

Major et al. (2001) stated that the historical eruptions of San Miguel have been mostly quiescent emplacements of lava flows

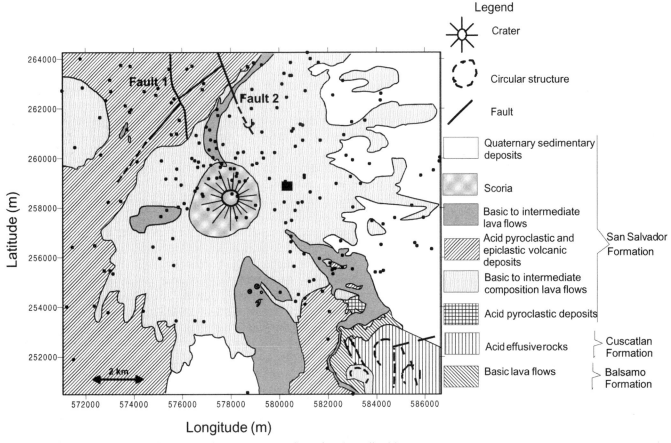

Figure 2. Geologic map of San Miguel volcano (after Geographical Institute of El Salvador, 1978) and sampling sites for the survey of soil gases.

and modest tephra falls. Considering the chemical composition of San Miguel's volcanic rocks and its historical relatively quiet eruptions, Major et al. (2001) suggested that the prehistoric activity of San Miguel was probably similar. However, the older rocks of San Miguel have not been investigated and a different past behavior cannot be excluded. A change in slope close to the summit suggests that San Miguel has suffered collapses in the past and that the summit region could have been destroyed by violent eruptions.

Pyroclastic deposits cover San Miguel volcano close to the summit area, and heavy rainfall produces small lahars. These deposits are thicker at the summit area and are a threat to the population, especially at the NW slope where the extension of fault 1 (Fig. 2) bisects the crater, forming a Y-shaped depression. Gases are emitted along this Y region, suggesting that volcanic gases propagate within the fault plane. It has been recognized that acidic volcanic fluids circulating along fault zones alter the surrounding rocks to clay minerals, facilitating rock movement when external triggers such as earthquakes, hurricanes, or strong winds act on the gravitationally unstable edifice (López and Williams, 1993). Unfortunately, this area of the volcano is difficult to access and a good distribution of soil gas samples could not be taken.

The occurrence of landslides and associated lahars has been recognized as one of the major hazards at San Miguel volcano. However, Major et al. (2001) have found that the distribution of lahar deposits at San Miguel suggests that most eruptions have not produced large landslides and lahars, and that large lahar deposits produced during noneruptive periods are not evident. Small lahar deposits (less than 1 million cubic meters) generated by rainfall and erosion processes of the minor tephra falls are common. According to Major et al. (2001), future eruptions of San Miguel are likely to be small and to produce relatively minor lahars.

Diffuse Degassing of CO_2, Radon, and Mercury

The term "diffuse degassing" at volcanic systems is used to identify the gases released from the magmatic system, incorporated to the hydrothermal envelope, and released to the atmosphere throughout the volcanic edifice (e.g., Hernández et al., 2001; Chiodini et al., 1996). In recent years, the importance of this type of degassing in volcanoes has been recognized. In 1990, at Mammoth Mountain in California, high emissions of CO_2 killed trees and burned vegetation, signaling unrest of the magmatic system (Sorey et al., 1998; Farrar et al., 1995; McGee and Gerlach, 1998). Chiodini et al. (1996) showed that diffuse emissions at the Fossa crater, Vulcano Island, are of the same order of magnitude as those released by the high-temperature fumaroles. Diffuse emissions of CO_2, methane, and ^3He from Teide volcano, Canary Islands, have been shown to correlate with geothermal anomalies and the distribution of faults and dikes of the three main volcanic rift zones surrounding the volcano (Hernández et al., 1998). More recently, the diffuse emissions of CO_2 at Usu volcano, Japan, were found to increase at the summit area (from 120 to 340 t d^{-1}) prior to the March 2000 eruption, and decrease (to 39 t d^{-1}) after the event (Hernández et al., 2001). Correlation between gas emissions and permeable faults has been identified at several sites (e.g., at Ustica Island, Etiope et al., 1999; at Ahuachapán Geothermal Field, Magaña et al., 2002). Diffuse degassing is also affected by seismic activity. Recent studies at San Vicente volcano, El Salvador, found an increase in CO_2 flux at a continuous monitoring station before an earthquake in May 2001 (an aftershock of the February 13, 2001, earthquake) (Salazar et al., 2002). Changes in the stress field of Earth's crust prior to earthquakes produce changes in gas pressure and emissions.

Studies of radon (^{222}Rn) concentrations and its isotope thoron (^{220}Rn) in other volcanic systems of the world have shown that radon emanations are related to volcanic and seismic activity. At Galeras volcano, Colombia, the onset of volcano-seismic activity was preceded by increases of radon at stations located on the flanks of the volcano (Heiligmann et al., 1997). During the 1995 eruption of Cerro Negro volcano, Nicaragua, a pulse of ^{222}Rn was identified in the soils of the volcano, suggesting that even movement of small volumes of magma can produce changes in radon concentrations (Connor et al., 1996). At Long Valley Caldera, a high radon and low mercury anomaly was identified at the zone of dike intrusion and associated seismicity, uplift, and ground deformation that started in 1978 (Williams, 1985). A similar correlation between high seismicity, high radon, and depletion of mercury was detected at Rabaul Caldera, Papua New Guinea (Pérez and Williams, 1990). Radon and mercury have also been used at Masaya Caldera Complex to identify fault traces (Crenshaw et al., 1982).

Magma, wallrock, and hydrothermal fluids can release ^{222}Rn gas that diffuses throughout the soils of volcanoes. Connor et al. (1996) indicated that if diffusion is the dominant transport mechanism, the mean depth from which ^{222}Rn escapes from soils is ~2 m. This short distance suggests that anomalous concentrations of radon in volcanic regions are produced by rapidly moving hydrothermal fluids and not by diffusion. Laboratory experiments with andesitic melts suggest that the exsolution of water and other major gases is responsible for the transport of trace gases like radon (and mercury) from the melt to the surrounding hydrothermal system (Gauthier et al., 1999).

METHODOLOGY

The gas survey at San Miguel volcano was done during December 1999 and January 2000. Samples were taken at 205 stations, whose locations are shown in Figure 2. Note that the southern slope of the volcano was not well covered by sampling sites. The reason is that this area is covered by low-permeability lava flows and lacks good soil coverage. A Pylon Model AB-5 Portable Radiation Monitor and a Pylon Model 300A Lucas Cell were used for the radon measurements. Gas samples at ~60 cm depth were taken at each sampling site. A vacuum pump was used to extract the gas sample through a sonde into the Lucas Cell. Before the gas sample was taken, the gas was purged three times to avoid atmospheric contamination. The radiation monitor detects the alpha particles produced during the decay process of radon and its daughter products. The alpha particles hit the walls of the Lucas Cell and produce light pulses that are amplified by a photomultiplier tube and converted into electronic pulses. Counting the number of pulses (proportional to the number of alpha particles) emitted per minute allows the calculation of the radon and thoron concentrations (Pylon Electronics, 1993).

Gas samples for carbon dioxide concentration, carbon isotopes, and mercury were extracted at ~60 cm depth using a sonde with a septum and a 60 cc syringe. Gas was purged from the soil three times to avoid atmospheric contamination. CO_2 concentrations were measured using a LI-COR Gas Hound CO_2 Analyzer Model LI-800. Gas samples for carbon isotopes were collected in a vacutainer and sent to the Laboratory for Earthquake Chemistry, University of Tokyo, for analysis of $\delta^{13}C$ in CO_2. CO_2 flux was measured using the accumulation chamber method (Baubron et al., 1990). In this method, the accumulation of CO_2 in a chamber collecting the emitted CO_2 from the soil produces a change in concentration in the chamber's air. The slope of concentration versus time is proportional to the flux of soil CO_2. The LI-COR analyzer was used to detect the change in concentration of CO_2 in the chamber through time.

Mercury was analyzed using the Jerome 431-X Mercury Vapor Analyzer. In this analyzer, mercury is adsorbed to a gold film, changing the electrical resistance of the film. The change in electrical resistance is proportional to the concentration of mercury in the circulating air, and measured using a Wheatstone bridge circuit (Arizona Instrument, 1996).

DISCUSSION OF RESULTS

The results of this investigation are listed in Table 1 and displayed in Figures 3, 4, and 5. The main findings for every one of the surveyed gases are discussed next.

Radon

Radon (^{222}Rn) concentrations range from 2 to 833 pCi/L (Table 1). Inflection points in plots of logged concentrations versus accumulative frequency distribution (Table 1 and Fig. 3A) allow the identification of three distinct populations (Tennant and White, 1959). Figure 4A displays a contour map of radon concentrations. Several anomalous regions within the study area can be identified: A, B, C, and D. Note that higher values occur at the flanks of the

TABLE 1. CONCENTRATIONS OF RADON (^{222}Rn), THORON (^{220}Rn), AND CO_2 SOIL GASES, δ^{13}C ISOTOPIC COMPOSITION OF SOIL CO_2, AND FLUX OF CO_2 AT SAN MIGUEL, VOLCANO, EL SALVADOR

Soil gas	Range and average	Range background population & average	Range threshold population & average	Range peak population & average
Radon (^{222}Rn)	2–833 pCi/L 110 pCi/L	2–52 pCi/L 28 pCi/L	52–72 pCi/L 62 pCi/L	72–833 pCi/L 187 pCi/L
Thoron (^{220}Rn)	20–2178 pCi/L 356 pCi/L	20–400 pCi/L 198 pCi/L	400–612 pCi/L 488 pCi/L	630–2178 pCi/L 849 pCi/L
Thoron/Radon	1.3–61.6 4.4	1.3–3.8 2.9	3.8–6.9 5.0	7.1–61.6 15.2
CO_2	85–1175 ppm 415 ppm			
CO_2 flux	<0.1–5.0 g m^{-2}day^{-1} 1.5 0 g m^{-2}day^{-1}			
δ^{13}C in CO_2	–32.8 to –21.4‰ –25.9‰			
Mercury	0–56 µg m^{-3} 2 µg m^{-3}			

Note: Date of Survey: December 1999–January 2000.

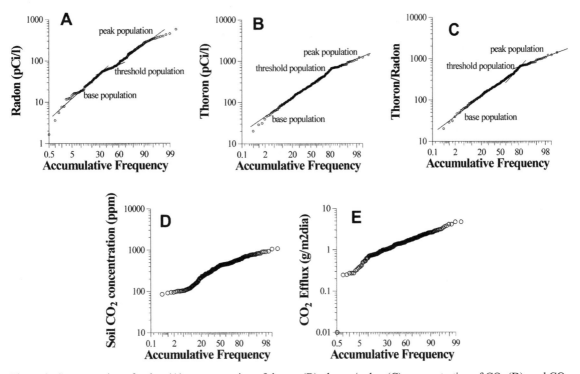

Figure 3. Concentration of radon (A), concentration of thoron (B), thoron/radon (C), concentration of CO_2 (D), and CO_2 flux (E) in soil gases versus accumulative frequency at San Miguel volcano. See text for discussion.

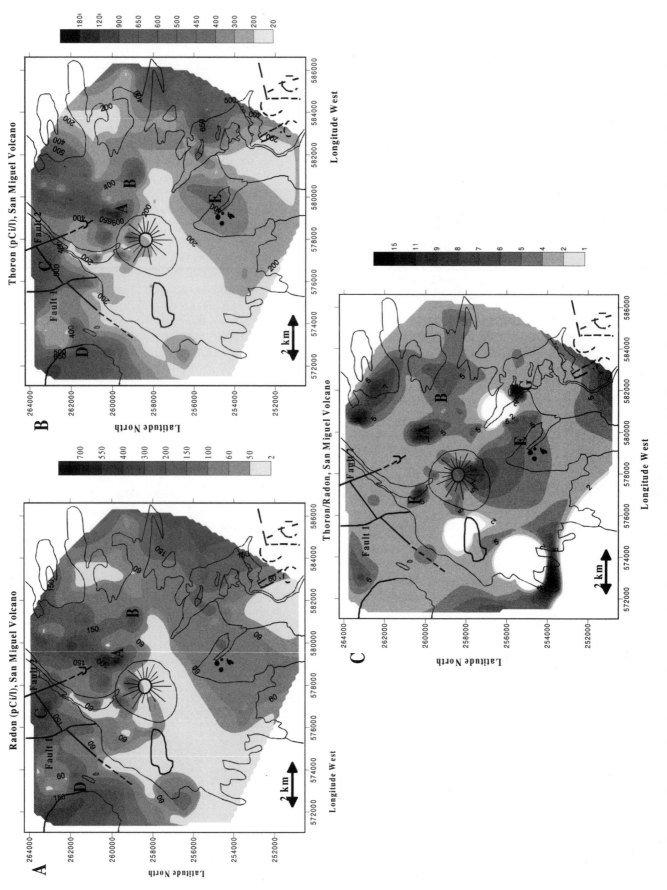

Figure 4. Concentration of radon (A), concentration of thoron (B), and thoron/radon (C) in soil gases at San Miguel volcano. Anomalies A, B, C, D, E, and F are shown.

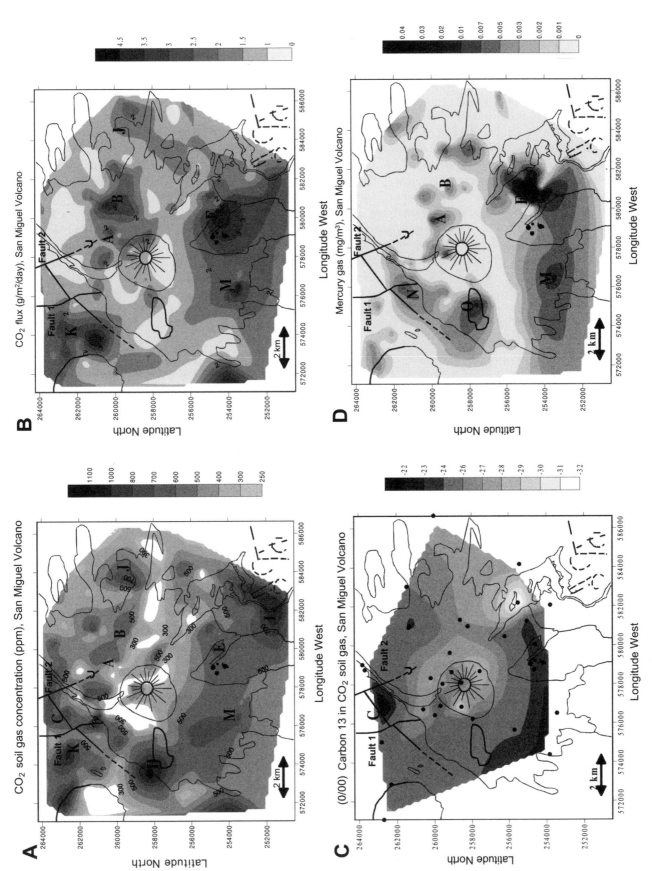

Figure 5. Concentration of CO_2 (A), CO_2 flux in soil gases (B), isotopic composition of CO_2 soil gases (C), and concentration of mercury gas (D) in soils of San Miguel volcano. Anomalies A to O are shown.

volcano, close to the contacts between different volcanic deposits, rather than close to the summit. Anomaly A occurs at the extension of fault 2, and anomaly C along a SW-NE–trending fault.

Thoron (^{220}Rn) concentrations (Table 1 and Fig. 3B) range from 20 to 2178 pCi/L. Three populations are identified. Figure 4B shows that the thoron anomalies are similar to the radon anomalies A–D. Another anomaly, E, is detected along the extension of fault 1. Thoron/radon ratios (Table 1 and Fig. 3C) range from 1.3 to 61.6. Three populations are identified (Table 1). Anomalies A, B, and E are found again in Figure 4C for the thoron/radon ratio. In addition, anomalies G and F should be noted. Anomalies A and G are oriented along the extension of fault 2, anomalies F and E along the extension of fault 1. The NE and SW regions of the study area also present relatively high values of the thoron/radon ratio. Along that direction we also see anomaly A, suggesting the possible existence of a buried fault, parallel to the fault that runs perpendicular to fault 1 and fault 2. These fault systems occur in all El Salvador (Weyl, 1980) and are the result of the tectonic stresses in the region. These results suggest that emissions of radon and thoron at San Miguel volcano are structurally controlled. Radon is released throughout permeable faults and contacts between different rock units.

The radon and thoron emissions from San Miguel volcano can be compared with those of other volcanic systems in El Salvador and the rest of the world. Similar radon surveys have been carried out in two other Salvadorian volcanoes. At San Salvador volcano a range of 1–284 pCi/L with a mean of 23 pCi/L has been measured (Pérez et al., this volume, Chapter 17). At Ilopango Caldera, radon concentrations ranged from 1 to 109 pCi/L with a mean of 14 pCi/L, and thoron concentrations ranged from 7 to 272 pCi/L with a mean of 50 pCi/L (López et al., this volume, Chapter 14). Peak values at San Miguel for radon and thoron are one order of magnitude higher than at San Salvador and Ilopango volcanoes. Radon emissions at San Miguel are similar to those of erupting volcanoes like Cerro Negro in Nicaragua, where a range of 7 to 720 pCi/L of soil radon was detected during the June 1995 eruption (Connor et al., 1996). It should be noted that emissions at Cerro Negro were one order of magnitude lower one year before this eruption. San Miguel's radon values are also similar to those found at Galeras volcano in Colombia, where values up to 1020 ± 60 pCi/L were measured close to active faults. Radon content in spring waters at Karymsky volcano, Kamchatka, was reported to increase from 50 to 350 pCi/L during the 1970 eruption (Gasparini and Mantovani, 1978). The concentrations of radon and thoron detected at San Miguel during the period of our experiment are therefore high and comparable with those of erupting volcanoes.

Carbon Dioxide

Carbon dioxide concentrations (Table 1 and Fig. 3D) range from 85 to 1175 ppm, with an average of 415 ppm. Two inflection points are observed in this graph at low concentration values, ~100 and 300 ppm. However, only six data points have values lower than 100 ppm and 81 have values between 100 and 350 ppm. These two populations make only 42% of the values, imply-

ing that the anomalous population is the low values rather than the high values. These low values are below the average value for the average atmospheric CO_2 concentration of 365 ppm. Diurnal and seasonal variations in the concentration of CO_2 in the atmosphere and in interstitial soil gases have been observed elsewhere, and a correlation with ambient temperature has been suggested (e.g., Maljanen et al., 2002; Ball et al., 1999; Granier et al., 2000). In addition, diurnal variations of CO_2 flux have been detected at San Vicente volcano in El Salvador (Salazar et al., 2002) as well as in a recently installed CO_2 flux monitoring station in San Miguel (Fig. 2) (N. Pérez, 2002, personal commun.). These variations in flux suggest variations in CO_2 concentrations.

CO_2 concentrations present several higher-value anomalies, the most important being A, B, C, E, H, I, J, K, and M (Fig. 5A). Higher soil CO_2 concentration values are observed at the lower elevations and to the N, NW, and SE regions. Anomalies are again frequently located close to contacts between different rock units. The region of faults 1 and 2 and the fault connecting them shows significantly high values of CO_2. Anomalies E and I seem to be aligned with fault 1.

CO_2 fluxes (Table 1 and Fig. 3E) at San Miguel are low. Values range from less than 0.1 to 5.0 g m^{-2} d^{-1}, with an average of 1.5 g m^{-2} d^{-1}. As for CO_2 concentrations, the two inflection points in Figure 3E suggest that the low values are the anomalous population, with only six data points with values lower than 0.3 g m^{-2} d^{-1} and 35 points between 0.3 and 0.9 g m^{-2} d^{-1}. These two populations make only the 20% of the total population and cannot be considered a background population. Similarly to ^{222}Rn, ^{220}Rn, and CO_2 concentrations, higher values are encountered at the lower elevations of the volcano (Fig. 5B) than close to the summit. Several high-value anomalies can be distinguished (Fig. 5B), including regions A, B, E, J, K, and M. All these anomalies are also present on the contour map of CO_2 concentrations.

Carbon isotopes (δ^{13}C in CO_2) were determined in 32 of the 205 sampling sites (Table 1 and Fig. 5C). The data range from −32.8 to −21.4‰, with an average of −25.9‰. According to Faure (1986), biogenic carbon has δ^{13}C values between −24 and −34‰ and most magmatic carbon ranges between −2 and −8‰. Other authors assign a range between −19 and −25‰ for forest soils (Farrar et al., 1995). Because the highest value at San Miguel is only −21.4%, the soil gases sampled at this volcano are predominantly biogenic. However, the spatial distribution of these values is compatible with the degassing of CO_2 and radon observed in the previous figures. Higher values of δ^{13}C are found at the lower elevations, and lower, more biogenic values, at the higher elevations. This behavior suggests that soil gases are receiving small contributions of volcanic CO_2 at the lower elevations of the volcano where other gases, like radon, probably also are discharged. Note the existence of a high anomaly at C, which is also observed in the other contour maps.

The CO_2 fluxes at San Miguel can be compared with those at other volcanoes in El Salvador and the world. Fluxes of CO_2 have been determined at San Salvador volcano, with a maximum value of 213 and a mean of 13 g m^{-2} d^{-1} (Pérez et al., this volume Chapter 17), at the Santa Ana–Izalco–Coatepeque volcanic complex, with a maximum value of 293 g m^{-2} d^{-1} (Salazar et al., this volume, Chapter 10), and at Ilopango caldera, with a maximum value of 3.9 g m^{-2} d^{-1} and a mean of 1.6 g m^{-2} d^{-1} (López et al., this volume, Chapter 14). The maximum diffuse fluxes at San Miguel are much lower than at San Salvador volcano and the Santa Ana–Izalco–Coatepeque volcanic complex. However, the San Miguel CO_2 fluxes are similar to the background emissions in these volcanoes and similar to the emissions at Ilopango caldera. It should be noted that the high fluxes in San Salvador and the Santa Ana complex are located in areas where convective heat and gas transfer are dominant (areas of fumaroles). In comparison, fumaroles or hot springs are not observed on the area of Ilopango caldera exposed to air and outside of the crater area of San Miguel volcano. Integration of the total area covered by this investigation and the flux of CO_2 values give a total emission of 284 t d^{-1} of CO_2 through the soils of San Miguel volcano. This value is much lower than the emissions from San Salvador volcano (4000 t d^{-1}, Pérez et al., this volume, Chapter 17), Ilopango caldera (between 644 and 1111 t d^{-1}, López et al., this volume, Chapter 14), and the Santa Ana–Izalco–Coatepeque volcanic complex (1400 t d^{-1}, Salazar et al., this volume Chapter 10). This low CO_2 diffuse flux compares with the low SO_2 flux of 200 t d^{-1} measured at the plume of San Miguel (W. Rose, 2003, personal commun.). The low fluxes of CO_2 are unexpected because they compare with fluxes at volcanoes like Teide in the Canary Islands, which are in a quiescent period (Hernández et al., 1998). Unlike them, San Miguel is not quiescent; it has had frequent eruptions and is passively degassing at the present time.

Mercury

Mercury soil gas concentrations (Table 1 and Fig. 5D) at San Miguel were also low. Values ranged from zero to 56 µg m^{-3}, with a mean of 2 µg m^{-3}. Anomalies observed in the mercury gas concentration were identified as A, B, E, M, N, and O. As in the case of radon and CO_2, higher values are found at lower elevations. Anomalies N and E and the small anomaly close to the crater are aligned with fault 1. The mercury soil gas concentrations at San Miguel are lower than those measured at the Masaya Caldera Complex (mean of 35.8 ppb or 41.5 mg m^{-3}) and Long Valley Caldera (peak population with a mean of 260 ppb or 302 mg m^{-3}, Williams, 1985), but they are similar to those found in nearby Ilopango Caldera (0 to 16 µg m^{-3}, with a mean of 2 µg m^{-3}, López et al., this volume, Chapter 14).

Of the different anomalies identified in Figures 4 and 5, A, B, and E are especially important. A and B occur in all the gas diagrams, and E occurs in all except the radon concentration maps. Anomalies A and B could be interesting targets for geochemical monitoring of gases at San Miguel. E is located close to vents that have erupted lava flows in the recent past (Fig. 2). Another anomaly of interest is M, which occurs for CO_2 concentration, CO_2 flux, and mercury concentration. Monitoring of these four points throughout frequent measurements of soil gases (at least radon and CO_2) could give important information about the soil gas concentrations and

fluxes and their relationship with the volcano behavior. After the 2001 earthquakes, a continuous monitoring geochemical station to measure fluxes of CO_2 was installed close to site A (Fig. 2).

CONCLUSIONS

At San Miguel volcano, emissions of radon seem to be related to the zones of high permeability along faults 1 and 2, with some anomalies located close to contacts between different rock units. Concentrations of radon are high and similar to soil radon measured in other eruptive volcanoes of the world. In comparison, fluxes of CO_2 are low and their isotopic composition suggests that the CO_2 is mostly biogenic, with a few sites close to faults and contacts emitting slightly higher fluxes. This suggests that the radon is probably released at the rock cover rather than being magmatic. The CO_2 fluxes emitted from the magma chamber and injected into the hydrothermal envelope are not high enough to perturb significantly the isotopic composition of biogenic gases emitted through the soils.

The low CO_2 fluxes and mercury concentrations at San Miguel volcano were not expected for an active volcano that is releasing a gas plume. However, the distribution of values observed in the maps on Figures 4 and 5—with higher values usually located at lower elevations on the volcanic edifice, along faults, and close to contacts—suggests that the reason for this spatial distribution is that the low-permeability volcanic products of the thicker parts of the volcano do not allow for high gas fluxes. Gases can migrate and discharge at the air-soil interface (1) close to permeable faults that are not completely covered at the lower elevations or (2) at contacts between different rock units. This behavior of the discharge of gases and concentrations in the soils is consistent with the lack of thermal springs at San Miguel volcano. Previous studies have shown that heat is transported advectively by groundwater from the recharge zone (at higher elevations) and from the core of mountains to the adjacent valleys, preferentially through permeable faults (Forster and Smith, 1989; López and Smith, 1995). Heat and gases are transported by similar transfer mechanisms. At San Miguel, a low-permeability thick cover in the recharge zone (upper elevations) is probably precluding the migration of gases, and the small fluxes that are migrating are discharged at the moderate-permeability zones of the volcano's flanks. The flux of CO_2 is low because there is an open conduit (volcanic pipe) that is allowing the release of gases to the atmosphere. Gases will migrate through the path of least resistance; in this case, the volcanic pipe. Only if the amount of gas released from the magmatic environment increases significantly could the increase in pressure produce the injection of higher fluxes of gases into the hydrothermal system and increase the soil fluxes. During the time of this investigation, the CO_2 released from San Miguel's magma chamber was not enough to inject high fluxes into the hydrothermal system. This behavior could change during periods of increased volcanic activity. These results are important to the monitoring of San Miguel and other volcanoes of the world. Consideration should be given to fluid flow, heat, and solute transfer in order to identify the areas of higher discharge of heat and gases and their interaction with the magmatic system.

ACKNOWLEDGMENTS

We thank K. Notsu for his help in this research. Fieldwork for this investigation was supported by a grant from Universidad de El Salvador. We deeply appreciate the changes suggested by reviewers William Rose and Lizzette A. Rodriguez.

REFERENCES CITED

Arizona Instrument, 1996, Jerome 431-X mercury vapor analyzer operation manual. Part Number SS-086, Doc # 6J21-0001, Rev C., Phoenix, Arizona, p. 60.

Ball, B.C., Scott, A., and Parker, J.P., 1999, Field N_2O, CO_2 and CH_4 fluxes in relation to tillage, compaction and soil quality in Scotland: Soil and Tillage Research, v. 53, p. 29–39.

Baubron, J.C., Allard, P., and Toutain, J.P., 1990, Diffusive volcanic emissions of carbon dioxide from Vulcano Island, Italy: Nature (Letters to Nature), v. 344, p. 51–53.

Chesner, C.A., Pullinger, C.R., and Escobar, C.D., 2004, Physical and chemical evolution of San Miguel volcano, El Salvador, *in* Rose, W.I., et al., eds., Natural hazards in El Salvador: Boulder, Colorado, Geological Society of America Special Paper 375, p. 213–226 (this volume).

Chiodini, G., Frondini, F., and Raco, B., 1996, Diffuse emission of CO_2 from the Fossa crater, Vulcano Island (Italy): Bulletine of Volcanology, v. 58, p. 41–50.

Connor, C., Hill, B., LaFemina, P., Navarro, M., and Conway, M., 1996, Soil ^{222}Rn pulse during the initial phase of June–August 1995 eruption of Cerro Negro, Nicaragua: Journal of Volcanology and Geothermal Research, v. 73, p. 119–127.

Crenshaw, W.B., Williams, S.N., and Stoiber, R.E., 1982, Fault location by radon and mercury detection at an active volcano in Nicaragua: Nature, v. 300, p. 345–346.

Etiope, G., Beneduce, P., Calcara, M., Favali, P., Frugoni, F., Schiattarella, M., and Smriglio, G., 1999, Structural pattern and CO_2-CH_4 degassing of Ustica Island, Southern Tyrrheanian basin: Journal of Volcanology and Geothermal Research, v. 88, p. 291–304.

Farrar, C.D., Sorey, M.L., Evans, W.C., Howie, F.F., Kerr, B.D., Kennedy, B.M., King, C.Y., and Southon, J.R., 1995, Forest-killing diffuse CO_2 emission at Mammoth Mountain as a sign of magmatic unrest: Nature, v. 37, p. 675–678.

Faure, G., 1986, Principles of isotope geology: New York, John Wiley & Sons, 589 p.

Forster, C., and Smith, L., 1989, The influence of groundwater flow on thermal regimes in mountainous terrain: A model study: Journal of Geophysical Research, v. 94, p. 9439–9451.

Gasparini, P., and Mantovani, M.S.M., 1978, Radon anomalies and volcanic eruptions: Journal of Volcanology and Geothermal Research, v. 3, p. 325–341.

Gauthier, P.J., Condomines, M., and Hammouda, T., 1999, An experimental investigation of radon diffusion in an anhydrous andesitic melt at atmospheric pressure: Implications for radon degassing from erupting magmas: Geochimica et Cosmochimica Acta, v. 63, no. 5, p. 645–656.

Geographical Institute of El Salvador, 1978, Geological map of El Salvador (1964–1970), scale 1:100,000, 6 sheets.

Giammanco, S., Gurrieri, S., and Valenza, M., 1995, Soil CO_2 degassing on Mt Etna (Sicily) during the period 1989–1993; discrimination between climatic and volcanic influences: Bulletin of Volcanology, v. 57, p. 52–60.

Granier, A., Ceschia, E., Damesin, C., Dufrene, E., Epron, D., Gross, P., Lebaube, S., Le Dantec, V., Le Goff, N., Lemoine, D., Lucot, E., Ottorini, J.M., Pontailler, J.Y., and Saugier, B., 2000, The carbon balance of a young Beech forest: Functional Ecology, v. 14, p. 312–325.

Heiligmann, M., Stix, J., Williams-Jones, G., Lollar, B.S., and Garzon, G., 1997, Distal degassing of radon and carbon dioxide on Galeras volcano, Colombia: Journal of Volcanology and Geothermal Research, v. 77, p. 267–283.

Hernández, P.A., Perez, N.M., Salazar, J.M., Nakai, S., Notsu, K., and Wakita, H., 1998, Diffuse emission of carbon dioxide, methane, and helium-3 from Teide volcano, Tenerife, Canary Islands: Geophysical Research Letters, v. 25. no. 17, p. 3311–3314.

Hernández, P.A., Mori, T., Natale, G., Notsu, K., Okada, H., Perez, N.M., Salazar, J.M., Sato, M., Shimoike, Y., and Virgili, G., 2001, Carbon dioxide degassing by active flow from Usu volcano, Japan: Science, v. 292, no. 5514, p. 83–86.

Johnson, D.M., Hooper, P.R., and Conrey, R.M., 1999, XRF analysis of rocks and minerals for major trace elements on a single low dilution Li-tetraborate fused bead: Advances in X-ray Analysis, v. 41, p. 843–867.

López, D.L., and Smith, L., 1995, Fluid flow in fault zones: Analysis of the interplay of convective circulation and topographically driven groundwater flow: Water Resources Research, v. 31, no. 6, p. 1489–1503.

López, D.L., and Williams, S.N., 1993, Catastrophic volcanic collapse; relation to hydrothermal processes: Science, v. 260, no. 5115, p. 1794–1796.

López, D.L., Ransom, L., Pérez, N.M., Hernández, P.A., and Monterrosa, J., 2004, Dynamics of diffuse degassing at Ilopango Caldera, El Salvador, *in* Rose, W.I., et al., eds., Natural hazards in El Salvador: Boulder, Colorado, Geological Society of America Special Paper 375, p. 191–202 (this volume).

Magaña, M.I., López, D.L., Tenorio, J., and Matus, A., 2002, Radon and carbon dioxide diffuse soil degassing at Ahuachapan Geothermal Field, El Salvador: Geothermal Resources Council Transactions, v. 26, p. 341–344.

Major, J.J., Schilling, S.P., Pullinger, C.R., Escobar, C.D., Chesner, C.A., and Howell, M.M., 2001, Lahar-hazard zonation for San Miguel Volcano, El Salvador: U.S. Geological Survey Open-File Report 01-395, 14 p.

Maljanen, M., Martikainen, P.J., Aaltonen, H., and Silvola, J., 2002, Short-term variation in fluxes of carbon dioxide, nitrous oxide and methane in cultivated and forested organic boreal soils: Soil Biology & Biochemistry, v. 34, p. 577–584.

McGee, K.A., and Gerlach, T.M., 1998, Annual cycle of magmatic CO_2 in a tree-kill soil at Mammoth Mountain, California: Implications for soil acidification: Geology, v. 26, p. 463–466.

Meyer-Abich, H., 1956, Los volcanes activos de Guatemala y El Salvador (America Central): Anales del Servicio Geológico Nacional de El Salvador, 102 p.

Molnar, P., and Sykes, L., 1969, Tectonics of the Caribbean and Middle America regions from focal mechanisms and seismicity: Geological Society of America Bulletin, v. 80, p. 16–1684.

Pérez, N.M., and Williams, S.N., 1990, Radon and elemental mercury distribution at Rabaul Caldera, Papua New Guinea: Eos (Transactions, American Geophysical Union), v. 71, p. 649.

Pérez, N.M., Salazar, J.M.L., Hernández, P.A., Soriano, T., López, D.L., and Notsu, K., 2004, Diffuse CO_2 and ^{222}Rn degassing from San Salvador volcano, El Salvador, Central America, *in* Rose, W.I., et al., eds., Natural hazards in El Salvador: Boulder, Colorado, Geological Society of America Special Paper 375, p. 227–236 (this volume).

Pylon Electronics, 1993, Pylon Model AB-5 Portable Radiation Monitor Instruction Manual, June 5, Manual Number A900024: Ottawa, Canada, Pylon Electronics, 69 p.

Salazar, J.M.I., Pérez, N.M., Hernández, P.A., Soriano, T., Barahona, F., Olmos, R., Cartagena, R., López, D.L., Lima, N., Melián, G., Castro, L., Galindo, I., and Notsu, K., 2002, Precursory diffuse carbon dioxide soil degassing signatures of recent earthquakes in El Salvador, Central America: Earth and Planetary Sciences Letters, v. 205, p. 81–89.

Salazar, J.M.L., Hernández, P.A., Pérez, N.M., Olmos, R., Barahona, F., Cartagena, R., Soriano, T., López, D.L., Sumino, H., and Notsu, K., 2004, Spatial and temporal variations of the diffuse CO_2 degassing at Santa Ana–Izalco–Coatepeque volcanic complex, El Salvador, Central America, *in* Rose, W.I., et al., eds., Natural hazards in El Salvador: Boulder, Colorado, Geological Society of America Special Paper 375, p. 135–146 (this volume).

Simkin, T., and Siebert, L., 2000, Appendix 2: Catalog of historically active volcanoes on Earth, *in* Sigurdsson, H., et al., eds., Encyclopedia of volcanoes: San Diego, California, Academic Press, p. 1365–1383.

Sorey, M.L., Evans, W.C., Kennedy, B.M., Farrar, L.J., Hainsworth, L.J., and Hausback, B., 1998, Carbon dioxide and helium emissions from a reservoir of magmatic gas beneath Mammoth Mountain, California: Journal of Geophysical Research, v. 103, no. B7, p. 15,303–15,323.

Tennant, C.B., and White, M.L., 1959, Study of the distribution of some geochemical data: Economic Geology, v. 54, p. 1281–1290.

Weyl, R., 1980, Geology of Central America: Berlin, Gebrüder Borntraeger, 371 p.

Williams, S.N., 1985, Soil radon and elemental mercury distribution and relation to magmatic resurgence at Long Valley Caldera: Science, v. 229, p. 551–553.

MANUSCRIPT ACCEPTED BY THE SOCIETY JUNE 16, 2003

Geological Society of America
Special Paper 375
2004

Physical and chemical evolution of San Miguel volcano, El Salvador

Craig A. Chesner*

Department of Geology/Geography, Eastern Illinois University, Charleston, Illinois 61920, USA

Carlos R. Pullinger
C. Demetrio Escobar
*Seccion Vulcanologia, Servicio Geologico de El Salvador, c/o Servicio Nacional de Estudios Territoriales,
Alameda Roosevelt y 55 Avenida Norte, Edificio Torre El Salvador, Quinta Planta, San Salvador, El Salvador*

ABSTRACT

San Miguel volcano in eastern El Salvador is a classic composite cone, symmetrical and concave upwards. Its summit crater exceeds 344 m in depth and consists of several nested craters with nearly vertical walls. The inner crater has grown in depth and size since it was first described in 1924. The dominant eruptive product at San Miguel has been lava flows. Spatter and scoria cones commonly occur at flank vents that erupted historic lava flows. Lava flows erupted from flank vents on several occasions between 1699 and 1867, traveling as far as 8 km from their vents. During this time interval, the elevation of flank vents increased. All subsequent activity has been minor Strombolian and ash eruptions in the summit crater. Occasionally, scoria fall deposits, pyroclastic flow deposits, and phreatomagmatic ashes have been produced, but no substantial explosive event has been positively linked to San Miguel. Few lahar deposits have been identified as a result of the preponderance of lavas exposed on the upper and middle slopes. Geochemical analyses of lavas, tephras, and block and ash flow deposits erupted from San Miguel indicate that the majority of activity has been mafic in character, ranging between 51 and 53 wt% SiO_2. Historic flank lavas plot at the mafic end of the chemical range and are basaltic. The most evolved flank lavas are basaltic andesites and comprise two chemically distinct subsets, distinguished mostly by the presence or absence of phenocrystic magnetite and their V and Al_2O_3 contents. They occur only on the eastern flank of the volcano and appear to represent some of the oldest exposed lavas. A stratigraphic sequence of 22 crater lavas has the most restricted compositional range and exhibits two chemical trends. These trends may represent rapid-fire eruptions, with little to no intervening reposes, from two distinct batches of magma. Overall, the development of San Miguel volcano is fairly simple and consists of two evolutionary stages. The first stage consisted of eruption of lavas from a central vent and growth of the symmetrical cone. Shallowing of the subvolcanic magma chamber may have been associated with this stage. This was followed by a significant change in the subvolcanic plumbing system characterized by flank eruptions and onset of the second growth stage. Magma draining laterally from the magma chamber to the flank vents led to the collapse of the summit region. Modification of the summit crater is an active process that continues today.

Keywords: geochemistry, mafic lavas, summit crater, historic activity.

---

*cfcac@eiu.edu

Chesner, C.A., Pullinger, C.R., and Escobar, C.D., 2004, Physical and chemical evolution of San Miguel volcano, El Salvador, *in* Rose, W.I., Bommer, J.J., López, D.L., Carr, M.J., and Major, J.J., eds., Natural hazards in El Salvador: Boulder, Colorado, Geological Society of America Special Paper 375, p. 213–226. For permission to copy, contact editing@geosociety.org. © 2004 Geological Society of America

INTRODUCTION

San Miguel volcano, known locally as Volcán Chaparrastique, is a symmetrical composite volcano located in eastern El Salvador at 13°26′ N, 88°16′ W (Fig. 1). The city of San Miguel, El Salvador's second largest city (~300,000) and the economic center of eastern El Salvador, is built upon the lowermost northeastern flank of the volcano, 2000 m below its summit. A few large towns are built on the west and southwestern flanks and include San Jorge, San Rafael Oriente, and El Transito. The Pan-American and Coastal highways cross the lowermost northern and southern flanks respectively. Coffee plantations cover a large portion of the northern flank, whereas the southern flank is mostly ranch land.

In the past 300 yr, San Miguel volcano has produced several mafic lava flows that have traveled to the base of the cone (Fig. 2). Most of these lava flows occurred in the 1700s and 1800s and erupted from flank vents (Meyer-Abich, 1956). Since 1867, the only eruptive activity has occurred within the summit crater (Simkin and Siebert, 1994). Based upon its past eruptive activity and its proximity to several population centers and transportation routes in eastern El Salvador, a future eruption at San Miguel could present several volcanic hazards to the local population and influence the economy of this region.

Fieldwork at San Miguel was conducted in January 2000 and January 2002 with the objective to characterize the physical and chemical evolution of the volcano. During these trips, the distribution and occurrence of lavas, tephras, lahar deposits, and pyroclastic flow deposits were noted and described. A total of 145 rock samples were collected from the volcano and its immediate vicinity (Fig. 2A). A stratigraphic sequence of 22 consecutive lavas sampled within the summit crater is among this suite.

This paper will describe the field occurrence of the various eruptive products from San Miguel, present preliminary evaluations of the petrography and geochemistry of the new rock suite, and propose a model for the physical and chemical evolution of the volcano.

GEOLOGIC SETTING

The Central American volcanic arc originates from northeastward subduction of the Cocos plate beneath the Caribbean plate. Stoiber and Carr (1973) divided the arc into eight distinct segments based upon variations in strike and offsets between adjacent volcanic centers. All nine large volcanic centers in El Salvador belong to the same segment, which strikes at about N74°W. San Miguel volcano is located in the eastern portion of this segment. The physical and chemical characteristics of the El Salvador segment have been described by Carr (1984). In general the Salvadoran segment has an intermediate crustal thickness (32–40 km) and subduction zone dip (45°–55°) when compared to Guatemala with the thickest crust (48 km) and gentlest dip (40°), and to Nicaragua with the thinnest crust (32 km) and the steepest-dipping subduction zone (75°) (Carr et al., 1990). The

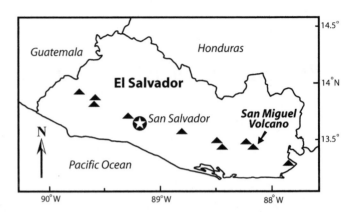

Figure 1. Location map of San Miguel volcano. Triangles represent Holocene stratovolcanoes.

volcanic arc in El Salvador was first described by Williams and Meyer-Abich (1955), and later mapped by Weber et al. (1978). Their work indicates that San Miguel, like all the active volcanic centers in El Salvador, lies within the "Median Trough." This ~2 m.y. old trough parallels the active volcanic arc and is faultbounded on both sides by active strike-slip faults. Many NNWtrending diagonal fractures cross the Median Trough and are thought to have guided fissure eruptions at San Miguel as well as Santa Ana, Boquerón, and El Playon volcanoes.

PHYSICAL VOLCANOLOGY

Morphology

Cone

San Miguel volcano is a composite volcano with a classic symmetrical, concave upwards, cone shape (Fig. 3). It rises 2130 m above sea level, and its southern and eastern flanks extend to the coastal plain to a base level of ~100 m. The northern flank reaches the Rio San Estaban ~100–150 m above sea level, whereas the western flank is buttressed against the Ojo de Agua (Cerro El Limbo) and Chinameca (Cerro El Pacayal) volcanoes at ~300–900 m in elevation. Although its radius generally ranges from 10 to 13 km, the northwestern flank has only a 5 km radius because of the buttressing effect of the nearby volcanoes.

Spatter and scoria cones commonly occur at the flank vents that erupted historic lava flows. Some are elongate in the downslope direction and could be classified as ramparts. They can reach heights of 100 m, although most are ~20 to 50 m in height. The majority of these vents occur in the southeastern and northwestern quadrants. Southeastern flank vents occur at elevations between ~200 and 800 m, whereas most northwestern vents are higher, 800–1200 m. One exception is an unusual spatter/scoria cone at ~1900 m on the southeastern flank. This vent produced an elongate flat-topped ridge of scoria fall to the north that parallels the contour of the cone. Viewed from the base of the volcano, this cone and its associated ridge of scoria appear

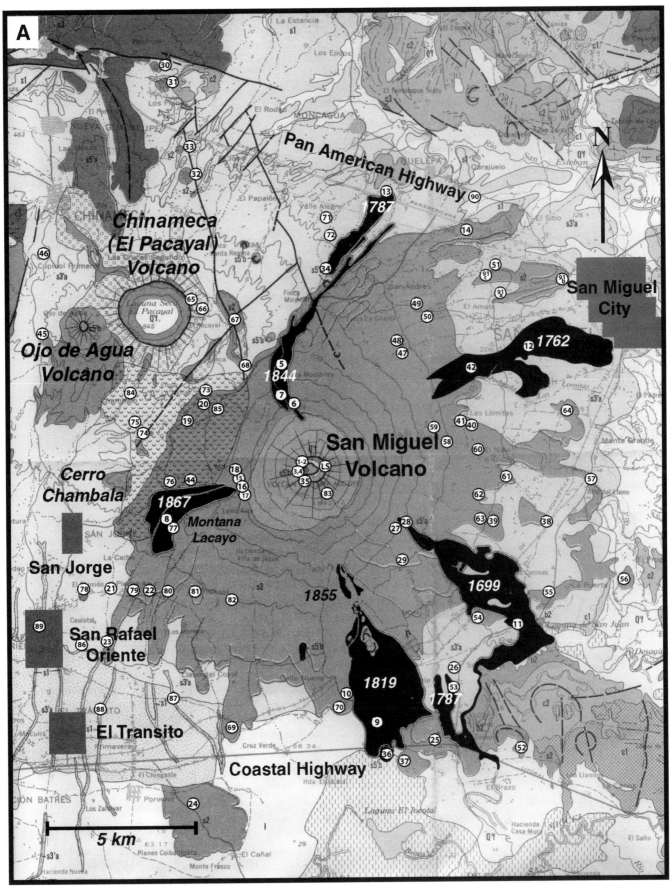

Figure 2 (*on this and following page*). A: Geologic map of San Miguel volcano and the nearby Chinameca, Ojo de Agua, and Cerro Chambala volcanoes. Historic lavas, sample locations, important cities, towns, and roads are also indicated. (Modified from Weber et al., 1978.) B: Sketch map of some historic and younger San Miguel lavas from Meyer-Abich (1956).

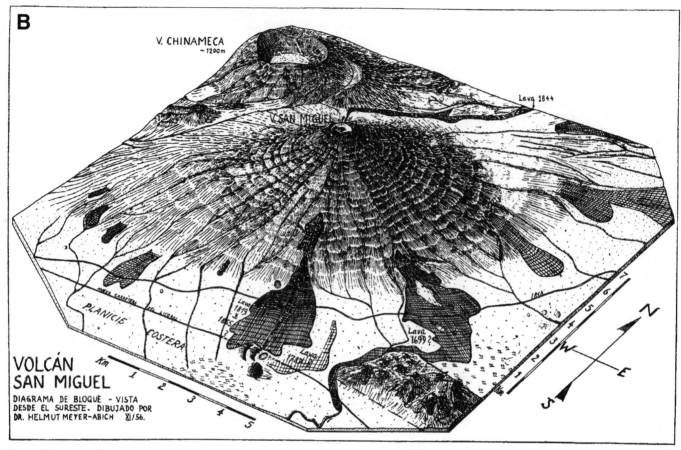

Figure 2 (*continued*).

as a "shoulder" on the main cone (Fig. 3B). The first historical accounts of San Miguel were made in 1586 and indicate that the main summit crater already existed at that time, although the volcano was previously known to have had a pointed top. Coupled with reports of a significant eruption in the early 1500s, Meyer-Abich (1956) has suggested that the summit crater may have formed as a result of an explosive eruption ~1510.

Summit Crater

San Miguel's summit crater is ~600 m in diameter and 344 m deep (Fig. 4). The outer crater walls are steep but allow access to a bench ~100 m below the outer crater rim at ~1990 m. A lower, inner bench occurs in the western part of the outer crater. The deep inner crater is ~250 m in diameter, has vertical walls, and consists of at least two nested craters separated by a small bench on the western side. The bottom of the inner crater cannot be viewed from anywhere on the outer crater rim. It has truncated a small pit crater on the east side of the upper bench. The inner bench and portions of the outer bench are littered with scoriaceous bombs from the 1976–1977 eruption. Surface coverage by these bombs declines rapidly away from the inner crater and indicates that bombs were probably not ejected outside the

outer crater. A water vapor–dominated gas plume is persistently emitted from the inner crater. Occasionally, this gas plume can carry some ash, which causes it to be distinctly brown in color and exhibits minor convective rise.

Several noticeable changes have taken place in the summit crater since Sapper mapped the summit crater in 1924 and aerial photographs were first taken in 1949. In 1924, there were three shallow inner craters dropping off from the 1990 m outer bench. The deepest of these craters was 80 m below the outer bench, and explosion vents from the 1921 eruption were visible (Sapper, 1925). By 1950, Williams and Meyer-Abich (1955) observed that the three shallow inner craters had been replaced by a single pit crater, although the upper bench remained essentially unchanged. The floor of the inner pit crater had subsided significantly to ~200 m beneath the rim of the outer bench. Scoriaceous bombs and blocks from the 1936 eruption littered the surface of the outer bench.

A comparison between aerial photographs taken in 1949, 1970, 1976 (during an eruption), 1979, and 1999 indicates that the inner pit crater has gradually enlarged to the east and north and increased in depth (Fig. 4). Furthermore, the floor of the inner pit crater has shown significant modifications related to

Figure 3. Photographs of San Miguel volcano taken from the west (A) and south (B). A small scoria cone can be seen on the southeast flank, below the summit in B.

Figure 4. Aerial photographs of San Miguel's summit crater from 1949, 1970, during the 1976 eruption, 1979, and 1999. North is to the top in all photographs.

eruptive activity and subsequent collapse. The photo taken in 1949 shows one main inner pit crater with an irregular floor. By 1970, a nested circular pit crater had formed in the eastern portion of the main inner pit crater. This new crater contained three small scoria/spatter cones aligned in a north-south direction. These cones presumably formed during the 1970 eruption, and each emitted a small gas plume. An oblique photo taken of the crater during the 1976 phase of the 1976–1977 eruption shows the inner crater nearly filled with lava. A circular vent in the center of the lava-covered inner crater was emitting a gas plume, and scoria/spatter covered the northern walls of the outer crater. In 1979, two small spatter/scoria cones, aligned in a northwesterly direction, were visible on the floor of the inner pit crater. These cones most likely formed during the 1977 phase of the 1976–1977 eruption. By 1999, two new nested pit craters had formed in the inner pit crater, and it had expanded toward the east. The 1977 spatter/scoria cones were no longer visible. In 1978, prior to the formation of the nested pit craters, the deepest part of the crater

was surveyed to be 1786 m (344 m below the summit) (Guzman, 1984). Thus, the 1999 and present crater depth is thought to exceed 344 m.

Stratigraphic sequences of lava flows are well exposed in both the inner and outer crater walls. The outer crater rim is dominated by hydrothermally altered dense lava flows, some of which can be several meters in thickness and probably represent lava flows that ponded in the summit crater. A distinct notch occurs in the eastern side of the outer crater. This notch affords access to the upper bench from where it was possible for us to collect a stratigraphic sequence of 22 lava flows (Fig. 5).

Lavas

Most of San Miguel's historic and prehistoric eruptive products are basaltic lava flows. These flows cover the middle and lower flanks of the volcano and are also common at the crater rim and exposed in the walls of the summit craters. Historic lavas

Figure 5. Photograph from within San Miguel's summit crater looking east. The prominent notch is in the center of the photograph, and the 22 sequential crater lavas are to its right.

have conspicuously blocky flow tops, but older lavas and summit lavas suggest that all flows have dense interiors.

Historic Lava Flows

According to Meyer-Abich (1956) and Martinez (1977), 12 lava flows have erupted from San Miguel since 1699. Nine of these have erupted from flank vents, and three have been confined to the summit crater. Lavas issued from vents on the southeastern flank in 1699, 1787, 1819, and 1855 (Fig. 2). The two largest flows were erupted in 1699 and 1819 from vents at ~400 m. The 1855 flow originated from a fissure at ~800 m above sea level. The location of the 1787 vent is unclear, perhaps being covered by the 1819 flow. The 1819 flow traveled ~4 km from its vent, whereas the 1699 lava flowed a total of 8 km downslope. Both flows crossed the future location of the Coastal Highway and descended to ~25 m above sea level. The greater runout distance of the 1699 flow can be attributed to channelization in a narrow paleo-valley. These lavas spread out to a maximum of ~2 km perpendicular to their flow direction as they descended the volcano. The 1855 flow only traveled a short distance (~1 km) from its source fissure higher on the southeastern flank.

The northern flank was also the site of vents for historic lava flows (Fig. 2). At the same time that lava was being erupted from the southeast flank in 1787, a separate flow issued from vents at ~500 m on the north flank. This flow eventually crossed the future path of the Pan-American Highway after traveling ~4 km. In 1844, lavas issued from 14 vents (Williams and Meyer-Abich, 1955) along a fissure between ~1000 and 1200 m, eventually producing a 7 km long lava flow. Sporadic explosions occurred at some of the fissure vents until 1848, when a small lava flow was erupted. The 1787 and 1844 lava flows are narrow, not exceeding ~0.5 km in width, as they were constrained by the flanks of San Miguel and Chinameca Volcano to the west.

Another important historic lava flow is the 1762 flow, which erupted from the northeast flank at ~400 m elevation (Fig. 2). This flow traveled ~6 km, descending to 120 m above sea level and ~2 km from the major river Rio Grande de San Miguel. The western rural edge of the city of San Miguel is built directly upon this lava flow and has spread ~1 km up its path. This flow was less confined than the northern flank flows and is up to 1.5 km wide. Another significant lava extrusion occurred in 1769, although its location is not documented. The lava flow located due east of the summit on Meyer-Abich's map (Fig. 2B) might represent this eruption. A scoria cone located at ~600 m may have been the source of this flow.

The youngest flank lava from San Miguel occurred in 1867 on the WSW flank (Fig. 2). This lava erupted from poorly defined vents at ~1000 m, flowed 2 km west toward the town of San Jorge, then turned south and flowed another 1.5 km. Directly to the southwest of the vent is the hilly, irregular terrain known as the Montaña Lacayo. The area is elongate in the southwesterly direction for ~2 km and is ~1 km wide. Several ridges parallel the long axis of Montaña Lacayo and consist entirely of thick accumulations (up to 100 m?) of highly vesicular scoria. This scoria is interpreted to represent fallout from tall fire fountains associated with the 1867 eruption. Thus, Montaña Lacayo owes its origin to downwind scoria fallout from Hawaiian-style fire fountaining associated with the 1867 eruption. This scoria also covered much of the surface of the 1867 lava flow, which explains why the youngest flank lava flow on the volcano is vegetated and even cultivated.

The older historic lava flows, 1699, 1762, 1787, and 1819, were all erupted from vents situated at ~400–500 m (1600–1700 m below the summit). Subsequent lava flows in 1844, 1848, 1855, and 1867 all had vents located between 800 and 1200 m. Lava flows confined to the floor of the summit crater were reported in 1884, 1976, and possibly in 1931. Thus, vent location has apparently migrated upward on the edifice with time.

Prehistoric Lava Flows

Radial drainages on the middle and lower flanks of the volcano commonly are floored by dense resistant lava flows. Because of thick vegetation and soil development between these drainages, it is impossible to map or extend these lava flows out of their respective narrow ravines. Several lava flow lobes, delineated by narrow rounded ridges, are evident on the lower slopes of the southern and northern flanks. The upper and middle slopes of the eastern flanks show well-defined lava flow levees. The greatest distance that prehistoric lavas occur from the summit (although this may not be their source) is 13 km. As described earlier, numerous lava flows are also exposed in the walls of the summit crater.

Tephras

Explosive tephra-producing eruptions have occurred at least 13 times since 1844. Most of these have been small eruptions that produced minor ashfalls that were rapidly erased from the geological record. Several coarser prehistoric and a few coarser historic pyroclastic fall deposits have been identified on the western flanks of the volcano, in the direction of the prevailing winds. On this flank, the upper portion of the cone is covered with a thick sequence of red scoriaceous fall deposits from prehistoric summit eruptions. The only possibly significant historic explosive event at San Miguel may have occurred ~70–80 yr prior to 1586 (Meyer-Abich, 1956). The frequent minor ashfalls and infrequent thicker scoria falls collectively indicate that tephras are volumetrically a minor eruptive product at San Miguel.

Historic Tephras

Mafic pyroclastic activity has occurred several times since 1586. The most significant historic accumulation of tephra occurred at Montaña Lacayo. Here, 100 m or more of highly vesicular tephra constructed elongated ridges and conspicuously hilly terrain. This tephra was apparently produced by tall fire fountains associated with the 1867 eruption and is very localized in the immediate area of Montaña Lacayo. Ash was produced during scoria/spatter cone growth associated with some of the flank lavas, especially the 1844–1848 and 1867 eruptions. Ashfall was reported 20 km northwest of the volcano during the 1844–1848 eruption. Ash was also emitted during several explosive eruptions in the summit crater that were not associated with lava flow emission, especially during the 1900s. The largest of these produced ashfall up to 10 km from the crater. During one of these events in 1931, ash fell on the city of San Miguel. Most of these pyroclastic fall events have been of short duration and small volume, affecting only a narrow zone immediately downwind from the vent. Ashfall from the 1976–1977 eruption reportedly caused some crop damage on the flanks of the volcano (Smithsonian Institution, 1976). Minor ashfalls were reported in 1995 (2–3 mm, 3 km southwest of the summit crater), 1997, and 2002.

Prehistoric Tephras

Dark gray to black, mafic tephras were exposed and sampled at 16 different localities near San Miguel. Fourteen exposures are located west of the summit crater, and the other two are to the northeast. About 6 km west of the summit, south of Ojo de Agua and Chinameca volcanoes, a thick tephra sequence is exposed in the Quebrada Joya Grande (sites 74 and 75). Here, up to 11 different mafic tephras are exposed. A thick tephra sequence consisting of nine different mafic tephras also outcrops on the east rim of Chinameca Volcano (site 65). In most instances, individual fall units are less than 50 cm thick. Physical volcanology suggests that several of the tephras in the Quebrada Joya Grande section may have come from the Cerro Chambala mafic center.

The thickest, most significant tephra unit outcrops in several locations centered between San Miguel, Chinameca, Ojo de Agua, and Cerro Chambala volcanoes (sites 19, 74, 75, 84, and 85). This fall deposit ranges from ~2 m to over 4 m thick, and contains scoriaceous fragments up to 10 cm in diameter. In addition, it is the youngest tephra unit encountered, typically overlain by less than 30 cm of soil. Should isopach data confirm that this

thick tephra unit was erupted from San Miguel, it would represent the sole significant explosive event in the exposed eruptive history of the volcano. Possibly, it could even correlate to the proposed eruption in 1510.

Mafic phreatomagmatic ashes with accretionary lapilli were noted at five separate locations on the southwestern flank. One exposure occurs in Quebrada La Quebradona ~3.5 km from the summit (site 44). The other exposures occur between 8 and 10 km from the summit in Quebradas La Ceibita o La Ceiba (sites 21 and 23), Aguacate (site 78), and San Jorge (site 89). Two distinct ashes separated by ~1.2 m of soil were found in Quebradas Ceibita o La Ceiba (site 23) and Aguacate. The younger deposit was ~35–40 cm thick, while the older deposit was ~50 cm thick. At another location in Quebrada Ceibita o La Ceiba (site 21), a 1.5 m thick phreatomagmatic ash was encountered. The most distal location, in Quebrada San Jorge on the west side of San Rafael Oriente, was 30 cm thick.

The only other mafic phreatomagmatic ash occurs in the summit crater on the west side of the upper bench and at the notch on the east side of the crater. The ash on the west side of the upper bench is ~2 m thick, well bedded, and demonstrates cross-beds and bomb sags. It is interpreted to represent base surge that originated during an eruption through a small lake in the inner crater and may be associated with one of the historic events. Phreatomagmatic ash near the notch is also well bedded, but finer grained than the ash on the west side of the crater. Scoria entrained in a matrix of mud constitutes an unusual deposit found at the notch.

Pyroclastic Flow Deposits

Mafic block and ash flow deposits were identified in two general areas on the southwestern and eastern flanks of San Miguel volcano. In the southwestern area, one deposit was sampled in Quebrada La Ceibita o La Ceiba (site 23), ~9 km from the summit where it is 40 cm thick. Another similar deposit occurs ~1 km to the west in Quebrada Aguacate (site 86). The largest block and ash flow deposit encountered in the vicinity of San Miguel is exposed in Quebrada San Jorge, on the west side of the town of San Rafael Oriente (site 89). Here, ~10 km from the summit of San Miguel, the deposit is ~3.5 m thick. Present topography suggests a more northerly source for this deposit, perhaps the Cerro Chambala center; however, it is possible that this deposit originated from San Miguel when drainage patterns were different. Preliminary observations (January 2003) that the base of this deposit immediately overlies and possibly even mixes with a debris avalanche deposit are potentially of great significance.

The other main area where block and ash flow deposits were found is on the east side of the volcano beneath the notch in the summit crater. Five different deposits outcrop ~4.5–5 km from the summit (sites 40, 41, 58, 60, and 62). Although some are in close proximity to one another, geochemistry suggests that they are distinct deposits. The youngest and best exposed of these deposits is at least 5 m thick and carries juvenile glassy bread-crust bombs up to 0.5 m in diameter (site 41). It is covered by 1 m of soil. Two older deposits are ~3 m thick and covered by 2 m of soil (sites 40, 60). Two other deposits were found in the area, but were highly weathered (sites 58 and 62). They were recognized as block and ash flow deposits by their quenched, breadcrust blocks. There are no reports of historic pyroclastic flows.

Debris Flow Deposits

Debris flows caused by erosion of the cone have occurred on the northwestern flank of San Miguel sporadically during periods of unusually heavy rainfall. A few prehistoric lahar deposits have also been identified; however, they too are considered infrequent events at San Miguel volcano. Major et al. (2001) have produced a lahar hazard zonation map for San Miguel volcano.

Historic Debris Flows

Occasionally heavy seasonal rains have mobilized scoria from high on the upper northwestern flank of the volcano. These scoria-dominant debris flows have descended the volcano in a NNW direction, damaging a few homes and/or crossing the main road to San Jorge between Canton El Volcan and La Placita. The source region for these debris flows appears to be an over-steepened, highly oxidized area of scoria located on the upper northwestern portion of the cone. Debris flows were reported in 1985, 1988, 1992, 1994, 1999, 2000, and 2001 (Major et al., 2001; Smithsonian Institution, 2002) These flows are not related to eruptive activity and have been relatively small. They have presented the local population with minor loss of property and are mostly a short-term inconvenience.

Prehistoric Debris Flows

Several lahar deposits were identified on the southwestern flanks of the volcano but nowhere else. Two well-indurated lahar deposits outcrop in the Quebrada La Quebradona on the west flank of San Miguel ~3.5 km from the summit (site 44). Farther downslope in Quebrada La Ceibita o La Ceiba, indurated lahar deposits are exposed 7.5 and 10 km from the summit (sites 21 and 23). The stratigraphic position of the lahars at these three localities, below several soils and tephras, suggests that they are prehistoric in age. The farthest outrunning of lahars occurs in Quebradas El Clavo (site 88) and El Llano (site 87), ~11 km from the summit, at elevations of 140 and 160 m respectively. All of these indurated lahar deposits appear to have originated from barrancas that incise the thick accumulations of scoria and ash on the western portion of the upper cone.

PETROLOGY

Mineralogy/Petrography

All lavas, tephras, and pyroclastic flow deposits collected from San Miguel volcano during this study are mafic in character.

They are dark gray to black in color and porphyritic, containing phenocrysts of plagioclase, olivine, and orthopyroxene. A few samples have up to 2% magnetite phenocrysts. Clinopyroxene occasionally occurs in glomeroporphyritic clumps but is usually restricted to the groundmass. All samples contain phenocrysts of olivine and can be described as basalts and basaltic andesites. Monomineralogic glomeroporphyritic clumps are common, but polymineralogic clumps can also occur. Orthopyroxene contents increase with increasing SiO_2 content. Groundmass textures are typically pilotaxitic and intergranular, although some interstitial glass is usually present. One of the north flank lavas (14) is unusually rich in olivine and clinopyroxene phenocrysts and could represent a cumulate magma tapped from the lower portion of a fractionating magma chamber.

Whole Rock Geochemistry

One hundred forty-five samples from San Miguel volcano, nearby volcanic features possibly associated with San Miguel, and tephra deposits were analyzed by XRF for major and trace elements at Washington State University (Johnson et al., 1999). These samples include 52 flank lava flows, 5 summit crater lavas, a stratigraphic sequence of 22 lavas sampled within the walls of the outer summit crater, blocks from 8 different block and ash flow deposits, 40 scoria fall deposits, 3 ash deposits, 13 samples from distal scoria cones, lava domes, and a maar, 1 lava from Chinameca Volcano, and a sample of the 1976 sco-riaceous bombs collected within the outer summit crater. Based upon location and geochemistry (Figs. 2A and 6A), 13 of these samples were interpreted to be part of other nearby volcanic systems, and were excluded from subsequent interpretations. They are all significantly more evolved than samples erupted from San Miguel and do not follow the same trends on variation diagrams. These samples include six distal scoria cones and domes (30, 31, 32, 33A, 33B, 72A) sampled to the northwest of San Miguel that are andesitic and presumed to belong to the Chinameca volcanic system. Three andesitic tephra samples collected to the north of San Miguel (71, 72B, and 90) may have also come from China-meca or perhaps Ojo de Agua or their flank vents. Other samples excluded from further analysis include a lava flow collected from Chinameca Volcano (67), an andesitic dome to the east of San Miguel (56), a sample from the Laguna de Aramuacu maar (43), and a rhyolitic ash collected from the southeastern flank of San Miguel (26) presumably associated with a significant explosive eruption from elsewhere in Central America.

Most samples attributed to the San Miguel volcanic system are calc-alkaline, medium-K basalts and basaltic andesites, typically containing between 51 and 53 wt% SiO_2 (Fig. 6). Only a few lavas and tephras have lower SiO_2 contents. Rock samples with higher SiO_2 contents consist of a small group of lavas and block and ash flow deposits collected from the eastern flank, and three scoria/lava cones near Laguna El Jocotal (36, 37, and 52). Conspicuously high Cu values, typically between 170 and 210 ppm and some as high as 250 ppm, are characteristic of the suite.

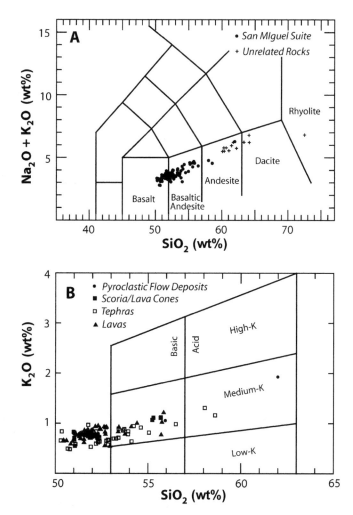

Figure 6. LeBas et al. (1986) and Gill (1981) volcanic and andesite rock classification diagrams. All 145 samples are plotted on the volcanic rock classification diagram (A). Only those samples attributed to San Miguel are shown on the andesite classification diagram (B).

Analyses of the historic lavas and representative crater lavas, flank lavas, block and ash flows, and tephras are given in Table 1.

Lavas

Historic flank lavas are all basalts and plot at the mafic end of the chemical range for all flank lavas (50.4–55.9 wt% SiO$_2$) (Fig. 7A). No systematic geochemical trends with age or vent elevation were evident. The most mafic flank lavas were the 1762 flow (12 and 42) that underlies the western portion of the city of San Miguel, an olivine/pyroxene-rich flow (14) that the Pan-American Highway crosses northeast of the summit, and two flows collected from the southwestern flank (69 and 82). Thus, the most mafic lavas appear to have been erupted from the northeastern and southwestern sectors of the volcano.

The most evolved lava flows are eight basaltic andesites (53–56 wt% SiO$_2$) that were collected from the eastern flank.

Five of these rocks belong to a low-V group (<260 ppm) that also consists of three east-flank lavas with nearly 53 wt% SiO$_2$ and three scoria/lava cones located near Laguna El Jocotal that have ~55 wt% SiO$_2$ (Fig. 7A). Petrographically, they are also distinct because they are the only San Miguel samples that contain phenocrystic magnetite. These observations strongly suggest that magnetite was fractionated prior to crystallization of these rocks. Furthermore, with the exception of one sample (59), all of these low-V samples (25, 36, 37, 38, 39A, 39B, 52, 55, 57, and 64) were collected from the most distal and geomorphologically oldest exposures associated with San Miguel volcano. Thus, this group of prehistoric samples may owe their distinctive geochemical signature to elevated fO_2 associated with an older and different magmatic plumbing system than all the subsequent lavas.

The three other evolved lavas (49, 61, and 63), plus a block and ash flow sample (40), have similar V contents (265–375 ppm) to the majority of San Miguel lavas, but can be distinguished from them by lower Al$_2$O$_3$ (<18 wt%, Fig. 7B) and CaO concentrations. These samples also have higher Cu contents than the samples in the low-V group (Fig. 7C). Eilenberg and Carr (1981) have shown that V and Cu contents in San Miguel lavas are a function of magnetite fractionation. Carr et al. (1981) have attributed high V and Cu contents in basaltic andesites from San Miguel to the fractionation of an anorthositic cumulate with very low proportions of magnetite and clinopyroxene. They suggest that fractionation of a gabbroic cumulate (with higher magnetite content) is more typical at San Miguel. Our data are consistent with fractionation of an anorthositic, magnetite-poor cumulate to produce the high-V-Cu and low-Al$_2$O$_3$ basaltic andesites. However, our data suggest that fractionation of a gabbroic cumulate with minimal magnetite would be required to produce the majority of San Miguel lavas, whereas fractionation of gabbro with a higher magnetite content is required to produce the low-V basaltic andesites.

Crater lavas are more restricted in their composition (51.0–52.8 wt% SiO$_2$) than the flank lavas (Fig. 7). Compared to the historic lavas that erupted exclusively from flank vents, the crater lavas have lower Sr, CaO (Fig. 8), and Al$_2$O$_3$ contents. Otherwise, they cannot be distinguished geochemically. Thus, magmas erupting at the summit apparently had fractionated slightly more plagioclase than those that erupted from the flanks. The stratigraphic sequence of 22 crater lavas has a narrow compositional range (51.6–52.3 wt% SiO$_2$) and exhibits two chemical trends when plotted against stratigraphy (Fig. 9). The lowermost 5–7 samples of the sequence have slightly lower TiO$_2$ and higher MgO contents than the lavas above them (Figs. 9A, B). These older lavas also become slightly more evolved up-section in terms of their MgO, Al$_2$O$_3$, and CaO contents. The uppermost 15–17 samples show virtually no chemical variation. This entire lava sequence may record a change in eruptive behavior from a period of growth punctuated by short reposes to a period of nearly continuous or rapid-fire eruptions with little to no repose between eruptions. Alternatively, because the chemical variation is so slight, the sequence might simply represent closely spaced eruptions from two slightly different magma batches.

TABLE 1. REPRESENTATIVE GEOCHEMICAL ANALYSES OF HISTORIC LAVA FLOWS, PYROCLASTIC FLOW AND FALL, CRATER LAVAS, AND OLDER FLANK LAVAS

	Historic flank lavas						Crater lavas		Older flank lavas					Pyroclastic flows			Tephras	
	11	42	13	9	1844	8	1	22	High-V group		Low-V group			41	40	60	19	44B
	1699	1762	1787	1819	1844	1867			49	63	59	38	39A					
SiO_2	51.52	50.39	51.18	51.64	51.59	50.79	51.74	51.51	55.66	53.30	55.75	53.38	52.76	51.72	55.45	61.95	51.61	50.57
TiO_2	1.03	.91	.98	.96	.96	.99	.95	.98	1.04	1.16	.83	.76	.76	.87	.96	.95	1.12	1.02
Al_2O_3	19.07	19.76	19.41	19.71	19.84	18.32	19.25	19.72	17.03	16.97	19.87	20.34	20.23	20.12	17.49	15.94	16.92	18.66
FeO	9.16	9.19	9.43	8.77	8.83	9.91	9.38	8.73	9.56	10.40	7.02	7.77	7.83	8.29	8.69	7.13	11.35	10.24
MnO	.18	.18	.18	.17	.17	.19	.18	.18	.20	.21	.15	.17	.16	.16	.19	.19	.21	.19
MgO	4.12	4.76	4.16	3.55	3.70	5.21	4.54	4.22	3.67	4.41	2.53	3.37	4.15	3.93	3.39	1.93	4.85	4.69
CaO	10.46	10.99	10.61	10.49	10.48	10.05	10.12	10.13	7.86	9.01	8.78	9.63	10.11	10.58	8.22	5.25	9.46	10.31
Na_2O	2.71	2.58	2.80	2.89	2.96	2.67	2.78	2.68	3.55	3.14	3.44	2.92	2.95	2.77	3.44	4.32	2.81	2.47
K_2O	.82	.65	.69	.86	.86	.76	.73	.76	1.02	.92	1.21	.78	.54	.71	1.04	1.92	.76	.82
P_2O_5	.20	.16	.17	.23	.23	.19	.19	.19	.24	.23	.23	.17	.14	.16	.21	.32	.20	.21
Total	99.27	99.57	99.60	99.26	99.62	99.08	99.85	99.10	99.82	99.75	99.81	99.29	99.63	99.31	99.07	99.90	99.28	99.18
Sc	40	39	37	34	36	35	37	35	37	45	28	29	31	28	32	26	48	34
V	316	310	305	282	293	319	307	338	266	377	169	199	236	274	267	100	371	301
Cr	28	29	25	27	26	22	23	20	12	25	9	11	16	19	11	—	30	30
Ni	4	47	4	4	4	12	11	6	—	5	3	6	7	7	—	3	4	7
Cu	197	180	212	203	199	191	203	169	198	224	168	104	51	172	170	67	233	223
Zn	83	83	84	82	84	93	91	89	101	100	81	77	75	75	86	96	101	91
Ga	19	18	19	22	18	20	21	20	16	19	19	17	17	18	19	21	18	20
Rb	15	9	12	14	13	14	11	12	15	12	20	13	6	10	17	37	11	15
Sr	475	485	487	489	487	447	456	460	455	421	466	537	493	508	446	390	425	444
Y	24	20	21	25	25	23	25	23	28	26	26	20	17	21	26	37	26	24
Zr	72	55	60	77	76	69	67	67	82	76	95	63	46	60	91	136	70	75
Ba	484	394	425	515	515	475	471	474	608	551	663	455	369	450	601	997	505	491
La	8	16	11	12	5	16	3	12	11	11	20	2	3	14	—	29	—	3
Ce	4	2	19	32	17	20	24	22	15	14	29	17	5	6	14	38	23	24

Note: All of the geochemical data is available from C.A. Chesner or in a Web-based GIS from D. Escobar. All Fe as FeO. Major elements (SiO_2 through P_2O_5) given in wt%; trace elements (Sc through Ce) given in ppm.

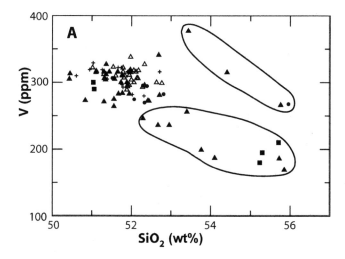

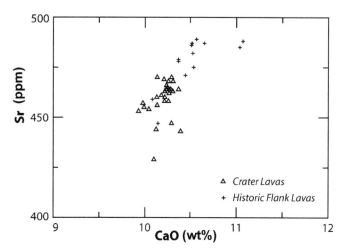

Figure 8. CaO vs. Sr variation diagram showing crater lavas and historic flank lavas.

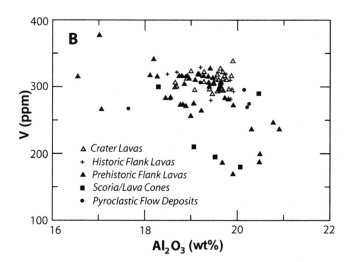

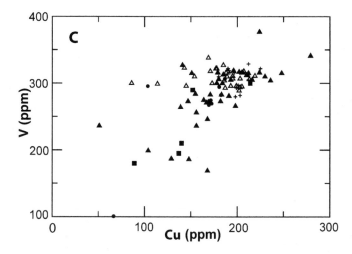

Figure 7. Variation diagrams plotted against V for all lavas, scoria cones, and pyroclastic flow deposits. Lavas have been plotted separately as historic flank lavas, prehistoric flank lavas, and crater lavas. The high-V and low-V basaltic andesites have been distinguished by enclosures on the SiO_2 vs. V plot (A).

Pyroclastic Flow Deposits

Individual blocks analyzed from eight separate block and ash flow deposits ranged from 51 to 62 wt% SiO_2. Six of these samples contain between 51 and 53 wt% SiO_2 (Fig. 7A). The two most silicic samples (40 and 60) were collected in close proximity to one another on the east flank of the cone and contain 56 and 62 wt% SiO_2 respectively. Sample 60 is the most silicic sample in the entire San Miguel suite. A block and ash flow deposit of particular interest is the thick deposit sampled in Quebrada San Jorge within the town of San Rafael Oriente (site 89). Although present topography suggests a more northerly source than San Miguel, geochemically it is very similar to other block and ash flow deposits and lavas that originated from San Miguel (Fig. 10). If this deposit was erupted from San Miguel, as the chemistry suggests, it represents the farthest known runout of a San Miguel pyroclastic flow.

Tephras

Tephra samples collected in the immediate vicinity of San Miguel have the greatest chemical range of any subgroup in this study (50–59 wt% SiO_2) (Fig. 6B). Most samples (75%) contained between 50 and 54 wt% SiO_2, and only a few samples had more than 56 wt% SiO_2. Like the lavas, there is also a low-Al_2O_3 group of tephras (Fig. 10A). The thick mafic tephra unit located between San Miguel, Chinameca, Cerro Chambala and Ojo de Agua can be clearly distinguished from other tephras by its high FeO (11.0–11.5 wt%) and Zn (100–110 ppm) contents (Fig. 10B). Its distinct chemistry allowed correlation of a young tephra on the west flank of Ojo de Agua (45) with samples collected farther east (16, 19, and 74A). However, geochemical correlation among other fall deposits from various exposures proved to be difficult.

One goal of this study was to characterize San Miguel tephras geochemically, thus providing a means to discriminate

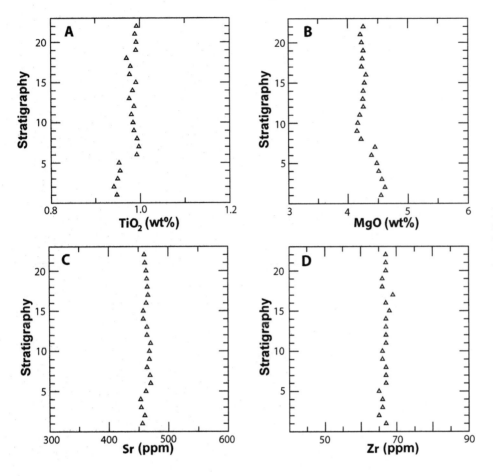

Figure 9. Selected major and trace elements plotted against stratigraphy for the sequence of 22 consecutive crater lavas. Sample 1 is the oldest and 22 is the youngest.

tephras from other sources. We believe that it is possible to distinguish the rather silicic tephras produced by Chinameca Volcano from other tephras in the sample set. However, until the other mafic centers in the area are chemically characterized, it will be difficult to unequivocally link many of the other tephras in this sample suite to San Miguel. Thus, geochemical fingerprinting alone cannot be used to correlate individual tephras with San Miguel, but should be coupled with isopach and isopleth data in order to make positive correlations.

EVOLUTION OF SAN MIGUEL VOLCANO

Based upon our fieldwork, the historic activity (1586–present), and geochemistry of 145 new samples, we have developed an evolutionary model for San Miguel volcano. This model consists of two distinct edifice growth stages. The first growth stage involved eruption from a central summit vent, with consequent increase in cone height. Lavas with at least three different chemical signatures were erupted during this growth stage. A shift in vent location from the summit to flank vents characterizes the second and present growth stage. It is during this stage that the summit crater formed by internal collapse. The summit crater has

continuously been modified by historic eruptions and additional internal collapses.

Stage I—Cone Growth from a Central Summit Vent

The majority of the symmetrical cone visible today probably was constructed from central vent eruptions at a well-established summit vent. During this growth stage, the low-V basaltic andesites were probably erupted earliest because of their generally older appearance and more distal locations. These were likely followed by eruption of the high-V, low-Al_2O_3 group of basaltic andesites, which are all located near the base of the cone and in geomorphologically younger positions than the low-V group. Collectively, these two geochemical groups represent the most evolved lavas sampled from the entire volcano. The transition from the low-V group to the high-V, low-Al_2O_3 group could be attributed to different fractionation assemblages caused by a shallowing of the subvolcanic magma chamber.

The third chemical group, represented by the majority of San Miguel lavas, then began to erupt, covering most of the lavas from the earlier chemical groups. This change toward more mafic lavas could also be attributed to changing fractionation

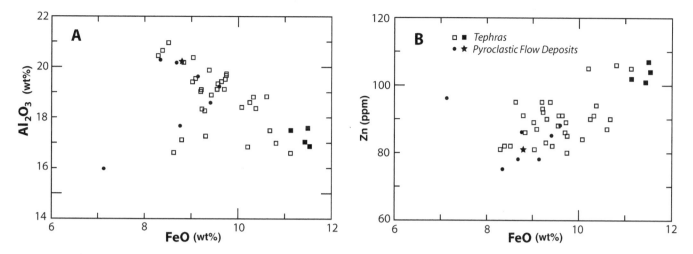

Figure 10. Variation diagrams showing only tephras and pyroclastic flow deposits. The thick tephra unit located between San Miguel, Chinameca, Cerro Chambala, and Ojo de Agua is represented by closed squares. The block and ash flow deposit located in San Rafael Oriente is highlighted by a filled star.

assemblages and magma chamber depth. Near the final stages of this growth period, nearly continuous eruptions were probably occurring as recorded by the 22 sequential lavas collected from the outer crater wall. Two chemical groups are apparent in the sequence and could represent eruptions from different magma batches. Slight chemical differences among the older lavas in the sequence suggests short reposes between eruptions, whereas little to no intervening reposes are indicated by the virtually identical lavas in the upper portion of the sequence. Such rapid-fire eruptions are consistent with a shallow magma chamber that eventually led to a change in eruption style.

Stage II—Flank Eruptions and Formation of the Summit Crater

After construction of the upper cone by central vent eruptions, a fundamental change in the plumbing system occurred. Instead of summit vent eruptions, flank vents at elevations ranging between 400 and 1200 m above sea level became established. These vents, located ~900–1700 m below the summit, became the preferred sites for lava flow venting. No change in lava chemistry from the youngest lavas of Stage I accompanied this shift. The dramatically reduced magma-static head may have promoted accelerated eruption rates, thus emptying the shallow magma chamber that had supplied the upper cone. Rapid draining of this chamber resulted in collapse of the summit region, forming the present summit crater, which exceeds 344 m in depth. Although it is likely that the crater formed in the early 1500s, a major explosive event is not required by this mechanism.

It is unclear whether the main summit crater formed from a single collapse event or by several piecemeal collapses. However, different benches and pit craters, and the dynamic changes reported and observed on aerial photographs, indicate that the

summit crater has been modified by numerous subsequent piecemeal collapses. Incremental collapses might have been directly associated in time with flank eruptions, as recently observed at Miyakejima Volcano, Japan, in 2000 (Geshi et al., 2002). Experimental studies of pit crater formation by Roche et al. (2001) indicate that terraced pit craters with piston-like subsidence develop when the thickness to width ratio of the subsiding roof block is ~1. No surface collapse occurs with thickness to width ratios of 2. Slightly deeper magma chambers with roof thickness to width ratios of ~1.1–1.2 develop craters with overhanging walls as a result of stoping and sudden collapse. The nested summit craters at San Miguel exhibit features indicative of both types of collapse. The outer pits have terraces, and the inner pits have vertical or overhanging walls. Applying the experimental results described above to San Miguel suggests that the magma chamber was ~600 m below the former summit when the outer crater formed. The magma chamber depth when the main inner pit crater formed would have been ~300 m below its rim, at ~1700 m elevation. Thus, the floor of the deepest inner nested pit crater is presently <90 m above the top of the recent magma chamber.

Periodically the summit crater has contained a lake as evidenced by at least two significant phreatomagmatic ash deposits on the southwestern flank and at least one within the summit crater. The high concentration of pyroclastic flow deposits downslope from the notch in the east wall of the summit crater suggests the likelihood that a few significant Vulcanian eruptions have occurred after formation of the summit crater. Thick crater-ponded lavas exposed in the upper walls of the summit crater, and the aerial photograph taken during the 1976 eruption, suggest that lava occasionally partly fills the summit crater. Since 1867, no large volumes of magma have been erupted from San Miguel. Activity has been restricted to minor ash and Strombolian explosions from the deep inner pit crater.

ACKNOWLEDGMENTS

We would like to thank the Sección Vulcanología, Servicio Geologico de El Salvador, and the Dept. de Investigaciones Sismologicas, Centro de Investigaciones Geotecnicas, El Salvador, for their support and assistance in the field. Jon Major helped collect the crater lava sequence and provided valuable field lessons on lahar deposits. The U.S. Geological Survey Cascade Volcano Observatory provided funding for geochemical analyses. Funding was provided by National Science Foundation grants INT 96-13647 and INT 01-18587. Cara Schiek assisted with sample preparation and preliminary assessment of the geochemical data. Paco provided friendly accommodations. Lastly, we would like to thank Bill Rose, who enthusiastically and graciously arranged and coordinated this study.

REFERENCES CITED

Carr, M.J., 1984, Symmetrical and segmented variation of physical and geochemical characteristics of the Central American volcanic front: Journal of Volcanology and Geothermal Research, v. 20, p. 231–252.

Carr, M.J., Mayfield, D.G., and Walker, J.A., 1981, Relation of lava compositions to volcano size and structure in El Salvador: Journal of Volcanology and Geothermal Research, v. 10, p. 35–48.

Carr, M.J., Feigenson, M.D., and Bennett, E.A., 1990, Incompatible element and isotopic evidence for tectonic control of source mixing and melt extraction along the Central American arc: Contributions to Mineralogy and Petrology, v. 105, p. 369–380.

Eilenberg, S., and Carr, M.J., 1981, Copper contents of lavas from active volcanoes in El Salvador and adjacent regions in Central America: Economic Geology, v. 76, p. 2246–2248.

Geshi, N., Shimano, T., Chiba, T., and Nakada, S., 2002, Caldera collapse during the 2000 eruption of Miyakejima Volcano, Japan: Bulletin of Volcanology, v. 64, p. 55–68.

Gill, J.B., 1981, Orogenic andesites and plate tectonics: Berlin, Springer Verlag, 389 p.

Guzman, P.A., 1984, Usulutan, El Salvador: Republica de El Salvador, Ministerio de Obras Publicas, Instituto Geographica Nacional, scale 1:50,000.

Johnson, D.M., Hooper, P.R., and Conrey, R.M., 1999, XRF analysis of rocks and minerals for major and trace elements on a single low dilution Litetraborate fused bead: Advances in X-ray Analysis, v. 41, p. 843–867.

LeBas, M.J., LeMaitre, R.W., Streckeisen, A., and Zanettin, B., 1986, A chemical classification of volcanic rocks based on the total alkali silica diagram: Journal of Petrology, v. 27, p. 745–750.

Major, J.J., Schilling, S.P., Pullinger, C.R., Escobar, D.C., Chesner, C.A., and Howell, M.M., 2001, Lahar-hazard zonation for San Miguel volcano, El Salvador: U.S. Geological Survey Open-File Report 01-395, 8 p. and map.

Martinez, H.M., 1977, Actividad historica del Volcan de San Miguel y ciertas consideratciones sobre su comportamiento futuro. Asociacion Salvadoreña de Ingeniero y Aarquitectos, El Salvador, Central America, no. 45, p. 9–16.

Meyer-Abich, H., 1956, Los Volcanes Activos de Guatemala y El Salvador (America Central). Anales del Servicio Geologico Nacional de El Salvador, Bd, 3, p. 49–62.

Roche, O., van Wyk de Vries, B., Druitt, T.H., 2001, Subsurface structures and collapse mechanisms of summit pit craters: Journal of Volcanology and Geothermal Research, v. 105, p. 1–18.

Sapper, K., 1925, Los Volcanes de la America Central. Estudios sobre America y Espana, Extraserie, Halle, 116 p.

Simkin, T., and Siebert, L., 1994, Volcanoes of the world: Tucson, Arizona, Geoscience Press, p. 133–134.

Smithsonian Institution, 1976, Bulletin of the Global Volcanism Network, December 1976, v. 1, no. 15.

Smithsonian Institution, 2002, Bulletin of the Global Volcanism Network, February 2002, v. 27, no. 2, p. 5.

Stoiber, R.E., and Carr, M.J., 1973, Quaternary volcanic and tectonic segmentation of Central America: Bulletin of Volcanology, v. 37, p. 304–325.

Weber, H.S., Wiesemann, G., Lorenz, W., and Schmidt-Thome, 1978, Mapa geologico de la Republica de El Salvador/America Central, 1:100,000. Bundesanstalt fur Geowissenschaften und Rhhstoffe, Hannover.

Williams, H., and Meyer-Abich, H., 1955, Volcanism in the southern part of El Salvador with particular reference to the collapse basins of Lakes Coatepeque and Ilopango: University of California Publications on Geological Sciences, v. 32, p. 1–64.

Manuscript Accepted by the Society June 16, 2003

Geological Society of America
Special Paper 375
2004

Diffuse CO_2 and ^{222}Rn degassing from San Salvador volcano, El Salvador, Central America

Nemesio M. Pérez
José M.L. Salazar
Pedro A. Hernández
Environmental Research Division, Instituto Tecnológico y de Energías Renovables (ITER), 38611 Granadilla, S/C de Tenerife, Spain

Tomás Soriano
Instituto de Ciencias de la Tierra, Universidad de El Salvador, San Salvador, El Salvador, Central América

Dina L. López
Department of Geological Sciences, 316 Clippinger Laboratories, Ohio University, Athens, Ohio 45701, USA

Kenji Notsu
Laboratory for Earthquake Chemistry, Graduate School of Science, University of Tokyo, Hongo, Bunkyo-Ku 113-0033, Tokyo, Japan

ABSTRACT

A diffuse CO_2 and ^{222}Rn degassing survey was carried out at San Salvador volcano in February 1999. The goal of this study was to evaluate the spatial distribution of diffuse degassing rates of both gas species and its utility for San Salvador volcano hazard forecasting. Three hundred eighty-eight measurements of CO_2 efflux and 380 of soil gas ^{222}Rn were performed covering most of San Salvador volcano, but taking into consideration geological and structural characteristics of the study area. CO_2 efflux and soil gas ^{222}Rn data showed a wide range of values up to 760 g m^{-2} d^{-1} and 284 pCi L^{-1}, respectively. Most of the study area showed CO_2 efflux background values (B = 12.6 g m^{-2} d^{-1}). Anomalously high CO_2 efflux data (> 9 × B) were mainly detected outside El Boquerón crater, suggesting a close spatial correlation with major faults and complex fracture/fault systems. Relatively high ^{222}Rn emissions (>140 pCi L^{-1}) were also detected outside El Boquerón crater and located parallel to the NW fissure related to the 1917 eruption. The total diffuse CO_2 output of the volcano was estimated at ~4000 t d^{-1} for an area of 266 km^2. Carbon isotope signatures of the soil CO_2 ranged from −6.9‰ to −35.4‰, suggesting a mixing between different carbon reservoirs. Spatial distribution of the carbon isotope ratios showed that most of the isotopically positive carbon emission is actually occurring at El Boquerón crater. These results suggest that monitoring spatial and temporal variations of diffuse degassing will be a potential geochemical tool for the San Salvador volcanic surveillance program.

Keywords: San Salvador, degassing, diffuse, carbon dioxide, radon.

Pérez, N.M., Salazar, J.M.L., Hernández, P.A., Soriano, T., López, D.L., and Notsu, K., 2004, Diffuse CO_2 and ^{222}Rn degassing from San Salvador volcano, El Salvador, Central America, *in* Rose, W.I., Bommer, J.J., López, D.L., Carr, M.J., and Major, J.J., eds., Natural hazards in El Salvador: Boulder, Colorado, Geological Society of America Special Paper 375, p. 227–236. For permission to copy, contact editing@geosociety.org. © 2003 Geological Society of America.

INTRODUCTION

Volcanoes pose a variety of geologic hazards, both during eruptions and in the absence of eruptive activity. Volcanic eruptions are certain to occur at San Salvador volcano in the future and can be neither prevented nor stopped, but actions can be taken to limit damage from them. Monitoring of volcanic precursors (increase of seismicity, ground deformation, increase of gas emission, etc.) generally can help to identify the locality of impending volcanic activity, even though it often does not indicate the nature or timing of an eruption, or even its certainty.

This article describes and discusses the potential use of diffuse degassing studies at San Salvador volcano for its surveillance program. Most volcano monitoring programs are performed by means of conventional geophysical methods and do not apply a multidisciplinary approach, which might include geophysical and geochemical monitoring tools, to improve and optimize the capabilities for detecting precursory signatures of magmatic reactivation. Scientists have long recognized that gases dissolved in magma provide the driving force of volcanic eruptions, but only recently new techniques have permitted routine measurement of different types of volcanic gases released into the atmosphere (Stoiber et al., 1983; Menyailov et al., 1985; Hirabayashi et al., 1986; Sato et al., 1988; Sano et al., 1991; Casadevall et al., 1994; Symonds et al., 1994; Francis et al., 1995; Giggenbach, 1996; Mori and Notsu, 1997; McGee and Gerlach, 1998; Horrocks et al., 1999; Hernández et al., 2001a). Significant amounts of gases are emitted to the atmosphere by active volcanoes through visible manifestations such as plumes, fumaroles, and hot springs, but invisible emanations through the surface environment, known as "diffuse degassing," are also a significant source of gases released to the atmosphere from depth by active volcanoes. Diffuse degassing studies are becoming a potential geochemical tool for volcano monitoring programs (Hernández et al., 2001a). The objectives of this article are (1) to evaluate the spatial distribution of diffuse degassing, and (2) to discuss the origin of these invisible emanations at San Salvador volcano.

Diffuse degassing at volcanoes is especially significant for major gas species such as CO_2, and other minor and trace gas components such as ^{222}Rn, both acting as useful indicators of gas flow. Diffuse degassing of CO_2 and ^{222}Rn have implications in monitoring volcanic activity and forecasting eruptions (Cox et al., 1980; Williams, 1985; Capaldi et al., 1988; Lombardi and Reimer, 1990; Allard et al., 1991; Baubron et al., 1991a; Caparezza and Diliberto, 1993; Farrar et al., 1995; Connor et al., 1996; Pérez et al., 1996; Gerlach et al., 1998; Hernández et al., 2001a, 2001b; Salazar et al., 2001).

Carbon dioxide is the major gas species after water vapor in both volcanic hydrothermal fluids and magmas, and it is an effective tracer of subsurface magma degassing due to its low solubility in silicate melts at low to moderate pressure, favoring its early exsolution (Gerlach and Graeber, 1985; Stolper and Holloway, 1988; Pan et al., 1991). For these reasons, carbon dioxide is one of the first gases to be exsolved and released from an ascending

magma. Therefore, increases in volcanic CO_2 emissions can be used as early indicators of rising magma (Baubron et al., 1991a; Bruno et al., 2001; Hernández et al., 2001a).

Radon (^{222}Rn) is produced in the subsurface by radioactive decay of U and Th in minerals. Its movement is limited by its half-life, which is 3.8 days. The parent isotopes of ^{222}Rn are often transported in groundwater, and their movement depends on several factors such as oxidation/reduction potentials, water pH, and presence of various anions (Smith et al., 1976). Radon can be brought to the surface also by fluid convection caused by high geothermal gradients, and radon transport occurs more effectively in areas of high permeability or fracturing (Fleischer and Mogro-Campero, 1978).

GEOLOGICAL SETTING

San Salvador volcano (Fig. 1) is part of the active Central American Volcanic Belt, which includes a 1100 km long chain of 41 active volcanoes from Guatemala to Panamá (Simkin and Siebert, 1994). It has a long history of hazardous volcanism and is located next to San Salvador, El Salvador's capital city with a population of ~2 million inhabitants. San Salvador volcano has not erupted for more than 80 years, but it has a long history of repeated, and sometimes violent, eruptions. This volcano erupted ash-rich tephra and pyroclastic flows 800 years ago and caused mudslides that would likely kill many thousands today.

San Salvador volcano is composed of remnants of multiple eruptive centers and numerous flank deposits (Meyer-Abich, 1956; Fig. 2). Several tephra deposits composed mainly of pumice indicate that some explosive eruptions were phreatomagmatic and involved interactions of magma and water (Major et al., 2001). The central part of the volcano, known as El Boquerón (maximum elevation 1893 m), is a well-defined edifice with a large circular crater (1600 m in diameter). El Picacho, the prominent peak of

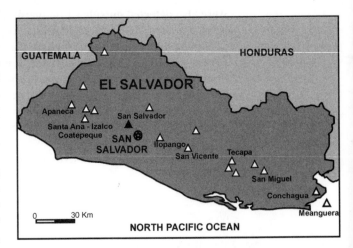

Figure 1. Location of the major volcanoes in the El Salvador Volcanic Belt showing the location of the San Salvador volcano.

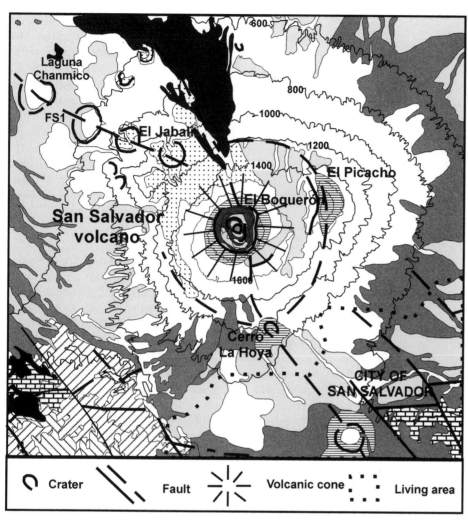

Figure 2. Main geological and structural features at San Salvador volcano (modified from topographic base, Instituto Geográfico de El Salvador, 1978). Two major fracture/fissure systems are pictured as FS1 and FS2.

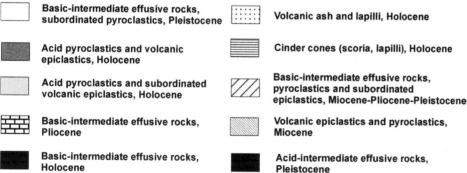

highest elevation (1959 m) to the northeast of the crater, and El Jabalí, the peak to the northwest of the crater, represent remnants of an older, larger edifice. Chemical analysis of lavas from the walls of the Boquerón crater and from historical lava flows on the northwest flank of the volcano were done by Fairbrothers et al. (1978) and classified the dominant rock types as calc-alkaline basalts and andesites to tholeiitic andesites.

Three eruptions have occurred at San Salvador volcano in the last 470 years. These events have consisted of strombolian flank eruptions and lava flows. The most recent eruption occurred in 1917, and at about the same time at least four fissures opened up along a N40W-trending fault on the NW flank of the volcano. After the 1917 eruption, fumarolic activity continued until the late 1970s (Fairbrothers et al., 1978), but this is absent today.

When eruptions occur again violently and explosively, substantial population and infrastructure will be at risk.

PROCEDURES AND METHODS

A diffuse degassing survey of 388 sampling sites was carried out in and around San Salvador volcano on February 1999, after taking into consideration geological and structural characteristics of the study area as well as accessibility. The sampling sites were selected in order to obtain a good spatial distribution. Soil CO_2 efflux measurements were carried out at each sampling site on the surface environment, together with soil gas collection and ground temperature readings at a depth of 30–40 cm. The survey was performed during the dry season, under sunny skies and low wind. These conditions ensured minimal soil gas variations from atmospheric changes, and minimized disturbances from soil moisture changes during the measurements.

Soil CO_2 efflux measurements were performed according to the accumulation chamber method (Baubron et al., 1991b; Chiodini et al., 1996), by using a nondispersive IR (NDIR) spectrophotometer, a cylindrical chamber open on one circular side, a datalogger, and a personal computer. Each CO_2 efflux measurement starts when the open side of the chamber is placed on the soil surface. A pump allows the air contained in the chamber to circulate through the IR spectrophotometer and then back into the chamber. Signals proportional to the CO_2 concentration in the chamber are acquired by the computer, which plots the CO_2 concentration versus time. The reproducibility for the range 10–20,000 g m^{-2} d^{-1} is ±10%. We assumed a random error of ±10% in the emission rates of CO_2, based on variability of the replicate measurements carried out on known CO_2 efflux rates in the laboratory. Values of CO_2 efflux (g m^{-2} d^{-1}) are estimated from the rate of concentration increase in the chamber at the observation site, accounting for changes of atmospheric pressure and temperature, and are converted from volumetric concentrations to mass concentrations.

Soil gas ^{222}Rn concentration was measured by emanometry using a Pylon Model AB-5. This instrument is equipped with a scintillation cell (Lucas cell), whose walls are coated with silver-activated zinc sulfide, ZnS(Ag). Radon in the soil gas, which is pumped into the scintillation cell, decays to ^{218}Po, giving off alpha particles. The impulses (scintillations) caused by adsorption of the alpha particles are amplified by a photomultiplier. The number of counts per minute caused by radon can be estimated roughly using the method described by Morse (1976).

Soil gas samples were collected at a depth of 30–40 cm using a metallic soil probe and 20 cc hypodermic syringes, and stored in evacuated 10 cc vacutainers for laboratory analysis of bulk and isotopic composition (Hinkle and Kilburn, 1979). Soil gas concentrations were measured using an in situ laboratory with a VARIAN microGC 2002P. To evaluate the origin of the carbon dioxide in the soil atmosphere, carbon isotope analyses were performed on 41 soil gas samples at the Laboratory for Earthquake Chemistry, University of Tokyo. Carbon isotope analyses were determined at the same laboratory with a Finnigan MAT Delta S mass spectrometer. The ^{13}C/^{12}C ratios are reported as δ^{13}C values (±0.1‰) with respect to V-PDB standard.

RESULTS AND DISCUSSION

Analytical results for all soil gas samples are summarized in Table 1. Soil CO_2 efflux values ranged from undetectable values up to 780 g m^{-2} d^{-1} with a median of 12.8 g m^{-2} d^{-1}. Carbon isotope ratios for soil gas CO_2 ranged from –35.4 to –6.9‰, suggesting different sources for the carbon dioxide in the soil atmosphere. Soil gas ^{222}Rn concentration ranged from 1.04 to 284 pCi L^{-1} with a median of 25.2 pCi L^{-1}. Both CO_2 efflux and ^{222}Rn data were used to construct contour maps of the summit and flank areas of San Salvador, using kriging with a linear variogram model as interpolation technique. Isotropic linear variograms with negligible nugget effect were fitted to the lognormal distributed data (Journel, 1988; Cressie, 1990).

The sampling frequency distribution of the diffuse CO_2 emission rates showed a polymodal shape (Fig. 3A), and the histogram of the lognormally distributed CO_2 efflux values showed a skewness and kurtosis of –0.37 and 5.95, respectively. To check whether the data come from a unimodal or a polymodal distribution, we applied the probability-plot technique (Tennant and White, 1959; Sinclair, 1974) to the entire CO_2 efflux data set (Fig. 3B). At least three distinct modes are found: normal I, normal II, and the mode between both distributions. These three distinct populations are known as background, peak, and intermediate mixed-population, respectively. The background population (normal I) accounts for 96.0% of the total data with a mean of 12.6 g m^{-2} d^{-1}. Peak population (normal II) represents 0.9% of the total data with a mean of 602 g m^{-2} d^{-1}.

The sampling frequency distribution of the soil ^{222}Rn values showed a unimodal shape. Soil gas ^{222}Rn data were also lognormally distributed (Fig. 4A), and the histogram of the lognormally distributed soil ^{222}Rn data showed a skewness and kurtosis of

TABLE 1. ANALYTICAL RESULTS OF SOIL GASES FROM STUDY AREAS AT SAN SALVADOR VOLCANO, EL SALVADOR

	CO_2 efflux (g m^{-2} d^{-1})	^{222}Rn (pCi L^{-1})	δ^{13}C-CO_2	pH	Org. Mat. (%)
Minimum	0.1	1.04	–35.4	3.22	2.6
Maximum	780	284	–6.9	8.68	69.3
Median	12.8	25.2	–27.4	6.32	25.6

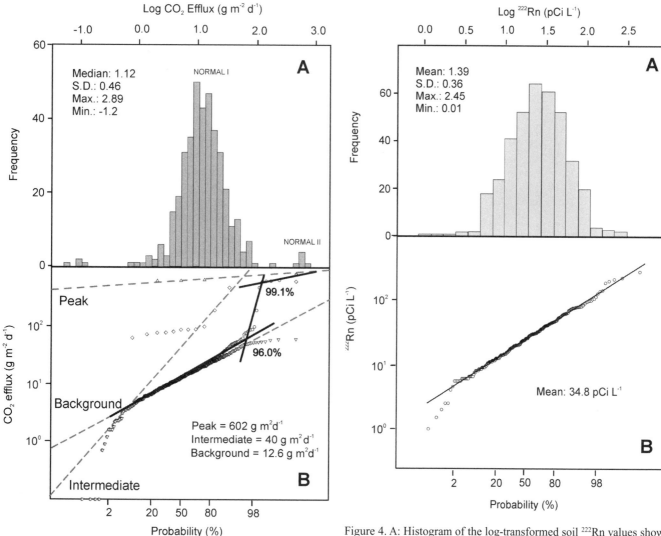

Figure 3. A: Histogram of the log-transformed soil CO_2 efflux values showing the existence of at least two modes within the sampling frequency distribution. B: Probability plot of the log-CO_2 efflux values observed in 388 sampling sites at San Salvador volcano.

Figure 4. A: Histogram of the log-transformed soil ^{222}Rn values showing the existence of one mode within the sampling frequency distribution. B: Probability plot of the ^{222}Rn values observed in 380 sampling sites at San Salvador volcano.

–0.19 and 0.62, respectively. A probability-plot technique was also applied to check whether the ^{222}Rn data come from a unimodal or a polymodal distribution (Fig. 4B). Only one normal distribution was identified using this log plot. This observation suggests that a large proportion of the ^{222}Rn values comes from a simple physical-chemical process.

The existence of three different geochemical populations for San Salvador's CO_2 efflux data suggests a deep perturbation of the surface environment degassing process by the volcanic hydrothermal system. Advection might be the responsible transport mechanism to explain the relatively high CO_2 efflux values, whereas background CO_2 efflux values could be mainly characterized by a

biogenic CO_2, which is mainly transported by diffusion (Chiodini et al., 2001). This observation suggests that there is influence of different physical-chemical processes acting at different scales on the diffuse CO_2 degassing at San Salvador volcano.

An inspection of the CO_2 efflux contour map (Fig. 5) shows that most of the study area displayed background values of diffuse degassing for carbon dioxide. Three clusters of relatively high soil CO_2 efflux values (>100 g m^{-2} d^{-1}) are observed (areas 1, 2 and 3). Area 1 is to the northeast of El Boquerón crater, within the 6 km wide caldera formed by collapse of the older Picacho and El Jabalí volcanoes, and situated along the intersection of two major fault systems (FS1 and FS2) where the most recent magmatic activity has occurred (June, 1917). Area 2 is located outside the 6 km wide caldera on the eastern flank of

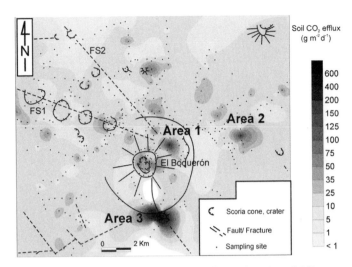

Figure 5. Contour map of the soil CO₂ efflux values (g m⁻² d⁻¹) measured at San Salvador volcano. Dots indicate sampling sites.

Picacho volcano. Area 3 is to the south of El Boquerón, at Cerro La Hoya, which is located outside the southern rim of the 6 km wide caldera, and is characterized by the highest observed CO_2 efflux. Diffuse CO_2 degassing anomalies suggest a close spatial correlation with major faults and complex fracture/fault systems (FS1 and FS2) in the study area. The lack of relatively high CO_2 efflux values observed along most of the FS2 fault of NW-SE direction might be related to the low-permeability cover formed by the recent lava flows. Similarly, most of the measurements carried out along the FS1 fault yielded background CO_2 efflux values. Most of the San Salvador volcanic edifice is covered with vegetation, which may contribute to the production of biogenic CO_2 in the soil atmosphere. The soils of the study area are strongly desaturated and developed over pyroclastic and ash volcanic deposits, with variable organic matter content. Relatively high soil gas CO_2 concentrations might be related to degradation of the organic matter. Degradation is enhanced by the acidity and permeability of the soils (Hinkle, 1994). To evaluate the influence of soil characteristics on the superficial production of CO_2, soil pH and organic matter content were also analyzed in soil samples from San Salvador volcano. CO_2 efflux anomalies were not spatially well correlated with soil pH and organic matter content (R^2 = 0.033 and 0.002, respectively). Therefore, relatively high CO_2 efflux values might show a contribution of deep-seated CO_2 as it was identified by means of statistics.

Radon distribution in the soil atmosphere showed a different pattern. Most of the study area showed relatively low values of soil gas ^{222}Rn, whereas relatively high ^{222}Rn concentrations (>140 pCi L⁻¹) were mainly identified outside the 6 km wide caldera formed by collapse of the older Picacho and Jabalí volcanoes, in the north and northeastern flanks of San Salvador volcano (Fig. 6). Spatial distribution of soil ^{222}Rn anomalies could be related to areas of greater permeability, and it is possible that

a buried or unidentified fault exists in this area, defined by the linear trend of the radon anomalies on the north and east volcano flanks. Note that the orientation of the anomalies is parallel to fault FS2. The shift in the linear trend of ^{222}Rn at the northern sector probably suggests a higher permeability of the deposits, or that relative displacement has occurred between the adjacent rocks, due to the recent seismic activity at San Salvador.

Diffuse CO_2 degassing rates at San Salvador were lower than those observed at other volcanoes, indicating either low permeability of the volcanic cover or low release of deep-seated CO_2 at the time of this survey. For comparison, Table 2 shows data related to soil CO_2 efflux mean peak populations, total diffuse CO_2 output, and the extension of study areas for diffuse CO_2 degassing surveys performed at Vulcano, Italy; Santorini, Greece; Cerro Negro and Masaya, Nicaragua; Teide, Canary Islands; Poás and Miravalles, Costa Rica; Usu and Miyakejima, Japan (Chiodini et al., 1998; Hernández et al., 1998, 2001a, 2001b; Pérez et al., 2000; Salazar et al., 2001; Melián et al., 2001). The low values measured for both CO_2 efflux and ^{222}Rn could be explained in terms of the characteristics of the deposits forming San Salvador volcano. Quite impermeable tephra deposits of reduced permeability and variable thickness related to the formation of Ilopango caldera are observed at the summit and flanks of San Salvador volcano. Low permeability at depth would decrease the gas diffusion to the upper layers, whereas high permeability of surface layers would enhance the infiltration of air. Subsurface sealing processes of fissures and the existence of numerous lava flows covering the surface could also contribute to the low diffuse degassing levels at San Salvador volcano.

To check the spatial and temporal variability of the CO_2 diffuse degassing indicated by the 1999 measurements at San Salvador volcano, 59 measurements of CO_2 efflux were carried out

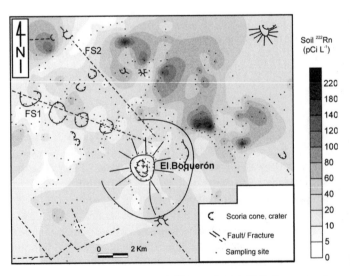

Figure 6. Contour map of the soil ^{222}Rn values (pCi L⁻¹) measured at San Salvador volcano. Dots indicate sampling sites.

TABLE 2. DIFFUSE CO₂ DEGASSING RATES OF SAN SALVADOR AND OTHER VOLCANIC SYSTEMS

Volcano	Year	Area (km^2)	Mean peak population CO$_2$ efflux (g m^{-2} d^{-1})	Total CO$_2$ output (t d^{-1})
San Salvador	1999	266	602	4,000
Solfatara of Pozzuoli	1994	0.09	5,530	133
Nea Kameni Santorini	1995	0.37	2,840	15.4
Cerro Negro	1999	0.58	>10,000	2,800
Masaya	1999	44	5,500	28,800
Teide	1996	0.53	–	380
Poás	2001	3.4	1,876	111
Miravalles	2000	65	1,158	12,700
Usu	1998	2.65	2,400	120
Usu	1999	2.65	7,000	340
Usu	2000	2	1,800	39
Miyakejima	1998 (May)	33.4	9,500	146
Miyakejima	1998 (September)	41.4	8,100	93

in March 2001 on a north-south transect across the San Salvador volcanic edifice. This additional survey yielded similar values to those measured in 1999. The mean value was 13.8 g m^{-2} d^{-1}, which is close to the background value estimated for the 1999 CO$_2$ efflux data set (12.6 g m^{-2} d^{-1}). This finding suggests that significant changes of the volcanic hydrothermal system of San Salvador volcano did not occur during the time period between the two surveys.

Diffuse degassing studies carried out at some subduction-related stratovolcanoes in Central America such as Arenal and Poás (Costa Rica), and at Popocatépetl (México), showed that negligible diffuse degassing of magmatic gas together with light δ¹³C–CO₂ occurred at the upper flanks and within the summit crater rims of these volcanoes (Williams-Jones et al., 2000; Varley and Armienta, 2001). At San Salvador volcano, carbon isotope ratios of the CO₂ from diffuse degassing (Table 1) suggest a clear biogenic origin for most of the study area. The heaviest δ¹³C–CO₂ values, –9.1‰ and –6.9‰, were identified inside El Boquerón's crater, indicating a mixing of biogenic and deep-seated carbon for the diffuse CO₂ degassing. In contrast, lighter carbon isotope ratios were observed along the two main faults located at the northwest flank of San Salvador volcano (Fig. 7). There are clear trends toward decreasing CO₂ efflux and lighter δ¹³C isotopic values with increasing distance from the central crater El Boquerón. This pattern has been observed at other volcanoes such as Fossa cone on Vulcano Island (Baubron et al., 1990), and it is associated with a higher magmatic contribution near the crater. The decreasing of both CO₂ efflux and δ¹³C isotopic values with distance at the San Salvador volcanic edifice might indicate a decrease of diffuse degassing due to a minor surface activity

and/or a mixing with an organic CO₂ component associated with higher humidity and biogenic activity. These results point out the importance of structural control and channeling on the CO₂ degassing, as has been observed at other volcanoes such as the Santa Ana–Izalco–Coatepeque volcanic complex (Salazar et al., this volume, Chapter 10).

The short half-life of ²²²Rn, 3.8 days, indicates that much of the ²²²Rn measured is derived from a shallow source unless it is

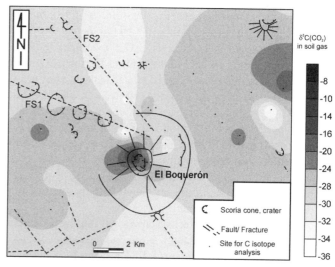

Figure 7. Contour map of δ¹³C(CO₂) in soil gases measured at San Salvador volcano. Dots indicate sampling sites.

transported quickly to the surface. Its limited migration indicates that a very significant portion of the radon must have a $^{226}Ra/^{238}U$ source in the aquifers and/or weathered bedrock of San Salvador volcano. No correlation was observed between CO_2 efflux and soil gas ^{222}Rn distribution at the study area. The very poor correlation observed between the log-CO_2 efflux and log-^{222}Rn values ($R^2 = 0.00024$), suggests that CO_2 does not act as a ^{222}Rn carrier to the surface. This indicates that the mechanisms controlling the transport of CO_2 and ^{222}Rn to the surface environment are independent.

$\delta^{13}C$–CO_2 values of soil gas samples suggest that the CO_2 emanating from the San Salvador volcanic edifice represents a mixing of different sources: (1) marine limestones including slab carbonate, which have an average $\delta^{13}C$ value near 0‰, (2) organic matter degradation affecting the sedimentary deposits of the volcanic basement (assuming that elemental and isotopic fractionation is minimal in the hydrothermal system), and (3) degassing of the magma chamber underneath the volcano. Deep-seated CO_2 is not released through the main faults and fissures located along the north and northwest flanks of San Salvador volcano and might be trapped by local meteoric water. The relatively high CO_2 efflux values measured at areas 1, 2, and 3 are higher than the contribution of diffusive CO_2 efflux by biogenic activity estimated for other volcanoes (7 g m^{-2} d^{-1} for Santa Ana volcanic complex, Salazar et al., this volume; 94 g m^{-2} d^{-1} for Miyake-jima volcano, Japan, Hernández et al., 1998). This observation together with the carbon isotope ratios measured suggests that deep-seated CO_2 contributes to the diffuse degassing phenomena at the surface environment of San Salvador volcano.

To estimate the total diffuse CO_2 output released from the San Salvador edifice, we considered the positive volume enclosed by the soil CO_2 efflux three-dimensional surface built up by the kriging in the contouring over the study area, yielding a total of 4000 t d^{-1} for a covered area of 266 km^2. Carbon dioxide output from other sources such as the CO_2-rich groundwater has not been considered in this study. This value (4000 t d^{-1}) is lower than the estimated value for other volcanoes (Chiodini et al., 1996, 1998; Hernández et al., 1998, 2001a, 2001b; Salazar et al., 2001) if we consider the ratio of total CO_2 output to area. This observation, together with the $\delta^{13}C$–CO_2 measured, indicates a small contribution of deep-seated CO_2 to the total CO_2 diffuse emission. This difference could be related to the level of volcanic activity. To constrain this possibility we have calculated the fraction of magmatic CO_2 emitted from this volcano. Based on the preliminary results of chemical and isotopic analysis of CO_2, we have considered CO_2 efflux corresponding to the peak (normal II) population as representative of magmatic CO_2. Our estimation indicates that ~40 t d^{-1} of magmatic CO_2 (0.9% of the total) are being released as diffuse emanation from the summit of San Salvador volcano at present.

With the aim of detecting changes in diffuse CO_2 degassing related to seismic activity at the San Salvador volcano, an automatic geochemical station (WEST Systems) was installed in November 19, 2001. This station continuously monitors diffuse CO_2 degassing at Finca El Espino, located in Cerro La Hoya at the south flank of San Salvador (1400 masl). Soil CO_2 efflux is also estimated according to the accumulation chamber method. Continuous measurements of CO_2 efflux together with soil temperature (30 cm depth) and barometric pressure carried out during almost seven months at San Salvador volcano show a general increase in the diffuse degassing during this period (increase started on December 28, 2001) with a mean value of 4000 g m^{-2} d^{-1} for the CO_2 degassing. This value is much higher than measured during the 1999 soil gas survey (760 g m^{-2} d^{-1}), suggesting that in the future it will be beneficial to carry out soil gas surveys together with continuous CO_2 efflux monitoring at San Salvador, in order to understand the relationship between the temporal variations in the total output of CO_2 and the volcanic activity.

CONCLUSIONS

This work shows that San Salvador volcano emits relatively low amounts of CO_2 in diffuse form compared with other volcanic systems. Most of the diffuse CO_2 degassing is concentrated (1) outside El Boquerón crater, to the northeast and within the 6 km wide caldera formed by collapse of the older Picacho and El Jabalí volcanoes, and situated along the intersection of two major faults (FFS1 and FFS2) where the most recent magmatic activity has occurred (June, 1917), (2) outside the 6 km wide caldera on the eastern flank of Picacho volcano, and (3) at Cerro La Hoya, which is located outside the southern rim of the 6 km wide caldera, and characterized by the highest observed CO_2 efflux. Soil ^{222}Rn distribution in interstitial soil gas showed a different pattern. Most of the study area showed relatively low values of soil gas ^{222}Rn, whereas relatively high ^{222}Rn concentrations were identified outside the 6 km wide caldera, on the north and northeastern flanks of San Salvador volcano.

Our results suggest that the main conduit of San Salvador volcano represents the preferential way for the gases to escape toward the surface, and that this is the best connection between the magma chamber and the summit crater. Low permeability of the buried volcanic deposits together with the high permeability of surface layers enhancing the infiltration of air would limit the escape of gases to the atmosphere. Sealing at depth of the fissures located at the northwest flank of the volcano, which are related to the last eruption of San Salvador volcano (1917), could explain the absence of significant emission of deep-seated gas to the surface, whereas an anomalous ^{222}Rn area near these parallel fissures seems to have a shallow origin. The poor correlation observed between the CO_2 efflux and ^{222}Rn values indicates that the mechanisms controlling the transport of CO_2 and ^{222}Rn to the surface environment are independent. The relatively low CO_2 efflux values together with light $\delta^{13}C(CO_2)$ measured in soil gases indicate that there is only a small contribution of mantle degassing proximal to or incorporated within the volcanic hydrothermal system of San Salvador volcano. The estimated emission rate of CO_2 for the San Salvador volcanic edifice is 4000 t d^{-1}.

This study emphasizes the importance of using geochemical monitoring, such as CO_2 efflux and soil ^{222}Rn surveys, to evaluate the relation between the level of the diffuse discharges and the state of a volcanic system. If CO$_2$ efflux surveys produce different temporal patterns before and after eruptions, they can well be adopted as a precursory tool possibly signaling volcanic destabilization, and be used to mitigate potential volcano hazards.

ACKNOWLEDGMENTS

We are grateful to the Agencia Española de Cooperación Internacional (AECI) in El Salvador, the Spanish Embassy in El Salvador, The Ministry of Environment and Natural Resources of the Government of El Salvador, and Geotérmica Salvadoreña (GESAL) for their assistance during the fieldwork. Helium isotopic analyses were done by H. Sumino, A. Shimizu, and professor K. Nagao at the Laboratory for Earthquake Chemistry, University of Tokyo, Japan. This research was mainly supported by the Cabildo Insular de Tenerife, Caja Canarias (Canary Islands, Spain), the European Union, and the University of El Salvador, but additional financial aid was provided by the Spanish Aid Agency (Agencia Española de Cooperación Internacional—AECI).

REFERENCES CITED

Allard, P., Carbonelle, J., Dajlevic, D., Bronce, J.Le., Morel, P., Robe, M.C., Maurenads, J.M., Faivre-Pierret, R., Martin, D., Sabroux, J.C., and Zettwoog, P., 1991, Eruptive and diffuse emissions of CO$_2$ from Mount Etna: Nature, v. 351, p. 387–391.

Baubron, J.C., Allard, P., and Toutain, J.P., 1990, Diffuse volcanic emissions of carbon dioxide from Vulcano Island, Italy: Nature, v. 344, p. 51–53.

Baubron, J.C., Allard, P., Sabroux, J.C., Tedesco, D., and Toutain, J.P., 1991a, Soil gas emanation as precursory indicators of volcanic eruptions: Geological Society [London] Journal, v. 148, p. 571–576.

Baubron, J.C., Mathieu, R., and Miele, G., 1991b, Measurement of gas flows from soil in volcanic areas: The accumulation method. Napoli'91 International Conference Active Volcanoes and Risk Mitigation, Abstracts, 27 August–1 September 1991, Napoli, Italy.

Bruno, N., Caltabiano, T., Gianmanco, S., and Romano, R., 2001, Degassing of SO$_2$ and CO$_2$ at Mount Etna (Sicily) as an indicator of pre-eruptive ascent and shallow emplacement of magma: Journal of Volcanology and Geothermal Research, v. 110, p. 137–153.

Capaldi, G., Gallucio, F., Pece, R., Veltri, C., and Williams, S.N., 1988, A radon survey in soil gases of the Campi Flegrei Caldera the year after the end of the 1982–1984 bradyseismic crisis: Rendiconti della Societa Italiana di Mineralogia e Petrologia, v. 43, p. 947–954.

Caparezza, M.L., and Diliberto, I.S., 1993, Vulcano: Helium and CO$_2$ soil degassing: Acta Vulcanologica, v. 3, p. 273–276.

Casadevall, T.J., Doukas, M.P., Neal, C.A., McGimsey, R.G., and Gardner, C.A., 1994, Emission rates of sulfur dioxide and carbon dioxide from Redoubt Volcano, Alaska during the 1989–1990 eruptions: Journal of Volcanology and Geothermal Research, v. 62, p. 519–530.

Chiodini, G., Frondini, F., and Raco, B., 1996, Diffuse emission of CO$_2$ from the Fossa crater, Vulcano Island (Italy): Bulletin of Volcanology, v. 58, p. 41–50.

Chiodini, G., Cioni, R., Guidi, M., Raco, B., and Marini, L., 1998, Soil CO$_2$ flux measurements in volcanic and geothermal areas: Applied Geochemistry, v. 13, p. 543–552.

Chiodini, G., Frondini, F., Cardellini, C., Granieri, D., Marini, L., and Ventura, G., 2001, CO$_2$ degassing and energy release at Solfatara volcano, Campi Flegrei, Italy: Journal of Geophysical Research, v. 106(B8), p. 16,213–16,221.

Connor, C., Hill, B., LaFemina, P., Navarro, M., and Conway, M., 1996, Soil ^{222}Rn pulse during the initial phase of the June–August 1995 eruption of Cerro Negro, Nicaragua: Journal of Volcanology and Geothermal Research, v. 73, p. 119–127.

Cox, M.E., Cuff, K.E., and Thomas, D.M., 1980, Variations of ground radon concentrations with activity of Kilauea volcano, Hawaii: Nature, v. 288, p. 74–76.

Cressie, N.A.C., 1990, The origings of kriging: Mathematical Geology, v. 22, p. 239.

Fairbrothers, G.E., Carr, M.J., and Mayfield, D.G., 1978, Temporal magmatic variation at Boqueron volcano, El Salvador: Contributions to Mineralogy and Petrology, v. 67(1), p. 1–9.

Farrar, C.D., Sorey, M.L., Evans, W.C., Howle, J.F., Kerr, B.D., Kennedy, B.M., King, Y., and Southon, J.R., 1995, Forest-killing diffuse CO$_2$ emission at Mammoth Mountain as a sign of magmatic unrest: Nature, v. 376, p. 675–678.

Fleischer, R.L., and Mogro-Campero, A., 1978, Mapping of integrated radon emanation for detection of long distance migration of gases within the Earth. Techniques and principles: Journal of Geophysical Research, v. 83, p. 3539–3549.

Francis, P., Maciejewski, A., Oppenheimer, C., Chaffin, C., and Caltabiano, T., 1995, SO$_2$:HCl ratios in the plumes from Mt. Etna and Vulcano by Fourier transform spectroscopy: Geophysical Research Letters, v. 22, p. 1717–1720.

Gerlach, T.M.J., and Graeber, E.J., 1985, Volatile budget of Kilauea volcano: Nature, v. 313, p. 273–277.

Gerlach, T.M.J., Doukas, M.K., McGee, K., and Kessler, R., 1998, Three-year decline of magmatic CO$_2$ emission from soils of a Mammoth Mountain tree kill: Horseshoe Lake, California, 1995–1997: Geophysical Research Letters, v. 25, p. 1947–1950.

Giggenbach, W., 1996, Chemical composition of volcanic gases, in Scarpa, R., and Tilling, R., eds., Monitoring and mitigation of volcanic hazards: Berlin, Springer-Verlag, p. 221–256.

Hernández, P.A., Pérez, N.M., Salazar, J.M., Notsu, K., and Wakita, H., 1998, Diffuse emission of carbon dioxide, methane, and helium-3 from Teide volcano, Tenerife, Canary Islands: Geophysical Research Letter, v. 25, p. 3311–3314.

Hernández, P.A., Notsu, K., Salazar, J.M., Mori, T., Natale, G., Okada, H., Virgili, G., Shimoike, Y., Sato, M., and Pérez, N.M., 2001a, Carbon dioxide degassing by advective flow from Usu volcano, Japan: Science, v. 292, p. 83–86.

Hernández, P.A., Salazar, J.M., Shimoike, Y., Mori, T., Notsu, K., and Pérez, N.M., 2001b, Diffuse emission of CO$_2$ from Miyakejima volcano, Japan: Chemical Geology, v. 177, p. 175–185.

Hinkle, M., 1994, Environmental conditions affecting concentrations of He, CO$_2$, O$_2$, and N$_2$ in soil gases: Applied Geochemistry, v. 9, p. 53–63.

Hinkle, M.E., and Kilburn, J.E., 1979, The use of vacutainer tubes for collection of soil gas samples for helium analysis: U.S. Geological Survey Open-File Report 79-1441.

Hirabayashi, J., Ossaka, J., and Ozawa, T., 1986, Geochemical study on volcanic gases at Sakurajima volcano, Japan: Journal of Geophysical Research, v. 91, p. 12,167–12,176.

Horrocks, L., Burton, M., Francis, P., and Oppenheimer, C., 1999, Stable gas plume composition measured by OP-FTIR spectroscopy at Masaya volcano, Nicaragua: Geophysical Research Letters, v. 26, p. 3479–3500.

Journel, A., 1988, Principles of environment sampling: Washington, D.C., American Chemical Society, p. 45–72.

Lombardi, S., and Reimer, G.M., 1990, Radon and helium in soil gases in the Phlegraean Fields, Central Italy: Geophysical Research Letters, v. 17(6), p. 849–852.

Major, J.J., Schilling, S.P., Sofield, D.J., Escobar, C.D., and Pullinger, C.R., 2001, Volcano hazards in the San Salvador region, El Salvador: U.S. Geological Survey Open-File Report 01-366, 30 p.

McGee, K.A., and Gerlach, T.M., 1998, Airborne volcanic plume measurements using a FTIR spectrometer, Kilauea volcano, Hawaii: Geophysical Research Letters, v. 25, no. 5, p. 615–618.

Melián, G., Galindo, I., Salazar, J.M.L., Hernández, P.A., Pérez, N.M., Ramirez, C., Fernández, M., and Notsu, K., 2001, Spatial and secular variations of diffuse CO2 degassing from Poás volcano, Costa Rica, Central América: Eos (Transactions, American Geophysical Union), v. 82, p. F1332.

Menyailov, I.A., Nikitina, L.P., and Zarpar, V.N., 1985, Results of geochemical monitoring of the activity of Ebeko volcano (Kurile Islands) used for eruption prediction: Journal of Geodynamics, v. 3, p. 259–274.

Meyer-Abich, H., 1956, Los Volcanes Activos de Guatemala y El Salvador: Anales del Servicio Geológico Nacional de El Salvador, Bd., 3, p. 49–62.

Mori, T., and Notsu, K., 1997, Remote CO, COS, CO$_2$, SO$_2$, HCl detection and temperature estimation of volcanic gas: Geophysical Research Letters, v. 24, p. 2047–2050.

Morse, R.H., 1976, Radon counters in uranium exploration: International Atomic Energy Agency, IAEA-SM-208/55.

Pan, V., Holloway, J.R., and Hervig, R.L., 1991, The pressure and temperature dependence of carbon dioxide solubility in tholeiitic basalts melts: Geochimica et Cosmochimica Acta, v. 55, p. 1587–1595.

Pérez, N.M., Wakita, H., Lolok, D., Patia, H., Talai, B., and McKee, C.O., 1996, Anomalous soil gas CO_2 concentrations and relation to seismic activity at Rabaul caldera, Papua New Guinea: Geogaceta, v. 20, p. 1000–1003.

Pérez, N.M., Melián, G., Salazar, J.M.L., Saballos, A., Alvarez, J., Segura, F., Hernández, P.A., and Notsu, K., 2000, Diffuse degassing of CO_2 from Masaya caldera, Nicaragua, Central America: Eos (Transactions, American Geophysical Union), v. 81, p. F1318.

Salazar, J.M., Hernández, P.A., Pérez, N.M., Melian, G., Alvarez, J., and Notsu, K., 2001, Diffuse volcanic emissions of carbon dioxide from Cerro Negro volcano, Nicaragua: Geophysical Research Letters, v. 28, p. 4275–4278.

Salazar, J.M.L., Hernández, P.A., Pérez, N.M., Olmos, R., Barahona, F., Cartagena, R., Soriano, T., López, D.L., Sumino, H., and Notsu, K., 2004, Spatial and temporal variations of diffuse CO_2 degassing at the Santa Ana–Izalco–Coatepeque volcanic complex, El Salvador, Central America, *in* Rose, W.I., et al., eds., Natural hazards in El Salvador: Boulder, Colorado, Geological Society of America Special Paper 375, p. 135–146 (this volume).

Sano, Y., Notsu, K., Ishibashi, J., Igarashi, G., and Wakita, H., 1991, Secular variations in helium isotope ratios in an active volcano: Eruption and plug hypothesis: Earth and Planetary Science Letters, v. 107, p. 95–100.

Sato, M., 1988, Continuous monitoring of hydrogen in volcanic areas: Petrological rationale and early experiments: Rendiconti della Societa Italiana di Mineralogia e Petrologia, v. 43, p. 1265–1281.

Simkin, T., and Siebert, L., 1994, Volcanoes of the world (2nd edition): Tucson, Arizona, Geoscience Press, 349 p.

Sinclair, A.J., 1974, Selection of thresholds in geochemical data using probability graphs: Journal of Geochemical Exploration, v. 3, p. 129–149.

Smith, A.Y., Barreto, P.M.C., and Pournis, S., 1976, Radon methods in uranium exploration: Symposium Proceedings, Exploration for Uranium Ore Deposits, IAEA (International Atomic Energy Agency) Vienna, p. 185–211.

Stoiber, R.E., Malinconico, J.L.L., and Williams, S.N., 1983, Use of correlation spectrometer at volcanoes, *in* Tazieff, H., and Sabroux, J.C., eds., Forecasting volcanic events: New York, Elsevier, p. 425–444.

Stolper, E., and Holloway, J.R., 1988, Experimental determination of the solubility of carbon dioxide in molten basalt at low pressure: Earth Planetary and Science Letters, v. 87, p. 397–408.

Symonds, R.B., Rose, W.I., Bluth, G.J.S., and Gerlach, T.M., 1994, Volcanic-gas studies: Methods, results, and applications, *in* Carroll, M.R., and Holloway, J.R., eds., Volatiles in magmas: Reviews in Mineralogy, v. 30, p. 1–60.

Tennant, C.B., and White, M.L., 1959, Study of the distribution of some geochemical data: Economic Geology, v. 54, p. 1281–1290.

Varley, N.R., and Armienta, M.A., 2001, The absence of diffuse degassing at Popocatépetl volcano, Mexico: Chemical Geology, v. 177, p. 157–173.

Williams, S.N., 1985, Soil radon and elemental mercury distribution and relation to magmatic resurgence at Long Valley caldera: Science, v. 229, p. 551–553.

Williams-Jones, G., Stix, J., Heiligmanm, M., Charland, A., Sherwood Lollar, B., Arner, N., Garzon, G., Barquero, J., and Fernandez, E., 2000, A model of diffuse degassing at three subduction-related volcanoes: Bulletin of Volcanology, v. 62, p. 130–142.

Manuscript Accepted by the Society June 16, 2003

Geological Society of America
Special Paper 375
2004

Lessons from the mud, lessons from the Maya:
Paleoecological records of the Tierra Blanca Joven eruption

Robert A. Dull*

Department of Geography, Texas A&M University, College Station, Texas 77843, USA

ABSTRACT

Stratigraphic studies of lake sediments were carried out in the Rio Paz Valley and Sierra de Apaneca of western El Salvador. Pollen, charcoal, and organic matter content of the sediments record the Holocene environmental history of the region. Reported here are records from four lake sites, Lagunas Cuzcachapa, El Trapiche, Llano, and Verde. All sites sampled in this region contain tephra from the Tierra Blanca Joven (TBJ) eruption of Volcán Ilopango, a major explosive event that occurred at ca. A.D. 430. This ubiquitous stratigraphic marker allows cross correlation of environmental conditions and human land use throughout the region before and after the eruption. Before the TBJ eruption, most of the study area was under cultivation by Preclassic and Early Classic period farmers; pollen from *Zea mays* (maize) and agricultural weeds are prominent components of the pollen assemblages from ca. 2000 B.C. until the A.D. 430 eruption. The post-eruption data record flooding and increased erosion, as well as secondary forest succession. Abandonment by human populations is clearly indicated in two of the three pollen records (Laguna Cuzcachapa and Laguna Verde), but data from the third pollen site (Laguna Llano) are inconclusive.

Keywords: Ilopango, volcanic eruption, hazards, paleoecology.

INTRODUCTION

Historical accounts of volcanic eruptions in El Salvador are rich sources of data on the nature of volcanic hazards in the country over the past several hundred years (Feldman, 1993; Lardé y Larín, 1978). These historical data on high-frequency, low-magnitude events help planners anticipate future volcanism in El Salvador and provide baseline scenarios for volcanic hazards mitigation. The postcolonial period (sixteenth century–present), however, has not witnessed any large (VEI 5+) explosive eruptions (Rose et al., 1999; Simkin and Siebert, 1994). The most recent such event in El Salvador was the caldera-forming eruption of Ilopango volcano in ca. A.D. 430, which is known as the Tierra Blanca Joven (TBJ) eruption (Dull et al., 2001; Hart and Steen-McIntyre, 1983). Understanding extreme volcanic events like the TBJ eruption can help in an effort to anticipate worst-

case scenarios while planning for future volcano-related hazards in El Salvador.

The TBJ eruption consisted of a violent series of base surge pyroclastic flows and tephra emissions that blanketed most of central and western El Salvador under meters of white silicic ash and highly vesiculated pumice (Hart and Steen-McIntyre, 1983) (Fig. 1). There were probably no survivors in the 100–200 km² area of greatest devastation—including the Basin of San Salvador (Valle de las Hamacas) and the eastern portion of the Zapotitán Valley. The people in this zone were most likely killed instantly by gaseous and fiery pyroclastic flows. Farther to the west and north, in areas such as the Rio Paz and upper Rio Lempa basins, the inhabitants fled and did not completely resettle the region until the end of the sixth century (Dull et al., 2001; Sheets, 1979).

The nature and extent of devastation caused by the TBJ eruption, and the responses of the fifth-century inhabitants of El Salvador, are still only partially understood. Any data that can be gleaned from geological, archaeological, and paleoecological studies on the subject should help advance our understanding of

*robdull@geog.tamu.edu

Dull, R.A., 2004, Lessons from the mud, lessons from the Maya: Paleoecological records of the Tierra Blanca Joven eruption, *in* Rose, W.I., Bommer, J.J., López, D.L., Carr, M.J., and Major, J.J., eds., Natural hazards in El Salvador: Boulder, Colorado, Geological Society of America Special Paper 375, p. 237–244. For permission to copy, contact editing@geosociety.org. © 2004 Geological Society of America

the full range of volcanic hazard potentialities in El Salvador. This paper reviews paleoecological data from western El Salvador in attempt to shed light on the TBJ eruption, its effect on the landscape, and its affect on the Maya. Finally, the paleoecological record of the TBJ disaster is considered as it relates to prospects for present-day hazards planning and mitigation in El Salvador.

THE TBJ TEPHRA AND LAKE SEDIMENT CROSS CORRELATION

The first detailed studies of the TBJ eruption were carried out by Payson Sheets and colleagues in the 1970s and 1980s (Sheets, 1971, 1979, 1983). Radiocarbon dating results at that time suggested a calendar date of ca. A.D. 260 ± 114 for the eruption (Sheets, 1983), but a reanalysis of those data combined with more recent AMS (accelerator mass spectrometry) results shows that the event occurred nearly two centuries later at ca. A.D. 430 (2 sigma = A.D. 408–536) (Dull et al., 2001). The refined dating of the TBJ event allows more precise chronological comparisons of archaeological and paleoecological sites throughout central and western El Salvador, western Honduras, and southeastern Guatemala. Although erosion and bioturbation have erased vestiges of the TBJ tephra in some western Salvadoran terrestrial sites, the conspicuous white stratum is as thick as 50 cm along El Salvador's western border with Guatemala (Fig. 2) (Dull et al., 2001). The TBJ horizon is present in the sediments of all western Salvadoran lakes thus far investigated by the author (Dull, 2001).

HISTORICAL ECOLOGY OF THE TBJ ERUPTION

Lake sediment cores were extracted from four lake sites in the Rio Paz drainage basin of western El Salvador (Fig. 2). Three of these sites—Lagunas Cuzcachapa, Trapiche, and Llano—are located at mid-elevations (~650–700 m) in the Rio Paz bottomlands near the towns of Chalchuapa and Ahuachapán. The fourth site, Laguna Verde, is a crater lake at the summit of a composite cone in the Apaneca highlands at ~1600 m. Cores from all sites were split and subsampled for several standard paleoecological analyses, including pollen, charcoal, and loss on ignition (LOI) (organic carbon content). Because of poor preservation of pollen and charcoal, the record from Laguna El Trapiche has only been partially analyzed and only the LOI data are reported here. Several (3–5) accelerator mass spectrometry radiocarbon dates were obtained for each core, but not enough to construct a detailed sub-millennial chronology at any of the sites. The TBJ tephra was found at each of the four sites and varied in thickness from ~15 cm at Laguna Verde to ~25 cm at Laguna Cuzcachapa. This paper focuses on the pre- and post-TBJ environments of the Rio Paz Valley and the Sierra de Apaneca. Complete core analyses and environmental reconstructions from Lagunas Cuzcachapa, Llano, and Verde are reported elsewhere (Dull, 2001).

Chalchuapa

The Chalchuapa archaeological zone was the preeminent center of cultural and economic development in western El Salvador from ca. 1000 B.C. until roughly A.D. 250 (Sharer, 1974; Sheets, 1984). The important role of Chalchuapa in the regional economy of the southern Maya realm was most likely solidified

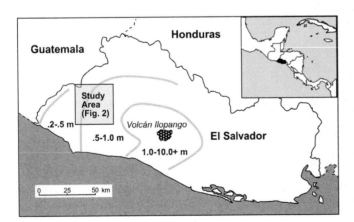

Figure 1. Location map showing study area in relation to Tierra Blanca Joven (TBJ) tephra isopachs (in meters). Adapted from Hart and Steen-McIntyre (1983), Dull et al. (2001).

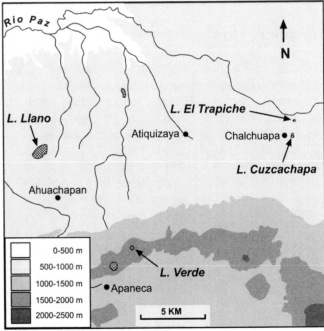

Figure 2. Study area topographic map including paleoecological study sites (boldfaced labeled lakes) and modern towns (black dots) in the region.

by its affiliation with the Olmec beginning ca. 1000 B.C. However, even as the Olmec influence waned during the latter Middle Preclassic period, Chalchuapa gained ground as a regional polity and early population center in the southeastern Maya "periphery" (Sharer, 1978). The economy of Preclassic Chalchuapa was based on the production and trade of Usulután ceramics and other regional commodities that might have included cacao and Ixtepeque obsidian (Bergmann, 1969; Demarest and Sharer, 1986; Neff et al., 1999; Sharer, 1974). Agricultural production in the region was not limited to cash crops such as cacao, but also included the intensive cultivation of subsistence crops. Chalchuapa's Preclassic florescence is amply recorded in the lake sediments at Laguna Cuzcachapa, where the effects of intensive agricultural land use are recorded.

Laguna Cuzcachapa

A 9.7-meter sediment core from Laguna Cuzcachapa, Chalchuapa, records agricultural and land-use history in the region both before and after the TBJ eruption (Fig. 3). Extremely high levels (over 6%) of *Zea mays* pollen are present in the deepest levels of the core, which date to ca. 1760 B.C. (3710 cal. yr B.P.). Agricultural weeds ("ruderal taxa") dominate the total pollen assemblage in the pre-TBJ levels, with the Chenopodiaceae/Amaranthaceae group comprising as much as 45% of the total pollen in the pre-TBJ levels. Other taxa in this group include *Ipomoea, Ambrosia*, Asteraceae, and *Iresine/Althernathera.*

It may be that Chalchuapa was in decline by the Early Classic Period (Earnest, 1999). The paleoecological record from Laguna Cuzcachapa supports this theory, recording gradual decreases in virtually all indices of human impact between ca. A.D. 250 and 430. The complete abandonment of the site, however, appears to have occurred only after the TBJ eruption. Indices of human occupation in the Laguna Cuzcachapa core ebb after the TBJ eruption. The lowest levels of agricultural weed pollen are recorded during this period and *Zea mays* pollen drops out altogether. Secondary forest taxa (e.g., *Piper, Celtis, Cecropia, Trema*) increase dramatically following the eruption, suggesting a period of succession uninterrupted by agricultural clearing and burning. Although the inhabitants of Chalchuapa evidently fled, erosion rates continued to rise after the eruption. Sedimentation rates reach a maximum of 1.06 cm/yr between 430 and 775 cal. yr A.D., an order of magnitude higher than the rate at the base of the core (.11 cm/yr), and 500% higher than the historic period sedimentation rate (.19 cm/yr). The increase in sediment flux into Laguna Cuzcachapa during this period suggests that unconsolidated tephra was probably being removed from the uplands by rain and wind, and accumulating in the Rio Paz Valley. Mass wasting events were probably common in the denuded landscape following the eruption, and would have contributed to the unusually high sedimentation rate.

The TBJ tephra thickness has been measured to 50+ cm in terrestrial sites around Chalchuapa and 25 cm in Laguna Cuzcachapa. Assuming that erosion, bioturbation, and compaction have thinned these deposits, the actual depth of the initial ash deposition at Chalchuapa was probably somewhat greater. It is assumed here that the actual tephra depth in Chalchuapa at the time of the eruption was 50–100 cm. The weight of the tephra would have collapsed most modern roofs (Baxter, 2000), which

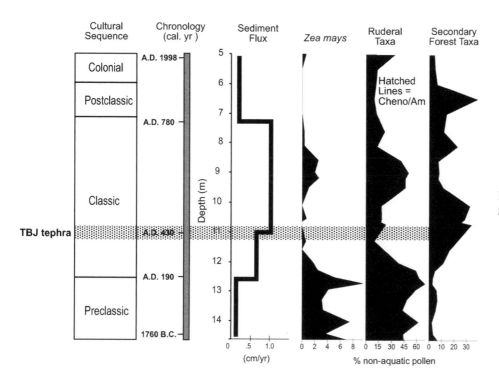

Figure 3. Summary diagram from Laguna Cuzcachapa, Chalchuapa.

are structurally stronger than the thatched roofs that presumably covered Chalchuapa's fifth-century dwellings. Crop failure was evidently widespread, and the reestablishment of agriculture would have been difficult for years following the event (Dull et al., 2001). Another potential effect over the first few years following the eruption may have been the reactivation of tephra in the rainy season. Debris flows and lahars may have been common as the thick mantle of tephra worked its way downslope and into the valley bottom.

Laguneta El Trapiche

Sediments from Laguneta El Trapiche, located less than 2 km north of Laguna Cuzcachapa (Fig. 2), also document an abrupt change in the sedimentation regime following the TBJ event (Fig. 4). Dating back to ca. 6720 cal. yr B.C., the El Trapiche core provides nearly a complete Holocene record of environmental change in the basin, including at least three volcanic eruptions in the area prior to the TBJ eruption. The physical stratigraphy from the 4.5 m El Trapiche core records a shift from emergent marsh to shallow lake following the TBJ eruption (peat to gyttja). The denuded landscape around the lake following the TBJ eruption would have been susceptible to higher rates of mass wasting, overland flow, and rill erosion. Indeed, the water-level rise at Laguna El Trapiche following the TBJ eruption is accompanied by an abrupt increase in sediment density. This increase in sediment density supports the conclusion that before the eruption, accreted marsh sediments accounted for most of the sedimentation—a regime dominated by autochthonous inputs. Following the eruption, however, most of the sediment deposited on the lake bottom apparently came from allochthonous sources (material transported to the lake by water and wind). Post-TBJ sedimentation rates cannot be accurately measured because the upper sediments of the lake were reportedly dredged in the early 1990s.

The Reoccupation of Chalchuapa

The duration of the abandonment period at Chalchuapa is still unknown. An attempt was made to date organic residue in the first level (10.66 m) containing *Zea mays* following the event, but insufficient carbon quantity resulted in a large margin of error (1440 ± 640 ^{14}C yr B.P.), thereby rendering the AMS radiocarbon result useless. Radiocarbon dates from the archaeological site of Tazumal provide a clearer picture of when monumental construction was resumed: ca. A.D. 590 (Dull et al., 2001; Stuiver and Deevey, 1961). The initial reoccupation of the region, however, was probably not carried out by a well-organized and stratified political entity (e.g., chiefdom-level). Considering the political and economic organization presumably required to undertake the construction of a ceremonial center, the Tazumal dates probably reflect activity occurring at some point after the initial reoccupation.

Ahuachapán

Laguna Llano

The paleoecological record from Laguna Llano records human occupation of the Ahuachapán savanna during the late Holocene and maize cultivation beginning ca. 960 cal. yr B.C. (Fig. 5). The TBJ stratum in the Laguna Llano core is ~20 cm thick, but unlike the Cuzcachapa and Verde records there is no clear evidence for depopulation immediately following the event. Pollen from *Zea mays* is present in the first post-TBJ level ana-

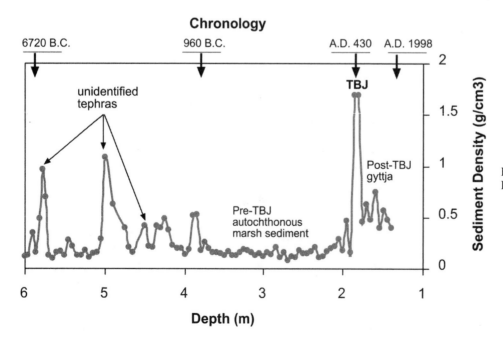

Figure 4. Sediment density graph from Laguneta El Trapiche, Chalchuapa.

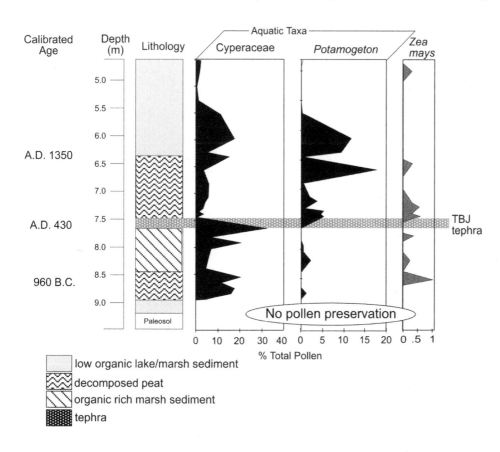

Figure 5. *Zea mays* and aquatic pollen taxa from Laguna Llano, Ahuachapán.

lyzed (7.45 m), including pollen in the diameter size range exclusive to cultivated maize: 90+ microns (Dull, 2001). One explanation for this result is that the TBJ tephra is found at a break in core sections. Ideally a one-meter core would be obtained that includes the TBJ tephra in the middle so high-resolution pre- and post-TBJ pollen samples could be extracted. Although no radiocarbon dates for the post-TBJ stratum have been obtained, the presence of maize pollen in this stratum may be a significant finding in relation to the post-TBJ habitability of western El Salvador. Still, as the post-TBJ pollen data reported here come from a different core section than the tephra itself, these post-TBJ pollen data should be viewed with caution. Replication of the data is absolutely necessary before drawing any firm conclusions.

The most dramatic effect attributable to the TBJ eruption observed in the Laguna Llano record is a seesaw shift in the pollen record of the aquatic taxa Cyperaceae and *Potamogeton*. Cyperaceae (sedge) drops from 35% of the total pollen before the eruption to <2% after it, while *Potamogeton* (pond weed) rises abruptly from zero before the eruption to over 5% immediately following it. This sequence implies an abrupt drowning of marginal emergent aquatic plant taxa. The deposition of 20–50 cm of tephra into the lake would explain such a rapid expansion of the lake surface area, and as a result the drowning of established emergent aquatic plant taxa. *Potamogeton* would have thrived in the expanded shallow water margins after the eruption.

The Sierra de Apaneca

Laguna Verde

Zea mays was first introduced in the Sierra de Apaneca at ca. 2500 B.C. (4440 cal. yr B.P.). The pollen and charcoal records at Laguna Verde record intermittent human impacts by highland farmers from 2500 B.C. until the TBJ eruption occurred at ca. A.D. 430 (Fig. 6). The highest influx levels of *Zea* in the core are found in the level immediately preceding the TBJ eruption, however not a single *Zea* pollen grain was found in the levels postdating the eruption. Charcoal and several weedy plant taxa typically found in association with human occupation and agriculture (*Ambrosia*, Chenopodiaceae/Amaranthaceae, and Asteraceae) also decrease significantly following the eruption. A decrease in *Osmunda* spore influx and an increase in the inorganic fraction of the sediments may indicate rising water levels and increased erosion after the eruption.

Unlike the lowlands of the Rio Paz basin and coastal plain, which were apparently reoccupied within 100–150 yr, people were slow to resettle the highlands of western El Salvador. This trend is understandable if the initial Preclassic settlement of the Apaneca highlands was at least partially driven by cultivable-land scarcity in the lowlands. When resettlement of western El Salvador commenced a century after the TBJ event, land scarcity was no longer a factor. There was probably abundant prime val-

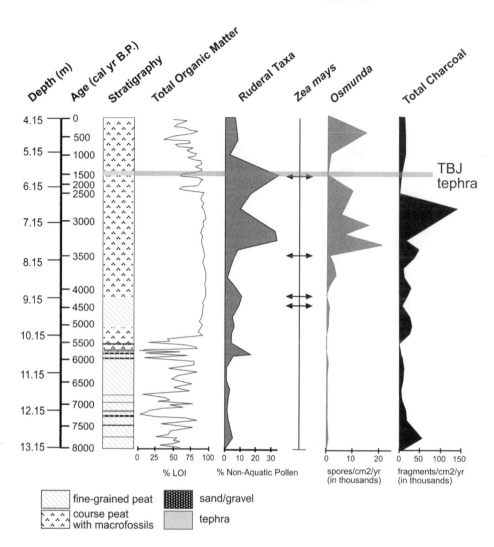

Figure 6. Summary diagram from Laguna Verde, Sierra de Apaneca.

ley farmland available to the pioneers of the sixth and seventh centuries A.D., obviating any need for settling steep highland regions that were more marginal for farming. This reinforces the conclusion that people did not return to their ancestral villages and traditional farming regions when they reoccupied western El Salvador. As has been argued elsewhere (Dull et al., 2001), it appears that the TBJ eruption was not merely a temporary setback or inconvenience for the people of western El Salvador. The bulk of the evidence (e.g., settlement patterns, ceramic affinities, paleo-ecological records) suggests that the TBJ eruption forced a whole-sale social and economic reconfiguration of western El Salvador and southeastern Guatemala. This cultural discontinuity is amply recorded in the Laguna Verde lake sediment record.

DISTAL HAZARDS: RECONSTRUCTING THE EFFECTS OF THE TBJ IN WESTERN EL SALVADOR

Stratigraphic records of Holocene environmental change in El Salvador can provide useful information about the nature of past volcanic eruptions. In the case of the TBJ eruption, reconstructed pre- and post-eruption ecologies can help us understand the effects of a large (VEI 6+) explosive eruption in distal hazard zones, outside the zone of complete destruction. TBJ tephra deposits in western El Salvador today range from 20 to 50 cm at terrestrial sites. Thickness measurements of the TBJ tephra in western Salvadoran lakes range from 15 to 25 cm.

Paleoenvironmental data from four lakes in the Rio Paz–Sierra de Apaneca region facilitate the reconstruction of the historical ecology of western El Salvador as it relates to the TBJ eruption. No detailed chronological sequences of events are reported here, however, as the number of AMS dates used to construct each site chronology is too low: 3–5 (Dull, 2001).

The pre-TBJ ecology of western El Salvador is marked by intensive land use, widespread cultivation, deforestation, burning, and low water levels in lakes and marshes. Following the TBJ eruption, water levels apparently rose at Lagunas Llano, El Trapiche, and Verde. With the resolution of the data we now have it is difficult to know how abruptly these areas were flooded.

Coeval with the flooding appears to be a trend of relatively high flux of inorganic material into the lakes from upland areas. This is likely related to post-eruption mass wasting and erosion on a landscape of greatly reduced vegetative cover. It is clear that tephra fall was significant in this area, and probably resulted in collapsing roofs and perhaps respiratory problems. The impacts on agricultural production appear to have been severe, and this may be fundamentally what led to the mass emigration.

The post-TBJ demography of western El Salvador cannot be clearly reconstructed with the data reported here, but some general observations can be offered. Chalchuapa was apparently abandoned following the eruption, as evidenced by the lack of *Zea* pollen in the strata immediately above the TBJ tephra. The length of the abandonment is still unclear, although chiefdom-level political organization was apparently reconstituted (as evidenced by construction at Tazumal) by the end of the sixth century. The evidence from Laguna Verde in the Sierra de Apaneca also suggests a wholesale abandonment of agricultural settlements in that area, with no obvious precolonial reoccupation following the event. The pollen record from Laguna Llano, Ahuachapán, does not record any clear hiatus in human occupation. This apparent continuity of occupation needs to be verified in future studies, as breaks in the core sections from this study may have resulted in a missing core stratum that would have recorded the depopulation signal in the sediments at Laguna Llano; alternatively, this finding may indicate that the Ahuachapán region was reoccupied much sooner than the rest of the Rio Paz basin and the Sierra de Apaneca.

Despite the puzzling data from Laguna Llano, a general picture emerges from western El Salvador of a region abandoned following the ca. A.D. 430 TBJ eruption. A period of seral succession accompanied this depopulation, as the regional flora recovered from the eruption and pre-TBJ anthropogenic impacts. A region formerly marked by forest clearance and intensive cultivation was transformed, however briefly, into a lush landscape of ecological regeneration. By the end of the sixth century the lowlands of western El Salvador had been completely resettled, setting the stage for a Late Classic and Early Postclassic resurgence of population growth and economic florescence.

THE TBJ EVENT AS A LESSON LEARNED: KNOWLEDGE, PREPAREDNESS, AND THE "SOCIAL PRODUCTION OF VULNERABILITY"

The TBJ experience of the fifth-century inhabitants of western El Salvador cannot be directly projected on twenty-first-century El Salvador from a cultural perspective. The Maya had no idea what was to befall them when the TBJ eruption hit sometime in the early fifth century A.D. They had no prior knowledge of volcanic events of such magnitude and could not have anticipated such a disaster. Planning for volcanic hazards today is obviously more viable, but significant hurdles still exist. As the 1998 Hurricane Mitch disaster illustrated, many developing countries are not adequately prepared to anticipate severe geophysical hazards, and aid from the developed countries is usually "too little, too late" (Comfort et al., 1999). Pre-disaster planning is crucial to ameliorating the effects of disasters like the TBJ eruption or even more recent ones like Hurricane Mitch. Understanding the geophysical processes that can result in loss of lives and livelihoods should be of paramount importance in this endeavor.

Although some recent hazards literature has concluded that the "social, economic, and political origins of the disaster remain as the root causes" (Blaikie et al., 1994), consideration of the "social production of vulnerability" should not be so emphasized as to neglect the real geophysical and environmental conditions that constitute the central threats of most natural disasters. This point has been concisely and elegantly argued by Harold Brookfield: "It is wrong to neglect geophysical change and attribute all blame to human forces, as has been done in a significant part of the modern social science literature" (Brookfield, 1999). Certainly social, political, and economic factors do contribute to the hazardousness of place and the vulnerability of people, especially in the developing world, but these must be considered *in concert with* a solid understanding of natural, recurring geophysical processes.

El Salvador is the most densely populated mainland country in the Americas (Wilkie and Ortega, 1997), and consequently the competition for basic resources remains high across the country (Daugherty, 1969; Durham, 1979; Utting, 1993). Because of the scarcity of land and resources in the country, people cannot readily move to a safer area or build with better construction materials in order to mitigate risk. Strict building codes and no-development hazards zones might be feasible in a wealthier, more sparsely populated country, but these are not realistic options in El Salvador. This makes the task of planning for any natural hazard substantially more difficult. No matter how much information has been generated about potentially hazardous zones and land-use practices in El Salvador, modes of living today appear not to have been adjusted in any measurable way so as to reduce human vulnerability to volcanic hazards.

Population pressure and uneven distribution of land resources are not the only factors hampering disaster planning in El Salvador. The ability of most developing countries to plan for and mitigate geophysical hazards is limited by a dearth of federal and local resources devoted to hazards planning (Tobin and Montz, 1997). This is certainly the case in El Salvador, where there is little prospect of a poor *Salvadoreño* changing his lifestyle in a way that reduces risk simply because he is encouraged to do so by either a governmental or nongovernmental organization. Whether living on the flanks of an active volcano or in a flood plain or at the base of an unstable slope, what matters most to El Salvador's economic underclass is not necessarily risk avoidance; paramount in their lives are food, shelter, and clothing. Hazards risk avoidance in the case of Salvador's poor is most often hindered by inadequate access to safe, viable alternatives—e.g., cultivable farmland, clean water, sturdy construction materials, etc.

The Ilopango eruption of ca. A.D. 430 forever changed the cultural and economic trajectory of the inhabitants of central and western El Salvador. Paleoecological records from western

El Salvador tell a tale of agricultural collapse, erosion, flooding, and widespread abandonment following the TBJ event. These sites are all 60–80 km from the volcano. Of course, closer to the vent, an area that includes modern-day San Salvador, the devastation was much worse. The prevention of future disasters like the TBJ is predicated on sound pre-disaster planning today. Hopefully this book can begin to raise awareness about natural hazards in El Salvador—the causes and consequences of disasters—and can be a first step in publicizing the specific risks of living in a country that is marked by such frequent and violent geophysical activity.

ACKNOWLEDGMENTS

This research was supported by the Tinker Foundation, the Stahl Endowment for Archaeological Research, the University of California at Berkeley Department of Geography, and the University of California at Berkeley Vice President for Research. I thank Payson Sheets, Karen Bruhns, and Roger Byrne for reviewing this manuscript and for helpful discussions about the TBJ eruption. Others who have contributed to my thinking on this topic include John Southon, Cameron McNeil, and Paul Amaroli. I thank the Escobar and Bowling families for hosting me in El Salvador and Gamaliel Garza for assisting with fieldwork.

REFERENCES CITED

Baxter, P.J., 2000, Impacts of eruptions on human health, *in* Sigurdsson, H., et al., eds., Encyclopedia of volcanoes: San Diego, California, Academic Press, p. 1035–1043.
Bergmann, J.F., 1969, The distribution of cacao cultivation in pre-Columbian America: Annals of the Association of American Geographers, v. 59, p. 85–96.
Blaikie, P., Cannon, T., Davis, I., and Wisner, B., 1994, At risk: Natural hazards, people's vulnerability, and disasters: London, Routledge, 284 p.
Brookfield, H., 1999, Environmental damage: Distinguishing human from geophysical causes: Environmental Hazards, v. 1, p. 3–11.
Comfort, L., Wisner, B., Cutter, S., Pulwarty, R., Hewitt, K., Oliver-Smith, A., Wiener, J., Fordham, M., Peacock, W., and Krimgold, F., 1999, Reframing disaster policy: The global evolution of vulnerable communities: Environmental Hazards, v. 1, p. 39–44.
Daugherty, H.E., 1969, Man-induced ecological change in El Salvador [Ph.D. thesis]: Los Angeles, University of California, 248 p.
Demarest, A.A., and Sharer, R.J., 1986, Late Preclassic ceramic spheres, culture areas, and cultural evolution in the southeastern highlands of Mesoamerica, *in* Urban, P.A., and Schortman, E.M., eds., The southeast Maya periphery: Austin, University of Texas Press, p. 194–223.
Dull, R.A., 2001, El bosque perdido: A cultural-ecological history of Holocene environmental change in western El Salvador [Ph.D. thesis]: Berkeley, University of California, 137 p.
Dull, R.A., Southon, J.R., and Sheets, P., 2001, Volcanism, ecology, and culture: A reassessment of the Volcán Ilopango TBJ eruption in the southern Maya realm: Latin American Antiquity, v. 12, p. 25–44.
Durham, W.H., 1979, Scarcity and survival in Central America: Ecological origins of the soccer war: Palo Alto, California, Stanford University Press, 209 p.
Earnest, H., 1999, A reappraisal of the Ilopango volcanic eruption in central El Salvador [Ph.D. thesis]: Cambridge, Massachusetts, Harvard University, 463 p.
Feldman, L.H., 1993, Mountains of fire, lands that shake: Earthquakes and volcanic eruptions in the historic past of Central America (1505–1899): Culver City, California, Labyrinthos, 295 p.
Hart, W., and Steen-McIntyre, V., 1983, Tierra Blanca Joven tephra from the A.D. 260 eruption of Ilopango caldera, *in* Sheets, P.D., ed., Archaeology and volcanism in Central America: Texas Pan-American Series: Austin, University of Texas Press, p. 14–34.
Lardé y Larin, J., 1978, El Salvador: inundaciones e incendios, erupciones y terremotos: San Salvador, Academia Salvadoreña de la Historia, 227 p.
Neff, H., Cogswell, J.W., Kosakowsky, L.J., Belli, F.E., and Bove, F.J., 1999, A new perspective on the relationships among cream paste ceramic traditions of southeastern Mesoamerica: Latin American Antiquity, v. 10, p. 281–299.
Rose, W.I., Conway, M.F., Pullinger, C.R., Deino, A., and McIntosh, W.C., 1999, An improved age framework for late Quaternary silicic eruptions in northern Central America: Bulletin of Volcanology, v. 61, p. 106–120.
Sharer, R.J., 1974, The prehistory of the southeastern Maya periphery: Current Anthropology, v. 15, p. 165–176.
Sharer, R.J., 1978, The prehistory of Chachuapa, El Salvador: Philadelphia, University of Pennsylvania, 226 p.
Sheets, P.D., 1971, An ancient natural disaster: Expedition, v. 13, p. 24–31.
Sheets, P.D., 1979, Environmental and cultural effects of the Ilopango eruption in Central America, *in* Sheets, P.D., and Grayson, D.K., eds., Volcanic activity and human ecology: New York, Academic Press, p. 525–564.
Sheets, P.D., 1983, Archaeology and volcanism in Central America: Austin, University of Texas Press, p. 307.
Sheets, P.D., 1984, The prehistory of El Salvador: An interpretive summary, *in* Lange, F.W., and Stone, D.Z., eds., The archaeology of lower Central America: Albuquerque, New Mexico Press, p. 85–112.
Simkin, T., and Siebert, L., 1994, Volcanoes of the world: Tuscon, Geoscience Press, 349 p.
Stuiver, M., and Deevey, E.S., 1961, Yale natural radiocarbon measurements VI: Radiocarbon, v. 3, p. 126–140.
Tobin, G.A., and Montz, B.E., 1997, Natural hazards: Explanation and integration: New York, Guilford Press, 388 p.
Utting, P.A., 1993, Trees, people, and power: London, Earthscan Publications, 206 p.
Wilkie, J.W., and Ortega, J.G., 1997, Statistical abstracts of Latin America, University of California, Los Angeles Latin American Center Publications, Volume 33: Los Angeles, The Regents of the University of California, p. 1051.

MANUSCRIPT ACCEPTED BY THE SOCIETY JUNE 16, 2003

Geological Society of America
Special Paper 375
2004

Hydrothermal eruptions in El Salvador: A review

Salvador Handal*
Luz Antonina Barrios
LaGeo, S.A. de C.V. 15 Avenida Sur, Col. Utila., Nueva San Salvador, El Salvador, Central América

ABSTRACT

Along the volcanic chain of El Salvador, hydrothermal systems and thermal manifestations coexist with communities, agriculture, and geothermal power generation. Hydrothermal eruptions occur in geothermal fields, compromising the safety of communities and infrastructure. We document hydrothermal eruption phenomena, focusing on the Ahuachapán geothermal field, where hydrothermal eruptions occurred at Agua Shuca in 1990 and at El Sauce in 2001. In addition, we include a summary of the thermal manifestations at TR-6 fumarole in Berlín that changed drastically at the time of the January 2001 El Salvador earthquake.

The Agua Shuca and El Sauce fumaroles discharge primary steam from underground dilute geothermal fluids. The dynamic behavior of Agua Shuca has not been influenced by the mass extraction from and pressure drop of the Ahuachapán geothermal reservoir. A radius of 200 m around Agua Shuca fumarole was established as a limit for the potential high-risk zone.

The January 13, 2001, earthquake caused increased steam emissions at TR-6 fumarole, which is separated from underground dilute geothermal fluids. Temperatures increased from 99.8 to 110 °C in 12 days after the earthquake. During the rainy season of 2001, subsidence occurred and water saturation was observed in TR-6 discharges. However, significant changes in gas content have not occurred since the earthquake. Protective measures taken at TR-6 include a fence, a channel to drain fluid flow, and installation of tubes to release pressure.

The information summarized in this paper should help to plan future studies, to legislate land use, and to design risk control programs in El Salvador.

Keywords: hydrothermal eruption, Ahuachapán geothermal field, Berlín geothermal field, hydrothermal manifestations, faults, subsidence.

INTRODUCTION

Understanding the physical and chemical characteristics and evolution of hydrothermal discharges in volcanic areas is important for the development of geothermal resources. Volcanoes provide fertile soil and thermal energy, but urban development and geothermal industry in volcanic areas are exposed to hazards such as landslides, gas emissions, seismic activity, volcanic eruptions, hydrothermal eruptions, and so on. This paper is a summary of the characteristics of two hydrothermal eruptions in the Ahuachapán geothermal field and earthquake-induced changes in hydrothermal features at the Berlín geothermal field in El Salvador. The purpose of this paper is to summarize hydrothermal eruption processes in developed geothermal fields. This information can be used for future risk assessment and land management in El Salvador.

Eruption Types

Normally, an eruption is the process by which materials are ejected into the atmosphere or onto the surface of the earth or sea bottom as a result of volcanic activity. Volcanologists have

*shandal@lageo.com.sv

Handal, S., and Barrios, L.A., 2004, Hydrothermal eruptions in El Salvador: A review, *in* Rose, W.I., Bommer, J.J., López, D.L., Carr, M.J., and Major, J.J., eds., Natural hazards in El Salvador: Boulder, Colorado, Geological Society of America Special Paper 375, p. 245–255. For permission to copy, contact editing@geosociety.org. © 2004 Geological Society of America

classified eruptions into several types (U.S. Geological Survey, 2002), including phreatomagmatic, phreatic, and hydrothermal eruptions. These types of eruptions apparently occur in connection with the influx of considerable quantities of water. Phreatomagmatic eruptions involve fresh magma along with water, steam, and brecciated country rock. Phreatic eruptions involve the transfer of magmatic heat to circulating groundwater and subsequent eruption of steam and country rock but, often, without the eruption of fresh magma. The U.S. Geological Survey (2002) and other researchers (e.g., Seach, 2002) indicate that phreatic eruptions involve heat transfer by intrusive or extrusive magma to circulating ground- or surface water. Once groundwater is expanded into steam, ash and fragments of country rock without magma are erupted onto the surface. Sometimes phreatic eruptions are precursors of volcanic activity. The duration of either phreatic or phreatomagmatic eruptions is directly related to the amount of water, which flashes into steam. Both types of events create craters and violent explosions.

Germanovich and Lowell (1995) establish that hydrothermal eruptions may be related to transient pressure changes in near-surface (<300 m) regions of hydrothermal systems. Hydrothermal eruptions are violent events driven by the expansion of gases and liquids from below the earth's surface as their pressures drop to atmospheric levels. According to Rynhart (1999), once a hydrothermal eruption is under way the mixture of gases and liquids increases in velocity as it rushes toward the surface and, as a result, rock particles, along with a two-phase mixture of superheated water and steam, are ejected onto the earth's surface. Hydrothermal eruptions have taken place all over the world, including countries like New Zealand and the United States. In 1983, a steam eruption occurred at the Craters of the Moon thermal area in Wairakei, New Zealand (Allis, 1984). Muffler et al. (1971) have documented hydrothermal explosion craters in Yellowstone National Park, USA.

Conceptual Model of a Hydrothermal Eruption

Smith and McKibbin (2000) presented a conceptual model of a boiling front that causes a hydrothermal eruption. Hot fluid, which lies below the surface at boiling-point conditions, suddenly loses pressure due to some triggering event, such as seismic activity, hydraulic fracturing, erosion, or climatological factors. The pressure reduction causes boiling, and the fluids expand. The continuation of boiling requires the presence of permeable conduits. If permeable conduits are not present, the fluid will not continue to boil. If permeable conditions prevail, the fluid will move toward regions of lower pressure and will move upward (Fig. 1). The upward fluid momentum pressurizes the rock above and, if this pressure is large enough to overcome the weight and cohesive stresses of the rock, rock fragments and fluid mix and are ejected upward. The fluid will continue to flash into steam as it rises, and the eruption column will probably have a greater steam fraction at the top than it does toward the bottom.

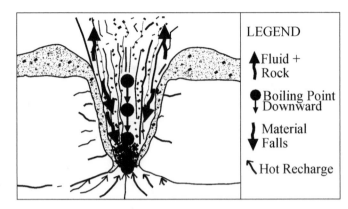

Figure 1. Hydrothermal eruption. After Smith and McKibbin (2000).

As fluid and rock are ejected upward, the depressurized propagation path progresses downward, causing the boiling front to continue downward, prompting more fluid and rock to be ejected. Much of the ejected material falls back into the vent and is re-ejected in the eruptive column (Fig. 1).

The continuing advancement of the boiling front is dependent on hot recharge or the inflow of heated water to the two-phase conduit. The boiling front will stop its advancement if it encounters a region in the rock matrix of negligible permeability. As the eruption continues, pressure reduction, cooling effects, and gravitational forces slowly dissipate the energy of the eruption and eventually stop it. Much of the erupted material falls back into the newly formed crater. Residual steam continues to rise from the crater floor (Fig. 2). Due to release of rock stresses, the walls of the crater will eventually begin to slump inward. The

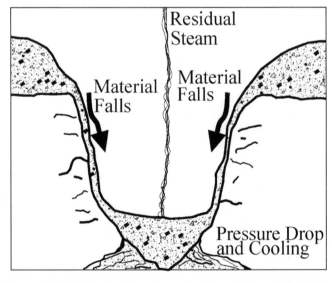

Figure 2. End of a hydrothermal eruption. After Smith and McKibbin (2000).

loss of significant amounts of material from under the surface may also cause the ground to subside locally.

HYDROTHERMAL ERUPTIONS IN EL SALVADOR

In 1990 and again in 2001, distinct hydrothermal eruptions occurred close to the Ahuachapán geothermal field, located in western El Salvador (Fig. 3). Simultaneous with the earthquake of January 13, 2001, an existing fumarole in the Berlín geothermal field, located in eastern El Salvador, increased its thermal activity, gas emissions, surface area, and subsidence features. However, an explosion has not occurred at this site yet.

Hydrothermal eruptions in El Salvador have occurred before recorded history. Even though hydrothermal eruptions are not well documented, morphological evidence suggests that these eruptions have occurred many times in the Ahuachapán geothermal field. Meyer-Abich (1953) wrote that Agua Shuca probably erupted in 1868 and that hydrothermal eruptions occurred at El Playón and La Labor (Fig. 4). Another example of a hydrothermal eruption, which is not well documented, can be found close to Apastepeque Lagoon near the city of San Vicente (Fig. 5). In spite of the occurrence of these eruptions in the last 140 yr, there are no recent studies to determine the frequency of these events.

The Agua Shuca Hydrothermal Eruption of 1990

Agua Shuca, meaning "dirty water," is one of the hydrothermal discharges encountered in the community of El Barro close to the Ahuachapán geothermal field, western El Salvador. The Agua Shuca hydrothermal eruption is the best-documented hydrothermal eruption in El Salvador. It occurred in October 1990, killing 25 people and injuring 15. Figure 6 shows a view of steam rising from the Agua Shuca crater after the eruption, and Figure 7 shows some of the social and environmental impact in the immediate surroundings. Escobar et al. (1992) gathered interesting information about this event and it is partially described next.

The eruption took place in a zone ~25 m in diameter with several thermal features: two hot water pools ~5 m in diameter, mud pools, mud cones as high as 0.5 m, and steaming ground. The eruption zone was surrounded by a group of small wooden houses, all of them built on private land. Neither physical changes nor seismic warning occurred prior to the explosion. On October 13, 1990, at 3:00 a.m. a sudden and strong mix of boiling water droplets, mud, and hydrothermally altered soil was erupted over the surrounding area. It covered an area almost 200 m in diameter. Some survivors indicated that the eruption lasted no more than half a minute and that the only warning was a thunder-like noise before the eruption.

The ejected material had a volume of ~1600 m³ and consisted of silicified clay-rich altered rocks and effusive material from the Las Ninfas volcanic group (Hernández, 1990). As a result of that event, a crater ~40 m in diameter and 5 m deep was formed at the place of the eruption. Hot water pools, mud pools, steaming ground, gas vents, and mud craters remained inside the crater (see Fig. 8).

As a consequence of the disaster, an area within a radius of 200 m around Agua Shuca was declared to have potentially high risk, and the area between 200 and 300 m radius, as an area of low risk (see Fig. 9). However, the designation of these areas did not correspond with a prevention law, because none existed. As

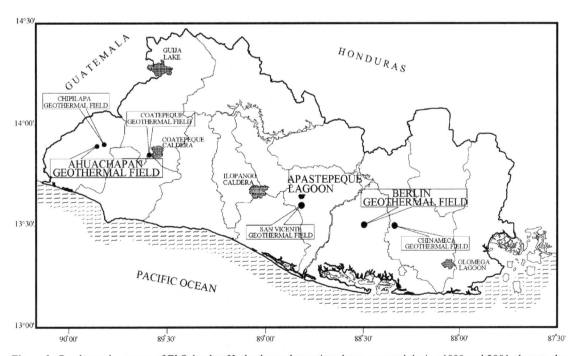

Figure 3. Geothermal systems of El Salvador. Hydrothermal eruptions have occurred during 1990 and 2001 close to the Ahuachapán geothermal field.

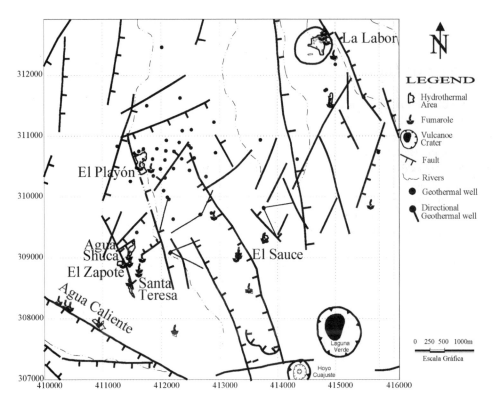

Figure 4. The main hydrothermal features in Ahuachapán geothermal field. After Montalvo et al. (2001).

Figure 5. Crater Hoyón de Caldera in Apastepeque, San Vicente. It is an example of an undocumented hydrothermal eruption.

Figure 6. East view of the Agua Shuca crater after the 1990 eruption (CEL, 1990).

a prevention measure, similar areas were designated for the land surrounding the La Labor hydrothermal discharges (see Fig. 4).

Historical Data of Agua Shuca Surface Hydrothermal Activities

Based on reports of several visitors to Agua Shuca, Meyer-Abich (1953) listed the changes that occurred in Agua Shuca from 1838 to 1888. A lake of 90 m of circumference with clear brown boiling water and bubbles up to 1.3 m high existed in 1838. In 1840, there was a pool 45 m in circumference, with large orifices from which steam rushed out with violence. In 1868, the area featured a boiling lake 12 m in diameter, and mud eruptions up to 5 m high. Both features had a maximum temperature of 98 °C. Meyer-Abich summarized reports from 1868 of other hydrothermal features nearby. There was a steam vent or fumarole with a temperature of 98.5 °C issuing from a rock fissure located ~300 m to the east of Agua Shuca. Furthermore, 150 m southeast of the foot of the mountain, there was a thermal area with temperatures

Figure 7. House and vegetation damage produced by the 1990 Agua Shuca eruption (CEL, 1990).

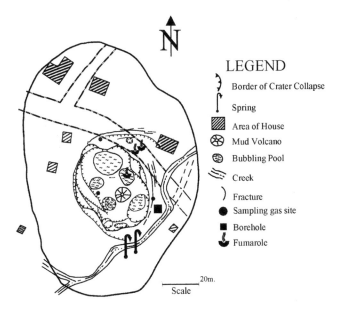

Figure 8. Agua Shuca crater and thermal features. After Escobar et al. (1992).

as high as 88 °C. This discharge coincided with the current location of the El Zapote hydrothermal site (see Figure 4).

In 1868, people living near Agua Shuca heard one blast. When people reached the area, they found steam discharging all over the place. This could indicate that a hydrothermal eruption had just occurred in the area.

Geochemistry of Agua Shuca Effusions

Based on the methodology of Stewart (1990) for the deuterium and ^{18}O isotopic composition of steam in a depressurizing system, Montalvo et al. (2002) established that Agua Shuca, El Playón, and Agua Caliente are fumaroles of primary steam from diluted geothermal fluids with temperatures between 210 and 190 °C; while Santa Teresa, El Zapote, and San Carlos are fumaroles of secondary steam separated from superheated local groundwater with temperatures between 180 and 110 °C. Figure 10 classifies the fumaroles based on 250 °C, $\delta^{18}O = -3.86‰$, and $\delta^2H = -42.5‰$ for deep reservoir conditions and 27 °C, $\delta^{18}O = -7.8‰$, and $\delta^2H = -54.06‰$ for surface water.

In Figure 11, carbon dioxide (CO_2) concentrations in fumarole steam at Agua Shuca show a possibly decreasing trend from 1975 to 1997 if the extreme data of 1979 and 1981 are neglected. In general, pressure drop in a geothermal reservoir results in increasing CO_2 concentration due to boiling, but the data of Figure 11 indicate that since 1975 the Agua Shuca fumarole has not been affected by more than a 15 bar pressure drop (Quijano, 1994) as a result of mass extraction from the Ahuachapán geothermal reservoir. If all the data in Figure 11 are taken into account, one obtains the linear equation, $CO_2 = 3555 - 27 \times year$, with a correlation coefficient of 0.11. The maximum reduction of CO_2 during 24 yr will be ~648 mmol/100 mol. This change in CO_2 is of the same order as the variability of the CO_2 measurements from 1989 to 2000. Therefore, it is suggested that mass

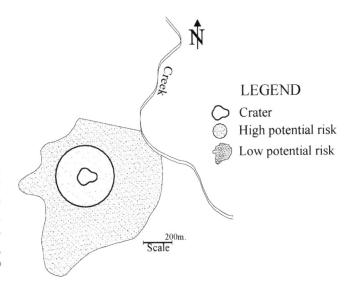

Figure 9. Areas of potential risk around Agua Shuca crater. After Escobar et al. (1992).

extraction from the Ahuachapán reservoir has not affected the dynamic behavior of the Agua Shuca fumarole.

El Sauce Hydrothermal Eruption

García et al. (2001) reported the occurrence of a small hydrothermal eruption in the fumarole El Sauce, which is located ~2 km to the east of the Agua Shuca. According to information

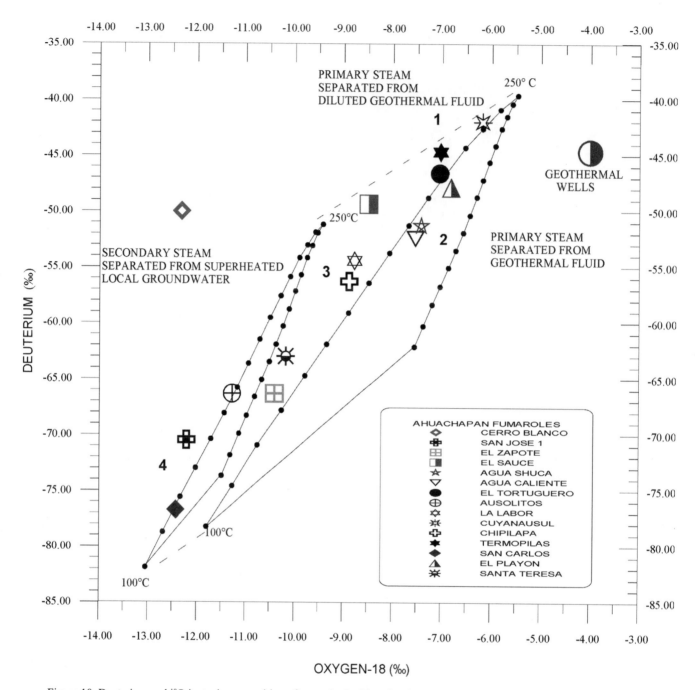

Figure 10. Deuterium and ^{18}O isotopic composition of steam in the Ahuachapán geothermal system. After Montalvo et al. (2002).

provided by local people, the date for the explosion was most probably October 7, 2001.

Before the eruption, the El Sauce hydrothermal manifestation mainly consisted of an altered area of ~5100 m², several small fumaroles, a mud crater, and a small pool at boiling conditions. During the explosion, the mud crater and the boiling pool were destroyed and only a small fumarole remained. This fumarole was located at the site of the boiling pool (Fig. 12).

Figure 13A shows pictures of a small geyser, boiling water, and fumarole thermal features that existed before the eruption. Figure 13B displays the remaining fumarole after the eruption. As a consequence of the eruption, a crater (Fig. 13C) 5 m in diameter and 2 m in depth was formed in the northwest side of the area.

The explosion ejected mud to the north, south, and west, covering an area of ~50 m in radius. According to the distance between the crater and some nearby trees affected by the eruption

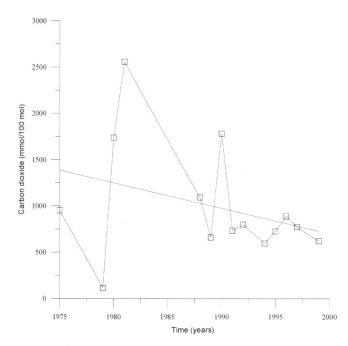

Figure 11. Evolution of CO_2 concentration in the Agua Shuca fumarole. $CO_2 = 3555 - 27 \times$ time (year), with a correlation coefficient of 0.11.

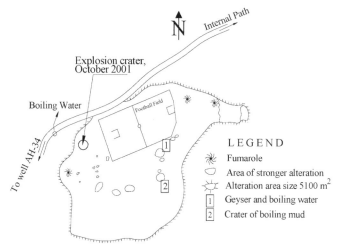

Figure 12. El Sauce hydrothermal manifestation before eruption. After García et al. (2001).

products, it is estimated that the mininum height reached by the mud column was 20 m. The El Sauce hydrothermal explosion area has not been studied as well as the Agua Shuca eruption. However, according to the position of El Sauce fumarole samples in a deuterium versus $\delta^{18}O$ diagram for steam from the Ahuachapán field (Montalvo et al., 2002), El Sauce is also a fumarole formed by primary steam separated from diluted geothermal fluids.

The El Sauce fumarole is included in the monitoring program for the Ahuachapán geothermal field. Monitoring at El Sauce, carried out at least once a year by LaGeo, consists mainly of studying the evolution of some chemical species (e.g.,

$NaCl$, HCO_3, SO_4, SiO_2, H_2S, Ar, CH_4, clay minerals), and stable isotopes in waters and gases. Direct geological observation of the different altered areas and stability of the ground are performed, along with microgravity, resistivity, and microseismicity measurements. Starting at the beginning of 2003, the monitoring program will include observations of gradients of ground temperature and heat losses.

Fumarole TR-6 in the Berlín Geothermal Field

In the Berlín geothermal field (Fig. 3), simultaneous with the earthquake of January 2001, an existing fumarole (fumarole TR-6) increased its thermal activity, gas emissions, surface area, and subsidence features. Because of these changes, TR-6 is now monitored to understand its evolution and to advise the local communities and geothermal industry of any imminent danger.

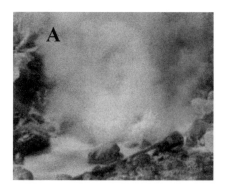
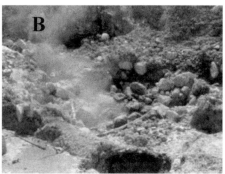
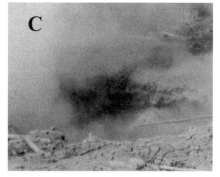

Figure 13.Thermal manifestations in El Sauce area. A: Before eruption. B: After eruption. C: Resulting crater due to explosion. Photos from Garcia et al. (2001).

This information is shared with SNET (Servicio Nacional de Estudios Territoriales), a national agency created in 2001 for the assessment of natural hazards and risks in El Salvador.

Fumarole TR-6 is located at the site of an abandoned well to the northeast of the Berlín power plant (Fig. 14). A steam cap was reached at ~144 m depth when this well was drilled at the end of 1991. The drill hole into the steam cap went out of control and, after trying to stabilize the subsurface, the drilling operation was stopped and the well abandoned. Since then, steam discharges remained in the TR-6 site and so it was called fumarole TR-6 (Fig. 15). Its activity remained practically unchanged until the January 13, 2001, earthquake. After the earthquake, its position and intensity changed (Fig. 16).

Morphology of TR-6 Fumarole after the 2001 Earthquake

According to a stress analysis made by Raymond (2001), the La Pila fault, trending NNW-SSE, has a direct control on the hot fluids that contribute to the formation of fumarole TR-6 and the generation of hydrothermal alteration zones. The perturbed area of the TR-6 fumarole is formed by a subvertical set of fractures, which form an ellipse of 10 m². The trend of the larger axis of the ellipse is 315 degrees. The northwest nose and the center of this ellipse collapsed with a vertical displacement of 40 cm and 20 cm, respectively (Fig. 17).

Earthquake-induced subsidence probably occurred as a consequence of water saturation combined with clay-rich hydro-thermal alteration. By May 2002, the fumarole had not changed its horizontal position or extension from that in January 2001. However, as a result of land flow during 2001, the subsidence at the fumarole had increased from 40 cm in the middle of 2001 to more than 1.5 m in December of 2001 (Fig. 18).

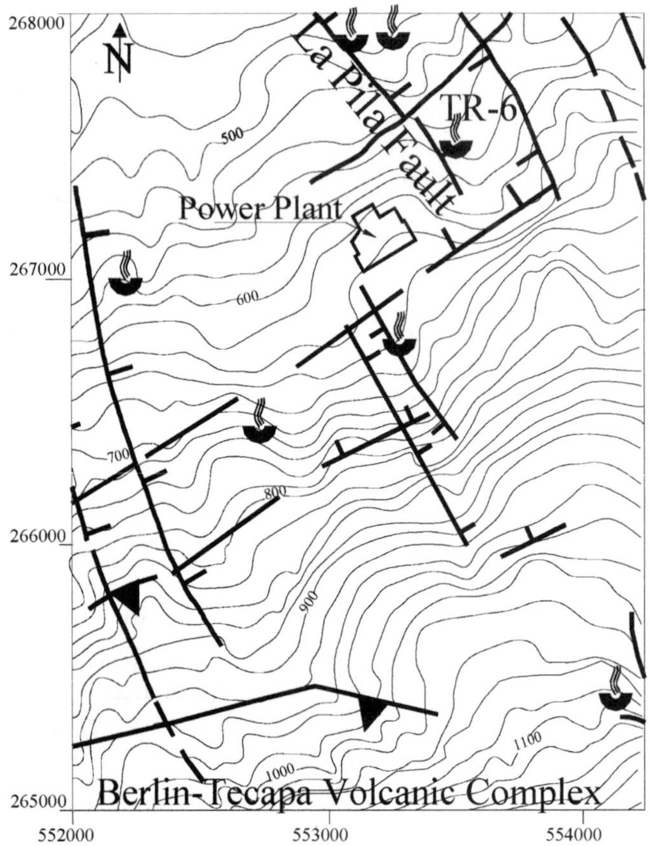

Figure 15. TR-6 fumarole before earthquake of January 2001. It was located 20 m to the SE of the ellipse described in the text and shown in Figure 16. Photo from Quezada and Matus (2001).

Figure 16. Fumarole TR-6 after earthquake of January 2001. Ellipse-shaped fumarole ~10 m² in size. Photo from Quezada and Matus (2001).

Figure 14. Some thermal manifestations and faults in the Berlín geothermal field.

Figure 17. Fumarole TR-6. Vertical displacement of 40 cm at the NW nose of the ellipse and 20 cm average vertical displacement at the center. After Quezada and Matus (2001).

Figure 18. Increase in subsidence of the collapsed zone of the ellipse from 40 cm to almost 2 m, from middle of 2001 to December 2001. Visual inspection in May 2002 did not show further increase in diameter. After Quezada (2002).

Alteration Zones at TR-6 Fumarole

There are two different alteration zones at TR-6 fumarole. The largest zone is distributed along the fractures of the ellipse. Silica and native sulfur have been deposited on these fractures. The second alteration zone is located at the older site of TR-6 fumarole (see Figs. 15 and 16), where native sulfur and minor silica have also precipitated. Field observations during August 2001 indicated that the deposits of native sulfur forming on the second alteration zone were almost disappearing. Also, clay minerals had sealed the fracturing in the former fumarole and the extension of the altered zone had decreased.

Thermal Manifestations at TR-6 Fumarole

TR-6 fumarole intensity varies with time, depending on meteorological conditions. Visual observations during 2001 showed that while fumarole vents increased their steam discharge during the rainy season, they decreased when rain ceased. The steam emissions had less intensity in May 2002, before the start of the rainy season. The water feeding the zone of the fumarole has several origins. According to a deuterium-$\delta^{18}O$ isotopic analysis, part of the water comes from precipitation and local recharge, and another portion is from diluted geothermal fluids. Also, gas analyses indicate that fumarole TR-6 did not show significant variation in gas content (Gerencia de Estudios, 2001). Figure 19 summarizes the stable isotope geochemistry of the Berlín field.

The temperature of the TR-6 fumarole in December of 2000 was 99.8 °C. On January 18 and 25, 2001, the temperatures were 107 and 110 °C, respectively. However, on June 18, 2001,

during a wet period, the temperature was 99 °C (M. Cubías, LaGeo, 2002, personal commun.). By August 2001, during the dry season, the activity manifestation in the TR-6 fumarole was very weak, and the temperature was 99 °C (Quezada and Matus, 2001). By May 2002 the temperature remained in the range of 99 to 100 °C, as it was before the January 13, 2001, earthquake.

Engineering Protection Measurements

Several engineering measures were taken to avoid future damage near the TR-6 fumarole. A channel was built around TR-6 to collect the flow from rain and slope wash and discharge it into a larger channel that goes into a nearby stream. A protective fence was built 35 m to the northwest of TR-6, to close access to the fumarole area and avoid any harm to the neighboring community. Tubes were installed around the fumaroles to decrease the pressure in the formation and allow the steam to escape from the system more easily.

CONCLUSIONS

A sudden pressure drop in a subsurface fluid under boiling conditions can cause a hydrothermal eruption. Decreasing pressure is a consequence of a triggering event like an earthquake, hydraulic fracturing, or meteorological conditions. If permeable paths are present, the fluid will move toward regions of lower pressure, and so it will usually move upward. The upward fluid momentum pressurizes the rock above, and if this pressure is large enough to overcome the weight and cohesive stresses of the rock, then a mixture of rock fragments and fluid is ejected

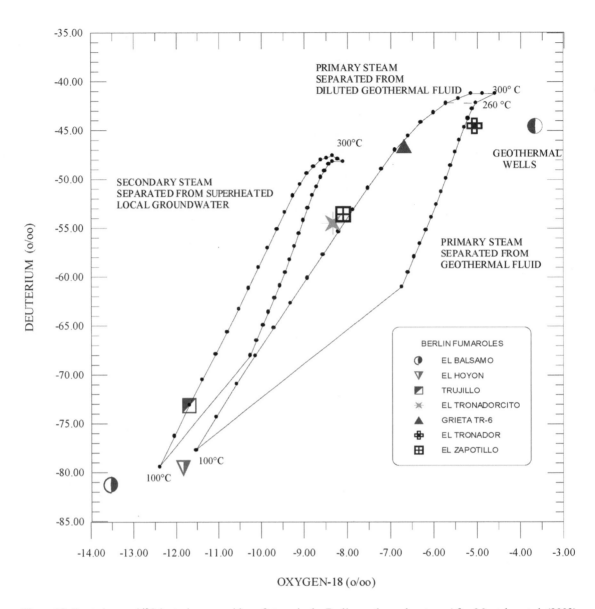

Figure 19. Deuterium and ¹⁸O isotopic composition of steam in the Berlín geothermal system. After Montalvo et al. (2002).

upward. As the eruption continues, pressure reduction, cooling effects, and gravitational forces slowly dissipate the energy of the eruption until it eventually stops.

Hydrothermal eruptions have occurred in El Salvador before and after the initiation of geothermal exploitation. According to literature, information from local people, and morphological evidence, several hydrothermal eruptions have occurred in the Ahuachapán geothermal field: at La Labor, El Playón, El Zapote, Agua Shuca, and El Sauce, including two documented eruptions and one probable eruption since 1868.

The Agua Shuca eruptive event in 1990 was probably produced by changes in the spatial distribution of both heat and mass

flow and changes in the self-sealing characteristics of the system, rather than the mass extraction from the geothermal reservoir. Based on their stable isotopic compositions, Agua Shuca, El Sauce, El Playón, and Agua Caliente are classified as fumaroles of primary steam, derived from diluted geothermal fluids with temperatures between 210 and 190 °C. Santa Teresa, El Zapote, and San Carlos are fumaroles of secondary steam separated from overheated local groundwater with temperatures between 180 and 110 °C.

At TR-6 fumarole in Berlín, an increment of almost 10 °C over the boiling point was recorded after the earthquake, in August 2001. However, by May 2002, during the dry season, the temperature in TR-6 fumaroles had returned to 99 °C, and gas analysis indicated no variation in gas content. Visual observations at TR-6 fumarole during 2001 showed that while fumarole vents increased their steam flow discharge during the rainy season, they decreased when rain ceased. According to deuterium-$\delta^{18}O$ correlation, part of the water comes from direct precipitation and local recharge, and another proportion from diluted geothermal fluid.

RECOMMENDATIONS

An inventory, classification, and dating of hydrothermal eruptions should be done along the extinct and active volcanic chain in order to estimate the volume of these eruptions, describe the erupted material, and define the eruption frequency and most probable areas of hazards. These results should be helpful for land use legislation and risk management.

Studies that should be included in investigating hydrothermal eruption are *(a)* evolution of chemical species and stable isotopes in water and gases, *(b)* monitoring of volatile gases like radon, helium, and carbon dioxide, *(c)* direct geological observation of the hydrothermal rock alteration, *(d)* stability of the ground, *(e)* temperature gradients, *(f)* heat losses, *(g)* natural electrical potential or self potential observations on the ground, *(h)* microseismicity monitoring, and *(i)* trajectory of the ejected material. Integrated computer modeling could then better define areas of potential risk and predict future eruptions.

ACKNOWLEDGMENTS

The authors thank LaGeo for the permission to publish some information they own. Specials thank to Dr. Dina Larios for an initial review and for making several important improvements to this paper. Also, the authors thank Marlene Serrano for drawing and modifying some of the figures in this article.

REFERENCES CITED

Allis, R.G., 1984, The 9 April 1983 Steam eruption at Craters of the Moon thermal area, Wairakei: New Zealand, Geophysics Division, Department of Scientific and Industrial Research, Report no. 196. 27 p.

CEL (Comisión Ejecutiva Hidroeléctrica del Río Lempa), 1990, La Explosion Hydrothermal de Agua Shuca, Internal reports: LaGeo, El Salvador.

Escobar, C., Burgos, J., and Mendoza, S., 1992, The Agua Shuca Hydrothermal Eruption: Geothermal Resources Council Bulletin.

García, O., Quezada, A., and Barrios, L., 2001, Informe Preliminar de Explosión Freática en la Fumarola El Sauce, Campo Geotérmico Ahuachapán, Unidad de Geología, Reporte Interno: El Salvador, GESAL (Geotérmica Salvadoreña), 8 p.

Gerencia de Estudios, 2001, Informe integral de las diferentes disciplinas de geociencias sobre el comportamiento de la nueva fumarola en la plataforma del TR-6, Reporte interno: El Salvador, GESAL (Geotérmica Salvadoreña), 12 p.

Germanovich, L.N., and Lowell, R.P., 1995, The mechanism of phreatic eruptions: Journal of Geophysical Research, v. 100, p. 8417–8434.

Hernández, W., 1990, Erupción Freática, Cantón El Barro, Ahuachapán; Depto. Exploración, Reporte Interno: Comisión Ejecutiva Hidroeléctrica del Río Lempa, El Salvador.

Meyer-Abich, H., 1953, Los Ausoles de El Salvador: Comisión Ejecutiva Hidroeléctrica del Río Lempa, El Salvador, Servicio Geológico Nacional, 51 p.

Montalvo, F., 2001, Gas monitoring of the Ahuachapán Geothermal Field, Internal report: El Salvador, GESAL (Geotérmica Salvadoreña), 41 p.

Montalvo, F., Matus, A., Magaña, M., Tenorio, J., and Guevara, W., 2002, Isotopic contribution for monitoring the exploitation and development of the Ahuachapán-Chipilapa and Berlín Geothermal fields, Internal report: El Salvador, GESAL (Geotérmica Salvadoreña), 54 p.

Muffler, L., White, D., and Truesdell, A., 1971, Hydrothermal explosion craters in Yellowstone National Park: Geological Society of American Bulletin, v. 82, p. 723–740.

Quezada, A., 2002, Estado de la fumarola del TR-6. Campo geotérmico de Berlin, Reporte interno mayo 22/2002: El Salvador, GESAL (Geotérmica Salvadoreña), 6 p.

Quezada, A., and Matus, A., 2001, Estado Actual de la fumarole del TR-6. Campo geotérmico de Berlìn, Reporte interno junio 21/2001: El Salvador, GESAL (Geotérmica Salvadoreña), 3 p.

Quijano, J., 1994, A revised conceptual model and analysis of production data for the Ahuachapán-Chipilapa geothermal field in El Salvador: Orkustofnum, Iceland, United Nations University, Geothermal Training Program, report 1994, no. 10, p. 237–266.

Raymond, J., 2001, Trabajo en la Fumarola del pozo TR-6. Campo Geotérmico de Berlín. Reporte Interno: El Salvador, GESAL (Geotérmica Salvadoreña), 11 p.

Rinhart, P., 1999, Hydrothermal eruptions: Institute of Fundamental Sciences, Massey University, New Zealand: http://www.massey.ac.nz/~prynhart/honours.pdf.

Seach, J., 2002, Phreatic eruptions: Volcano Live Pty Ltd. Australian: http://www.volcanolive.com/phreatic.html (June 2003).

Smith, T., and McKibbin, R., 2000, An investigation of boiling processes in hydrothermal eruption: Proceedings of the World Geothermal Congress, International Geothermal Association, Kyushu-Tohuku, Japan.

Stewart, M.K., 1990, Environmental isotope study of Ahuachapán Geothermal Area, A reassessment Mission Report: New Zealand, Department of Scientific and Industrial Research, Institute of Nuclear Sciences, 23 p.

U.S. Geological Survey, 2002, Photo glossary of volcano terms; Phreatic eruption: California, USA, USGS Volcano Hazards Program, http://volcanoes.usgs.gov/Products/Pglossary/HydroVolcEruption.html.

MANUSCRIPT ACCEPTED BY THE SOCIETY JUNE 16, 2003

Geological Society of America
Special Paper 375
2004

Seismograph networks and seismic observation in El Salvador and Central America

Mario Fernández*
Central America Seismological Center (CASC), Apdo. 214-2060, San José, Costa Rica

C. Demetrio Escobar
*Servicio Nacional de Estudios Territoriales de El Salvador (SNET), Avenida Las Mercedes,
Contiguo Parque de Pelota, San Salvador, El Salvador*

Carlos A. Redondo
Escuela Centroamericana de Geología, Universidad de Costa Rica, Apdo. 214-2060, San José, Costa Rica

ABSTRACT

Central America is one of the most tectonically active areas in the world, and its high seismicity rate demands the establishment of seismic instrumentation to monitor earthquake occurrence. The first seismological equipment installed in the region dates from 1882. In El Salvador, seismological observations began in the late nineteenth century when 15 Wiechert seismographs arrived in the country. A network of 12 short-period telemetric seismic stations became fully operational in El Salvador in late 1984; this network currently consists of 18 seismographs.

Earthquake monitoring in the other countries of the region by permanent networks has operated since the 1970s. In the 1990s, all the countries in the region acquired SEISLOG data acquisition systems and new equipment, including broadband stations, which were donated to improve their seismic networks. In 1998, the Central America Seismological Center (CASC) was established in Costa Rica, marking a new stage in seismological observation in the region. The goal of this center is to locate regional earthquakes and compile seismic databases to be used in the estimation of seismic hazard. This paper summarizes the progress on seismic monitoring in El Salvador and reviews the scientific achievements of the CASC after three years of operation.

Keywords: seismicity, earthquakes, Central America, networks, monitoring.

INTRODUCTION

During the last 500 years, Central America has experienced destructive earthquakes and suffered damage from other processes associated with seismic events. As in almost any area of the world, the region is presently undergoing a period of accelerated growth and economic development. Consequently, the impact of earthquakes on human activity can be extremely large, and therefore preparedness and mitigation are needed. The recent (January 2001) earthquake disaster in El Salvador

is a good example of the vulnerability of the population to the seismic hazard. This catastrophe brought out new ideas on the seismic process in El Salvador and indicated that more accurate information on earthquakes and associated processes is required to identify the vulnerability and areas at risk, and to make precise estimates of the seismic hazard of the region.

Seismic equipment has been operating in the region since the nineteenth century, for monitoring seismic activity, studying the seismic hazards, and for understanding tectonics in more detail. The seismic potential of the region can then be estimated from knowledge of the historic and recent records; for evaluating earthquake hazards in the region, it is necessary to know details

*mefernan@cariari.ucr.ac.cr

Fernández, M., Escobar, C.D., and Redondo, C.A., 2004, Seismograph networks and seismic observation in El Salvador and Central America, *in* Rose, W.I., Bommer, J.J., López, D.L., Carr, M.J., and Major, J.J., eds., Natural hazards in El Salvador: Boulder, Colorado, Geological Society of America Special Paper 375, p. 257–267.

of the seismic sources. Therefore, an accurate and homogeneous earthquake catalog is needed. Such a catalog is also useful for testing statistical methods that may be applicable in earthquake prediction. Earthquake catalogs have been widely used for hazard assessment, seismicity studies, long-term earthquake prediction and forecasting, and detailed studies of plate tectonics.

Seismic observation with high-efficiency networks is a prerequisite today for obtaining high-quality hypocenter data and the preparation of complete and reliable catalogs. Until the last few decades, instrumental monitoring of seismicity within the region chiefly relied on worldwide seismographic WWSSN (World Wide Standardized Seismographic Network) coverage, which, however, had a fairly poor coverage of Central America. Through the efforts of all the countries of Central America, more seismic stations and accelerometers have been installed in response to the need for improving local and regional seismic coverage. All of the countries in the region have made concerted attempts to improve their local networks; in 1998, a regional seismological center, the CASC, was established at the Geology Department of the University of Costa Rica.

More than 100 years after the installation of the first seismic equipment in El Salvador and four years after the CASC opened, it is an appropriate time to review what has been achieved in seismic monitoring in El Salvador and Central America. It became apparent in the first years of the CASC's operation that this center gives the detailed and accurate location of those earthquakes located near the borders of the countries where the national networks have problems obtaining reliable solutions. In this paper, we describe both the development of the seismological studies in El Salvador and the role of the CASC in the reduction of the seismic disaster in the region. Of principal concern in this paper is the history of seismology in El Salvador because this volume focuses on the natural hazards of that country.

HISTORY OF THE SEISMIC NETWORK OF EL SALVADOR

The repeated seismic disasters in El Salvador have made it necessary for government and scientific groups to install equipment to understand seismic sources better. The intentions have been the best, but the aims have not always been achieved due to reasons beyond the scientists' control. Some seismic stations of the seismic network of El Salvador have operated only intermittently during the network's history because of severe civil conflict and vandalism. The network has been revitalized and reorganized, but the problems in keeping it operational persist. In the following, we describe the evolution of this network since the installation of its first seismometer, based on Schulz (1969), Alvarez (1980), and Gálvez (2002).

The Earliest Seismographs

The history of seismograph station operations in El Salvador began in the middle of nineteenth century when Dr. Dario González initiated the first astronomical and meteorological observations at the National Institute. In 1889, the Minister of Public Instruction, Dr. Hermogenes Alvarado, ordered the acquisition of the first seismic equipment from Europe, and in October 1890 the Astronomical and Meteorological Observatory (AMO) was founded by official decree, marking the beginning of the earliest seismological observations in El Salvador. The honor of being the first director of the AMO fell to Dr. Alberto Sanchez. The earliest seismographs arrived to the country in 1896; they consisted of 15 Ewing seismometers acquired and operated by AMO.

Single Stations with On-Site Recording

On June 7, 1917, there was a destructive earthquake in El Salvador with an eruption of the San Salvador volcano. The National Seismological Observatory (NSO) was founded by official decree following the earthquake of February 25, 1918. This observatory was the first devoted to seismology in El Salvador; its foundation marked the beginning of a period of growth in seismology in the country. The eminent Dr. Jorge Larde was the first director of the new observatory; he immediately tried to get more seismic instruments.

The AMO and NSO were merged on September 1, 1928, giving birth to the National Observatory of El Salvador. At the end of 1930, two Wiechert instruments started recording at the National Observatory. (They are still operating at the Seismology Department of the Center for Geotechnical Research [CIG] of El Salvador, which is now SNET.) Another new seismic station was acquired in 1930, a three-component Wiechert seismograph, which, despite its early arrival, was not permanently installed until 1953 under the name Estación Sismológica San Salvador (SSA).

On May 6, 1951, a large earthquake destroyed the cities of Jucuapa, San Buena Ventura, Nueva Guadalupe, and Chinameca. The German seismologist Rudolf Schulz arrived in El Salvador in 1953 (Schulz, 1969) to assume responsibility for the seismological investigations in the country. Two years later, the National Geological Service (NGS) of El Salvador was established and, in 1956, the Santiago de María station (SDMS) was installed in the Province of Usulután; this station recorded on site.

The following years saw several significant contributions to seismology in El Salvador. In 1957, SDMS started continuous recording. In 1961, another station, built in El Salvador by Schulz and technicians from El Salvador, was installed in the city of Ahuachapán (Estación Sismológica de Ahuachapán [AHS]), and the U.S. Geological Survey placed WWSSN station #95 in the city of La Palma (Chalatenango) in 1962; this station was designated Estación Sismológica la Palma (LPS) and was used to record local and regional earthquakes. The long-period seismometers of this station were the first of this type installed in El Salvador.

Development of the Seismograph Network

The Center for Geotechnical Research (CIG from its acronym in Spanish) was founded by decree in January 1964, and since then the seismic equipment of El Salvador has been part of the Section for Seismological Research of this center. That very year, the government of the Federal Republic of Germany donated a Stuttgart seismograph to El Salvador, which was installed at the El Salvador station. On May 3, 1965, a strong earthquake destroyed part of San Salvador, damaging the mechanical seismic equipment and making it impossible to have a full record of the event. This fact revealed the fragility and vulnerability of the seismic instruments at that time. New studies were then started to modernize and expand the seismic network of El Salvador. As a result, a new effort began in 1975 to install a telemetric network of national coverage. Four years later, the CIG took steps to expand the seismic network of four stations. Soon after, CIG, with its own resources, acquired 17 telemetric seismographs to monitor the local seismic and volcanic activity of El Salvador. The seismic signals were recorded at CIG on thermo-sensitive paper and also on photographic records. Table 1 shows the original distribution of stations.

Despite the fact that the equipment was received in 1980, it could not be installed immediately because the selected sites were unsafe for both the technical group and the instruments due to the civil conflict that started that year. The original configuration of the network was modified, and new, safer sites were selected. The capability to locate earthquakes effectively using instrumental data did not come until 1983, however, when seven stations had been installed. Effective seismic monitoring began ca. 1984 (12 stations, Table 2), with fairly complete seismic coverage of the country by the end of that year. The densest part of the network lay in the province of San Salvador and its surroundings, covering an area ~65 km by 43 km, with an average station spacing of 30 km. Data from the 12 stations were telemetered to CIG via radio. For earthquake location within the study area, arrival time data from the entire network were routinely integrated and processed at CIG.

In 1985, three of the stations were lost (Cauta, Coatepeque, and Montecristo), and the stations at Volcán de San Miguel, La Fuente, and Tecomasuche suffered irreparable damage. These six stations were replaced with equipment waiting to be installed elsewhere. Consequently, the network could not be expanded as was required for the new project.

The monitoring problems and the need to monitor specific areas like San Salvador and Lago de Ilopango were therefore evident. These areas could not be adequately monitored with the national network designed to detect and locate earthquakes over the entire country. To solve the problem, 11 telemetric analog seismic stations were set up around San Salvador in 1991 to monitor the tectonic and volcanic activity. Digital acquisition data (DATASEIS) was used for the information processing. Tables 3 and 4 list the stations of the two new seismic networks.

Besides the 11 stations, a digital accelerometer and Earthquake Analysis Software (SEISAN [Havskov, 1997]) were obtained through the Centro de Coordinación para la Prevención de los Desastres Naturales en America Central (CEPREDENAC) and the Norwegian Agency for International Cooperation (NORAD). This improvement greatly facilitated the collection, processing, and storage of seismic information.

TABLE 1. STATIONS OF THE PLANNED TELEMETRIC SEISMIC NETWORK OF EL SALVADOR IN 1980

No.	Name	Code	Location
1	Montecristo	MTO	Parque Nacional Montecristo,
2	Chingo	CHI	Volcán El Chingo, Santa Ana
3	Coatepeque	COA	Cantón Potrerillos, Santa Ana
4	Cauta	CTA	Jujutla, Ahuachapán
5	Nanahuazin	NAN	Sacacoyo, La Libertad
6	Tecomasuch	TEM	Tacachico, La Libertad
7	Chachacaste	CHA	Cerro Chachacaste, Cuscatlán
8	Guazapa	GUA	Cerro Guazapa, Cuscatlán
9	Victoria	VIC	Villa Victoria, Cabañas
10	San Jacinto	SJA	Cerro San Jacinto, San Salvador
11	Quezalapa	QZA	Rosario de La Paz, La Paz
12	Sihuatepequ	SHE	Cerro Sihuatepeque, San Vicente
13	Jucuarán	JCR	Villa Jucuarán, Usulután
14	Polorós	POL	Polorós, La Unión
15	Jocoro	JOC	Jocoro, La Unión
16	Conchagua	VCH	Volcán Conchagua, La Unión
17	Volcancillo	VOL	Cerro Volcancillo, Morazán

TABLE 2. STATIONS INSTALLED IN LATE 1984

No.	Name	Code	Location
1	Montecristo	MTO	Parque Nacional Montecristo, Metapán
2	Yupe	YPE	Cerro El Yupe, El Paste, Santa Ana
3	Cusmapa	CUS	Tacuba, Ahuachapán
4	Coatepeque	COA	Cantón Potrerillos, Santa Ana
5	Cauta	CTA	Jujutla, Ahuachapán
6	Nanahuazín	NAN	Sacacoyo, La Libertad
7	Tecomasuche	TME	Tacachico, La Libertad
8	Volcán de San Salvador	VSS	Volcán de San Salvador, San Salvador
9	San Jacinto	SJA	Cerro San Jacinto, San Salvador
10	La Fuente	LFU	Cerro La Fuente, San Salvador
11	Quezalapa	QZA	Cantón Barahona, Rosario de la Paz
12	Volcán de San Miguel	VSM	Cantón Placitas, San Miguel

TABLE 3. TELEMETERED NETWORK TO DETECT TECTONIC SEISMICITY AROUND THE METROPOLITAN AREA OF SAN SALVADOR

No.	Name	Code	Location
1	San Jacinto	SJA1	Cerro San Jacinto, San Salvador
2	El Huehuecho	HUE2	Cerro El Huehuecho, San Vicente
3	El Angel	ANG3	Cerro El Angel, San Salvador
4	El Ojo de Agua	OJO4	Cerro Ojo de Agua, La Libertad
5	Santa Adelaida	ADE5	Comasagua, La Libertad

TABLE 4. TELEMETERED NETWORK TO DETECT VOLCANIC
SEISMICITY AROUND THE EL SALVADOR VOLCANO

No.	Name	Code	Location
1	El Faro	LFR1	San Francisco Chinameca, La Paz
2	La Ceiba	LCB2	San Emigdio, La Paz
3	Las Brisas	LBR3	San Martín, San Salvador
4	El Picacho	PIC4	Volcán de San Salvador
5	Las Granadillas	GRD5	Quezaltepeque, La Libertad
6	El Boquerón	BOQ6	Volcán de San Salvador

In 1993, there were 20 telemetric stations operating continuously in El Salvador, four with recording on site, as well as the digital acceleration signals. Unfortunately, between the years 1994 and 1995 the stations HUE2, NAN, GRD5, VSS, SJA, SJA1, and El ANG3 (see Tables 2, 3, and 4) were closed because of vandalism. In the years 1996 to 1998, three more stations were also lost (OJO4, ADE5, and TEM), and LPS and AHS were closed for maintenance problems.

Scientific and technological developments during the last decade have provided encouragement to increase the effort toward solving the problem of earthquake location. In this sense, CIG began a new attempt to extend the coverage of stations to eastern El Salvador in 1997. To do this, it made an alliance with the Commission Hidroeléctrica del Río Lempa (CEL) to incorporate CEL's four seismic stations, located at the Berlín Geothermal Field, into the national network. These seismographs (Table 5) would be used to record seismic activity near the Rio Lempa and Berlín. Another significant advance in seismic observations in El Salvador during 1997 was the acquisition of the SEISLOG data acquisition system (Utheim and Havskov, 1997).

In 1998, El Salvador put a broadband station with long-period seismometers into operation at the closed La Palma Station; this station would be used for regional purposes. In early 1999, equipment from the closed Tecomasuche station was used to install a new telemetric station at Parque Nacional Montecristo in Metapán (Montecristo 2). This augmented the coverage of stations toward the north. By 1999, the seismic network of El Salvador had 11 stations of its own and four more stations from CEL (Table 6). This was enough to monitor the seismic activity of the entire country.

Early in 2000, the DATASEIS program failed with the change of millennium (Y2K fault). SEISLOG then became the

only program used to record seismic events. In July of the same year, a severe electrical discharge disabled part of the network, damaging the SEISLOG and the receivers of Las Brisas, La Fuente, El Picacho, and El Boquerón seismic stations, as well as the system to update and correct universal time. To solve the problem, the radios from the undamaged stations at Berlín and El Faro were redistributed, recovering most of the lost seismic signals. In spite of this, LFR1, LCB2, and LBR3 remained out of operation. During the El Salvador earthquakes of January and February 2001, only six seismic stations were in operation, which made it difficult to locate the events.

The present-day network (Fig. 1) includes 18 one-component, short-period, analog stations. El Salvador intends to install nine new stations in the near future, showing that the seismological group is determined to have a high-resolution network with continuous recording and efficient maintenance.

THE CENTRAL AMERICA SEISMOLOGICAL CENTER (CASC)

History

Central America is a densely populated region located near the active convergent margin formed by the interaction of the Cocos and Caribbean plates (Fig. 2). The stress generated by the

TABLE 6. LIST OF THE TELEMETERED STATIONS IN 1999

No.	Name	Code	Location
1	Cusmapa	CUS	Tacuba, Ahuachapán
2	El Yupe	YPE	Chalchuapa, Santa Ana
3	Montecristo	MTO2	Parque Nacional Montecristo, Santa Ana
4	La Fuente	LFU	Tonacatepeque, San Salvador
5	Quezalapa	QZA	Rosario de La Paz, La Paz
6	Las Brisas	LBR3	San Martín, San Salvador
7	La Ceiba	LCB2	San Emigdio, La Paz
8	El Faro	LFR1	San Francisco Chinameca, La Paz
9	El Boquerón	BOQ6	Parque Nacional El Boquerón, San Salvador
10	El Picacho	PIC4	Volcán de San Salvador, El Salvador
11	Volcán de San Miguel	VSM	Volcán de San Miguel, San Miguel
12	Santa Julia	SJUZ	Berlín, Usulután
13	Loma Alta	LALZ	Berlín, Usulután
14	Santiago de María	SDMZ	Santiago de María, Usulután
15	Las Palmas	LPAZ	Berlín, Usulután

TABLE 5. TELEMETERED STATIONS OF THE BERLIN
GEOTHERMAL FIELD

No.	Name	Code	Location
1	Santa Julia	SJUZ	Cerro Santa Julia, Usulután
2	Loma Alta	LALZ	Cerro Loma Alta, Usulután
3	Santiago de María	SDMZ	Santiago de María, Usulután
4	Las Palmas	LPAZ	Cerro Las Palmas, Usulután

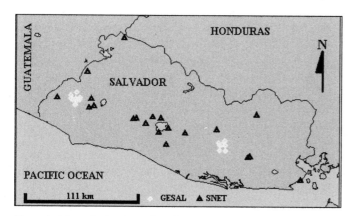

Figure 1. Map of seismic stations in El Salvador showing the location of the seismographs. Black triangles represent stations of the seismic network of El Salvador operated by the Servicio Nacional de Estudios Territoriales (SNET). Gray diamonds are stations of the El Salvador Electricity Company (GESAL).

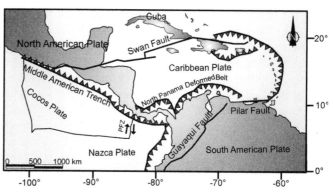

Figure 2. Tectonic features of Central America. The Caribbean Plate is bordered by the Cocos, North American, Nazca, and South American plates. The tectonic boundaries between these plates are the Middle American Trench, the Panama Fracture Zone (PFZ), the Polochic-Motagua–Swan fault system, and the North Panama Deformed Belt (NPDB).

collision of these plates extends hundreds of kilometers toward the continent and results in many earthquakes that sometimes strike population centers, causing tremendous damage and great disasters. The most recent seismic catastrophes are those of Nicaragua, 1972; Guatemala, 1976; El Salvador, 1986; Costa Rica, 1991; and El Salvador, 2001.

The first seismological instruments in Central America were deployed in Panama in 1882. In Costa Rica, El Salvador, Guatemala, and Nicaragua, the first installations took place in 1888, 1896, 1925, and 1966, respectively. Local seismic monitoring of Central America has been performed since the 1970s through the operation of separate local networks. The networks have been installed for monitoring seismicity in every country of the region. The detection/location threshold of coda magnitude is 2 for most of the networks. The first network of the region was installed in Costa Rica in 1974, and the latest was that in Panama, which started in 1991. The Nicaragua, Guatemala, and El Salvador networks began in 1974, 1976, and 1984, respectively. Honduras has only three seismic stations. At the present, most countries have more than one network.

These local networks face some difficulties in monitoring the regional seismicity. First, they are rather linear and do not provide accurate locations for earthquakes near the border regions of the countries. Stability in earthquake detection and location capability is difficult to establish because of the changing configurations and procedures as the networks develop. Catalogs will therefore contain reporting rate changes caused by the changes in seismograph networks, gain setting, computer codes for analysis, personnel, and extra work loads due to aftershock sequences, resulting in inhomogeneous catalogs. These aspects pose serious problems that affect the quality of hazard estimations in the region.

Fortunately, the Center for Prevention of Natural Disasters in Central America (Centro para la Prevencion de los Desastres Naturales en America Central [CEPREDENAC]) was founded

in 1988. In 1991, the Reduction of Natural Disasters in Central America Earthquake Preparedness and Hazard Mitigation project was initiated, coordinated by CEPREDENAC, funded by NORAD, and operated by the University of Bergen and NORSAR in cooperation with seismological agencies in all of the member countries. In the first phase of the project (1991–1994), the goal was to improve the seismic networks of the region and to conduct regionally based research, particularly hazard estimation. The second phase of the project was more concerned with local problems, such as microzonation, while still maintaining its attention on the national data centers. Once the networks had high earthquake detection and location capability, a regional seismological center would be created.

From 1990 to 1996, every country of the region received seismological equipment, especially data acquisition systems. Accelerometers were also provided to different institutions and more professionals from the field of seismology were involved at post-graduate level. In 1998, the Central America Seismological Center (CASC) was opened at the Geology Department of the University of Costa Rica (Alvarenga et al., 1998).

The CASC was created to provide reliable detection and location of earthquakes based on data from short-period and broadband stations and to offer a sufficiently large data recording and storage capacity for seismic data in Central America. The principal reasons for establishing the CASC network are to evaluate the potential for seismic and tsunami hazards in the region, to understand the tectonics in more detail, and hopefully, to be able to reduce the impact of future earthquakes and tsunamis in the region.

The CASC Information Center

A combined network of nearly 150 short-period and eight permanent broadband (or near-broadband) stations is now operating in Central America (Fig. 3, Table 7), providing regional coverage of the area. From this total, all the broadband records and most of the short-period ones must be used in the quasi-real time location of regional earthquakes. Every country operates local analog networks whose data are sent to the recording centers by radio or telephone. The SEISLOG data acquisition system is used to input the signals from the sensors and declare events based on the number of stations having detections in a given time window.

Physically, the CASC network is built around two interconnecting systems, SEISLOG and SEISNET (Ottemöller, 1998). While SEISLOG connects the stations to the local networks, including the broadband ones, SEISNET is a data collection system used by the main microprocessor located at the CASC to communicate with the SEISLOG systems in the region. Microprocessors oversee the operation of collecting the information and locate the events while others handle the storage of selected data.

Using these interconnected systems, the CASC automatically collects data for large events (M > 4) from the regional stations via modem or Internet to determine the location and magnitude of the events. SEISNET downloads detected information if there are more than two triggers within a time window of five minutes. The event parameters provide sufficient information for approximate event location and magnitude assessment. Further processing will

be done manually to give a reliable epicentral location and magnitude. SEISNET also collects data from Incorporated Research Institutions for Seismology (IRIS) stations located in Central America and also from the National Earthquake Information Center (NEIC) data center. Once an event is located, the information is stored in the database of the CASC. The main problem with the automatic data collection is the low Internet speed in the region. Because of this, many of the seismological agencies of the region are not contributing to real-time location of the earthquakes, and therefore the CASC has to locate the events with fewer stations than otherwise could have been used.

CASC Products

The Catalog of Historical Earthquakes

For Central America, important fragmentary historical earthquake information is scattered in many ancient and recent documents. There was, therefore, an urgent need to establish an historical earthquake catalog for the whole region so as to integrate the historical data with the instrumental observations (Rojas et al., 1993).

The historical catalog of Central American earthquakes dates from the beginning of the sixteenth century, and consists

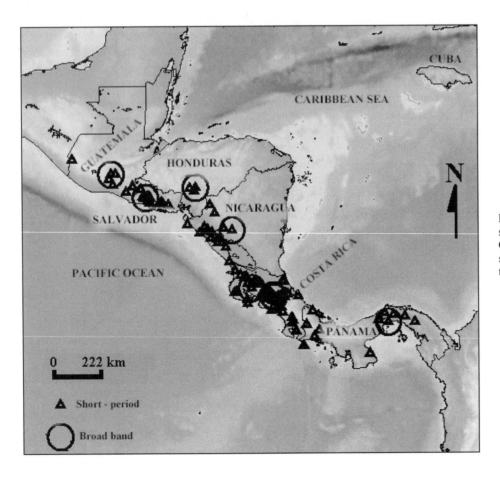

Figure 3. Map showing the location of seismic stations currently operational in Central America. Triangles represent the short-period stations; open circles show the location of broadband stations.

TABLE 7. SEISMIC STATIONS AND ACCELEROMETERS IN CENTRAL AMERICA

Country	Short-period stations	Broadband stations	Accelerometers
Guatemala	12	1	8
El Salvador	36	1	15
Honduras	3	1	1
Nicaragua	36	1	20
Costa Rica	54	3	53
Panama	13	1	10
Total	154	8	107

of ~20,000 events recorded between 1505 and 2000. The largest events (M > 5.4) in the catalog and the associated casualties are listed in Table 8.

The data can be divided in two sets: macroseismic information from 1505 to 1901 and both macroseismic and instrumental information after 1902. Maximum intensities have been assigned to the events on the basis of damage reports; for pre-instrumental earthquakes, magnitudes have been taken from publications. A magnitude threshold of M or Ms 6.0 was chosen for the instrumentally recorded events, but shallow crustal events with magnitudes below 6.0 have also been included in the catalog. The catalog is considered complete for earthquakes with magnitude Ms > 5.5 in the period from 1900–2002; for Ms > 7.0, the completeness spans the period of 1820–2000. The catalog also lists the date, intensity, magnitudes, epicentral coordinates, depth, distance indicator, quality indicator, fault-plane solutions, reporting agency, and descriptions of the damage or felt effects. The primary source of indexed material was the writers' personal collections of seismological reports. Most of the information was scattered in secondary sources.

The catalog follows the so-called "Extended Nordic Format" (Rojas, 1993), which made it possible to include a variety of macroseismic data, associated information such as type of event, cultural aspects, references, and instrumental data. In this catalog, it is possible to differentiate between shallow upper-plate and subduction zone events. The catalog was designed to provide

basic statistics so the user could differentiate between located and non-located earthquakes, events with magnitude, events with intensity, etc. This earthquake database contains a wealth of magnitude data of different kinds, often with many estimates for the same event. Also, associations between the different magnitudes can be found in the catalog.

The catalog has been used for a preliminary estimate of the seismic hazard of the region (Montero et al., 1997). Future applications of the catalog could be the study of seismogenic potential and the development of earthquake recurrence models.

Catalog of Instrumental and Recent Earthquakes

The catalog contains instrumental seismicity data for the entire region from 1992–2001. The goal of the CASC is to locate events as soon as possible, yielding monthly bulletins of seismicity and more data for the catalog. The catalog will hopefully serve as a useful index of important Central American earthquakes and as a tool for seismicity studies and hazard analyses. The database of instrumentally recorded earthquakes contains over 130,000 earthquakes. As the mission of the CASC is to develop a database of M > 4.0 events, the center has re-determined the locations of earthquakes from the 1992–1997 period. Since 1998, the CASC has been routinely locating seismic events from the region. Summing the located events, we find that there are more than 10,000 M > 4.0 events in the database for the 1992–2001 period. Besides the locations, there are ~20,000 waveforms in the database, and ~10,000 more are expected to be part of the database shortly.

Table 9 gives the general statistics of the catalog. The entries include earthquake origin time, hypocentral coordinates, magnitude (several), RMS error, and type of source mechanisms. The present threshold for the detection and location of earthquakes is magnitude 4.0.

Figure 4 shows the distribution of epicenters that have been located by the CASC from 1992–2001. Within the studied area, the majority of earthquakes have been located in a rectangular area that extends from Nicaragua to El Salvador, suggesting that this is currently the most active segment of the Middle American Trench. The largest earthquakes are concentrated in Guatemala, El Salvador, Nicaragua, and the Panama Fracture Zone. Costa

TABLE 8. NUMBER OF LARGEST EARTHQUAKES AND DEATH TOLL

Country	Events	Deaths
Guatemala	518	28,088
El Salvador	252	4935
Honduras	90	3
Nicaragua	386	14,150
Costa Rica	256	871
Panama	371	124
Total	1873	48,171

TABLE 9. STATISTICS FROM THE INSTRUMENTAL EARTHQUAKES CATALOG

Year	Events	Required waveforms	Available waveforms
1992	446	693	583
1993	2245	5365	1721
1994	1302	3129	1981
1995	864	2165	1001
1996	853	2102	1393
1997	1240	2826	2273
1998	931	3162	2757
1999	2207	7203	6239
Total	9988	26,645	17,948

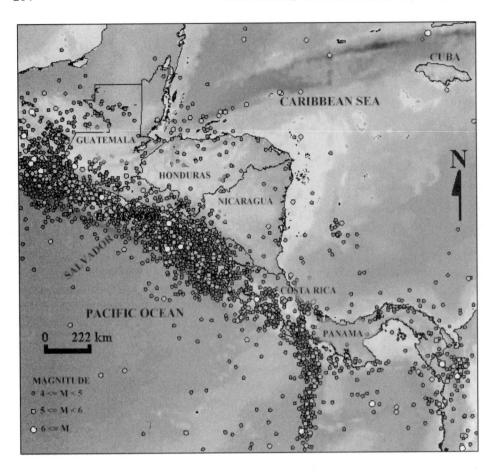

Figure 4. Spatial distribution of seismicity (M > 4.0) in Central America and surrounding area. Seismicity spans the entire distance between Guatemala and Panama along the Middle American Trench. The map is for the period 1992–2001.

Rica and most of the Panamanian territory had few strong earthquakes during the period 1992–2001. Some earthquakes have also been recorded by the CASC along the Polochic-Motagua Fault System and the Panama Deformed Belt. Of particular interest are some moderate magnitude earthquakes located east of Nicaragua (in the Caribbean Sea, probably associated with the Hess Scarp) and the cluster of larger earthquakes in Guatemala and Nicaragua.

A striking feature of the map is the excellent definition of the Middle American Trench (MAT) and the Panama Fracture Zone, which confirms the good relationship of this data to the known plate movements. The pattern of the new map suggests that moderate earthquakes are uniformly distributed along the MAT, and no conspicuous gaps are observed in Figure 4.

Catalog of Fault-plane Solutions

During the past 50 years or so, numerous fault-plane solutions of earthquakes have been published. Since 1960, the number of such solutions has increased rapidly, primarily as a consequence of the installation of the World Wide Standardized Seismographic Network (WWSSN). A catalog of fault-plane solutions has been established for the region (Redondo et al., 1993), comprising 650 solutions for the period of 1934–2001, 537 from individual events, and 13 composite focal mecha-

nisms. The information was largely taken from previous publications, and only one of the two possible nodal planes is tabulated. The database is divided into two groups: shallow upper-crustal events (depth < 50 km) and the intra-slab events that occurred at greater depths.

The solutions are plotted in Figures 5 and 6, where they are also classified according to the triangular faulting mode representation of Frohlich and Apperson (1992). The deeper earthquakes clearly show a strong dominance of reverse faulting with little strike-slip faulting, while the shallow earthquakes show a predominance of strike-slip faulting. This information is available to all researchers interested in more detailed studies of the earthquake sources in the region.

Catalog of Tsunamis

Central America has been subjected to damaging tsunamis, some generated in the region and others approaching from regions surrounding the Pacific basin. The September 2, 1992, Nicaraguan tsunami is the most recent example of a destructive tsunami in the region (Fernández et al., 2000). To face this natural hazard, an exhaustive investigation on historical tsunamis was carried out, yielding a tsunami catalog that can serve as a basis for further work related to the reduction of the tsunami impact in the coastal areas of Central America. Work on the tsunami

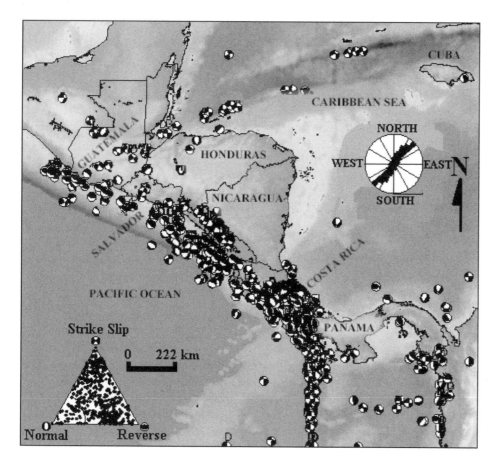

Figure 5. Distribution, triangle representation, and rose diagram of all fault-plane solutions for shallow (< 50 km) earthquakes.

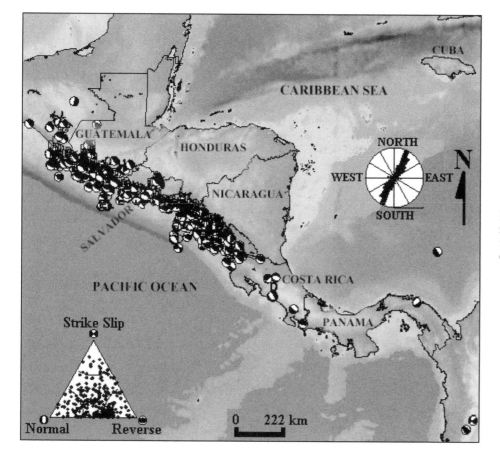

Figure 6. Distribution, triangle representation, and rose diagram of all fault-plane solutions for deep (> 50 km) earthquakes.

catalog was begun in 1997 by Enrique Molina (INSIVUMEH, Guatemala) and was concluded by Mario Fernández (University of Costa Rica), who also wrote a proposal to establish a tsunami warning in the region (Fernández, 1998). Molina (1997) did the preliminary editing of the catalog, and the publication of the catalog was carried out by Fernández et al. (2000).

The catalog records 49 tsunamis for the period 1539–2001, 37 from the Pacific and 12 from the Caribbean. They have struck all the countries of Central America, but the most affected countries have been Panama, Nicaragua, and El Salvador, where almost 500 people have been killed by destructive tsunamis (~100 in Panama, ~170 in Nicaragua, and ~185 in El Salvador). On the Caribbean coast, the tsunamis are concentrated in two tsunamigenic areas: the Gulf of Honduras and the coastal segment between Panama and Costa Rica. On the Pacific coast, the tsunamis are well distributed along the entire coast.

The current regional tsunami information is being used to study the tsunami hazards of the region in detail (Fernández and Rojas, 2000; Ortiz et al. 2001; Fernández et al., this volume, Chapter 32). Our goal is to identify the areas at maximum risk and make a map of inundation. Besides these catalogs, the CASC has almost 400 publications related to the seismicity and tsunamis of Central America. We hope to collect more publications now scattered in different journals, books, and other documents.

CONCLUSIONS

Seismological observations began in El Salvador in the late nineteenth century. After 1896, routine earthquake recording in El Salvador was done with Wiechert seismographs. Seismological research began in 1955 and continued through the 1960s, although relatively few publications emerged from this period. The principal figure among the first seismologists in El Salvador was Dr. Rudolf Schulz, whose influence on seismological investigations in El Salvador was considerable.

The telemetric network in El Salvador has operated almost continuously since 1980, but some stations have operated only intermittently. The reason for this is that they are stolen or damaged, and the substitution or necessary repairs cannot be made because of the lack of both security and funds. In such cases, the network will run with partial coverage until repairs can be made. This condition is undesirable from the point of view of continuous and homogeneous monitoring. A more detailed monitoring of microearthquake activity began in the 1990s with the expansion of the network and the acquisition of modern SEISLOG acquisition data system. The current network consists of 36 telemetric stations.

The establishment of the CASC in 1998 marked a significant acceleration of the effort to monitor seismic activity by providing more dense and standardized instrumental coverage in the region. Through the CASC, earthquakes from the region are subjected to systematic and detailed analyses. This center is important because it monitors the seismic activity of a region with both high seismicity and a large population. It provides useful information on tectonics and helps to better estimate the seismic potential of Central America.

With the adoption of new techniques for collecting and processing seismic station observations, the CASC has created a seismic catalog that contains almost 10,000 well-located events and 17,000 waveforms. The CASC also has three more catalogs: a historical earthquake catalog with ~2000 events of magnitude up to 5.4, a database of fault-plane solutions including 650 focal mechanisms, and a tsunami catalog with 37 tsunamis, most of them from the Pacific. Furthermore, the CASC has almost 400 publications on regional seismicity and tsunamis. These databases are now the most important sources of the numerous and high-quality earthquake data from the region.

RECOMMENDATIONS

Seismic data are potentially very important for assessing both local and regional seismic hazard, and a close cooperation between the countries of Central America is therefore essential if they want to truly mitigate seismic disasters. Estimates of seismic hazard also depend on knowledge of local site conditions. We thus need to consider other ground-motion parameters in addition to the peak ground accelerations to make seismic hazard estimation more reliable, precise, and pertinent to the user. Continued work to identify active faults is particularly important.

ACKNOWLEDGMENTS

Two authors thank CEPREDENAC and NORAD for the financial support to do this investigation. We also thank the staff of all the seismological laboratories of Central America for their supply of instrumental data. Special thanks go to Griselda Marroquin from the Servicio Nacional de Estudios Territoriales (SNET) in El Salvador and Wilfredo Rojas from Red Sismologica Nacional de Costa Rica (RSN: ICE-UCR) for their contribution to this work. We are grateful to Robin Adams and Hilmar Bungum for comments on an early version of the paper, which led us to substantially improve it.

REFERENCES CITED

Alvarenga, E., Barquero, R., Boschini, I., Escobar, J., Fernández, M., Mayol, P., Havskov, J., Galvez, N., Hernandez, Z., Ottomoller, L., Pacheco, J., Redondo, C., Rojas, W., Vega, F., Talavera, E., Taylor, W., Tapia, A., Tenorio, C., and Toral, J., 1998, Central American Seismic Center (CASC): Seismological Research Letters, v. 59, no. 5, p. 394–399.

Alvarez, S., 1980, Una solución para mejorar el sistema de la red sismográfica en El Salvador: San José, Costa Rica, Centro de Investigaciones Geotécnicas de El Salvador, 45 p.

Fernández, M., 1998, Proposal to establish a tsunami warning system in Central America: Bergen, Norway, Institute of Solid Earth Physics, University of Bergen, Technical Report no. II 1–10, 26 p.

Fernández, M., Molina, E., Havskov, J., and Atakan, K., 2000, Tsunamis and tsunami hazard in Central America: Natural Hazards, v. 22, p. 91–116.

Fernández, M., Ortiz-Figueroa, M., and Mora, R., 2004, Tsunami Hazards in El Salvador, *in* Rose, W.I., Bommer, J.J., López, D.L., Carr, M.J., and Major, J.J., eds., Natural Hazards in El Salvador: Boulder, Colorado, Geological Society of America Special Paper 375, p. 435–444 (this volume).

Fernández, M., Rojas, W., 2000, Amenaza sísmica y por tsunamis, *in* Denyer, P., Kussmaul, S., eds., Geología de Costa Rica: Editorial Tecnológica de Costa Rica, Instituto Tenologico de Costa Rica, 515 p.

Frohlich, C., and Apperson, K., 1992, Earthquake Focal Mechanisms, Moment tensors, and the Consistency of Seismic Activity near Plate Boundaries: Tectonics, v. 11, no. 2, p. 279–296.

Gálvez, N., 2002, Historia de la red telemétrica sísmica de El Salvador: San José, Costa Rica, Servicio Nacional de Estudios Territoriales de El Salvador, 10 p.

Havskov, J., 1997, The SEISAN earthquake analysis software for the IBM, PC and Sun version 6.0 manual: Bergen, Norway, Institute of Solid Earth Physics, University of Bergen, 236 p.

Molina, E., 1997, Tsunami Catalog for Central America 1539–1996: Bergen, Norway, Institute of Solid Earth Physics, University of Bergen, Technical Report no. II 1–4, 87 p.

Montero, W., Peraldo, G., and Rojas, W., 1997, Proyecto de amenaza sísmica de América Central: Ottawa, Canada, Instituto Panamericano de Geografía e Historia (IPGH), 79 p.

Ortiz, M., Fernández, M., and Rojas, W., 2001, Análisis de riesgo de inundación por tsunamis en Puntarenas, Costa Rica: Boletín de la Unión Geofísica Mexicana (GEOS), v. 21, no. 2, p. 108–113.

Ottemöller, L., 1998, SEISNET user manual, version 1.1: Bergen, Norway, Institute of Solid Earth Physics, University of Bergen, 30 p.

Redondo, C., Lindholm, C., and Bungum, H., 1993, Earthquakes Focal Mechanisms in Central America: Oslo, Norway, NORSAR, Technical Report, 21 p.

Rojas, W., 1993, A Catalog of Historical and Recent Earthquakes in Central America: San José, Costa Rica, Escuela Centroamericana de Geología, Universidad de Costa Rica, Tesis de Licenciatura, 77 p.

Rojas, W., Bungun, H., and Lindholm, C., 1993, A catalog of historical and recent earthquakes in Central America: Oslo, Norway, NORSAR, Technical Report, 78 p.

Schulz, R., 1969, Die Entwicklung des salvadorenischen Erdbebendienstes: Stuttgart, Germany, Festschrift Wilhelm Hiller, Institut für Geophysik der Universität Stuttgart, p. 84–95.

Utheim, T., and Havskov, J., 1997, The SEISLOG data acquisition system version 7.0 user manual: Bergen, Norway, Institute of Solid Earth Physics, University of Bergen, 101 p.

MANUSCRIPT ACCEPTED BY THE SOCIETY JUNE 16, 2003

Geological Society of America
Special Paper 375
2004

Seismic vulnerability of the healthcare system in El Salvador and recovery after the 2001 earthquakes

Ruben L. Boroschek*

World Health Organization/Pan-American Health Organization Collaborating Centre for Disaster Mitigation in Health Facilities, University of Chile, Blanco Encalada 2120, Piso 4, Santiago, Chile

ABSTRACT

During the January and February 2001 earthquakes, the national healthcare system in El Salvador suffered infrastructure damage that caused loss of functionality and the evacuation of the main hospitals in the country. These evacuations produced a strong effect on the capacity to care for the wounded and to continue offering the normal daily health assistance. The loss of functionality was due to damage in nonstructural elements, limited damage in engineered structural elements, and loss of confidence of the staff and patients in the safety of hospital buildings. Many hospitals had to operate totally or partially outside their premises with temporary shelters for two years after the events. A review of the hospital infrastructure indicates that the seismic design objective used, if any, only intended to protect the main structural elements. Only limited seismic design was made to protect nonstructural components and the functionality or the hospitals' investment. Additionally, construction practices contributed to the damage observed. Due to massive damage to the health infrastructure, a recovery strategy is needed in which new standards and performance objectives are considered. However, they are limited by economic resources and the need for a fast and massive reconstruction. Initially, a few strategically located hospitals will be reinforced structurally and nonstructurally to establish hospitals with a low probability of earthquake-induced functional disruptions. Other hospitals will primarily be repaired to recover their preexisting capacity and others will be reinforced to obtain a higher level of structural safety. This strategy will ensure a minimum health response capacity and a reduction in infrastructure and economic losses.

Keywords: hospital, structural damage, health, functionality loss.

THE PUBLIC HEALTHCARE SYSTEM IN EL SALVADOR

El Salvador is one of the most densely populated countries in Latin America, with a density of 280 inhabitants per square kilometer and a growth rate of ~2.2% per year. Forty-five percent of the population is located in urban areas, with a life expectancy of 73 and 66 yr for women and men, respectively. Child mortality is 38.9 deaths per each thousand born alive, and the country has a literacy rate of 73%. The population by department for year 2000

is shown in Figure 1. Two main health facilities networks exist in El Salvador: the public healthcare system administered by the Health Ministry and the social security system administered by Salvadorian Social Security Institute.

The Health Ministry hospital system is the institution that has the major population coverage, reaching the total national population (~6,350,000 inhabitants). It is organized on three basic levels of care as described by the National Health Regulation: Level I, with the following types of centers: health homes (with a national total of 161), rural nutrition centers (51), and health units (361); Level II, with peripheral general hospitals (11) and central general hospitals (14); and Level III, with special-

*rborosch@ing.uchile.cl

Boroschek, R.L., 2004, Seismic vulnerability of the healthcare system in El Salvador and recovery after the 2001 earthquakes, *in* Rose, W.I., Bommer, J.J., López, D.L., Carr, M.J., and Major, J.J., eds., Natural hazards in El Salvador: Boulder, Colorado, Geological Society of America Special Paper 375, p. 269–279.

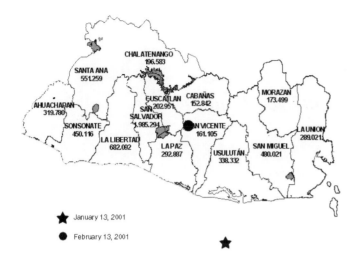

Figure 1. Political division and population distribution of El Salvador. Approximate epicenter locations of the January and February events.

ized national hospitals (5). Some hospitals' characteristics are presented in Table 1. In general terms, one health unit is created for 5000 to 40,000 inhabitants, one peripheral hospital for 50,000 to 100,000 inhabitants, and one central hospital for each country department. In addition to the population density, variables such as distance and geographical difficulties are considered in the development and location of health centers.

The geographic distribution of the 30 hospitals that make up the network of the Health Ministry is displayed in Figure 2. There are 18 health departments, 13 of them corresponding to the national political division, and the remaining five health departments are located in San Salvador.

According to the Pan-American Health Organization (PAHO/WHO, 2000), Health Ministry (2000b), and Boroschek (2001), 57% of the hospitals have buildings that are 46 years old or more, 29% between 16 and 30 years old, and 14% between 1 and 15 years old. The average construction age, by hospital category, is as follows: specialized national hospitals, 80 years, general central, 73 years, and peripheral national, 28 years. Less than three-story-high construction is present in 77% of the hospitals. Their main structures are based on adobe, brick walls, or steel structures with steel or concrete infills (Rosales and Santa Ana hospitals). The remaining 23% have reinforced concrete frame-type construction with infilled brick walls. Most of these frames utilize structural details that provide limited ductility capacity. Just a few medium-rise buildings have structural walls, and if they exist, they are mainly located in elevator shafts (Rosales Surgery Building, Zacatecoluca, and San Pedro de Usulután main structures). Modern structural control techniques, such as base isolation, are not present in the Salvadorian hospital system.

The total constructed area of Health Ministry hospitals is ~296,500 m². By category, 33.4% corresponds to specialized

hospitals, 48.4% to central hospitals, and 18.1% to peripheral national hospitals. These hospitals had an occupancy rate of 86% in the first half of the year 2000 (Health Ministry, 2000b).

Considering that there are a total of 4677 registered hospital beds, there is one bed for every 1358 inhabitants and for each 63 m² of hospital construction. Also, 35.7% of the beds are assigned to specialized hospitals. Of these hospitals, the Rosales Hospital, a 98-year-old building, and Santa Ana Hospital, a 155-year-old building, have 14% of the square meters and 21% of the country's beds. The next hospital in number of beds is the San Juan de Dios de San Miguel Hospital, with 8.3% of the beds. That is to say, the older hospitals show the highest number of beds and have been assigned 53% of the total country's population.

Specialties and Human Resources

According to PAHO/WHO (2000), all hospitals have the four basic areas: gynecology, obstetrics, pediatrics, and surgery and internal medicine, with the exception of the specialized Benjamin Bloom Hospital (pediatrics) and the Maternity Hospital. The most complex hospital is Rosales Hospital with 22 specialties, followed by Benjamin Bloom with 21, San Miguel with 18, Santa Ana with 17, San Rafael with 14, and Zacamil with 10 specialties. Notably, specialized hospitals make up 22.4% of the significant surgical clinical procedures, the central general hospitals 48.7%, and the peripheral hospitals 28.9%. One important feature is that intensive care units are present only in five hospitals (Rosales, Benjamin Bloom, Maternity, Santa Ana, and San Miguel), requiring a quick transfer system between locations and hospitals for critical patients.

Because the size of the country is rather small, critical cases and the need for specialized treatment is solved by transferring patients from local hospitals to the greater San Salvador area. In normal times, it is possible to cross the longer axis of the country by car in 4 hours 30 minutes, and the shorter axis, with connections to Honduras and the Pacific Ocean, in 2 hours 30 minutes.

In addition to the hospitals, the Health Ministry has two national laboratories: the Dr. Max Bloch Central Laboratory and the Biological Laboratory, both located in greater San Salvador. The former certifies medicines and food and makes diagnoses for other laboratories. In fact, it is also dedicated to its own diagnoses, and it is an important center for national referrals. Other laboratories are located in the San Miguel, Santa Ana, and Rosales Hospitals, and they rely on Dr. Max Bloch Central Laboratory for specialized analysis.

The Biological Laboratory is committed to developing simple vaccines against hydrophobia. Also, the national medicine storage center is within its facilities.

Because of these characteristics, the Health Ministry services show a great concentration of infrastructure toward the coastal sector and in greater San Salvador. All specialized hospitals are located in greater San Salvador; 83% of these hospitals, which by regulation are specialized, actually render general services.

TABLE 1. HEALTH MINISTRY HOSPITALS: CHARACTERISTICS AND DAMAGE

Hospital	Age of Main Structure	Construction M²	# Beds	Ref Population	Surgery Rooms	Intensive Care	# Stories >3	# Stories <3	Permanently Lost Beds % Jan.	Feb.	Damage
Specialized National											
Rosales	98	25.000	531	National	15	X		X	0	0	Severe non-structural damage in Surgery Building: collapse of ceilings, damaged partitions, elevator derailment, excessive motion and collapse of ducts and piping.
Benjamin Bloom	72	21.752	325	National	6	X	X		0	0	No damage—temporarily evacuated.
Maternidad	46	17.792	308	National	10	X	X		100	100	Severe non-structural damage, Fertility Building. 100% evacuated several times. Damage to partitions, ceilings, and ducts.
Psiquiatria/General	105/2	23.002	216	284,003	3			X	0	0	Minor non-structural damage. Damage to slab stucco.
Neumologico/General	81	11.500	292	209,747	2			X	37	0	
General Central											
S. Juan Dios Santa Ana	155	15.750	473	319,252	13	X		X	0	0	Minor non-structural damage. Partition cracking.
S. Juan Dios San Miguel	176	37.122	390	321,418	15	X	X		36	0	Minor structural damage. Severe non-structural damage: collapse of partitions, ceiling, light fixtures, piping and duct. Elevator derailment.
Zacamil	7	11.985	255	1,212,551	7			X	0	0	
San Rafael	140/ND	1.766	222	682,092	7		X		100	100	Severe adobe structural damage. Moderate structural and severe non-structural damage in newer structure: heavy cracking, partitions and facades.
Sonsonate	96	18.800	189	450,116	2			X	0	0	Minor non-structural damage.
Zacatecoluca	ND	ND	155	292,887	5		X		100	100	January: Moderate non-structural damage: partition cracking. February: Moderate structural damage and severe non-structural damage: column shear failure, collapsed equipment, ceilings, piping, ducts.
Francisco Menendez	117	14.227	157	319,780	5			X	0	0	Minor non-structural damage.
Santa Gertrudis	153	8.387	126	161,105	2			X	0	100	Moderate structural and non-structural damage.
San Pedro	27	4.345	130	161,243	2		X		100	0	Minor structural damage. Severe non-structural damage: collapsed ceilings, rupture of piping, elevator derailment.
Dr. Luis e. Vazquez	28	6.000	94	110,466	5		X		0	0	Moderate structural damage, new structure: wall and partition cracking.
Cojutepeque	127	3.600	78	187,205	2			X	0	30	Severe damage to old structures.
La union	49	3.000	64	150,388	3			X	0	0	
Francisco Gotera	21	12.852	60	173,499	2			X	0	0	
Sensuntepeque	54	5.729	60	73,975	3			X	0	0	
General Peripheral											
Chalchuapa	20	4.990	78	162,405	1			X	0	0	
San Bartolo	21	4.436	65	262,374	2			X	0	0	
Ilobasco	12	5.300	59	78,867	2			X	0	0	
Nueva Guadalupe	12	3.800	55	90,078	1			X	0	0	
Jiquilisco	12	7.500	50	73,190	2			X	0	0	
Santiago de Maria	49	9.153	44	103,899	1			X	100	100	Infrastructure damage.
Metapan	47	2.000	41	69,602	2			X	0	0	
Santa Rosa de Lima	20	4.003	47	138,633	2			X	0	0	
Ciudad Barrios	20	4.662	41	68,525	2			X	0	0	Infrastructure damage.
Nueva Concepcion	19	5.278	40	86,289	2			X	0	0	
Suchitoto	60	2.650	32	15,746	1			X	0	0	

Note: Source: MSPAS Gestión en Salud. Bulletin N 1. 2000.
OPS Encuesta de actividades y recursos de los hospitales públicos de El Salvador, 2000.
COEN 2001

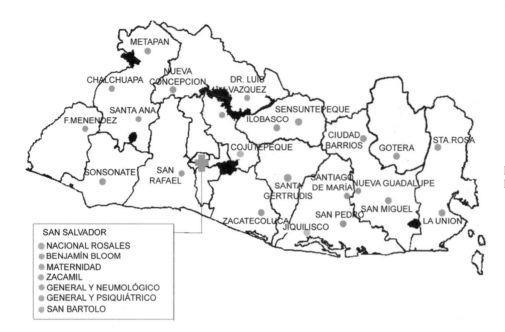

Figure 2. Geographic distribution of Health Ministry hospitals.

The Health Ministry health network is the largest in the country and has a close relationship with the Salvadorian Social Security Institute hospital network. This relationship implies mutual support and sharing of the infrastructure, generally owned by the Health Ministry.

SALVADORIAN SOCIAL SECURITY INSTITUTE HEALTH NETWORK

The network of the Salvadorian Social Security Institute covers 963,950 inhabitants, which represent 15.1% of the population and 21.2% of the economically active population in the country.

The Salvadorian Social Security Institute network has the following hierarchical structure: Level I, with communal clinics (31), company clinics (149), and medical units (35); and Levels II and III, with hospitals (18). The more complex hospitals, Level III, are Médico Quirúrgico (surgical-medical), Especialidades (specialties), Oncología (Oncology), and 1° de Mayo. Partial bed distribution is indicated in Table 2 and Figure 3 (SSSI, 2000a, 2000b). The Salvadorian Social Security Institute has 81.4% of its beds located in the greater San Salvador area. Communal and company clinics do not have beds and have a small resolution capacity: general medicine, epidemiology, and education.

HOSPITAL DESIGN CRITERIA IN EL SALVADOR

The Salvadorian seismic design code, like most of Latin American countries, is based mainly on U.S. codes and on its national seismic experience. Latin American codes, in general, were developed in the first half of the twentieth century. Before this date, engineering and earthquake-resistant design was not specified in codes, but applied by some experienced individuals. Countries such as Chile created their first seismic codes in the 1920s, but in that period of earthquake-resistant design, a clear objective of protection did not exist. From the 1960s, and after some destructive seismic events, engineers began to develop codes with the basic aim of protecting lives in severe events and allowing controlled damage in relatively moderate events. This philosophy of seismic design is currently accepted worldwide.

In these codes, hospitals were considered special structures, increasing the seismic loads with the hope that in this indirect manner the expectations of protecting the functionality required by the users would be fulfilled.

Until the 1990s, Latin American earthquake design codes were not dedicated to the protection of the investment or the functionality of hospitals. Therefore, it is unlikely that these objectives were accomplished in the infrastructure constructed before this date, or even today, if these aspects have been only considered at the design stage and not during the construction and operating stages (maintenance).

The first earthquake-resistant design regulation of El Salvador was published in 1966 and the second one in 1989, after the 1965 and 1986 earthquakes, respectively. The current seismic design code of 1994 (Ministry of Public Works, 1994) belongs to the groups of codes mainly oriented to protect the life of the inhabitants and not to protect the hospital's investment or function. Nevertheless, it has clauses that, if applied, could indirectly contribute to increasing the functionality protection.

The 1994 code is based on the 1991 U.S. Uniform Building Code and introduces the demands and requirements for the analysis of structural and nonstructural elements. This code establishes the seismic demands for El Salvador in two seismic zones (effective acceleration of 0.4 g and 0.3 g, respectively), which in rela-

TABLE 2. SALVADORIAN SOCIAL SECURITY INSTITUTE HOSPITALS (PARTIAL NETWORK): CHARACTERISTICS AND DAMAGE DURING THE JANUARY 13 EVENT

#	Department	Hospital	Beds	Damage during January 13, 2001
1	San Salvador	Medico Quirurgico	255	
2	San Salvador	Materno Infantil 1º de Mayo	239	Minor damage in secondary beams and moderate damage in facades and interior partitions.
3	San Salvador	Psiquiatrico	112	Minor damage in walls and non-structural elements and slab.
4	San Salvador	Especialidades	274	Minor damage in walls and non-structural elements and slab.
5	San Salvador	Oncologia	50	Moderate damage in structural walls. Severe damage in ceilings in 4th floor.
6	San Salvador	Neumologico	95	Minor damage in non structural elements (potable water).
7	San Salvador	Amatapec	165	Moderate damage in walls and electrical system. Severe damage in partitions and ceiling, 4th floor.
8	San Salvador	Roma	70	Damage in roof steel structure and elevators.
9	La Libertad	San Rafael	70	Severe non-structural damage.
10	La Paz	Santa Teresa Zacatecoluca	16	Moderate non-structural damage.
11	San Miguel	San Juan de Dios de San Miguel	105	Severe non-structural damage (walls, partitions, ceilings, elevators, water supply, sewer).
12	La Unión	La Unión	14	Minor structural damage. Small cracks in columns and spalling.
13	Usulután	Santiago de Maria	10	Moderate damage in structural walls and severe damage to non-structural walls.
14	Usulután	Usulután	24	Severe non-structural damage and moderate structural damage.
15	Santa Ana	Santa Ana	135	
16	Sonsonate	Sonsonate	70	Moderate structural and non-structural damage.

Note: Sources: SSSI Department of Infrastructure, 2001, and Boroschek, 2001.

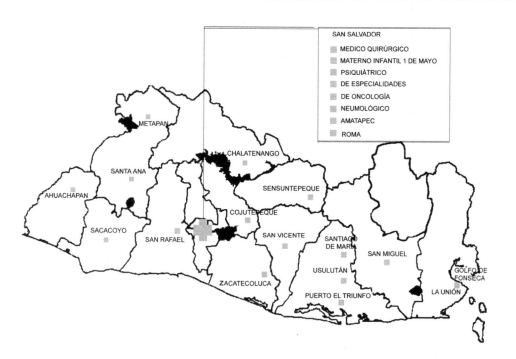

Figure 3. Geographic distribution of Salvadorian Social Security Institute hospitals.

tive terms, describes the whole country as highly seismic (Fig. 4). Additionally, for hospitals, the code requires a factor of 1.5 times the requirements established for structures with low occupancy, and additionally limits the relative drift between adjacent floors to a maximum of 1% of the floor height.

Due to the population distribution, 80% of hospitals (92% of the beds) and 100% of intensive care units are located in the highest-risk seismic zone. Most of the healthcare network is relatively old and was constructed with limited seismic design criteria. When it does exist, the seismic safety level is to protect

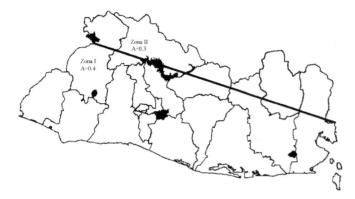

Figure 4. Seismic zones according to El Salvador seismic design code.

life and not to protect the infrastructure or functionality. This is the main reason for the observed damage of the healthcare system and imposes a big challenge to overcome years of design and construction that use a philosophy that nowadays is considered inadequate for critical facilities.

JANUARY 13 EARTHQUAKE

The earthquake of January 13, 2001, caused serious human casualties and infrastructure damage. The number of victims was 844, mainly concentrated in Las Colinas, Department of La Libertad, due to a landslide (Table 3).

As a result of its magnitude, the event severely affected an important number of hospitals (San Juan de Dios, San Miguel, San Pedro de Usulután, San Rafael, Maternity, General and Neumológico, Zacatecoluca, and Santiago de Maria, belonging to the Health Ministry; San Rafael, 1º de Mayo, Oncología, Amatepec, Roma, Usulután, San Miguel, and Sonsonate, from the Salvadorian Social Security Institute). Tables 1 and 2 describe the damage and functional consequences. This implied a temporary loss of 34% of the beds from the Health Ministry national total, necessitating the use of temporary shelters (Figs. 5 and 6). Detailed descriptions of the damages are given in Boroschek and Retamales (2001), Boroschek (2001), and Baroni and Quaglia (2001). In addition to the damage in hospitals, there was damage to 20% of the other components of the Health Ministry system. According to Ruales and Ayala (2002), 17 health units suffered severe damage, 21 suffered moderate damage, and 77 suffered minor damage.

The structural damage observed can be classified as severe for structures with low seismic capacity, such as adobe construction (San Rafael Hospital and others). In general, the main hospitals structured with moment resisting frames and frame-wall buildings suffered relative minor to moderate structural damage and no collapse (Fig. 7). On the other hand, nonstructural damage was generalized and severe. A visit to the main hospitals of the country indicated a complete absence of seismic design for architectural elements, medical and industrial equipment, and distribution systems such as piping, ducts, etc. Because of the severity of the earthquake and the relatively small size of the country, most of the hospitals in the coastal and central areas were

TABLE 3. CASUALTIES IN THE JANUARY AND FEBRUARY EVENTS

Department	Total population	Affected population			Wounded			Dead		
		January 13 data 02/12/01	February 13 data 02/21/01	Total	January 13 data 02/12/01	February 13 data 02/21/01	Total	January 13 data 02/12/01	February 13 data 02/21/01	Total
Ahuachapán	319,780	71,086	0	71,086	247	0	247	0	0	0
Cabañas	152,842	2997	2368	5635	7	0	7	0	0	0
Chalatenango	196,583	1250	0	1250	4	0	4	0	0	0
Cuscatlán	202,951	38,119	106,120	144,239	43	1372	1415	20	165	185
La Libertad	682,092	143,215	0	143,215	1364	0	1364	585	0	585
La Paz	292,887	227,034	75,821	302,855	157	806	963	44	58	102
La Unión	289,021	15,062	0	15,062	8	0	8	1	0	1
Morazán	173,499	498	0	498	3	0	3	0	1	1
San Miguel	480,021	76,665	230	76,895	43	0	43	19	0	19
San Salvador	1,985,294	107,083	1370	108,453	386	0	386	24	4	28
San Vicente	161,105	92,395	66,443	158,838	53	1220	1273	29	87	116
Santa Ana	551,259	112,561	0	112,561	327	0	327	47	0	47
Sonsonate	450,116	101,487	0	101,487	1295	0	1295	48	0	48
Usulután	338,332	340,354	0	340,354	786	1	787	27	0	27
Total	6,275,782	1,329,806	252,622	1428	4723	3399	8122	844	315	1159

Note: Sources: OPS El Salvador and COEN, 2001; Estimated population, Dirección General de Estadística y Censos (DIGESTYC). From PAHO.

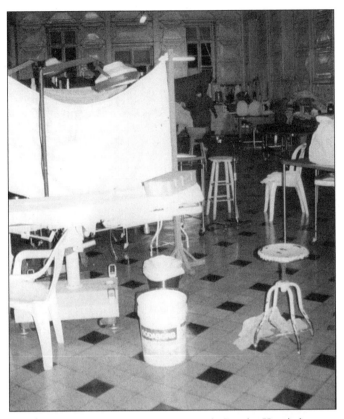

Figure 5. Temporary surgical rooms in Rosales Hospital.

Figure 6. Neonatology in outside shelter structure. 1° Mayo Hospital.

Figure 7. Typical structural system based on frame and infilled wall system.

affected. Peak ground acceleration was recorded in several areas, so attenuation curves can be developed and damage correlated with earthquake intensity. The damage observed is attributed to an absence of seismic design of the nonstructural elements and an inappropriate structural system, with excessive flexibility, for a hospital. Damage can be divided and described as follows:

1. Severe damage and collapse of suspended ceilings and lighting fixtures (Fig. 8) due to absence of restraints, inadequate use of railings and supporting systems, and lack of structural expansion joints.

2. Severe damage and collapse of infilled walls and facades (Fig. 9) due to lack of expansion joints, inadequate expansion joints, and inadequate out-of-plane support.

3. Severe damage to elevator systems due to complete absence of seismic design for elevator systems and damage to elevator shafts (Fig. 10).

4. Severe damage to piping and ductwork due to lack of restraints, no consideration for structural expansion joints, and poor maintenance, shown previously in Figure 8.

5. Severe damage to medical equipment and supplies due to inadequate furniture and supports (Fig. 11).

6. Structure pounding between buildings due to inadequate separation or maintenance of expansion joints.

This nonstructural damage caused total or partial evacuation and lost of functionality in the main hospitals at San Miguel, Usulután, Zacatecoluca, La Libertad, and San Salvador.

Figure 8. Damage to suspended ceiling, lighting fixtures, and air conditioning. Fertility building in Maternity Hospital.

Figure 9. Damage to partition and facades due to inadequate isolation. San Rafael Hospital.

The extensive damage to hospitals in the coastal and central area and the damage to the road system due to landslides considerably limited the possibility of transferring critical patients from local hospitals to San Salvador or to undamaged units. For example, the Santa Ana hospital suffered no damage and had a high resolution capacity, but due to its location and a landslide on the main route, was not able to adequately respond to health demands.

Immediately after the event, the Health Ministry and the Salvadorian Social Security Institute organized evaluation groups to survey damage to the hospitals and to the central offices. Additionally, the Salvadorian Association of Engineers and Architects

Figure 10. Counterweight derailment in elevator system. Usulután Hospital.

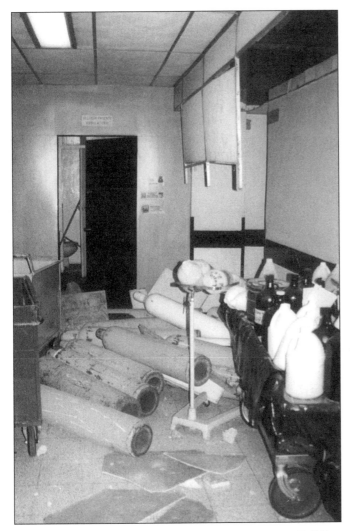

Figure 11. Medical equipment and supply damage. Surgery Building, Rosales Hospital.

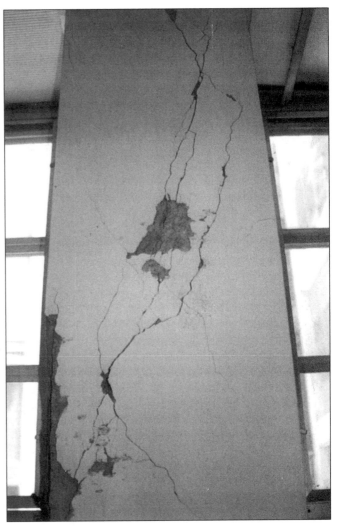

Figure 12. Shear failure in column due the presence of partial-height infilled walls. Zacatecoluca Hospital.

(ASIA), as well as international institutions such as the Pan-American Health Organization through the Collaborating Centre for Disaster Mitigation in Health Facilities, carried out evaluations. These evaluations allowed the establishment of safety areas for immediate use and a long-term recovery strategy.

FEBRUARY 13 EARTHQUAKE

On February 13, the second strong earthquake occurred. This event increased the fear of the population and caused damages in the Zacatecoluca, Santa Gertrudis of San Vicente, and Cojutepeque hospitals (Tables 1 and 2; Fig. 12).

Because this event had a shallow focus, the affected zone was very limited, and damage was not as widespread as the January earthquake. This allowed a relatively fast health response. Table 3 presents the casualties. As happened during the January event, landslides in the roads occurred, causing difficulties with transporting patients, especially between the San Vicente and Cojutepeque hospitals and San Salvador.

EVACUATION AND TEMPORARY HOSPITALS

The events caused an immediate evacuation of most of the main hospitals in the affected region, producing a negative impact on the functionality of the healthcare system. Many evacuations after the earthquakes were made hastily, without any technical criteria. This behavior can be associated with:

1. The employees' perception of a possible structural collapse, considering the experience of the 1986 earthquake. During the 1986 event, a significant number of deaths occurred due to the collapse of structures in San Salvador. At that time, the health sector had complete functional loss of the General Hospital (Salvadorian Social Security Institute); partial collapse of the Benjamin Bloom Hospital in an aftershock (100% evacuated); damage to

the Maternity Hospital (Health Ministry) in the surgery and laundry rooms (100% evacuated); and damage to the Rosales Hospital surgical rooms. In addition, the San Rafael (Health Ministry), Maternity 1° de Mayo (Salvadorian Social Security Institute), and the Military Hospital of San Salvador were evacuated.

2. The lack of contingency plans that clearly established the actions to be taken during an emergency, especially procedures on how to decide whether or not to evacuate.

3. The absolute absence of seismic design along with poor maintenance for nonstructural elements: ceiling, walls, facades, windows, plasters, lighting fixtures, medical and industrial equipment, furniture, and lifelines. Moderate and severe nonstructural damage caused chaos and the partial loss of the functional capacity, affecting the staff and response capacities. It is not possible for a health employee to establish the difference between nonstructural and structural damage and it is even less likely for a health employee to establish if it is a life-threatening situation.

4. The use of structural systems with little ability to tolerate damage. As presented before, the main Health Ministry hospitals (Rosales and Santa Ana) are more than 80 years old; nevertheless, their older sections suffered no structural damage. Newer structures built after the 1960s are mainly based on moment resisting frames with relatively large floor flexibility and low ductility capacity.

These factors created a great insecurity in the hospitals' staff and made the decision to evacuate easier without knowing the real condition of the facilities and without considering the consequences on the health care capacity after the evacuation.

It is possible to conclude that the healthcare system of El Salvador was seriously affected by the January and February earthquakes due to the low seismic and functional safety of its centers (infrastructure and organization), the high concentration of its specialized services, the high dependence on highways to fulfill the specialties' treatments, and the limited organization for disasters considering its vulnerabilities. According to Ruales and Ayala (2002) the total loss in the health sector was U.S.$200 million.

Due to the evacuation of the most important hospitals, health care was performed outdoors. The operation required a large number of tents and temporary housing that was organized in the vicinity of existing structures (Figs. 5, 6, and 13).

A high percentage of hospitals remained evacuated for more than two years, so several field hospitals were installed in Santa Gertrudis, San Pedro de Usulután, Maternidad, San Rafael, Zacatecoluca, and San Juan de Dios de San Miguel. In some cases, existing room functions were reassigned, as in 1° Mayo, Neumológico, Rosales, and Cojutepeque. Others, such as Oncología, stopped functioning and had no alternatives.

Some hospitals had evacuated temporarily and worked outdoors. Later they returned to undamaged or partially damaged structures, but the tents were held in reserve in case of other earthquakes. In some cases, semipermanent shelters had even been constructed in case another evacuation was considered necessary.

A year after the earthquake, 28% of the total existing beds remained in field hospitals; the reasons, among others, are the following:

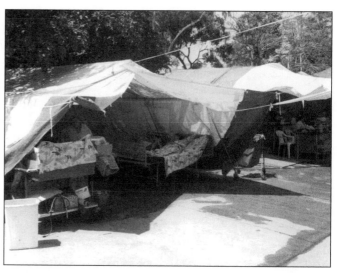

Figure 13. Temporary shelters for long-term patients.

1. Existence of structural damage or real risk of damage in at least the following hospitals: San Rafael, San Pedro, Zacatecoluca, and San Juan de Dios of San Miguel.

2. Severe nonstructural damage requiring long-term recovery procedures.

3. Absence of previous knowledge about the condition and real risk of the structures. Response decisions require information that takes a relatively long time to gather.

INFRASTRUCTURE RECOVERY PLAN

Due to the extensive damage, the reconstruction and recovery of the health system care capacity is taking a long time. The extensive loss of the infrastructure puts the health suppliers in the dilemma of either recovering themselves to the condition that existed before the seismic events of 2001, or trying to establish better conditions. These decisions have to be made considering that an important number of hospitals have an obsolete infrastructure. Therefore, the variables depend on the level of structural and nonstructural damage, the infrastructure obsolescence, the need to rebuild an urgently needed health network, the lack of redundancy in the system, the severe limitation of economic resources, and the lack of design and construction standards that allow a speedy process of reconstruction and that ensure that the seismic vulnerability that existed previously will not be reconstructed.

Initially, the Health Ministry has two recovery strategies. Under the first strategy, hospitals will be repaired to restore their present structure and service lines. These facilities are San Pedro de Usulután, Rosales, and Psiquiátrico Hospitals, and the Max Bloch and Biological Laboratories. In the second strategy, hospitals will receive a complete upgrade (structural and nonstructural). These are San Juan de Dios of San Miguel, Zacatecoluca, and San Rafael. Also, new structures will be considered for some of these hospitals.

Before the seismic events, the Health Ministry, with Pan-American Health Organization support, was developing a new scheme that decentralizes care while maintaining effective levels of response. The plan establishes two types of care: outpatient and inpatient care. For the outpatient case, three levels of care are established: Level I includes the health units, the health homes and the rural nutrition centers; Level II includes the health units with basic specialties (delivery care and basic emergencies); and Level III includes specialized institutes, central laboratories, and health units with specialties. For the inpatient case, the plan establishes Level II with peripheral hospitals and general hospitals of the department, Level III with regional referral hospitals, and Level IV with national specialized hospitals.

Regional referral hospitals will allow decentralization of healthcare, reducing the dependency on greater San Salvador facilities. The preliminary proposal recommends one hospital in the west, one in the east, one in the central coastal sector, and one in the central border region.

El Salvador's healthcare system is confronted with the huge task of recovering its service capacities. The consequences of the damages could have been foreseen by vulnerability studies of the hospitals. This is the approach that other Central American countries have decided to take. During 2001 and 2002, after Hurricane Mitch, Honduras, El Salvador, Guatemala, and Nicaragua started to evaluate their health networks considering natural disasters. Costa Rica is the most advanced country in the area in this matter because since 1985, they have performed several vulnerability studies and structurally reinforced several of their hospitals. This mitigation strategy allowed the healthcare system of Costa Rica to survive the 1991 Limon earthquake.

The main problem of health infrastructure in developing countries is that most existing hospitals were constructed under design codes that do not protect the investment or its functionality. And even in the case that this protection was initially established, the maintenance budget is very small, so this protection becomes lost with time.

In order to overcome this situation, countries such as Chile, in South America, have established and applied regulations that require a vulnerability study of the infrastructure if more than 20% of the hospital has been affected by modification or repair. For new hospitals these regulations require design procedures for the protection of the investment and functionality, depending on their complexity and role in the healthcare system network.

CONCLUSIONS

Two major earthquakes that caused a large number of human losses and injuries affected El Salvador. The country's main hospitals suffered severe damage, and after three years, the healthcare system network still experiences a substantial loss of capacity. This loss can be attributed to several causes but especially to the high concentration of health services in San Salvador, the complete dependency on vulnerable lifelines, and a hospital seismic design philosophy whose only objective is to protect the life of the user and does not consider the protection of the health infrastructure investment, its functionality, and a fast and low-cost recovery of the system after a moderate or severe event.

Damage cannot be attributed to a limited application of the codes or use of poor materials. The damage observed is associated with an inappropriate design concept, which nowadays is no longer acceptable.

ACKNOWLEDGMENTS

We would like to extend our thanks for the support for this work to the Minister of Health of El Salvador, J.F. Lopéz; and to Dr. de Nieto, Arq. De Rivera and Eng. Vargas from the Health Ministry; Eng. Baltazar Mejía from SSSI; and Dr. Ruales and Dr. Jenkings from PAHO and the Civil Engineering Department of the University of Chile.

REFERENCES CITED

Baroni, M., and Quaglia, J., 2001, República de El Salvador. Análisis Físico Funcional de los Hospitales. Propuesta de Acciones ("Republic of El Salvador. Physical and Functional Analysis of Hospitals. Proposal for Actions"), Pan-American Health Organization El Salvador Report (in Spanish).

Boroschek, R., 2001, Los terremotos del 13 de enero y 13 de febrero y el sistema de salud de El Salvador ("The January and February 13 earthquakes and the health system in El Salvador"), WHO/PAHO Collaborating Centre for Disaster Mitigation in Health Facilities, University of Chile, Report (in Spanish), 56 p.

Boroschek, R., and Retamales, R., 2001, Daños Observados en los Hospitales de la Red Asistencial de Salud de El Salvador en el Terremoto del 13 de Enero de 2001 ("Damages Observed in Hospitals of the Public Health Network of El Salvador in the January 13th, 2001 Earthquake"), WHO/PAHO Collaborating Centre for Disaster Mitigation in Health Facilities, University of Chile, Report (in Spanish), 50 p.

Health Ministry, 2000a, Gestión en Salud ("Health Management"), Newsletter no. 1, October, El Salvador (in Spanish), 8 p.

Health Ministry, 2000b, Sistemas Básicos de Salud Integral (SIBASI), Definición Geográfica y Poblaciones ("Basic Systems of Comprehensive Health (SIBASI), Geographic and Population Definition"), October (in Spanish), 83 p.

Ministry of Public Works, 1994, Norma Técnica para Diseño por Sismo ("Technical Standard for Seismic Design"), El Salvador (in Spanish).

PAHO/WHO (Pan-American Health Organization/World Health Organization), 2000, Encuesta de actividades y recurso de los hospitales públicos de El Salvador ("Survey of Activities and Resources of El Salvador Public Hospitals"), El Salvador (in Spanish).

Ruales, J., and Ayala, G., 2002, Proyecto: Preparativos para los desastres y reducción de la vulnerabilidad de sector salud en los países afectados por el Huracán Mitch ("Disaster and mitigation in the health sector in countries affected by Hurricane Mitch"), Pan-American Health Organization El Salvador Report, April 2002 (in Spanish).

Salvadorian Social Security Institute (SSSI), 2000a, Estadísticas 99 ("Statistics 99"), Unit of Actuarial Advising of Statistics Department, San Salvador, April (in Spanish), 127 p.

Salvadorian Social Security Institute (SSSI), 2000b, Red de Centros de Atención del ISSS por Niveles de Atención ("Network of Care Centers of the SSSI, sorted by Levels of Care"), October 11th, Internal Report (in Spanish).

Salvadorian Social Security Institute, 2001, Estado Preliminar de la Infraestructura Física de los Centros de Atención del ISSS ("Preliminary State of Physical Infrastructure of the SSSI Care Centers"), Infrastructure and Conservation Unit Report, February 28th (in Spanish), 13 p.

MANUSCRIPT ACCEPTED BY THE SOCIETY JUNE 16, 2003

Geological Society of America
Special Paper 375
2004

Adobe housing in El Salvador:
Earthquake performance and seismic improvement

D.M. Dowling*

Centre for Built Infrastructure Research, Faculty of Engineering,
University of Technology, P.O. Box 123 Broadway, Sydney NSW 2007, Australia

ABSTRACT

Adobe is the predominant housing material in rural El Salvador, due mostly to economic advantages and ease of construction. The high seismicity of El Salvador has repeatedly exposed the vulnerability of traditional adobe housing to the forces of earthquakes, as spectacularly demonstrated in the severe earthquakes of 2001. This paper presents the features of traditional adobe housing in El Salvador, including construction techniques and distribution, followed by a discussion of the performance of adobe buildings in recent earthquakes in El Salvador. The impact of the 2001 earthquakes is demonstrated by statistical data, which also reveal the severe housing deficit in El Salvador. Common damage patterns evident in earthquake-affected adobe buildings are detailed, with emphasis on the failure mechanisms and exacerbation features. These aspects are then linked to a presentation of improved seismic design and construction techniques; seismic retrofitting and damage repair systems for adobe structures are also considered. Finally, the obstacles to reducing damage to adobe houses are presented, and some key recommendations for adobe strengthening in El Salvador are discussed, which involve both social and technical solutions.

Keywords: adobe, housing, El Salvador, earthquake, performance, failure, resistance, mitigation, seismic improvement.

INTRODUCTION

The force of earthquakes can cause devastation and destruction to both infrastructure and lifestyle. The housing sector in developing countries is particularly vulnerable due to resource limitations and poor construction quality. Adobe (mudbrick) houses are one of the most severely affected types of building because of their extensive use, general poor quality, and inherently brittle nature. A history of severe earthquakes in El Salvador has exposed the deficiencies in traditional adobe housing, particularly in rural communities. Other devastating earthquakes in Peru, India, Afghanistan, Iran, and México in 2001 and 2002 have confirmed that the loss from earthquakes can be frequent, unpredictable, and global. Despite this key limitation, there is little doubt that adobe will continue to be the chosen construction material for a significant proportion of the population who simply cannot afford any alternative.

The capacity of an adobe house to resist earthquakes is dependent on individual adobe block characteristics, building location and design, and the quality of construction and maintenance. These factors are mutually dependent and should be considered when undertaking an adobe construction project. In recent times there has been an increased emphasis on the seismic improvement of adobe, and research has revealed various methods to improve the seismic resistance of simple adobe dwellings. Despite this ever-expanding body of technical knowledge, there are certain obstacles that are preventing the widespread application of this information. These obstacles relate to deficiencies in the promotion and support of improved adobe, as well as a shortage of skills and resources to facilitate improved construction. A combination of social and technical solutions is required to reduce the vulnerability of adobe structures, and local and international collaboration is necessary to realize these goals.

*dominic.m.dowling@uts.edu.au

Dowling, D.M., 2004, Adobe housing in El Salvador: Earthquake performance and seismic improvement, *in* Rose, W.I., Bommer, J.J., López, D.L., Carr, M.J., and Major, J.J., eds., Natural hazards in El Salvador: Boulder, Colorado, Geological Society of America Special Paper 375, p. 281–300. For permission to copy, contact editing@geosociety.org. © 2004 Geological Society of America

The author of this paper is a Ph.D. candidate who is researching appropriate means of improving the earthquake resistance of adobe dwellings. His focus is on low-cost, low-tech solutions for developing countries and includes static and dynamic testing and analysis of adobe specimens. The author has been involved in various earthquake relief, rehabilitation, research, and reconstruction projects in post-earthquake El Salvador. In early 2002 he undertook an Earthquake Engineering Research Institute–sponsored, Lessons Learned Over Time evaluation of the rehabilitation and reconstruction response to the 2001 El Salvador earthquakes, focusing on adobe housing. In the second half of 2002, the author coordinated the design and construction of an improved adobe child-care center in rural El Salvador, which was a skills-building program in collaboration with the local community, Imperial College, London, and the Catholic Agency for Overseas Development (CAFOD) UK, among others.

ADOBE IN EL SALVADOR

History

Moreira and Rosales (1998) reported that in preconquest times, the indigenous population mainly used light materials, such as palm, straw, and reeds, although there is also evidence of the use of adobe in the region. Colonial constructions generally consisted of stone foundations and thick adobe walls (up to 1.5 m thick), which were reinforced with pilasters and buttresses (external columns that strengthen a wall). Lime was often added to the mud mortar and the walls were rendered. The roofs were built with timber sections covered with tiles. The lack of confidence in the capacity of such structures to resist earthquakes, coupled with the introduction of "modern" construction materials (such as cement, corrugated iron, and brick and stone masonry, followed in recent times by precast concrete, steel, aluminum, plywood, asbestos-cement, and concrete blocks), generated changes in the usage patterns and attitudes toward traditional materials. Despite these changes, adobe is still extensively used in poor rural communities and urban areas outside San Salvador.

Distribution

In 1999, it was estimated that ~1.6 million people (26% of the population) lived in adobe homes, with 70% of these located in rural areas, 26% in urban areas outside San Salvador, and less than 4% in the San Salvador metropolitan area (DIGESTYC, 1999). Table 1 describes the common housing types in El Salvador. Figure 1 shows the housing distribution in El Salvador, considering the whole country, urban areas, and rural areas.

Traditional Adobe Construction in El Salvador

The following generalized aspects relate to the traditional form of adobe construction in El Salvador. Some of the more serious limitations of the current practice are presented, although

TABLE 1. COMMON HOUSING TYPES IN EL SALVADOR

Material	Description
Concrete	For example, concrete block, concrete panel walls, reinforced concrete, etc.
Mixto	System of confined masonry, consisting of lightly reinforced beams and columns with infill brick masonry.
Adobe	Sun-dried mudbrick.
Bahareque	Matrix of vertical and horizontal timber or cane elements confining mud or stones, also known as "wattle-and-daub."
Timber	Any type of timber construction.
Lamina	Metal lamina sheeting, typically corrugated, which is supported by a timber frame.
Other	For example, plastic sheeting, palm fronds, cardboard, "waste" materials.

it should be noted that these shortcomings are not representative of all adobe houses. Some houses feature many deficiencies, whereas others are well built and maintained. Later sections detail practical methods of overcoming some of these limitations.

Land ownership is a key factor influencing the current housing situation in El Salvador. Wisner (2001) suggests that the roots of disaster vulnerability lie in the imbalance of land ownership and the consequent violent struggles that have become a feature of El Salvador's history. The process of land redistribution, a key part of the 1992 Peace Accords, has been bureaucratic and slow, and as such, a large proportion of families do not possess official land titles. This has created uncertainty, with a general reluctance to commit effort and funds into the adequate construction of a house, when the legal title to the land is in question.

Site selection is linked with land ownership and rarely takes into account local hazards. Most families simply utilize land that is available, and consequently, houses are often built in high-risk areas, which may be subject to flooding, volcanic activity, and slope or soil instability.

Labor is usually family or community based and is normally directed by the family head or someone with experience in construction. Formal training is rarely available, and techniques (both good and bad) are passed on informally.

Adobe blocks are generally made with the locally available soil and may contain unsuitable proportions of sand, silt, and clay, as well as undesirable organic material, large particles, and foreign matter. The mud is often inadequately mixed and placed into molds that are poorly constructed, and the blocks are commonly cured in direct sunlight or exposed to rain. These factors often result in severe cracking, erosion, or deformation of the blocks.

Construction. Crude foundations consist of stones held by a weak cement or mud mortar without reinforcement. Foundations are often shallow and are not raised above the natural ground level, such that the first course of blocks is susceptible to water ingress from the ground and excess rainwater. A moisture-proof course or layer is generally not used. Often, the floor consists of

Total country (urban + rural): 1,383,145 houses

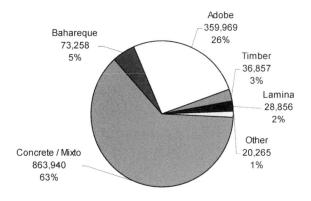

Urban: 860,082 houses

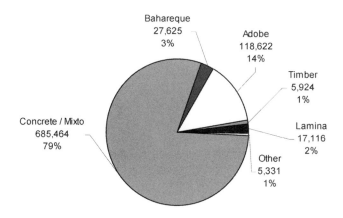

Rural: 523,063 houses

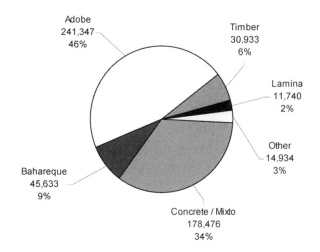

Figure 1. Housing distribution in El Salvador (DIGESTYC, 1999).

bare earth, which has been compacted and smoothed over time. In other cases, a thin (10–30 mm) concrete layer is applied or tiles are laid. Walls are generally thin (150–250 mm), and mortar joints tend to be thick (>30 mm) and in some cases are thicker than the blocks themselves (Fig. 2). Common features also include unlevel horizontal courses, insufficient horizontal block overlap, and vertical joints running several courses high. Foreign material, such as ceramic, metal, timber, and glass is frequently embedded in the walls, which are often out of plumb. Ring beams (also known as collar-, bond-, or tie-beams), pilasters/buttresses, and reinforcement, either vertical or horizontal, are hardly ever used. The walls may be unprotected or rendered with cement, lime, or mud mortar. The majority of houses are one story high.

The roof structure is normally made from heavy rough-hewn timbers that support a heavy tile covering (Fig. 2). The structure generally rests directly upon the walls, with little or no attachment. The tiles are not normally attached to the frame, or to each other.

Other houses consist of crude timber columns supporting the roof structure, with infill adobe walls (Fig. 3). The walls are seldom attached to the columns and behave as unrestrained cantilever walls, which are highly susceptible to collapse during seismic events. The advantage of this system is that the roof is less prone to collapse if the columns and roof frame have sufficient strength.

Many buildings have been built in stages or repaired using different materials and techniques. Such construction generally lacks continuity, homogeneity, and uniformity. During an earthquake, the response of each component is different, which often causes damage to the adjoining components due to pounding.

Maintenance of adobe buildings is generally poorly done. Many of the buildings that failed during the earthquakes of early 2001 were older buildings that were poorly maintained. The attack of natural agents, such as water, wind, animals, insects, and plants, is often left unattended and reduces the structural integrity of the building (Fig. 4).

Organization and Training

Prior to the earthquakes of 2001, the focus on adobe construction and training programs was limited. Notable exceptions include the activities of a local NGO (nongovernmental organization), Fundacion Salvadoreña de Desarrollo y Vivienda Minima (FUNDASAL), which has been involved in community adobe construction projects since 1979, and the Universidad Centroamericana José Simeón Cañas (UCA), which conducted some training programs with support from institutions from Germany, France, and Peru. Other small-scale initiatives include projects by Alain Hays and Unidad Ecológica Salvadoreña (UNES), among others. The performance of these buildings is discussed later.

Since the earthquakes of 2001, there has been a renewed and concentrated focus on construction projects, training programs, and publications aimed at promoting the principles of improved seismic resistance of adobe buildings. Encouragingly, some over-

Figure 2. Typical adobe house, La Paz, El Salvador, 2002.

Figure 3. Timber columns with unrestrained adobe walls, Usulután, El Salvador, 2002.

seas aid has been provided to support these endeavors (although not on the scale of projects using "modern" materials, such as concrete block). Despite such efforts, the shear magnitude and wide dispersion of need means that many people have been left unsupported and are either living in temporary shelters or rebuilding in the traditional manner because they are unaware of simple techniques that can be adopted to improve seismic resistance.

Throughout this paper, reference is made to the adobe supplement of the El Salvador building code, RESESCO (*Reglamento para la Seguridad Estructural de las Construcciones*). The code itself was published in 1994, and the 1997 supplement provides a set of recommendations for adobe construction. Because of the widespread and unregulated manner of adobe construction

in El Salvador, it is understandable that the adobe supplement does not form part of the legal building code of El Salvador.

PERFORMANCE OF ADOBE BUILDINGS IN EARTHQUAKES IN EL SALVADOR

2001

Various estimates as to the extent of damage have been produced by the United Nations (UN) and different departments within the Government of El Salvador (GOES). For this paper, data from the *Dirección General de Estadísticas y Censos* (DIGESTYC, 1999, 2001), which is part of the GOES Ministry of the Economy, have been adopted because a correlation can be drawn between official pre-earthquake housing statistics (1999) and official post-earthquake damage statistics (2001) to provide an overall picture of the extent of damage relative to preexisting housing stocks. Analysis of these data sets reveals the following features:

- 20% (276,594) of all houses in El Salvador were affected, consisting of 166,529 (12%) uninhabitable (destroyed) and 110,065 (8%) repairable (damaged).
- 37% of the affected houses were in urban areas and 63% were in rural areas.
- Of the affected houses, 57% (157,070) were made of adobe; of the houses destroyed, 68% (113,469) were made of adobe; and of the houses damaged, 40% (43,601) were made of adobe.
- 8.3% (71,444) of all houses constructed of "modern" materials (concrete and *mixto*) were affected, consisting of 20,432 (2.4%) destroyed and 51,012 (5.9%) damaged.
- 39.5% (205,150) of all houses constructed of "low-cost" materials (adobe, *bahareque*, lamina, and discarded materials) were affected, consisting of 146,097 (28.1%) destroyed and 59,053 (11.4%) damaged.
- 44% (157,070) of all houses constructed of adobe were affected, consisting of 113,469 (32%) destroyed and 43,601 (12%) damaged. The damage percentages for houses of *bahareque* were similar (44% affected: 34% destroyed and 10% damaged), although prior to the earthquakes only 5% of houses were made of this material.

Based on these statistical features, combined with field observations, the following comments can be made:

1. Houses constructed of "low-cost" materials were more vulnerable than those constructed of "modern" materials.

2. Adobe houses were more prone to complete destruction, rather than repairable damage (Fig. 5). This suggests the high susceptibility of traditional adobe houses to sudden and catastrophic failure, thus increasing the risk of fatalities, injury, and loss of possessions. This tendency toward destruction also meant that there was a disproportionately smaller number of damaged adobe houses available for assessment of damage patterns and failure modes.

3. Contrary to widespread local opinion, houses constructed of "modern" materials (concrete and *mixto*) were not "immune" to damage, reaffirming that the type of building material is not the only factor that influences seismic performance.

4. Damage in rural areas was more extensive, reflecting greater use of traditional/low-cost materials, reduced access to resources (materials, tools, technical support), and consequent lower standards of quality.

It should be noted that these statistics cover the whole country and that some departments, such as Morazán, La Unión, and Chalatenango, were practically unaffected, while others, notably Cuscatlán, La Paz, San Vicente, and Usulután were severely affected. In the municipality of San Agustín, Usulután, it was

Figure 4. Water-damaged adobe wall (ruler for scale, 30 cm).

Figure 5. Catastrophic failure of adobe house, Cuscatlán, El Salvador, 2001 (Menjivar, EERI, 2001).

reported that 93.5% of all houses were affected in some way, with 83% of all houses destroyed (DIGESTYC, 2001).

Even prior to the destructive earthquakes of 2001, there was a severe housing deficit in El Salvador. The government housing ministry, the *Vice-Ministerio de Viviendas y Desarrollo Urbano* (VMVDU, 2001), suggested that in 1999 this deficit totaled 554,324 houses and was divided into qualitative and quantitative categories. The quantitative deficit accounts for homes that are deficient in all six of the basic housing requirements (safe/secure walls, safe/secure roof, hygienic floor, adequate sanitation facilities, access to potable water, and access to electricity). In 1999, the quantitative deficit was 42,817. The qualitative deficit accounts for homes that are deficient in up to five of the six basic housing features and was 511,507 in 1999.

VMVDU data suggest that the total post-earthquake housing deficit was 571,250, consisting of a quantitative deficit of 208,136 and a qualitative deficit of 363,114. This implies an overall deficit increase of only 3.1%, suggesting that "adequate" housing was largely unaffected by the earthquakes. As expected, the increase in the quantitative deficit was predominantly due to previously classed qualitative-deficit houses being redesignated as quantitative-deficit houses.

In the aftermath of the earthquakes, many families utilized both a "day house" and a "night house." The "day house" typically consists of a damaged adobe or *bahareque* house that has been partially repaired and is used for daytime activities, such as cooking, relaxing, and meeting (Fig. 6). The "night house" is generally made of lamina sheeting that has been constructed as temporary accommodation by the Government of El Salvador (GOES), aid agencies, or the residents themselves (Fig. 7). These shelters have been dubbed *hornos* (ovens) by the local folk, due to the oppressive internal buildup of heat during the day, and as such, they are only used for sleeping at night. The families are distrustful of the "day houses," due to the risk of failure in an earthquake, with a greater risk of injury if the occupants are sleeping within.

It is expected that many of the houses that were affected by the 2001 earthquakes were structurally weakened by the 1986 earthquake, as well as other seismic activity in recent years. Bommer et al. (2002) noted the occurrence of a significant seismic swarm in 1999 in the San Vicente area and suggested it was very likely that the damage levels experienced in the area in 2001 were "exacerbated by the damage inflicted during the 1999 swarm."

1986

The M_w 5.7 earthquake of October 10, 1986, caused severe damage in the San Salvador area. Considerable damage to non-engineered buildings occurred but was largely unreported, due to the more dramatic impacts to engineered buildings, many of which suffered total collapse.

The United Nations Economic Commission for Latin America and the Caribbean (ECLA/CEPAL) estimated that 22,800 dwellings were either totally destroyed or needed to be demolished, and a further 29,800 homes required repair. It was estimated that the earthquake added 40,000 families to the homeless population (Durkin and Hopkins, 1987). It is expected that nonengineered buildings, including adobe, made up a large proportion of these affected houses.

A 1986 reconnaissance team from the Earthquake Engineering Research Institute (EERI), in reference to low-cost housing, stated: "Improving the connection between the rafters and walls, attaching the tiles to the rafters, and ensuring that the roof can function as a diaphragm to transfer lateral wall reactions to shear walls/cross walls are important considerations when using this type of construction" (Anderson, 1987). The same recommendations are being reiterated in the aftermath of the earthquakes of 2001, which indicates the presence of limita-

Figure 6. Damaged adobe "day house," Usulután, El Salvador, 2002.

Figure 7. "Night house" temporary shelter, Usulután, El Salvador, 2002.

tions in the transfer of relevant information and/or the capacity of the rural housing sector to take appropriate action. These limitations and various action strategies are discussed later in this paper.

DAMAGE PATTERNS AND FAILURE MODES

This section details the structural response and resultant damage patterns that are commonly observed in traditional adobe housing subjected to earthquakes. A brief description of the damage pattern is given, followed by an analysis of the failure mechanisms that cause the damage and a description of the exacerbation features that increase the structural response and stresses that generate failure.

Vertical Cracking at Corners

Failure Mechanism

Large relative displacement between orthogonal walls. Shear walls (subject to in-plane forces) have a low response (displacement). Transverse walls (subject to out-of-plane forces) have little resistance to bending and overturning action, resulting in a larger response. The large relative displacement between shear and transverse walls induces stresses at the connection of the walls (highest stresses at the top). This type of failure is very common because relative response is largest at the wall-wall interface. Cracking occurs when the material strength is exceeded in either shear (Fig. 8) or tearing (Fig. 9). Oblique seismic forces will induce a combination of both shear and tearing stresses (Fig. 10) Vertical corner cracking may lead to the post-failure overturning of the wall panel, as detailed below.

Exacerbation Features

1. Large roof mass, which is transferred to the walls. The larger mass generates larger inertial forces, which in turn cause larger response (displacement).

2. Poor block arrangement (block-mortar interfaces are planes of weakness whose seismic resistance is lowered by inadequate overlap, thick mortar joints, and the presence of vertical joints).

Figure 10. Typical damage patterns: vertical cracking, corner dislocation, spalling of render, San Vicente, El Salvador, 2002.

3. Thin walls, which have a lower relative area of resistance.

4. Long walls, which attract greater out-of-plane response about the vertical axis due to bending, which induces a splitting-crushing cycle at the corners, thus reducing the shear area.

Vertical Cracking and Overturning of Upper Part of Wall Panel

Failure Mechanism

Out-of-plane seismic forces inducing bending about the vertical axis (dominantly) and the horizontal axis. Bending about the vertical axis causes a splitting-crushing cycle generating vertical cracks in the upper part of the wall (Fig. 10). These vertical cracks reduce the resistance to bending about the horizontal axis in the damaged panel, which may result in overturning (Figs. 11 and 12).

Exacerbation Features

1. Poor roof anchoring, where roof beams and trusses often rest directly on the wall, creating zones of high stress

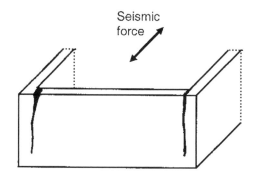

Figure 8. Vertical corner cracking due to shear failure.

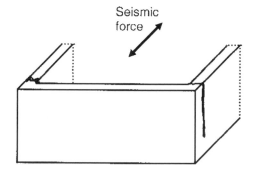

Figure 9. Vertical corner cracking due to tearing failure.

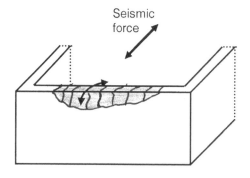

Figure 11. Vertical cracking and overturning of upper part of wall.

Figure 12. Typical damage patterns: inclined cracking, overturning of wall panels, vertical corner cracking, displacement of roof tiles, Cuscatlán, El Salvador, 2001 (Menjivar, EERI, 2001).

concentration (exacerbated by large roof mass). Both localized shear (dominant) and localized bending stresses are generated in these zones.

2. Poor block arrangement (inadequate overlap, thick mortar joints, vertical joints).

3. Long, thin, and slender walls, which are more susceptible to bending about the vertical and horizontal axes.

Overturning of Wall Panel

Failure Mechanism

Out-of-plane seismic force acting on a wall panel with lack of edge restraint on three sides (wall-wall corner connection, wall-roof connection). The lack of fixity of the wall-wall connections may be due to vertical cracking at the corners (Figs. 8 and 9) or the presence of timber columns at the corners that are inadequately attached to the walls (Fig. 3). In these cases, the wall-foundation interface behaves as a pin connection, which has little resistance to overturning when an out-of-plane force is applied (Fig. 13). This type of failure is particularly common for long walls without intermediate lateral restraint, such as boundary walls and garden walls. For buildings, the wall panel will generally overturn outward, due to the outward force exerted by the roof in the absence of an adequate roof diaphragm. This type of failure frequently results in the total collapse of the building, as commonly observed in El Salvador in 2001.

Exacerbation Features

1. Same features as for vertical cracking.

2. Poor conditions at the base of the wall, due to moisture damage and/or poor mortar-block bonding, which have little rotational restraint.

Inclined Cracking in Walls

Failure Mechanism

1. Out-of-plane deformation due to bending, causing "bulging," which generates "X" pattern cracking (Fig. 14). Adequate three- or four-sided restraint is required for this condition.

2. Very large in-plane shear forces, which generate tensile stresses at ~45 degrees, thus causing single-direction inclined cracking, or "X" pattern cracking due to cyclic loading (Fig. 15).

Exacerbation Features

1. Poor block arrangement.

2. Thin walls, which have a smaller shear area and are more susceptible to bending about the vertical axis.

3. Slender walls (large height-thickness ratio), which are more susceptible to bending about the horizontal axis.

4. Long walls, which attract greater out-of-plane response due to bending.

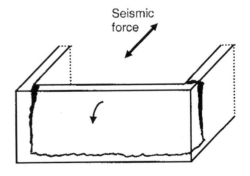

Figure 13. Overturning of wall panel.

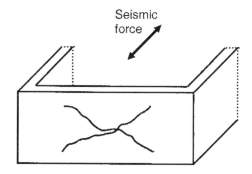

Figure 14. Inclined cracking in wall due to "bulging."

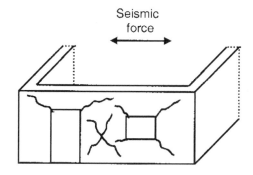

Figure 15. Inclined cracking in wall due to in-plane shear.

5. Openings (doors and windows), which induce high stress concentrations, as well as reducing the effective area of the walls. The wall panel between openings acts as a slender column and as such is subject to greater shear stresses, as well as greater compressive stresses due to the weight of the wall above. The influence of openings is further exacerbated when they are located close to other zones of high stress concentration, such as corners and other openings (Figs. 12 and 15).

Dislocation of Corner

Failure Mechanism

Initial failure is due to vertical corner cracking induced by shear or tearing stresses, as described above. The lack of fixity at the corners allows greater out-of-plane displacement of the wall panels, which generates a pounding impact with the orthogonal wall. The top of the wall has a greater response, which causes a greater pounding impact, thus inducing greater stresses that lead to failure (Figs. 10 and 16). An oblique seismic force will cause both orthogonal walls to respond, and the pounding impact will be greater.

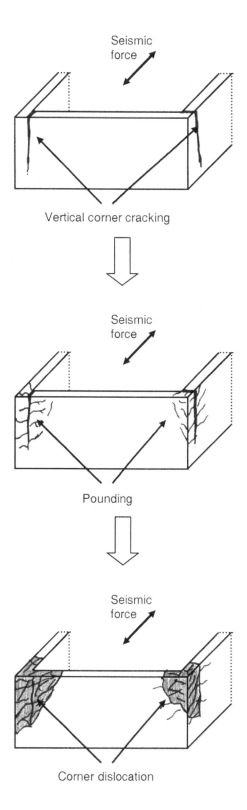

Figure 16. Sequence leading to corner dislocation.

Exacerbation Features

1. Same features as for vertical corner cracking.

2. High stress concentrations due to poor roof anchoring and lack of uniform distribution of roof loads.

Horizontal Cracking in Upper Section of Wall Panel

Failure Mechanism

Out-of-plane or in-plane shear failure, which generally occurs when there is a ring beam or the roof is securely attached to the top of the wall. A seismic force creates a relative difference in lateral movement between the wall and roof, which induces high shear stresses leading to shear cracking when the shear resistance is exceeded (Fig. 17).

Exacerbation Features

1. Lack of adequate fixity of ring beam to wall.

2. Poorly anchored roof structure, which creates zones of high stress concentration.

3. Poor mortar-block bonding, creating a horizontal plane of weakness. The resistance to shear due to friction is greatly reduced at the top of the wall because there is less applied weight force.

Displacement of Roof Structure

Failure Mechanism

Relative displacement between different components of the roof structure and the walls. The worst-case scenario of this type of failure is the total collapse of the roof structure.

Exacerbation Features

1. Inadequate three-dimensional roof diaphragm, with a lack of triangulation and diagonal bracing.

2. Poor connection to walls and between different roof components. In many cases, the only connection between roof components is provided by nails, which only provide limited resistance to shear.

3. Heavy roof mass, which attracts greater seismic force due to inertia.

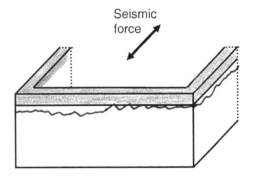

Figure 17. Horizontal cracking in upper section of wall panel.

Falling Roof Tiles

Failure Mechanism

1. No attachment between tiles and roof structure. The roof tiles are simply held by friction, which is easily overcome during seismic events, resulting in the roof tiles sliding off the roof (Fig. 12).

2. The failure of purlins, which support the tiles, resulting in tiles falling inside the building.

Exacerbation Features

1. Deteriorated purlins, which easily fail when an additional force is applied.

Linking Simplified Analysis to Improved Design and Construction

It should be stressed that in the above assessments the seismic force is generally considered as acting in a single plane. This should not alter an appreciation that during a seismic event any given wall in a particular structure will act as both a transverse wall and a shear wall, being subjected to components of both in-plane and out-of-plane forces (in alternately opposite directions). A wall will influence and be influenced by other structural members and experience bending, tensile, compressive, shear, and tearing stresses. The resulting loads and stress conditions are a complex blend of these factors, and combinations of the above damage patterns are common. Despite this limitation, assessment of simplified scenarios allows a greater understanding of the performance and response of structural elements, which supports the development of concepts of improved seismic design and construction.

IMPROVED SEISMIC DESIGN AND CONSTRUCTION OF ADOBE BUILDINGS

Design Concepts

There are many design concepts that can greatly improve the seismic resistance of a building. Some of these design concepts are simple and inexpensive, while others are more complex and resource-dependent. The design concepts discussed below have been divided into separate structural components and configurations for ease and clarity. This should not alter the appreciation that a structure will behave as an integrated unit under earthquake loading. In order to take full advantage of this integration, the structure must be securely tied together, thus promoting redundancy. Furthermore, a focus on improved seismic design should not diminish the importance of other design factors, such as design for wind loading.

It should be remembered that earthquakes are unpredictable, variable, and dynamic in nature. The recommendations below do not guarantee that an improved building will be unaffected by an earthquake. It is hoped that the use of these suggestions will, at the very least, reduce the loss of life during seismic events, and hopefully minimize the damage and destruction to common housing.

Site Conditions

The main consideration when choosing a site for construction is the stability of the ground. This takes the form of slope stability and soil stability. Sloping ground is highly susceptible to slipping or sliding during earthquakes, and it is recommended that buildings not be built on or near excessively sloping ground.

Certain soils exhibit unstable characteristics when shaken or wetted, as occurs during an earthquake. Settlement, compaction, loss of shear resistance, and liquefaction are some of the unstable results of ground movement and inundation. The soils that are most vulnerable to such instability are fill materials, very loose sands, and volcanic ash deposits. The predominance of the volcanic ash *tierra blanca* in El Salvador has been linked to the large number of landslides that occurred during the 2001 earthquakes (Bommer et al., 2002).

The ideal site is flat, firm, and dry, with good drainage. For the majority of the rural poor, however, access to such ideal sites is limited because land is expensive and terrain is variable. This is particularly the case in El Salvador, where "the roots of the civil war were in land ownership conflicts" (J. Bommer, 2003, personal commun.). Various remediation measures, such as leveling, drainage, stabilization, and revegetation, are available, which may improve site conditions.

Foundations

Foundations are key elements of a building, which serve to evenly distribute the wall load on the ground, thus minimizing zones of high stress concentration. If the foundation fails, then the superstructure will almost inevitably fail also. It is recommended that some form of continuous horizontal reinforcement (e.g., steel bar, bamboo) be included in the foundations, which will provide resistance to the bending, tensile, and shear stresses that are generated during a seismic event. Obviously, the addition of reinforcement will increase the cost and complexity of the construction, although it is suggested that even simple reinforcement will provide significant benefits. Figure 18 shows the foundation configuration as outlined in the RESESCO (1997) adobe supplement.

The plinth is a concrete "top foundation," which elevates the adobe wall above the ground, thus reducing potential moisture damage.

Adobe Soil

Various investigations and regulations recommend that a certain mix of the components of the soil (sand, silt, and clay) will improve the structural integrity of the adobe blocks, although there is no universal agreement as to the most appropriate mix. The volcanic ash *tierra blanca* is widely used in adobe block fabrication in

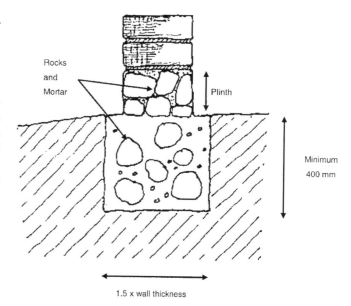

Figure 18. Foundation configuration (RESESCO, 1997).

El Salvador, and it has been suggested that it possesses pozzolanic properties (P. Meyer, 2002, personal commun.). Table 2 is a compilation of soil mix proportions recommended by various sources.

Despite these recommendations, it should be noted that the average rural homebuilder will generally not have access to various soil types or to accurate means of testing the soil. The most effective manner of assessing the suitability of the soil is to make several trial bricks (using combinations of the available soils) and cure these under different conditions (direct sun, shade, with additives, etc.). It is suggested that at the end of the curing period blocks should "be able to be handled without crumbling or being easily damaged; and not have developed any crack longer than 75 mm and wider than 3 mm or deeper, irrespective of length or width, than 10 mm" (Middleton, 1987).

Adobe Block Dimensions

There is no firm agreement relating to the ideal dimensions of adobe blocks, although several factors should be considered,

TABLE 2. SOIL MIX PROPORTIONS FOR ADOBE BLOCKS

Material	RESESCO (1997)	Reglamento Nacional, Peru (2000)	Pérez (2001)	Equipo Maiz (2001)
Sand	55–75%	55–70%	50–70%	40%
Silt	10–28%	15–25%	5–20%	40%
Clay	15–18%	10–20%	15–30%	20%

including wall thickness, horizontal overlapping, block weight, and provision for reinforcement. In general, a larger block is recommended, although this will be limited by a manageable block weight. A 300 × 300 × 100 mm block will weigh ~15 kg. Larger blocks will be more difficult to handle.

Wall Dimensions

General recommendations for acceptable wall dimensions include:

1. Long walls without intermediate support should be avoided. The RESESCO (1997) adobe supplement recommends that the length of unsupported wall should be less than ten (10) times the thickness of the wall. IAEE (1986) recommends that the length of unsupported wall "should not be greater than 10 times the wall thickness t or greater than $64t^2/h$ where h is the height of the wall." If longer walls are desired, they should be restrained by intermediate vertical walls or pilasters, as shown in Figure 19.

2. The height of a wall should be less than eight (8) times its thickness (IAEE, 1986; RESESCO, 1997; Minke, 2000).

3. Thicker walls are more resistant to seismic loading, although they attract greater seismic force. Thicker walls also require more materials and greater labor energy to construct.

The above suggestions relating to suitable wall dimensions are assumed to refer to traditional, unreinforced adobe walls, which do not employ improved seismic resistance systems (e.g., reinforcement, ring beam, pilasters, roof diaphragm, etc.). The provision of improvement systems is expected to alter these recommendations, although further research is required to ascertain the degree of such changes.

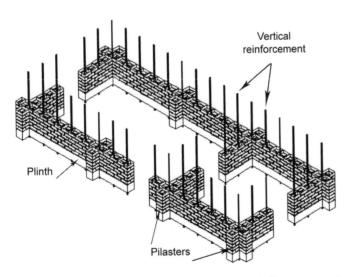

Figure 19. Sample placement of plinth, vertical reinforcement, and pilasters.

Mortar Joints

Mortar-block interfaces are planes of weakness in an adobe wall. Effective mortar bonding between blocks increases the shear resistance and can be achieved by:

1. Selecting a suitable mortar (ideally the same as the soil used for the bricks). Mortars with high clay contents are subject to excessive shrinkage, which causes mortar cracking.

2. Roughening the bonding surfaces of blocks prior to laying (increases the friction coefficient of the blocks).

3. Wetting the blocks prior to laying (promotes better adhesion of mortar).

4. Maintaining a mortar thickness of between 15 and 25 mm (RESESCO, 1997).

There should be sufficient horizontal overlap of bricks and continuous vertical joints should be avoided, as these provide very little resistance to shear and bending stresses. This is particularly important at the intersection of orthogonal walls, where high stress concentrations exist.

Reinforcement

Horizontal and vertical reinforcement may take the form of any ductile material, including bamboo, reeds, cane, vines, burlap, rope, timber, chicken wire mesh, welded mesh, barbed wire, and steel bars. A local cane, *vara de castilla* (*Gynerium sagittatum*, Hays, 2001) is widely used in the construction of *bahareque* houses and has been used as internal vertical and horizontal reinforcement in improved adobe construction projects in El Salvador. The durability, availability, and expense of each material should be considered when selecting the most effective material within resource constraints. As described above, a complex array of responses and stresses are generated in a structure subject to seismic forces, and the specific location and structural properties of the reinforcement should be carefully considered. The shear, tensile, and bending capacity of the mortar-reinforcement interface is an important factor in the structural performance of the system, and this aspect requires greater research. All reinforcement should be securely tied together and to the other structural elements (foundations, ring beam, roof). This attachment provides a stable matrix, which is inherently stronger than the individual components.

Table 3 provides a generalized assessment of the benefits of key improvement systems (ring beam, reinforcement, and pilasters) in reducing the common damage in adobe housing, as described above. This assessment assumes ideal integrity of the systems, particularly the connections.

Internal (within the wall) vertical reinforcement transfers lateral forces to the ring beam and the foundation beam, thus restraining in-plane and out-of-plane shear (in a doweling action), as well as providing restraint to out-of-plane response due to bending. Internal vertical reinforcement provides a moderate to high contribution in the mitigation of vertical corner cracking, overturning of wall panels, inclined cracking, corner dislocation,

TABLE 3. CONTRIBUTIONS OF IMPROVEMENT SYSTEMS IN REDUCING COMMON DAMAGE

Damage Patterns	Improvement Systems						
	Ring Beam	Vertical Reinforcement (internal)	Horizontal Reinforcement (internal)	Vertical Reinforcement/Mesh (External)	Horizontal Reinforcement/Mesh (External)	Pilasters (intermediate)	Pilasters (corners)
Vertical corner cracking (due to shear failure) (Fig. 8)	3	2	2	1	2	2	0
Vertical corner cracking (due to tearing failure) (Fig. 9)	3	2	2	1	2	2	3
Vertical cracking and overturning of upper part of wall panel (Fig. 11)	3	2	2	1	2	3	0
Overturning of wall panel (post shear failure corner cracking) (Fig. 13)	3	2	2	1	3	3	0
Overturning of wall panel (post tearing failure corner cracking) (Fig. 13)	3	2	2	1	3	3	2
Inclined cracking in walls (due to out-of-plane "bulging") (Fig. 14)	0	1	1	2	2	2	0
Inclined cracking in walls (due to in-plane shear) (Fig. 15)	1	3	3	2	2	2	0
Dislocation of corner (post failure) (Fig. 16)	3	2	3	2	2	1	3
Horizontal cracking in upper section of wall panel (Fig. 17)	0	3	0	2	0	1	0

Note: Contribution Key: 0—Insignificant contribution; 1—Low contribution; 2—Moderate contribution; 3—High contribution

and horizontal cracking (Table 3). Internal vertical reinforcement is protected from exposure to the environment, but is difficult to place and align, as well as requiring blocks with voids that need to be specifically made and positioned (Fig. 19).

Horizontal reinforcement helps to transmit the bending and inertia forces in transverse walls to the orthogonal shear walls, as well as increasing the shear resistance capacity at the wall-wall interface and minimizing vertical crack propagation. In the case of out-of-plane shear forces causing corner cracking (Fig. 8), the shear stress concentration is generally largest at the mid-thickness of the wall (and negligible at the face), so internal reinforcement is more effective at minimizing crack formation than external reinforcement. In the case of in-plane shear forces causing inclined cracking in wall panels (Fig. 15), the shear forces are uniformly distributed across the wall thickness, and therefore wide reinforcement (such as mesh or burlap) would be more effective than thin reinforcement (such as barbed wire). Table 3 shows the contribution of internal horizontal reinforcement in the mitigation of common damage patterns. Internal horizontal reinforcement is easy to position along horizontal mortar joints.

External (outside the wall) mesh reinforcement provides restraint to the out-of-plane forces due to bending about both vertical and horizontal axes. A strong mesh and an effective connection to the wall are critical in achieving restraint. Connectors must have sufficient tensile, shear, and axial capacity (for both inward and outward motion). Ties attached to internal vertical reinforcement or external mesh (on the other side of the wall) would provide adequate fixity. Nails and staples are less effective connectors (due to a lack of axial resistance and minimal shear capacity due to poor interaction with adobe).

External mesh reinforcement also provides some restraint to in-plane shear forces. The horizontal component of the external mesh binds orthogonal walls, thus providing some resistance to the vertical cracking induced by shear and tensile stresses at the corners. It also provides resistance to overturning and corner dislocation after the formation of vertical cracks, as well as restraining the response due to shear and bending stresses that cause inclined cracking (Table 3). The vertical component of the external mesh serves to maintain the form of the mesh, as well as acting to restrain the shear and bending stresses that generate inclined and horizontal cracking (Table 3). External reinforcement is easy to place, but is vulnerable to deterioration due to exposure to the environment.

Pilasters

Pilasters can be positioned at critical parts of a structure to increase stability and stress resistance. A pilaster is a projection from a wall, which is interconnected by masonry bonding such that both components respond as a single unit during seismic events (Fig. 19). The term "buttress" is also often used to describe this type of element, although technically speaking, buttresses are not connected to the wall and act independently when subject to seismic forces (Roselund,

1995). Pilasters are counter supports that act to restrain out-of-plane response, and are most effective at intermediate locations in long walls. Intermediate pilasters offer resistance to the formation of common damage patterns, such as vertical cracking, overturning, and bulging (Table 3). Corner pilasters provide some resistance to the out-of-plane response that generates tearing stresses in the corners. Corner pilasters, however, do not offer restraint to the shear stresses that form vertical corner cracks because the critical section remains in the same location and has the same contact area. In this case, the pilasters may actually attract greater seismic force due to the increased mass at the corners.

Several organizations in El Salvador are building improved adobe houses with pilasters, with the depth of the pilasters varying from 160 to 420 mm. Deeper pilasters provide greater restraint but are more costly and attract greater seismic force. Further research is required to determine optimal pilaster size and ideal pilaster location.

The inclusion of pilasters requires a considerable amount of additional skill, labor, and expense in the preparation and construction of foundations, plinths, pilasters, and roof system. The inclusion of 300 mm deep pilasters on a 6 m by 8 m building with 600 mm roof overhang results in an increase of 20% in the plan area of the roof, and various elements of the roof structure must be designed and built to adequately support cantilever actions of up to one meter to provide this overhang. This represents a significant increase in resource requirements. The additional skills, labor, and expense associated with using pilasters should be balanced with the structural benefits provided, and should be the subject of further research.

Openings

Openings are locations of high stress concentration and reduce the effective area of a wall. The RESESCO (1997) adobe supplement suggests several design concepts that will reduce the risk of failure, as shown in Figure 20.

External diagonal reinforcement around the corners of openings would also provide restraint to the stresses that induce cracking from these locations of high stress concentration. This diagonal reinforcement should be firmly attached to the walls in a similar fashion to external wire mesh, as described above.

Ring Beam

A ring beam (also known as a collar-, bond-, or tie-beam) is a continuous horizontal beam that encircles a building (Fig. 21). In the case of single-story buildings, a ring beam should be located at the top of the wall to form part of both the wall and roof structures. The ring beam allows a more uniform distribution of the roof load onto the walls, thus minimizing zones of high stress concentration. The ring beam also provides fixity for vertical reinforcement and transfers forces from vulnerable transverse walls to stiffer shear walls, thus restraining

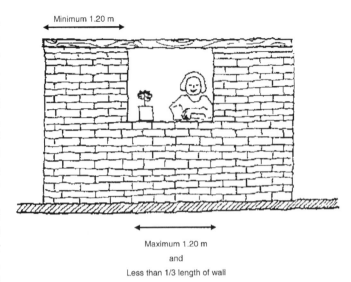

Minimum 1.20 m

Maximum 1.20 m
and
Less than 1/3 length of wall

Figure 20. Configuration of openings (RESESCO, 1997).

Figure 21. Ring beam construction, Sonsonate, El Salvador, 2002 (UNES project) (ruler for scale, 30 cm).

out-of-plane response. The ring beam is a very effective seismic resistance system, which significantly contributes to the mitigation of vertical corner cracking, overturning of wall panels, and corner dislocation (Table 3).

The ring beam should exhibit good tensile and shear strength (e.g., reinforced concrete, reinforced soil-cement, timber) and should be continuous. A soil-cement beam with bamboo or *vara de castilla* reinforcing is often used as a low-cost alternative. The effectiveness of the ring beam depends on the wall–ring beam and ring beam–roof connections. Connections must have the capacity to withstand strong shear, tensile, and bending stresses induced by seismic and wind forces. A strong connection between the ring beam and the wall can be achieved by attachment to vertical reinforcement, which provides restraint to shear stresses, combined with strapping, which utilizes the large wall mass to restrain uplift of the roof due to wind loading. Strapping can take the form of steel straps or heavy-gauge wire or nylon.

Roofing

The two main parts of the roof are the structure (trusses, beams, rafters, purlins, braces) and the covering (sheeting, tiles, natural material). If a reinforced concrete or soil-cement ring beam is used, an effective connection between the roof structure and ring beam can be achieved by placing a longitudinal timber beam atop the ring beam and attaching it to the ring beam and wall using strapping. A securely attached and strong roof diaphragm will transfer seismic forces to orthogonal walls, thus reducing the relative displacement between the walls and the roof structure. Diagonal bracing provides significant additional restraint to torsional response. Roof structure connections should adequately resist shear, tensile, and bending stresses. For many connections, nails do not provide sufficient restraint, so strapping or bolts should be used. Where possible, roof beams and rafters should not be positioned above openings, as this increases the bending stresses in the wall panel above the opening. If this is unavoidable, the lintels should be reinforced.

A lighter roof will attract less seismic force, although it is subject to greater response during wind loading. Corrugated iron sheets are light, relatively cheap, and widely available, although they are less heat resistive than traditional tile roof covers. Iron roof sheeting can be covered by natural materials, such as grasses or palm fronds, to improve the thermal properties. More expensive, lightweight roofing materials are also available, including steel sheeting with a zinc-aluminum coating, fibrocement sheeting, fiberglass sheeting, and microcement tiles. A light roof cover also allows resource savings in the roof structure, which supports a smaller roof mass.

Building Configuration

The above considerations relate to the individual components of a structure. These considerations should also be viewed in the broader context of total building design. The following recommendations relate to overall building configuration:

1. Buildings should be regular and symmetrical. Simple regular shapes (e.g., rectangles) will perform better than shapes with many projections. Asymmetry will promote detrimental torsional stresses in a building. Symmetry in each axis, as well as in openings (size and location) will reduce the effects of torsion (IAEE, 1986; Equipo Maíz, 2001).

2. Building height should be restricted to one floor (plus attic) in earthquake zones (IAEE, 1986; Norton, 1986; Equipo Maíz, 2001).

3. The length of a building should be no greater than three times the width (Norton, 1986).

Erosion and Moisture Control

The structural integrity of an adobe building is severely reduced if erosion and moisture damage are not controlled. Some of the recommendations to control erosion and moisture include:

1. Provision of adequate drainage that allows water to be cleared from the vicinity of the wall.

2. Construction of a plinth, or "top foundation," which raises the adobe wall from the ground (Fig. 18). The plinth may be constructed of stabilized blocks or concrete and should be raised at least 200 mm above the ground level. A moisture-proof course (e.g., plastic sheeting) should be laid between the plinth and adobe wall to prevent water penetration due to capillary action.

3. An overhanging roof means that the area of wall susceptible to direct wetting by rain is reduced. An overhanging roof does, however, increase the complexity and expense of the roof construction.

4. Application of a protective coating is the most economical and practical way to reduce the vulnerability of superstructure walls. Woodward (1996) suggests that impervious membranes should be avoided. He argues that impervious membranes crack and deteriorate over time, which allows the penetration of water that is not easily evaporated. Woodward suggests that a combination of linseed oil and mineral turpentine performs particularly well. The linseed oil–turpentine mix is a clear finish, so the natural appearance of the adobe is demonstrated. Mineral turpentine is widely available in El Salvador, whereas linseed oil is difficult to obtain outside San Salvador. A simple, but valuable, research investigation would be to assess the use and effectiveness of locally available oils (e.g., corn oil, cotton seed oil) as a protective coating.

Because adobe is often associated with poverty and "cheap" housing, local homeowners would generally prefer to have a "non-adobe-looking house." Rendering of the walls not only provides protection and durability, but it also improves the appearance and increases the impression of "solidity" of the structure. Common renders are made using various combinations of the following materials: soil, cement, sand, lime, bitumen, straw, cow dung, and cactus "glue."

5. Use of stabilizers. Addition of stabilizing agents (such as cement, lime, asphalt emulsion, organic or chemical compounds)

is designed to improve the resistive capacity of a block. Improvements in the soil cohesion result in increased durability and may also increase compressive, tensile, and shear strengths. The use of stabilizers, however, should be undertaken with caution. Stabilizers increase the costs of production, and each stabilizer responds uniquely to different soil components and proportions and requires specific preparation and curing conditions. For this reason, it is important that tests be carried out to determine the correct procedure and quantities to be used. The fabrication of trial blocks is highly recommended to ensure satisfactory results.

Examples of Improved Adobe Construction in El Salvador

The effectiveness of improved adobe construction has been demonstrated by the performance of several improved adobe buildings that were constructed prior to the 2001 earthquakes. These include:

1. Three hundred houses constructed around the country since 1977 by the Salvadoran NGO FUNDASAL, which have "passed the test of two earthquakes" (Moreno et al., 2001).

2. A series of houses constructed in 1994 in San Francisco de Ayutuxtepeque, San Salvador, and a house constructed in 1994 in San Juan de Letrán, Usulután, under the coordination of Alain Hays, which suffered no significant damage (Hays, 2001).

3. A community hall constructed in 1997 in San José de La Ceiba, San Vicente, by FUNDASAL, which suffered some minor cracking that has been repaired.

4. A small house constructed in 1999 in La Palomera, Santa Ana, by the Salvadoran NGO Unidad Ecológica Salvadoreña (UNES), which suffered no damage (Fig. 22).

These buildings exhibited excellent performance during the earthquakes, despite the damage and destruction of nearby houses made of *mixto*, traditional adobe, and *bahareque*. It is, however, difficult to provide a strong statistical confirmation of the success of improved adobe based on a relatively small sample size (due to the scarcity of such buildings prior to the earthquakes of 2001). Nevertheless, such examples indicate that improvement is possible, and they have also proved to be extremely valuable tools in the promotion of improved adobe to a generally skeptical population.

SEISMIC RETROFITTING AND DAMAGE REPAIR OF ADOBE STRUCTURES

Seismic retrofitting can be undertaken on adobe buildings to reduce the likelihood of damage due to future earthquake events. Many retrofitting systems are also suitable for the repair of damaged adobe buildings. The option to repair a damaged building is often overlooked, as minor cracking is often perceived as serious damage, and such buildings are commonly abandoned or demolished. In other cases, minor damage is ignored, which increases the risk of collapse during future seismic events. There are widespread needs and opportunities for programs aimed at training in seismic retrofitting and damage repair in El Salvador. Current institution-supported adobe projects tend to focus on new construction rather than rehabilitation and improvement of existing dwellings.

In each retrofit or repair case, the connection between the existing structure and the supplementary system is extremely important. If the connection does not adequately resist the imposed shear, tensile, and bending stresses, then the system will be less effective. The systems described below attempt to address the common structural deficiencies of traditional or damaged adobe houses, as described earlier.

Chicken Wire Mesh or Welded Mesh

Mesh is placed over the entire wall or in 45 cm wide horizontal and vertical strips at the corners and upper part of each wall, simulating beams and columns. Experiments undertaken by the Pontificia Universidad Católica del Perú (Zegarra et al., 2000) and Centro Nacional de Prevencion de Desastres (CENAPRED), México (Flores et al., 2001), have revealed this to be a very effective method of reducing the damage to adobe specimens. They report that the mesh is held with bottle tops and nails that are attached to the wall at 250–300 mm spacings and then covered with render.

The mesh provides restraint to out-of-plane response due to bending that generates vertical, horizontal, and inclined cracking, as well as overturning of wall panels. The mesh also provides restraint to in-plane shear cracking. Unless the mesh is connected to the foundations, it is unlikely to provide base restraint to overturning, although the restraint at the corners should minimize this overturning motion. It is recommended that some form of through-wall cross tie that resists tensile and shear stresses be used, because nails and staples have little resistance to out-of-plane movement. There is also some concern that the use of nails and staples may cause local cracking in the adobe. These aspects should be subjects of further research.

Figure 22. Undamaged adobe house, Santa Ana, El Salvador, 2002 (UNES project).

Strapping

Shake table testing undertaken at Stanford University, California, and University SS Cyril and Methodius, Republic of Macedonia, revealed that the addition of vertical and horizontal straps "reduced the amount of severe damage and dramatically decreased the risk of collapse" (Tolles et al., 2000). In these experiments, nylon straps were looped horizontally and vertically around the building and walls and securely connected by ductile cross ties that passed through holes drilled in the walls. Wire mesh was placed under the straps at the corners to reduce local stress concentrations. The mechanisms of restraint are similar to those for external wire mesh, described above. It is expected that other ductile materials, such as wire, steel strapping, or rope, would also be effective in this system, provided that adequate tension in the material can be maintained and local failure of the adobe at the connections is restricted.

Rope

Rope is placed in a horizontal channel formed in the wall, tightened, and then rendered. Zegarra et al. (2000) report that this system performs better than unreinforced specimens under earthquake simulations but is not as effective as the application of mesh as described above. The rope loop provides restraint to overturning due to out-of-plane seismic forces and may also provide some restraint to the tearing stresses that induce vertical corner cracking.

Buttresses

Buttresses can be built adjacent to walls to restrain out-of-plane motion. Unattached buttresses will provide limited restraint to motion toward the buttress only, although the presence of an adequate roof diaphragm will restrain motion away from the buttress. If the buttress can be securely attached to the wall, using cross ties, then greater restraint can be achieved for motion both toward and away from the buttress.

Ring Beam

A ring beam can be added to a building, although the roof will need to be removed to undertake this process. The ring beam should be securely attached to the walls, which can be achieved using external strapping and/or inserting dowels into the walls. In all cases, great care must be taken to prevent damage to the adobe wall during this process. The mechanisms of restraint achieved by ring beams are detailed above.

Other Systems

Other retrofit and repair systems include the insertion of fiberglass or steel center-core rods, which are epoxy-grouted in place (Tolles et al., 2000), and injection of cracks with a mix of soil, cement, expansive additive, and synthetic adhesive emulsion, which is under development at the Universidad Centroamericana José Simeón Cañas (R. Castellanos, 2002, personal commun.). Despite these advanced systems being more complex and expensive, they are regarded as being an important alternative in the conservation of historical and culturally significant adobe buildings. It is hoped that this technology can be applied to common housing in the future.

OBSTACLES TO REDUCING DAMAGE TO ADOBE HOUSES

Key obstacles to the reduction of earthquake damage to adobe houses include:

1. Lack of confidence in adobe as a construction material.
2. Lack of widespread promotion and support of improved adobe, due to limited resources, traditional attitudes, and vast populations.
3. Lack of experience and skill in seismically resistant design and construction.
4. Lack of personal financial resources, which restrict skills development and the use of better materials and techniques.

Recommendations to overcome these obstacles are discussed below.

RECOMMENDATIONS FOR ADOBE STRENGTHENING IN EL SALVADOR

A combination of social and technical solutions is required to improve the current housing situation in El Salvador and other developing countries. The key components are research and development, and implementation. The implementation component consists of three interconnected elements; namely, promotion, skills, and resources.

Promotion

General promotion is first required to demonstrate the existence and effectiveness of improved adobe construction techniques to local communities, government institutions, and nongovernmental organizations. In local communities, visual presentations tend to be more effective and should include examples of the seismic performance of improved adobe houses, as well as general aspects of improved seismic design and construction. Where possible, interested community members could be invited to visit nearby examples of improved adobe housing, thus providing a tangible demonstration of the improved system. Effective promotion increases the interest, awareness, and acceptance of improved adobe, which are prerequisites for any skills-building training and community construction projects.

Skills

"Skills" relates to both skills building and provision of adequately skilled personnel to support improved construction.

Skill building involves the training of interested and capable people in techniques of improved adobe construction, considering both theoretical and practical components. In order to thoroughly cover all aspects of the design and construction process, the most effective training program involves the construction of a complete building (Fig. 23). This process requires a significant period of time (upwards of two or three months) and, as such, requires the full commitment and interest of all stakeholders. The program should be organized such that participants are not unduly burdened by the lengthy process, with options for worksite rotation rosters and payment (cash or food) for work. Naturally, the initial assessment of the project and community should include such discussion and should focus on the needs, interests, and availability of the participants. The training program should be complemented by a workbook or construction manual that details each step of the preparation and construction process.

Current improved adobe techniques have placed adobe construction in the domain of skilled artisans. It is unlikely that local people with little or no construction experience or skill will be capable of applying the full range of improvement suggestions in the construction of their homes, even after participation in a skills-building project. Nevertheless, it is hoped that some of the simpler techniques will be adopted by unskilled owner-builders, and at the very least there will be a greater awareness that improved adobe houses can be constructed. The real benefit of the training programs lies in the improved capacity of local artisans to construct safer adobe homes.

Skilled personnel are required to coordinate adobe promotion activities, facilitate training programs, and provide post-training support for community construction. Skilled personnel are also sought for large-scale improved adobe construction projects, such as those implemented in the aftermath of the 2001 earthquakes, including those by Asociación Salvadoreña de Desarrollo Integral (ASDI), Asociación Bálsamo, Atlas Logistique, FUNDASAL, Trocaire, and Unidad Ecológica Salvadoreña (UNES), among others. Improved adobe is increasingly being seen as a viable and appropriate alternative for agency-coordinated housing projects, which have tended to focus on concrete block construction.

Resources

Resources are required to support improved adobe promotion and skills-building programs, as well as support community construction. These resources may be sourced from local and international governmental and nongovernmental agencies in the form of micro-credit or finance schemes, direct payment, or provision of personnel, materials, or equipment. A significant example of the increased support for improved adobe was shown by the European Commission Humanitarian Aid Office (ECHO), which provided 528,000 € (US$475,000 at that time) for improved adobe training programs as part of the post-earthquake relief operations in El Salvador. Other resources have come from Australia, Canada, Cuba, France, Germany, Ireland, Netherlands, Peru, Spain, the United Kingdom, and the United States. Resource burdens may

Figure 23. Improved adobe model house, San Miguel, El Salvador, 2002 (FUNDASAL project).

also be relieved by promoting recycling and reuse of available materials. Resources are also required to support the essential ongoing research and development aspects of improved adobe.

Research and Development

The research and development component involves the improvement of the current state of technical knowledge, while maintaining strong social links to ensure the investigation of appropriate solutions. Further research is required to improve the statistical depth of knowledge, leading to greater confidence in analysis and design. Research may focus on individual adobe blocks (soil composition, additives, mechanical properties, etc.), adobe masonry components (block-mortar bonding, reinforcement-block interaction, mechanical properties, etc.), and complete building response (wall length-width-height ratios, pilasters, reinforcement, ring beams, connections, computer modeling, etc.). Research and development also relates to the social aspects of implementation, such that lessons are learned and effective programs are developed, thus increasing the capacity of affected communities to respond appropriately.

There is great value in research undertaken in El Salvador that assesses the performance of local materials and building techniques. The relative lack of testing facilities in El Salvador, particularly a shake table, is seen as a key obstacle to further research in-country. It is expected that key findings from research conducted around the world will increase the profile of improved adobe and the confidence in its use, both in El Salvador and in other countries where adobe is commonly used.

Research findings combined with active promotion are required to increase the acceptance of improved adobe by local and international agencies. Government support for adobe in El Salvador is still tentative, and in a global context funding for development projects is extremely competitive. If the advantages of adobe can be shown in some quantifiable measure, then support for improved adobe is expected to increase.

CONCLUSIONS

Adobe (mudbrick) is one of the most widely used construction materials in the world. Two factors are responsible for the predominant use of adobe in El Salvador and other developing countries. These are (1) low cost, and (2) minimal specialist skills required for traditional adobe construction. Other advantages of adobe include durability, energy efficiency, ecological sustainability, and thermal capacity. The main disadvantages of adobe include longer fabrication and construction time, susceptibility to water damage, "poor" image, and vulnerability to damage by earthquakes. This final feature has been tragically observed in earthquake-prone areas around the world, particularly El Salvador.

The vulnerability of adobe buildings to damage from earthquakes and an assessment of damage patterns and failure modes have led to various recommendations for improved seismic resistance. A number of low-tech and low-cost amendments can be made to existing construction practices to improve the seismic resistance of an adobe structure. These include tying the structure together, use of vertical and horizontal reinforcement, inclusion of pilasters, and construction of wide walls with good horizontal overlap of bricks. Other factors relate to foundations, floors, walls, openings, ring beams, roofing, building configuration, erosion and moisture control, and construction and maintenance quality. There is large scope for further research in this field.

There are several challenges associated with the introduction of change to traditional building practices. These challenges focus on overcoming resource limitations and the lack of awareness, experience, and skill in improved seismic design and construction. Any solution must incorporate both social and technical aspects and requires the involvement of local, national, and international institutions to promote and support improved adobe design and construction systems. Advances in the development and application of improved adobe are essential to reduce the vulnerability of common housing to the effects of earthquakes. El Salvador has both the need and the opportunity to lead this advance.

ACKNOWLEDGMENTS

Special appreciation is extended to the many organizations and individuals that have provided invaluable information and assistance, including Asociación Salvadoreña de Desarrolo Integral (ASDI), Asociación Bálsamo, Atlas Logistique, Centro Nacional de Prevencion de Desastres (CENAPRED) México, Ing. Rafael Colindres, Equipo Maíz, Expedition El Salvador, Fundación Böll, Fundacion Salvadoreña de Desarrollo y Vivienda Minima (FUNDASAL), Instituto Salvadoreño de Formación Profesional (INSAFORP), Goal, Las Melidas, United Nations Development Program (PNUD/UNDP) El Salvador, Ing. Gabriel Pons, Christian Rademaker, Melvin Tebbutt, Trocaire, Universidad Centroamericana José Simeón Cañas (UCA), Universidad de El Salvador (UES), Unidad Ecológica Salvadoreña (UNES), and Vice-Ministerio de Viviendas y Desarrollo Urbano (VMVDU). Many thanks to the Earthquake Engineering Research Institute (EERI) and the University of Technology, Sydney, for funding, resource, and logistical support, without which the field trips would not have been possible. The contribution and input of Ing. Patricia Méndez de Hasbun and Ing. Ricardo Castellanos of the Universidad Centroamericana José Simeón Cañas are gratefully acknowledged. Particular thanks are due to Dr. Julian Bommer of Imperial College, London, for his invaluable support and assistance with field investigations, information transfer, and establishing contacts with many of the organizations and individuals mentioned above. Thanks are also directed to Professor Bijan Samali and Mr. Joko Widjaja of the University of Technology, Sydney, for their patience, support, and guidance. And the sincerest gratitude and respect to the people of El Salvador, for opening their homes and sharing their reality.

REFERENCES CITED

Anderson, R.W., 1987, The San Salvador Earthquake of October 10, 1986: Review of building damage: Earthquake Engineering Research Institute, EERI Earthquake Spectra 1987, v. 3(3), p. 497–541.

Bommer, J.J., Benito, M.B., Ciudad Real, M., Lemoine, A., Lopez Menjivar, M.A., Madariaga, R., Mankelow, J., Mendez de Hasbun, P., Murphy, W., Nieto Lovo, M., Rodriguez Pineda, C.E., and Rosa, H., 2002, The El Salvador earthquakes of January and February 2001: Context, characteristics, and implications for seismic risk: Soil Dynamics and Earthquake Engineering, v. 22, p. 389–418.

DIGESTYC (Dirección General de Estadística y Censos), 1999, Encuesta de hogares de propósitos multiples: Ministerio de Economía, El Salvador, internal report.

DIGESTYC (Dirección General de Estadística y Censos), 2001, Censo de viviendas afectadas por la actividad sísmica del año 2001: Ministerio de Economía, El Salvador, internal report.

Durkin, M., and Hopkins, J., 1987, The San Salvador Earthquake of October 10, 1986: Architecture and urban planning: Earthquake Engineering Research Institute, EERI Earthquake Spectra 1987, v. 3(3), p. 609–620.

EERI (Earthquake Engineering Research Institute), 2001, Preliminary reports and annotated images from the El Salvador earthquakes of January 13th and February 13th, 2001: Oakland, California, Earthquake Engineering Research Institute, CD-ROM.

Equipo Maíz, 2001, La casa de adobe sismorresistente: Asociación Equipo Maíz, El Salvador, 91 p.

Flores, L., Pacheco, M., and Reyes, C., 2001, Algunos estudios sobre el comportamiento y rehabilitación de la vivienda rural de adobe: CENAPRED (Centro Nacional de Prevencion de Desastres), México, Informe IEG/03/01, 123 p.

Hays, A., 2001, Geo-Arquitectura en El Salvador: Efectos de los terremotos del 2001 en el patrimonio arquitectónico tradicional Salvadoreño y perspectivas de restauración y reconstrucción con geomateriales: GEOdomus Internacional, 170 p.

IAEE (International Association for Earthquake Engineering), 1986, Guidelines for earthquake resistant non-engineered construction: Tokyo, IAEE, 158 p.

Middleton, G.F., 1987, Bulletin 5: Earth-wall construction (4th edition): Chatswood, NSW, Australia, National Building Technology Centre, 65 p.

Minke, G., 2000, Earth construction handbook: Southampton, UK, WIT Press, 206 p.

Moreira, R.A., and Rosales, J.A., 1998, Diseño, construcción y control de calidad de estructuras de adobe para vivienda rural [graduation thesis]: El Salvador, Universidad Centroamericana José Simeón Cañas, 369 p.

Moreno, F., Cáceres, M., Santos, M., Aviles, C., and Sanabria, A., 2001, Hacia un adobe seguro: El Diario de Hoy, 18th March 2001, p. 3–7.

Norton, J., 1986, Building with earth: A handbook: Rugby, UK, Intermediate Technology Development Group, 68 p.

Pérez, A.H., 2001, Manual técnica para la producción y construcción con adobe natural: Habitat–Cuba, 74 p.

Reglamento Nacional de Construcciones, 2000, Norma técnica de edificación (NTE) 080 adobe: Ministro de Transportes, Comunicaciones, Viviendas y Construcción, Lima, Perú, 35 p.

RESESCO (Reglamento para la Seguridad Estructural de las Construcciones), 1997, Folleto complementario adobe: Asociación Salvadoreña de Ingenieros y Arquitectos, Ministerio de Obras Publicas, El Salvador, 36 p.

Roselund, N., 1995, Buttresses, pilasters and adobe wall stability, Workshop on the seismic retrofit of historic adobe buildings, March 10, 1995: Lafayette, California, Earthen Building Technologies, p. 89–90.

Tolles, E.L., Kimbro, E.E., Webster, F.A., and Ginell, W.S., 2000, Seismic stabilization of historic adobe structures: Final report of the Getty Seismic Adobe Project: Los Angeles, USA, The J. Paul Getty Trust, 158 p.

VMVDU (Vice-Ministerio de Viviendas y Desarrollo Urbano), 2001, Riesgo sísmico—reconstrucción habitacional: la experiencia Salvadoreña: 44 slide presentation, San Salvador, El Salvador.

Wisner, B., 2001, Risk and the neoliberal state: Why post-Mitch lessons didn't reduce El Salvador's earthquake losses: Disasters 2001, v. 25(3), p. 251–68.

Woodward, B., 1996, Mudbrick notes (2nd edition): Wollombi, Australia, Earthways, 48 p.

Zegarra, L., Quiun, D., San Bartolomé, A., and Giesecke, A., 2000, Reforzamiento de viviendas de adobe existents: Pontificia Universidad Católica del Perú, Perú, 10 p.

MANUSCRIPT ACCEPTED BY THE SOCIETY JUNE 16, 2003

Geological Society of America
Special Paper 375
2004

Seismic hazard assessments, seismic design codes, and earthquake engineering in El Salvador

Manuel López

Escuela de Ingeniería Civil, Universidad de El Salvador, Apartado Postal 740, San Salvador, El Salvador

Julian J. Bommer*

Department of Civil & Environmental Engineering, Imperial College London, SW7 2AZ, UK

Rui Pinho

European School for Advanced Studies in Reduction of Seismic Risk (ROSE School), Pavia, Italy

ABSTRACT

Seismic hazard is very high in parts of El Salvador, with destructive earthquakes occurring on average at least once per decade. Probabilistic estimates of the hazard published by various researchers differ appreciably, in large part due to the uncertainties associated with the earthquake catalog and the sparse database of strong-motion accelerograms. Proposals have been put forward to create two seismic zonation maps, one for upper crustal events and another for earthquakes in the subducted Cocos plate. Observed differences in the spatial distribution and specific characteristics of ground shaking caused by the two types of earthquakes support this proposal, but it is clear that site effects are also of great importance, hence seismic microzonation is also required for urban areas. The high level of seismic hazard in El Salvador makes an appropriate code for the earthquake-resistant design of buildings a vital tool for risk mitigation. Seismic codes have been published in 1966, 1989, and 1994, the most recent being a reasonably comprehensive and technically robust set of guidelines. The effectiveness of the current code to reduce seismic risk in El Salvador is low, but not because of technical deficiencies in the requirements for structural design. The most serious deficiency is clearly defined as the complete lack of any effective mechanism for the enforcement of the code requirements for minimum levels of seismic resistance. Another important element of the risk in El Salvador, which is due to the frequency of destructive events, is lack of adequate repair and strengthening of damaged buildings, an issue not addressed by the code.

Keywords: El Salvador, seismic hazard assessment, seismic design codes, earthquake engineering.

INTRODUCTION

The republic of El Salvador is struck by a destructive earthquake, or earthquake sequence, once per decade on average. With a geographical extension of a little more than 20,000 km², the frequency of damaging earthquakes clearly demonstrates that El Salvador is a country of very high seismic hazard. Indeed, the capital city, San Salvador, is probably the city in the Americas that has been most frequently damaged by earthquakes.

Despite the fact that there are many pressing demands on El Salvador's very limited economic resources, a concerted program of measures to reduce earthquake risk is clearly an

*Corresponding author: Department of Civil & Environmental Engineering, Imperial College London, South Kensington Campus, London SW7 2AZ, UK, j.bommer@imperial.ac.uk.

López, M., Bommer, J.J., and Pinho, R., 2004, Seismic hazard assessments, seismic design codes, and earthquake engineering in El Salvador, *in* Rose, W.I., Bommer, J.J., López, D.L., Carr, M.J., and Major, J.J., eds., Natural hazards in El Salvador: Boulder, Colorado, Geological Society of America Special Paper 375, p. 301–320. For permission to copy, contact editing@geosociety.org. © 2004 Geological Society of America.

indispensable element of any realistic plan for sustainable development. The purpose of this paper is to provide an overview and critical assessment of the work carried out to date in three key areas that together may be considered as the basis of an effective earthquake risk mitigation plan: quantitative assessment of seismic hazard, development and implementation of seismic design regulations for structures, and training of engineers and architects in earthquake engineering. The objective of performing such a review is neither to praise the noble efforts of many dedicated individuals who have worked very hard to promote earthquake engineering in El Salvador, nor to gratuitously find fault where there have been shortcomings and even failures. Rather the paper attempts to review the work carried out in and for El Salvador within the context of the country's turbulent history in recent decades and the opportunities and perspectives presented by the relatively stable conditions that now exist. In particular, the paper aims to identify gaps in the current knowledge and weaknesses in the risk mitigation measures being applied in El Salvador. On the basis of these observations, recommendations and suggestions are made for future work and for improvements in seismic risk mitigation, building on recent progress and the increase in external aid that has arisen following Hurricane Mitch and the 2001 earthquakes.

SEISMIC HAZARD ASSESSMENTS FOR EL SALVADOR

The quantitative assessment of seismic hazard is a fundamentally important tool for any effective long-term seismic risk mitigation program.

Earthquakes in El Salvador

The seismicity and tectonics of El Salvador are discussed in detail by Dewey et al. (this volume, Chapter 27), but a brief overview is provided here in terms of the sources of destructive earthquakes.

Large earthquakes (M > 7) generated directly by the subduction of the Cocos plate below the Caribbean plate in the Middle America Trench caused widespread but moderate damage in El Salvador on 7 September 1915, 21 May 1932, and 19 June 1982. Although reports differ in their estimates, only a few deaths resulted from each of these events (Ambraseys and Adams, 2001; Lara, 1983).

Shallow crustal earthquakes, of smaller magnitude (M ~6) have caused very intense damage over limited areas at several locations along the line of Quaternary volcanoes that is situated in the southern side of the Great Interior Valley that runs through El Salvador, parallel to the coast. White (1991) proposed that these earthquakes, invariably caused by strike-slip fault ruptures along the volcanic axis, are the result of a right-lateral shear zone caused by an oblique component of the Cocos-Caribbean collision. This forearc sliver model has recently been confirmed by GPS measurements of plate velocities (DeMets, 2001).

The capital city, San Salvador, was badly damaged by such local earthquakes on 8 June 1917 and again on 28 April 1919, the former preceding an eruption 80 minutes later of the San Salvador volcano. The death tolls in San Salvador from these two earthquakes are 40 and between 70 and 100, respectively (White and Harlow, 1993; Ambraseys and Adams, 2001). The city of San Vicente was struck by an earthquake on 20 December 1936, leaving 100–200 dead (Levin, 1940; White and Harlow, 1993). An earthquake on 26 December 1937 close to the border with Guatemala caused damage in the towns of Ahuachapán and Atíquizaya (White and Harlow, 1993; Ambraseys and Adams, 2001). The towns of Jucuapa and Chinameca in the eastern provinces of Usulután and San Miguel were destroyed by a series of three earthquakes on 6–7 May 1951, with 400 fatalities (Ambraseys et al., 2001).

Moderate-magnitude local earthquakes again struck San Salvador on 3 May 1965 (Lomnitz and Schulz, 1966) and on 10 October 1986 (EERI, 1987; Bommer and Ledbetter, 1987). The 1965 earthquake left 120 dead, whereas the death toll in 1986 was 1500, a significant number of the victims in the second earthquake being killed by the collapse of engineered structures that had been weakened in the 1965 event. The most notorious example was the collapse of the five-story Rubén Darío building, which had been condemned following the 1965 earthquake, where 300 perished (Bommer and Ledbetter, 1987; Anderson, 1987).

At the turn of the new century, El Salvador was hit by two major earthquakes within the space of one month, the first a normal-faulting event in the subduction zone on 13 January 2001, the second a strike-slip crustal earthquake to the east of San Salvador (Bommer et al., 2002). The 13 January earthquake, of magnitude M_w 7.7, was of similar depth and source mechanism as the M_w 7.3 1982 subduction earthquake, but had a far greater impact on the country. About 850 people were killed by this earthquake, the vast majority buried by landslides triggered by the ground shaking (Jibson et al., this volume, Chapter 6), most of the remaining victims being due to the collapse of *adobe* houses. The earthquake had surprisingly little impact on engineered structures and there was not a single case of major structural collapse due to this event. In particular, despite widespread damage to vernacular houses and triggering of landslides both east and west of the capital, San Salvador was practically unscathed by the earthquake. The 13 February earthquake, of magnitude M_w 6.6, occurred in approximately the same location as the 1936 earthquake (M_s 6.1) and similarly affected the city of San Vicente. The death toll due to the 13 February earthquake was a little over 300, again due to landslides and the collapse of *adobe* and *bahareque* (wattle-and-daub) houses. Despite the relatively shallow focus of the second earthquake, estimated at ~15 km, damage to engineered structures was less severe than could have been expected.

The damage patterns in the 2001 earthquakes in El Salvador could be interpreted as implying that shaking damage to engineered structures is not an important element of risk. The vulnerability of vernacular building systems, particularly *adobe* and to a lesser extent *bahareque*, is a very important element of seismic risk in El Salvador. Similarly, the hazard posed by earthquake-triggered landslides is also a major component of the risk and a feature in all earthquakes: At least 200 of the victims of the 1986 San Salvador earthquake were directly due to slope failures. However, the apparently low importance of direct damage to engineered structures observed in the 2001 earthquakes is misleading. Firstly, subduction zone earthquakes in the region seem generally to cause only limited damage to engineered structures, which has been at least partially explained by the relatively low rate of energy input by the associated ground motions (Bommer and Martínez-Pereira, 1999). The Arias intensity—a measure of the energy content—in the ground motions recorded in San Salvador during the 1982 and 1986 earthquakes was very similar, but in the latter case the energy was imparted to structures almost an order of magnitude more rapidly. The second point to note is that the 13 February earthquake occurred in the area of San Vicente, whereas large concentrations of medium- and high-rise engineered structures in the country are confined to San Salvador. Therefore, the performance of engineered structures in San Salvador in the 2001 earthquakes is no guarantee against heavy damage and potentially high loss of life in future local earthquakes near the capital, which occur with average recurrence intervals on the order of 30 yr (Harlow et al., 1993). Indeed, if a large part of the destruction in San Salvador in 1986 resulted from the collapse of engineered structures weakened in 1965—and very possibly further debilitated in 1982—then there is now the worrying scenario of the collapse, in the next San Salvador earthquake, of a far larger number of engineered structures adversely affected in 1986, and possibly further deteriorated by the earthquakes of January and February 2001.

Seismic Hazard Studies for El Salvador

Quantitative assessment of the geographical distribution and degree of hazard from seismic shaking in El Salvador is fundamental to seismic design of new and existing structures, and also to land-use planning. Reliable assessment of the ground-shaking hazard is also a requirement for the quantification of other hazards, in particular that due to earthquake-induced landslides.

The maps of peak ground acceleration (PGA) and velocity with a return period of 200 yr presented for large parts of the world by Hattori (1979) indicate a PGA value of 0.2 g in San Salvador. During the 1970s, seismic hazard maps were produced by the John A. Blume Earthquake Engineering Center at Stanford University for most Central America countries, presenting contours of PGA for return periods of 100, 500, and 1000 yr. These studies were produced for the countries neighboring El Salvador: Nicaragua (Shah et al., 1975), Guatemala (Kiremidjian et al., 1977), and Honduras (Kiremidjian and Shah, 1979). All three of these studies made use of the attenuation equation derived by Esteva (for which no reference is given), although the hazard assessment for Honduras also used an equation derived by Woodward-Clyde Consultants in 1978. The 500-year hazard levels in El Salvador indicated by these studies were PGA values of 0.5 g along the northern border with Honduras and a maximum of 0.3 g on the western border with Guatemala.

More recent studies have been carried out for El Salvador's neighbors, including a hazard assessment for Nicaragua by Larsson and Mattson (1987). This study made use of the same attenuation equation employed in the Stanford University reports and an earlier equation by Esteva that was presented by Cornell (1968). Ligorría et al. (1995) carried out a seismic hazard study for Guatemala, in which they used an attenuation relationship for PGA derived by Climent et al. (1994) using strong-motion data from Central America and Mexico. The contours for the 500-year PGA (at soil sites) in El Salvador indicate levels of acceleration of 0.45 g. The equation of Climent et al. (1994), subsequently presented by Dahle et al. (1995), was also used for a later hazard assessment of Guatemala City (Villagran et al., 1997). Cáceres and Kulhánek (2000) presented hazard maps for Honduras and surrounding areas, including all of El Salvador, where the 475-year hazard on hard rock is shown to increase from 0.8 g along the northern border with Honduras to 1.3 g between the volcanic chain and the coastline. Cáceres and Kulhánek (2000) considered four PGA attenuation equations (Climent et al., 1994; McGuire, 1976; Boore et al., 1997; Schnabel and Seed, 1973) in a logic-tree formulation, finally opting to use the outdated equation of Schnabel and Seed (1973). Tavakoli and Monterroso (this volume, Chapter 31) have developed new seismic hazard maps for El Salvador and neighboring countries.

Lindholm et al. (1995) presented a seismic hazard map in terms of PGA for the entire Central American isthmus, in which the 500-year hazard in El Salvador is almost constant, increasing from 0.25 g at the northern border to just above 0.3 g along the volcanic chain. For a country as small as El Salvador, it is inevitable that regional seismic hazard maps will indicate almost constant hazard throughout the national territory. The seismic hazard map of the Americas produced as part of the Global Seismic Hazard Assessment Program (GSHAP) indicates almost constant hazard throughout El Salvador, with 475-year PGA on the order of 0.5 g (Shedlock, 1999). The Central American hazard maps generated as part of a regional earthquake loss estimation model by Chen et al. (2002) indicate a 475-year PGA level of 0.2–0.4 g in northern El Salvador and 0.4–0.8 g in the southern and western parts of the country. While it can be argued that there is no geophysical justification for separating El Salvador from neighboring areas of Central America, land-use planning and building regulations are necessarily confined in their application to the national territory of the country, and hence there is a need to develop seismic hazard maps for the area within El Salvador's borders.

Following the 1986 San Salvador earthquake, three seismic hazard studies were published for El Salvador, which have been reviewed in detail by Bommer et al. (1996). The first study was produced by the U.S. Geological Survey (Algermissen et al., 1988) and made use of an attenuation relationship for PGA derived from a suite of accelerograms recorded in and around San Salvador. The second study was produced by the John A.

Blume Center at Stanford University (Alfaro et al., 1990) and used two attenuation equations, one for crustal earthquakes, the other for subduction events, using just 20 accelerograms for each. The third study was produced by the Universidad Nacional Autónoma de México (Singh et al., 1993) and was commissioned as part of the project to develop the 1994 seismic design code for El Salvador, which is discussed in greater detail below. Singh et al. (1993) derived an attenuation equation for PGA, using a data set of Central American accelerograms from which they determined the vectorial resolution of the two horizontal components. Despite the fact that the three studies had essentially the same input data at their disposal, differences in the definition of seismic source zones, attenuation relationships, and method of hazard calculation resulted in very different hazard maps. The 475-year PGA values at the northern border of El Salvador were shown to be 0.3 g, 0.5 g, and 0.7 g in the maps of Algermissen et al. (1988), Alfaro et al. (1990), and Singh et al. (1993), respectively; the PGA values for San Salvador were 0.5 g, 0.8 g, and 1.0 g respectively in the three studies.

Seismic Zonation Maps for El Salvador

Three seismic zonation maps of El Salvador have been produced for the three seismic design codes discussed in the next section. The zonations, shown in Figure 1, have always divided the country into two zones, the zone of higher hazard containing all of the Great Interior Valley and the coastal mountain ranges and coastal plains.

The first zonation was proposed by Rosenblueth and Prince (1966), who clearly identified the map as being tentative. The modified frontier between the two zones presented in the 1989 seismic design code zonation was based on a map of observed intensities in El Salvador due to local and subduction earthquakes as well as the 4 February 1976 earthquake in Guatemala. The 1994 zonation, based on the hazard map of Singh et al. (1993), is almost identical to the first zonation from 1966. The boundary in the current map between zones I and II approximately follows the contour of 0.8 g from the hazard map for a return period of 500 yr, although it was actually drawn to follow the 1.0 g contour on the 1000-year return period map.

One of the reasons that Rosenblueth and Prince (1966) classified their crude zonation map as tentative was that they proposed that for a more definitive seismic code than the one introduced in 1966 as a response to the 1965 San Salvador earthquake, two zonation maps should be produced. One zonation map would correspond to the hazard represented by shallow crustal earthquakes generated along the volcanic chain, and the other to the hazard from subduction zone earthquakes. Rosenblueth and Prince (1966) suggested that the map for the shallow-focus earthquake hazard would indicate a maximum zone along the volcanic axis and parallel zones of lower hazard on either side, whereas the hazard map for subduction earthquakes would consist of broad zones, approximately parallel to the coast, with the levels of ground motion reducing with distance from the coast.

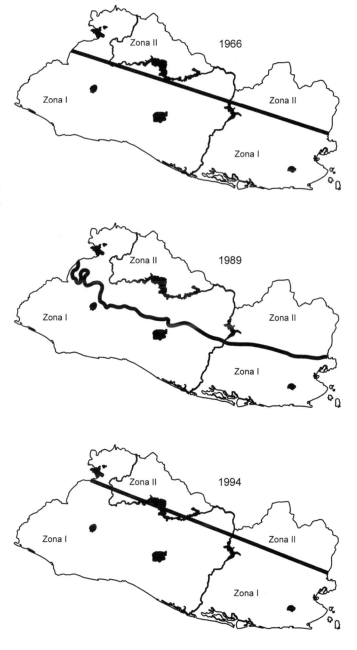

Figure 1. Zonation maps from El Salvadoran seismic design codes (Bommer et al., 1996).

This proposal has never been considered in the drafting of subsequent building codes published in 1989 and 1994. One objection that could be raised is that the use of two zonation maps, each with a corresponding response spectrum, would not be consistent with the concept of the uniform hazard spectrum (UHS). However, the appropriateness of the UHS as the basis for engineering design has been questioned, since in many cases its use is equivalent to designing for the simultaneous occurrence of two earthquakes (e.g., Reiter, 1990; Bommer et al., 2000). The seismic design codes of both Portugal and China include separate spectra for near sources of moderate-magnitude earthquakes and for more distant sources of large-magnitude events, so there is a precedent for such a departure from the use of the UHS.

There are good reasons for separating the hazard from upper-crustal and subduction seismicity in El Salvador, since the ground motions are very different in terms of duration and root-mean-square amplitude (Bommer et al., 1997). Moreover, the spatial distributions of the two types of hazard are different, with the highest hazard being concentrated around the volcanic centers where destructive earthquakes have been frequently generated (Bommer et al., 1998). In this respect, unlike the proposal of Rosenblueth and Prince (1966), the highest-hazard zone would not be a continuous band along the southern side of the Great Interior Valley, but rather a series of areas surrounding the seismic sources. In fact, these could be approximately drawn from a map showing maximum observed values of intensity.

A zonation map, or pair of maps, that reflected the spatial variation of the seismic hazard, and in particular the limited area where Modified Mercalli intensities of VIII and more can be expected, would be very useful for land-use planning. It would also mean that the penalty for building in the highest-hazard zone (which in the west of the country is severely overpopulated) in terms of design criteria and insurance premiums could help to reverse the trends that have led to three-quarters of the population living in just one-third of the national territory. Bommer (1999) presented elements for the development of seismic design criteria based on separation of crustal and subduction seismicity in terms of zonation, elastic spectral shapes, and spectral reduction factors. An alternative form of seismic hazard mapping, which allows all features of the ground motion to be estimated without simplification, has been proposed on the basis of the same idea of separating the hazard from different sources of seismicity (Bommer, 2000). The basis of this alternative mapping procedure is to define, for each site, an earthquake scenario compatible with the seismic hazard, and then map the magnitude and distance values that define these scenarios to create two zonation maps. Bommer and White (2001) developed preliminary maps of this type for El Salvador. A further refinement that could be considered is to differentiate between interface and intraslab subduction earthquakes (Bent and Evans, this volume, Chapter 29).

The proposal for two separate seismic zonation maps is based on differences in the ground motions from the two types of earthquakes due to source and path characteristics. A question that then arises is the relative importance of these differences when compared to the influence of local site effects on the ground motion, since the latter would make microzonation maps more useful than separate macrozonation maps. Figures 2A and 2B show the ratios of spectral shapes (normalized to PGA) from the 13 January and 13 February 2001 earthquakes at eight digital strong-motion accelerograph stations (Bommer et al., 2002). In simplistic terms, if the ratios were equal to unity it would imply that the nature of the motion is entirely dominated by site effects. This assumption would need to be qualified by consideration

of the effects of soil nonlinearity, since at some of the stations (Armenia, La Libertad, and Santa Tecla) the absolute amplitudes of motion recorded in the February earthquake were more than an order of magnitude lower than in January. At Armenia, where the ratio of absolute amplitudes between the two events was highest, the ratio of spectral shapes is appreciably less than unity at short response periods. The two stations at which the absolute amplitudes, in terms of PGA, were closest in the two earthquakes are Panchimalco and San Bartolo; in both cases, the spectral ratio is close to unity for one component, and not the same component in both cases, but significantly higher for other component. A third station at which comparable levels of PGA were recorded in both earthquakes (slightly higher in the second event) is Zacatecoluca, where the spectral ratios are quite close to unity at most periods. The Zacatecoluca records indicate a predominance of site effects, whereas many other stations strongly indicate source and path effects, suggesting that all three effects are likely to be significant, and hence the definition of seismic design loads should take both factors into account. Exact definition of the relative influences will require detailed geotechnical characterization of the strong-motion recording sites, which is not currently available.

SEISMIC DESIGN CODES IN EL SALVADOR

Historical Background

In the colonial period, a construction system named *calicanto* was introduced in El Salvador, consisting mainly of unreinforced stone masonry bonded by lime mortar. However, the shortage of this construction material, coupled with the need for building stronger structures capable of withstanding the earthquakes that frequently hit the region, led, around the eighteenth century, to a gradual change from stonework to *adobe* and *bahareque* types of construction.

After the 1873 and 1917 earthquakes in San Salvador (Harlow et al., 1993), however, another conversion in building practice took place. This time, a structural system composed of wood or metal frames (covered by thin metal sheets) was used to rebuild the city. This system had the important benefit of possessing low mass, thus allowing, for the first time in El Salvador, the construction of buildings with more than one story. The oldest buildings that can currently be found in San Salvador date from the beginning of the twentieth century and make use of this particular structural system.

Prior to 1942, no building above three stories existed. It was only during the period between 1942 and 1965 that the first series of high-rise buildings, up to eight stories high, were erected, using reinforced concrete frames and cast-in-place slabs (Lara, 1987). The building design did not take into account irregularities or torsional effects affecting the seismic behavior of the structure, and the majority of the infill panels were built with full interaction with the frames. Soil investigations (introduced in El Salvador ~1950) became mandatory, for structures above three stories, only around the year of 1960 (Lara, 1987).

A

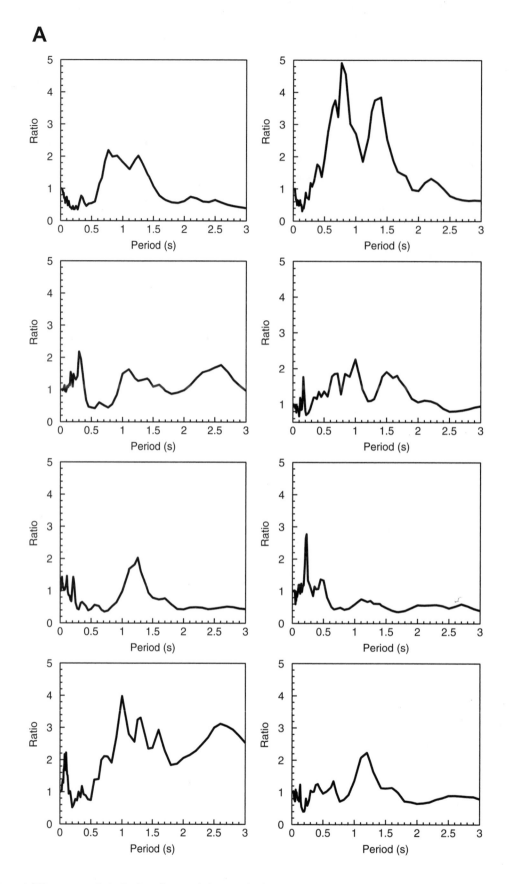

Figure 2 (*on this and following page*). A: Ratios of spectral shapes (absolute acceleration for 5% of critical damping) normalized to PGA of the NS components (left) and EW components (right) of accelerograms recorded in the 13 January and 13 February 2001 earthquakes. Recording stations from top to bottom: Armenia, San Salvador (Externado de San José), La Libertad, and Panchimalco. B: Ratios of spectral shapes as in A for recording stations (from top to bottom): Santa Tecla (Hospital San Rafael), San Bartolo, Tonacatepeque, and Zacatecoluca.

B

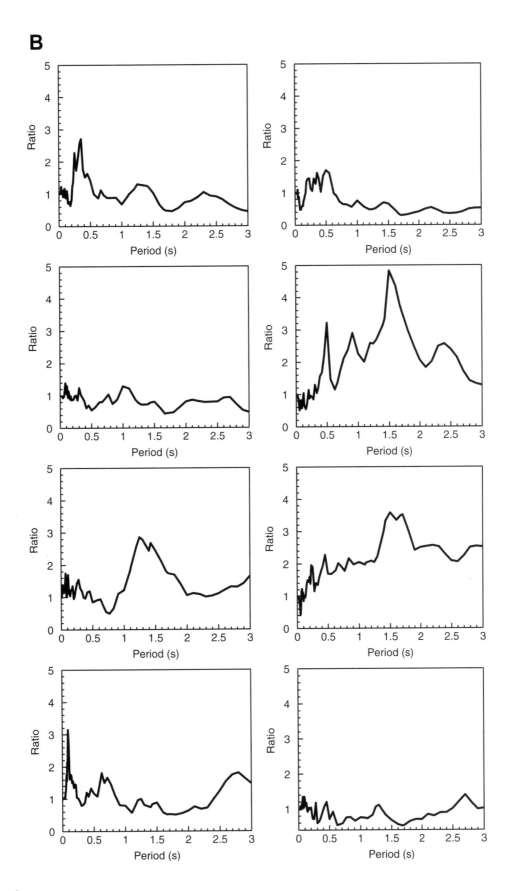

Figure 2 (*continued*).

As far as building regulations are concerned, there was no national code (and no foreign structural codes were systematically applied) up until the mid-1940s. In May 1946, the introduction of a national seismic design code was abandoned (Rosenblueth and Prince, 1966), giving way to the adoption of foreign regulations, such as the 1956 San Francisco code or the 1958 Uniform Building Code. The employment of the latter was coupled with the adoption of a very low design base shear (~3% of the seismic weight), either uniformly or linearly distributed along the height of the structure. In the structural design of steel and reinforced concrete buildings, the German DIN code and ACI-318-1956 were used instead (Rosenblueth and Prince, 1966).

By May 1965, when a magnitude M_s 6.0 earthquake hit San Salvador (Harlow et al., 1993), the tallest structure in this city was a bell tower, 70 m in height, followed by a 13-story shear-wall building and a few 11-story reinforced concrete frame structures (with masonry infills). It is also noteworthy that amongst a relatively large number of 6- to 10-story buildings, in only one was steel employed as the main structural material (Rosenblueth and Prince, 1966).

Following the 1965 San Salvador earthquake, a national seismic design code, based on the code of Acapulco (Guerrero, Mexico) was introduced in 1966 by the Ministry of Public Works (MOP, 1966). This code did consider torsional effects and provided some guidance and recommendations on the use of dynamic analysis for seismic design. However, it did not take into account soil conditions in the derivation of the design spectrum, a significant drawback given the terrain characteristics in El Salvador. Nonetheless, by the mid-1970s, a few structural engineers in El Salvador were already aware of the importance of considering site effects in the response of structures, and were estimating the thickness of volcanic ash layers, the characteristic site period, and effects of soil-structure interaction in the design of their structures. In fact, as reported by Lara (1987), the few structures where design had indeed accounted for soil effects performed significantly better during the 1986 earthquake.

As mentioned previously, another moderate-magnitude earthquake hit the city of San Salvador in the year of 1986. In the aftermath of this event, an Emergency Seismic Design Code (MOP, 1989), drafted by a technical committee appointed by the Salvadoran Association of Engineers and Architects (ASIA), was promptly developed and implemented three years later (Alfaro, 1994). The most significant modifications to the previous regulations were the increase of elastic acceleration spectral coordinates, reflecting the accelerograms recorded during the 1986 earthquake, and the consideration of geometric nonlinearities (P-Δ effects). However, and somewhat disappointingly, soil properties were still not included in the formulation of the seismic coefficient and no detailed specifications on dynamic analysis were provided.

In 1994, the Ministry of Public Works released a set of regulations for building practice in which the current seismic code is included (MOP, 1994). The whole project was financed through a World Bank loan, and was carried out by a private company with input from expert consultants, including a prestigious team from the Universidad Nacional Autónoma de México (UNAM). This code has the merits of not having been issued as an ill-matured post-earthquake measure and of including a probabilistic seismic hazard assessment of the country. Moreover, the subsoil conditions were finally accounted for in the base shear computation procedure.

Keeping in line with the focus of the present paper, this section addresses mainly those issues of the code that are somehow hazard-related, with only minor references to topics on structural design and detailing. In describing the current code requirements, a brief comparison with the corresponding prerequisites of previous regulations will also be given, so as to highlight the developments achieved so far. Table 1 summarizes the definition of the static base shear coefficient for the three Salvadoran codes.

Design Peak Ground Acceleration

In the 1966 and 1989 editions of the El Salvador seismic code, the value of peak ground acceleration (PGA) was not explicitly provided. Indeed, this factor was instead embedded within the coefficient C, which would simultaneously account for the seismic zone being considered (two were proposed in these codes) and for the response modification factor. This approach was evidently very disadvantageous, since it not only forced a "blind" design approach with reduced knowledge of the actual parameters involved, but it also prevented more in-depth studies, including comparison with other codes or design methodologies.

The present regulation, released in 1994, did eliminate this troublesome feature by introducing independent values for PGA and response modification factors, which are now explicitly used in the calculation of the base shear coefficient. The PGA is obtained through the zonation map of the country shown in Figure 1. The design values of PGA are 0.4 g and 0.3 g in zones I and II, respectively.

Importance Categories and Soil Types

In comparison to the 1966 version, the 1989 and 1994 seismic codes feature an additional importance category, referring to essential buildings such as medical facilities, communication centers, and fire and police stations, required for emergency response. The importance categories, as currently defined in the 1994 code, are described in Table 2.

Furthermore, and as mentioned above, the 1994 code also features the overdue introduction of soil classes, used in the determination of the response spectrum. This is achieved by means of site coefficients, C_o and T_o, which vary from 2.5 to 3.0 and 0.3 to 0.9, respectively, depending on the soil category considered; S1—rock, S2—stiff soil, S3—soft soil, S4—very soft soil (shear wave velocity below 150 m/s). The use of site classes referred to as S1–S4 is reminiscent of previous versions of the Uniform Building Code (UBC), but the classification criteria in the Salvadoran code are quite different; for example, the

TABLE 1. DEFINITIONS OF THE STATIC BASE SHEAR COEFFICIENT IN THE SEISMIC DESIGN CODES OF 1966, 1989, AND 1994

	1966	1989	1994
Base shear coefficient	$V/W = ICD$	$V/W = ICD$	$V/W = \left[\dfrac{AIC_o}{R}\left(\dfrac{T_0}{T}\right)^{2/3}\right]$
Importance factor	$I = 1.3, 1.0, 0.0$	$I = 1.5, 1.3, 1.0, 0.2$	$I = 1.5, 1.2, 1.0$
Zonation	Implicit in factor C Zone I, PSA = 1.0 Zone II, PSA = 0.5	Implicit in factor C Zone I, PSA=1.0 Zone II, PSA=0.5	Considered as A Zone I, PGA = 0.4 Zone II, PGA = 0.3
Response spectrum factor	$D = \sqrt{C/X_C}$, $0.6 \le D \le 1$; where, X_c = Displacement of the center of gravity of the building, in cm	$D = \dfrac{0.72}{T^{2/3}} \le 1.0$	Included in the base shear formula; however, there are limits by using: $T_o < T < 6T_o$
Soil characteristics	Not considered	Not considered	$C_o = 2.5, 2.75, 3.0, 3.0$ $T_o = 0.3, 0.5, 0.6, 0.9$
Structural factor	Implicit in factor C	Implicit in Factor C	R, selected depending on the structural type
Fundamental period	Not used in the computation of the static base shear	The fundamental period can be computed using any of the following methods: Rayleigh's Formula. For all buildings For framed buildings: $T = C_t h_n^{3/4}$, where: $C_t = 0.085$ steel frames $C_t = 0.073$ R/C structures h_n: height of the structure For other buildings: $T = \dfrac{0.09h_n}{\sqrt{Le}}$, $Le = L_{s\,max}\left[\sum \dfrac{L_s^2}{L_{s\,max}^2}\right]$	The fundamental period can be computed using any of the two following methods: For all buildings: $T_A = C_t h_n^{3/4}$, where: $C_t = 0.085$ steel frames $C_t = 0.073$ R/C structures $C_t = 0.049$ for other structural systems h_n: height of the structure Rayleigh's Formula For all buildings, where: T_B = fundamental period; however, $T_B \le 1.4 T_A$

TABLE 2. DESCRIPTIONS OF IMPORTANCE CATEGORIES IN THE 1994 SEISMIC DESIGN CODE

Category	Type of occupancy	Factor
I	Dangerous or essential structures: Important facilities necessary for life, care, and safety after an earthquake, such as hospitals, fire and police stations, military compounds, etc.	1.5
II	Special occupancy buildings: Facilities with high occupancy or non-permanent high occupancy, necessary for continued operation immediately after an earthquake (e.g., government buildings, day-care centers, markets, transportation terminals, etc.).	1.2
III	Normal occupancy buildings: Structures with low occupancy including structures to be used as family residences, offices, commercial stores, hotels, industrial facilities, etc.	1.0

shear wave velocity to distinguish rock sites in the latter code is 500 m/s, compared with 760 m/s in the UBC. Although the authors have not been able to establish the basis for the site classifications in the current Salvadoran code, it would appear to have been calibrated to conditions encountered in El Salvador.

In Figure 3 the 1994 code spectra for each importance category and each soil class are plotted. The differences and variations for each of the parameters are obvious and require little explanation. Hereafter in this paper, an importance category III, representing normal-occupancy buildings, is adopted in all subsequent spectra plots, together with a soil class S3, which best matches the characteristics of the young volcanic ash deposits that are predominant in the most populated parts of El Salvador (Rolo et al., this volume, Chapter 5).

Elastic Response Spectra

The shape of the elastic code spectra has been revised and updated in every edition of the three aforementioned earthquake-resistant design regulations in El Salvador. Indeed, following the initial proposal of 1966, the code spectral shape was then modified in order to reflect better the characteristics of the strong motion recorded during the 1986 earthquake (Shakal et al., 1987). This meant that the constant acceleration plateau was made wider and that the rate at which response acceleration decreases thereafter was reduced.

Then, in 1994, the code spectral shape was modified once more, adding a further degree of conservatism, as is observed in Figure 4, where the three elastic code spectra previously discussed are plotted, together with the response spectra (enveloping the two horizontal components) of two records from the 1982 and 1986 earthquakes in El Salvador, obtained at the OBS and IGN stations, respectively.

It is perhaps important to note that since, as described earlier, the values of PGA were not stated in the 1966 and 1989 codes, it was necessary for the authors to infer such values. A value of 0.4 g was thus adopted as the PGA, assuming that the maximum elastic spectral acceleration implied in these previous codes was 1.0 g and that the maximum spectral amplification ratio was 2.5, according to standard code conventions. In this way, the three code spectra could be normalized at the origin, and their qualitative features compared.

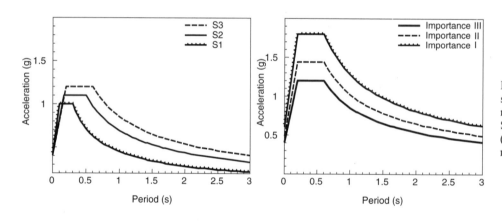

Figure 3. Design spectra from the 1994 seismic code for different soil types and normal occupancy (left) and for soil S3 and different importance categories (right). Importance category III is normal occupancy.

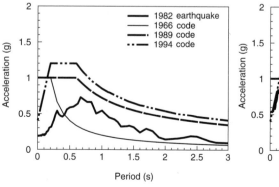
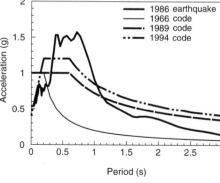

Figure 4. Elastic design response spectra from the 1966, 1989, and 1994 seismic codes compared to the elastic response spectra of the 1982 and 1986 earthquakes.

Response Modification Factor

As mentioned above, in the 1966 and 1989 codes, the "response modification factor" was not explicitly considered, but rather implicitly included in the coefficient C. This had the important drawback of leaving the designer with little knowledge of the degree of inelasticity that was being adopted in the design, since the ratio of design (inelastic) to elastic base shear was unknown.

This limitation was removed from the 1994 code edition, which for the first time explicitly defined a response modification factor R, based on the structural type and the building material used. However, this factor R, used to scale down the elastic response spectrum into the inelastic design spectrum, does not have a direct physical meaning. In effect, if the response modification factor, as defined in the current code, represented the actual ratio between the elastic and design base shears, this would effectively mean that a ductile moment-resisting frame, for instance, had to be designed for an extremely demanding overall structural ductility of 12.

This apparent idiosyncrasy results from the need of the response modification factor to account for the fact that a load combination for gravity load design is used to calculate the inertial weight of the structure, from which the base shear is derived. Since it is unlikely that a building will be at its maximum gravitational load capacity during the occurrence of an earthquake, the vertical loading adopted in a seismic combination case may be reduced. Hence, the R factor used in the El Salvador code constitutes in fact a ratio between a "gravity-load-based" elastic base shear and a "seismic-load-based" inelastic base shear, thus requiring higher values than those expected in a response modification factor that would account only for the capacity of the structure to respond inelastically to seismic loading.

In order to estimate the actual ductility demand implicit within a given response modification factor, one can compare the values of R with those of the response modification factor, q, in Eurocode 8 (CEN, 2002). The latter can be directly related to structural ductility μ, through the use of well-known expressions available in the literature (e.g., Chopra, 2001). Taking also into account the load factoring and load combination employed in both codes ($1.4G_K + 1.4Q_K + 1.4E_K$ in the El Salvador code and $G_K + 0.3Q_K + E_K$ in Eurocode 8), it is straightforward to conclude that R is two to three times larger than its equivalent counterpart in Eurocode 8, q.

Therefore, a high-ductility R value of 12 may be assumed as equivalent to a q value of 5, which is in fact the maximum value of a high-ductility moment-resisting frame. Hence it is also now possible to determine the corresponding ductility values: For intermediate- and long-period structures $\mu = q$ (equal-displacement approach), while for the case of short-period systems $\mu = (q^2 + 1)/2$ (equal-energy hypothesis).

Design Response Spectra

In Figure 5, the inelastic design spectra from the three editions of the Salvadoran building code are plotted, and again compared against the two aforementioned records from the 1982 and 1986 earthquakes. Observation of these plots renders evident the fact that all three versions of the design code spectra consistently, and significantly, underpredict the inelastic structural response below 1.0 s.

This deficiency comes as a result of a conspicuous failure of these codes to acknowledge, and account for, the variation of the response modification factor with the vibration period, particularly within the short-period range. In the limit, the inelastic spectrum of the current code suggests that ground acceleration is dependent on structural ductility, scaling it down by a factor of up to 12. This constitutes a gross misinterpretation of seismic ground-motion mechanisms and theory of structural dynamics, which clearly states that ductility demands on very short-period systems may be very large even if their strength is only slightly below that required for the system to remain elastic (Chopra, 2001). In fact, because of the latter, many seismic design regulations, such as the Mexico Federal District and the European (Eurocode 8) codes, feature inelastic spectral amplification factors that depend on both the response modification factor and the period of the structure.

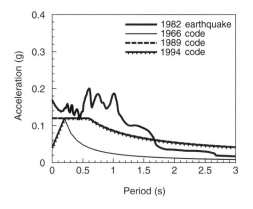
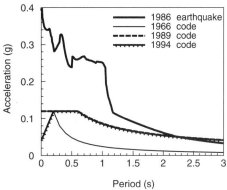

Figure 5. Inelastic design response spectra from the 1966, 1989, and 1994 seismic codes compared to inelastic response spectra of the 1982 and 1986 earthquakes.

It is important to note that since the code design spectra are derived for a constant modification factor, a valid comparison with records from previous events at El Salvador can be made only if the corresponding inelastic spectra are also computed for an invariable value of R. Since, for a given response modification factor, the value of ductility varies with period range, it results that the records' inelastic spectra must be derived for a variable ductility factor. In the current work, and for an R factor of 12 (approximately equivalent to $q = 5$, as discussed above), the authors chose to employ ductility values of $\mu = 10$, $\mu = 8$, and $\mu = 5$ for the period ranges of $0 < T < 0.3$, $0.3 < T < 0.5$, and $T > 0.5$, respectively. While the first and third period limits correspond to short- and intermediate/long-period ranges, for which the expression discussed earlier can be employed in the determination of μ, the second range has been arbitrarily selected to provide a transition between the two. This applies to Figure 5 and to all subsequent plots of inelastic spectra.

Verification Criteria

Similar to the philosophy behind the two previous editions, the 1994 code requires the verification of three main limit states for structures subjected to earthquake action, so as to ensure that structures have sufficient capacity to resist small earthquakes with no damage, moderate earthquakes with nonstructural or minor damage, and strong earthquakes without collapse. However, the code also assumes that verification of structural response under a single seismic design level (collapse prevention, corresponding to a ground motion with a return period of 475 yr) is sufficient to guarantee adequate performance at all possible levels of earthquake loading.

This performance verification is carried out by means of checks on deformation and strength capacity. The former, however, tend to control the design of framed structures in El Salvador, since the criteria used (inter-story drift limit) have always been very severe. Indeed, in the 1966 code, the verification criterion consisted of a very stringent drift limit of 0.2% of story height, independently of building type and ductility level. In 1989, the same story drift limit was increased up to a much more realistic value of 1.5%, although still independent of building type and level of ductility. It is noteworthy that, in the latter code, this limit was to be compared to the inelastic deformation of the structure, obtained as the product of the elastic displacement under the design seismic load with a displacement amplification coefficient C_d, specified in the code.

With the 1994 code, three drift limits (1%, 1.5% and 2%) were instead provided, depending both on the building type and on the importance of the structure. In general, a 1.5% drift limit is to be adopted for a framed building of normal importance (category III). This value does seem to be in tune with the current guidelines in a number of codes and guidance documents, such as Eurocode 8 (CEN, 2002) and FEMA-273 (ATC, 1997).

Regarding the verification of strength capacity, all three editions of the code have systematically prescribed an allow-

able stress design methodology, carried out within a force-based framework. A detailed description of such procedure is, however, well beyond the scope of the current work.

Brief Assessment of 1994 Code Requirements

One of the most effective ways of evaluating the merits and/or weaknesses of a seismic code is to carefully assess how structures designed according to such regulations perform under the action of an earthquake. In the particular case of El Salvador, this pragmatic code-assessing methodology can indeed be employed, since, as discussed earlier, earthquakes occur with relatively high frequency. Therefore, in the following subsection, the effects that the two recent events of January and February 2001 had on structures designed according to the 1994 code will be briefly reviewed.

Additionally, the adequacy of the code prerequisites can equally be assessed through a comparison between the seismic input prescribed by the code, and the characteristics of the strong motion that have been effectively recorded at the affected areas. Hence, the aforementioned review on structural response in El Salvador will be preceded by a brief overview of code input requirements and strong-motion characteristics of previous events.

Seismic Input

In Figure 6, the code elastic response spectrum for "normal" structures (importance category III) is compared to the response spectra of a number of records from the June 1982, October 1986, January 2001, and February 2001 earthquakes.

It is observed that the current code spectrum does fail to fully enclose the acceleration spectral characteristics of past events, albeit by a not too significant margin, especially taking into account that the value of PGA is reasonably close to the recorded values in most cases.

The above scenario would become gloomier if one were to consider those limited cases where local characteristics might cause a record to exceed by up to a threefold factor the code-specified response, as was observed in the recording station at La Libertad during the 13 January earthquake in El Salvador, for instance. However, as shown by Bommer et al. (2002), such peculiar situations are usually associated with specific local site conditions, and should therefore not be extrapolated to the more generalized case.

In Figure 7, the code inelastic design response spectrum for "normal" high-ductility (R = 12) structures is compared to the inelastic response spectra of the aforementioned recordings from the June 1982, October 1986, January 2001, and February 2001 earthquakes. It is clear that the current code design spectrum severely underestimates the values of spectral response observed during past events. The main reason for this deficiency stems from the gross underprediction of structural inelastic response in the short-period range, already discussed above.

It is clear from the above that a review of the design spectra currently prescribed in the 1994 seismic design code is called for.

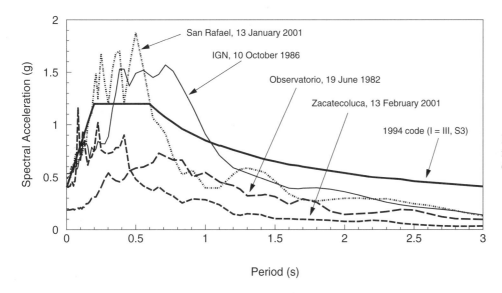

Figure 6. Elastic response spectra for records from the 1982, 1986, and 2001 earthquakes.

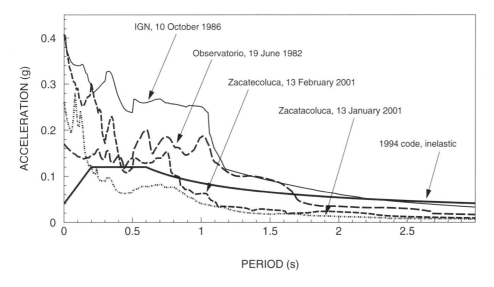

Figure 7. Inelastic response spectra for records from the 1982, 1986, and 2001 earthquakes.

This should be carried out in tandem with other improvements and modifications, suggested in a later section.

Structural Performance

The most common building practice in El Salvador involves the employment of reinforced concrete moment-resisting frames, the design of which is mainly drift-controlled, as pointed out before. Most of the damage observed after the 2001 earthquakes in recently constructed buildings was to nonstructural components, such as infill panels and wall mortar plaster. Structural components of such buildings remained, by and large, undamaged, but for reasons that are explained subsequently this cannot

be interpreted as demonstrating the efficiency of the current code in protecting structural members in this type of construction.

Relatively new buildings constructed using *adobe* and *bahareque*, on the other hand, were heavily damaged, despite the presence in the code of an annex that does cover adobe construction. This can probably be attributed to the fact that such regulations have never had proper divulgation in areas where adobe construction dominates, outside the main city centers (Dowling, this volume, Chapter 22).

Finally, it was also observed that most of the masonry and reinforced concrete buildings damaged or destroyed by the two earthquakes of 2001 had been designed and erected more than 25 years ago, at a time when the present regulations were obviously not yet in place. Furthermore, in many cases these structures had already been weakened by previous large seismic events, such as those in 1982 and 1986, as well as by swarms of smaller earthquakes in 1975, 1985, and 1999.

In summary, it does seem that the seismic code released in 1994, to the extent that it is actually applied, could contribute to a reduction of structural vulnerability in El Salvador. However, it is also important to note that there is still a large percentage of buildings that have been designed using the 1989 code, the 1966 code, some other regulations, or none at all, neglecting seismic design considerations completely. Therefore, although the regulations of the 1994 code may seem to cover future seismic demands in a satisfactory manner, with minor adjustments, there is still going to be a large percentage of structures whose seismic performance is unknown.

Future Developments for the Seismic Code in El Salvador

The 1994 code established a Technical Committee of Structural Safety, serving as an advisory group to the Ministry of Public Works (MOP), and which has the responsibility of studying and proposing changes to the code in order to incorporate new technological and scientific developments in the earthquake engineering field. This Committee constitutes therefore a commendable source of incentive for a constant updating of current seismic regulations at El Salvador. In this section, an attempt is made to somehow contribute to this objective, particularly taking into account the fact that the 1994 seismic code has to undergo revision every ten years.

Developments in Terms of Seismic Input

It is known that using a single value, the PGA, to define the amplitude of a design spectrum produces a spectrum that does not have uniform return periods in all its ordinates. Some codes, such as the 1997 edition of the Uniform Building Code, UBC97 (ICBO, 1997), attempt to overcome such inconsistency by using two parameters, one related to short-period spectral ordinates, the other to the spectral ordinates at intermediate response periods. Similarly, the Canadian code provides one zoning map in terms of PGA and another one as a function of PGV, the latter being employed to define the intermediate ordinates of the spectrum, while the former is used for the short-period range. It is believed by the authors that a similar approach should be employed in the case of El Salvador.

An important characteristic of the seismic activity is the near-source rupture directivity effect (Somerville et al., 1997). Indeed, the highly destructive effects of the ground motions close to the causative fault rupture are now well recognized. However, UBC97 and IBC 2000 (ICBO, 2000) are the only seismic design codes that explicitly account for near-source effects through the application of spectral amplification factors, N_a and N_v. These factors are used to increase the spectral ordinates at short and intermediate periods, respectively, for sites located less than

15 km from an active fault capable of producing a 6.5 magnitude earthquake or higher. Again, it is the opinion of the authors that this feature should be included in all seismic codes, including that of El Salvador, especially since there is evidence that the destructive motions of the 10 October 1986 earthquake in San Salvador, despite its relatively small magnitude (M_w 5.7), were in part due to near-source directivity effects (Bommer et al., 2001).

Another important feature of ground motion exclusive to sites near to the causative fault is high vertical accelerations. Indeed, vertical motion can contribute significantly to the level of damage, either directly or through interaction with the horizontal shaking. A few seismic codes consider this component of ground movement by suggesting the use of a vertical response spectrum that is obtained by multiplying the corresponding horizontal component by a constant factor of 2/3. This approach, however, ignores the fact that the two spectral shapes are completely different and that in the near field, the vertical spectral ordinates can even exceed the horizontal ordinates. Hence, for the case of El Salvador, the authors would rather suggest the employment of a methodology similar to that introduced in the new edition of Eurocode 8, where an independent vertical spectral shape is defined, featuring a maximum vertical ordinate of 0.9 g (Sabetta and Bommer, 2002).

Regulations and Guidelines for Existing Structures

It was stated earlier that many new structures, a few of which were designed according to the current code, seem to have performed satisfactorily, from a structural safety point of view, in the recent events of 2001. However, and although these new structures did not approach their collapse point, it is probable that they have indeed incurred some level of damage, even if not directly visible to the naked eye. In fact, and as described earlier, history in El Salvador has shown that the structural weakening effect that a previous earthquake has on apparently undamaged structures might lead to disastrous consequences when a new event hits the area.

Furthermore, the current seismic code provides no guidance on how to assess the remaining capacity of a building after an earthquake. In addition, this lack of structural assessment guidelines is complemented by an equally conspicuous absence of guidance on methods for structural retrofitting, be it for repair of damaged structures or for the strengthening of capacity-deficient ones.

The above becomes even more significant if one takes into account the fact that the majority of the building stock in El Salvador has been designed according to outdated regulations (or following none at all), and has been subjected to a large number of events during its lifetime. Therefore, it is imperative that structural assessment and redesign methodologies are introduced in future versions of the El Salvador seismic code.

Design Philosophy

It is now understood, and widely accepted, that the damaged inflicted by an earthquake on a structure is more related to the displacement experienced by the construction than to the forces imposed on the structure. Therefore, it is almost certain that the new generation of seismic codes will consider displacements, as opposed to acceleration or forces, to represent seismic input.

In addition, recent developments in earthquake engineering research and application have highlighted the importance of the adoption of an explicit performance-based seismic design of structures. Although, conceptually, this does not constitute a novelty on its own, its current implicit implementation in seismic codes such as the Salvadoran one, whereby only one level of design is prescribed and deemed capable of guaranteeing verification of three performance targets, has proven not to be effective. Therefore, a complete shift into full performance-based design, whereby up to four limit states, related to different return periods and different criteria of structural behavior, need to be verified during the design stage, is also called for.

Finally, it is noteworthy that the above would also imply a shift from the current allowable stress design philosophy to an ultimate capacity verification requirement, as currently implemented in the vast majority of modern seismic design codes.

Code Enforcement

The effectiveness of a seismic design code in contributing to the long-term reduction of earthquake risk, provided it is free of major technical deficiencies, is primarily dependent on the degree to which its implementation is adopted by engineers and enforced by the appropriate authorities. Arguably, the most effective building codes have been those whose main purpose has been to apportion responsibility for the performance of the building, the clearest example being the first known legal requirements for buildings included in the code of laws enacted by the Babylonian king Hammurabi ~2100 B.C. This ancient code afforded almost no guidance on house construction but imposed penalties on the builder for poor structural performance: In case of collapse of a house killing the occupants, the penalty was death (Berg, 1982). While such drastic measures may no longer be acceptable, there is a valuable lesson to be learned in terms of code objectives, in addition to providing guidance on design: that of making the engineer responsible if these guidelines are ignored.

Lara (1987) reports that the 1966 code for earthquake-resistant design was rarely applied in El Salvador and much less enforced. There is little evidence to suggest that the situation is significantly different with respect to the 1989 and 1994 codes. The lack of enforcement is symptomatic of the general weakness of both central and municipal governments in El Salvador, which has been termed "institutional vulnerability" (Bommer et al., 2002). The weakness of municipal governments is the main obstacle to effective code enforcement, since within the legal framework of El Salvador, the responsibility for enforcement of regulations such as building codes ultimately resides with the local authorities. Municipalities in El Salvador, with the possible exception of those that collectively form the Oficina de Planificación del Area Metropolitana de San Salvador (OPAMSS) in

the capital city, are unable to carry out this task not only because of lack of resources in general, but specifically because of the almost complete lack of adequately trained technical personnel. Furthermore, it has been pointed out that to enforce building regulations, and for that matter land-use policy, it is also necessary to possess strong legal teams to take on the lawyers representing the major construction companies and developers, another area in which the municipalities are extremely weak (Wisner, 2001).

The lack of seriousness in addressing this issue is reflected in the proposed reforms to the Law of Urbanism and Construction, enacted as part of the process of introducing the new seismic code in 1994. This law formed the Technical Commission for Structural Safety (CTSE), mentioned previously, consisting of one representative from each of the following organizations:

- Salvadoran Association of Engineers and Architects (ASIA)
- College of Architects of El Salvador (CADES)
- Salvadoran Chamber of the Construction Industry (CASALCO)
- Salvadoran Society for Earthquake Engineering (SSIS)
- Three universities: Universidad Centroamericana "José Simeón Cañas" (UCA), Universidad de El Salvador (UES), and Universidad Albert Einstein (UAE) (see "Earthquake Engineering Education" below)
- The Ministry of Public Works (MOP)

The CTSE, which is coordinated by ASIA and operates without any material or financial support, is supposed to meet at least once a month, and its duties include (1) proposing changes to the building regulations to incorporate scientific and technological advances; (2) evaluating any innovative materials or systems of construction that are to be introduced; (3) verifying the fulfillment of the code and the technical norms in those projects that the commission, through their own work, through reports, or accusations, judge to be in need of inspection; and (4) collaborating in the evaluation of structural damage caused by any catastrophes. No small undertaking for eight individuals, without technical, logistical, or even secretarial support, to carry out on an effectively voluntary basis in addition to their normal daily commitments.

Further evidence of the extent to which MOP has delegated all responsibility to ASIA is the fact the regulations for structural safety and the technical norms that accompany them have actually been published by ASIA on behalf of MOP. Furthermore, ASIA is the sole organization that distributes these documents, at costs that are not insignificant to Salvadorans: The regulations and the norms for wind and earthquake loading, documents of 27 and 36 pages respectively, cost a total of U.S.$35.

EARTHQUAKE ENGINEERING IN EL SALVADOR

In many civil engineering degree courses around the world, elective courses are offered on engineering seismology, earthquake engineering, and seismic design. In a country like El Salvador, where it can almost be guaranteed that all civil engineering works will be subjected to at least one earthquake during their useful lives, it is hard to conceive of earthquake-resistant design as an option. The long-term benefits to seismic risk reduction of incorporating earthquake engineering as an integral part of civil engineering education are obvious; this section reviews the work that is being done toward this goal.

Earthquake Engineering Education

There are a disproportionately high number of universities in El Salvador. The national university, Universidad de El Salvador (UES), was founded in 1841 and has in the past enjoyed a reputation as one of the foremost seats of learning in Central America in some disciplines, particularly medicine. The School of Civil Engineering at the UES was founded in 1954. Civil engineering is also taught at a number of private universities including the prestigious Universidad Centroamericana "José Simeón Cañas" (UCA), established and run by the Society of Jesus, as well as by Universidad Politécnica (UPES), Universidad Albert Einstein (UAE), Universidad Tecnológica Latinoamericana (UTLA) in Santa Tecla, Universidad Católica de Occidente (UNICO) in Santa Ana, and Universidad Gerardo Barrios in San Miguel.

The UCA, UPES, and UAE have all existed for several decades, but the proliferation of universities in El Salvador, which at its peak reached 43 until legislation by the Ministry of Education forced eight of these to close due to lack of academic credentials, can be traced back to a series of military occupations of the UES. The university was closed from 1970 to 1972 during the dictatorship of General Molina, and in 1975 was again closed for about six months. The greatest damage to the university occurred at the beginning of the civil war (1980–1992) when the university was occupied by the armed forces from 1980 to 1983. During this time many academics were killed or exiled and the university lost most of its resources due to vandalism and theft. During this extended closure, while UES academics continued to give classes outside the university in whatever way could be arranged, tens of new private universities, some with extremely weak academic credentials, appeared. Some of these universities have flourished, while the UES, even after reopening, continued to suffer from chronic lack of funding and the effects of the damage inflicted during the military occupation. The UES suffered a further major blow in 1986 when the 10 October earthquake caused the collapse of several of the university buildings and extensive damage to others.

The civil war inevitably took a toll on education in general and was disruptive to other universities as well. A particularly infamous example was the murder of six leading academics at the UCA, including the rector, by the armed forces in November 1989. After ten years of peace, the universities have begun to recover, and many positive initiatives have been put in place as a result of increased funding, greater freedom of operation, and in particular through increasing links with overseas universities.

In terms of the integration of seismic design principles into civil engineering training, some progress has been made. At the UCA, structural dynamics has always formed a core module in the civil engineering degree course, and for the last ten years, engineering seismology and earthquake-resistant design have also been compulsory modules in the final years of study.

At the UES, elective courses related to earthquake engineering have been offered at different times, depending upon the availability of lecturers with specialist training in the subject. In 1979, an elective module on earthquake engineering was introduced by Héctor David Hernández Flores, who had been awarded a Ph.D. in this field at the University of Illinois; this module, however, ceased to run more than a decade ago. In 1994, the second author of this paper introduced another elective course on fundamentals of engineering seismology and earthquakes engineering. In 1998, a new curriculum was introduced, which includes elective modules in the following subjects: seismology, dynamics of structure, earthquake engineering, and advanced structural analysis. All of the modules listed were introduced by academic staff members after having studied postgraduate courses overseas.

To the knowledge of the authors of this paper, subjects related to engineering seismology, structural dynamics, and earthquake engineering are not taught as part of the civil engineering degree courses at any other university in El Salvador.

Paradoxically, although teaching related to seismic risk reduction measures is still very limited at the undergraduate level, there have been some important recent initiatives regarding postgraduate training. In the past, a few engineers from El Salvador have studied for masters and doctoral degrees in seismology and earthquake engineering overseas, and some of those who have subsequently returned to El Salvador have made significant contributions to developing these fields within the country. The most common destination of Salvadorans studying seismology and earthquake engineering overseas has been the International Institute of Seismology and Earthquake Engineering in Tsukuba, Japan. During the decade of the 1990s, programs of assistance in these disciplines from Scandinavian countries resulted in many Salvadorans being trained, mainly in earth sciences, at the University of Bergen and NORSAR in Norway and at the Royal Institute of Technology (Kungliga Tekniska Högskolan or KTH) and Uppsala University in Sweden. These training periods have varied from short visits of a few months' duration to master's and Ph.D. courses.

In 2001, a master's course in Geo-environmental Technology was launched at UPES in collaboration with the Universidad Politécnica de Madrid (UPM) in Spain. The master's course is made up of three subject areas: geo-resources, environmental science, and geological risks, the last of these including earthquakes, tsunamis, landslides, and techniques of natural disaster mitigation amongst the topics covered. The courses are mainly taught by professors from Madrid, and the degrees will be awarded by UPM rather than by UPES. A number of faculty members from other Salvadoran universities were enrolled amongst the first cohort of master's students.

Another significant new development is the Graduate School on Natural Disaster Mitigation in Central America (NADIMCA), which planned to begin offering master's degrees in 2003. The course is coordinated by Uppsala University (Sweden) and the

Universidad Nacional de Costa Rica (UNA) with support from the Swedish International Development Agency. The master's courses will be run over two years, with all students based in Costa Rica for the first year, taking common modules related to natural hazards and disaster mitigation. During the second year of the degree, students will study at one of the Central American universities in the NADIMCA network, according to the particular topic in which they choose to specialize. The UCA in El Salvador will operate as the regional center for landslides under the technical coordination of Ing. José Mauricio Cepeda, a graduate of the M.S. program in Soil Mechanics and Engineering Seismology at Imperial College London. The NADIMCA project will also include the opportunity for students to study for doctoral degrees in Sweden.

Earthquake Engineering Professional Training

The lack of compulsory modules related to earthquake engineering in most undergraduate civil engineering degrees in El Salvador, plus the continual advance of earthquake engineering technology and practice, make training in this area for practicing engineers another important tool in the risk mitigation arsenal. The universities, particularly UES and UCA, offer many seminars and lectures open to members of the public, and in this way contribute to this continuing professional development. Two other organizations exist that would also be expected to play an important role in this respect: SSIS and ASIA. The Salvadoran Society of Earthquake Engineering (SSIS) was formed after the 1986 earthquake to promote the development of earthquake engineering in El Salvador, and during its first years of existence was quite active, particularly in collecting information related to the effects of the earthquake. In 1987, SSIS organized a regional seminar on the seismic design and performance of reinforced masonry. Nowadays, SSIS is fairly dormant, with membership having dropped from ~50 at the time of its foundation to just over ten. The Society is associated with ASIA but has no legal status and produces no publications, nor does it operate a Web site. Curiously, the SSIS has never been formally recognized as the national association for earthquake engineering within the International Association for Earthquake Engineering (IAEE), despite the approval of El Salvador's participation in the preparatory committee for the foundation of IAEE in 1963. In April 1963 the National Committee on Earthquake Engineering in El Salvador formed and was affiliated to the IAEE; the president of this national committee was one A.G. Prieto and the national delegate was Dr. Robert Schulz, a seismologist working at the Center for Geotechnical Investigations (CIG) of MOP, who held this post until 1976. Two other Salvadorans have held the post of national delegate to IAEE, both employees of CIG and neither connected to SSIS. The last appointment expired in 1996, and El Salvador is not currently represented within the international association.

The Salvadoran Association of Engineers and Architects (ASIA) could be considered as the most important professional association in the country with regard to earthquake engineering.

In addition to its important roles as the publisher and distributor of the current seismic codes, and coordinator of the CTSE, ASIA also operates, again on behalf of MOP, the Committee for the Evaluation of Damages. This committee has operated after both the 1986 and 2001 earthquakes, organizing its members to carry out preliminary assessments of affected structures, to tag them, and to build up an inventory of damages. The evaluations are carried out using a standard six-page form that classifies each building by its structural characteristics and function, finally classifying the building as green, yellow, orange, or red tag, depending on the severity of damages and the need to carry out repair, strengthening, or demolition. Sadly, the comprehensive and critically important inventory of damages submitted by the committee to MOP following the 1986 earthquake is no longer available, and repeated attempts to obtain copies of the document from MOP have met with failure, the explanation given being that the document has been lost.

ASIA also contributes to professional development through its publications and through courses and seminars. ASIA publishes a technical journal, *Revista ASIA*, several times a year with articles, written by members of the association and also by foreign specialists, on many topics of relevance to civil and environmental engineering, including many on themes related to earthquake engineering. Special issues on earthquake engineering have been published after both the 1986 and 2001 earthquakes. ASIA has also organized innumerable courses related to earthquake engineering and seismic design, ranging from single presentations to courses held over periods of one or more weeks. Lecturers on these courses have included both national and foreign experts in various aspects of geotechnical, structural, and earthquake engineering. Courses of particular note are those held in 1997 and 1998 to introduce and explain the 1994 code of construction as a means of disseminating the new technical norms and encouraging their adoption in engineering practice. ASIA has also hosted on two occasions, in 1995 and 2000, the UNESCO-sponsored Roaming International Course on Earthquake-Resistant Engineering. These courses, organized by Dr. José Grases of Venezuela on behalf of RELACIS (Red Latinoamericana y del Caribe de Centros de Ingeniería Sísmica), have also been held in several other countries in Latin America and the Caribbean.

Although ASIA clearly plays an important role as a focal point for the development of earthquake engineering amongst the civil engineering profession in El Salvador, it has been criticized for operating on an excessively commercial basis, with the attendance fees for many of its courses being set quite high and frequently out of the reach of many interested individuals.

Earthquake Engineering Research

A great deal of research has been, and continues to be, carried out in the field of earthquake engineering and related subjects in El Salvador. Tens of graduation theses in the civil engineering departments of the universities have been carried out, looking at topics such as the seismicity and seismic hazard of the country, microzonation of the capital, and many aspects of structural response and seismic upgrade in both vernacular and engineered structures. However, sustained research programs in this field have almost exclusively been carried out with foreign assistance, both in terms of technical expertise and, perhaps more importantly, financial support. Projects related to seismology and earthquake engineering have been funded by the European Union, the Norwegian Agency for Development Cooperation (NORAD), the United States Agency for International Development (USAID), the Danish International Development Agency (DANIDA), the Swedish International Development Cooperation Agency (SIDA), the Spanish Association for International Cooperation (AECI) and the Japanese Agency for International Cooperation (JICA), amongst others. The projects have been channeled through government agencies, such as CIG (and now through its successor SNET), through nongovernmental organizations, and through the universities.

There are many obstacles to creating a sustained culture of scientific and technical investigation carried out independently by academics and professionals within the country, but the lack of research funding must be amongst the most serious. In the early 1990s the government formed the National Council of Science and Technology (CONACYT) to promote science and technology in El Salvador. Similar government bodies in neighboring countries, most notably Mexico, provide research grants and scholarships for postgraduate studies overseas, but CONACYT in El Salvador provides relatively little research funding and no scholarships for students. In 2002 the Universidad de El Salvador (UES) received a grant of U.S.$500,000 from the Ministry of Education for research projects to be funded in the university. This modest sum, distributed amongst all faculties and departments, provides just U.S.$15,000 for each project, but this is at least a clear indication that research is beginning to be valued and supported: Prior to this grant, the annual allowance for the UES included a nominal sum simply to keep open the budget item for research—the sum was just over one dollar for the entire university.

The formation of the National Service for Territorial Studies (SNET) under the auspices of the Ministry of Environmental and Natural Resources (MARN) is a positive move toward increasing and coordinating research in natural hazards (Rose et al., this volume, Chapter 1). The focus of SNET is almost exclusively on the monitoring and characterization of natural hazards in El Salvador, and in general there has been a tendency for research to be focused more on the earth sciences rather than on actual risk mitigation (e.g., Salazar, 1999). There is clearly a need for the creation of one or more agencies that take on as their responsibility, preferably in collaboration with the universities, the promotion of research in earthquake engineering as well as engineering seismology in order to extend the work from characterization of seismic hazard to quantification and reduction of seismic risk. Alternatively, as it consolidates its organization and expands its activities, these aspects of research could be incorporated into the remit of SNET, which is forging strong research links with the universities.

CONCLUSIONS

El Salvador is a country of very high seismic hazard and consequently high seismic risk. Ongoing research, in particular an expansion of the strong-motion database, is refining the understanding of the hazard, although models of the hazard have to some extent been excessively based on procedures developed in and for regions with very different seismic characteristics. Further studies to characterize expected ground motions, including local site effects, and to quantify their destructive potential, are clearly needed. As in nearly all countries, there is also a need to continually update the representation of the seismic hazard in the seismic design code, to reflect the improved understanding of the seismicity and of the characteristics of the ground motion. The spectral reduction factors and the verification criteria have also been shown to be in need of revision. However, these refinements to the seismic code will have only a very minor impact on the seismic risk, if any at all. The most important task to increase seismic safety in El Salvador is to create a procedure for the rigorous enforcement of the seismic design code in all building and construction practice. Equally important are the promotion of improved earthquake resistance in vernacular housing and improved land-use control to avoid, or otherwise mitigate, the risk associated with the hazard of earthquake- and rainfall-induced landslides, topics dealt with in other papers in this volume.

The post–civil war era in El Salvador, despite the acute social, economic, and environmental problems that currently beset the country, has created the possibility of expanding and improving applied research in engineering seismology and earthquake engineering. With foreign technical and financial support, and limited government support, the first fruits of this new situation are already visible. There will be a need to sustain this support and to provide conditions to prevent a "brain drain" of the most able professionals trained in this field.

The effective mitigation of earthquake risk in El Salvador clearly requires a concerted program over many years and at many different levels. To some it may seem like an impossible task, and perhaps it is true that until fundamental social and health issues in El Salvador are resolved, it is unreasonable to expect earthquake risk mitigation programs comparable to those in countries such as the United States, Japan, and New Zealand, or even those in Mexico and Chile. Nonetheless, it is important to the sustainability of its development that El Salvador makes earthquake risk mitigation an integral component in the expansion of its housing and infrastructure.

ACKNOWLEDGMENTS

The authors are grateful to many individuals in El Salvador who provided information for this article, including Ing. Julio Bonilla, MSc José Mauricio Cepeda, Dr. Héctor David Hernández, Ing. Patricia Méndez de Hasbun, Ing. Mario Nieto Lovo, Ing. Rodolfo Nosiglia, Lic. José Pérez, and MSc Carlos Pullinger. The second author also wishes to acknowledge the generous support of the Royal Academy of Engineering for funding the fieldwork carried out to gather information for this paper.

REFERENCES CITED

Alfaro, C., 1994, El Salvador, *in* Paz, M., ed., International handbook of earthquake engineering: Codes, programs and examples: New York, Chapman and Hall, p. 205–214.

Alfaro, C.S., Kiremidjian, A.S., and White, R.A., 1990, Seismic zoning and ground motion parameters for El Salvador: The John A. Blume Earthquake Engineering Center Report No. 93, Palo Alto, California, Stanford University, 157 p.

Algermissen, S.T., Hansen, S.L., and Thenhaus, P.C., 1988, Seismic hazard evaluation for El Salvador: Report for the U.S. Agency for International Development, 21 p.

Ambraseys, N.N., and Adams, R.D., 2001, The seismicity of Central America—A descriptive catalogue 1898–1995: London, Imperial College Press, 309 p.

Ambraseys, N.N., Bommer, J.J., Buforn, E., and Udías, A., 2001, The earthquake sequence of May 1951 at Jucuapa, El Salvador: Journal of Seismology, v. 5, no. 1, p. 23–39.

Anderson, R.W., 1987, The San Salvador earthquake of October 10, 1986—Review of building damage: Earthquake Spectra, v. 3, no. 3, p. 497–541.

ATC (Applied Technology Council), 1997, NHERP guidelines for the seismic rehabilitation of buildings: Federal Emergency Management Agency Report No. 273, Washington, D.C., 407 p.

Bent, A.L., and Evans, S.G., 2004, The M$_W$ 7.6 El Salvador earthquake of 13 January 2001 and implications for seismic hazard in El Salvador, *in* Rose, W.I., et al., eds., Natural hazards in El Salvador: Boulder, Colorado, Geological Society of America Special Paper 375, p. 397–404 (this volume).

Berg, G.V., 1982, Seismic design codes and procedures: Oakland, California, Earthquake Engineering Research Institute, 119 p.

Bommer, J., 1999, Cargas sísmicas para el diseño estructural en El Salvador: Revista ASIA, no. 132, p. 15–25.

Bommer, J.J., 2000, Seismic zonation for comprehensive definition of earthquake actions, *in* Proceedings, International Conference on Seismic Zonation, 6th, Oakland, California: Earthquake Engineering Research Institute, 6 p.

Bommer, J., and Ledbetter, S., 1987, The San Salvador earthquake of 10th October 1986: Disasters, v. 11, no. 2, p. 83–95.

Bommer, J.J., and Martínez-Pereira, A., 1999, The effective duration of earthquake strong motion: Journal of Earthquake Engineering, v. 3, no. 2, p. 127–172.

Bommer, J.J., and White, N., 2001, Una propuesta para un método alternativo de zonificación sísmica en los países de Iberoamérica, *in* Proceedings, Congreso Iberoamericano de Ingeniería Sísmica, Madrid: Asociación Española de Ingeniería Sísmica, p. 80–89.

Bommer, J.J., Hernández, D.A., Naverrete, J.A., and Salazar, W.M., 1996, Seismic hazard assessments for El Salvador: Geofísica Internacional, v. 35, no. 3, p. 227–244.

Bommer, J.J., Udías, A., Cepeda, J.M., Hasbun, J.C., Salazar, W.M., Suárez, A., Ambraseys, N.N., Buforn, E., Cortina, J., Madariaga, R., Méndez, P., Mezcua, J., and Papastamatiou, D., 1997, A new digital accelerograph network for El Salvador: Seismological Research Letters, v. 68, no. 3, p. 426–437.

Bommer, J., McQueen, C., Salazar, W., Scott, S., and Woo, G., 1998, A case study of the spatial distribution of seismic hazard (El Salvador): Natural Hazards, v. 18, p. 145–166.

Bommer, J.J., Scott, S.G., and Sarma, S.K., 2000, Hazard-consistent earthquake scenarios: Soil Dynamics and Earthquake Engineering, v. 19, p. 219–231.

Bommer, J.J., Georgallides, G., and Tromans, I.J., 2001, Is there a near-field for small-to-moderate magnitude earthquakes?: Journal of Earthquake Engineering, v. 5, no. 3, p. 395–423.

Bommer, J.J., Benito, M.B., Ciudad-Real, M., Lemoine, A., López-Menjívar, M.A., Madariaga, R., Mankelow, J., Méndez de Hasbun, P., Murphy, W., Nieto-Lovo, M., Rodríguez-Pineda, C.E., and Rosa, H., 2002, The El Salvador earthquakes of January and February 2001: Context, characteristics, and implications for seismic risk: Soil Dynamics and Earthquake Engineering, v. 22, no. 5, p. 389–418.

Boore, D.M., Joyner, W.B., and Fumal, T.E., 1997, Equations for estimating horizontal response spectra and peak acceleration from western North American earthquakes: A summary of recent work: Seismological Research Letters, v. 68, no. 1, p. 128–153.

Cáceres, D., and Kulhánek, O., 2000, Seismic hazard of Honduras: Natural Hazards, v. 22, p. 49–69.

CEN (Comité Europée de Normalisation) 2002, Eurocode 8: Design of structures for earthquake resistance, Part 1: General rules, seismic actions and rules for buildings, draft no. 5, Brussels: European Committee for Standardisation, Doc CEN/TC250/SC8/N317, 197 p.

Chen Yong, Chen Ling, Güendel, F., Kulhánek, O., and Li Juan, 2002, Seismic hazard and loss estimation for Central America: Natural Hazards, v. 25, p. 161–175.

Chopra, A., 2001, Dynamics of structures: Upper Saddle River, New Jersey, Prentice Hall, 844 p.

Climent, A., Taylor, W., Ciudad Real, M., Strauch, W., Santana, G., Villagran, M., Dahle, A., and Bungum, H., 1994, Spectral strong motion attenuation in Central America: Technical Report No. 2-17, NORSAR, 46 p.

Cornell, C.A., 1968, Engineering seismic risk analysis: Bulletin of the Seismological Society of America, v. 58, p. 1583–1606.

Dahle, A., Taylor, W., Ciudad Real, M., Strauch, W., Santana, G., Villagran, M., and Bungum, H., 1995, New spectral strong motion attenuation models for Central America, *in* Proceedings, International Conference on Seismic Zonation, 5th, Nice, v. 2, p. 1005–1012.

DeMets, C., 2001, A new estimate for present-day Cocos-Caribbean plate motion: Implications for slip along the Central American volcanic arc: Geophysical Research Letters, v. 28, no. 21, p. 4043–4046.

Dewey, J.W., White, R.A., and Hernández, D.A., 2004, Seismicity and tectonics of El Salvador, *in* Rose, W.I., et al., eds., Natural hazards in El Salvador: Boulder, Colorado, Geological Society of America Special Paper 375, p. 363–378 (this volume).

Dowling, D.M., 2004, Adobe housing in El Salvador: Earthquake performance and seismic improvement, *in* Rose, W.I., et al., eds., Natural hazards in El Salvador: Boulder, Colorado, Geological Society of America Special Paper 375, p. 281–300 (this volume).

EERI (Earthquake Engineering Research Institute), 1987, The San Salvador earthquake of October 10, 1986: Earthquake Spectra, v. 3, no. 3, p. 415–634.

Harlow, D.H., White, R.A., Rymer, M.J., and Alvarez, G.S., 1993, The San Salvador earthquake of 10 October 1986 and its historical context: Bulletin of the Seismological Society of America, v. 83, no. 4, p. 1143–1154.

Hattori, S., 1979, Seismic risk maps in the world (maximum acceleration and maximum particle velocity) (II)—Balkan, Middle East, Southeast Asia, Central America, South America and others: Bulletin of the International Institute of Seismology and Earthquake Engineering, v. 17, p. 33–96.

ICBO (International Conference of Building Officials), 1997, Uniform Building Code 1997: Whittier, California, ICBO, v. 2, 416 p.

ICBO (International Conference of Building Officials), 2000, International Building Code 2000: Whittier, California, ICBO, 756 p.

Jibson, R.W., Crone, A.J., Harp, E.L., Baum, R.L., Major, J.J., Pullinger, C.R., Escobar, C.D., Martínez, M., and Smith, M.E., 2004, Landslides triggered by the 13 January and 13 February 2001 earthquakes in El Salvador, *in* Rose, W.I., et al., eds., Natural hazards in El Salvador: Boulder, Colorado, Geological Society of America Special Paper 375, p. 69–88 (this volume).

Kiremidjian, A.S., and Shah, H.C., 1979, Seismic hazard analysis of Honduras: The John A. Blume Earthquake Engineering Center Report No. 38, Palo Alto, California, Stanford University, 294 p.

Kiremidjian, A.S., Shah, H.C., and Lubetkin, L., 1977, Seismic hazard mapping for Guatemala: The John A. Blume Earthquake Engineering Center Report No. 26, Palo Alto, California, Stanford University, p. 195.

Lara, M.A., 1983, The El Salvador earthquake of June 19, 1982: Earthquake Engineering Research Institute, EERI Newsletter, v. 17, no. 1, p. 87–96.

Lara, M., 1987, The San Salvador earthquake of October 10, 1986—History of construction practices in San Salvador: Earthquake Spectra, v. 3, no. 3, p. 491–496.

Larsson, T., and Mattson, C., 1987, Seismic hazard analysis in Nicaragua: Examensarbete No. 3:87, Inst. För Jord-och Bergemaknik, Kungl. Tekniska Högskola, Stockholm, 211 p.

Levin, S.B., 1940, The Salvador earthquakes of December, 1936: Bulletin of the Seismological Society of America, v. 30, p. 1–45.

Ligorría, J.P., Lindholm, C., Bungum, H., and Dahle, A., 1995, Seismic hazard for Guatemala: Technical Report No. 2-21, NORSAR, 47 p.

Lindholm, C., Rojas, W., Bungum, H., Dahle, A., Camacho, E., Cowan, H., and Laporte, M., 1995, New regional seismic zonation for Central America, *in* Proceedings, International Conference on Seismic Zonation, 5th, Nice, v. 1, p. 437–444.

Lomnitz, C., and Schulz, R., 1966, The San Salvador earthquake of May 3, 1965: Bulletin of the Seismological Society of America, v. 56, p. 561–575.

McGuire, R.K., 1976, FORTRAN computer program for seismic risk analysis: U.S. Geological Survey Open-File Report 76-67, 90 p.

Ministerio de Obras Públicas (MOP), 1966, Regulaciones para el Diseño Sísmico de la República de El Salvador: San Salvador, Official Gazette, 21 January 1966, v. 210, no. 12.

Ministerio de Obras Públicas (MOP), 1989, Reglamento de Emergencia de Diseño Sísmico de la República de El Salvador: San Salvador, Official Gazette, 14 August 1989, v. 304, no. 14.

Ministerio de Obras Públicas (MOP), 1994, Norma Técnica para Diseño por Sismo, *in* Reglamento para la Seguridad Estructural de las Construcciones: San Salvador, Ministry of Public Works of El Salvador, 24 p.

Reiter, L., 1990, Earthquake hazard analysis—Issue and insights: New York, Columbia University Press, 254 p.

Rolo, R., Bommer, J.J., Houghton, B.F., Vallance, J.W., Berdousis, P., Mavrommati, C., and Murphy, W., 2004, Geologic and engineering characterization of Tierra Blanca pyroclastic ash deposits, *in* Rose, W.I., et al., eds., Natural Hazards in El Salvador: Boulder, Colorado, Geological Society of America Special Paper 375, p. 55–67 (this volume).

Rose, W., Bommer, J.J., and Sandoval, C., 2004, Natural hazards and risk mitigation in El Salvador: An introduction, *in* Rose, W.I., et al., eds., Natural hazards in El Salvador: Boulder, Colorado, Geological Society of America Special Paper 375, p. 1–4 (this volume).

Rosenblueth, E., and Prince, J., 1966, El temblor de San Salvador, 3 de mayo de 1965; Ingeniería sísmica: Ingeniería, v. 36, no. 1, p. 31–58.

Sabetta, F., and Bommer, J.J., 2002, Modification of the spectral shapes and subsoil conditions in Eurocode 8, *in* Proceedings, European Conference on Earthquake Engineering, 12th: London, Elsevier Science, CD-Rom, paper no. 518.

Salazar, W., 1999, Research activities on seismology and earthquake engineering in El Salvador, Central America: Research reports on earthquake engineering, Japan: Tokyo Institute of Technology, Japan, no. 70, p. 1–17.

Schnabel, P.B., and Seed, H.B., 1973, Accelerations in rock for earthquakes in the western United States: Bulletin of the Seismological Society of America, v. 63, no. 2, p. 501–516.

Shah, H.C., Mortgat, C.P., Kiremidjian, A.S., and Zsutty, T.C., 1975, A study of seismic risk for Nicaragua—Part 1: The John A. Blume Earthquake Engineering Center Report No. 11, Palo Alto, California, Stanford University, p. 187.

Shakal, A.F., Huang, M.J., and Linares, R., 1987, The San Salvador earthquake of October 10, 1986—Processed strong motion data: Earthquake Spectra, v. 3, no. 3, p. 465–481.

Shedlock, K.M., 1999, Seismic hazard mapping of North and Central America and the Caribbean: Annali di Geofisica, v. 42, no. 6, p. 977–997.

Singh, S.K., Gutiérrez, C., Arboleda, J., and Ordaz, M., 1993, Peligro sísmico en El Salvador: Universidad Nacional Autónoma de México, 74 p.

Somerville, P.G., Smith, N.F., Graves, R.W., and Abrahamson, N.A., 1997, Modification of empirical strong ground motion attenuation relations to include the amplitude and duration effects of rupture directivity: Seismological Research Letters, v. 68, no. 1, p. 199–222.

Tavakoli, B., and Monterroso, D., 2004, Monte Carlo seismic hazard maps for northern Central America, covering El Salvador and surrounding area, *in* Rose, W.I., et al., eds., Natural hazards in El Salvador: Boulder, Colorado, Geological Society of America Special Paper 375, p. 423–434 (this volume).

Villagran, M., Lindholm, C., Dahle, A., Cowan, H., and Bungum, H., 1997, Seismic hazard assessment for Guatemala City: Natural Hazards, v. 14, p. 189–205.

White, R.A., 1991, Tectonic implications of upper-crustal seismicity in Central America, *in* Slemmons, D.B., et al., eds., Neotectonics of North America: Boulder, Colorado, Geological Society of America, Decade Map, v. 1, p. 323–338.

White, R.A., and Harlow, D.H., 1993, Destructive upper-crustal earthquakes of Central America since 1900: Bulletin of the Seismological Society of America, v. 83, no. 4, p. 1115–1142.

Wisner, B., 2001, Risk and the neo-liberal state: Why post-Mitch lessons didn't reduce El Salvador's earthquake losses: Disasters, v. 25, no. 3, p. 251–268.

MANUSCRIPT ACCEPTED BY THE SOCIETY JUNE 16, 2003

Geological Society of America
Special Paper 375
2004

Local site effects on microtremors, weak and strong ground motion in San Salvador, El Salvador

Kuvvet Atakan
Department of Earth Science, University of Bergen, Allégt.41, N-5007 Bergen, Norway

Mauricio Ciudad Real
Kinemetrics Inc., 222 Vista Ave., Pasadena, California 91107, USA

Rodolfo Torres
Universidad Centroamericana, San Salvador, El Salvador

ABSTRACT

The San Salvador area lies along the axis of the Central American volcanic chain, which is the product of the northeast-directed subduction of the Cocos plate beneath the Caribbean plate. The destructive earthquakes that occurred in El Salvador during the last few decades are either shallow, moderate-sized earthquakes with epicenters in the volcanic zone, or large earthquakes from the subduction zone. The earthquakes of 13 January and 13 February 2001 are the most recent examples of both types of events. A considerable part of the damage related to these earthquakes comes from the local site effects. The present study describes the local site effects in the metropolitan area of San Salvador based on the spectral ratio technique through a comparison between three data sets from microtremors, weak motion, and strong motion.

Data from four sites within the metropolitan area are used to investigate the local site response in San Salvador. Spectral ratios of the horizontal components from the four sedimentary sites with respect to a reference site are calculated and compared between microtremors, earthquakes from the subduction zone (weak motion), and the earthquake of October 10, 1986 (strong motion). Results indicate site amplification factors ranging from 4 to 10 at all sites on all three data sets within the frequency band 1.0–10.0 Hz. However, within this band, no correlation is found at any dominant frequency. This is probably due to the vertical and lateral variations in the thickness and the properties of the sedimentary layers. Average amplification factors obtained from the three data sets systematically show a decreasing trend from microtremors to weak motion to strong motion. Estimates on all four sites indicate higher values for microtremors compared to the weak-motion data, whereas strong-motion data show the lowest. In order to obtain reliable estimates of site amplification, nonlinear sediment response has to be taken into account. Unless calibrated with the strong-motion data, caution is recommended in using microtremors.

Keywords: local site effects, microtremors, weak-motion, strong ground motion, earthquakes.

*Kuvvet.Atakan@geo.uib.no

Atakan, K., Ciudad Real, M., and Torres, R., 2004, Local site effects on microtremors, weak and strong ground motion in San Salvador, El Salvador, *in* Rose, W.I., Bommer, J.J., López, D.L., Carr, M.J., and Major, J.J., eds., Natural hazards in El Salvador: Boulder, Colorado, Geological Society of America Special Paper 375, p. 321–337. For permission to copy, contact editing@geosociety.org. © 2004 Geological Society of America

INTRODUCTION

Any reliable estimation of seismic hazard is largely dependent on detailed information of the local site response. This phenomenon has been recognized for some time (e.g., Milne, 1898; Takahashi and Hirano, 1941; Gutenberg, 1957). The methods used to determine the site response can be categorized into two major groups, the theoretical (analytical) and the empirical methods. The theoretical calculation of the site response, mainly based on inversion techniques, requires a very good knowledge of the geotechnical parameters to constrain the results. Depending on the type of inversion used, several assumptions have to be made regarding the shape of the source spectrum or the site response at the selected stations. Empirical methods are somehow more effective in the sense that they are based on calculating the frequency spectrum directly from the recorded ground motion. Among the empirical methods, the spectral ratio of a sedimentary site with respect to a bedrock reference site is a widely used technique (e.g., Borcherdt, 1970; Borcherdt and Gibbs, 1976; Rogers et al., 1984). Effective use of this technique is demonstrated at different sites and geological conditions following large destructive earthquakes (e.g., Singh et al., 1988; Lermo et al., 1988; Borcherdt et al., 1989; Hough et al., 1990). Some review papers discuss the effectiveness of the different methods used in the site response estimate (e.g., Aki, 1988; Hartzell, 1992; Aki, 1993; Bard, 1994; Atakan, 1995; Field and Jacob, 1995). Other methods include H/V spectral ratios (horizontal vs. vertical components) using single-station recordings (Nakamura, 1989; Lermo and Chávez-García, 1993; Field and Jacob, 1993; Lachet and Bard, 1994), and the cross-spectrum estimate (Safak, 1991; Field et al., 1992).

The present study intends to demonstrate the applicability and the limitations of the different techniques used in estimating the local site response, through a comparison between three sets of data from microtremors, weak motion, and strong motion. In this study, we have used the spectral ratio method on data from the selected five sites in the metropolitan area of San Salvador, where there were permanent accelerograph stations when the October 10, 1986, earthquake occurred, although a more recent accelerograph network does exist (see Bommer et al., 1997). These are the Centro de Investigaciones Geotécnicas (CIG), Instituto Geográfico Nacional (IGN), Universidad Centroamericana (UCA), Hotel Camino Real (HCR), and Hotel Sheraton (SHE) (Fig. 1). In addition, two other sites at the western flank of the San Jacinto Hill (SJA and SOC) were used as bedrock reference sites. The spectral ratios from the microtremors were compared with the data collected from the subduction zone earthquakes recorded at the same sites. These two data sets were then correlated with the strong-motion records of the October 10, 1986, earthquake.

GEOLOGICAL SETTING

The San Salvador area lies along the axis of the Central American volcanic chain, which is the product of the northeast-directed subduction of the Cocos plate beneath the Caribbean plate with an absolute plate motion of 8.0 cm/yr (Rymer, 1987; Weinberg, 1992). The plate-tectonic framework of Central America and the Caribbean was previously reviewed by several authors (e.g., Molnar and Sykes, 1969; Malfait and Dinkelmann, 1972; Burke et al., 1984; Mann and Burke, 1984; Dengo, 1985). The generalized geological overview of El Salvador was discussed by Wiesemann (1975), and later the geological map was published by Weber et al. (1978).

The metropolitan area of San Salvador is underlain by middle and upper Cenozoic volcanic and volcaniclastic deposits of different origins, mainly derived from the adjacent volcanoes (Weber et al., 1978). The area lies on the gentle eastern slope of the San Salvador volcano, dissected by deeply incised streams and rivers (Fig. 2A), and is surrounded by four major volcanic features. These are the San Salvador volcano, with a minor cinder cone in its center to the west-northwest; the Cerros de Mariona to the north; a collapsed caldera, Lake Ilopango, to the east; and an inactive volcanic dome with a minor parasitic cone, Cerro San Jacinto, to the south-southeast.

Relevant to the local site response, there are two major deposits forming part of the San Salvador Formation that cover the metropolitan area. These are (1) the younger Holocene volcanic ash, locally named "tierra blanca" (Rolo et al., this volume, Chapter 5) because of its light color, and (2) the underlying Pleistocene, brown-colored tuffs (Schmidt-Thomé, 1975). The latter unit is as thick as 25 m and is derived from the San Salvador volcano, whereas the younger unit is ~25 m thick immediately underneath the city and may reach up to 50 m toward Lake Ilopango (Rymer, 1987). This younger unit represents multiple eruptions from the former Ilopango volcano. The characteristics of the uppermost 10 to 30 m of these sediments are revealed in either geotechnical investigations or in cross sections outcropping along the riverbanks. The clast compositions, grain sizes, and thicknesses vary considerably within short distances. The degree of consolidation is another characteristic feature varying considerably, as indicated by the standard penetration tests (Fig. 2B).

There are three major fault trends in the area, in NNE-SSW (or N-S), NE-SW, and NW-SE directions (Durr et al., 1960). In addition, the fourth and the oldest trend can be inferred from the general strike direction of the volcanic chain in the WNW-ESE (or E-W) direction (Rymer, 1987). Among the other three, the north-south–trending faults are less distinct and occur only locally (Rymer, 1987). The youngest faults represent the dominant trend in the NE-SW and the conjugate set in the NW-SE directions. All faults seem to have steep dip angles generally larger than 65° and appear active (Schmidt-Thomé, 1975). Although the individual fault trends may have developed at different times, they have been repeatedly reactivated. The reactivation of the NNE-SSW–trending fault during the October 10, 1986, earthquake is such an example (White et al., 1987; Harlow et al., 1993).

SEISMICITY

The city of San Salvador has experienced several destructive earthquakes within the last century. The most damaging were

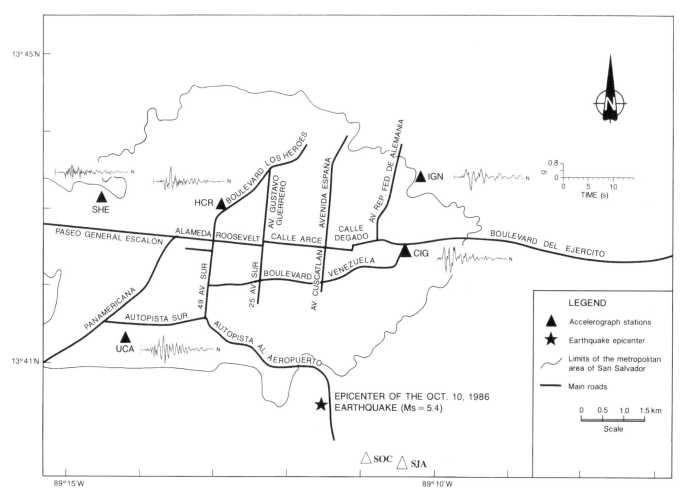

Figure 1. The locations of the five accelerograph stations used in this study. At each site, the N-S component of the acceleration recordings from the October 10, 1986, earthquake are shown for comparison. Note the lower amplitudes at SHE. The accelerograms are in the same scale, where the vertical axes show ground motion amplitudes in g (see Table 2 for the maximum values of the amplitudes), and the horizontal axes show time in seconds (the total length of each signal is 14 seconds). The major roads are shown for orientation.

(1) the June 8, 1917, earthquake, where two almost consecutive events with magnitudes 6.4 and 6.3 (M_s) occurred, respectively, followed by an eruption half an hour later; (2) the April 27, 1919, earthquake (M_s = 6.0); and (3) the May 3, 1965, earthquake (M_s = 6.0), which was preceded by an unusual number of shocks (more than 8000) (White et al., 1987). Although the epicenter of the latter is close to the October 10, 1986, earthquake, the casualties were much lower due to the high number of foreshocks, which alerted the population.

The destructive earthquakes experienced in El Salvador during the last few decades originate either from shallow, moderate-sized earthquakes within the volcanic zone or from large subduction zone earthquakes. The October 10, 1986, earthquake (M_s = 5.4) was an example for the first case. The main shock of the October 10, 1986, earthquake reached intensities as high as IX on the Modified Mercalli intensity scale mainly in the area along

the fault plane striking NNE-SSW (Alvarez, 1987) (Fig. 3). The highest intensities also correspond to the central and densely populated part of the city. Figure 3 shows the distribution of damage with increasing degree of intensity, as well as the macroseismic intensity contours. The shallow depth of the main shock has resulted in high damage concentrated in a relatively narrow zone. This, combined with the fact that most of the damage to buildings was the result of poor construction practices, led to a large number of casualties, reaching 1500 with more than 10,000 injured (Olson, 1987). The material losses reached U.S.$1.5 billion. Much of this damage was related to the local site conditions (Atakan and Torres, 1993). The most recent destructive earthquake from the subduction zone occurred on January 13, 2001 (M_W = 7.6). It claimed nearly 1000 lives and left ~100,000 homeless, both through direct shaking and through numerous landslides (NORSAR et al., 2001).

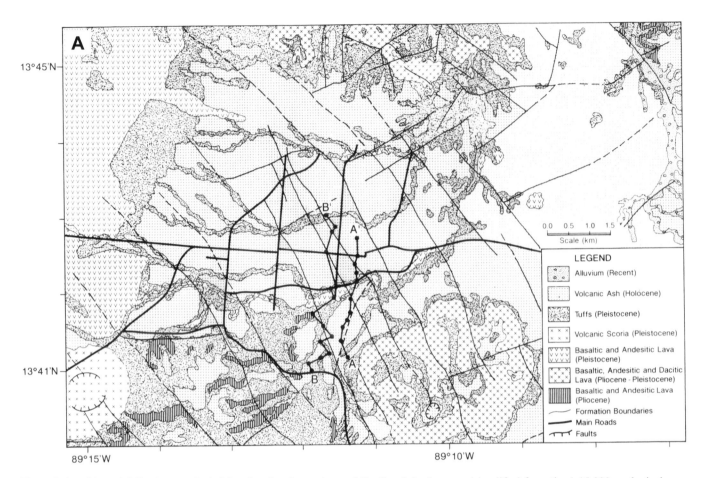

Figure 2 (*on this and following page*). A: Map showing the geology of the San Salvador area (simplified from the 1:15,000 geological map prepared by Centro de Investigaciones Geotécnicas, 1987). Major fault trends (shown as solid lines with hatches indicating normal faults and without hatches indicating strike-slip faults), as well as the main roads are shown. B: The parallel N-S cross sections, A–A' and B–B', showing the lithology and thickness variations. The numbers at each site indicate the results from the standard penetration tests.

DATA SETS

In the spectral ratio analyses three different data sets were used, collected from the four sedimentary sites CIG, IGN, UCA, and HCR. These are the microtremors, weak motion from the subduction zone earthquakes, and strong motion from the October 10, 1986, earthquake.

The strong-motion data set was collected by the permanent accelerograph network operated by Centro de Investigaciones Geotécnicas (CIG). Data from CIG, IGN, UCA and HCR, recorded by Kinemetrics SMA-1 instruments, were available. Recordings from the SHE station were used as the reference site, because of its proximity to the bedrock outcrops. For the other data sets, the reference site SJA, located on the southern flank of the San Jacinto dome, was used. In addition, another reference site, SOC (close to SJA), was used in some of the earthquake recordings. SJA station was used as a standard reference, and the other reference sites were corrected accordingly.

The weak-motion data were chosen from the 16 events recorded for this study. Only those events that had similar locations and magnitudes were used in the analysis. The epicenter locations of these eight events are shown in Figure 4 (see also Table 1). Because of the limitations in the number of instruments (Kinemetrics SSR-1 digital recording systems) used for the microtremor and the weak-motion recordings, only a pair of recordings at a time, simultaneously taken at a sedimentary and a bedrock reference site, were collected. The strong-motion records, on the other hand, were taken simultaneously at all five sites.

Spectral ratios of the four sedimentary sites with respect to the bedrock reference site were calculated for all three data sets. For each pair of recordings, the spectra were computed on windows with same duration (i.e., same number of samples) chosen at the S-wave part of the signal, including the S-wave coda (believed to give similar amplifications: Gutierrez and Singh, 1992). The signals were cosine tapered and Fourier transformed (with fast Fourier transforms); the spectral amplitudes were then smoothed with a one-third-octave band filter.

B

PROFILE A

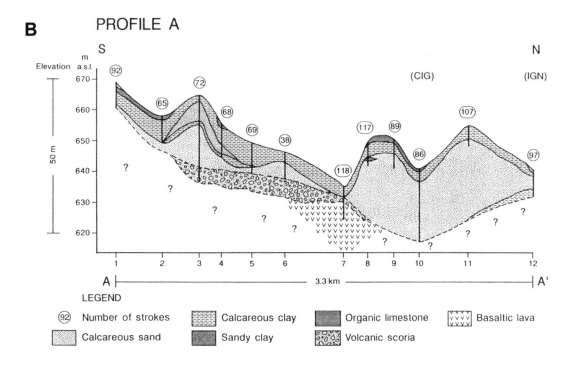

LEGEND

(92) Number of strokes	Calcareous clay	Organic limestone	Basaltic lava
Calcareous sand	Sandy clay	Volcanic scoria	

PROFILE B

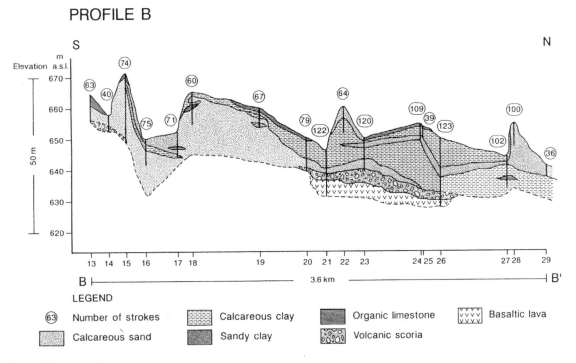

LEGEND

(63) Number of strokes	Calcareous clay	Organic limestone	Basaltic lava
Calcareous sand	Sandy clay	Volcanic scoria	

Figure 2 (*continued*).

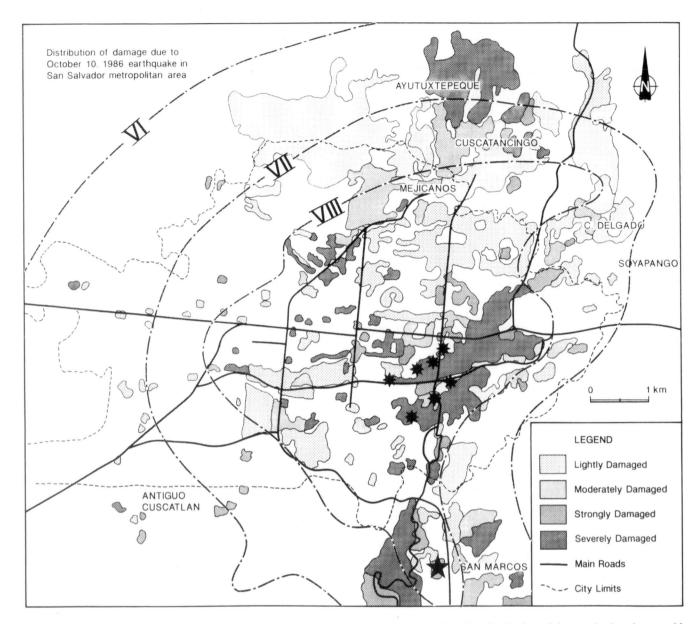

Figure 3. The macroseismic intensity (Modified Mercalli) contour map (from Alvarez, 1987). The distribution of damage is also shown, with darker patterns indicating higher damage (modified from Centro de Investigaciones Geotécnicas, 1987). The stars indicate the areas where total collapse of buildings occurred. The epicenter of the October 10, 1986, earthquake is also shown by a larger star. Dashed line indicates the city limits. Note that the distribution is largely along a NNE-SSW trend.

SPECTRAL RATIOS

Microtremor data sets were gathered from the pre-event windows of 30 seconds duration from the earthquake records used for the weak-motion data set. The results are presented in Figures 5 and 6 for the N-S and E-W components, respectively. At all sites, spectral amplifications are observed within a broad range of frequencies, but especially between 1.0 to 10.0 Hz. For the N-S components, average spectral ratios that show amplification factors are 10.4 for CIG, 11.1 for IGN, 6.0 for UCA, and 5.5 for HCR. The E-W components show somewhat lower values of amplifications. The average spectral ratios that show amplification factors for the same stations for the E-W components are 7.8 for CIG, 10.1 for IGN, and 6.2 for HCR. Here, the records from UCA are missing due to instrumental problems. Mean values of N-S and E-W components are also calculated for the average amplification factors. These are 9.1 for CIG, 10.6 for IGN, and 5.9 for HCR.

Weak-motion records are chosen from the 16 earthquakes located along the subduction zone in the south. The eight events

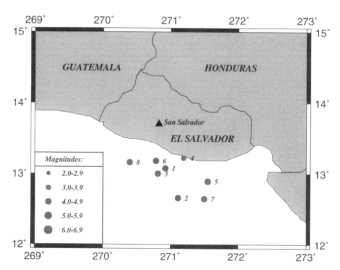

Figure 4. Epicenter distribution of the earthquakes from the subduction zone that are used in the weak-motion data set (see Table 1).

sedimentary sites are calculated with respect to the reference site SHE situated on the eastern flank of the San Salvador volcano. Amplifications occur in general on lower frequencies (less than 4 Hz), when compared to the weak-motion and the microtremor data. For the N-S components, the average values indicate 4.1 for CIG, 6.1 for IGN, 3.3 for UCA, and 5.2 for HCR (Fig. 9). The E-W components give somewhat lower values (Fig. 10); the average values are 3.1 for CIG, 2.9 for IGN, 3.4 for UCA, and 1.4 for HCR. The mean values of the two horizontal components indicate 3.6 for CIG, 4.5 for IGN, 3.4 for UCA, and 3.3 for HCR.

The spectral amplifications observed on strong-motion records may be biased by the short source-receiver distances (Table 2), especially since two of the stations, IGN and CIG, lie along the direction where fault rupture propagated (Bommer et al., 2001). This issue, together with the problems related to the radiation pattern, was discussed in earlier studies (Atakan and Figueroa, 1993; Atakan and Torres, 1993). Although the present results may include some effects of the fault directivity, radiation pattern, etc., the amplifications related to the local sediments seem to be dominant and overshadow the other possible influences. A simple test was performed to see the effect of radiation pattern, using forward waveform modeling (Herrmann, 1985). The relative comparison of the accelerograms indicates that the observed differences in the S-wave amplitudes (i.e., giving higher amplifications on the sedimentary sites with respect to that of the reference site at SHE) are not the result of the radiation pattern. A comparison between the sites SHE, CIG, and IGN, indicates in fact an opposite effect on the synthetic accelerograms (i.e., higher amplitudes at SHE when compared to CIG and IGN). Furthermore, similar spectral amplifications in terms of the frequency ranges involved are observed also on the weak-motion records from the subduction zone, where the assumptions related to the spectral ratio method are considered valid (i.e., far-field conditions where source to receiver distances are large enough to provide similar source and radiation pattern effects at all sites).

used in the analysis are listed in Table 1. In general, spectral amplifications are observed on all records within a wide range of frequencies. At higher frequencies (usually above 10.0 Hz), the phenomenon is reversed, giving deamplifications. For the N-S components (Fig. 7), the spectral ratios of four sedimentary sites with respect to the bedrock reference site indicate average amplification factors of 4.9 for CIG, 7.3 for IGN, 4.6 for UCA, and 7.0 for HCR. The E-W components (Fig. 8) indicate average amplification factors of 4.0 for CIG, 6.3 for IGN, and 3.5 for HCR (record from UCA is missing due to instrumental problems). The mean values of the two horizontal components indicate amplification factors of 4.5 for CIG, 6.8 for IGN, and 3.5 for HCR.

The strong-motion records are taken from the October 10, 1986, earthquake, which occurred close to the metropolitan area of San Salvador. The spectral amplifications of the four

TABLE 1. LIST OF EARTHQUAKES FROM THE SUBDUCTION ZONE, USED IN THE WEAK-MOTION DATA SET

No.	Year	Date (mm/dd)	Time (hh:mm)	Lat (N)	Long (W)	Depth (km)	Number of stations	Rms (sec)	Mc
1	1992	10/25	09:54	13.077	89.078	50.0	17	0.1	4.1
2	1992	10/29	06:40	12.660	88.866	11.0	14	0.1	4.1
3	1992	11/05	08:52	13.001	89.190	37.0	18	0.2	3.8
4	1992	12/09	19:28	13.224	88.804	72.2	18	0.1	4.6
5	1993	04/27	05:56	12.893	88.445	46.8	16	0.2	4.2
6	1993	04/30	10:01	13.184	89.218	51.6	16	0.1	4.3
7	1993	07/01	05:55	12.649	88.498	50.0	15	0.4	4.4
8	1993	07/22	18:31	13.163	89.607	35.2	18	0.2	4.6

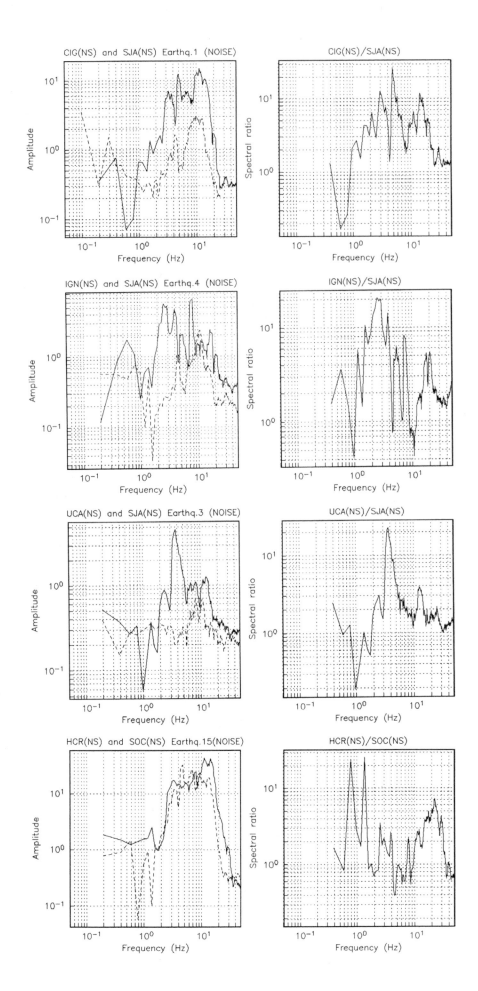

Figure 5. The spectral plots and the spectral ratios (right-hand side) of N-S components of the pre-event microtremor records from four sedimentary sites with respect to the bedrock reference sites SJA and SOC. The solid lines are the spectra corresponding to the sedimentary sites, whereas the dashed lines indicate the spectra of the bedrock reference site. Amplitudes are in cm/s^2 (valid also for all following figures).

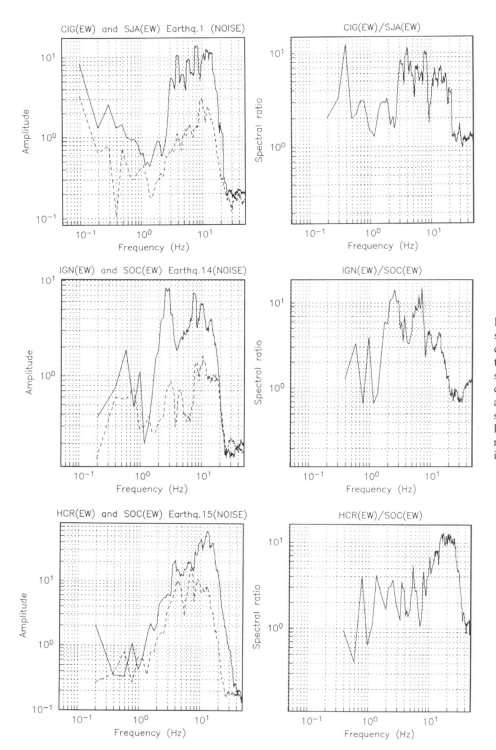

Figure 6. The spectral plots and the spectral ratios (right-hand side) of E-W components of the pre-event micro-tremor records from three sedimentary sites with respect to the bedrock reference sites SJA and SOC. The solid lines are the spectra corresponding to the sedimentary sites, whereas the dashed lines indicate the spectra of the bedrock reference site. Data from UCA are missing due to instrumental problems.

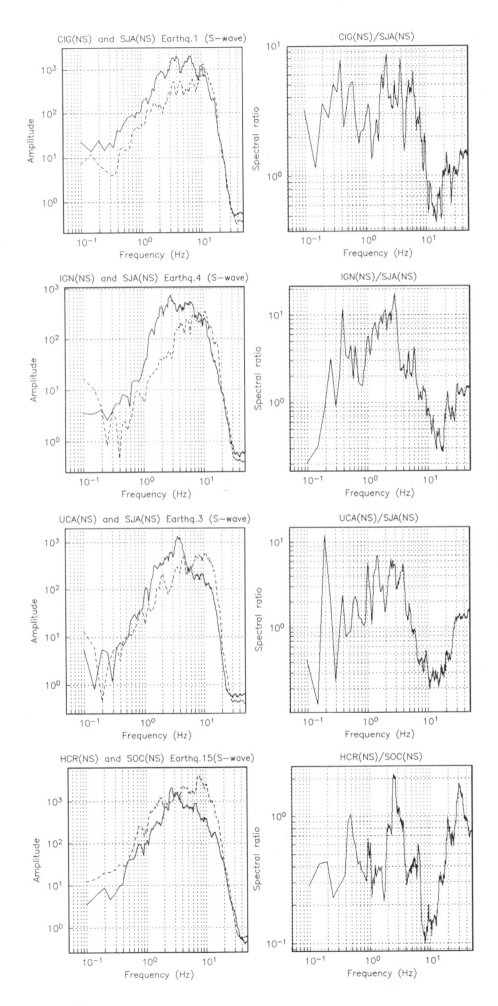

Figure 7. The spectral plots and the spectral ratios (right-hand side) of N-S components of the weak-motion records from four sedimentary sites with respect to the bedrock reference sites SJA and SOC. The solid lines are the spectra corresponding to the sedimentary sites, whereas the dashed lines indicate the spectra of the bedrock reference site.

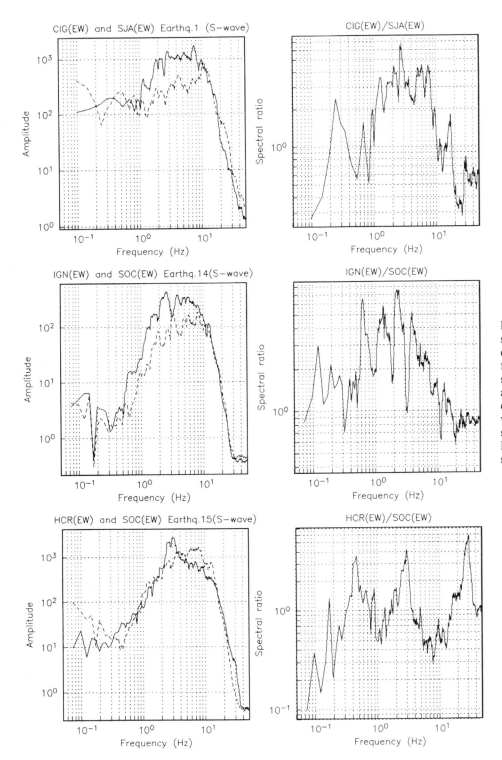

Figure 8. The spectral plots and the spectral ratios (right-hand side) of E-W components of the weak-motion records from three sedimentary sites with respect to the bedrock reference sites SJA and SOC. The solid lines are the spectra corresponding to the sedimentary sites, whereas the dashed lines indicate the spectra of the bedrock reference site. Data from UCA are missing due to instrumental problems.

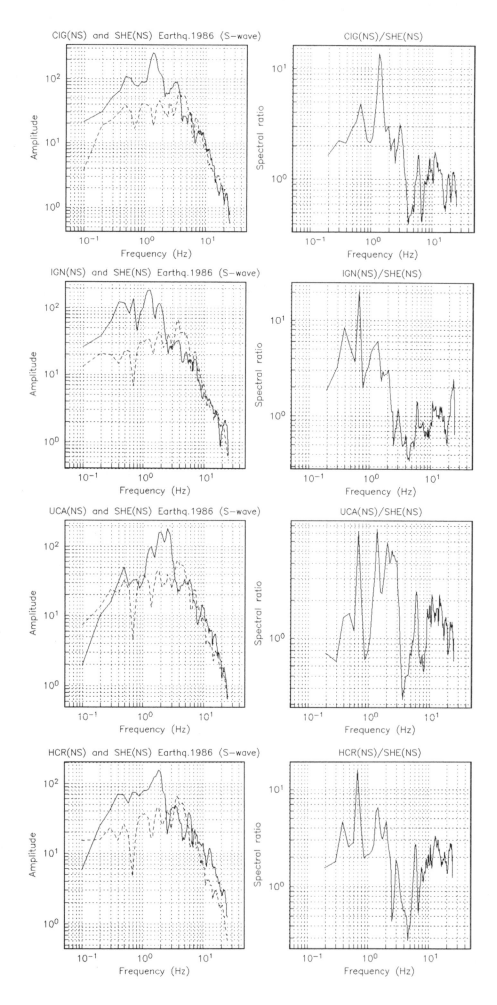

Figure 9. The spectral plots and the spectral ratios (right-hand side) of N-S components of the strong-motion records from four sedimentary sites with respect to the bedrock reference site SHE. The solid lines are the spectra corresponding to the sedimentary sites, whereas the dashed lines indicate the spectra of the bedrock reference site.

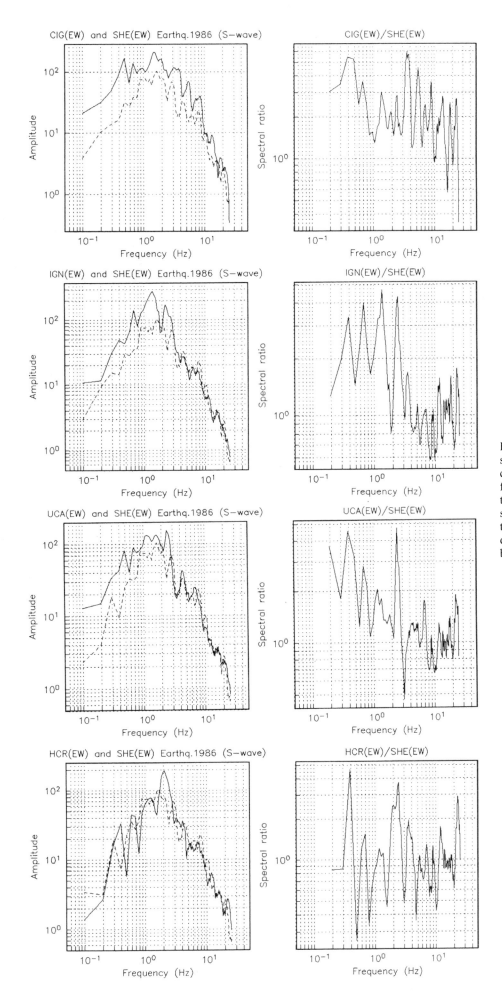

Figure 10. The spectral plots and the spectral ratios (right-hand side) of E-W components of the strong-motion records from four sedimentary sites with respect to the bedrock reference site SHE. The solid lines are the spectra corresponding to the sedimentary sites, whereas the dashed lines indicate the spectra of the bedrock reference site.

TABLE 2. SITE LOCATIONS AND THE MAXIMUM
ACCELERATIONS RECORDED

Station	Lat (N)	Long (W)	Maximum acceleration (g)			Distance (km)
			N-S	E-W	Z	
CIG	13.700	89.175	0.42	0.71	0.40	4.3
IGN	13.714	89.171	0.40	0.54	0.46	5.7
UCA	13.683	89.237	0.39	0.43	0.24	3.8
HCR	13.712	89.215	0.47	0.34	0.26	4.5
SHE	13.713	89.243	0.22	0.32	0.15	6.2

Note: From Shakal et al., 1986.

DISCUSSION

Comparisons are made on the spectral ratios of the three different data sets from the four sedimentary sites (CIG, IGN, UCA, and HCR), and are shown in Table 3. On both horizontal components, the average amplifications give systematically higher estimates at all sites for the microtremors. Weak-motion records give slightly lower estimates but still higher than the strong-motion records.

The general lack of correlation between the frequencies at which the maximum amplifications occurred at different sites and the appearance of several peaks in the spectral ratios at different frequencies imply significant lateral and vertical variations in the sedimentary layers. The surface geology (Fig. 2A), and the standard penetration test results from two parallel cross sections only a few hundred meters apart (Fig. 2B), confirm that these variations are substantial. A possible fundamental resonant frequency for the basin as a whole is therefore obscured by these variations. In this sense, the situation in San Salvador differs from the well-known example of the case of Mexico City, where the site amplifications are related to a predominant frequency and give much higher amplification factors (Singh et al., 1988).

Previous studies of the site effects using analytical methods (Faccioli et al., 1987) attributed the highest damage to sediment amplifications in the period range of 0.2 to 0.5 seconds, combined with the poor construction practices. It is also mentioned that increased vulnerability due to the partial damage from pre-

vious earthquakes is an important factor that contributed to the damage. There seems to be good agreement between the analytical and empirical results (Atakan and Figueroa, 1993). The present seismic design code (ASIA, 1997) takes into account the local site effects associated with the sediments following the recommendations from these previous studies (López et al., this volume, Chapter 23).

The spectra and the spectral ratios from the three data sets indicate that the site response is highly frequency dependent. Yu et al. (1992), in their study on nonlinear sediment response, have demonstrated that in the Fourier-transform domain the sediment response can be separated into three frequency bands: first, the lowest frequency range, in which spectral amplitude is not affected by the nonlinearity; second, the central frequency range, in which the spectral amplitudes are decreased; and third, the frequency range above a crossover frequency, where a reversal in the amplifications occur. The transition frequencies between these three bands are expected to shift depending on the geotechnical properties and the thickness of the sediments, as well as the amplitude and the frequency content of the input signal. At lower frequencies (lower than the crossover frequencies), the low-impedance effect of the sedimentary layers is dominant and produces the site amplification, whereas at higher frequencies the absorption effect becomes more dominant and results in deamplifications.

In our observations, all three frequency bands discussed above are present on the weak- and strong-motion data sets. On the microtremor data, however, the transitions are not so clear, and it is not possible to distinguish these three frequency bands. On both the weak- and the strong-motion data sets, we observe a reversal of the amplifications on the spectral ratios. The crossover frequencies above which deamplification occurs at individual sites differ, varying from 2.5 to 3.5 Hz in the case of strong motion, and 7.0–9.0 Hz in the case of weak motion. These transition frequencies for the strong motion are systematically higher than the weak motion with a difference of 5–6 Hz. This is consistent with the results obtained for central California (Chin and Aki, 1991; Aki, 1993), where the crossover frequency was at 5 Hz on strong motion and 12 Hz on weak motion. The slight difference between the two cases, the results from San Salvador being slightly smaller than from California, may be due to the differences in the magnitude levels of the strong ground motions used, the differences

TABLE 3. COMPARISON OF THE AVERAGE SPECTRAL AMPLIFICATIONS BETWEEN
MICROTREMORS, WEAK-, AND STRONG-MOTION DATA SETS

Station	Microtremors			Weak-motion			Strong-motion		
	N-S	E-W	Mean	N-S	E-W	Mean	N-S	E-W	Mean
CIG	10.4	7.8	9.1	4.9	4.0	4.5	4.1	3.1	3.6
IGN	11.1	10.1	10.6	7.3	6.3	6.8	6.1	2.9	4.5
UCA	6.0	–	–	4.6	–	–	3.3	3.4	3.4
HCR	5.5	6.2	5.9	7.0	3.5	5.2	5.2	1.4	3.3

in the epicentral distances, and more likely the differences in the thickness of the sediments (i.e., thicker sediments would cause a shift toward the lower frequencies: Yu et al., 1992).

The two separate observations of *(a)* the amplification factors being systematically overestimated by both microtremor and weak-motion data with respect to the strong ground motion, and *(b)* the frequencies on which the reversal of amplifications occur on the strong-motion records being significantly lower than the weak-motion data indicate a clear nonlinear response of the sediment layers.

Nonlinear sediment response will reduce the difference in amplification between the sediment and bedrock sites as the ground motion increases. Based on this assumption and the observations from California, Aki (1993) proposed that the peak ground accelerations (PGA) within epicentral distances less than 50 km were not affected by the site conditions. In San Salvador, PGA values from the October 10, 1986, earthquake indicate a variation in the range 0.2–0.7 g between the sites within distances of 4–5 km (Shakal et al., 1986). The difference can be due to several factors. The combined effects of source directivity and radiation pattern may cause azimuthal variation (Joyner and Boore, 1988). The differences in the magnitude levels of the strong motions used for the two cases may also explain the differences observed in the PGA values in San Salvador at short epicentral distances.

CONCLUSIONS

Quaternary and especially Holocene volcanic activity provides the necessary sediment supply to deposition centers within the volcanic chain, in which thick volcaniclastic sediments accumulate. This process creates a vulnerable zone prone to amplification of seismic waves originating from shallow destructive earthquakes along the volcanic chain. In this respect, the situation at San Salvador is no exception to other parts of Central America, where the majority of the population centers are usually located in similar sediment-filled valleys parallel to the Pacific coast. The impact of local site effects on ground motion, therefore, becomes extremely important in estimating earthquake hazard, and hence in risk mitigation. The results of this study demonstrate the importance of local site amplifications in San Salvador and compare different data sets that can be used in estimating these. The conclusions and the recommendations related to the spectral ratios given here should be treated carefully, bearing in mind the assumptions involved. However, it is considered important that these are used as guidelines in any seismic hazard estimate, as they provide a clear indication of site amplifications and a rough estimate in the absolute levels for the San Salvador metropolitan area. Here, the following conclusions are deduced.

1. All three data sets—microtremors, weak ground motion, and strong ground motion records—give spectral amplifications related to the local site conditions. Amplifications were observed in a range of frequencies of engineering interest (as defined by the resonant frequencies of one- to ten-story buildings).

2. Within the frequency range of concern, no preferred trend is observed on the maximum amplifications. The average amplification factors (computed over the frequency range of concern), on the other hand, provide a more stable estimate of the spectral amplifications, and a preferred trend of decreasing spectral amplification factors are observed starting from the microtremors, to weak motion, to strong motion. This seems to be an indication of nonlinearity and is considered important.

3. On both weak and strong ground motion data, above a crossover frequency, reversal of amplifications was observed. This reversal occurs at lower frequencies (2.5–3.5 Hz) on the strong-motion data when compared with the weak motion.

4. Regarding the comparison between the data sets, the most realistic estimates are undoubtedly obtained from the strong-motion records, as they represent the real target events. As to the strong-motion records used in this study, there are a few considerations related to the radiation pattern, fault directivity, etc., which may pose problems in the results obtained. However, as discussed earlier, the spectral amplifications related to the sedimentary layers overshadow these effects. The amplifications that can be expected from a local shallow source, similar to the October 10, 1986, earthquake, are in the range of a factor of 3 to 4.

5. When compared to strong motion, the poorest results are obtained from the microtremors, where they overestimate the amplification factors. Unless calibrated with weak or strong motion, these records can only be used as a relative indication of spectral amplifications at different sites.

Increased knowledge about the geological and geotechnical characteristics of the selected sites used in this study will naturally provide a better foundation for correlating with the spectral amplifications. In this respect, detailed studies of sediment lithology, lateral and vertical distribution of the different units, etc., are needed. Specific investigations, such as simultaneous downhole and surface recordings of microtremors and/or earthquakes (e.g., Tables 3 and 4 in Bommer et al., 2002), may provide a good calibration for the results obtained in this study. Finally, similar studies from other major urban areas in El Salvador are needed, in order to have a more complete knowledge of the local site amplifications on population centers vulnerable to earthquake damage.

Previous comparisons based on empirical data from other population centers in Central America show similar results, where similar local site conditions exist (Atakan, 1997). In general, the bulk of the data indicate amplifications on the order of a factor 2 to 6 over a fairly broad range of frequencies (~1.0 to 8.0 Hz). This is in good agreement with the results based on the weak-motion data (partly also the microtremor data as well) from the present study. However, possible effects of nonlinearity should be taken into account, as strong ground motion data in general exist only for few cases. In this sense, the results based on the strong-motion data from the October 10, 1986, earthquake as shown in this study, as well as other existing or future strong-motion data, are important to provide valuable constraints in the estimates of the local site effects due to near-surface local geology in Central America.

ACKNOWLEDGMENTS

We thank Centro de Investigaciones Geotécnicas, El Salvador, for providing the digitized accelerograph recordings and the logistic support during the fieldwork. The geotechnical and geological information is largely based on the unpublished work by the staff at the Dept. of Geology, Centro de Investigaciones Geotécnicas, El Salvador. K. Atakan's fieldwork was supported by the project "Reduction of Natural Disasters in Central America: Earthquake Preparedness and Hazard Mitigation" sponsored by the Norwegian Agency for Development Cooperation (NORAD). We thank Sri K. Singh, Mario Ordaz and Jens Havskov for reading the earlier version of the manuscript and for valuable comments. We appreciate the detailed reviews made by Drs David Boore and John Douglas. Their comments and suggestions have helped improve the manuscript significantly. Jane K. Ellingsen has kindly helped with the drawings. The computer program used in calculating the spectral ratios (DEGTRA) was kindly provided by Mario Ordaz.

REFERENCES CITED

Aki, K., 1988, Local site effects on strong ground motion: Proceedings, Earthquake Engineering and Soil Dynamics II, Park City, Utah, 27–30 June, American Society of Civil Engineers p. 103–155.

Aki, K., 1993, Local site effects on weak and strong ground motion, *in* Lund, F., ed., New horizons in strong motion: Seismic studies and engineering practice: Tectonophysics, v. 218, p. 93–111.

Alvarez, S., 1987, Informe técnico-sismológico del terremoto de San Salvador del 10 de Octubre de 1986: (Internal Report) Ministerio de Obras Públicas, Centro de Investigaciones Geotécnicas, Departamento de Sismología, San Salvador, April 1987, 83 p.

ASIA, 1997, Reglamento para la seguridad estructural de las construcciones. Normas tecnicas para diseño por sismo: Asociación Salvadoreña de Ingenieros y Arquitectos (ASIA), Ministerio de Obras Publicas, Republica El Salvador, 19 p.

Atakan, K., 1995, A review of the type of data and the techniques used in empirical estimation of local site response: Proceedings, Fifth International Conference on Seismic Zonation, Nice, France, October 17–19, 1995, p. 1451–1460.

Atakan, K., 1997, Empirical site response studies in Central America: Present status, *in* Proceedings, Seminar on Evaluation and Mitigation of Seismic Risk in the Central American Area, Universidad Centroamericana "Jose Simeón Cañas" San Salvador, El Salvador, 22–26 Sept. 1997, p. 77–88.

Atakan, K., and Figueroa, J.C., 1993, Local site response in San Salvador: Comparison between the synthetics and the observed ground motion of the October 10, 1986, earthquake: Report No. 8 under the project "Reduction of Natural Disasters in Central America: Earthquake Preparedness and Hazard Mitigation," Institute of Solid Earth Physics, University of Bergen, Norway, 52 p.

Atakan, K., and Torres, R., 1993, Local site response in San Salvador based on the earthquake of October 10, 1986: Report No. 6 under the project "Reduction of Natural Disasters in Central America: Earthquake Preparedness and Hazard Mitigation," Institute of Solid Earth Physics, University of Bergen, Norway, 35 p.

Bard, P-Y., 1994, Effects of surface geology on ground motion: recent results and remaining issues: Proceedings, Tenth European Conference on Earthquake Engineering, Vienna, Austria, August 1994, 19 p.

Bommer, J.J., Udías, A., Cepeda, J.M., Hasbun, J.C., Salazar, W.M., Suárez, A., Ambraseys, N.N., Buforn, E., Cortina, J., Madariaga, R., Méndez, P., Mezcua, J., and Papastimatiou, D., 1997, A new digital accelerograph network for El Salvador: Seismological Research Letters, v. 68, no. 3, p. 426–437.

Bommer, J.J., Georgallides, G., and Tromans, I.J., 2001, Is there a near-field for small-to-moderate magnitude earthquakes?: Journal of Earthquake Engineering, v. 5, no. 3, p. 395–423.

Bommer, J.J., Benito, M.B., Ciudad-Real, M., Lemoine, A., López-Menjívar, M.A., Madariaga, R., Mankelow, J., Mendéz de Hasbun, P., Murphy, W., Nieto-Lovo, M., Rodriguez-Pineda, C.E., and Rosa, H., 2002, The El Salvador earthquakes of January and February 2001: Context, characteristics, and implications for seismic risk: Soil Dynamics and Earthquake Engineering, v. 22, p. 389–418.

Borcherdt, R.D., 1970, Effects of local geology on ground motion near San Francisco Bay: Bulletin of the Seismological Society of America, v. 60, p. 29–61.

Borcherdt, R.D., and Gibbs, J.F., 1976, Effects of local geological conditions in the region on ground motions and the intensities of the 1906 earthquakes: Bulletin of the Seismological Society of America, v. 66, p. 467–500.

Borcherdt, R.D., Glassmoyer, G., Der Kiureghian, A., and Cranswick, E., 1989, Results and data from seismologic and geologic studies following earthquakes of December 7, 1988 near Spitak, Armenia, S.S.R.: U.S. Geological Survey Open-File Report 89-163A.

Burke, K., Cooper, C., Dewey, J.F., Mann, P., and Pindell, J.L., 1984, Caribbean tectonics and relative plate motions, *in* Bonini, W.E., Hargraves, R.B., and Shagam, R., eds., The Caribbean–South American plate: Boulder, Colorado, Geological Society of America Memoir 162, p. 31–63.

Centro de Investigaciones Geotécnicas, 1987, Mapa geológico-tectónico del área San Salvador y alrededores, scale 1:15,000.

Chin, B-H., and Aki, K., 1991, Simultaneous determination of source, path and recording site effects on strong ground motion during the Loma Prieta earthquake—A preliminary result on pervasive nonlinear site effect: Bulletin of the Seismological Society of America, v. 81, p. 1859–1884.

Dengo, G., 1985, Mid-America: Tectonic setting for the Pacific margin from southern Mexico to northwestern Colombia, *in* Naim, H.E.M., et al., eds., The ocean basins and margins: New York, Plenum, p. 123–180.

Durr, F., et al., 1960, Energía Geotérmica: Informe No. 1, Servicio Geológico Nacional, San Salvador, 268 p.

Faccioli, E., Battistella, C., Alemani, P., and Tibaldi, A., 1987, Seismic microzoning investigations in the metropolitan area of San Salvador, El Salvador, following the destructive earthquake of October 10, 1986, *in* Schuëller, G.I., et al., eds., Proceedings, International Seminar on Earthquake Engineering, Report 23-88, Universität Innsbruck, Institutt für Mechanik, Innsbruck, Austria, 1988, p. 28–65.

Field, E.H., and Jacob, K.H., 1993, The theoretical response of sedimentary layers to ambient seismic noise: Geophysical Research Letters, v. 20, p. 2925–2928.

Field, E.H., and Jacob, K.H., 1995, A comparison and test of various site response estimation techniques, including three that are not reference-site dependent: Bulletin of the Seismological Society of America, v. 85, p. 1127–1143.

Field, E.H., Jacob, K.H., and Hough, S.E., 1992, Earthquake site response estimation: A weak motion case study: Bulletin of the Seismological Society of America, v. 82, p. 2283–2307.

Gutenberg, B., 1957, Effects of ground on earthquake motion: Bulletin of the Seismological Society of America, v. 47, p. 221–250.

Gutierrez, C., and Singh, S.K., 1992, A site effect study in Acapulco, Guerrero, Mexico; a comparison of results from strong-motion and microtremor data: Bulletin of the Seismological Society of America, v. 82, p. 642–659.

Harlow, D.H., White, R.A., Rymer, M.J., and Alvarez, G.S., 1993, The San Salvador earthquake of 10 October 1986 and its historical context: Bulletin of the Seismological Society of America, v. 83, p. 1143–1154.

Hartzell, S.H., 1992, Site response estimation from earthquake data: Bulletin of the Seismological Society of America, v. 82, p. 2303–2327.

Herrmann, R.B., 1985, Computer programs in earthquake seismology, vol. VI: Regional seismograms: Wavenumber integration: Saint Louis University, Missouri, USA.

Hough, S.E., Borcherdt, R.D., Friberg, P.A., Busby, R., Field, E., and Jacob, K.H., 1990, The role of sediment-induced amplification in the collapse of the Nimitz freeway during the October 17, 1989 Loma Prieta earthquake: Nature, v. 344, p. 853–855.

Joyner, W.B., and Boore, D.M., 1988, Measurement, characterization, and prediction of strong motion, *in* Proceedings, Earthquake Engineering and Soil Dynamics, Park City, Utah, 27–30 June, American Society of Civil Engineers, p. 43–102.

Lachet, C., and Bard, P-Y., 1994, Numerical and theoretical investigations on the possibilities and limitations of Nakamura's technique: Journal of Physics of the Earth, v. 42, p. 337–397.

Lermo, J.F., and Chávez-García, F.J., 1993, Site effect evaluation using spectral ratios with only one station: Bulletin of the Seismological Society of America, v. 83, p. 1574–1594.

Lermo, J.F., Rodriguez, M., and Singh, S.K., 1988, The Mexico earthquake of September 19, 1985: Natural period of sites in the valley of Mexico from microtremor measurements and from strong motion data: Earthquake Spectra, v. 4, p. 805–814.

López, M., Bommer, J.J., and Pinho, R., 2004, Seismic hazard assessments, seismic design codes, and earthquake engineering in El Salvador, *in* Rose, W.I., et al., eds., Natural hazards in El Salvador: Boulder, Colorado, Geological Society of America Special Paper 375, p. 301–320 (this volume).

Malfait, B.T., and Dinkelmann, M.G., 1972, Circum-Caribbean tectonic and igneous activity and the evolution of the Caribbean plate: Geological Society of America Bulletin, v. 83, p. 251–271.

Mann, P., and Burke, K., 1984, Neotectonics of the Caribbean: Revisions of Geophysics, v. 22, p. 309–362.

Milne, J., 1898, Seismology (1st edition): London, Kegan Paul, Trench, Truber.

Molnar, P., and Sykes, L.R., 1969, Tectonics of the Caribbean and the Middle America regions from focal mechanisms and seismicity: Geological Society of America Bulletin, v. 89, p. 1639–1684.

Nakamura, Y., 1989, A method for dynamic characteristics estimation of subsurface using microtremor on the ground surface: Quarterly Report of Railway Technical Research Institute, v. 30, no. 1, p. 25–33.

NORSAR, NGI, and NTNU (Norwegian Seismic Array, Norges Geoteknisk Institutt, Norsk Naturvitenskapelig Teknisk Universitet, in cooperation with E. Camacho, V. Schmidt, G. Marroquin and G. Cruz), 2001, Technical mission to El Salvador, following the January 13 earthquake, Report prepared for CEPREDENAC (Center for the Prevention of Natural Disaster in Central America), 33 p.

Olson, R.A., 1987, The San Salvador earthquake of October 10, 1986: Overview and context: Earthquake Spectra, v. 3, p. 415–435.

Rogers, A.M., Borcherdt, R.D., Covington, P.A., and Perkins, D.M., 1984, A comparative ground response study near Los Angeles using recordings of Nevada nuclear tests and the 1971 San Fernando earthquake: Bulletin of the Seismological Society of America, v. 74, p. 1925–1949.

Rolo, R., Bommer, J.J., Houghton, B.F., Vallance, J.W., Berdousis, P., Mavrommati, C., and Murphy, W., 2004, Geologic and engineering characterization of Tierra Blanca pyroclastic ash deposits, *in* Rose, W.I., et al., eds., Natural hazards in El Salvador: Boulder, Colorado, Geological Society of America Special Paper 375, p. 55–67 (this volume).

Rymer, M.J., 1987, The San Salvador earthquake of October 10, 1986—Geological aspects: Earthquake Spectra, v. 3, p. 436–462.

Safak, E., 1991, Problems with using spectral ratios to estimate site amplification, *in* Proceedings, Fourth International Conference on Seismic Zonation, Stanford, California, II, p. 277–284.

Schmidt-Thomé, M., 1975, The geology in the San Salvador area (El Salvador, Central America), a basis for city development and planning: Geologisches Jahrbuch, v. 13, p. 207–228.

Shakal, A.F., Huang, M.J., Parke, D.L., and Linares, R., 1986, Summary of the processed strong motion data, San Salvador earthquake of October 10, 1986, *in* Proceedings, San Salvador Earthquake Briefing, November 24, 1986, Washington, D.C., USA.

Singh, S.K., Lermo, J., Dominguez, T., Ordaz, M., Espinoza, J.M., Mena, F., and Quass, R., 1988, The Mexico earthquake of September 19, 1985: A study of amplification of seismic waves in the Valley of Mexico with respect to a hill zone site: Earthquake Spectra, v. 4, p. 653–673.

Takahashi, R., and Hirano, K., 1941, Seismic vibrations of soft ground: Bulletin of the Earthquake Research Institute, Tokyo University, v. 19, p. 534–543.

Weber, H.S., Weisemann, G., Lorenz, H., and Schmidt-Thomé, M., 1978, Mapa geológico de la República de El Salvador/América Central: Bundesanstalt für Geowissenschaften und Roshtoffe, Hannover, Germany, scale 1:100,000.

Weinberg, R.F., 1992, Neotectonic development of western Nicaragua: Tectonics, v. 11, p. 1010–1017.

Weisemann, G., 1975, Remarks on the geologic structure of the Republic of El Salvador, Central America: Mitteilungen Geologisch–Palaontologischen Institut, University of Hamburg, v. 44, p. 557–574.

White, R.A., Harlow, D.H., and Alvarez, S., 1987, The San Salvador earthquake of October 10, 1986—Seismological aspects and other recent local seismicity: Earthquake Spectra, v. 3, p. 419–434.

Yu, G., Anderson, J., and Siddharthan, R., 1992, On the characteristics of the nonlinear soil response: Bulletin of the Seismological Society of America, v. 83, p. 218–244.

MANUSCRIPT ACCEPTED BY THE SOCIETY JUNE 16, 2003

Geological Society of America
Special Paper 375
2004

Analysis of the spatial and temporal distribution of the 2001 earthquakes in El Salvador

B. Benito

*Escuela Universitaria de Ingeniería Topográfica, Universidad Politécnica de Madrid, Campus Sur,
Autovía de Valencia km 7, 28031 Madrid, Spain*

J.M. Cepeda

Universidad Centroamericana Simeón Cañas, Boulevard Los Procéres, San Salvador, El Salvador

J.J. Martínez Diaz

Departamento de Geodinámica, Universidad Complutense, Madrid, Spain

ABSTRACT

This paper presents a study of the spatial and temporal distribution of the large destructive earthquakes that occurred in El Salvador during January and February 2001, together with the static stress transfer after each main shock, associated with their respective rupture processes. The sequence began with the magnitude M_w 7.7 earthquake of 13 January, located off the western Pacific Coast in the subduction zone between the Cocos and Caribbean plates. One month later, a second destructive earthquake of M_w 6.6 occurred in the Caribbean plate farther inland, the epicenter of which was located near San Pedro Nonualco. This shock was linked to the local faults beneath the volcanic arc and also produced significant damage. The two main shocks and their aftershock sequences, together with other minor events that followed successively, produced unusually intense activity in the zone, in a short interval of time. The aims of this study are to document the spatial and temporal evolution of each seismic sequence and also to understand the possible interaction between the different events. We have inferred that some events with M > 5 triggered other shocks with the same or different origin (subduction zone or local crustal faults). The Coulomb stress transfer has been studied, and some models developed, using the rupture parameters derived from the geometric distribution of aftershocks. These results suggest the existence of a dynamic interaction, since the 13 February event occurred in a zone where the Coulomb stress increased following the January 13 event. Subsequently, some further events with magnitude around M_w 5 in turn were located in other zones of increased stress associated with the two previous large earthquakes.

Keywords: seismicity of Central America, subduction, aftershock distributions, Coulomb stress transfer, triggering mechanism.

INTRODUCTION

On January 13, 2001, a destructive earthquake of M_w 7.7 (U.S. Geological Survey) struck El Salvador, Central America. The earthquake was centered at 12.80° N, 88.79° W, with a focal depth of 40 km, in the subduction zone between the Cocos and Caribbean plates. This earthquake was followed by numerous aftershocks with the same origin; ~540 events with M > 2 occurred in the first month, and 4000 in the first six months, nearly half of which were larger than M 3.0.

Just one month later, on 13 February, a second major earthquake of M_w 6.6 occurred, this time located farther inland

*ma_ben@euitto.upm.es

Benito, B., Cepeda, J.M., and Martínez Diaz, J.J., 2004, Analysis of the spatial and temporal distribution of the 2001 earthquakes in El Salvador, *in* Rose, W.I., Bommer, J.J., López, D.L., Carr, M.J., and Major, J.J., eds., Natural hazards in El Salvador: Boulder, Colorado, Geological Society of America Special Paper 375, p. 339–356. For permission to copy, contact editing@geosociety.org. © 2004 Geological Society of America

(13.64° N, 88.94° W) and with a shallower focal depth of ~15 km. This earthquake, located near San Pedro Nonualco (30 km from San Salvador), was associated with the local fault system aligned with the Central American volcanic arc that bisects El Salvador from east to west. This shock was preceded by numerous local events, ~100 events between 13 January and 13 February (M > 2), and followed by numerous aftershocks, 685 during the first month and 1300 in the first six months.

A third moderate-magnitude event (m_b = 5.1, CIG) occurred four days later, on 17 February, located south of metropolitan San Salvador (12.90° N, 89.10° W), but also associated with faulting along the volcanic axis.

The seismicity map with the epicenters of the events of 2001 in El Salvador and surrounding areas, recorded by the Center for Geotechnical Investigations (CIG), is shown in Figure 1.

Besides the three principal earthquakes described above, other events with magnitude close to and larger than M 5 followed the first shock until September 2001, alternating between events of the subduction zone and those of the volcanic arc.

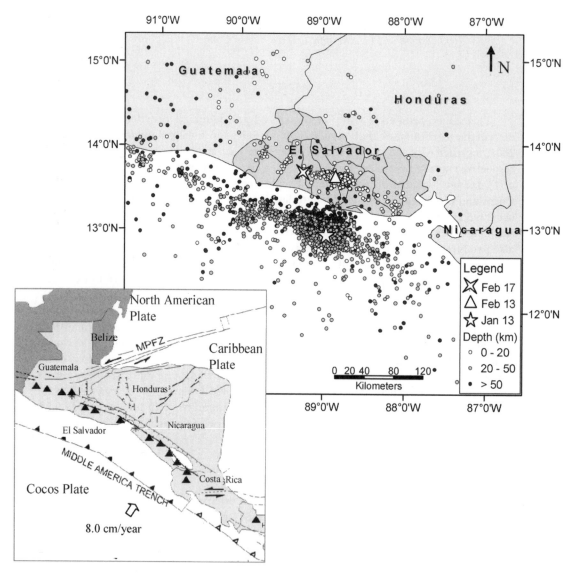

Figure 1. Map showing the distribution of seismicity during 2001, recorded and relocated by the Salvadoran Short-Period Network of the Center for Geotechnical Investigations (CIG). Estimation of the parameters was carried out with SEISAN system (Earthquake Analysis Software, 2000). Inset shows Regional Tectonics of Central America (after Rojas et al., 1993). Solid and open triangles indicate thrust faulting at subduction and collision zones, respectively. Large open arrows are plate motion vectors; half arrows indicate sense of movement across strike-slip faults, ticks indicate downthrown side of normal faults. Large solid triangles are Quaternary volcanoes. MPFZ—Motagua-Polochic fracture zone.

The two large shocks of 13 January and 13 February, together with events of lesser magnitude and their respective aftershock sequences, produced an intense period of seismic activity during a short time interval. This activity at a certain moment did not appear to decrease in time and frequency, according to the laws known. The study of the spatial and time distribution of these series, related to the tectonic environment and the stress transfer evolution, is the main purpose of this paper. The historical seismicity in El Salvador shows that large subduction events are commonly followed by crustal earthquakes in a time interval of four to five years (White, 1991). The main question we address here is whether the M 6.6 February 13 earthquake was in some way triggered by the larger subduction earthquake a month earlier. This behavior would be important for future seismic hazard in the area.

SEISMOTECTONIC ENVIRONMENT

The seismic activity registered in 2001 is framed in the particular seismotectonic context of El Salvador, and of Central America at a regional scale, which has been already described by many authors (Dewey and Suarez, 1991; Ambraseys and Adams, 2001; White and Harlow, 1993; Bommer et al., 2002).

The El Salvador earthquakes of 2001 are associated with the two principal seismic sources that define the seismotectonic structure of Pacific Central America. The largest earthquakes are generated in the subducted Cocos plate and its interface with the Caribbean plate beneath the Middle America trench (Dewey and Suarez, 1991). Relative plate motion of 8 cm/yr produces frequent earthquakes extending to intermediate depths (~200 km), beneath the Pacific Coast of El Salvador. Some earthquakes in this zone in the past century include those of 7 September 1915 ($M_S = 7.7$), 28 March 1921 ($M_S = 7.4$), 21 May 1932 ($M_S = 7.1$), and 19 June 1982 (M = 7.3) (Ambraseys and Adams, 2001).

A second major source of seismicity is related to a local system of faults that extends from west to east along the volcanic chain. These upper-crustal earthquakes have a tectonic origin, but are often called "volcanic chain events" due to their proximity to the volcanic axis. The majority of the events in this source have moderate magnitudes (5.5 < M < 6.8) and shallow depths (h < 20 km). These events contribute significantly to the seismic hazard and risk in the region, and historically have caused more deaths and damage than large earthquakes in the subduction zone (White and Harlow, 1993). During the twentieth century such earthquakes struck El Salvador on at least seven occasions, sometimes occurring in clusters of two or three similar events with a time difference of minutes or hours (White and Harlow, 1993) (Table 1 and Fig. 2).

The 2001 earthquakes of 13 January and 13 February are recent examples of the seismic potential of the subduction zone and the volcanic chain. Specifically, the event of 13 January is similar to that of 19 June 1982 in terms of mechanism, focal depth, and of the damage pattern in the southwest of the country. This earthquake was followed by the crustal event in 1986. On the other hand, the location of the 13 February event is similar to that which occurred in 1936, which was preceded by the subduction event in 1932.

Regarding the focal mechanism for both types of earthquakes, different authors give solutions for the subduction events of 1982 and of 13 January 2001, corresponding to normal faulting with horizontal extension in NE-SW direction. The shallow crustal events in 1965, 1986, and February 2001 present a strike-slip mechanism, with vertical fault planes oriented in NS and EW directions.

A tectonic interpretation of the region given by Harlow and White (1985) suggested that the relative motion between the Caribbean and Cocos plates is slightly oblique and decoupled into two components: a larger normal component, manifest as thrust faulting along the Middle America trench, and another smaller

TABLE 1. SOURCE PARAMETERS OF THE DESTRUCTIVE EARTHQUAKES IN EL SALVADOR THROUGHOUT THE TWENTIETH CENTURY

Year	Month	Day	Hour	Latitude (°)	Longitude (°)	MS	Depth (km)	Intensity (MM)	Source
1915	09	07	01:20	13.90	−89.60	7.7	60	IX	Subduction
1917	06	08	00:51	13.82	−89.31	6.7	10	VIII	Local
1917	06	08	01:30	13.77	−89.50	5.4	10	VIII	Local
1919	04	28	06:45	13.69	−89.19	5.9	10	X	Local
1930	07	14	22:40	14.12	−90.25	6.9	30	VII	Local
1932	05	21	10:12	12.80	−88.00	7.1	150	VIII	Subduction
1936	12	20	02:45	13.72	−88.93	6.1	10	VIII	Local
1937	12	27	00:43	13.93	−89.78	5.9	10	VII–VIII	Local
1951	05	06	23:03	13.52	−88.40	5.9	10	VIII	Local
1965	05	03	10:01	13.70	−89.17	6.3	15	VIII	Local
1982	06	19	06:21	13.30	−89.40	7.3	80	VII	Subduction
1986	10	10	17:49	13.67	−89.18	5.4	10	VIII–IX	Local
2001	11	03	17:33	13.05	−88.66	7.8	60	VIII	Subduction
2001	02	13	14:22	13.67	−88.94	6.5	10	VIII	Local

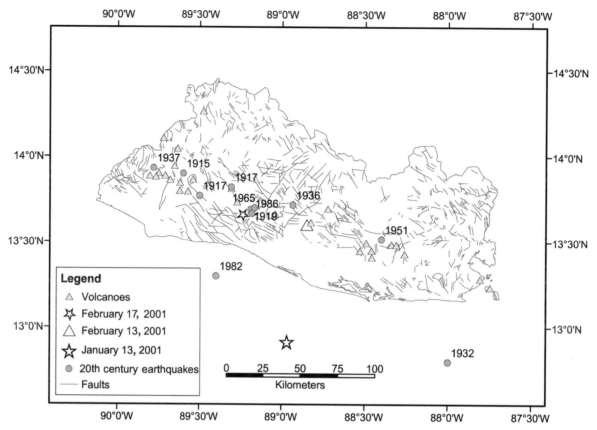

Figure 2. Local tectonic map of El Salvador with locations of the main shocks of the twentieth century (epicenters represented by circles) and the beginning of 2001 (special symbols) and the active volcanoes (triangles). A list with the parameters of these earthquakes is included in Table 1.

longitudinal component, manifest as right-lateral shear along the volcanic chain. The focal mechanisms and geologic features along the volcanic zone are compatible with the interpretation of this zone as a strike-slip right-lateral shear zone caused by an oblique component of Cocos-Caribbean collision (White, 1991). The existence of right-lateral slip faults within the volcanic arc of El Salvador and adjacent regions of Guatemala and Nicaragua (Weinberg, 1992), and the clustering of earthquakes along these faults, is consistent with the trench-parallel component of motion concentrated along the volcanic chain (DeMets, 2001). The rate of the strike-slip motion along the arc is estimated to be 8 mm/ yr (Guzman-Speciale, 2001), the motion being parallel to the trench. In both El Salvador and Guatemala this rate is predicted to be slower than in Nicaragua, due to the extension east of the forearc (DeMets, 2001). Recent geodetic observations in the region not only support the model of strain partitioning proposed by Harlow and White (1985) but also constrain the rate of forearc slip (DeMets, 2001).

In this tectonic environment, normal-faulting subduction earthquakes are usually followed within four or five years by large thrust events or by shallow intraplate earthquakes. This behavior has been observed in other regions where the tectonic regime involves a subduction limit offshore and a volcanic axis inside the continent, such as Mexico (Lomnitz and Rodríguez, 2001). This inference may be explained because the stress transfer due to relaxation in one area leads to heightened tectonic stress in adjacent areas. The present study shows that a similar pattern may exist in El Salvador.

SPATIAL AND TEMPORAL CHARACTER OF THE 13 JANUARY AND 13 FEBRUARY 2001 EARTHQUAKES SEQUENCE

We focus on the 13 January and 13 February 2001 events, which were the largest shocks of that year, and on their respective aftershock series. Our study is intended to shed light on a possible interaction between both types of events.

Correlation between Magnitude Scales

The study aims to characterize the evolution of the seismicity recorded in El Salvador during 2001, taking into account the magnitude of the earthquakes that followed subsequently. For this purpose, a homogeneous magnitude is required, so we have calculated moment magnitude (M_W) for all the significant events included in the available catalog.

The El Salvador seismic catalog for 2001 compiled by the Servicio Nacional de Estudios Territoriales (SNET) contains 3755 events with magnitudes ranging from 2.1 to 7.8. The magnitude scales used are coda magnitude (M_C), local magnitude (M_L), surface wave magnitude (M_S), and moment magnitude (M_W).

Analyses of earthquake recurrence and strong-motion attenuation use magnitude in terms of M_S or M_W, to avoid saturation of local magnitude scales, such as M_C and M_L, for earthquakes larger than about M 7. In order to allow for comparisons of our analyses with other studies, a regression of the data was performed to obtain an M_W-M_C conversion relationship (Rojas et al., 1993).

A subset was created from the catalog, selecting events containing both magnitude scales. The subset was fitted to a second degree polynomial, which produced the best solution compared to the linear, logarithmic, power, and exponential forms. The resulting relation is given by:

$$M_w = -0.0155M_c^2 + 0.9731M_c + 0.3719 \qquad (1)$$

The correlation coefficient is 0.9. Figure 3 shows a plot of the M_W-M_C distribution from which the moment magnitude M_W has been estimated for all the events in the catalog.

Source Parameters

The source parameters of the studied shocks together with the focal mechanism, given by different agencies and authors, are shown in Tables 2A and 2B.

The source time function for the 13 January event indicates two subevents: the first with higher amplitude and 22 seconds duration, and a second one of 24 seconds (Bommer et al., 2002). The seismic moment release is 5.54×10^{20} Nm with no apparent directivity effects. The earthquake with intermediate depth occurred inside the down-going Cocos plate, its mechanism being a normal fault with subvertical fault plane and a tension (T-axis) subparallel to the dip direction of the descending slab.

For the 13 February event, located in the upper plate at the volcanic chain, the fault plane solution is a strike-slip event. This event occurred at a depth of 14 km, with a seismic moment of 6.05×10^{18} Nm and a total duration of 12 s (Bommer et al., 2002). The aftershock distribution delineates a rupture plane subparallel to the volcanic chain and thus subparallel to the trench.

Spatial Distribution of Aftershocks

The map depicted in Figure 1 shows the total distribution of events in 2001, relocated by the SNET. In that figure, it is possible to observe some clusters corresponding to the aftershock sequences of the different main shocks, with the largest clusters located around the epicenters of 13 January and 13 February events. Our purpose is to identify the aftershocks associated with both events, as well as their rupture surfaces.

To obtain an overview of the seismicity pattern and associate events with each series, we examined the distribution of the aftershocks week by week within the time period January 13 to June 7, 2001. The results are included in Figures 4A and 4B.

During the first week following the January 13 earthquake, different clusters of local events occurred in the upper plate inland, together with one offshore cluster in the southwest part of the main-shock rupture. In the second week, from 21 to 28 January, overall activity decreased and in the third and fourth weeks ceased altogether beneath the volcanic chain. There is a quiescent period of 15 days before the M_W 6.6 volcanic chain event of 13 February, following which seismicity increased along a system of faults located parallel and perpendicular to the coast and included the M 5 event of 17 February. At the same time, the subduction activity increased again during the week from 13 to 20 February, as if it had been reactivated by the two volcanic chain events.

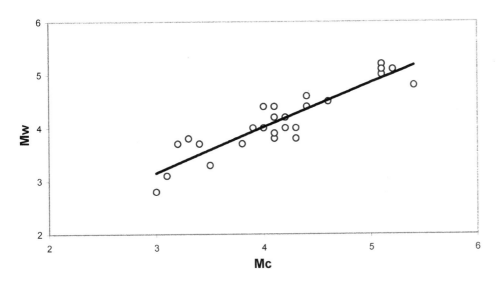

Figure 3. Relationship between coda and moment magnitude of events obtained from the subset used in the study. The agency source for magnitude is the Servicio Nacional de Estudios Territoriales (SNET), with the exception of the 13 January main shock, where the sources are the U.S. Geological Survey (USGS) and Central American Seismic Centre (CASC).

TABLE 2A. SOURCE PARAMETERS AND FOCAL MECHANISM FOR THE 2001 JANUARY 13 EVENT, ACCORDING TO DIFFERENT AGENCIES AND AUTHORS

Harvard (CMT)	USGS (CMT)	Buforn et al., 2001 inversion P wave	CIG	CASC	Bommer et al., 2002
M_0: $4.6*10^{20}$ Nm	M_0: $2.9*10^{20}$ Nm	M_0: $5.54*10^{20}$ Nm	M_0:	M_0:	M_0: $5.54*10^{20}$ Nm
Mw: 7.7	Mw: 7.6	Mw: 7.8	Mw: 7.6	Mw: 7.7	Mw: 7.7
Time: 17:33:45.80	Time: 17:33:32.38	Time: 17:33:46	Local time: 11:34	Time: 17:33:30	
Epicentral location:	Epicentral location:	Epicentral location:	Epicentral location:	Epicentral location:	Epicentral location:
Lat 12.97° N	Lat 13.05° N	Lat 12.77° N	Lat 12.92° N	Lat 12.87° N	
Long 89.13°	Long. N 88.66° W	Long. 88.83° W	Long. 88.97° W	Long. 88.77° W	
Depth: 56 km	Depth: 36 km	Depth: 49 km	Depth: 32 km	Depth: 60 km	Depth: 50 km
		STF: 46 s			
p1: STK: 121°; DP: 35°; slip: −95°	p1: STK: 121°; DP: 37°; slip: −95°	p1: STK: 128°; DP: 63°; slip: −98°	Localization: 45 km south, los Blancos Beach, Dpto. La Paz		2 sub-events of 22 and 24 s
p2: 307°; 56°; −86°	p2: 307°; 54°; −86°	p2: ??			
		Rupture duration: 46 s			Fault plane: subvertical normal fault. Tension axis subparallel to dip direction of the descending slab.
Axis T, N Y P	Axis P, T Y N				
1. (T) VAL = 4.58; PL = 10; AZM = 35	1. (T) VAL = 3.17; PL = 9; AZM = 34	Complex rupture: most of released energy during the first 20 s.			
2. (N) −0.03; 3; 125	2. (N) −0.47; 3; 125				
3. (P) −4.55; 79; 231	3. (P) −2.70; 81; 233				

TABLE 2B. SOURCE PARAMETERS AND FOCAL MECHANISM FOR THE 2001 FEBRUARY 13 EVENT, ACCORDING TO DIFFERENT AGENCIES AND AUTHORS

Harvard (CMT)	USGS (CMT)	Buforn et al., 2001 inversion P wave	CIG	CASC	Bommer et al., 2002
M_0: $8.1*10^{18}$ Nm	M_0: $6.2*10^{18}$ Nm	M_0: $6.05*10^{18}$ Nm	M_0:	M_0:	M_0: $6.05*10^{18}$ Nm
Mw: 6.6	Mw: 6.5	Mw: 6.5	Mw: 6.6	Mw: 6.6 (mc = 6.7)	Mw:
Time: 14:22:16.40	Time: 14:22:05.82	Time: 14:22:16	Local time: 08:22	Time: 14:22:09.80	Time:
Epicentral location:	Epicentral location:	Epicentral location:	Epicentral location:	Epicentral location:	Epicentral location:
Lat 13.98° N	Lat 13.67° N	Lat 13.60° N	Lat 13.60° N	Lat 13.44° N	
Long 88.97° W	Long. N 88.94° W	Long. 88.96° W	Long. 88.85° W	Long. 88.94° W	
Depth: 15 km	Depth: 15 km	Depth: 14 km	Depth: 8 km	Depth: 13.9 km	Depth: 14 km
		STF: 12 s			Duration: 12 s
p1: STK: 276°; DP: 74°; slip: −175°	p1: STK: 96°; DP: 81°; slip: −178°	p1: STK: 90°; DP: 91°; slip: −180°	Localization: San Pedro Nonualco. Dpto. La Paz.		Strike-slip. Fault plate subparallel to the subduction trench.
p2: 7°; 86°; 16°	p2: 6°; 88°; −9°	p2: 90°; 101°			
Axis T, N Y P	Axis P, T Y N				
1. (T) VAL = 8.50; PL = 14; AZM = 233	1. (T) VAL = 6.17; PL = 5; AZM = 51				
2. (N) −0.83; 73; 22	2. (N) −0.02; 81; 173				
3. (P) −7.66; 8; 141	3. (P) −6.14; 7; 321				

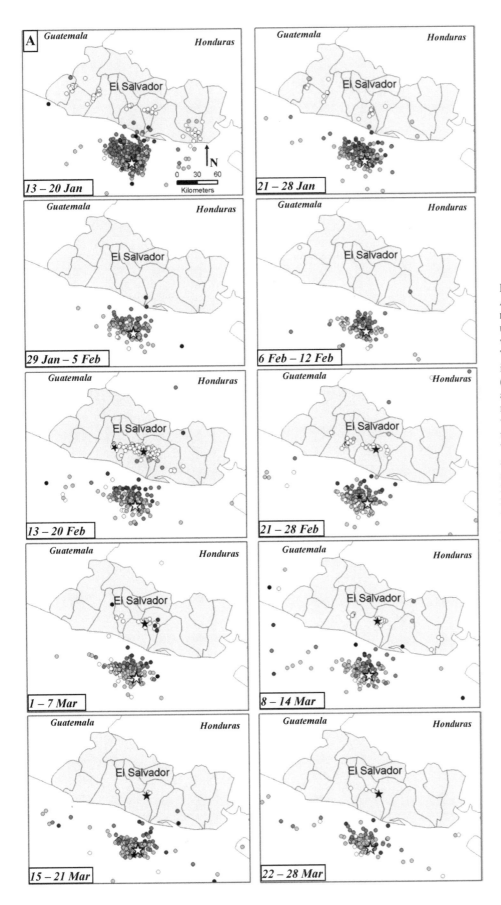

Figure 4 (*on this and following page*). A: Seismicity of El Salvador after the main shock of January 13 (white star) until March 28, for intervals of one-week duration (magnitude $M_w \geq 3.0$). The epicenter of the February 13 event is also represented after its occurrence (black star). The other events identified as changing the seismicity rates are also represented in their corresponding time windows (M 5.6, 28 February, subduction; M 5.7, 16 March, subduction). The locations of the remaining events are represented by circles, whose color shows the range of depths (white—h < 20 km, gray—20 < h < 50 km, black—h > 50 km). Representation has been done using a geographic information system, Arc-Info 8.0. B: Seismicity for the period from March 29 until June 7, with the same representation criteria as in Figure 4A. New symbols of epicenters in some time windows correspond to events that act as triggers (M 4.9, 10 April, volcanic chain; M 4.6, 10 April, subduction; and M 4.6, May 8 and 9, volcanic chain).

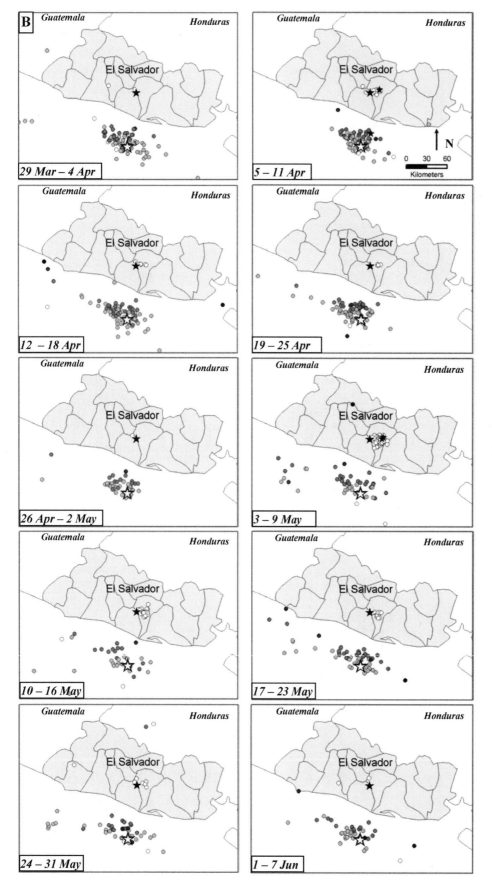

Figure 4 (*continued*).

Analysis of subsequent weeks suggests further changes in the seismicity rate following events with moderate magnitude: 28 February (M 5.6, subduction), 16 March (M 5.7, subduction), 10 April (M 4.9, volcanic chain), 8 and 9 May (series of three shocks, M 4.8, 4.6, and 4.6, volcanic chain). Activity began to subside in July, even though seven more events of M > 4.5 took place until the end of the year in both scenarios.

To sum up, the seismic activity as far as the subduction and the local strike-slip faults are concerned are alternately increasing and decreasing, with some weeks in which both types of seismicity seem to alternate. This tendency becomes evident as, coincidentally to the occurrence of new shocks with magnitude close to 5.0, the activity around their epicenters is again triggered. Therefore, such events can act as triggers of new shocks, inducing in turn other events with similar magnitude and different source. Due to the importance of these events in the global activity registered, which may be at the same time the cause and effect of itself, it is worthwhile to analyze them in detail. Table 3 shows the parameters of all the events registered during 2001 with magnitude M ≥ 4.5.

In summary, the temporal superposition of these series produced an unusual activity during the first six months of 2001. It appears that the 13 January earthquake triggered one or several local faults of the volcanic chain, and these in turn affected seismicity in the subduction area. One must consider whether the local events would have occurred in the absence of the 13 January earthquake. A reasonable supposition is that they would occur but perhaps not until much later. The historical seismicity of El Salvador shows that approximately every 20 yr a destructive volcanic earthquake occurs, and the last one took place in 1986 (White and Harlow, 1993). It is probable then that the fault on which the 13 February earthquake was originated had accumulated sufficient strain so that, although it might have not been released by itself at that time, the additional loading caused by the event of 13 January could have acted as a trigger.

Modeling of Rupture Surfaces

After studying the spatial distribution of the aftershock sequences, we now attempt to model the rupture planes associated with the earthquakes on 13 January and 13 February, starting with the distribution of the aftershocks for the first three days following the main shocks. In theory, the distribution for each aftershock series must define a plane obtained by the fit of the hypocenters, which agrees with the solutions given by the focal mechanism. However, when examining the solutions given by different agencies (Tables 2A and 2B), we can see that there are some differences among the values of azimuth and dip. Therefore, we intend to obtain more information from the aftershock areas, which allows us to confirm some of these solutions.

We have tested different orientations of faulting, centered in the aftershock cloud, according to the focal mechanism given in Tables 2A and 2B. Then we have looked for the best solution, as that which represents higher coherence between the mechanism and aftershock distribution.

For the January 13 event, our best-fit solution is a fault plane dipping 60° to the NE with a strike of N 128° E, subparallel to the Middle America trench (Fig. 5A). The fault has a length of 67 km and a rupture area of 2532 km², indicated by the spatial distribution of the aftershocks.

For the crustal earthquake of 13 February, the best solution indicates a plane of 471 km² striking N 94° E and dipping 70° SW (Fig. 5B).

Temporal Distribution of Aftershocks

We studied the temporal evolution of each aftershock sequence independently for the 13 January and 13 February events to assess the possible interaction between both series. In the second sequence, the foreshocks have also been included in the analysis. The discrimination of shocks associated with each series was made by taking the events associated with each surface rupture previously determined but extending the time interval through 2001. Figure 6 illustrates the number of events as a function of time for the two sequences and reveals alternating increases and decreases of activity in the respective source zones.

A significant feature is the occurrence of volcanic chain events from 13 to 25 January, which we interpret as local activity triggered by the previous subduction earthquake. The absence of local events during the following two weeks until February 13, when the main shock of this series took place, is also apparent.

These observations seem to corroborate the interaction between the sequences of different source, which can also be illustrated by superposing both series logarithmically with the same reference origin time (on 13 January), revealing strong irregularity for the two sequences (Fig. 7).

We also studied the decay of the aftershock activity analytically, with respect to Omori's Law (Omori, 1884), which in logarithmic form corresponds to the expression: log N(t) = a – b log t; N(t) being the number of events by day and t the time in days from the main shock. Figure 8 shows the fit for each series in two time intervals. For the subduction sequence, a first fit is made for the whole series (six months) and a second fit taking only the events between 13 January and 13 February. The aftershocks can be seen to have decayed approximately according to Omori's law in the period up to 13 February, but they were gradually dying out when the second earthquake occurred (Fig. 8A). The fit is better if we only consider the aftershocks prior to the second major event (equation 3; Fig. 8B) rather than with the complete sequence (equation 2).

$$\log N(t) = 2.7 - 0.8 \log t, \text{ with } R^2 = 0.7 \tag{2}$$

$$\log N(t) = 2.4 - 0.7 \log t, \text{ with } R^2 = 0.8 \tag{3}$$

For the February 13 main shock and aftershock sequence an initial fit was made using the total set (Fig. 8C), in all the time intervals, where it is possible to appreciate that the decreasing

TABLE 3. SUMMARY OF THE SOURCE PARAMETERS FOR THE EVENTS THAT OCCURRED IN 2001 WITH MAGNITUDE M ≥4.5

Year	Mo	Day	Hr	Min	Sec	Lat	Long	Depth (km)	Type	Magnitude	Agency
2001	1	13	17	35	52.7	12.92	−88.97	32.1	Subduction	$M_w = 7.6$	USGS
2001	1	16	8	23	24.3	13.33	−88.71	54	Subduction	$M_c = 4.6$	SAL
2001	1	16	10	59	32.2	13.04	−88.83	56.8	Subduction	$M_c = 4.9$	SAL
2001	1	16	11	25	48.0	13.00	−88.96	57.7	Subduction	$M_c = 4.5$	SAL
2001	1	17	1	36	18.7	12.93	−89.16	33.8	Subduction	$M_c = 4.6$	SAL
2001	1	17	21	5	40.5	12.93	−88.99	63.5	Subduction	$M_c = 4.6$	SAL
2001	1	25	10	28	51.8	12.79	−88.77	37.5	Subduction	$M_c = 4.8$	SAL
2001	2	2	8	10	43.7	12.89	−89.29	26.1	Subduction	$M_c = 5.1$	SAL
2001	2	7	10	23	11.3	13.03	−89.08	47	Subduction	$M_c = 5.1$	SAL
2001	2	13	2	50	57.0	12.00	−88.34	68	Subduction	$M_c = 4.5$	SAL
2001	2	13	14	22	05.8	13.60	−88.85	11.1	Volcanic chain	$M_w = 6.6$	SAL
2001	2	17	1	17	31.6	12.67	−88.96	50	Subduction	$M_c = 5.1$	SAL
2001	2	17	20	25	15.9	13.68	−89.25	5	Volcanic chain	$M_L = 5.1$	SAL
2001	2	21	6	51	28.1	12.98	−88.94	64.4	Subduction	$M_c = 4.7$	SAL
2001	2	23	18	40	56.8	13.04	−88.85	43.9	Subduction	$M_c = 4.5$	SAL
2001	2	24	3	46	52.1	13.49	−88.67	13.8	Volcanic chain	$M_c = 4.5$	SAL
2001	2	25	8	28	14.6	13.69	−89.23	8.5	Volcanic chain	$M_L = 4.6$	SAL
2001	2	26	19	51	10.2	12.64	−89.24	28.8	Subduction	$M_c = 4.7$	SAL
2001	2	28	18	50	14.5	13.00	−89.08	51.4	Subduction	$M_L = 5.6$	SAL
2001	3	16	0	1	19.6	12.84	−89.02	50	Subduction	$M_L = 5.7$	SAL
2001	3	18	15	43	23.1	12.58	−87.92	24	Subduction	$M_c = 5.2$	SAL
2001	3	29	6	54	31.6	13.01	−88.94	61.7	Subduction	$M_L = 5.4$	SAL
2001	4	3	1	7	14.2	12.83	−88.71	32.9	Subduction	$M_c = 4.5$	SAL
2001	4	10	3	16	53.9	13.08	−88.84	46.4	Subduction	$M_L = 4.9$	SAL
2001	4	10	21	46	58.9	13.64	−88.72	7.6	Volcanic chain	$M_c = 4.4$	SAL
2001	4	14	19	38	33.9	12.40	−88.74	17.9	Subduction	$M_c = 4.6$	SAL
2001	5	2	7	5	54.8	12.97	−89.09	40.4	Subduction	$M_c = 4.6$	SAL
2001	5	8	18	2	17.3	13.62	−88.71	9	Volcanic chain	$M_L = 4.8$	SAL
2001	5	8	18	15	47.0	13.61	−88.68	10.5	Volcanic chain	$M_L = 4.6$	SAL
2001	5	9	7	23	57.4	13.63	−88.67	18	Volcanic chain	$M_c = 4.6$	SAL
2001	6	2	19	36	36.4	12.83	−88.25	65.6	Subduction	$M_c = 4.7$	SAL
2001	6	6	19	27	46.9	12.27	−88.29	79.4	Subduction	$M_c = 4.5$	SAL
2001	7	17	20	1	52.3	12.62	−87.53	77.5	Subduction	$M_c = 4.8$	SAL
2001	9	6	22	59	53.5	12.34	−89.08	50	Subduction	$M_c = 4.5$	SAL
2001	9	18	14	50	58.2	13.04	−89.05	33.4	Subduction	$M_L = 5.0$	SAL
2001	11	9	0	49	37.9	13.32	−88.32	26.7	Subduction	$M_c = 4.6$	SAL
2001	12	1	9	0	04.9	12.72	−88.62	34.6	Subduction	$M_c = 4.6$	SAL

Note: USGS—United States Geological Survey; SAL—Salvadorian Local Network.

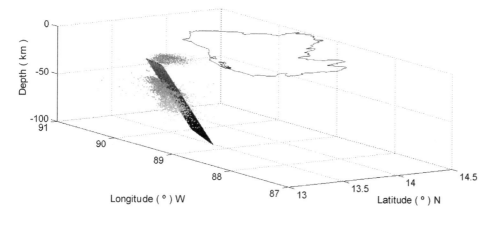

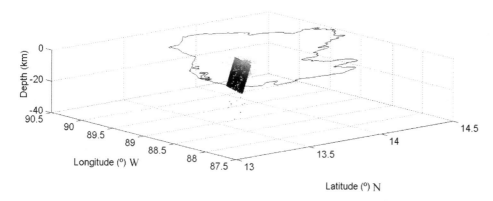

Figure 5. Fault plane of main shocks fitted to aftershocks of the first three days, together with the hypocenters and epicenters of these events and the outline of El Salvador. A: Representation for the 13 January M_w 7.7 event. The fault trace is subparallel to the coast and the uppermost part of the rupture reaches a depth of 20 km. B: Representation for the 13 February M_w 6.6 event. The uppermost part of the rupture reaches a depth of 5 km, without breaking the surface, and most of the hypocenters are constrained to a depth less than 15 km.

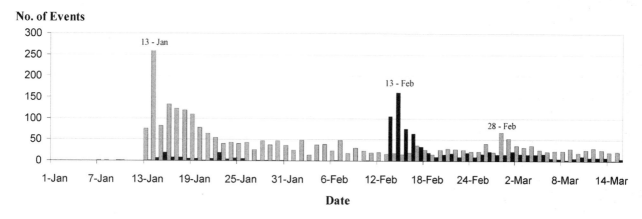

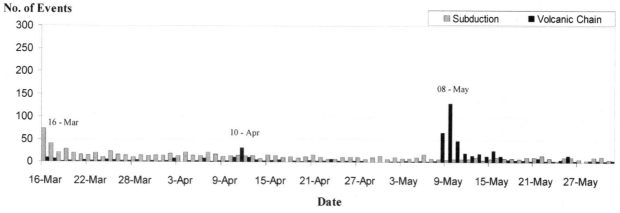

Figure 6. Histogram with the number of events as a function of time for the period January13–May 31, identifying with a different color those associated with 13 January and 13 February. It makes appreciable the occurrence of volcanic chain events from 13 January to 25 January, which can considered as local activity triggered by the previous subduction earthquake, and the total lack of local events during the following two weeks until February 13. Also remarkable is the alternating increase and decrease of subduction and volcanic chain events. The events that induce new activity are easily identified.

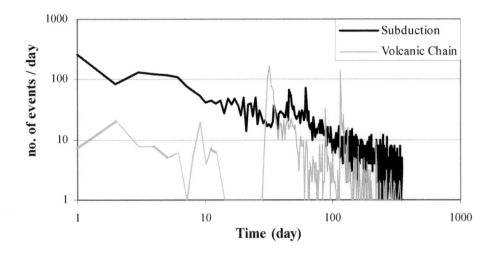

Figure 7. Joint temporal distribution of the aftershock series for January 13 and February 13, with reference to the origin time of the first main shock (13 January), in logarithmic scale. A strong irregularity for both series, in the temporal interval in which they are superposed, is observed.

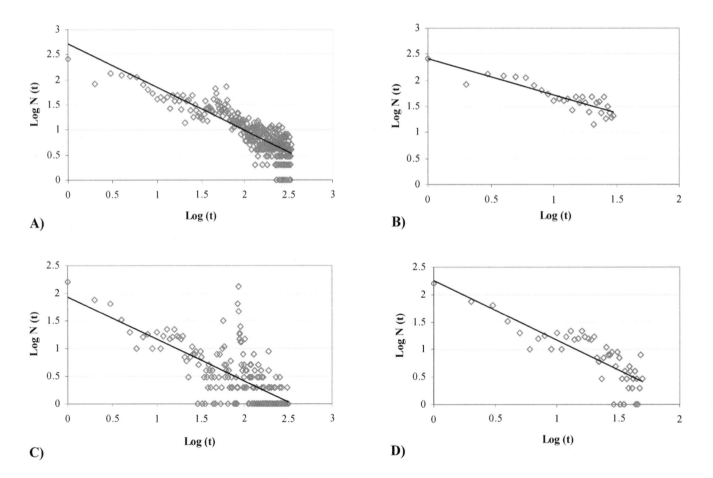

A)

B)

C)

D)

Figure 8. A: Fit to Omori's law for the total aftershock series of January 13 (six months). B: Fit to Omori's law only for the period between January 13 and February 13, prior to the second main shock. In this case the fitting is better than in the previous one. C: Fit to Omori's Law for the total aftershock series of February 13. D: Same as C for the period between February 13 and May 8, previous to the three local shocks with M > 4.6. A better fit is also found with regard to the total series.

exponential rate is lost after approximately 100 days (log t = 2), when the events of May 8 and 9 took place. A second test, considering only the events until this date (Fig. 8D), shows a better fit. The expressions derived for these two intervals are given respectively in equations 4 and 5:

$$\log N(t) = 1.9 - 0.7 \log t, \text{ with } R^2 = 0.5 \quad\quad (4)$$

$$\log N(t) = 2.3 - 1.1 \log t, \text{ with } R^2 = 0.7 \quad\quad (5)$$

The results confirm a perturbation to the lineal fit of aftershock decay when another earthquake with moderate magnitude occurred. A similar result was found in the analysis of the aftershock sequence of the M_W 7.6 August 17, 1999, Izmit, Turkey, earthquake. The time distribution of aftershocks follows Omori's law, except for the perturbations due to the activity following events of magnitude ~5, in particular the M 5 Marmara Island shock of September 20 (Polat et al., 2002).

Magnitude-Frequency-Depth Distribution of Aftershocks

The magnitude–frequency distribution of earthquakes commonly follows a power law of the Gutenberg-Richter form: $\log N(m) = a - bm$; where "m" is a threshold magnitude and N(m) a cumulative number of earthquakes with $M \geq m$. We computed parameters for the 13 January and 13 February series and obtained the equations 6 (subduction) and 7 (volcanic chain), as follows:

$$\log N(m) = 6.2 - 1.0 \, m, \text{ with } R^2 = 0.9 \quad\quad (6)$$

$$\log N(m) = 5.6 - 1.1 \, m, \text{ with } R^2 = 0.9 \quad\quad (7)$$

In both cases the parameter b was close to unity, with a high correlation ($R^2 = 0.9$). A lower value for parameter a was obtained for the volcanic chain events (13 February) and indicates less activity for the crustal events than for the subduction zone, as could be expected, given the source dimensions.

Finally, magnitude versus depth was also assessed for the two sequences of aftershocks (Fig. 9). The lesser magnitude and depth for the volcanic chain is clearly observed. The larger aftershocks (M > 5) of the subduction series are deeper than 25 km. The higher stress of rocks in deeper levels of the seismogenic crust produce higher stress drops in deeper faults and may explain the observed difference in magnitude (Scholz, 1990).

MODELING OF STATIC COULOMB STRESS TRANSFER: TRIGGERING MECHANISM

The space-time relationship between the two main earthquakes and their aftershock series invites the study of possible causal relations between both events resulting from dynamic and from static changes in the state of stress. Such changes may advance or retard the failure of faults in the region, as proposed in other seismically active regions, for instance the North Anatolian fault zone (Stein, 1999), and in particular the Marmara area after the Izmit earthquake (Parsons et al., 2000).

In the historical period, several large (M > 7) subduction earthquakes along the Cocos and Caribbean plate boundary have been succeeded by shallower crustal earthquakes in the volcanic chain in time intervals of years or months (White and Harlow, 1993; Bommer et al., 2002). This suggests the existence of a dynamic interaction between the faults of these principal seismic source zones. We have modeled the stress transfer produced by the main shocks on January 13 and February 13

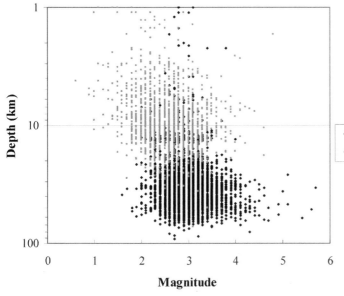

Figure 9. Magnitude versus depth for the two sequences of 13 January and 13 February. Most of the hypocenters of subduction events have depths in the range (20–80 km), while this range for local events is 3–20 km. The larger aftershocks (M > 5) of the subduction series have depths >25 km.

using the parameters of the rupture surfaces consistent with the regional tectonics and the seismological data.

It is known that the stress drop on a fault plane due to the occurrence of an earthquake produces an increase of effective shear stress around the rupture area (Chinnery, 1963). This transfer of the static stress may explain the generation and location of aftershocks and other main shocks at large distances from the fault, even at tens of kilometers, in those zones where the increase of the Coulomb failure stress (CFS) is ~1 bar. This fact has been recognized in numerous works in different geodynamic frameworks since 1992 (e.g., Jaumé and Sykes, 1992; King et al., 1994; Toda et al., 1998). The peculiarities of the El Salvador seismic sequences give special interest to this kind of analysis.

Methodology

During the last ten years, observations of seismicity from different seismogenic settings and magnitudes have indicated that variations in static stress less than 1 bar are able to induce the reactivation of nearby faults that are close to failure, either as aftershock activity or as larger earthquakes. This phenomenon has been described as a triggering process (King et al., 1994; Harris et al., 1995). It has also been observed that the triggering process may involve not only the generation of aftershocks or major shocks, but also changes of seismicity rate in a certain zone, increasing or decreasing for several months after a main shock (Stein and Lisowski, 1983; Reasenberg and Simpson, 1992; Stein, 1999).

The triggering effect is attributed to changes in Coulomb failure stress (CFS):

$$CFS = \tau_\beta - \mu\,(\sigma_\beta - p) \qquad (8)$$

where τ_β is the shear stress over the fault plane, σ_β is the normal stress, p is the fluid pressure, and μ is the frictional coefficient.

For the seismic series of 2001 in El Salvador, we have estimated the change in the static Coulomb failure stress by the expression given in equation 9:

$$\Delta CFS = \Delta\tau_\beta - \mu'\,\Delta\sigma_\beta \qquad (9)$$

where $\Delta\tau_\beta$ is considered positive in the direction of the slip fault, and $\Delta\sigma_\beta$ is also positive in a compressional regime. μ' is the apparent coefficient of friction and includes the effects of pore fluid as well as the material properties of the fault zone (see Harris, 1998, for a deeper explanation of this parameter). The positive values for ΔCFS are interpreted as promoting faulting, while negative values inhibit the activity.

We have estimated the stress change in an elastic half-space following the Okada (1992) method, taking for the shear modulus a value of 3.2×10^{10} N m^{-2} and for the Poisson coefficient a value of 0.25. The apparent friction coefficient is taken as 0.4, which is an acceptable value as proposed by Deng and Sykes (1997) from the study of 10 yr of seismicity in southern California. The intro-

duction of different values for the apparent frictional coefficient, ranging from 0.2 to 0.6, does not produce significant changes in the obtained results.

Models of stress transfer have been constructed for the ruptures associated with the January 13 and February 13 earthquakes, respectively. The dimension and orientation of the surface ruptures are those derived earlier, taking into account the focal mechanisms published in previous studies.

The surface rupture estimated for the January earthquake (M 7.7) is ~2500 km^2. The focal mechanism calculated by Harvard University, U.S. Geological Survey, Buforn et al. (2001), and Bommer et al. (2002) using different approaches (CMT [Centroid Moment Tensor] and wave polarities) (see Table 2A) gives practically the same orientation for the fault plane solution, between N 120° E and N 129° E. This direction agrees with the orientation of the horizontal axis of the ellipse fitted with the aftershock sequence. A bigger discrepancy is found for the dip of the fault, ranging from 48° NE to 63° NE. Taking into account the spatial distribution of the aftershocks, our model has been built for a plane oriented N 128° E dipping 60° NE, in agreement with the rupture solution presented earlier. The rake of the slip vector used is 98°, following the focal mechanism of Buforn et al. (2001), which corresponds with a normal fault. The aftershock sequence delineates a rupture extending between 15 and 78 km in depth.

In the case of the February event (M 6.6), the aftershock distribution, as well as the focal mechanism estimated by the U.S. Geological Survey and Buforn et al. (2001), indicate a steeply dipping, dextral strike-slip rupture plane, oriented N 94° E dipping 70° SW. The rupture area previously estimated from the aftershock distribution is 471 km^2. This result is consistent with the empirical magnitude/rupture area relationships of Wells and Coppersmith (1994).

Results and Interpretation

We obtained a model of Coulomb failure stress change for the January 13 event, which is included in Figure 10. Figure 10A represents a map view of the model for the January M 7.7 earthquake made for a horizontal plane at 14 km depth, which is the focal depth of the 13 February, M 6.6 earthquake. The color scale represents the different values in bars of the static Coulomb stress change generated by the rupture on planes parallel to the local fault reactivated on February 13 (N 94° E, 70° S). The epicenters of the main shock and the aftershocks produced 48 hours after the two main shocks are also projected. Figure 10B represents a cross section of the same model and shows that the February sequence occurs in an area where the January event produced an increase of CFS.

The stress change produced by the February 13 event is in general lower, but the shallower depth of the rupture produces strong effects in the surrounding area. Figure 11A shows a map view of the stress change produced by this strike-slip event across planes parallel to the January rupture plane, calculated for a 5 km

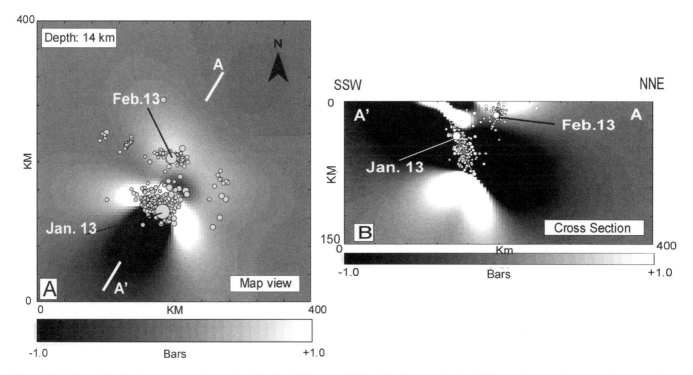

Figure 10. Inferred Coulomb stress transfer produced by the 13 January 2001 subduction earthquake. White and gray colors show the areas of predicted stress increase, while black represents the areas of predicted stress decrease. A: Map view for a horizontal plane, 14 km depth. B: NE-SW cross section view. The epicenters and hypocenters of the aftershocks that occurred within 48 hours of the two main shocks (13 January and 13 February) are shown. The location of the February sequence seems to be controlled by the lobe of increased stress produced by the January 13 event.

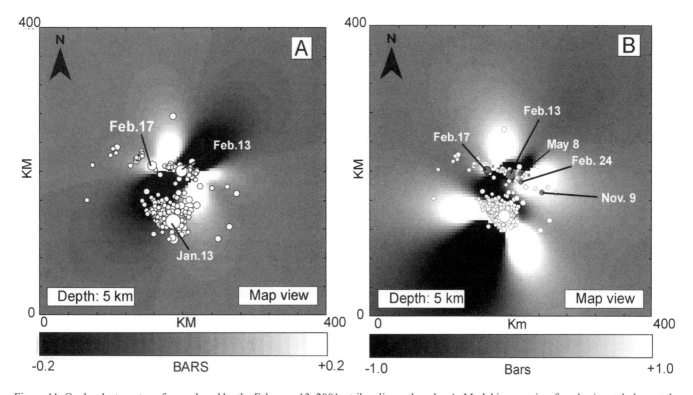

Figure 11. Coulomb stress transfer produced by the February 13, 2001, strike-slip earthquake. A: Model in map view for a horizontal plane at the focal depth of the 17 February event (5 km). This event occurred in a lobe of predicted stress increase. B: Stress transfer model for the two main shocks together in map view for a horizontal plane 5 km depth. The gray circles are the aftershocks of the volcanic chain with magnitude higher than 4.5, which occurred after the 13 February event.

depth horizontal plane (focal depth of the 17 February event). The February 17 event occurred on a lobe where CFS increased more than 0.8 bars following the February 13 event. We also observe that the aftershock area of the January event suffered either relative increase or decrease of CFS. Figure 11B represents the model of CFS change produced by the two main ruptures (M 7.7 and 6.6) on planes parallel to the February plane of rupture. After this event, significant areas of the volcanic chain are affected by increase of CFS higher than 0.4 bars. The aftershocks with magnitude higher than 4.5 of February 17, February 24, and November 11 occurred in areas of stress increase (Fig. 11B). However, the two aftershocks of May 8 happened in an area of reduced stress. Nevertheless, these two aftershocks are close to the rupture area of the February 13 event, where the development of static stress may be more complex.

In summary, we can conclude that the stress transfer generated by the January 13 event induced an increase of stress higher than 0.8 bars in the hypocentral zone of the February 13 event. Most of the aftershocks that occurred during the 48 hours after the February main shock are located in an area of increased CFS, and most of the aftershocks that delineate the rupture surface are located in the area with an increase higher than 1.5 bars. In turn, the February 13 shock increased the CFS by 0.4 bars in the hypocentral zone of the February 17 event (Fig. 11A).

On the other hand, the evolution of the aftershock rate for the January sequence seems to show a complex short-term dynamic evolution in the aftershock area. The change of CFS produced by the February 13 strike-slip event induced an increase of CFS up to 0.2 bars in the western part of the January rupture area and a decrease of CFS up to 0.18 bars in the eastern part. This process, repeated for all the local events with M > 4.5, may induce alternating stress increases and decreases either in time or in space, thus generating the observed complexity in the aftershock rate.

The correlation between CFS increases and observed seismicity in 2001, together with the historical pattern of subduction earthquakes followed by volcanic chain earthquakes, suggests that static stress transfer may be an important mechanism for this region. The events bigger than M 7 generated in the subducted Cocos plate are responsible for reactivating strike-slip faults along the volcanic chain on the Caribbean plate.

DISCUSSION

The study of the historical seismicity in El Salvador shows that large subduction earthquakes were often followed by shallow earthquakes along the volcanic chain in a time interval of 4 or 5 yr. The question we pose now is whether the 13 February 2001 earthquake was in some way triggered by the large subduction earthquake a month earlier.

One possible explanation is that the second event would have occurred anyway, without being triggered. A destructive volcanic chain earthquake has occurred in El Salvador approximately every 20 yr throughout the twentieth century, the last one in 1986. The 13 February event could simply have been the latest

volcanic chain event in that series and thus could have occurred in the absence of the 13 January event.

However, the results of our study suggest that the 13 January earthquake triggered one or several local faults, and at the same time they were activated reciprocally and new events were induced in the area of the subduction event. The fault where the 13 February earthquake occurred probably had sufficient energy accumulated, and the stress storage derived from the adjustment of the tractions after 13 January acted as a trigger, in other words, "the straw that breaks the camel's back."

Anyway, many events in El Salvador have occurred in compound subduction–volcanic axis sequences throughout history. Of special interest should be the study of the time delay from the subduction to the continental events and also the study of the time interval between major subduction events. Figure 12 shows the time correlation between the main volcanic chain and subduction events from 1900 until 2001. Subduction events occur less frequently, that is, they have longer recurrence intervals than the volcanic chain events, but they also have larger magnitudes. A delay of three to four years for the continental events following the ones of subduction is also appreciated, with the exception of the two events of 2001. The analysis of these delays combined with the long-term stress loading derived from plate convergence can provide new insights into the mechanical coherence of a systematic triggering behavior.

CONCLUSIONS

A study of the spatial and temporal distribution of the earthquakes that occurred in El Salvador in 2001 has been carried out, with different purposes: first, to identify the aftershocks linked to each main shock; second, to model the corresponding rupture surfaces; and third, to know the evolution of the activity and the stress transfer associated with each rupture process. The results indicate that the M_W 7.7 event of 13 January in the subduction zone triggered later events associated with a system of crustal faults along the volcanic chain farther inland. The second destructive earthquake of M_W 6.6 on February 13 was located on one of these faults, near San Pedro de Nonualco. The superposition in such a short interval of time of both main shocks, together with the respective aftershock series, produced an intense period of activity that did not decay according to known laws, such as Omori's.

Our analysis of the ruptures and aftershock distribution leads us to the conclusion that the observed activity can be explained by interaction between the respective earthquakes' sources (subduction and local faults), whose aftershocks could have induced each other. Some events with a magnitude ~5 could be acting as triggers of other events with the same or different origin. Such events are, at the same time, the cause and the effect of the intense activity recorded.

On the other hand, the stress transfer after the two main shocks leads us to conclude that the 13 February event occurred in a zone where the Coulomb stress had increased by more than 0.8 bar following the January 13 event. A similar pattern may

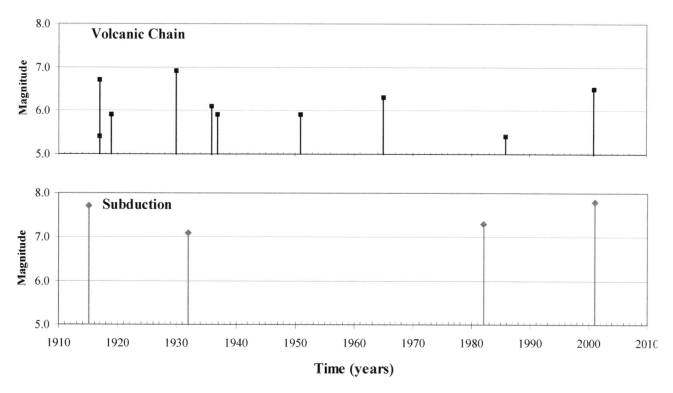

Figure 12. Temporal evolution of the main shocks in El Salvador from 1900 until 2001, showing magnitude versus time for volcanic chain events (top) and subduction events (bottom). The frequency of the last ones is lower and the magnitude higher than that for the local events. The delay of three to four years for the continental events following the ones of subduction is observed, with the exception of January 13 and February 13, 2001.

be inferred related to further events that occurred in the volcanic chain faults on February 17, due to the stress changes induced by the two previous shocks. The stress change also seems to have influenced the aftershock rate associated with the process.

Finally it is worth emphasizing the importance of the behavior of certain events as triggers of other events with a different origin in the seismic hazard of the region, and in other zones with a similar tectonic regime. A challenge for future study will be to model the conditions under which a subduction event may interact with events of the volcanic chain, and to repeat processes such as the one studied in this paper. If the triggering mechanism can be modeled systematically, it may lead to improved estimates of earthquake recurrence and seismic hazard in El Salvador.

ACKNOWLEDGMENTS

The research has been developed within the framework of two projects financed by the Spanish Agency of International Cooperation (AECI) and the Spanish Ministry of Science and Technology (Ren2001-0266-C02-02). The stress modeling is part of the project UCM PR78/02-11013. We are grateful to R. Robinson at the Institute of Geological and Nuclear Sciences in New Zealand for his comments and permission to use the software GNStress 5.1 to perform the stress change modeling. We would also like to thank the technicians at the seismologi-cal center of CIG (Center for Geotechnical Investigations), who carried out the epicenter relocations, Griselda Marroquin for her useful comments as well as Miryam Bravo, Moisés Contreras, and Erick Burgos, who have helped to obtain some of the results, with the use of GIS, and the rupture planes. Finally, the authors should like to express our appreciation for Dr. Hugh Cowan's support and contribution. In his capacity as a referee, he has put forward important suggestions and corrections that have greatly contributed to the clarity of this paper. We also thank another anonymous referee for his corrections, as well as Dr. Julian Bommer for his outstanding help in the final revision of the paper.

REFERENCES CITED

Ambraseys, N.N., and Adams, R.D., 2001, The seismicity of Central America: A descriptive Catalogue1898–1995: London, Imperial College Press.

Bommer, J., Benito, B., Ciudad-Real, M., Lemoine, A., López-Menjivar, M., Madariaga, R., Mankelow, J., Mendez-Hasbun, P., Murphy, W., Nieto-Lovo, M., Rodríguez, C., and Rosa, H., 2002, The Salvador earthquakes of January and February 2001: Context, characteristics, and implications for seismic risk: Soil Dynamics and Earthquake Engineering, v. 22, no. 5, p. 389–418.

Buforn, E., Lemoine, A., Udias, A., and Madariaga, R., 2001, Mecanismo focal de los terremotos de El Salvador: 2° Congreso Iberoamericano de Ingeniería Sísmica, Octubre 2001, Madrid, p. 65–72.

Chinnery, M.A., 1963, The stress changes that accompany strike-slip faulting: Bulletin of the Seismological Society of America, v. 53, p. 921–932.

DeMets, C., 2001, A new estimate for present-day Cocos-Caribbean plate motion: Implications for slip along the Central American volcanic arc: Geophysical Research Letters, v. 28, no. 21, p. 4043–4046.

Deng, J., and Sykes, L.R., 1997, Stress evolution in southern California and triggering of moderate-, small-, and micro-sized earthquakes: Journal Geophysical Research, v. 102, no. 24, p. 411–435.

Dewey, J.W., and Suarez, G., 1991, Seismotectonics of Middle America, *in* Slemmons, D.B., et al., eds., Neotectonics of North America: Boulder Colorado, Geological Society of America, Decade Map, v. 1, p. 309–321.

Guzman-Speciale, M., 2001, Active seismic deformation in the grabens of northern Central America and its relationship to the relative motion of the North America–Caribbean plate boundary: Tectonophysics, v. 337, p. 39–51.

Harlow, D.H., and White, R.A., 1985, Shallow earthquakes along the volcanic chain in Central America: Evidence for oblique subduction: Earthquake notes, v. 55, p. 28.

Harris, R.A., 1998, Introduction to special section: Stress triggers, stress shadows, and implications for seismic hazard: Journal of Geophysical Research, v. 103, no. 24, p. 347–358.

Harris, R.A., Simpson, R.W., and Reasenberg, P.A., 1995, Influence of static stress changes on earthquake locations in southern California: Nature, v. 375, p. 221–224.

Jaumé, S.C., and Sykes, L.R., 1992, Change in the state of stress on the southern San Andreas fault resulting from the California earthquake of April to June 1992: Science, v. 258, p. 1325–1328.

King, G.C.P., Stein, R.S., and Lin, J., 1994, Static stress changes and the triggering of earthquakes: Bulletin of the Seismological Society of America, v. 84, p. 935–953.

Lomnitz, C., and Rodríguez, S., 2001, El Salvador 2001: Earthquake disaster and disaster preparedness in a tropical volcanic environment: Seismological Research Letters, v. 72, no. 3, p. 346–351.

Okada, Y., 1992, Internal deformation due to shear and tensile faults in a half space: Bulletin of the Seismological Society of America, v. 82, p. 1018–1040.

Omori, F., 1884, On the aftershocks of earthquakes: Journal College of Sciences, Imperial University, v. 7, p. 111–200.

Parsons, T., Toda, S., Stein, R., Barka, A., and Dieterich, J., 2000, Heightened odds of large earthquakes near Istanbul: An interaction based probability calculation: Science, v. 288, p. 661–665.

Polat, O., Eyidogan, H., Haessler, H., Cisternas, A., and Phillip, H., 2002, Analysis and interpretation of the aftershock sequence of the August 17, 1999, Izmit (Turkey) earthquake: Journal of Seismology, v. 6, p. 287–306.

Reasenberg, P., and Simpson, R., 1992, Response of regional seismicity to the static stress change produced by Loma Prieta Earthquake: Science, v. 255, p. 1687–1690.

Rojas, W., Bungum, H., and Lindholm, C.D., 1993, A catalogue of historical and recents earthquakes in Central America: Report NORSAR, 77 p.

Scholz, C., 1990, The mechanics of earthquakes and faulting: Cambridge, UK, Cambridge University Press, 439 p.

Earthquake Analysis Software, 2000, SEISAN for the IBM PC and SUN, Version 7.1, May 2000: Institute of Solid Earth Physics, University of Bergen, Norway.

Stein, R.S., 1999, The role of stress transfer in earthquake recurrence: Nature, v. 402, p. 605–609.

Stein, R.S., and Lisowski, M., 1983, The Homestead Valley earthquake sequence, California: Control of aftershocks and postseismic deformations: Journal Geophysical Research, v. 88, p. 6477–6490.

Toda, S., Stein, R.S., Reasenberg, P.A., and Dieterich, J.H., 1998, Stress transferred by the $M_W = 6.5$ Kobe, Japan, shock: Effect on aftershocks and future earthquake probabilities: Journal of Geophysical Research, v. 103, no. 24, p. 543–565.

Weinberg, R.F., 1992, Neotectonic development of western Nicaragua: Tectonics, v. 11, p. 1010–1017.

Wells, D.L., and Coppersmith, K.J., 1994, New empirical relations among magnitude, rupture length, rupture width, rupture area and surface displacement: Bulletin of the Seismological Society of America, v. 85, no. 1, p. 1–16.

White, R., 1991, Tectonic implications of upper-crustal seismicity in Central America, *in* Slemmons, D.B., et al., eds., Neotectonics of North America: Boulder, Colorado, Geological Society of America, Decade Map, v. 1, p. 323–338.

White, R.A., and Harlow, D.H., 1993, Destructive upper-crustal earthquakes of Central America since 1900: Bulletin of the Seismological Society of America, v. 83, p. 1115–1142.

MANUSCRIPT ACCEPTED BY THE SOCIETY JUNE 16, 2003

Geological Society of America
Special Paper 375
2004

Two earthquake databases for Central America

Conrad Daniel Lindholm*
NORSAR, P.O. Box 53, 2027 Kjeller, Norway

Carlos A. Redondo
Escuela Centroamericana de Geologia, Universidad de Costa Rica,
Centro Sismologico de America Central (CASC) San Jose, Costa Rica Aptdo: 214-2060

Hilmar Bungum
NORSAR, P.O. Box 53, 2027 Kjeller, Norway

ABSTRACT

The 1990s was a period of improved cooperation between the Central American countries, including the field of seismology. As part of the regional cooperation and data integration, two seismological databases for the region were established: (1) a database of all available earthquake focal mechanisms for the region, and (2) a database in which all data from national seismological observatories were integrated, leading to a published regional earthquake bulletin. These databases are potentially valuable for a wide range of analyses, albeit with some limitations. The objective of this paper is to attract attention to the existence of these databases and to encourage the wider use of them.

Keywords: Regional earthquake data merging, earthquake database, focal mechanisms, direction of horizontal stress.

INTRODUCTION

Worldwide seismological databases are generated and maintained by several agencies like the International Seismological Center (ISC) and the U.S. Geological Survey's National Earthquake Information Center (USGS/NEIC). On the national level, many countries (including those in Central America) have established, and continue to maintain, earthquake databases, but due to logistic reasons it may sometimes be difficult to obtain copies of these national databases. Regional databases, which may be of significance for regions with several smaller countries, are more rare on the global scene. Central America is a tectonically very active region where most of the countries are large enough to maintain national seismic networks, but when earthquakes occur near the national boundaries, the integration of phase and waveform data from neighboring countries is quite essential for improving location estimates and focal mechanism solutions.

In 1991, a regional seismology project funded by the Norwegian Agency for International Cooperation (NORAD) was implemented through the Center for the Prevention of Natural Disaster in Central America (CEPREDENAC), involving Guatemala, El Salvador, Honduras, Nicaragua, Costa Rica, and Panama. The project, involving Central American and Norwegian seismological observatories (NORSAR and the University of Bergen), was initiated with two distinct goals: (1) to improve national seismological networks and data exchange between the six countries, and (2) to improve the basis for carrying out seismic zonation and earthquake hazard assessment. As a result of this, new data are now continuously acquired by the national agencies and merged into a regional database that also publishes regional bulletins.

The present paper reports on two seismological databases with important potentials for the scientific community: (1) a database of earthquake focal mechanisms, and (2) a database of all regional phases and hypocenter solutions based on all available data from Central American seismological observatories.

---
*conrad@norsar.no

Lindholm, C.D., Redondo, C.A., and Bungum, H., 2004, Two earthquake databases for Central America, *in* Rose, W.I., Bommer, J.J., López, D.L., Carr, M.J., and Major, J.J., eds., Natural hazards in El Salvador: Boulder, Colorado, Geological Society of America Special Paper 375, p. 357–362. For permission to copy, contact editing@geosociety.org. © 2004 Geological Society of America

THE EARTHQUAKE FOCAL MECHANISM DATABASE

The work on establishing the focal mechanism database was started by Redondo et al. (1993) under the above-mentioned cooperation project between Central America and Norway. The goal was to establish a database that on an empirical and quantitative basis would contribute to enhanced understanding of local and regional tectonics.

The focal mechanism database presently (2003) consists of 1,942 reports; however, in particular for larger earthquakes, there are multiple reports (875 of the reports are focal mechanisms computed for the same earthquake from alternative sources). A significant part of the database reports are USGS and Harvard Centroid-Moment Tensor (CMT) solutions, and another large part is collected from the open literature. Currently, only 202 reports come from the national agencies in Central America (of which few have been published), although it is the hope and expectation that this proportion will change as focal mechanisms based on local data become available in larger amounts.

The focal mechanisms have varying degrees of quality, reflecting the time and size of the earthquake and the number of stations that recorded it. A four-step quality ranking from A (best) to D (poor) has been used whenever information for estimating the

quality was available. However, many of the older reports in particular were published without any information of this kind.

The early part of the earthquake focal mechanism database was established by collecting solutions from published sources, beginning with an M_S 7.5 earthquake on July 18, 1934, which caused widespread damage and a tsunami in the Gulf of Chiriqui, Panama. While the older reports (prior to 1990) are taken from a wide range of publications (e.g., Molnar and Sykes, 1969; Dean and Drake, 1978; White and Harlow, 1993), the later reports are largely those published by the USGS and national seismological agencies in Central America (INSIVUMEH in Guatemala, Centro de Investigaciones Geotécnicas in El Salvador, INETER in Nicaragua, Universidad de Costa Rica in Costa Rica, and Universidad de Panamá in Panama). The focal mechanism database was established for the region 4°–19° N and 77°–93° W and is maintained so that new mechanisms for the region are included as they become available.

There are currently 1,067 earthquakes in the Central American region for which focal mechanisms have been established, and it represents a rich source for the understanding of regional and local deformation processes. Figure 1 shows the complete database together with a rose diagram of the azimuth of the principal stress direction and a triangle plot demonstrating how

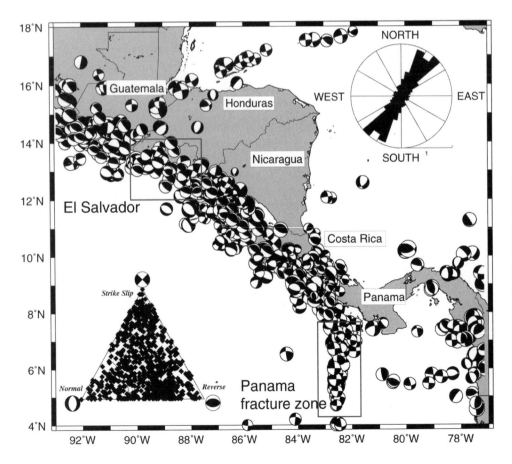

Figure 1. The focal mechanism database for Central America. All solutions are plotted in lower hemisphere plots with black areas representing areas of compressions. Upper inset: Rose diagram of azimuth of main compressional axis. Lower left inset: Distribution of mode of faulting in the database.

deformation is relatively distributed between the three principal faulting modes, strike-slip, normal, and reverse faulting. The overall direction of horizontal compression is N30°E, in good agreement with regional tectonics.

DATA ANALYSIS EXAMPLES

Stress Direction and Types of Faulting

The earthquake focal mechanism database leaves more to be desired with respect to quality assessment of the different reports; however, in particular for the older part of the database, information that can be used to assess the quality is missing. Nevertheless, the database represents a source of seismotectonic information that lends itself to various types of analyses, of which Figures 2 and 3 are brief examples. Figure 2 shows, for two regions (Panama fracture zone and El Salvador), an analysis for stress direction and deformation type. From Figure 2 it is seen

that the Panama fracture zone is dominated by shallow strike-slip earthquakes, and both the inverted stress directions and the type of faulting are so uniform that they can be used as a textbook example. The El Salvador area is more complex, with the coastal areas dominated by earthquakes associated with the subduction process (including the January 13, 2001, M_w 7.6 [NEIC] earthquake), and the volcanic chain exhibiting a variety of faulting types. As seen in Figure 2, all faulting types are present; however, the stress tensor inversion based on the same data is remarkably well constrained, indicating slab pull to be an important driving force (a more detailed analysis would probably indicate a systematic deformation difference between the volcanic chain events and the subduction zone earthquakes).

More generally, the focal mechanism data show some characteristic features of the coseismic deformation of Central America. When deep and shallow earthquakes for the whole region are separated, the difference in faulting mechanisms is distinct. The deeper earthquakes show a stronger dominance of reverse fault-

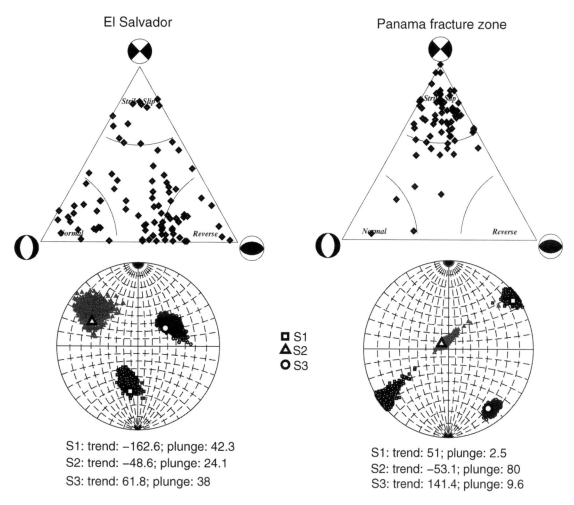

Figure 2. Triangle plots following the Frohlich and Apperson (1992) representation (upper plots) and stress inversions following the Michael (1984) method as implemented by Wiemer (2001) for two regions indicated in Figure 1. Left: A region around El Salvador. Right: A region encompassing the Panama fracture zone. Sx denotes the trend and plunge of the main stress vectors in the deviatoric stress tensor.

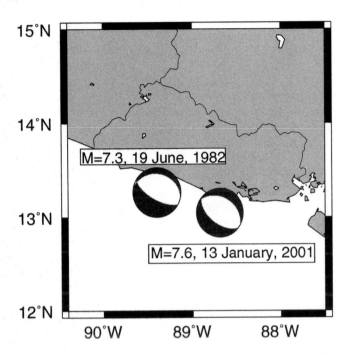

Figure 3. Focal mechanisms for two large, subduction zone earthquakes in El Salvador. One nodal plane of the 1982 earthquake focal mechanism had strike, dip, and slip of 102, 25, and −106 degrees, respectively, compared to 121, 37, and −95 degrees for the 2001 earthquake.

ing with hardly any strike-slip faulting, whereas for the shallow-focus earthquake, mechanisms are predominantly strike-slip.

When looking at the stress directions from the different areas, there is a clockwise rotation in the maximum stress direction (σ_1) when moving from northwest to southeast. This applies to both the shallow crustal and the subduction zone earthquakes, and reflects the different angles that the subducting Cocos plate has with the continent areas.

Comparative Studies: Similarities between the 19 June 1982 and 13 January 2001 El Salvador Earthquakes

There were several indicators that the in-slab earthquake in 2001 caused very high ground motions and corresponding damage (see other contributions in this volume). This may be related to the more general observation that in-slab earthquakes, rupturing the oceanic crust, often generate very high ground motions (e.g., Iglesias et al., 2002; Adams and Halchuk, 2000). The 19 June 1982 M_w 7.3 earthquake ruptured a segment to the northwest of the 2001 earthquake. Its focal depth was estimated to be ~80 km (ISC), which is somewhat deeper than the 2001 earthquake (60 km), and Figure 3 demonstrates the similarity of the two normal-faulting focal mechanisms. The similarities in depth, size, and focal mechanism between the 1982 and 2001 earthquakes are substantiated through similarities also in the inflicted damage (NORSAR, 2001; Ambraseys and Adams,

2001). Notwithstanding that there are several factors other than the earthquake source that may be responsible for the damage patterns (distance, local amplification, etc.), the similarities between the two earthquakes are striking, and may indicate that El Salvador has experienced two in-slab earthquakes over the last 20 yr. If this is correct, it has implications for the seismic hazard estimates.

THE CASC EARTHQUAKE DATABASE

As part of the Central American seismological data integration, a regional bulletin is regularly published based on phase readings from all stations in Central America. These data are collected at the Central American Seismological Center (CASC) and used in creating new hypocenter solutions and magnitudes (see also Alvarenga et al., 1998). CASC is physically hosted for a limited time at the different national observatories, and the responsibility for the publication has been shared by the national institutions. The database currently spans the time period from 1992 through 2000, with more than 11,000 located earthquakes, as shown in Figure 4. Figure 5 shows a smaller part of the database, covering the El Salvador region.

Much effort is put into the location procedures for the CASC database, in particular by applying calibrated velocity models (Alvarenga et al., 1998; Matumoto et al., 1977) in the location procedure. Even so, it has proved difficult to establish reliable locations for small earthquakes, in particular with respect to hypocentral depth. This is a well-known problem that is directly related to station density and distribution, and to the complicated crustal and subcrustal velocity structure. It has been claimed that location uncertainty in the volcanic chain is ~5 km, with depth uncertainties in the same range. For subduction earthquakes, a location uncertainty of 10 km has been claimed. An in-depth study of the location uncertainties in the CASC databases is pending, but the opinion of the present authors is that the above values are overly optimistic. The location accuracy problems are particularly evident in regions where there are reasons to expect that the seismicity follows the subducting plate to large depths (such as under El Salvador), in which case the one-dimensional velocity model used in the location inversion is clearly very unsuitable. The seismic activity following the subducting slab is only poorly mapped in the CASC earthquake catalog (as is the case also for many national catalogs).

SUMMARY AND REMARKS ON FUTURE CHALLENGES

The CASC databases of earthquake locations and focal mechanisms described above are complementary to the databases available through ISC, USGS/NEIC, or other international agencies. In a time when international data integration is widely recognized as a necessity, the establishment of these regional databases should be evaluated also under the perspective of cross-border cooperation.

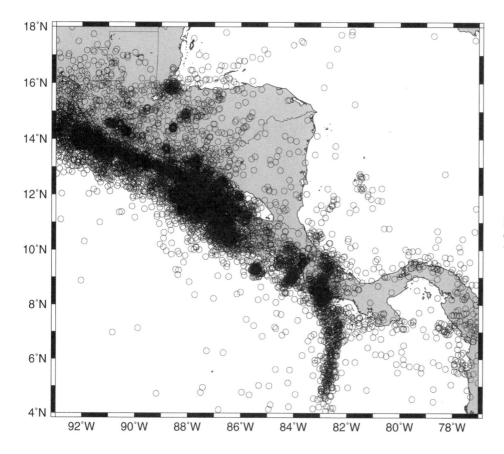

Figure 4. The complete CASC earthquake database from 1992 through November 2000 for Central America.

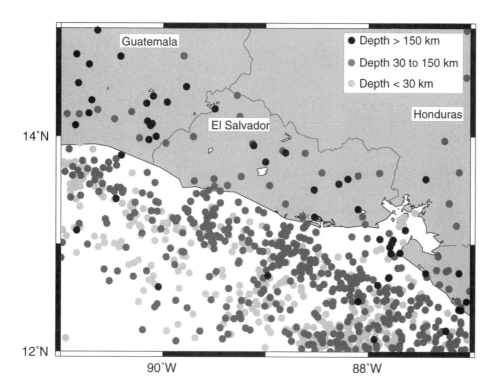

Figure 5. The CASC earthquake database covering the El Salvador area. Only earthquakes with $M_C > 4.0$ are included

Some features of the databases are summarized below:

1. The CASC database as described above is useful not only for the wider seismological community. The main advantage is possibly that scientists in the region, through redistribution of the data, receive immediate access to both phase readings and waveform data from all seismic stations throughout Central America.

2. All countries in the region, except Honduras, publish national seismic bulletins. Data from Honduras are only obtainable through the CASC database.

3. For the regions close to the national boundaries, the earthquake parameters are better constrained with data from both sides of the border.

4. The CASC database lends itself to studies of seismic hazard, attenuation (particularly through the redistribution of all waveforms), magnitude calibrations, crustal structure investigations, etc.

5. To the best knowledge of the present authors, the focal mechanism database is the only comprehensive database of its kind that has been compiled for this region. Many of the focal mechanisms have not been published and are only accessible through this database.

6. The focal mechanism solutions that are obtained for smaller earthquakes are of particular value in seismotectonic investigations.

However, the merging of data from different agencies has also revealed some hitherto unresolved challenges:

1. Phase identification practices may still deviate between the national observatories. This situation may gradually improve if the New Manual of Seismological Observatory Practice (Bormann, 2002) is followed by the seismological observatories.

2. The one-dimensional velocity model used in national and joint solutions is generally ill-conditioned for locating subduction zone earthquakes with precision. This is particularly evident for deep subduction events when these are recorded only on a few stations (as often happens in El Salvador). This situation entails that many of the deeper subduction zone earthquakes are particularly poorly resolved.

3. A thorough evaluation of location uncertainties in the CASC database for different regions is still pending. Figure 5 appears to indicate that some very deep events occur in areas dominated by shallow activity. A study focusing also on depth uncertainty would be particularly beneficial for the future seismic hazard studies in the region.

The regional seismological databases as described above are potentially useful for the general seismological community and in particular for the practicing seismologist in Central America. We have attempted to demonstrate briefly some of their potentials, and we encourage the further use of these catalogs, which are freely available on request. Moreover, the extensive use of such databases will stimulate the further expansion of the databases both in terms of quality and quantity. The existence and future elaboration of databases of the types described are par-

ticularly useful also for the participating countries and research within these countries.

The CASC homepage is http://www.cepredenac.org/08_cnc/casc/, and data can be obtained from Carlos Redondo Chavarria (credondo@geologia.ucr.ac.cr).

ACKNOWLEDGMENTS

This work has been supported by the Norwegian Agency for International Cooperation (NORAD) through the project "Reduction of Natural Disasters in Central America, Earthquake Preparedness and Hazard Mitigation, Seismic Zonation, and Earthquake Hazard Assessment." Dr. R.D. Adams and an anonymous reviewer are thanked for thoughtful comments and suggestions.

REFERENCES CITED

Adams, J., and Halchuk, S., 2000, Knowledge of in-slab earthquakes needed to improve seismic hazard estimates for southwestern British Columbia: http://www.seismo.nrcan.gc.ca/hazards/inslab2000/index.php.

Alvarenga, E., Barquero, R., Boschini, I., Escobar, J., Fernández, M., Mayol, P., Havskov, J., Galvez, N., Hernandez, Z., Ottemoeller, L., Pacheco, J., Redondo, C., Rojas, W., Vega, F., Talavera, E., Taylor, W., Tapia, A., Tenorio, C., and Toral, J., 1998, Central American Seismic Center (CASC): Seismological Research Letters, v. 69, p. 394–399.

Ambraseys, N., and Adams, R.D., 2001, The seismicity of Central America. A descriptive catalogue 1898–1995: London, Imperial College Press, ISBN 1-86094-244-X, 309 p.

Bormann, P., editor, 2002, IASPEI new manual of seismological observatory practice: Potsdam, Germany, GeoForschungsZentrum, 2 Volumes, ISBN 3-9808780-0-7, 1252 p. plus 1 CD-ROM.

Dean, H.W., and Drake, C.L., 1978, Focal mechanisms and tectonics of the Middle America arc: Journal of Geology, v. 86, p. 329–338.

Frohlich, C., and Apperson, K.D., 1992, Earthquake focal mechanisms, moment tensors, and the consistency of seismic activity near plate boundaries: Tectonics, v. 11, no. 2, p. 279–296.

Iglesias, A., Singh, S.K., Pacheco, J., and Ordaz, M., 2002, A source and wave propagation study of the Copalillo, Mexico, earthquake of 21 July 2000 (M 5.9): Implications for seismic hazard in Mexico City from inslab earthquakes: Bulletin of the Seismological Society of America, v. 92, p. 1060–1071.

Matumoto, T., Othake, M., Lathan, G., and Umana, J., 1977, Crustal structure of southern Central America: Bulletin of the Seismological Society of America, v. 67, p. 121–134.

Michael, A.J., 1984, Determination of stress from slip data: Faults and folds: Journal of Geophysical Research, v. 89, p. 11,517–11,526.

Molnar, P., and Sykes, L.R., 1969, Tectonics of the Caribbean and Middle America regions from focal mechanisms and seismicity: Bulletin of the Seismological Society of America, v. 80, p. 1639–1684.

NORSAR, 2001, Technical mission to El Salvador, following the January 13 earthquake: Report for CEPREDENAC (Center for the Prevention of Natural Disaster in Central America), 33 p.

Redondo, C., Lindholm, C., and Bungum, H., 1993, Earthquake focal mechanisms in Central America: NORSAR–CEPREDENAC report, 22 p.

White, R.A., and Harlow, H.D., 1993, Destructive upper crustal earthquakes of Central America since 1900: Bulletin of the Seismological Society of America, v. 83, p. 1115–1142.

Wiemer, S., 2001, A software package to analyze seismicity: ZMAP. Electronic seismologist: Seismological Research Letters, v. 72, p. 373–382.

MANUSCRIPT ACCEPTED BY THE SOCIETY JUNE 16. 2003

Geological Society of America
Special Paper 375
2004

Seismicity and tectonics of El Salvador

James W. Dewey*
U.S. Geological Survey, MS 966, Box 25046, Denver, Colorado 80225, USA

Randall A. White
U.S. Geological Survey, 345 Middlefield Road, Menlo Park, California 94025, USA

Douglas A. Hernández
Servicio Nacional de Estudios Territoriales (SNET), San Salvador, El Salvador

ABSTRACT

The large-scale plate-tectonics framework of El Salvador was defined in the "plate-tectonics revolution" of the 1960s and 1970s, but important issues related to seismic hazards depend on details that have been only recently, or are not yet, understood. Present evidence suggests that coupling across the interface-thrust zone beneath coastal El Salvador is sufficient to produce occasional interface-thrust earthquakes as large as M ~8. The rate of such earthquakes is determined by the percentage of relative plate motion that is accumulated as elastic strain on the thrust-fault interface between the Cocos and Caribbean plates, which appears to be lower than in many other subduction zones, but is not well established. Earthquakes in the interior of the Cocos plate, such as the El Salvador earthquake of January 13, 2001, account for a significant percentage of Wadati-Benioff zone earthquakes. Separate consideration of the seismic hazard posed by, respectively, Cocos intraplate earthquakes and interface-thrust earthquakes is complicated by the difficulty of separating interface-thrust and Cocos intraplate events in earthquake catalogs. Earthquakes such as the San Vicente–San Salvador sequence of February 13–25, 2001, probably result from the motion of the Central American forearc northwestward with respect to the interior of the Caribbean plate; the geometry of the fault systems that accommodate the motion remains to be worked out. Understanding of this tectonic complexity and associated seismic hazards will be facilitated greatly by the long-term operation of high-sensitivity local seismograph networks, such as that operated by, and currently being upgraded by, the Servicio Nacional de Estudios Territoriales (SNET) of El Salvador.

Keywords: El Salvador, Central America, earthquake, earthquakes, seismicity, seismotectonics, seismic hazards.

INTRODUCTION

The high level of earthquake activity in El Salvador is a consequence of its position at the boundary of two major tectonic plates, the Cocos plate and the Caribbean plate (Fig. 1; Molnar and Sykes, 1969; Dewey and Suárez, 1991). El Salvador

---

*dewey@usgs.gov

lies on the Caribbean plate. The subduction of the Cocos plate beneath the Caribbean plate along the Pacific coast of Central America produces several classes of damaging earthquakes that are distinguished by their positions within the tectonic plates or by their focal mechanisms (Figs. 1–5). Beneath the Pacific coast and offshore, thrust-fault earthquakes occur on the interface between the Cocos and Caribbean plates, and significant activity also occurs below the interface within the interior of the Cocos

Dewey, J.W., White, R.A., and Hernández, D.A., 2004, Seismicity and tectonics of El Salvador, *in* Rose, W.I., Bommer, J.J., López, D.L., Carr, M.J., and Major, J.J., eds., Natural hazards in El Salvador: Boulder, Colorado, Geological Society of America Special Paper 375, p. 363–378. For permission to copy, contact editing@geosociety.org. © 2004 Geological Society of America

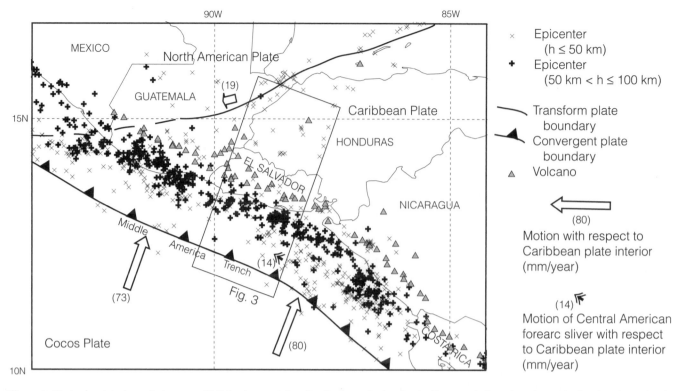

Figure 1. Tectonic plate boundaries near El Salvador, and the distribution of teleseismically recorded earthquakes calculated to have depths of 100 km and less in 1964–2000. Plate motions and motion of the Central American forearc sliver are from DeMets (2001). See Figure 2 for the location of the Central American forearc sliver. Volcanoes are from Siebert and Simkin (2002). Earthquake epicenters and focal depths are computed by the method of Engdahl et al. (1998) and were provided to the authors by E.R. Engdahl (2002, personal commun.; see text). Ninety percent of the earthquakes plotted had magnitudes of 4.2 or larger.

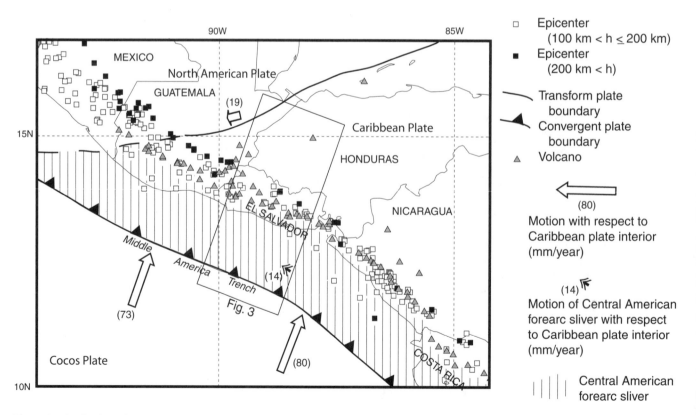

Figure 2. Distribution of earthquakes calculated to have depths greater than 100 km, for the period 1964–2000. Sources of earthquake data and information on plate boundaries, plate motions, and volcanoes are as in Figure 1.

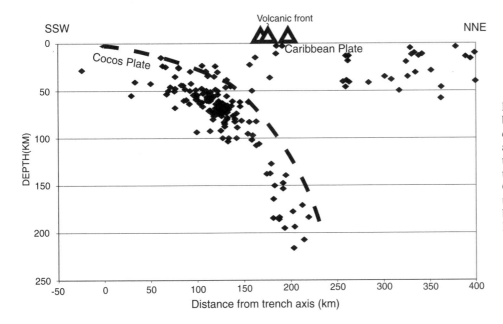

Figure 3. Cross section of seismicity beneath El Salvador, 1964–2000. Hypocenters plotted are those from Figures 1 and 2 lying within the "Fig. 3" box of those figures. Dashed curve represents the approximate location of the Cocos-Caribbean plate interface (see also Figure 5). The zone of hypocenters near and beneath the plate interface is the Wadati-Benioff zone.

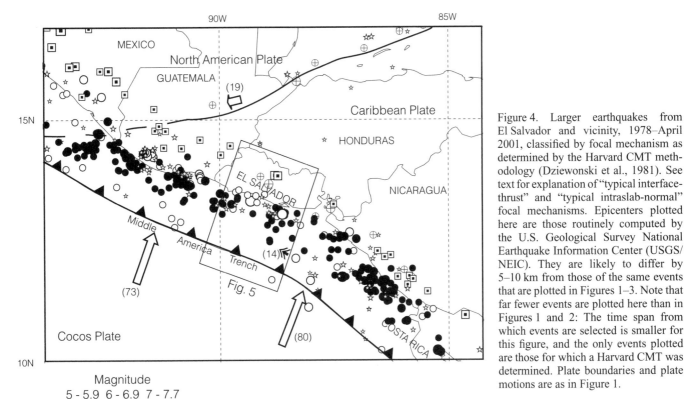

Figure 4. Larger earthquakes from El Salvador and vicinity, 1978–April 2001, classified by focal mechanism as determined by the Harvard CMT methodology (Dziewonski et al., 1981). See text for explanation of "typical interface-thrust" and "typical intraslab-normal" focal mechanisms. Epicenters plotted here are those routinely computed by the U.S. Geological Survey National Earthquake Information Center (USGS/NEIC). They are likely to differ by 5–10 km from those of the same events that are plotted in Figures 1–3. Note that far fewer events are plotted here than in Figures 1 and 2: The time span from which events are selected is smaller for this figure, and the only events plotted are those for which a Harvard CMT was determined. Plate boundaries and plate motions are as in Figure 1.

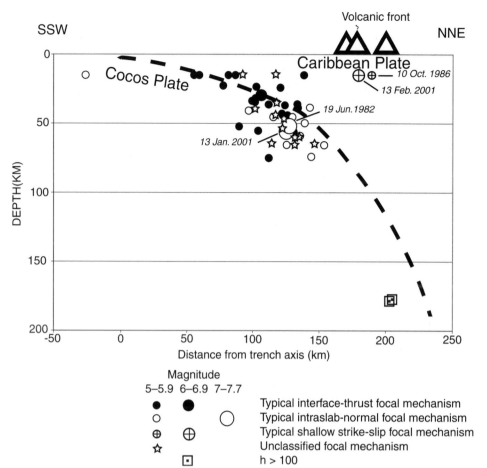

Figure 5. Cross section of earthquake hypocenters, 1978–April 2001, classified by focal mechanism as determined by the Harvard CMT methodology (Dziewonski et al., 1981). Events are those that lie within the box labeled "Fig. 5" on Figure 4. Hypocenters of the most destructive El Salvadoran earthquakes since 1978 are labeled by their dates. Shallow-focus earthquakes for which the hypocentroid cannot be accurately determined by Harvard CMT methodology are assigned default depths of 15 km.

plate. Inland from the coastline, intraplate earthquakes within the subducted Cocos plate occur to depths of more than 200 km beneath El Salvador (Fig. 3) and to more than 250 km elsewhere in Central America. A societally very significant class of earthquakes occurs at shallow depths in the overriding Caribbean plate, in the vicinity of the volcanic chain. The largest known earthquakes from El Salvador have had magnitudes of ~7.7. Shallow-focus earthquakes with magnitudes (m_b) less than 4.5 have been damaging when they have occurred directly beneath population centers.

SOURCES OF DATA ON EL SALVADOR EARTHQUAKES

Seismologists' current understanding of the seismicity of El Salvador is based on a record of earthquake locations, sizes, and focal mechanisms that becomes generally more complete and accurate as one approaches the present. Knowledge of earthquakes occurring before the twentieth century must be based on the macroseismic effects produced by the earthquakes (e.g., White et al., this volume, Chapter 28). Seismographic data contribute usefully to knowledge of Central American seismic-

ity from early in the twentieth century, becoming steadily more informative with the continued spread of seismographs (Ambraseys and Adams, 2001). In the early 1960s, a major improvement in the completeness of Central American earthquake catalogs resulted from the introduction of computer determinations of earthquake hypocenters and magnitudes as well as from the installation of the World-Wide Standardized Seismograph Network (WWSSN). Hypocenters and magnitudes have been routinely determined by computer since 1964 by the U.S. Geological Survey National Earthquake Information Center (USGS/NEIC) and the International Seismological Center (ISC), using data from seismographs distributed globally. Data from Salvadoran seismographic stations at San Salvador (international abbreviation SSS), La Palma (LPS), Nueva Concepcion (NCS), and Santiago de María (SDM) were able to contribute importantly to the locating of earthquakes in El Salvador and elsewhere in Central America by the USGS/NEIC and the ISC. Catalogs of earthquakes computed by the USGS/NEIC and the ISC are available online, at http://neic.usgs.gov/ and http://www.isc.ac.uk/, respectively. The installation of the WWSSN also made possible the determination of reliable focal mechanisms for larger Central American earthquakes by analysis of the geographic distribution of P-wave first motions (Molnar and Sykes, 1969; Dean and Drake, 1978). A dramatic increase in the number of digitally recording long-period and broadband seismographs in the 1970s led to the development of more sophisticated methods of describing the earthquake focal mechanism and accurately locating the earthquake source. Moment tensors of larger Central American earthquakes are independently determined and made available online by Harvard University (Dziewonski et al., 1981; http://www.seismology.harvard.edu/CMTsearch.html) and the USGS (Sipkin, 1982; http://neic.usgs.gov/neis/sopar/).

Teleseismic Hypocenters

In Figures 1–3, we have represented the seismicity of El Salvador and vicinity by use of hypocenters for earthquakes occurring in 1964–2000 that were calculated with the methodology of Engdahl et al. (1998) and provided to us by Engdahl (2002, personal commun.). The particular set of hypocenters plotted represents a subset of those determined by Engdahl and his colleagues, those for which the largest teleseismic "secondary open azimuth" (azimuthal spread between reporting seismographs that contains only one other reporting seismograph) is less than 130°. Approximately 25% of all teleseismically recorded earthquakes occurring since 1964 that are cataloged by the USGS/NEIC and ISC meet this criterion. Based on comparison of individual epicenters from Figures 1 and 3 with independently known epicenters of the same earthquakes, we think that the epicenters (but not the focal depths) of most shallow-focus inland earthquakes represented in Figures 1 and 3 are probably accurate to within 20 km. A significant fraction of location errors would be due to bias arising from the use of one-dimensional velocity models in an earth with three-dimensional velocity structure (e.g., Dewey and Algermissen, 1974). Epicenters within different parts of the Wadati-Benioff zone may also be biased by tens of kilometers, though we are not able to check for the presence of these biases without reliable independently determined epicenters. Uncertainties in focal depths shown in Figure 3 are likely to be several tens of kilometers. We do not, for example, attach tectonic significance to focal depths calculated to be greater than 25 km in the Caribbean plate at distances of 250–400 km inland from the Middle America trench (Fig. 3), although such focal depths would require a tectonic explanation if they were considered reliable. In spite of the uncertainties in the hypocenters of Figures 1–3, these hypocenters should still be on average more accurate than those routinely computed by the USGS/NEIC or the ISC (Engdahl et al., 1998). The imposition of a 130° largest teleseismic secondary open azimuth for Figures 1–3 dramatically reduces the scatter of hypocenters for earthquakes smaller than magnitude 5.5, compared to what would be obtained by plotting all USGS/NEIC hypocenters for the regions of Figures 1–3.

In Figures 4 and 5, we have represented the seismicity of El Salvador and vicinity by plotting hypocenters of earthquakes for which Harvard CMT (Centroid-Moment Tensor) focal mechanisms were calculated by the methodology of Dziewonski et al.

(1981). The epicenters in Figures 4 and 5 are those computed by the USGS/NEIC. The focal depths plotted in Figure 5, however, are the hypocentroid focal depths that are computed from long-period seismograph waveforms along with the Harvard CMT mechanisms. In the case of many earthquakes for which the routine Harvard CMT methodology would calculate a focal depth shallower than 15 km, the Harvard CMT convention is to fix the depth at 15 km, in order to avoid instability in the CMT computation (Dziewonski et al., 1987). For most events, the Harvard CMT focal depths should be more reliable than the focal depths computed by the USGS/NEIC from times of first-arriving P-waves. In the absence of errors, the hypocenters computed by the methodology of Engdahl et al. (1998) (Figs. 1–3) and the epicenters computed by the USGS/NEIC would represent the points at which the earthquakes nucleated. The Harvard CMT hypocentroid depths, by contrast, correspond to the centroid of moment release (Dziewonski et al., 1981).

Teleseismic Magnitudes

Teleseismically determined long-period magnitudes, such as M_S or M_W, are currently the most useful earthquake magnitudes for seismic hazard studies in Central America (Ambraseys and Adams, 2001). Instrumentally determined M_S are available for large Central American earthquakes since the beginning of the twentieth century (Ambraseys and Adams, 2001), and M_W have been systematically determined for larger earthquakes since the late 1970s. Since 1964, teleseismic short-period body-wave magnitudes (m_b) have been assigned to earthquakes worldwide by the USGS/NEIC and the ISC. The m_b are important because they can be determined for smaller earthquakes than can M_S and M_W; they are the only instrumental measure of earthquake size for many moderate-sized earthquakes. Teleseismic m_b cannot, however, be automatically assumed to be identical to other types of magnitude (Geller, 1976). Effective use of teleseismic m_b jointly with long-period magnitudes for seismic hazard studies of Central America may require development of special, source-region specific, scaling relationships between m_b and the long-period magnitudes. Teleseismic m_b values assigned to destructive shallow-focus earthquakes along the Central American volcanic front seem disproportionately small with respect to other types of earthquake magnitude, more so than would be expected from worldwide scaling relationships (e.g., Geller, 1976) between different types of magnitude. For example, the destructive El Salvador earthquake of October 10, 1986 [M_S(USGS) = 5.4; M_W(HRV) = 5.7], had m_b(USGS) = 5.0, and the earthquake of February 13, 2001 [M_S(USGS) = 6.5; M_W(HRV) = 6.6], had m_b(USGS) = 5.5.

Local Earthquake Monitoring

Currently, monitoring of small and moderate earthquakes within the overall territory of El Salvador is conducted by seismographs run by the Servicio Nacional de Estudios Territoriales (SNET) (Fig. 6). The network of Salvadoran stations is capable of

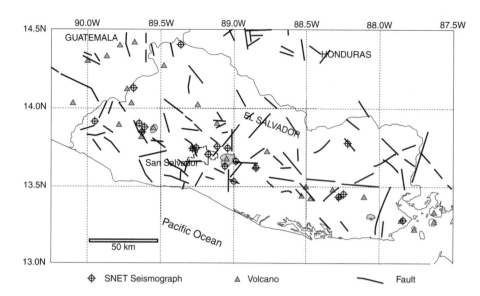

Figure 6. Seismographs operated in mid-2002 by the Servicio Nacional de Estudios Territoriales (SNET) for the monitoring of the seismicity of El Salvador, together with volcanoes (source as in Figure 1) and geologically mapped faults (Case and Holcombe, 1980).

locating earthquakes of magnitude 3.8 from throughout the area of El Salvador, and substantially smaller shocks in areas of dense instrumentation. At locations in El Salvador that are surrounded by nearby stations of the SNET network, epicenter accuracies of several kilometers can be claimed, with focal depth accuracies being somewhat poorer. Accuracies of hypocenters lying outside of the network would generally be much poorer. Expansion of the network is currently being undertaken to extend the region within which small earthquakes can be detected and earthquake hypocenters can be calculated to high accuracy. Accurate location of earthquakes on the borders of El Salvador will depend on observations from seismographs situated in neighboring countries, such as the Nicaraguan seismic network (Segura and Havskov, 1994). In addition to the network of sensitive stations used to monitor small earthquakes, several groups operate strong-motion networks in El Salvador (Shakal et al., 1987; Bommer et al., 1997). Although installed primarily to study the characteristics of the damaging ground shaking, data from the strong-motion seismographs can also provide important constraints on the locations of strong local earthquakes (e.g., White et al., 1987).

THE CENTRAL AMERICAN WADATI-BENIOFF ZONE

Earthquakes occurring in the Wadati-Benioff zone along the coast of Central America pose a significant hazard to El Salvador and other Central American countries. Earthquakes in this coastal Wadati-Benioff zone occur mostly at depths of ~100 km and less (Figs. 1–3), and most large shocks occur at depths shallower than 70 km (Fig. 5). In the twentieth century, earthquakes as large as M_S 7.9 have produced Modified Mercalli intensities as high as IX in Central America north of Costa Rica (Ambraseys and Adams, 2001). At least one coastal Wadati-Benioff zone earthquake produced a destructive tsunami.

Classification of Wadati-Benioff Zone Earthquakes by Focal Mechanism

The seismicity of the Central American coastal Wadati-Benioff zone comprises earthquakes occurring on the thrust interface between the Cocos and Caribbean plates and earthquakes occurring beneath the interface in the interior of the Cocos plate. In Figures 4 and 5 we have classified events occurring in 1978–April 2001 on the basis of their Harvard CMT focal mechanism (Dziewonski et al., 1981). Events plotted as "typical interface-thrust focal mechanism" earthquakes in Figures 4 and 5 had focal mechanisms consistent with their occurring as reverse motion on shallowly dipping planes, with slip being approximately parallel to the 30° azimuth at which slip occurs between the Cocos plate and the Caribbean forearc sliver (DeMets, 2001): These events had T-axes that plunge more than 45°, and they had one slip vector with an azimuth between 10° and 50° and a plunge less than 45°. Events plotted as "typical intraslab-normal focal mechanism" earthquakes in Figures 4 and 5 had focal mechanisms consistent with their occurring by normal faulting caused by extension approximately parallel to the downdip direction of the Wadati-Benioff zone: These events had P-axes that plunge more than 45° and T-axis azimuths between 0° and 50°. Most events in Figures 4 and 5 that are classified as "typical" interface-thrust and intraslab-normal earthquakes on the basis of their focal mechanisms had locations consistent with locations of interface-thrust and intraslab, normal-fault mechanisms in subduction zones around the world (e.g., Astiz et al., 1988), and we have defined the likely position of the Cocos-Caribbean thrust interface beneath El Salvador in agreement with the global pattern (Fig. 5).

Eventually, seismic hazard studies for El Salvador and for Central America in general will want to distinguish between the hazard caused by the coastal Wadati-Benioff zone interface-thrust and Cocos intraplate earthquakes, respectively (Bent and Evans, this volume, Chapter 29). A methodology for assessing seismic hazard that depends on physical models of strain accumulation will need to distinguish between earthquakes that release strain accumulated through different mechanisms. In addition, some studies (e.g., Youngs et al., 1997) suggest that shallow intraplate earthquakes tend to produce stronger shaking at a given epicentral distance than interface-thrust earthquakes of the same magnitude. These differences might reflect systematic differences in stress-drops between the two types of earthquakes, differences in radiation patterns from the differently oriented faults, or different focusing and attenuation along the slightly different paths from source to site. Finally, physical models for estimating the tsunami hazard to the Pacific coast of Central America may need to distinguish between interface-thrust earthquakes and Cocos intraplate earthquakes. At present, a separate accounting for interface-thrust earthquakes and Cocos intraplate earthquakes in seismic hazard studies is complicated by the difficulty of separating the two types of earthquakes in earthquake catalogs. Many of the Cocos intraplate earthquakes occur near the seismically active part of the interplate thrust interface, and it is commonly not possible to determine whether an earthquake was a Cocos intraplate earthquake or an interplate thrust-interface earthquake solely on the basis of its teleseismically determined hypocenter. Large earthquakes occurring in decades since the 1960s can be classified on the basis of their seismographically determined focal mechanism, but instrumental focal mechanisms have not been determined for most large earthquakes occurring before the early 1960s.

Seismic Gaps and Variations in Coupling across Subduction Zones

An example of a seismic-hazard assessment methodology that depends on a physical model of strain accumulation is the "seismic gap" methodology as used by McCann et al. (1979) and Nishenko (1991). In its general form, this methodology is intended to assess the probabilities of great earthquakes on plate interfaces only. The methodology is based on the assumptions that (1) strain on an interface accumulates at a rate that is proportional to the rate of relative plate motion and (2) the likelihood of a great earthquake on a particular segment of a plate interface increases with the time that has elapsed since strain was locally released by the last great earthquake on that segment. The methodology requires a catalog of past earthquakes on the interface to identify the times of the most recent great earthquakes on each segment of the plate interface and to estimate the proportion of strain on the plate boundary that is released seismically. Using the seismic gap methodology, Nishenko (1991) characterized the interface-thrust zone offshore of central and eastern El Salvador and western Nicaragua as having "significant but unknown seismic potential," and he characterized the interface-thrust fault offshore of western El Salvador and southeastern Guatemala as having a rather high probability for the recurrence of a magnitude 7.5 earthquake in upcoming decades. These characteriza-

tions were, however, based on poorly understood earthquakes of the early twentieth century and earlier. The earthquakes' focal mechanisms have not been instrumentally determined, and their hypocenters are not well enough determined that they can be definitely ascribed to rupture on the plate interface, instead of rupture in the interior of the Cocos plate. If some of the early earthquakes, instead of being interface-thrust earthquakes, were Cocos intraplate earthquakes, they would not be appropriate for use in the seismic gap methodology to assess the likelihood of future interface-thrust earthquakes.

Application of the seismic gap methodology to the Central American subduction zone is complicated by studies suggesting that interface-thrust zones worldwide differ in the sizes and frequencies of destructive earthquakes that they produce, corresponding to differences in the proportion of relative plate motion that is released respectively seismically and aseismically in the different zones. The differences in seismic behavior are broadly correlated with differences in the large-scale geotectonic characteristics of the subduction zones and with differences in the morphology of their Wadati-Benioff zones. The seismotectonic characteristics of a given subduction zone may be viewed against the characteristics of two "end-member" subduction zones—the "Chilean-type" and "Marianas-type" subduction zones (Uyeda and Kanamori, 1979; Kanamori, 1986; Jarrard, 1986). The Chilean type of subduction zone is characterized by a shallowly dipping Wadati-Benioff zone and an overriding plate that is strongly compressional; the Marianas-type subduction zone is characterized by a steeply dipping Wadati-Benioff zone and an overriding plate that is strongly extensional. The adjacent plates are tightly coupled across the interface thrusts of Chilean-type subduction zone; accumulated strain is mostly released in great thrust-fault earthquakes. The adjacent plates are loosely coupled across the interface thrusts of a Marianas-type subduction zone; relative motion between the plates is mostly accommodated aseismically, and great thrust-fault earthquakes do not occur along the margin.

The Central American subduction zone is intermediate in large-scale geotectonic style between the Chilean type and the Marianas type, and its characteristics change along strike. The part of the Central American Wadati-Benioff zone beneath Guatemala, El Salvador, and Nicaragua has a much steeper dip than Chilean-type subduction zones, with the dip slightly shallowing from south to north (Burbach and Frohlich, 1986). On a scale of 1 (active backarc spreading) to 7 (very strongly compressional), Jarrard (1986) classifies the overriding plate of the Nicaraguan section of the Central American subduction zone as a 3 (mildly tensional). The Costa Rica section of the Central American subduction zone has a tectonic style that is more Chilean-type than the rest of the Central American subduction zone; it is characterized by a much more compressive geotectonic environment and a shallower, less steeply dipping Wadati-Benioff zone (Burbach et al., 1984; Protti et al., 1994). Subduction zones that are intermediate in tectonic character and Wadati-Benioff zone morphology, between the Chilean type and the Marianas type, are expected to have seismic behavior that is in some sense intermediate between behaviors characteristic of the Chilean-type and Marianas-type zones, although the seismic behavior of such intermediate-style zones differs widely, according to the sizes and distribution of asperities (patches across which the plates are tightly coupled) on the interface thrust of each zone (Lay et al., 1982; Pacheco et al., 1993). McNally and Minster (1981) and Pacheco et al. (1993) have noted that, for the Central American subduction zone from Guatemala through Costa Rica, the historically observed rate of great earthquakes is substantially smaller than would be expected in an efficiently coupled subduction zone that accommodates ~80 mm/yr relative plate motion.

Interface-Thrust Earthquakes

Considering earthquakes occurring in 1978–April 2001 in Central America beneath or offshore of Guatemala, El Salvador, and Nicaragua, of depth less than 100 km and epicenter between 50 km and 170 km from the trench axis, we find that "typical interface-thrust focal mechanism" earthquakes accounted for

~60% of earthquakes of M_W 6 or greater (Fig. 7A). Only one of the typical thrust-mechanism earthquakes since 1978 had a magnitude approaching M ~8, consistent with the historical trend noted by McNally and Minster (1981). Even on a generally weakly coupled interface thrust, however, it is plausible that there would exist some asperities that could accumulate strain sufficient to produce interface-thrust earthquakes in the M ~8 range, and it is also possible that there could be along-strike variations in seismic coupling along the Guatemala-Nicaragua section of the Central American subduction zone. In the historical record, a good candidate for a Salvadoran M ~8 interface-thrust earthquake would be the shock of September 7, 1915, for which Ambraseys and Adams (2001) calculated a hypocenter beneath western El Salvador and an M_s of 7.7. White et al. (this volume, Chapter 28) show that P-wave first motions recorded worldwide on the sparsely distributed and relatively insensitive seismographs of 1915 are consistent with the earthquake having a thrust-faulting mechanism. The shock was followed by a number of large aftershocks (Ambraseys and Adams, 2001), a pattern more typi-

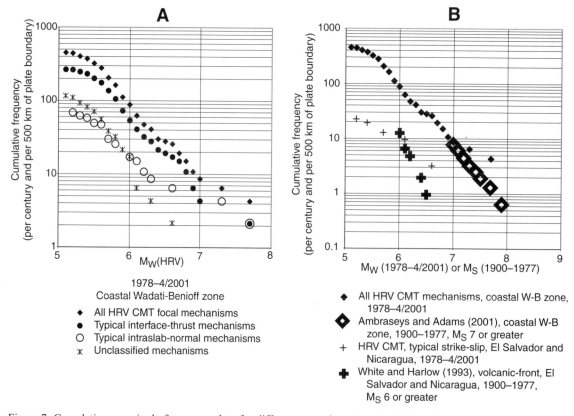

Figure 7. Cumulative magnitude-frequency plots for different sets of earthquakes distinguished by tectonic context or time of occurrence. For each set of earthquakes, the vertical axis shows the observed frequency of earthquakes with magnitudes equal to or greater than the magnitude shown on the horizontal axis. Symbols are plotted for a magnitude only if an earthquake of that magnitude is cataloged. A: Plots comparing the frequencies of different types of focal mechanism for the coastal Wadati-Benioff zone of Guatemala, El Salvador, and Nicaragua, 1978–April 2001. Events are those for which Harvard CMT focal mechanisms were determined. B: Plots comparing the frequencies for coastal Wadati-Benioff zone earthquakes and shallow-focus volcanic-front earthquakes, each determined with data from two different catalogs and time periods.

cal of interface-thrust earthquakes than slightly deeper intraplate normal earthquakes (see, however, discussion later in this paper on aftershocks to the intraplate normal earthquake of January 13, 2001). The 1915 earthquake produced Modified Mercalli intensities of VIII and higher at several locations in western El Salvador and adjacent parts of Guatemala (Ambraseys and Adams, 2001).

The largest (in terms of M_w) Central American interface-thrust earthquake in the past half-century was only lightly felt but produced a locally destructive tsunami. The Nicaraguan earthquake of 00:16 GMT, September 2, 1992, occurred as the result of slip at shallow depth on the Central American thrust interface (Satake, 1994). The earthquake nucleated in a seismic gap that had been identified by Harlow et al. (1981), but the earthquake rupture propagated well beyond the boundaries of the gap that had been identified. The local time of origin was 6:16 p.m., September 1. The shock was felt by only about half of the people in the coastal regions that were to be affected by the tsunami (Satake et al., 1993). The tsunami arrived on the coast at 8:00 p.m., local time; runup heights exceeded 2 m along ~300 km of the Nicaraguan coastline, and the maximum runup exceeded 9 m (Satake et al., 1993; Baptista et al., 1993). An estimated 168 people died, and 13,000 were made homeless. Compared with tsunamis produced by other thrust-interface earthquakes worldwide, the tsunami produced by the Nicaraguan earthquake was disproportionately large for the earthquake's magnitude. Tsunami runup values were those typical of an M_S 8.0 earthquake, whereas the observed M_S was 7.2 (U.S. Geological Survey, 1992) and the observed M_w (HRV) was 7.7. The efficiency of tsunami generation for the Nicaraguan earthquake is attributed to the earthquake having produced unusually large displacements of the seafloor and having had an unusually long faulting duration for its M_S or M_w: These characteristics are in turn attributed to the earthquake having occurred at an unusually shallow depth, with much of the rupture process taking place in low-velocity sediments (Satake, 1994; Polet and Kanamori, 2000).

At the present state of knowledge, we have to consider that earthquakes similar to the 1992 Nicaraguan earthquake might occur elsewhere along the trenchward edge of the Central American interface thrust, and that the El Salvador coast might be subject to destructive tsunamis from this type of source. Development of procedures for issuing tsunami warnings along the Central American coast would have to consider that the light shaking produced on land from the 1992 Nicaraguan earthquake might be typical of this specific type of tsunami earthquake.

Cocos Intraplate Earthquakes

The destructive El Salvador earthquakes of June 19, 1982 [M_w(HRV) = 7.3] (Lara, 1983), and January 13, 2001 [M_w(HRV) = 7.7] (Paulson and Bommer, 2001; Lomnitz and Elizarrarás, 2001), were normal-faulting intraplate earthquakes within the Cocos plate (Bent and Evans, this volume, Chapter 29). "Typical intraslab-normal mechanism" earthquakes, as defined in this paper, account for ~20% percent of M_w 6 or greater Cen-

tral American Wadati-Benioff zone earthquakes of Guatemala, El Salvador, and Nicaragua for which Harvard CMT mechanisms have been determined (Fig. 7A). Many of the "unclassified mechanisms" represented in Figures 4, 5, and 7A probably also correspond to Cocos intraplate earthquakes, with faulting styles or fault orientations differing from those of typical intraslab-normal earthquakes. Intraslab earthquakes immediately downdip of the seismogenically active interface thrust have been attributed to stresses arising from a variety of causes, such as extensional forces exerted by the deeper subducted slab or stresses caused by the distortion of the slab (e.g., Astiz et al., 1988; Bevis, 1988). We are not aware of a simple theoretical model of intraslab seismic-strain accumulation that would enable a prediction of the rate of such earthquakes that would be useful in seismic hazard studies.

The M_w 7.7 Cocos intraplate earthquake of January 13, 2001, was followed within one week by 28 aftershocks of m_b (USGS) 4.0 or greater, most of which were felt in San Salvador and the largest of which had M_w of 5.8. By comparison, the M_w 7.3 Cocos intraplate earthquake of June 19, 1982, was followed by only two aftershocks of m_b 4.0 or greater within one week of the main shock. In a global context, the 2001 aftershock sequence was exceptionally strong for an intraslab earthquake occurring well inland of the trench axis (Astiz et al., 1988). We have earlier cited the strong aftershock sequence of the earthquake of September 7, 1915, as evidence in favor of that earthquake being an interface-thrust shock rather than a Cocos intraplate shock. The aftershock sequence associated with the intraslab January 13, 2001, earthquake suggests that the presence of a significant aftershock sequence is not a conclusive basis for postulating that a historical coastal Wadati-Benioff zone earthquake occurred on the interface thrust.

Most seismic hazard associated with the Central American Wadati-Benioff zone is associated with shocks of depth less than 100 km occurring near the coast. Some large deeper shocks may also cause damage. Ambraseys and Adams (2001) favor a focal depth of ~150 km for the M_S 7.1 earthquake of May 21, 1932, which was damaging in southeastern El Salvador. Cocos intraplate normal-fault earthquakes occur near the axis of the Middle America trench, seaward of the coastal Wadati-Benioff zone (Figs. 4 and 5). Worldwide, uncommon great trench normal-fault earthquakes, such as the 1933 Sanriku, Japan, earthquake (Kanamori, 1971) and the Sumba, Indonesia, earthquake of 1977 (Spence, 1986; Lynnes and Lay, 1988), have produced devastating tsunamis on nearby coasts.

UPPER-CRUSTAL SEISMICITY IN THE CARIBBEAN PLATE

Historically, the most destructive earthquakes in El Salvador have been earthquakes occurring at upper-crustal focal depths (depth < 15 km) within ~10 km of the volcanic front in the overriding Caribbean plate (White, 1991; White and Harlow, 1993). More than half the length of the volcanic front in El Salvador experienced destructive upper-crustal earthquakes in the twentieth century—some communities more than once (Fig. 8). The largest of these earthquakes had magnitudes in the 6.5–6.7 range (White and Harlow, 1993; Ambraseys and Adams, 2001; U.S. Geological Survey, 2001), but smaller shocks have also been deadly. The M_w 5.7 San Salvador earthquake of October 10, 1986, for example, killed ~1500 people, left 100,000–150,000 homeless, and produced losses of about $1.5 billion (Bommer and Ledbetter, 1987; Olson, 1987). Earthquakes of m_b 4.5 or less, too small to be well recorded by the global network of seismographs, may cause significant alarm and minor damage (Ambraseys and Adams, 2001). Similar destructive upper-crustal earthquakes occur beneath the volcanic fronts of Guatemala, Nicaragua, and Costa Rica.

The majority of upper-crustal, volcanic-front earthquakes of Central America that have been well recorded teleseismically, and hence could have their focal mechanisms reliably determined by inversion of their seismic-wave radiation patterns, have had strike-slip focal mechanisms (White and Harlow, 1993). In Figures 4 and 5, we have plotted as "typical shallow strike-slip focal mechanism" earthquakes those for which both the P- and T-axes of the Harvard CMT solutions have plunges of less than 20° and for which the Harvard CMT depth is less than 50 km. The mechanisms of shallow strike-slip earthquakes of El Salvador and Nicaragua are consistent with the earthquakes occurring as slip on right-lateral faults that are approximately parallel to the main volcanic chain, or as slip on left-lateral faults that are approximately perpendicular to the trend of the volcanic chain.

For only a few Central American upper-crustal, volcanic-front earthquakes is there a strong basis from independent data (distribution of accurately located aftershocks, or observed fault displacement on the earth's surface) for establishing which of the two possible fault planes in the earthquake focal mechanism

actually corresponds to the causative fault (White and Harlow, 1993). For additional earthquakes with two possible fault planes suggested by instrumental data, one can argue in favor of one of the two planes on the basis of observed seismic intensities, although intensities are influenced by many factors in addition to proximity to the fault plane. On the basis of its focal mechanism and aftershock distribution, the San Salvador earthquake of October 10, 1986 (Figs. 5, 8, and 9), likely occurred as the result of left-lateral faulting oriented at a high angle to the regional trend of the principal volcanic chain (Harlow et al., 1993). As we shall discuss below, the distribution of aftershock epicenters and damage from the El Salvador earthquake of February 13, 2001, seems most easily explained in terms of right-lateral slip on a fault approximately parallel to the volcanic chain. From the distribution of their intensities and focal mechanisms, the largest of the Jucuapa earthquakes of 1951 (1951a in Fig. 8) is inferred to have occurred on a left-lateral fault approximately perpendicular to the volcanic chain (Ambraseys et al., 2001), and the San Salvador earthquake of 1965 is inferred to have occurred on a right-lateral fault subparallel to the volcanic chain (White and Harlow, 1993).

Upper-crustal, volcanic-front earthquakes account for a relatively small fraction of the overall earthquake activity of the Central American subduction zone: The destructiveness of these earthquakes is a consequence of their occurring at shallow focal depth, commonly beneath heavily populated regions. In Figure 7B, we plot the average frequency of such earthquakes for El Salvador and Nicaragua, as determined by two independent catalogs spanning nonoverlapping time periods. The catalogs are the 1900–1977 portion of White and Harlow's (1993) catalog and the Harvard CMT catalog, shallow, strike-slip events only, for 1978 through April 2001. Both catalogs imply an average

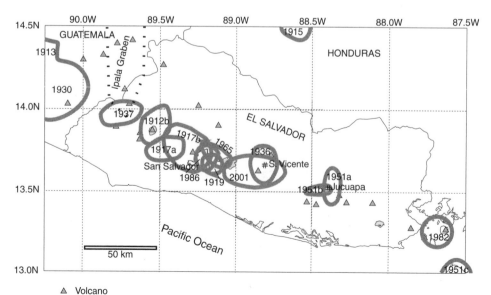

Figure 8. Regions of damage (Modified Mercalli intensities of VII or greater) from upper-crustal, volcanic-front earthquakes, twentieth century, El Salvador and vicinity. Pre-2001 isoseismals are from White and Harlow (1993) and are labeled as by White and Harlow. The Intensity VII isoseismal from the 13 February 2001 earthquake (Alvarenga et al., 2002) has been added. The boundaries of the Ipala graben are as defined by White (1991). Volcanoes are as in Figure 1.

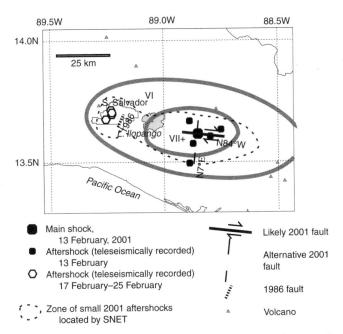

Figure 9. Sources of destructive El Salvador, volcanic-front earthquakes of October 1986 and February 2001. Location and trend of the fault that produced the earthquake of October 10, 1986, are from Harlow et al. (1993). "Likely 2001 fault" and "Alternative 2001 fault" correspond to possible fault planes of the main shock of February 13, 2001, that are implied by the Harvard CMT focal mechanism of the earthquake. They have strikes implied by the CMT mechanism and are positioned so as to pass through the main-shock epicenter and to come near to as many as possible of the teleseismically recorded aftershocks of February 13. Intensity VI and VII isoseismals are those of Alvarenga et al. (2002) for the main shock of February 13, 2001. Epicenters are those calculated by SNET using exclusively local data for earthquakes that occurred in February 2001 and that were large enough to also have been recorded at teleseismic distances. The "Zone of small 2001 aftershocks located by SNET" is generalized from maps published on the SNET Web site (http://www.snet.gob.sv/Geologia/Sismologia/sismi2001.htm) and represents the overall zone of seismicity that was activated by the February 13 main shock, from February 13 to the end of 2001.

frequency of ~10 earthquakes of magnitude greater than 6 per century per 500 km of length along the volcanic arc in El Salvador and Nicaragua. This is substantially less than the nearly 100 magnitude 6 or greater Wadati-Benioff zone earthquakes, per century per 500 km of arc-length, that is implied by the Harvard CMT catalog for 1978 through April 2001 and that is also consistent with the frequencies of large earthquakes cataloged by Ambraseys and Adams (2001) for 1900–1977 (Fig. 7B).

We note that upper-crustal, volcanic-front earthquakes of Guatemala and Costa Rica, although not represented in Figure 7B, also pose a significant hazard to the populations of those nations (White, 1991; White and Harlow, 1993). We represent earthquakes exclusively from El Salvador and Nicaragua in Figure 7B in order to estimate a regional frequency-of-occurrence of

upper-crustal, volcanic-front earthquakes that is not contaminated by strike-slip events originating on the Caribbean/North American plate boundary in Guatemala, nor by strike-slip events that might be caused by stresses associated with the shallow subduction of the Cocos Ridge beneath Costa Rica. Shallow strike-slip earthquakes along the Caribbean/North American plate boundary in eastern Guatemala (Fig. 4), for example, are occurring on the faults of that transform boundary and may play a different role in regional plate tectonics than that played by the upper-crustal, volcanic-front earthquakes of El Salvador. The destructive Guatemala earthquake of February 4, 1976 (Plafker, 1976; Espinosa, 1976), occurred on the Caribbean/North American plate boundary and would not be classified as a volcanic-front earthquake within the Caribbean plate.

Tectonic Origin of Upper-Crustal Seismicity

In a broad sense, the upper-crustal volcanic-front seismicity probably reflects the northwestward motion of the forearc of the Caribbean plate, south and west of the volcanic chain, with respect to the Caribbean plate that lies to the north and east of the volcanic chain (Plafker, 1976; Harlow and White, 1985; White, 1991). The importance of motion of "forearc slivers" parallel to plate boundaries has been recognized in a number of subduction zone environments (Fitch, 1972; Jarrard, 1986). DeMets (2001) reports evidence that slip vectors of Central American interface-thrust earthquakes are systematically rotated with respect to a well-constrained independent estimate of the motion of the Cocos plate with respect to the core of the Caribbean plate; he interprets this rotation in terms of a Central American forearc sliver that is moving northwest with respect to the core of the Caribbean plate with a velocity of ~14 mm/yr (Fig. 2).

In detail, the geometry of the structures that accommodate the right-lateral motion of the Central American forearc with respect to the interior of the Caribbean plate has not been defined. The currently seismogenic faults are likely to be a subset of the faults that have defined the distribution of Quaternary volcanoes (Carr and Stoiber, 1977). Figure 6 shows mapped faults in El Salvador as compiled by Case and Holcombe (1980). Not all faults in Figure 6 are likely to be potentially seismogenic at present, however, and potentially seismogenic faults remain unmapped. White (1991) and White and Harlow (1993) have argued that the most important faults for accommodating the motion of the forearc sliver are right-lateral strike-slip faults that are parallel to the volcanic chain. La Femina et al. (2002), in contrast, emphasize the possible importance of bookshelf faulting, in which much seismogenic slip occurs on NNE-trending left-lateral faults that are within a broad WNW-trending right-lateral shear zone. Aseismic deformation may also play an important role in accommodating the relative motion of the Central American forearc sliver with respect to the interior of the Caribbean plate. In many other subduction zones, forearc slivers are separated from the interiors of overriding continental plates by geologically conspicuous regional strike-slip faults that are parallel to the trends of the arcs, but the Central American forearc sliver is not separated from the interior of the Caribbean plate by such a regionally prominent fault (Jarrard, 1986).

A better understanding of the configuration of seismogenic faults and of the role of aseismic deformation would help estimate the likelihood of future strong earthquakes in those areas of the volcanic chain in El Salvador (such as much of eastern El Salvador) that did not experience damaging upper-crustal earthquakes in the twentieth century (Fig. 8). Such knowledge would also help anticipate the maximum magnitude of upper-crustal earthquakes that are likely to occur at a given location. In general, the maximum size of earthquakes in a region will be determined by the lengths of fault segments that can rupture in a single earthquake. Considering Central America from Costa Rica through Guatemala, the maximum observed magnitude of upper-crustal, volcanic-front earthquakes is ~6.9 (White and Harlow, 1993). White and Harlow (1993) postulate that individual segments of a regional, arc-parallel, right-lateral shear zone are demarcated by the volcanic centers at their ends; the maximum sizes of earthquakes along a given section of the Central American volcanic arc would therefore be determined by the distance between the volcanic centers that bound that section of the arc. Under the model that motion of the forearc sliver is accommodated by bookshelf faulting (La Femina et al., 2002), the sizes of earthquakes would be determined by the width of the overall WNW-trending right-lateral shear zone within which the individual NNE-trending seismogenic faults are situated.

Earthquake of February 13, 2001

The El Salvador earthquake of February 13, 2001 (M_w = 6.6), and its aftershocks illustrate several important characteristics of the upper-crustal seismicity. The main shock was assigned a Harvard CMT focal depth of 15 km (U.S. Geological Survey, 2001), the default focal depth for shallow-focus shocks (Fig. 5), and locally recorded aftershocks were overwhelmingly in the uppermost 15 km of Earth's crust (http://www.snet.gob.sv/Geologia/Sismologia/sismi2001.htm). The earthquake directly or indirectly caused the deaths of over 300 people and injuries to more than 3000 (U.S. Geological Survey, 2001). The earthquake caused shaking of Modified Mercalli intensity VI or higher in an area of ~3000 km^2 (Fig. 9; Alvarenga et al., 2002). The moment magnitude (M_w) of the earthquake, together with the Wells and Coppersmith (1994) equations relating rupture length to moment magnitude for strike-slip earthquakes, suggests that the causative fault was ~20 to 30 km long. The Harvard CMT focal mechanism solution indicates that the causative fault plane was either a NNE-striking (N7°E) left-lateral strike-slip fault or a WNW-striking (N84°W) right-lateral strike-slip fault (http://www.seismology.harvard.edu/CMTsearch.html). The February 13 main shock triggered aftershocks over an ~75 km long, WNW-trending zone that includes the region of the main-shock epicenter (Fig. 9), but that is far larger than the zone of highest intensity and is far longer than the length of the main-shock rupture that would be suggested by

the main shock's seismic moment. The overall 75 km long zone therefore likely reflects slip on a number of different faults that were activated by stress changes caused by the February 13 main shock. The WNW orientation of the overall aftershock zone supports the hypothesis that the February 13 earthquake occurred in a regional WNW-trending zone of right-lateral shear. The WNW elongation of the intensity VII isoseismal (Fig. 9) is most consistent with the main shock having occurred on a WNW-striking fault plane within the broader WNW-trending shear zone, and the WNW-trending plane is accordingly identified as the "likely fault" in Figure 9. The aftershock zone is, however, wide enough to leave open the possibility that the individual fault segment that produced the February 13 main shock may have been a NNE-trending left-lateral fault within the broader WNW-trending right-lateral shear zone. The NNE-trending nodal plane of the Harvard CMT focal mechanism is therefore indicated as the "alternative 2001 fault" in Figure 9.

The San Salvador Sequence of February 17–25

The western San Salvador earthquake sequence of February 17–25 may be cited as an example of a Central American upper-crustal earthquake sequence that consists of only small and moderate earthquakes but that causes great public concern. The sequence also illustrates the importance of local networks in locating such earthquakes. The sequence on February 17–25 occurred west-northwest of the source region of the February 13 earthquake. The source of the February 17–25 sequence lies on the trend of the inferred N84°W-striking fault of the February 13 main shock (Fig. 9), but it is well outside the region of strongest shaking in the February 13 main shock. The February 17–25 sequence therefore must have occurred on a different fault or fault segment than the fault segment that ruptured in the February 13 main shock. In February 17–25, 65 of these shocks were felt locally (http://www.snet.gob.sv/Geologia/Sismologia/sismi2001.htm), and several of the largest were recorded teleseismically. The first and largest earthquake, that of February 17, 20:25 GMT, had a body-wave magnitude, m_b (USGS), of 4.1 and a local magnitude of 5.1. One person was reported killed by the earthquake, and three injured (U.S. Geological Survey, 2001). The earthquakes were too small to be well recorded by the global seismographic network. Locations calculated for several of the shocks by the U.S. Geological Survey National Earthquake Information Center, using only data obtained at stations far from El Salvador, were mislocated by more than 50 km (Fig. 10). If the earthquakes had been slightly smaller, or the global seismographic network slightly less sensitive, any listing of the earthquakes in internationally compiled earthquake catalogs would have depended entirely on data from local seismographs.

Clustering of Upper-Crustal Seismicity

Upper-crustal, volcanic-front earthquakes of Central America are commonly preceded by foreshocks or occur in clusters of several damaging events closely spaced in time (White and Harlow, 1993). For example, the largest San Vicente earthquake of late 1936, that of December 20, was preceded by several weeks of smaller earthquakes from approximately the same source (Levin, 1940), and the San Salvador earthquake of May 3, 1965, was immediately preceded by foreshocks and occurred in a period of rather high activity that had begun with an earthquake swarm three months earlier (Lomnitz and Schulz, 1966). White and Harlow (1993, p. 1137) describe the clustering tendency of damaging upper-crustal, volcanic-front earthquakes as follows: "When a destructive earthquake occurs along the volcanic front after several years of calm, 43% of the time it is followed within one month by another destructive earthquake less than 60 km away." The strong 1917 earthquakes, centered just west of San Salvador (Fig. 8), occurred within 35 minutes of each other (White and Harlow, 1993), and three earthquakes of M ~6 occurred near Jucuapa on May 6–7, 1951 (Meyer-Abich, 1952; Ambraseys et al., 2001). The earthquake of February 13, 2001, was not preceded by an obvious foreshock sequence. The sequence of February 17–25, 2001, may correspond to the type of triggered seismicity that has been represented at other times and locations of the volcanic chain by temporal clusters of damaging earthquakes, although its largest event was much smaller than the main shock of February 13.

In addition to their tendency to be temporally clustered, upper-crustal volcanic-front earthquakes have over a period of decades tended to concentrate in spatial clusters. The epicentral region of the February 13, 2001, earthquake has a history of earlier destructive, shallow earthquakes. The San Vicente earthquake of December 20, 1936, M ~6.1 (White and Harlow, 1993),

for example, was destructive in several of the towns and villages that were heavily damaged in the 2001 earthquake (Levin, 1940), and the region had experienced damaging earthquakes in each of the eighteenth and nineteenth centuries (Harlow et al., 1993). San Salvador has repeatedly experienced damaging earthquakes, with at least nine earthquakes having produced intensity VII effects within the city since 1700 (Harlow et al., 1993). In addition to the shocks of 1951, the region of Jucuapa had experienced two damaging earthquakes in the nineteenth century (Meyer-Abich, 1952; Ambraseys et al., 2001).

Volcano Seismicity

Most Central American earthquakes of $M_S > 5$ are not obviously associated with volcanism (White and Harlow, 1993). Their concentration near the volcanoes of El Salvador suggests that their locations may be defined in part by magmatic processes or that both seismic and magmatic phenomena may be localized by a fundamental tectonic process, but most of the earthquakes don't have obvious links to specific volcanic eruptions. An example of an exceptional, strong, upper-crustal El Salvador earthquake that was directly associated with an eruption is the M_S 6.4 earthquake that occurred on June 8, 1917, and was followed 30 minutes later by the eruption of Volcán San Salvador. The earthquake damage was centered on the northwest flank of the volcano.

Although most damaging Central American earthquakes are not associated with volcanism, volcanic eruptions are invariably preceded and accompanied by seismic activity. A dramatic local example was the eruption of Volcán Pacaya in Guatemala during 1981 and 1982 that was preceded by at least 100,000 small to

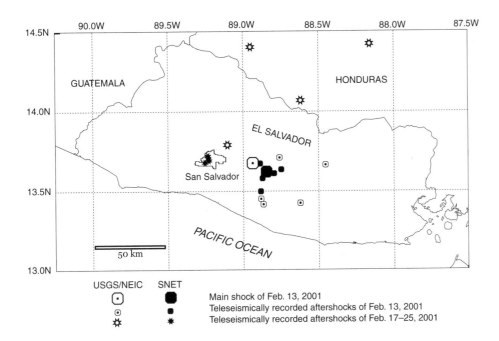

Figure 10. Comparison of epicenters calculated by the USGS/NEIC and SNET for teleseismically recorded, volcanic-front earthquakes of February 2001, illustrating the importance of locally recorded seismographic data for the accurate location of small and moderate El Salvador earthquakes. The SNET epicenters were determined with data obtained from seismographs recording in El Salvador and are accurate to within several kilometers. With the exception of the epicenter of the main shock, however, all USGS/NEIC epicenters were determined with data recorded from a limited range of azimuths by stations situated outside of El Salvador, and most are badly mislocated.

moderate earthquakes along faults southeast of the volcano (White et al., 1980). The largest of those earthquakes was the M_S 5 shock of October 9, 1979, that destroyed more than 150 houses. White and Power (2001) have shown that such earthquakes "have often been observed beneath distal regions of volcanoes weeks to years prior to eruptions." Such events, called volcano-tectonic earthquakes, are high frequency and broadband and appear to result from brittle failure along faults, resulting from magma-induced changes in stress or fault strength. However, volcano-tectonic events are often distinguishable from tectonic seismicity by their occurrence in spatial and temporal clusters, and by a tendency for the rate of earthquake activity and the maximum magnitudes of earthquakes within a cluster to increase with time. White and Power (2001) show that volcano-tectonic seismicity can sometimes be used to forecast the type and size of an eruption.

In addition to volcano-tectonic earthquakes, the other principal type of seismicity generally associated with magmatic and geothermal processes is long-period volcano seismicity. Long-period volcano seismicity is narrowband and has a dominant frequency less than 5 Hz. Long-period volcano seismicity occurs both as individual events and as extended periods of tremor lasting minutes to many days. Much of this type of seismicity is inferred to result from degassing of magma and/or boiling of the hydrothermal system (Chouet et al., 1994). Like volcano-tectonic seismicity, long-period volcano seismicity can also sometimes be used to forecast the type and size of an eruption (Chouet et al., 1994; Chouet, 1996). Such long-period volcano seismicity was used to successfully forecast the cataclysmic 1991 eruption of Mount Pinatubo in the Philippines (Harlow et al., 1996). Seismic signals are also caused by various other surface or near-surface volcanic phenomena, such as explosive eruptions, pyroclastic flows, lahars, dome collapse, and lava flows: These signals are generally of the long-period type.

Caribbean Plate Northeast of the Volcanic Front

The northwest corner of the Caribbean plate inland of the volcanic front, encompassing northern El Salvador, southeastern Guatemala, and western Honduras, has in recent decades been characterized by a moderate level of shallow-focus earthquake activity (Figs. 1 and 4), but the region has a history of occasional larger shocks. White et al. (this volume, Chapter 28) document the occurrence of a large (M ~7.5) earthquake in the northern Ipala graben of Guatemala in 1765. The Ipala graben extends into northwestern El Salvador (Fig. 8). The Honduras earthquake of 1915, whose intensity VII isoseismal extends into the region covered by Figure 8, had a magnitude (M_S) of 6.4 (White and Harlow, 1993). The relatively few Honduras events for which Harvard CMT mechanisms have been determined (Fig. 4) were characterized by predominantly normal-faulting focal mechanisms and have T-axes oriented E-W or ESE-WNW, consistent with geologic evidence for E-W extension in Honduras south of the North American/Caribbean plate boundary (Plafker, 1976). As noted in the section entitled "Sources of Data on El Salvador

Earthquakes," calculated focal depths at the north-northeast end of the profile in Figure 3 are uncertain by several tens of kilometers. We are not aware of strong independent evidence that shocks within the northwest corner of the Caribbean plate have focal depths approaching 50 km, as is suggested by Figure 3. Earthquakes this deep would be unusual, though not unprecedented (Wong et al., 1984), in a region of extensional tectonism. The relatively few Harvard CMT solutions in this region (Fig. 4) have hypocentroid depths less than 25 km.

CONCLUSIONS

This paper is being written at a time when the seismographic network run by El Salvador's Servicio Nacional de Estudios Territoriales (SNET) is being significantly upgraded. The existence of a high-quality network of local seismographs, together with a network staff of local scientists who collectively have acquired experience in the interpretation of seismographic data from earthquakes in different parts of El Salvador, will greatly facilitate addressing earthquake hazard issues. Solutions of most problems will also benefit from the systematic interchange of data between SNET and other Central American seismological organizations (Lindholm et al., this volume, Chapter 26), and from the systematic interchange of data between SNET and international agencies such as the USGS/NEIC and the ISC (data shared with the USGS/NEIC are automatically forwarded to the ISC). Let us consider, for example, seismological data that would help address issues identified in this paper.

1. Removal of bias in teleseismically determined epicenters and focal depths that are cataloged by international agencies (see section entitled "Sources of Data on El Salvador Earthquakes") could be accomplished by means of calibration events accurately located with local data.

2. Calibration of the teleseismic m_b scale for use in seismic hazard studies ("Sources of Data on El Salvador Earthquakes") could be accomplished by means of local magnitudes of small and moderate teleseismically recorded shocks.

3. Accurate calculation of earthquake epicenters on the borders of El Salvador ("Sources of Data on El Salvador Earthquakes") requires exchange of data with agencies in other Central American countries.

4. Distinguishing between interface-thrust earthquakes and Cocos intraplate earthquakes (see section entitled "The Central American Wadati-Benioff Zone") on the basis of accurate hypocenters would be facilitated by the transmittal of locally recorded arrival times to international agencies.

5. Distinguishing between interface-thrust earthquakes and Cocos intraplate earthquakes ("The Central American Wadati-Benioff Zone") on the basis of their focal mechanism, for earthquakes too small to have focal mechanisms reliably determined by the Harvard CMT methodology, may be accomplished with on-scale, broadband, local recording of these earthquakes.

6. An understanding of the long-term seismic potential of the interface thrust between the Cocos and Caribbean plates

("The Central American Wadati-Benioff Zone") may require mapping of asperities on the interface using locally recorded small earthquakes.

7. Issuing tsunami warnings for offshore Central American earthquakes ("The Central American Wadati-Benioff Zone") requires exchange of data with agencies in other Central American countries and exchange of data with Pacific-wide international tsunami warning agencies.

8. Seismographic definition of seismogenic faults in and near the volcanic front, for use together with geologic observations in seismic zoning (see section entitled "Upper-Crustal Seismicity in the Caribbean Plate"), requires precisely determined, locally recorded hypocenters.

9. Ensuring the reliable location and cataloging of potentially damaging earthquakes with M < 5.5 by international seismological agencies ("Upper-Crustal Seismicity in the Caribbean Plate") requires the sharing of local data with these agencies.

10. Evaluating the temporal and spatial clustering tendencies of upper-crustal, volcanic-front earthquakes ("Upper-Crustal Seismicity in the Caribbean Plate"), and developing ways to usefully characterize this clustering for the benefit of emergency management agencies and the El Salvadoran community, requires local seismological monitoring that is sustained over a long time.

11. Recording of volcano-tectonic and long-period seismicity for use in forecasting volcanic eruptions ("Upper-Crustal Seismicity in the Caribbean Plate") requires the collection of seismological data close to potentially active volcanoes.

ACKNOWLEDGMENTS

We thank Bob Engdahl for providing us with the hypocenters that are plotted in Figures 1–3. We appreciate the helpful reviews of Hugh Cowan, Paul Mann, George Choy, and Mark Petersen.

REFERENCES CITED

Alvarenga, E.R., Hernández, D.A., and Hernández Flores, D.A., 2002, Mapas preliminares de isosistas correspondientes a los sismos del 13 de Enero y 13 de Febrero de 2001, http://www.snet.gob.sv/Geologia/Sismologia/1deptsis.htm (June 2003).

Ambraseys, N.N., and Adams, R.D., 2001, The seismicity of Central America: A descriptive catalogue 1898–1995: London, Imperial College Press, 309 p.

Ambraseys, N.N., Bommer, J.J., Buforn, E., and Udias, A., 2001, The earthquake sequence of May 1951 at Jucuapa (El Salvador): Journal of Seismology, v. 5, p. 23–39.

Astiz, L., Lay, T., and Kanamori, H., 1988, Large intermediate-depth earthquakes and the subduction process: Physics of the Earth and Planetary Interiors, v. 53, p. 80–166.

Baptista, A.M., Priest, G.R., and Murty, T.S., 1993, Field survey of the 1992 Nicaragua tsunami: Marine Geodesy, v. 16, p. 169–203.

Bent, A.L., and Evans, S.G., 2004, The M_W 7.6 El Salvador earthquake of 13 January 2001 and implications for seismic hazard in El Salvador, *in* Rose, W.I., et al., eds., Natural hazards in El Salvador: Boulder, Colorado, Geological Society of America Special Paper 375, p. 397–404 (this volume).

Bevis, M., 1988, Seismic slip and down-dip strain rates in Wadati-Benioff zones: Science, v. 240, p. 1317–1319.

Bommer, J., and Ledbetter, S., 1987, The San Salvador earthquake of 10th October, 1986: Disasters, v. 11, p. 83–95.

Bommer, J.J., Udias, A., Cepeda, J.M., Hasbun, J.C., Salazar, W.M., Suarez, A., Ambraseys, N.N., Buforn, E., Cortina, J., Madariaga, P., Mendez, P.,

Mezcua, J., and Papastamatiou, D., 1997, A new digital accelerograph network for El Salvador: Seismological Research Letters, June 1997, v. 68, no. 3, p. 426–437.

Burbach, G.V., and Frohlich, C., 1986, Intermediate and deep seismicity and lateral structure of subducted lithosphere in the Circum-Pacific region: Reviews of Geophysics, v. 24, p. 833–874.

Burbach, G.V., Frohlich, C., Pennington, W.D., and Matumoto, T., 1984, Seismicity and tectonics of the subducted Cocos plate: Journal of Geophysical Research, v. 89, p. 7719–7735.

Carr, M.J., and Stoiber, R.E., 1977, Geologic setting of some destructive earthquakes in Central America: Geological Society of America Bulletin, v. 88, p. 151–156.

Case, J.E., and Holcombe, T.L., 1980, Geologic-tectonic map of the Caribbean region: U.S. Geological Survey Miscellaneous Investigations Map I-1000, scale 1:2,500,000, 2 sheets.

Chouet, B.A., 1996, Long-period volcano seismicity: Its source and use in eruption forecasting: Nature, v. 380, p. 309–316.

Chouet, B.A., Page, R.A., Stephens, C.D., Lahr, J.C., and Power, J.A., 1994, Precursory swarms of long-period events at Redoubt Volcano (1989–1990), Alaska: Their origin and use as a forecasting tool: Journal of Volcanology and Geothermal Research, v. 62, p. 95–135.

Dean, B.W., and Drake, C.L., 1978, Focal mechanism solutions and tectonics of the Middle America arc: Journal of Geology, v. 86, p. 111–128.

DeMets, C., 2001, A new estimate for present-day Cocos-Caribbean plate motion: Implications for slip along the Central American arc: Geophysical Research Letters, v. 28, p. 4043–4046.

Dewey, J.W., and Algermissen, S.T., 1974, Seismicity of the Middle America arc-trench system near Managua, Nicaragua: Bulletin of the Seismological Society of America, v. 64, p. 1033–1048.

Dewey, J.W., and Suárez, G., 1991, Seismotectonics of Middle America, *in* Slemmons, D.B., et al., eds., Neotectonics of North America: Boulder, Colorado, Geological Society of America, Decade Map, v. 1, p. 309–321.

Dziewonski, A.M., Chou, A.T., and Woodhouse, J.H., 1981, Determination of earthquake source parameters from waveform data for studies of global regional seismicity: Journal of Geophysical Research, v. 86, p. 2825–2852.

Dziewonski, A.M., Ekström, G., Franzen, J.E., and Woodhouse, J.H., 1987, Centroid-moment tensor solutions for January–March 1986: Physics of the Earth and Planetary Interiors, v. 45, p. 1–10.

Engdahl, E.R., van der Hilst, R., and Buland, R., 1998, Global teleseismic earthquake relocation with improved travel times and procedures for depth determination: Bulletin of the Seismological Society of America, v. 88, p. 722–743.

Espinosa, A.F., editor, 1976, The Guatemalan earthquake of February 4, 1976: A preliminary report: U.S. Geological Survey Professional Paper 1002, 90 p.

Fitch, T.J., 1972, Plate convergence, transcurrent faults, and internal deformation adjacent to Southeast Asia and the western Pacific: Journal of Geophysical Research, v. 77, p. 4432–4460.

Geller, R.J., 1976, Scaling relations for earthquake source parameters and magnitudes: Bulletin of the Seismological Society of America, v. 66, p. 1501–1523.

Harlow, D.H., and White, R.A., 1985, Shallow earthquakes along the volcanic chain of Central America: Evidence for oblique subduction [abs.]: Earthquake Notes, v. 56, no. 1, p. 28.

Harlow, D.H., White, R.A., Cifuentes, I.L., and Aburto, Q., A., 1981, Quiet zone within a seismic gap near western Nicaragua: possible location of a future large earthquake: Science, v. 213, p. 648–651.

Harlow, D.H., White, R.A., Rymer, M.J., and Alvarez, G.S., 1993, The San Salvador earthquake of 10 October 1986 and its historical context: Bulletin of the Seismological Society of America, v. 83, p. 1143–1154.

Harlow, D.H., Power, J.A., Laguerta, E.P., Ambubuyog, G., White, R.A., and Hoblitt, R.P., 1996, Precursory seismicity and forecasting of the June 15, 1991, eruption of Mount Pinatubo, *in* Newhall, C.G., and Punongbayan, R., eds., Fire and mud: Eruptions and lahars of Mount Pinatubo, Philippines: Seattle, University of Washington Press, p. 285–306.

Jarrard, R.D., 1986, Relations among subduction parameters: Reviews of Geophysics, v. 24, p. 217–284.

Kanamori, H., 1971, Seismological evidence for lithospheric normal faulting—The Sanriku earthquake of 1933: Physics of the Earth and Planetary Interiors, v. 4, p. 289–300.

Kanamori, H., 1986, Rupture process of subduction zone earthquakes: Annual Reviews of Earth and Planetary Sciences, v. 14, p. 293–322.

La Femina, P.C., Dixon, T.H., and Strauch, W., 2002, Bookshelf faulting in Nicaragua: Geology, v. 30, p. 751–754.

J.W. Dewey, R.A. White, and D.A. Hernández

Lara, G.M.A., 1983, El Salvador earthquake of June 19, 1982: Newsletter Earthquake Engineering Research Institute, v. 17, p. 87–96.

Lay, T., Kanamori, H., and Ruff, L., 1982, The asperity model and the nature of large subduction zone earthquakes: Earthquake Prediction Research, v. 1, p. 3–71.

Levin, S.B., 1940, The Salvador earthquakes of December 1936: Bulletin of the Seismological Society of America, v. 30, p. 377–407.

Lindholm, C.D., Redondo, C.A., and Bungum, H., 2004, Two earthquake databases for Central America, *in* Rose, W.I., et al., eds., Natural hazards in El Salvador: Boulder, Colorado, Geological Society of America Special Paper 375, p. 357–362 (this volume).

Lomnitz, C., and Elizarrarás, S.R., 2001, El Salvador 2001: Earthquake disaster and disaster preparedness in a tropical volcanic environment: Seismological Research Letters, v. 72, p. 346–351.

Lomnitz, C., and Schulz, R., 1966, The San Salvador earthquakes of May 3, 1965: Bulletin of the Seismological Society of America, v. 56, p. 561–575.

Lynnes, C.S., and Lay, T., 1988, Source process of the Great 1977 Sumba earthquake: Journal of Geophysical Research, v. 93, p. 13,407–13,420.

McCann, W.R., Nishenko, S.P., Sykes, L.R., and Krause, J., 1979, Seismic gaps and plate tectonics: Seismic potential for major boundaries: Pure and Applied Geophysics, v. 117, p. 1082–1147.

McNally, K.C., and Minster, J.B., 1981, Nonuniform seismic slip rates along the Middle America Trench: Journal of Geophysical Research, v. 86, p. 4949–4954.

Meyer-Abich, H., 1952, Das Erdbeben von Jucuapa in El Salvador (Zentralamerika) vom 6 und 7 Mai 1951, Neues Jahrbuch Geologie und Palaeontologie Abhandlungen, v. 95, p. 331–336.

Molnar, P., and Sykes, L., 1969, Tectonics of the Caribbean and Middle American regions from focal mechanisms and seismicity: Geological Society of America Bulletin, v. 80, p. 1639–1684.

Nishenko, S.P., 1991, Circum-Pacific seismic potential: 1989–1999: Pure and Applied Geophysics (PAGEOPH), v. 135, p. 169–259.

Olson, R.A., 1987, The San Salvador earthquake of October 10, 1986, overview and context: Earthquake Spectra, v. 3, p. 415–418.

Pacheco, J.F., Sykes, L.R., and Scholz, C.H., 1993, Nature of seismic coupling along simple plate boundaries of the subduction type: Journal of Geophysical Research, v. 98, p. 14,133–14,159.

Paulson, C., and Bommer, J., editors, 2001, Preliminary observations on the El Salvador earthquakes of January 13 and February 13, 2001: Earthquake Engineering Research Institute Special Earthquake Report—July 2001, p. 1–12.

Plafker, G., 1976, Tectonic aspects of the Guatemala earthquake of 4 February 1976: Science, v. 93, p. 1201–1208.

Polet, J., and Kanamori, H., 2000, Shallow subduction earthquakes and their tsunamigenic potential: Geophysical Journal International, v. 142, p. 684–702.

Protti, M., Guendel, F., and McNally, K., 1994, The geometry of the Wadati-Benioff zone under southern Central America and its tectonic significance: Results from a high resolution local seismographic network: Physics of the Earth and Planetary Sciences, v. 84, p. 271–287.

Satake, K., 1994, Mechanism of the 1992 Nicaragua tsunami earthquake: Geophysical Research Letters, v. 21, p. 2519–2522.

Satake, K., Bourgeois, J., Abe, Kuniaki, Abe, Katsuyuki, Tsuji, Y., Imamura, F., Iio, Y., Katao, H., Noguera, E., and Estrada, F., 1993, Tsunami field survey of the 1992 Nicaragua earthquake: Eos (Transactions, American Geophysical Union), v. 74, p. 145, 156, and 157.

Segura, F., and Havskov, J., 1994, The new Nicaraguan seismic network: Geofísica Internacional, v. 33, no. 2., p. 223–233.

Shakal, A.F., Huang, M.-J., and Lineares, R., 1987, The San Salvador earthquake of October 10, 1986, processed strong motion data: Earthquake Spectra, v. 3, p. 465–481.

Siebert, L., and Simkin, T., 2002, Volcanoes of the world: An illustrated catalog of Holocene volcanoes and their eruptions: Smithsonian Institution Global Volcanism Program Digital Information Series, GVP-3 (http://www.volcano.si.edu/gvp.world/).

Sipkin, S.A., 1982, Estimation of earthquake source parameters by the inversion of waveform data; synthetic waveforms: Physics of the Earth and Planetary Interiors, v. 30, p. 242–259.

Spence, W., 1986, The 1977 Sumba earthquake series: Evidence for slab pull force acting at a subduction zone: Journal of Geophysical Research, v. 91, p. 7225–7239.

U.S. Geological Survey, 1992, Preliminary Determination of Epicenters, Monthly Listing, September 1992, 35 p.

U.S. Geological Survey, 2001, Preliminary Determination of Epicenters, Monthly Listing, February 2001, 40 p.

Uyeda, S., and Kanamori, H., 1979, Back-arc opening and the mode of subduction: Journal of Geophysical Research, v. 84, p. 1049–1061.

Wells, D.L., and Coppersmith, K.J., 1994, New empirical relationships among magnitude, rupture length, rupture width, rupture area, and surface displacement: Bulletin of the Seismological Society of America, v. 84, p. 974–1002.

White, R.A., 1991, Tectonic implications of upper-crustal seismicity in Central America, *in* Slemmons, D.B., et al., eds., Neotectonics of North America: Boulder, Colorado, Geological Society of America, Decade Map, v. 1, p. 323–338.

White, R.A., and Harlow, D.H., 1993, Destructive upper-crustal earthquakes of Central America since 1900: Bulletin of the Seismological Society of America, v. 83, p. 1115–1142.

White, R.A., and Power, J.A., 2001, Distal volcano-tectonic earthquakes: Diagnosis and use in eruption forecasting: American Geophysical Union 2001 Fall Meeting Program, p. 183.

White, R.A., Sanchez, E., Cifuentes, I.L., and Harlow, D.H., 1980, Preliminary Report on the on-going earthquake swarm in the Department of Santa Rosa, Guatemala: U.S. Geological Survey Open-File Report 80–800, 23 p.

White, R.A., Harlow, D.H., and Alvarez, S., 1987, The San Salvador earthquake of October 10, 1986—Seismological aspects and other recent local seismicity: Earthquake Spectra, v. 3, p. 419–434.

White, R.A., Ligorría, J.P., and Cifuentes, I.L., 2004, Seismic history of the Middle America subduction zone along El Salvador, Guatemala, and Chiapas, Mexico, 1526–2000, *in* Rose, W.I., et al., eds., Natural hazards in El Salvador: Boulder, Colorado, Geological Society of America Special Paper 375, p. 379–396 (this volume).

Wong, I.G., Cash, D.J., and Jaksha, L.H., 1984, The Crownpoint, New Mexico, earthquakes of 1976 and 1977: Bulletin of the Seismological Society of America, v. 74, p. 2435–2449.

Youngs, R.R., Chiou, S.-J., Silva, W.J., and Humphrey, J.R., 1997, Strong ground motion attenuation relationships for subduction zone earthquakes: Seismological Research Letters, v. 68, p. 58–73.

MANUSCRIPT ACCEPTED BY THE SOCIETY JUNE 16, 2003

Geological Society of America
Special Paper 375
2004

Seismic history of the Middle America subduction zone along El Salvador, Guatemala, and Chiapas, Mexico: 1526–2000

Randall A. White*

U.S. Geological Survey, MS 910, 345 Middlefield Road, Menlo Park, California 94025, USA

Juan Pablo Ligorría

Coordinadora Nacional para la Reducción de Desastres, Guatemala City, Guatemala

Ines Lucia Cifuentes

Carnegie Institution of Washington, 1630 P Street NW, Washington, D.C. 20015, USA

ABSTRACT

We present a catalog of subduction zone earthquakes along the Pacific coast from central El Salvador to eastern Chiapas, Mexico, from 1526 to 2000. We estimate that the catalog is complete since 1690 for $M_S \geq 7.4$ thrust events and $M \geq 7.4$ normal-faulting events within the upper 60 km of the down-going slab. New intensity maps were constructed for the 27 earthquakes since 1690, using mostly primary data sources. By calibrating with recent events we find that the long axis of the (MM) VII intensity contour for such large earthquakes well approximates the length and location of rupture along the subduction zone and can thus be used to estimate the locations and magnitudes of older events.

The section from western El Salvador to Chiapas appears to have ruptured completely in a series of four to five earthquakes during each of the periods 1902–1915, 1743–1776, and possibly 1565–1577. Earthquakes of M_W 7.75 ± 0.3 have caused major damage along the 200 km long section from San Salvador to Guatemala City every 71 ± 17 yr, apparently since at least 1575. Although the January 2001 El Salvador earthquake caused damage within part of this zone, no major thrust earthquake has occurred there since at least 1915. We find that much of this section has been relatively quiescent for moderate earthquakes shallower that 50 km since at least 1963. The conditional probability that an earthquake of M_W 7.75 ± 0.3 will occur at this location in the next 20 yr is estimated at 50% (±30%).

Keywords: Central America, tectonics, historical seismicity, slip deficit, seismic gap.

INTRODUCTION

Instrumental catalogs of earthquakes rarely include more than one cycle of large and great earthquakes at a particular section of a plate boundary. A complete seismic history of large and great earthquakes extending over several cycles is essential for estimating repeat times and for long-term prediction in a given

region. Existing catalogs of earthquakes in Central America prior to 1900, which included Chiapas, Mexico, during the colonial era, do not contain enough information to estimate the locations or magnitudes of earthquakes, and are therefore of limited value for prediction and hazard analysis. For excellent catalogs for seismicity since ~1900, we refer the reader to Ambraseys and Adams (1996, 2001). We have examined primary sources, including those from several colonial archives, to compile as complete a seismicity catalog for the region as possible from which to make

*rwhite@usgs.gov

White, R.A., Ligorría, J.P., and Cifuentes, I.L., 2004, Seismic history of the Middle America subduction zone along El Salvador, Guatemala, and Chiapas, Mexico: 1526–2000, *in* Rose, W.I., Bommer, J.J., López, D.L., Carr, M.J., and Major, J.J., eds., Natural hazards in El Salvador: Boulder, Colorado, Geological Society of America Special Paper 375, p. 379–396. For permission to copy, contact editing@geosociety.org. © 2004 Geological Society of America

intensity maps. Most of the raw data for this catalog, including abstracted town-by-town descriptions of damage and their sources, have been published in Feldman (1993) through a project funded by the U.S. Geological Survey for this study. Other results of this project include catalogs of shallow earthquakes along the volcanic chain (White, 1991; White and Harlow, 1993) and along the Caribbean–North American plate boundary in Central America (White, 1984).

TECTONIC SETTING

The Cocos plate subducts beneath the North American and Caribbean plates at the Middle America Trench, along the Pacific coasts of Mexico and Central America (Fig. 1). At the surface, the triple junction formed by the three plates extends over a broad region from ~90.5° W to ~94° W. Significant differences are observed in the tectonics and characteristics of subduction at the Cocos–North American plate margin, northwest of the triple junction, compared with the Cocos-Caribbean plate margin to the southeast.

At the Cocos–North American plate boundary, the Cocos plate is composed of young sea floor (4–15 Ma) with rough topography (Klitgord and Mammerickx, 1982). A major topographic high, the Tehuantepec Ridge, is apparently being subducted near the southern end of the trench segment (e.g., LeFevre and McNally, 1985; Ponce et. al., 1992). The Cocos plate subducts at a shallow angle (10° to 20°) (e.g., Molnar and Sykes, 1969; LeFevre and McNally, 1985; Pardo and Suárez, 1995) and accretes sediments to the hanging wall (e.g., Karig et al., 1978; Moore et al., 1979). A forearc basin is absent (e.g., Ross and Shor, 1965; Karig et al., 1978) while a broad, relatively inactive volcanic chain lies several hundreds of kilometers inland from the trench and strikes obliquely to it (e.g., Nixon, 1982; Robin, 1982). Interplate thrust events are generally found down to typical depths of 40 km (e.g., Chael and Stewart, 1982; LeFevre and McNally, 1985; Pardo and Suárez, 1995) and have magnitudes as large as that of the recent Michoacan earthquake (M_w 8.5). The section of the trench where the Tehuantepec Ridge is subducting is apparently anomalous, having no known historical earthquakes (e.g., Kelleher et. al., 1973; McNally and Minster, 1981; Singh et al., 1981). This section is designated the Tehuantepec Seismic Gap (Ponce et. al., 1992; Guzmán-Speziale et al., 1989). Previous studies have detailed the seismicity and seismic history of the Cocos–North American plate boundary north of the Tehuantepec Gap and propose recurrence intervals of 30 to 56 yr for great earthquakes there (McNally and Minster, 1981; Singh et. al., 1981).

Along the Cocos-Caribbean section of the Middle America Trench, the Cocos plate is composed of moderate-age seafloor (20 to 40 Ma) of smooth topography (e.g., Fischer, 1961; Truchan and Larson, 1973; von Huene et al., 1980). At shallow-to-intermediate depths, the subducting plate dips at a moderate angle (40° to 55°), resembling the geometry of the Rivera plate subduction at the northernmost part of the Middle America Trench (e.g., Dewey and Algermissen, 1974; Burbach et al., 1984; DeMets and

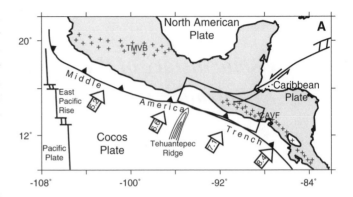

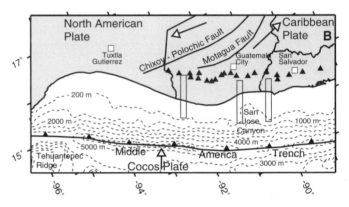

Figure 1. Regional tectonics surrounding the Middle America subduction zone. A: General plate tectonics of the region. The study area is enclosed in the box. The open arrows indicate the convergence rate (cm/yr) of the Cocos plate toward the North American and Caribbean plates (DeMets et al., 1990). TMVB—Trans-Mexican Volcanic Belt, CAVF—Central American Volcanic Front. B: The open arrows indicate the relative motions of the major plate segments. Open rectangles are the proposed segment boundaries of Stoiber and Carr (1973). The black triangles represent the position of some of the volcanoes with Quaternary activity. Bathymetric contours are labeled every 1000 m, and the 200-meter contour is also shown to represent the edge of continental margin.

Stein, 1990; Dewey and Suárez, 1991). There is no accretion of sediments (e.g., von Huene et al., 1980; Aubouin et al., 1982), but there is a well-developed forearc basin (e.g., Ladd et al., 1978; Seely, 1979). About 160 to 180 km landward of the trench, and parallel to it, lies a narrow, very active volcanic chain (e.g., Carr et al., 1982; Carr and Stoiber, 1990). The seismicity of this section of the Middle America Trench, from the Tehuantepec Gap to central El Salvador, has not been studied in any detail, and little has been published on the seismic history prior to 1900. Interplate thrust events occur down to depths of ~70 km (e.g., LeFevre and McNally, 1985; Ambraseys and Adams, 1996) and have maximum magnitudes of 7.8 to 7.9 and rarely 8.1. The average recurrence intervals for very large (M_s ~7.5) earthquakes at the Middle America Trench are ~68 yr off western El Salvador, as shown in this study, to 75 to 100 yr off Nicaragua (Harlow et al., 1981).

Coincident with the steepening in dip of the subducting slab southeast of the Tehuantepec gap (Ponce et al., 1992), there is a 10–25 Ma increase in age of the subducted plate and a gradual increase of the rate of convergence with distance from the Tehuantepec Ridge. In the region of our study, this convergence rate (see Fig. 1) is between 6.6 and 7.6 cm/yr (DeMets et al., 1990).

The crustal thickness of the overriding Caribbean plate in Guatemala differs from neighboring continental blocks. Ligorría and Molina (1997) defined a depth to Moho of ~46 km, as compared to a crustal thickness of 37 km obtained for the Tehuantepec Isthmus (Ligorría and Ponce, 1993). This regional difference was also identified by Couch and Woodcock (1981) from gravity data. Carr (1984), also through gravimetric data, estimated a thinner crust (~35 km) for the neighboring continental segment to the southeast in El Salvador. The effect of these variations in crustal thickness on the overriding plate motion in the Middle America Trench subduction zone along the Cocos-Caribbean border is difficult to ascertain, mainly because of the unknown influence of the North American–Caribbean plate interactions along the Motagua-Polochic fault system (e.g., Kanamori and Stewart, 1978; Guzmán-Speziale et al., 1989; Ligorría et al., 1996).

SOURCES FOR HISTORICAL SEISMICITY

Reports of historical earthquakes date from as early as 1526, when the conquerors, upon their return to Ciudad Vieja from the conquest of El Salvador, experienced an earthquake of such intensity that they were "unable to remain standing" (Diaz del Castillo, 1575). One of the first catalogs of earthquakes was compiled by Bustillo (1774) in a report on the disastrous earthquake of 1773 that caused widespread damage and prompted the relocation of the capital from Antigua to Guatemala City. Montessus de Ballore (1888) made the first attempt at an exhaustive catalog, which is more complete than earlier works. It contains most of the previously published data as well as data from primary sources; the majority of the information is anecdotal and primarily concerns the largest cities. Catalogs by Diaz (1930), Vassaux (1969), and Martinez (1978) contain data from primary sources only for the ~30 to 50 yr prior to their publication and rely extensively on data, sometimes erroneous, from earlier catalogs. In contrast to those catalogs, Larde (1960) contains considerable primary information and details the damage to outlying towns, but only for relatively recent events in El Salvador. The most noteworthy compilation of published secondary and tertiary sources is that by Grases (1974), which contains the complete quotations in the original language of publication from all of the catalogs mentioned above, as well as accounts of earthquakes from a few other sources. Recently, the works by Ambraseys (1995) and Ambraseys and Adams (1996, 2001) reviewed macroseismic and instrumental information for the twentieth century. The latter work overlaps with our study and we use it as a calibration for our magnitude estimations of pre-instrumental (older than one century) earthquakes.

An archival search was conducted for primary accounts of historical earthquakes in El Salvador, Guatemala, and Chiapas,

Mexico, which was part of Guatemala throughout the colonial era. More than 1000 documents from the colonial era, which we estimate accounts for at least 80% of the relevant primary documents still existing, were located and inspected. The vast majority of these documents are petitions from local priests and mayors asking for tax relief or reconstruction funds. After 1830 most of the data are from newspaper files and reports of the national observatories of Guatemala and El Salvador. There may yet remain some documents unseen by us (1) at the Archivo General de Centro America (Guatemala City) concerning earthquakes in Chiapas, Mexico, and (2) at the National Archives of Honduras concerning earthquakes in El Salvador. Pertinent documents may also exist at town halls and/or churches of a few individual towns. Table 1a summarizes the searched archives that were found to contain relevant documents, and Table 1b enumerates those archives searched that contain limited information for our purposes.

Antigua, Guatemala, was the capital of colonial Central America, which included Chiapas, Mexico, during most of the colonial era. The Spanish first arrived near Antigua in 1524 and established it as the center of power in 1526. Although Antigua contained churches and monasteries of almost every order, the influence of the Catholic Church, as the primary westernizing influence, was somewhat slow to spread throughout Central America. The likelihood that a given town would file an earthquake damage report with the King depended primarily on its size and its distance from Antigua, and much later, from San Salvador.

Reporting sources for earthquake damage were essentially limited to those towns with more than 1000 taxpayers. Figure 2 shows the geographical distribution of source towns in this region during three historical periods—1575, 1700, and 1830 (Feldman, 1993). For the sixteenth and seventeenth centuries we were not able to find information from a sufficient distribution of towns to produce reliable intensity maps. By 1690 the number of taxpayers in most towns had more than doubled and many additional towns supported at least 1000 taxpayers, enough that we could produce fairly reliable intensity maps, though during the early decades of that period the sizes of damage areas may be somewhat underestimated. By 1830 many towns of Guatemala and El Salvador contained more than 5000 eligible taxpayers, yet the coastal plain of both Guatemala and southeastern Chiapas remained little influenced by the Spanish. In 1830 Central America broke up into individual countries essentially as they exist today, and Chiapas had become part of Mexico. At this time national "gazettes" began to appear, and with them, better reports of newsworthy earthquake damage. By 1840, both Guatemala and El Salvador had informal geophysical observatories, and by the 1880s, were producing intermittent formal bulletins. The catalogs of Larde (1960), Vassaux (1969), and Martinez (1978) were produced by workers at these observatories.

ASSIGNMENT OF INTENSITIES

Intensities were assigned according to the Modified Mercalli (MM) scale of Richter (1958). We use this scale primarily

TABLE 1: CONSULTED SOURCES OF INFORMATION

Name of Archive	Location
a) Archives with significant relevant information	
Archivo General de Centro America	Guatemala City, Guatemala
Archivo Eclesiastico de Guatemala	Guatemala City, Guatemala
Archivo General de las Indias (AGI)	Seville, Spain
Cathedral Archives of San Cristobal las Casas	Chiapas, Mexico
Chiapas State Archives at Tuxtla Gutierrez	Chiapas, Mexico
National Archives of El Salvador	San Salvador, El Salvador
Bancroft Library	University of California at Berkeley
b) Archives with little relevant information	
Hacienda de Sevilla	Seville, Spain
Biblioteca Universitaria	Seville, Spain
Biblioteca de Circulo de Amistad	Cordoba, Spain
Biblioteca Publica Provincial de Cordoba	Cordoba, Spain
Biblioteca Publica de Granada	Granada, Spain
Biblioteca General del Universitario	Granada, Spain
Archivo de la Real Cancilleria	Granada, Spain
Archivo Historico Nacional	Madrid, Spain
Archivo del Servicio Historico Militar	Madrid, Spain
Archivo General de Simancas	Valladolid, Spain
Vatican files	St. Louis University, Missouri
U.S. National Archives	Washington, D.C.

because Brazee (1979) has evaluated the significance of each element of the scale, and we avoid those elements with significance <0.4. The great majority of the data concerns damage to churches, the construction of which is probably more homogeneous than that of private dwellings. We have, however, assigned both private dwellings and churches to the Masonry D category of the Modified Mercalli scale classification.

Some documents state the exact nature of the damage to various parts of a church, and in these cases assignment of intensity is simple. In most cases, however, the documents are not very specific as to the details of the damage to a church or other structure, but merely use a qualifying adjective. We have compared some of the qualifying adjectives used with the specific details of the account when available, and estimate the intensity that corresponds to the qualifying adjective (see Table 2).

HISTORICAL SEISMICITY CATALOG

Many of the damaging seismic events described in historical documents are local in their effects and originated on surface faults,

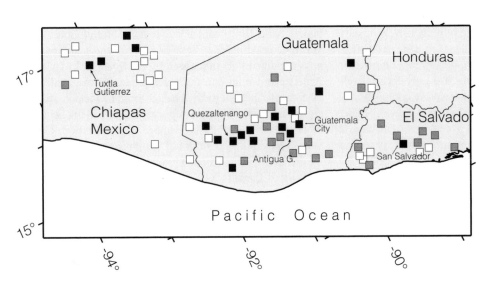

Figure 2. Spread of Spanish influence with time. The black, gray, and white square symbols represent towns that governed more than 1000 taxpayers in 1575, 1700, and 1830, respectively. Note the scarcity of significant towns in the Mexico portion of the study area.

TABLE 2. COMMON PHRASES ENCOUNTERED IN
CONSULTED DOCUMENTS AND INTENSITY ASSIGNMENT

Spanish Phrase	English Translation	Modified Mercalli Intensity
población asustada	town frightened	IV to VI
rajada, rajadura	cracked, cracks	VI ±
maltratada	badly affected	VI to VII
dañado, arruinado	damaged, ruined	VII ±
inútiles, muy maltratada	unusable, very badly affected	VII ±
casi enteramente arruinada	almost entirely ruined	VII to VIII
enteramente arruinada	entirely ruined	VIII
destruida, se cayó	destroyed, collapsed	VIII to IX
completamente en el suelo	total collapse	IX

predominantly along the volcanic chain. White and Harlow (1993) studied 47 of the largest such events during the twentieth century and found that they are confined to the upper crust. Upper-crustal events along the volcanic chain have surface-wave magnitudes of M_S 6.5 or less except in southeastern Guatemala where magnitudes have reached M_S 6.9. Damage from most upper-crustal events and from most large subduction zone events is centered along the volcanic axis, and the peak MM intensity produced by both sources is also similar (VII to IX). We can, however, distinguish damage produced by M ≥ 7.4 subduction zone events from damage produced by upper-crustal events along the volcanic axis by simply comparing the areas within the MM intensity VII contour (A_{VII}): For upper-crustal events, $A_{VII} < 600$ km², whereas for $M_W ≥ 7.4$ subduction zone events, $A_{VII} > 10,000$ km². Therefore to remove possible upper-crustal events, any events having $A_{VII} < 10,000$ km² and centered along the volcanic axis were excluded.

Downdip intraplate events apparently occurred on February 4, 1921 (M 7.2), March 28, 1921 (M 7.4), February 8, 1926 (M 7.1), May 21, 1932 (M 7.1), and January 26, 1947 (M 6.7), based on calculated depths (Ambraseys and Adams, 1996). Only the 1932 earthquake caused damage as serious as MM VIII. Two more recent events are well constrained to be downdip intraplate, those of June 19, 1982, and September 10, 1993, by Centroid-Moment Tensor solutions (Harvard CMT Solution Catalog). We use these two events as calibration events in an attempt at discriminating events within the subducting slab.

The seismogenic thickness of the young down-going slab is small, ~15 km at most, and the radius of curvature is relatively tight in this region (Burbach et. al., 1984). The tight curvature may cause the relatively frequent down-going-slab events, while the small seismogenic thickness may restrict the peak magnitude. The effect of generally greater depth for events within the down-going slab apparently helps restrict the damage produced. Down-going-slab normal-faulting events as large and shallow as the January 2001 El Salvador earthquake (Bommer et al., 2002) would certainly not have escaped our detection, however, so we believe our catalog may be complete for both M ≥ 7.4 thrust events and for M ≥ 7.4 down-going-slab normal-faulting events shallower than ~60 km since 1690.

The catalog of significant historical earthquakes along the Middle America subduction zone from El Salvador to Chiapas, Mexico, is presented in Tables 3 to 5. Table 3 summarizes all known earthquakes from 1526 to 1710, from all sources. The table presents the event date, estimated peak MM intensity at a few locations, a brief description of known damage, and, if known, the felt duration. Where original data are vague, we tried to preserve the vagueness in the table. Table 4 presents only those earthquakes from 1710 to 1900 that are judged to have originated along the Middle America subduction zone and that produced damage of MM ≥ VII. Table 4 presents the same elements as Table 3, plus estimated coordinates of the center of maximum intensity (which may approximate the location of the moment centroid), our estimate of the surface-wave magnitude, here designated M_I, based on intensities (discussed below), and the approximate along-arc lengths of the MM VII and MM VIII contours in km (i.e., the major axes measured parallel to the trench, designated L_{VII} and L_{VIII} respectively).

We attempt to scale the macroseismic magnitude of the historical earthquakes to surface-wave magnitude, which is a reasonably good estimate of moment magnitude within the magnitude range of the largest historical events in this region, i.e., from M 7.4 to M 8.2. We believe this is a good approximation for the great majority of events presented in this paper, owing to the fact that the events are likely of shallow origin (depths <60 km) based on the large areas of MM VII and VIII for most of them. The macroseismic magnitude, M_I, is estimated using chi-square (X^2) regressions, which take into account the errors of intensity contours (Press et al., 1996) (see Fig. 3):

$$M_I = 0.0024 L_{VII} + 7.06 \qquad (1)$$

and/or

$$M_I = 0.008 L_{VIII} + 7.02 \qquad (2)$$

Errors in M_I may vary by ±0.4 for $M_I ≥ 7.4$. The effect of depth is greater for smaller events so that errors in M_I may vary a bit more for values of $M_I < 7.4$.

TABLE 3: EARTHQUAKES ALONG CHIAPAS, MEXICO, TO EL SALVADOR (ALL SOURCES) FROM 1525 TO 1710

Year	mo/day	Modified Mercalli Intensity: Locations	Description [duration of felt shaking in brackets]
1526	7/20	VIII: Antigua; VII?: V. Fuego (G), Santa Ana (S), San Salvador (S); VI–?: Tecpan (G)	(or possibly 8/15?) earthquake knocked down soldiers; affected; affected; apparently unaffected ["long" duration]
1530	5/29	Int?: Ciudad Vieja, Quezaltenango, San Marcos (G)	great earthquakes
1533	early	Int?: Guatemala (G), Tehuantepec (M)	it shook often; events stronger in Guatemala than Tehuantepec, Mexico
1556		VI–VII: San Salvador (S)	minor damage from several earthquakes
1565	8/30	VII+:Escuintla, Chimaltenango, VIII: Antigua, Comalapa, Almolonga, Solola (G), las Casas (M)	there was a great earthquake; many houses fell; extaordinary violence [duration: "3 psalms" = ±90? s]
1566	5	VIII: Quezaltenango, Taxisco (G), Sonsonate (S)	houses collapsed; aftershocks for 9 days
1575		VIII: Chiapas (M) to Nicaragua, Antigua (G)	great destruction in all of Chiapas province, Mexico, to that of Nicaragua
1575	6/02	VIII: San Salvador, Sierra Texacuangos towns (S)	(5/23?) 1st major earthquake here; all houses fell
1576		VII: Antigua (G)	earthquakes; considerable damage
1577	11/29	VII: Antigua, San Marcos, Solola (G)	earthquake at midnight destroyed many buildings [duration "3 hours"]
1581		VIII: San Salvador (S)	total destruction; church ruined
1585	1/16	VII+: Antigua, Solola (G)	began 2 years of earthquakes less than 9 days apart; many dead
1586	12/05	VII: Antigua; VI: Solola (G)	earthquake destroyed some houses
1586	12/23	VII–VIII: Antigua (G)	church buildings and ~60 poorer houses ruined
1591	3/14	VIII: Las Casas (M)	cathedral tower and main chapel destroyed
1594	4/21	VII–VIII: San Salvador (S)	13 dead; 1 great shake for an instant [duration: "an instant"] (upper crustal?)
1607	10/09	VII: Antigua (G)	buildings cracked/ruined; 29 dead; aftershocks for 4–6 months
1631	2/18	VII: Antigua; VI+: Mixco (G)	hospital damaged; earthquakes (aftershocks?) for 9 days
1650		VIII: San Salvador (S)	great earthquakes; church destroyed
1651	2/18	VII–VIII?: Antigua, Amatitlan (G)	all buildings badly damaged; aftershocks to mid-April
1658 (or 1656?)		VIII: San Salvador and Sonsonate provs. (S)	destroyed buildings; worst in 70 years [duration: "1 creed" = ~30 sec]
1659		V?: Antigua (G)	earthquakes alarmed population
1663	5/01	VII–?: Antigua (G)	partial ruin; series of earthquakes, worst on 5/01
1666		VII–?: Antigua (G)	partial ruin
1671	8/16	VIII: San Salvador (S)	all buildings fell/unusable; greatest earthquake here so far
1676±		VI–VII?: San Andreas Itzapa, Antigua (G)	great earthquake (no damage mentioned)
1679	3/04	VII+: Comalapa; VI+: Antigua (G)	church very badly damaged by earthquakes
1681	7/22	MM?: Antigua (G)	series of earthquakes, worst on 7/22
1683	5	MM?: Antigua (G)	series of earthquakes
1684	8	MM?: Antigua (G)	series of earthquakes
1687±		VII?: las Casas (M)	damage
1687	9/10	MM?: Antigua (G)	earthquakes
1688	10/10	VI?: Mexico, Guatemala, (VII+: Chiapas?)	earthquake felt
1689	2/12	VIII: San Pedro las Huertas, Antigua (G)	all major buildings destroyed
1689	2/23	VIII: Antigua (G)	stronger than event of 2/12/1689
1691		VII?: Antigua (G)	Damage
1693		MM?: San Miguel (S)	earthquake and thundering sounds
1702	8/14	VII+: Ahuachapan (S)	church ruined
1707		MM?: San Salvador (S)	possible earthquake

Note: "±" following a year is uncertain within a few years. The month and day are shown if known. "+" or "–" following a value indicates the value may be one unit greater or less than stated, respectively. MM?—intensity unknown. Locations: (G)—Guatemala, (M)—Mexico, (S)—El Salvador.

TABLE 4. SUBDUCTION EARTHQUAKES ALONG CHIAPAS, MEXICO TO EL SALVADOR, FROM 1710 TO 1900

Year	mo/day	Modified Mercalli Intensity: Locations	Description [duration of felt shaking]	Macroseismic Information			
				Lat (N)–Long(W)	M	L_{VII} (km)	L_{VIII} (km)
1711		VII: Comasagua (S)	continuous quakes; church about to fall	13°40–89°25	7.0 ± 0.3	>5	0?
1712	12/14	VIII: Costuma, San Pedro Nonualco to San Vicente (S)	churches destroyed, aftershocks for several days	13°35–88°50	7.1 ± 0.2	>45?	23
1717	9/29	VIII+: Patzicia to Guazacapan; VII–: Mazatenango to Jilotepeque to Palin (G)	3000+ houses, all churches ruined or destroyed; didn't produce ruin here; aftershocks to 10/31	14°30–90°45	7.4 ± 0.2	85– / 140+	>3
1717	10/03	VI–VII: Amatitlan, Guazacapan, Itzapa (G)	some additional damage (aftershock?)	14°20–90°30	7.0 ± 0.2	20	0
1719	03/05	VIII: Ystepeque, Zacatecoluca, Cojutepeque, San Miguel Perulapan (S); VII+: San Salvador, San Vincente, San Miguel, Santa Ana, Apastepeque (S)	churches destroyed; 150+ earthquakes (aftershocks?); churches destroyed; many dead left in miserable state	13°40–89°00	7.6 ± 0.2	>100	>30
1720		VII: Siquinala, Antigua?, Santa Apalonia (G)	earthquake caused new damage	14°40–91°00	7.0 ± 0.2	20	0
1733	5	VII+: Patzicia; VII––: Alotenango, San Andres (G)	church ruined; churches suffered	14°40–91°05	7.0 ± 0.2	20	0?
1736	05/06	VIII: Tecoluca; VII: Lago .Ilopango to Soyapango (S)	churches destroyed; churches ruined	13°35–88°55	7.2 ± 0.1	~60	>1
1742 (or 1743?)		VIII: Taculula (G)	church collapsed from earthquake	14°00–90°05	7.1 ± 0.2	>1	>1
1742		VIII––: Comasagua, Ilopango, Tonacatepeque (S); VII: Zacatecoluca, Ostuma (S)	churches on the verge of collapse; churches ruined	13°35–89°05	7.2 ± 0.1	~65	>1
1742	08/10	VIII––: San Antonio Suchitepequez, Cuyotengo, Villa Seca, San Francisco Zapotitlan, Santiago Sambo (G)	churches on the verge of collapse; churches ruined	14°30–91°30	7.1 ± 0.2	25	>1
1743	05/30	VIII: Tuxtla, las Casas, Xitala, Tapala (M); VII?: Nuestra Senora Del Rosario (M)	churches collapsed, on the verge of collapse; church damaged (on day of San Fernando)	16°45–92°45	7.4–8.2	>140	>60
1747	10/13	VIII: Cuyotenango: VII+: Mazatenango, San Bernadino, Quezaltenango, Retalhuleu (G)	church demolished; church ruined; churches damaged	14°40–91°30	7.1 ± 0.2	>35	>1
1748	03/13	VIII: Olocuitla (S); VII+: Aculhuaca, Ilobasco, San Miguel Perulapan, Cojutepeque, San Miguel Obispo (S)	church destroyed; churches ruined	13°40–89°05	7.1 ± 0.2	>35	>1
1751	03/04	VIII: Compala, Antigua (G); VII+: las Huertas, San Francisco de Jesus, Zaragoza, Guazacapan (G)	chapel collapsed/churches ruined [duration 1 min] most buildings fell; numerous aftershocks	14°40–90°50	7.3 ± 0.1	75–130	>35
1757	10/04	VII––: Alotenango, Antigua (G)	churches damaged; less serious than that of 1751	14°30–90°45	6.8 ± 0.3	10?	0
1764		Int?: Quezaltenango (G)	possible earthquakes (shallow?)	14°50–91°50	6.5–7.1	?	
1765	10/24	VIII+: Ostuncalco to Tacana (G); VII+: San Marcos (G); VII: Chiapas (M) to Quezaltenango (G)	buildings fell; aftershocks to 11/01; churches badly damaged or ruined [duration 7–8 min]	15°00–91°55	7.6–8.2	140–200	>65
1772	06/15	VII+: S Antonio Suchitepequez, Zapotitlan, Mazatenango (G)	churches ruined; all houses damaged	14°35–91°30	7.0 ± 0.1	18	0
1773	06/11	VII: Antigua (G)	at 0530: damage; at 1700: worse damage, many aftershocks				0?
1773	07/29	VIII+: Antigua to L. Atitlan (G); VII: Ostuncalco, Mazatenango to Escuintla and Guatemala City area (G)	5–600 died immediately; 600+ died later; destruction in first 2 sec; violent foreshock; all churches within this area ruined [duration 1 min]	14°40–91°10	7.5 ± 0.1	125–140	67–90
1773	09/07	VII?: Antigua (G)	substantial additional damage	14°30–90°45	6.8 ± 0.3	5?	0
1773	12/13	VII+?: Antigua (G); VII?: Mazatenango (G)	3 great earthquakes like on 7/29; aftershocks to 1/04 [long]	14°35–91°30	7.1 ± 0.2	30–60	0?

(continued)

TABLE 4. SUBDUCTION EARTHQUAKES ALONG CHIAPAS, MEXICO TO EL SALVADOR, FROM 1710 TO 1900 (continued)

Year	mo/day	Modified Mercalli Intensity: Locations	Description [duration of felt shaking]	Macroseismic Information			
				N Lat–W Lon	M_l	L_{VII} (km)	L_{VIII} (km)
1776	03/30?	VII–VIII: Cuilapa, Chiquimulilla (G); VII+?: Zinacantan to Moyuta (G)	great earthquake; churches very badly damaged; churches very badly damaged	14°15–90°15	7.2 ± 0.1	>70	>1
1776	05/30	VII–VIII: Santa Ana and Ahuachapan provinces (S) to western La Paz and southwestern Cuscatlan provinces (S)	All churches ruined, mostly collapsed; churches totally ruined	13°50–89°50	7.9 ± 0.2	>160	36–80
1776	07/06	VI–VII: San Salvador (S)	more damage to churches and houses	13°35–89°00	6.8 ± 0.3	5?	0
1776	11/15	VII+?: San Salvador (S)	finished the ruin; more collapse	13°35–89°00	7.0 ± 0.3	>10	0?
1804		VII: las Casas (M)	cathedral damaged (upper crustal?)	16°30–92°40	6.8 ± 0.3	>5	0
1829	12/17	VII+: Chiquimulilla (G)	church threatening ruin (upper crustal?)	14°10–90°15	7.1 ± 0.1	>5	0?
1831	02/07	VII: Jalpatagua (G), San Salvador, Comasagua, Jayaque, Armenia, Cacaluta, Izalco (S)	churches left in bad state; suffered notable damage	13°50–89°35	7.1 ± 0.2	55–95	0?
1853	02/09	VI–VII: Quezaltenango, Cantel, Zunil (G)	several buildings damaged [duration "very prolonged"]	14°50–91°30	6.8 ± 0.3	10?	0
1859	08/25	VII?: La Union (S), Amapala, La Brea, San Diego (H)	damage, tsunami, aftershocks to 9/03	13°15–87°45	7.1 ± 0.1	>35	0?
1859	12/08	VIII: Ahuachapan and Santa Ana provinces (S); VII: San Salvador, Libertad, Sonsonate (S)	destruction/major damage; tsunami [duration 1 min] major buildings badly damaged; aftershocks to 12/15	13°50–89°00	7.3 ± 0.3	>90	>26
1860	01/19	VIII: San Jose Obispo; VI+: Escuintla; V: Guatemala City	single earthquake; church fell; houses damaged [1 min]	14°40–90°45	7.1 ± 0.1	20	>1
1860	12/03	VIII: Talnique to Cuscatancingo (S); VII: Opico, Ateos to Santo Tomas (S)	destruction; aftershocks to 12/11 [1 min]; some damage to major buildings	13°45–89°15	7.1 ± 0.1	40	21
1861	08/23	VII+?: Jalpatagua; VII: Conguaco (S)	continuous earthquakes: buildings heavily damaged. ["long"]	14°05–90°00	6.7 ± 0.4	10	0?
1862	04/01	VII?: Canales hills; VI: Guatemala City (G)	many earthquakes since February, continuous through December 1862	14°35–90°30	5.5–6.8	>1?	0
1862	12/19	VIII: Santa Ana, Ahuachapan, Sonsonate, La Libertad (S); VII: Escuintla, Chimaltenango (G) to Ahuachapan (S)	especially great to great damage [2+ min]; considerable damage, aftershocks continuous through December 31, 1862, plus January 4, 8, 10, 15, 20, and 24, 1863	13°50–89°35	8.1 ± 0.1	250–285	135
1867	06/30	VII: San Salvador and Cuscatlan provinces (S)	damage	13°40–89°05	7.1 ± 0.1	50	0?
1870	06/12	VII: Chiquimulilla;	completely destroyed; damaged; earthquakes began 4/12, continued past 7/15 [15 sec]	14°10–90°10	7.2 ± 0.1	83	>1?
1873	03/19	VII: Santa Rosa, Jutiapa (G), western Sonsonate (S); VIII: 10 km radius around San Salvador (S);	total destruction, some killed, damage; increasing foreshocks since 2/22	13°40–89°10	7.1 ± 0.1	35	15
1874	09/03	VII: 10 to 17 km around Lake Ilopango (S); VIII: Patzicia, Chimaltenango, to Itzapa (G);	destruction; earthquakes from 8/15/1874 to 1/16/1875;	14°40–90°50	7.1 ± 0.1	60	25
1891	09/09	VII: Tecpan, Escuintla, Amtitlan (G); VII: Analquito to Comasagua (S)	some buildings damaged [25–30 secs] semi-ruin	13°37–89°10	7.1 ± 0.1	50	0

Note: Locations: (G)—Guatemala, (M)—Mexico, (S)—El Salvador, (H)—Honduras. "+" following a value indicates the value may be somewhat greater than stated. M_l is our estimate of the surface wave magnitude calculated from L_{VII} and L_{VIII} using equations 1 and/or 2. L_{VII} and L_{VIII} are the major axes of the intensity VII and VIII contours respectively.

TABLE 5: SUBDUCTION EARTHQUAKES OF MS ≥ 7 ALONG CHIAPAS, MEXICO TO EL SALVADOR FROM 1900 TO 2000

Year	mo/day	Modified Mercalli Intensity: Locations	Macroseismic N Lat–W Long	Description [duration of felt shaking]	M_s	Instrumental N Lat–W Long–Depth (km)
1902	01/18	VII: Costa Cuca, San Francisco Zapotitlan (G)	14°40–91°40	partly destroyed; continuous earthquakes to 4/18	(6.4–7.0)	none
1902	02/26	III–?: Acajutla (S), Puerto San Jose (G)	N.D.	tsunami run-up height 2 meters, no felt reports	M_t 8.0	none
1902	04/18	VIII: San Marcos, to Quezaltenango and western Chimaltenango provinces (G); VII: Soconusco (M) and from southern Huehuetenango to Escuintla province (G)	14°50–91°55	all nearby churches collapsed or destroyed [1–2 min]; churches damaged; 8–900 dead; aftershocks frequent to 5/5	7.5 (7.9)	14.9°–91.5°–25
1902	09/23	VIII: central Chiapas (M)	16°35–92°35	great damage (map from Figueroa) [65 s]	7.6 (8.2)	16.5°–92.5°–25
1903	09/23	VII+: Soconusco (M)	15°00–93°00	Damage (map from Nishenko & Singh) [54–90]	7.6 (8.1)	15.0°–93.0°–25
1915	09/06	VIII–: northern Soconusco (S); VII: western La Paz and San Vincente provinces (S) to eastern Escuintla (G)	13°45–89°40	houses fell; some walls fell; 12 aftershocks at about one half hour intervals [1+ min]	7.7 (M7.9) 7.75 PAS	13.9°–89.6°–60
1915	10/15	VII+: Ahuachapan, Ataco, Apaneca (S)	13°55–89°50	ruined (aftershock?)	(6.4–7.0)	none
1919	04/17	VI–: Quezaltenango (G)	14°50–91°30	a statue fell, light aftershocks	M7.0	14.5°–91.7°–?
1919	06/29	VII: San Miguel, La Union (S); VI: Chinandega(N)	13°30–88°00	damaged; alarmed or affected	6.8DBN	14.5°–86.0°–?
1921	02/04	VII+: Tehuantepec (M)	16°?–95°?	damage; no damage in Guatemala	M7.5	15.0°–95.0°–120
1921	03/28	VI–VII: Gulf of Fonseca to Usulutan (S)	13°10–88°00	some houses damaged or destroyed	7.4 (7.4)	12.9°–88.7°–100
1929	01/24	VI–?: Guatemala, El Salvador	N.D.	no damage reported	7.1 (7.1)	12.8°–91.0°–?
1932	05/21	VIII: Zacatecolucas (S); VII: Santa Tecla to Usulutan (S)	13°30–88°50	complete ruin, 5 killed; some damage	7.1	12.8°–88.0°–150
1935	12/14	VI+: San Marcos to Quezaltenango(G)	14°55–91°40	some walls fell	(6.9 PAS) 7.3 (7.4)	14.8°–91.3°–?
1936	11/19	VI–?: Guatemala, El Salvador	N.D.	no damage reported	(7.4)	13.5°–90.8°–?
1942	08/06	VIII: Chimaltenango, Totonicapan (G) to San Marcos, and Retaluleu (G); VII: Chiapas (M); aftershocks felt in San Marcos to Escuintla provinces (G)	14°45–91°10	most buildings fell [± 1 min]; great damage to houses and buildings, some fell 18 felt aftershocks through 8/20, plus a few to 9/28	7.9(7.9) (7.7 mb)	14.8°–91.3°–40
1944	06/28	VII?: Soconusco (M)	N.D.	no damage reported in Guatemala	7.2	14.3°–91.8°–?
1947	01/26	VIII: La Union, V: Conchagua (S)	13°20–87°50	towns in rubble, buildings fell	6.7 (M7.2)	12.5°–86.3°–170
1950	10/23	VIII: San Marcos, La Esperanza (G); VII: Solola (G) to Mexico border	14°55–91°40	80% of houses seriously damaged [70 s]; buildings damaged; 12 aftershocks to 10/28 plus one on 11/05	7.3 (7.2)	14.3°–91.8°–100
1953	11/17	V: Huehuetenango; IV: San Marcos to Chimaltenango (G)	14°00–92°00	felt strongly; reported felt	6.9 (6.9)	13.6°–92.2°–56
1970	04/29	VII: Soconusco (M) along Guatemala border	15°00–92°20	Damage (map from Figueroa)	7.3 (7.1)	14.6°–92.6°–42
1982	06/19	VII: southern La Libertad province (S)	13°35–89°22	few homes fell; 40 dead; 8 aftershocks to 7/06 [25 s]	7.3(7.0 mb)	13.3°–89.4°–80
1993	09/10	VII: near coast of Chiapas (M)	15°00–92°30	Damage in parts of Mexico and southwestern Guatemala	7.3	14.7°–92.7°–30

Note: "+" or "–" following a value indicates the value may be one unit greater or less than stated. Locations: (G)—Guatemala, (M)—Mexico, (S) —El Salvador, (N)—Nicaragua. N.D. indicates insufficient intensity data to estimate location. M_s is instrumentally determined surface wave magnitude, taken from Ambraseys and Adams (1996). M_s values in parenthesis are from Abe (1981), unless m_b—body wave magnitude or M_t—tsunami magnitude, following Abe (1979) is specified, or, if preceded by M, indicates a "unified magnitude" from Gutenberg and Richter (1954) and/or followed a three letter station code. Instrumental epicenter locations are from Ambraseys and Adams (1996) and depths are either from USGS-PDE hypocenter files or from Gutenberg and Richter (1954).

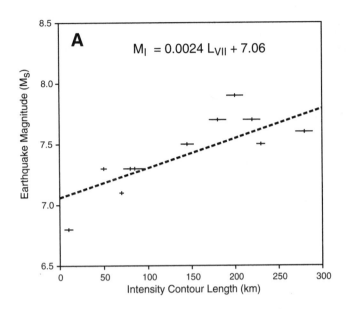

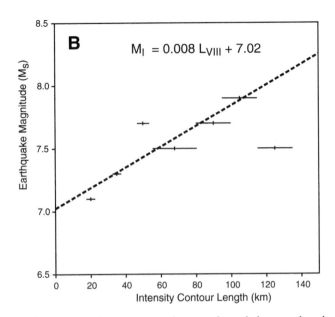

Figure 3. Magnitude versus along-arc length of Modified Mercalli intensity (MM) isoseismal contours. Maximum estimated along-arc lengths MM VII and VIII are shown versus instrumentally calculated magnitude. Where more than one value has been published for magnitudes, the values are connected by a straight line. Where values are less certain, the range is shown as a dashed line.

Table 5 presents earthquakes with $M_S \geq 7.0$ that occurred from 1900 to 2000 along the Middle America subduction zone. Elements of Table 5 are the same as for Table 4 except that instrumental surface-wave magnitude (M_S), epicenter, and depth (if known) are listed instead of M_I, L_{VII}, and L_{VIII}. A comprehensive listing of the damage descriptions from which intensities were taken and their sources may be found in Feldman (1993) and Peraldo and Montero (1999).

Approximate MM intensity VII and VIII contours for the 27 events with largest A_{VII} since 1690 are shown in Figures 4 and 5. The method used to draw these contours was to simply encircle values of a particular intensity, but where two adjacent sites differ by one intensity level, to draw the contour halfway between the sites. Similarly, where major towns or cities did not report major damage but are adjacent to sites that reported MM intensity VII, we drew the contour halfway between the sites. Recall that prior to 1800 the damage reports are less complete, and therefore the actual areas of MM intensity VII and VIII may be somewhat underestimated. Eight additional events, for which $A_{VII} < 600$ km^2, are shown as stars. These latter events are judged to have originated along the subduction zone, and not from the upper crust along the volcanic axis, because the events produced low peak intensities and have a slow attenuation of intensity with distance. Instrumentally located events similar to the latter have magnitudes up to 7.2 and have occurred mostly within the down-going slab. We estimate that this catalog is complete since 1690 for all interplate events of $M_S \geq 7.4$ and for down-going slab intraplate events of $M_S \geq 7.4$ shallower than 60 km, from central Chiapas to central El Salvador. A brief discussion of the largest ($M \geq 7.4$) events since 1717 is presented in the Appendix.

RUPTURE AREAS AND RELATIONSHIP TO TECTONIC FEATURES

The quality of teleseismic earthquake locations depends critically on the quantity and distribution of stations reporting data. Because larger events are more widely recorded than small events, and because the number and quality of stations has

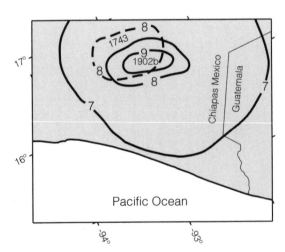

Figure 4. Comparison of 1743 and 1902 central Chiapas earthquakes. The MM intensity VIII contour of the 1743 events is similar in size and location to that of the September 23, 1902, event (M_S 7.6). For the 1743 earthquake, too few reports were found to construct a reliable MM VII intensity contour. Contours for 1902 are slightly modified from Figueroa (1973).

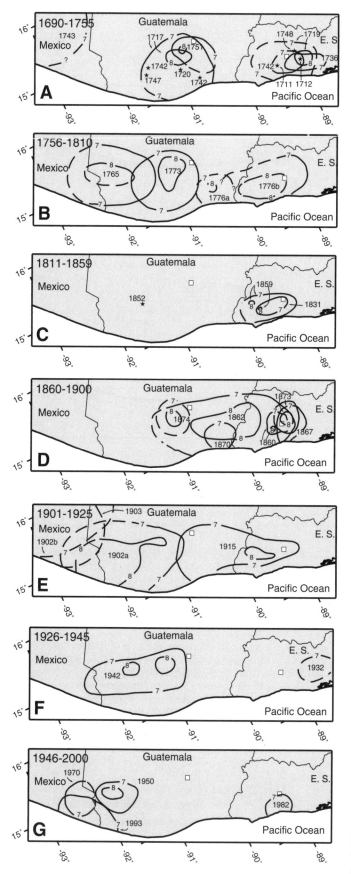

Figure 5. Areas of damage for earthquakes since 1690. Seven consecutive (A–G) though unequal time periods are shown. Contours are solid lines where known and dashed where inferred, enclosing areas of both MM intensity VII and VIII (where known) for all subduction zone earthquakes that caused damage over more than 600 km². Stars indicate isolated, usually minor, damage from earthquakes that were probably either smaller or deeper. The cities of Guatemala and San Salvador are shown as open squares.

gradually improved since the turn of the century, locations are generally better for larger events and improve with time since 1900. The largest instrumentally located events in the study area, which occurred during the first half of the twentieth century, have magnitudes no larger than about M 8, with aftershocks no larger than about M 6. The location errors for these aftershocks are therefore large with respect to the size of the rupture zones. A rupture zone estimated using a hundred or more locally recorded aftershocks (as for the 1982 event discussed above) is likely to be much more accurate than one approximated using but a few aftershocks recorded at teleseismic distances. Therefore, we conclude that MM VII contours provide reasonable approximations of the rupture zones in this region and, especially for older events, are probably much better approximations than are teleseismically determined aftershock zones.

To estimate the rupture areas of great earthquakes of the twentieth century, most workers have assumed that the rupture area corresponds to the aftershock area of the earthquake. Kelleher et al. (1973) used this technique to estimate the rupture areas of large earthquakes within our study area since 1930. In order to extend the data set to earlier centuries, we have used intensity data to estimate the location and size of rupture areas. In Figure 6, we compare the MM intensity VII contours with aftershock zones for four recent earthquakes in Guatemala and El Salvador. Figure 6A compares the aftershock zone for the intraplate earthquake of October 23, 1950 (M$_S$ 7.2), as delineated by Kelleher et al. (1973) from teleseismic data, with similar data from the June 19, 1982, and the September 10, 1993, intraplate earthquakes (Table 5). It can be seen that, for the 1950 event, the VII contour is considerably smaller than the aftershock zone and lies almost entirely within it, whereas for each of the two more recent events, which have more reliable aftershock area estimations, there is no overlap between the VII contour and the aftershock zone.

In the case of the 1942 earthquake, shown in Figure 6B, there is almost no overlap of the aftershock zone with the VII contour. Since a good location of the rupture area of the 1942 earthquake is necessary in order to estimate the distribution of current slip deficit along the Middle America Trench, we examined records of the main shock from a three-component Wiechert seismograph located in Guatemala City. The first motion from the horizontal seismograms shows compression on the east-west component and dilatation of the north-south component. This indicates a mainshock source to the west-northwest of Guatemala City, as shown in

Figure 6B, in agreement with the location of damage and with the estimated location of Ambraseys and Adams (1996) and not with a source 100 km to the southwest as determined by the International Seismological Center (ISC) and Kelleher et al. (1973). In addition, the first motions and S-P interval for an aftershock that occurred on August 8 are consistent with an epicenter ~160 km WNW of Guatemala City, where the town of Tectitlan (only) reported damage from the event, whereas Kelleher et al. (1973) located this aftershock ~120 km southwest of Guatemala City. We believe this demonstrates that macroseismically determined earthquake locations are at least as accurate and are often much more accurate than instrumentally determined locations for events prior to the expansion of the worldwide seismic network in the early 1960s.

Examination of Figure 5 shows that the MM VII contours for the earthquakes of 1765, 1773, and 1776 nearly abut but slightly overlap. Likewise, the MM VII contours for earthquakes in 1859, 1870, 1874, 1902, and 1903 seem to approximately abut, as do the 1915 and 1942 earthquakes, with slight overlap. This may be evidence that these events each represent thrust-type ruptures and would suggest that where data are most complete, the MM VII contour slightly overestimates the rupture zone. It might also be evidence that the catalog is complete for thrust-type earthquakes of $M_s \geq 7.4$. And if the catalog is complete for such large thrust earthquakes, this suggests it may be complete for earthquakes of the same size ($M \geq 7.4$) within the down-going slab in the same depth range, i.e., above ~60 km.

Figure 7B shows the rupture lengths along arc versus time for earthquakes in this catalog. Although data are too sparse to prepare intensity maps for earthquakes during the period from 1526 to 1700, reports show that several very large earthquakes occurred during this time, and these earthquakes are included on the figure. Rupture lengths are shown as solid lines where the data are good and as dashed lines where rupture length is uncertain.

It has been suggested that structural features transverse to the trench, or subducting tectonic features recognizable in the bathymetry and surface geology, may control rupture lengths of very large earthquakes (e.g., Mogi, 1969; LeFevre and McNally, 1985; McCann, 1985). Figure 1B shows the main bathymetric features in the Middle America Trench (i.e., the smooth seafloor seaward of the trench and the large submarine San Jose Canyon south of Guatemala City). The proposed tectonic segment boundaries of Stoiber and Carr (1973), thought to be tears in the down-going slab, are also shown. Villagrán et al. (1997) also suggested a segmentation boundary south of the Guatemala–El Salvador border. Figure 7B shows rupture lengths of the historical earthquakes versus time, and one can see that the April 1902 and 1915 earthquakes clearly ruptured across the San Jose Canyon and three of the proposed segment boundaries. With the possible exception of the Tehuantepec Ridge, it can be seen that the bathymetric features and proposed segment boundaries are ineffective as barriers to rupture in very large earthquakes. In fact, these results are taken as evidence against any relationship between segmentation of the down-going slab and seismogenic coupling, as suggested by Ligorría et al. (1995).

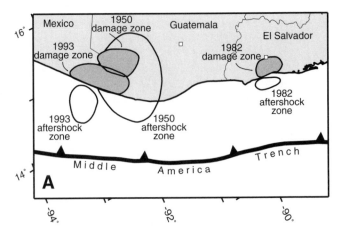

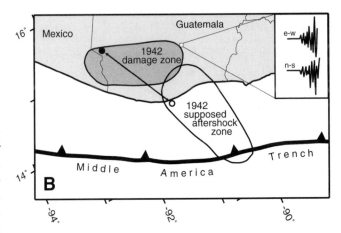

Figure 6. MM intensity VII contour vs. aftershock zones. The dark gray areas represent damage zones and the open areas indicate aftershock zones. A: The June 19, 1982, October 23, 1950, and September 10, 1993, intraplate earthquakes are shown. The aftershock zone for the 1982 event is from Harlow (2002, personal commun.), enclosing epicenters of 100 aftershocks that occurred within the first 48 hours following the earthquake, and located using data from 15 local high-gain stations in El Salvador and Guatemala. Due in part to poor quality of teleseismically located aftershocks, the apparent extent of the aftershock zone of the 1950 event is more than twice the area of the zone of damage. B: The aftershock area encircles epicenters from Kelleher et al. (1973) that pointed out the poor quality of locations of the offshore epicenters. The aftershock area and the zone of damage do not overlap, probably due to poor quality of teleseismically located aftershocks. Arrow connects the supposed epicenter (open circle) of the August 8 aftershock with the location of known damage (solid circle) from the aftershock, ~200 km away. The inset shows the first arrivals recorded at the horizontal components of the Wiechert seismograph located in Guatemala City (open square) at the time of the earthquake.

The seafloor beneath the study area becomes progressively rougher offshore from Chiapas as one approaches the Tehuantepec Ridge. Earthquakes near the Tehuantepec Ridge are less frequent than in the Middle America Trench section off Guatemala and El Salvador, where the sea floor is quite smooth and has no obvious ridges or seamounts that might act as asperities during subduction.

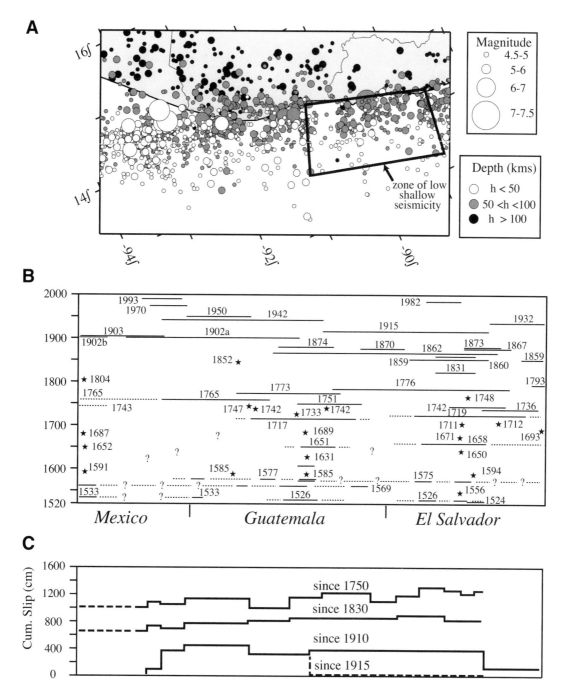

Figure 7. Seismicity since 1963, rupture lengths, and cumulative slip versus distance for the study area. A: Epicenters are shown for earthquakes from 1963 through 2000 with magnitudes $M_S \geq 4.5$ for depths between 11 and 50 km (white circles), 51 and 100 km (gray circles) and deeper than 100 km (black circles). Epicenters are from the USGS-PDE catalog. Note the zone off the coast of eastern Guatemala and western El Salvador that lacks any significant, shallow seismicity of $M \geq 6$ since at least 1963. B: The rupture lengths of the major earthquakes analyzed in this study vs. time. Solid lines represent the rupture lengths as approximated by the along-arc lengths of the MM intensity VII contours in Figures 4 and 5, measured parallel to the trench. Reliable magnitude estimates since 1900 are presented in Table 5. Dashed lines indicate uncertainty in the extent of rupture and/or suspected ruptures during the sixteenth and seventeenth centuries. Note the great earthquake ruptures across the Guatemala–El Salvador border (San Jose Canyon in Figure 1B). Absence of shocks in several areas indicated by question marks before 1700 reflects the poor historical record for those areas. C: Cumulative coseismic slip through 2000 vs. distance along the arc for time intervals since 1750, 1830, and 1910 if all events in B are assumed to have occurred along the megathrust zone. We used 1.5 cm of coseismic slip per km of rupture length to estimate the coseismic slip. If the 1915 earthquake was a thrust event, as we suspect, then the cumulative slip for the southeastern half of the study area since 1915 would be nearly zero, as shown by the dashed line at bottom right. If so, then relative to the northwestern section of the study area, there is a current deficit equal to one M_W 7.75 earthquake along the Guatemala–El Salvador border.

This last argument could suggest that smaller earthquakes should be expected at locations such as the Guatemala–El Salvador section, with a relative lack of topographic complications at the interface (e.g., Okubo and Dietrich, 1984; Scholz, 1990).

RECURRENCE RATES OF MAJOR EARTHQUAKES AND THE CURRENT SEISMIC SLIP DEFICIT

We next calculate the repeat times for damage from very large (M ≥ 7.4) earthquakes for three locations along the subduction zone. We wish to emphasize that, because we cannot distinguish historical earthquake source types, this repeat time is the average interval between damage from an earthquake of any type, whether interplate or intraplate, and damage from the next earthquake of any type. Major earthquake damage occurred in central Chiapas in 1743 and 1902 and probably in 1565 or 1591, giving an average repeat time for major damage of 155–169 ± 10 yr. In southwestern Guatemala, major earthquakes occurred in 1765, 1902, and 1942, and probably also in 1565 and 1651 or 1689, giving an average repeat time for major damage of 94 ± 54 yr. In western El Salvador, major earthquakes occurred in 1719, 1776, 1859, 1862 and 1915, and in 2001 as well as in 1575 and 1658 apparently. Aside from the unusual three-year interval between the 1859 and 1862 earthquake, the average repeat-times are 71 ± 17 yr for western El Salvador. All of the above intervals are much longer than the intervals of 33 to 50 yr suggested by Kelleher et al. (1973), McNally (1981), and Papadimitriou (1993), who all used much shorter data sets than this study.

It seems difficult to explain the apparent regularity of major earthquake damage in western El Salvador every 54–88 yr since at least 1575. Prior to the occurrence of the January 2001 El Salvador earthquake, these authors considered the simple explanation to be that all but the 1862 event originated along the same source fault, the thrust interface. With the occurrence of the 2001 earthquake within the down-going slab (Bommer et al., 2002), that explanation is less satisfactory. It is clear that there exist two sources of major earthquake damage for western El Salvador. It may well be one is a rather stably repeating source, e.g., the thrust interface, and the other is a much more randomly repeating source, e.g., within the down-going slab, but our historical data do not permit any such conclusions. We simply wish to draw the reader's attention to the fact that, although the 2001 earthquake occurred within the typical 54–88 year time window since the last major earthquake in 1915, no major thrust-interface earthquake has yet occurred here for at least 86 yr. If the thrust interface is the stably repeating source of major damage every 54–88 yr, as suggested above, then major damage from this source is apparently pending during the next few decades.

As a means of investigating the current seismic hazard from both inter- and intraplate sources of damage combined, we estimate seismic slip totals since 1743 for the combined sources along the central 75% of the study area where the historical coverage is most complete. Lacking good calibration events from within the study area, we apply the regional approach of Ambra-

seys and Adams (1996), adding data from our historic catalog. Figure 7C shows the cumulative slip versus distance along the trench for three progressively longer time intervals.

Note the large slip deficit after 1915 offshore from central El Salvador to Guatemala City. If this slip deficit is for the thrust interface and still exists, then was the 1915 earthquake such a thrust-interface event? From the relatively small surface-wave amplitudes and from P and S arrival times, Gutenberg and Richter (1954) assigned a focal depth of 80 km. Based on the arrival times of the maximum surface-wave amplitudes and the abundance of strong aftershocks, Ambraseys and Adams (1996) assigned a depth of 60 km. We obtained seismograms for the earthquake from Ottawa and Uppsala and find that the intervals of pP-P and sS-S at Ottawa and sS-S at Uppsala are most compatible with a depth of 60 km. Figure 8 shows the onset of the P wave train at ten stations. Assuming the nodal planes should strike subparallel with the trench axis, the first motions are compatible with a thrust mechanism along a plane dipping at ~55° to the northeast and are incompatible with a normal mechanism.

Thus, we believe that the 1915 earthquake may well have occurred on the thrust interface. As such, it would represent the last major thrust event along the section of subduction zone between San Salvador and Guatemala City. If not, then the last major thrust event in this area occurred at least as long ago as 1859 or 1862. For the thrust interface, the slip deficit since 1915 would be essentially eliminated by an earthquake of the approximate size of the 1915 earthquake, that is, an earthquake of about

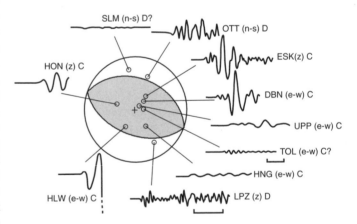

Figure 8. Focal mechanism of the 1915 earthquake. P-wave first motions are shown at ten stations. The arrivals are compatible with thrust along a plane dipping ~55° to the northeast, in accordance with the regional stress regime (Redondo et. al., 1993). The component of the record is indicated in parenthesis for each trace: east-west (e-w), north-south (n-s), and vertical (z). First motions are indicated by C for compression and D for dilatation. SLM—St. Louis, Missouri; OTT—Ottawa, Canada; ESK—Eskdalemuir, Scotland; DBN—DeBilt, Netherlands; UPP—Uppsala, Sweden; TOL—Toledo, Spain; HNG—Hongo, Japan; LPZ—La Paz, Bolivia; HLW—Helwan, Egypt; HON—Honolulu, Hawaii. The traces for HLW and HNG were actually recorded from right to left, but we have reversed them here for display purposes.

magnitude M 7.75 along this section of the subduction zone. This is approximately the same size as prior earthquakes along this section in 1776 (M_I 7.9) and in 1859 (M_I 7.6). If one disputes our assertion that the 1915 event represents the last major rupture of the thrust interface along this section, then the current slip deficit would be nearly double the above estimate.

During the several to 20 years preceding some large shallow earthquakes, seismic quiescence has been observed within the area of rupture (e.g., Kanamori, 1981; Ohtake et al., 1981; Lomnitz, 1994), while seismic activity markedly increases within the region downdip from the impending rupture and the outer rise region (e.g., Mogi, 1973; Dmowska et al., 1996). Figure 7A shows the seismicity of $M_s \geq 4.5$, separated in depth ranges of h < 50 km, 50 < h < 100 km, and h ≥ 100 km, along Guatemala and El Salvador from 1963 through 2000 from the U.S. Geological Survey–Preliminary Determination of Epicenters (USGS-PDE) catalog. Although local networks have operated in both Guatemala and El Salvador during small portions of this time span, those networks have been too intermittent and sparse for use here, whereas the USGS-PDE catalog is believed to be complete for M ≥ 4.5 for this period. Figure 7 shows that moderately shallow (h < 50 km) seismicity has been somewhat lower within the area from San Salvador to eastern Guatemala (within the box), than in the surrounding region. Within this area only a few M ≥ 5.0 earthquakes have occurred since 1963, while many such earthquakes have occurred within the surrounding region, especially since 1979. We also note that since 1979, seismicity has been noticeably greater along the deeper limits of the quiescent zone, at depths of 50 to 100 km. Between 1979 and the end of 2000, three events of M ≥ 6.5 have occurred in the subducted plate downdip from the thrust interface, beneath the low-seismicity zone, compared to none during the preceding 15 yr. More importantly, the January 2001 El Salvador earthquake (M_w 7.7) occurred downdip from this same zone (Bommer et al., 2002). It has been suggested that this downdip seismicity beneath a seismic gap may result from stress fluctuation associated with forthcoming underthrust rupture (Astiz et al., 1988; Lay et al., 1989; Dmowska et al., 1996). Also, the reduced thrust-zone seismicity coinciding with the San Jose Canyon resembles the seismicity around the Adak Canyon in the Andreanof Islands, where little seismic activity has been observed between major earthquakes that straddle the bathymetric expression of the canyon (Kisslinger and Kikuchi, 1997).

Although such a forecasting approach may seem simplistic, it has been previously applied successfully to other segments of the Middle America Trench having similar characteristics (e.g., Harlow et al., 1981; Lomnitz, 1994). In particular, Harlow et al. (1981) found a similarly long interval since the last major earthquake along the Nicaraguan section of the Middle America Trench and found that section to also be deficient in events of M > 5 shallow earthquakes instrumentally well located by the worldwide network. In addition, Harlow et al. (1981) had the benefit of having six years of local data from the Nicaraguan national seismograph network, which verified that this same section was

apparently quiet down to M > 3 for at least the previous 5 yr. They forecast an earthquake of M 7.5 based primarily on the size and location of the quiet zone in the local data and the time since the last major earthquake in that particular area. An earthquake of M 7.5 did occur at that precise location just 11 yr later, filling that quiet zone. At the current time, the seismograph network run by El Salvador's Servicio Nacional de Estudios Territoriales (SNET) is being significantly upgraded. It will be of great importance to use this local network data to investigate whether or not the aforementioned section of thrust interface off western El Salvador is similarly quiet for seismicity down to M > 3, as was the particular section of the Middle America Trench off Nicaragua.

Table 6 shows both the cumulative conditional probability assuming a lognormal distribution (Nishenko and Buland, 1987) and the distribution-free probability, both calculated according to Savage (1991), that a major thrust event will occur along the section of subduction zone between San Salvador and Guatemala City within the next 5, 10, and 20 yr. The mean probabilities indicate about a 50% chance of a magnitude 7.75 ± 0.3 earthquake occurring along the Pacific coast from Guatemala City to San Salvador by the year 2023.

CONCLUSIONS

Earthquake damage from events of magnitude M 7.5–8.0 occurred at intervals of 155–169 ± 10 yr in central Chiapas, in 1743 and 1902 and possibly also in 1565 or 1591. In southwestern Guatemala, earthquakes of that size occurred at intervals of 94 ± 54 yr in 1765, 1902, and 1942, and possibly also in 1565 and 1651 or 1689. In western El Salvador, damage from earthquakes of M 7.5–8.0 apparently occurred much more frequently, at intervals of 71 ± 17 yr, in 1719, 1776, 1859, 1862 and 1915, and in 2001 and possibly also in 1575 and 1658. All of the above intervals are much longer than the intervals of 33 to 50 yr suggested by Kelleher et al. (1973), McNally (1981), and Papadimitriou (1993).

Many other authors have previously pointed out the high seismogenic potential of this section of the Middle America subduction zone off El Salvador to Chiapas (e.g., Kelleher et al., 1973; McNally and Minster, 1981; Astiz and Kanamori, 1984; Lay et al., 1989; Papadimitriou, 1993; Lomnitz, 1996). We wish to point out that a significant slip deficit, equivalent to an earthquake of about magnitude 7.75, currently exists along one specific section of the

TABLE 6. CONDITIONAL PROBABILITY OF A MAGNITUDE 7.75 ± 0.3 EARTHQUAKE OCCURRING OFF THE GUATEMALA–EL SALVADOR COAST DURING THE NEXT 5, 10, AND 20 YEARS

Earthquake during next	Mean probability (and 90% confidence interval) Calculated from Savage (1994)	
	Probability assuming lognormal distribution	Distribution-free probability
5 years	16% (±13%)	50% (±30%)
10 years	29% (±20%)	50% (±30%)
20 years	47% (±25%)	50% (±30%)

thrust interface, the section from San Salvador to Guatemala City. If past major earthquakes in this area in 1575, 1658, 1719, 1776, 1859 or 1862, and 1915 were produced by slip along this thrust interface, then the recurrence interval for such earthquakes is apparently as frequent as 68 ± 15 yr. Based on these recent event intervals, the next such earthquake should have occurred between 1969 and 2001, and so seems overdue. The lack of moderate seismicity along this section of thrust interface since at least 1963 and the increased seismicity within the slab beneath this same section of thrust interface also appear to argue for an impending major earthquake along this locked section of thrust interface.

ACKNOWLEDGMENTS

We thank L. Feldman for reviewing pertinent handwritten documents at archives in Guatemala, El Salvador, and Spain. Also helping were S. Cristol at archives in Chiapas, B. Borg, T. Majewski, and I. Morales at archives in Guatemala, and S. Alvarez at archives in San Salvador. We thank N. Ambraseys, C. Ammon, B. Ellsworth, F. Klein, A. Lindh, W. Thatcher, C. Stephens, S. Kirby and J Havskov for helpful comments and reviews. Jim Savage calculated the conditional probabilities. Most of the historical work was carried out under OFDA/AID PASA IGT-5542-P-IC-3161-00 (PIO/T 361532), NSF EAR84-16635, and USGS 14-08-0001-19748.

APPENDIX. DISCUSSION OF MACROSEISMIC DESCRIPTIONS FOR THE LARGEST (M ≥ 7.4) EVENTS SINCE 1690 THAT ORIGINATED FROM EL SALVADOR TO CHIAPAS, MEXICO

September 29, 1717: The 1717 earthquake essentially destroyed Antigua, Guatemala, the colonial capital of Central America, where intensity reached at least MM IX. All churches, monasteries, and other major buildings along with at least 3000 houses were destroyed. A study was commissioned to consider relocating the capital to someplace less susceptible to earthquakes. M_I 7.4 ± 0.2 was estimated from both L_{VII} and L_{VIII}.

March 5, 1719: The 1719 earthquake produced intensity MM VII to VIII over a large portion of western El Salvador. Many churches were badly damaged to destroyed, and many people were killed. At least 150 smaller earthquakes were felt, though it is not clear if all were aftershocks or if some were foreshocks. We estimate a magnitude M_I 7.4 from the extent of actual reported damage, but if the damage was underreported, M_I may be as large as 7.7.

May 30, 1743: The 1743 earthquake initiated a 33-year sequence of very large events that apparently ruptured the entire length of the subduction zone from the Tehuantepec Gap to central El Salvador. Little information could be found concerning the extent of intensity MM VII for this event. However, churches collapsed or were on the verge of collapse over a large region of central Chiapas. The size and location of the area of intensity MM VIII were very similar to that for the September 1902 (M_S 7.6) earthquake, so we estimate an M_I of 7.6. These two events are compared in Figure 4.

October 24, 1765: The 1765 earthquake destroyed buildings over much of western Guatemala. At least 4000 people perished in the Ostuncalco area from a plague that apparently resulted from the destruction. Damage was reported only from Las Casas (MM VII) in central Chiapas, an area where reports are often few. The reported duration of felt shaking of 7 to 8 min is exceptionally long, and is compatible with rupture that

may have extended to central Chiapas. Alternatively, this extremely long duration may be interpreted as a sequence of immediate aftershocks. Damage in Guatemala was similar to that produced in April 1902, but if rupture extended to central Chiapas, the 1765 event may have been similar to the combined April 1902 and January 1903 events. Estimated M_I is 7.6 to 8.2, depending on extent of rupture into central Chiapas.

July 29, 1773: The 1773 earthquake was probably the single most traumatic event in Central American history, destroying the political, religious, and cultural heart of the region. All of the churches in south-central Guatemala were severely damaged or destroyed. The duration was reported to be nearly one minute, though in Antigua, Guatemala, the buildings collapsed during the first two seconds. Several severe aftershocks occurred through December. In Antigua alone, 500 to 600 people died in the earthquake, and at least 600 more died from the resulting poverty and disease. The King's engineer was sent from Spain to survey and report on the destruction. A study was again commissioned to find a new location for the capital of Central America, which was subsequently reestablished at what is currently Guatemala City. Fairly tight limits on the extent of intensity MM VII and MM VIII damage constrain the magnitude at M_I 7.5 ± 0.1.

May 30, 1776: On May 30, 1776, an earthquake collapsed or severely damaged buildings from central El Salvador well into southeastern Guatemala. In several nineteenth-century texts, the May event is reported as having occurred in 1774, having been confused with a very destructive storm, probably a hurricane, in the same area that year. A rather well-defined L_{VII} and less well defined L_{VIII} constrain the magnitude at M_I 7.9 ± 0.2.

December 8, 1859: The 1859 event caused destruction in western El Salvador and eastern Guatemala. Shaking is reported to have lasted a little over one minute and aftershocks lasted for one week. This is the first earthquake for which a tsunami was reported. From the L_{VII} and L_{VIII} shown, M_I is estimated at 7.6 ± 0.3. This event may represent thrust-zone rupture, while the following 1862 event may have been an intraplate normal-faulting event.

December 19, 1862: The 1862 earthquake destroyed much of western El Salvador and caused damage of MM VII from central Guatemala to central El Salvador. An unusually high level of destructive seismicity preceded this earthquake—at least ten damaging earthquakes occurred in this same area during the preceding ten years from both subduction zone and shallow volcanic axis sources. The shaking reportedly lasted for about two minutes. Aftershocks were continuous through December 31, 1862, and at least six others occurred through January 24, 1863. The limits of L_{VII} and L_{VIII} are well constrained and give an M_I of 8.0 ± 0.2. This is the largest event to occur along Guatemala or El Salvador from 1690 to the present.

February 26, 1902: A tsunami with a runup height of 2 m occurred along the coast of central El Salvador and possibly eastern Guatemala. This corresponds to a tsunami magnitude (M_t) of ~8.0 according to the relationships of Abe (1979). On this date, however, no large earthquake is known to have occurred locally or anywhere else in the circum-Pacific region. Since this is the only location for which the tsunami was reported, this event may represent a spontaneous underwater landslide or flow along the upper thrust zone off western El Salvador.

April 18, 1902: This earthquake (1902a in Figures 5E and 7B) was preceded by three months of continuous felt events. The earthquake was felt for one to two minutes, during which nearly all churches collapsed in western Guatemala and easternmost Chiapas. Between 800 and 900 people were killed. Aftershocks were frequent for 17 days. Gutenberg and Richter (1954) assigned a depth of 25 km and Ambraseys and Adams (1996) assigned a magnitude M_S 7.5.

September 23, 1902: This earthquake (1902b in Figures 4 and 5E) destroyed a large part of central Chiapas and produced damage of intensity VII over all of Chiapas and perhaps the border region of Guatemala. The intensity contours shown in Figure 4 are essentially from Figueroa (1973). Felt shaking was reported as 65 seconds. A depth of

25 km was assigned by Gutenberg and Richter (1954), and M_S 7.6 was assigned by Ambraseys and Adams (1996).

January 13, 1903: This earthquake, originally located in central Mexico by Gutenberg and Richter (1954) from instrumental data, has been relocated by Nishenko and Singh (1987). Though no damage was reported, the central area in southeasternmost Chiapas could well have suffered. Reports of shaking duration vary from 54 to 90 seconds. The event was assigned a depth of 25 km by Gutenberg and Richter (1954) and magnitude M_S 7.6 by Ambraseys and Adams (1996). The extent of damage from the event seems anomalously low for an earthquake of this magnitude, which may indicate possible mislocation of an event from the outer rise close to the Gulf of Tehuantepec.

September 7, 1915: This earthquake caused destruction (MM VIII) over much of western El Salvador and damage of MM VII from central Guatemala to central El Salvador. The felt duration was reportedly just over one minute. Strong aftershocks were felt at about half-hour intervals for six hours, and then tapered off. Abe (1981) assigned the 1915 earthquake a body-wave magnitude (m_b) of 7.4, and Richter (1958) assigned it a unified magnitude (M) of 7.9. Ambraseys and Adams (1996) assigned an M_S magnitude of 7.7. From the relatively small surface-wave amplitudes and from P and S arrival times, Richter assigned this event a focal depth of 80 km. Based on the arrival times of the maximum surface-wave amplitudes and abundance of strong aftershocks, Ambraseys and Adams (1996) assigned a depth of 60 km. Based on pP-P and sS-S at Ottawa and sS-S at Uppsala, we estimate the depth at 60 km. Figure 8 shows the onset of the P wave train at ten stations. Assuming the nodal planes strike subparallel with the trench axis, the first motions are compatible with a typical thrust mechanism along a plane dipping at ~55° to the northeast.

August 6, 1942: This is the most recent great earthquake in this catalog. The earthquake produced destruction in much of western Guatemala, similar to that produced in 1765 and 1902. It clearly did not produce any damage in south-central Guatemala at the location of the aftershock zone, as determined by Kelleher et al. (1973). The duration of felt shaking was about one minute. Aftershocks were frequent through August 20, with a few more through September 28. The earthquake was assigned a depth of 50 km by Gutenberg and Richter (1954) and a magnitude of M_S 7.9 by Ambraseys and Adams (1996).

REFERENCES CITED

Abe, K., 1979, Size of great earthquakes of 1837–1974 inferred from tsunami data: Journal of Geophysical Research, v. 84, p. 1561–1568.

Abe, K., 1981, Magnitudes of large shallow earthquakes from 1904 to 1980: Physics of the Earth and Planetary Interiors, v. 27, p. 72–92.

Ambraseys, N.N., 1995, Magnitudes of Central American earthquakes: 1898–1930: Geophysical Journal International, v. 121, p. 545–556.

Ambraseys, N.N., and Adams, R.D., 1996, Large-magnitude Central American earthquakes, 1898–1994: Geophysical Journal International, v. 127, p. 665–692.

Ambraseys, N.N., and Adams, R.D., 2001, Seismicity of Central America: A descriptive catalog, 1898–1995: London, Imperial College Press, 309 p.

Astiz, L., and Kanamori, H., 1984, An earthquake doublet in Ometepec, Guerrero, Mexico: Physics of the Earth and Planetary Interiors, v. 34, p. 24–45.

Astiz, L., Lay, T., and Kanamori, H., 1988, Large intermediate-depth earthquakes and the subduction process: Physics of the Earth and Planetary Interiors, v. 53, p. 80–166.

Aubouin, J., Stephan, J.F., Roump, J., and Renard, V., 1982, The Middle America Trench as an example of a subduction zone: Tectonophysics, v. 86, p. 113–132.

Bommer, J.J., Benito, M.B., Ciudad-Real, M., Lemoine, A., Lopez-Menjivar, M.A., Madariaga, R., Mankelow, J., Mendez de Hasbung, P., Murphy, W., Nieto-Lovo, M., Rodriguez-Pinedai, C.E., and Rosa, H., 2002, The El Salvador earthquakes of January and February 2001: Context, characteristics, and implications for seismic risk: Soil Dynamics and Earthquake Engineering, v. 22, p. 389–418.

Brazee, R.J., 1979, Reevaluation of the Modified Mercalli scale for earthquakes using distance as a determinant: Bulletin of the Seismological Society of America, v. 69, p. 911–924.

Burbach, G.V., Frohlich, C., Pennington, W.D., and Matumoto, T., 1984, Seismicity and tectonics of the subducted Cocos plate: Journal of Geophysical Research, v. 89, p. 7719–7735.

Bustillo, J.G., 1774, Razon puntual de los estragos mas memorables y de los estragos y daños que ha padecido la Ciudad de Guatemala y su vecindario desde que se fundo: Mixco, Guatemala, 55 p. (Found in the Archivo Géneral de Centro-América, Guatemala City.)

Carr, M.J., 1984, Symmetrical and segmented variation of physical and geochemical characteristics of the Central American volcanic front: Journal of Volcanology and Geothermal Research, v. 20, p. 231–252.

Carr, M.J., and Stoiber, R.E., 1990, Volcanism, in Dengo, G., and Case, J.E., eds., The Caribbean region: Boulder, Colorado, Geological Society of America, Geology of North America, v. H., p. 375–391.

Carr, M.J., Rose, W.I., Jr., and Stoiber, R.E., 1982, Regional distribution and character of active andesitic volcanism: Central America, in Thorpe, R.S., ed., Andesites: Orogenic andesites and related rocks: New York, John Wiley, p. 149–166.

Chael, E.P., and Stewart, G.S., 1982, Recent large earthquakes along the Middle America Trench and their implications for the subduction process: Journal of Geophysical Research, v. 87, p. 329–338.

Couch, R., and Woodcock, S., 1981, Gravity structure of the continental margins of southwestern Mexico and northwestern Guatemala: Journal of Geophysical Research, v. 86, p. 1829–1840.

DeMets, C., and Stein, S., 1990, Present-day kinematics of the Rivera plate and implications for tectonics in southern Mexico: Journal of Geophysical Research, v. 95, p. 21,931–21,948.

DeMets, C., Gordon, R.G., Argus, D.F., and Stein, S., 1990, Current plate motions: Geophysical Journal International, v. 101, p. 425–478.

Dewey, J.W., and Algermissen, S.T., 1974, Seismicity of the Middle America Arc Trench System near Managua, Nicaragua: Bulletin of the Seismological Society of America, v. 64, p. 1033–1048.

Dewey, J.W., and Suárez, G., 1991, Seismotectonics of Middle America, in Slemmons, D.B., et al., eds., Neotectonics of North America: Boulder, Colorado, Geological Society of America, Decade Map, v. 1, p. 309–321.

Diaz, V.M., 1930, Conmociones Terrestres en la America Central, 1469–1930: Guatemala City, Tipografia el Santuario, 268 p.

Diaz del Castillo, B., 1837, Historia verdadera de la Conquista de la Nueva España, v. 4, Paris (1st edition, 1575), 444 p.

Dmowska, R., Zheng, G., and Rice, J.R., 1996, Seismicity and deformation at convergent margins due to heterogeneous coupling: Journal of Geophysical Research, v. 101, p. 3015–3029.

Feldman, L.H., 1993, Mountains of fire, lands that shake: Earthquakes and volcanic eruptions in the historic past of Central America (1505–1899): Culver City, California, Labyrinthos Press, 288 p.

Figueroa, J., 1973, Sismicidad en Chiapas, Report no. 316: Instituto de Ingeniería, Universidad Nacional Autónoma de México, Mexico, 50 p.

Fisher, R.L., 1961, Middle America Trench, topography and structure: Geological Society of America Bulletin, v. 72, p. 703–720.

Grases, J., 1974, Relacion cronologica de los sismos destructores ocurridos en America Central desde 1525 hasta 1900: Sismicidad de la region Centro-Americana a la cadena volcanica del cuaternario, v. 2, Caracas, 253 p.

Gutenberg, B., and Richter, C.E., 1954, Seismicity of the Earth and associated phenomena (2nd edition): Princeton, New Jersey, Princeton University Press, 310 p.

Guzmán-Speziale, M., Pennington, W.D., and Matumoto, T., 1989, The triple junction of the North America, Cocos, and Caribbean plates: Seismicity and tectonics: Tectonics, v. 8, p. 981–998.

Harlow, D.H., White, R.A., Cifuentes, I.L., and Aburto Q., A., 1981, Quiet zone within a seismic gap near western Nicaragua: possible location of a future large earthquake: Science v. 213, p. 648–651.

Kanamori, H., 1981, The nature of seismicity patterns before large earthquakes, in Talwani, M., and Pittman, W.C., III, eds., Earthquake prediction: An international review: Washington, D.C., Maurice Ewing Series 4, American Geophysical Union, p. 1–19.

Kanamori, H., and Stewart, G.S., 1978, Seismological aspects of the Guatemala earthquake of February 4, 1976: Journal of Geophysical Research, v. 83, p. 3427–3434.

Karig, D.E., Cardwell, R.K., Moore, G.F., and Moore, D.G., 1978, Late Cenozoic subduction and continental-margin truncation along the northern

Middle America Trench: Geological Society of America Bulletin, v. 89, p. 265–276.

Kelleher, J.A., Sykes, L.R., and Oliver, J., 1973, Possible criteria for predicting earthquake locations and their applications to major plate boundaries of the Pacific and Caribbean: J. Geophysical Research Letters, v. 78, p. 2547–2585.

Kisslinger, C., and Kikuchi, M., 1997, Aftershocks of the Andreanof Islands of June 10, 1996, and local seismotectonics: Geophysical Research Letters, v. 24, p. 1883–1886.

Klitgord, K.D., and Mammerickx, J., 1982, North East Pacific Rise: Magnetic anomaly and bathymetric framework: Journal of Geophysical Research, v. 87, p. 6725–6750.

Ladd, J.W., Ibrahim, A.K., McMillen, K.J., Latham, G.V., Worzel, J.L., von Huene, R.E., Watkins, J.S., and Moore, J.C., 1978, Tectonics of the Middle America Trench offshore Guatemala, *in* Dengo, G., ed., Proceedings of the international symposium on the February 4th, 1976 Guatemalan earthquake and the reconstruction process. Volume I. Guatemala City, paper 5, 18 p.

Larde, J., 1960, Historia Sismica y Erupcio-volcanica de El Salvador, 1, San Salvador: San Salvador, Museo Nacional "David Guzman," p. 451–576.

Lay, T., Astiz, L., Kanamori, H., and Christensen, D.H., 1989, Temporal variation of large intraplate earthquakes in coupled subduction zones: Physics of the Earth and Planetary Interiors, v. 54, p. 258–312.

LeFevre, V.L., and McNally, K.C., 1985, Stress distribution and subduction of aseismic ridges in the Middle America subduction zone: Journal of Geophysical Research, v. 90, p. 4495–4510.

Ligorría, J.P., and Molina, E., 1997, Crustal velocity structure of southern Guatemala using refracted and Sp converted waves: Geofisica Internacional, v. 36, p. 9–19.

Ligorría, J.P., and Ponce, L., 1993, Estructura Cortical en el Istmo de Tehuantepec, Mexico, Usando ondas convertidas: Geofisica Internacional, v. 32, p. 89–98.

Ligorría, J.P., Lindholm, C., Bungum, H., and Dahle, A., 1995, Seismic hazard assessment for Guatemala: NORSAR, Technical Report No. 2-21, 47 p.

Ligorría, J.P., Lindholm, C., and Ammon, C.J., 1996, Regional seismic hazard and seismogenic coupling of the Middle America subduction zone in Guatemala: Eos (Transactions, American Geophysical Union), v. 77, p. F507 S31C-8.

Lomnitz, C., 1994, Fundamentals of earthquake prediction: New York, John Wiley & Sons, p. 88–127.

Lomnitz, C., 1996, Predicting earthquakes with the MRI algorithm: Seismological Research Letters, v. 67, p. 40–46.

Martinez, H.M., 1978, Cronología sismica y eruptiva de la Republica de El Salvador a partir de 1520: San Salvador, Ministerio de Obras Públicas, Centro de Investigaciones Geotecnicas, 40 p.

McCann, W.R., 1985, On the earthquake hazards of Puerto Rico and the Virgin Islands: Bulletin of the Seismological Society of America, v. 75, p. 251–262.

McNally, K.C., 1981, Plate subduction and prediction of earthquakes along the Middle America trench, *in* Talwani, M., and Pittman, W.C., III, eds., Earthquake prediction: An international review: Washington, D.C., Maurice Ewing Series 4, American Geophysical Union, p. 63–72.

McNally, K.C., and Minster, J.B., 1981, Nonuniform seismic slip rates along the Middle America Trench: Journal of Geophysical Research, v. 86, p. 4949–4959.

Mogi, K., 1969, Relationship between occurrence of great earthquakes and tectonic structures: Bulletin of the Earthquake Research Institute, Tokyo University, v. 47, p. 429–451.

Mogi, K., 1973, Relationship between shallow and deep seismicity in the western Pacific region: Tectonophysics, v. 17, p. 1–22.

Molnar, P., and Sykes, L.R., 1969, Tectonics of the Caribbean and Middle America regions from focal mechanisms and seismicity: Geological Society of America Bulletin, v. 80, p. 1639–1684.

Montessus de Ballore, F., 1888, Tremblements de terre et eruptions volcaniques au Centre Amerique: Depuis la conquete espagnole jusqu'a nos jours: E. Jobard, Dijon, Société des Sciences Naturelles de Saone-et-Loire, 281 p.

Moore, J.C., Watkins, J.S., Bachman, S.B., Beghtel, F.W., Butt, A., Didyk, B.M., Leggett, J.K., Lundberg, N., McMillen, K.J., Niitsuma, N., Shepard, L.E., Shipley, T.H., Stephan, J.F., and Stradner, H., 1979, Progressive accretion in the Middle America Trench, southern Mexico, Results from Leg 66 DSDP: Nature, v. 281, p. 638–642.

Nishenko, S.P., and Buland, R., 1987, A generic recurrence interval distribution for earthquake forecasting: Bulletin of the Seismological Society of America, v. 77, p. 1382–1399.

Nishenko, S.P., and Singh, S.K., 1987, Relocation of the great Mexican earthquake of 14 January 1903: Bulletin of the Seismological Society of America, v. 77, p. 256–259.

Nixon, G.T., 1982, The relationship between Quaternary volcanism in central Mexico and the seismicity and structure of subducted ocean lithosphere: Geological Society of America Bulletin, v. 93, p. 514–523.

Ohtake, M., Matumoto, T., and Latham, G.V., 1981, Evaluation of the forecast of the 1978 Oaxaca, southern Mexico earthquake based on a precursory seismic quiescence, *in* Talwani, M., and Pittman, W.C., III, eds., Earthquake prediction: An international review: Washington, D.C., Maurice Ewing Series 4, American Geophysical Union, p. 53–61.

Okubo, P.G., and Dietrich, J.H., 1984, Effects of physical properties on frictional instabilities produced on simulated faults: Journal of Geophysical Research, v. 89, p. 5817–5827.

Papadimitriou, E.E., 1993, Long-term earthquake prediction along the western coast of South and Central America based on a time-predictable model: PAGEOPH, v. 140, p. 301–316.

Pardo, M., and Suárez, G., 1995, Shape of the subducted Rivera and Cocos plates in southern Mexico: Seismic and tectonic implications: Journal of Geophysical Research, v. 100, p. 12,357–12,373.

Peraldo, G.H., and Montero P., W., 1999, Sismologia Historica de America Central: Instituto Panamericano de Geografia e Historia, Publication No. 513, Mexico, 347 p.

Ponce, L., Gaulon, R., Suárez, G., and Lomas, E., 1992, Geometry and state of stress of the downgoing Cocos plate in the Isthmus of Tehuantepec, Mexico: Geophysical Research Letters, v. 19, p. 773–776.

Press, W.H., Teukolsky, S.A., Vetterling, W.T., and Flannery, B.P., 1996, Numerical recipes in Fortran 77: The art of scientific computing (2nd edition): Cambridge, UK, Cambridge University Press, p. 653–664.

Redondo, C., Lindholm, C., and Bungum, H., 1993, Earthquake focal mechanisms in Central America. Technical Report. NORSAR, 21 p.

Richter, C.F., 1958, Elementary seismology: San Francisco, Freeman, 768 p.

Robin, C., 1982, Regional distribution and character of andesitic volcanism: Mexico, *in* Thorpe, R.S., ed., Andesites: Orogenic andesites and related rocks: New York, John Wiley, p. 137–148.

Ross, D.A., and Shor, G.G., 1965, Reflection profiles across the Middle America Trench: Geological Society of America Bulletin, v. 75, p. 5551–5572.

Savage, J.C., 1991, Criticism of some forecasts of the National Earthquake Prediction Council: Bulletin of the Seismological Society of America, v. 81, p. 862–881.

Scholz, C.H., 1990, The mechanics of earthquakes and faulting: Cambridge University Press, 439 p.

Seely, D., 1979, Geophysical investigations of continental slopes and rises, *in* Watkins, J.S., and Montadert, L., eds., Geological and geophysical investigation of continental margins: Tulsa, Oklahoma, American Association of Petroleum Geologists Memoir 51, p. 245–260.

Singh, S.K., Astiz, L., and Havskov, J., 1981, Seismic gaps and recurrence periods of large earthquakes along the Mexican subduction zone: A reexamination: Bulletin of the Seismological Society of America, v. 71, p. 827–843.

Stoiber, R.E., and Carr, M.J., 1973, Quaternary volcanic and tectonic segmentation of Central America: Bulletin of Volcanology, v. 37, p. 304–325.

Truchan, M., and Larson, R.L., 1973, Tectonic lineaments on the Cocos plate: Earth and Planetary Science Letters, v. 17, p. 426–432.

Vassaux, J., 1969, Cincuenta años de Sismología en Guatemala, Observatorio Nacional, Ministerio de Agricultura, Guatemala City, 98 p.

Villagrán, M., Lindholm, C., Dahle, A., Cowan, H., and Bungum, H., 1997, Seismic hazard assessment for Guatemala City: Natural Hazards, v. 14, p. 189–205.

von Huene, R., Aubouin, J., Azema, J., Blackinton, G., Coulbourn, W.T., Cowan, D.S., Curiale, J.A., Dengo, C.A., Faas, R.W., Harrison, W., Hesse, R., Hussong, D.M., Laad, J.W., Muzylov, N., Shiki, T., Thompson, P.R., Westberg, J., and Carter, J.A., 1980, Leg 67, DSDP Mid-America Trench transect off Guatemala: Geological Society of America Bulletin, v. 91, p. 421–432.

White, R.A., 1984, Catalog of historical earthquakes of back-arc Guatemala: U.S. Geological Survey Open-File Report 84-88, 34 p.

White, R.A., 1991, Tectonic implications of upper-crustal seismicity in Central America, *in* Slemmons, D.B., et al., eds., Neotectonics of North America: Boulder, Colorado, Geological Society of America, Decade Map, v. 1, p. 323–338.

White, R.A., and Harlow, D., 1993, Destructive upper-crustal earthquakes of Central America since 1900: Bulletin of the Seismological Society of America, v. 83, p. 1115–1142.

MANUSCRIPT ACCEPTED BY THE SOCIETY JUNE 16, 2003

Geological Society of America
Special Paper 375
2004

The M_W 7.6 El Salvador earthquake of 13 January 2001 and implications for seismic hazard in El Salvador

Allison L. Bent*

*National Earthquake Hazards Program, Geological Survey of Canada,
7 Observatory Crescent, Ottawa, Ontario K1A 0Y3, Canada*

Stephen G. Evans

Department of Earth Sciences, University of Waterloo, 200 University Avenue West, Waterloo, Ontario N2L 3G1, Canada

ABSTRACT

The 13 January 2001 earthquake (M_W 7.6) that occurred off the coast of El Salvador within the subducting slab of the Cocos plate caused nearly 1000 deaths, most of which were due to the hundreds of landslides triggered by the earthquake. The earthquake was well recorded by strong-motion stations within El Salvador and seismograph stations worldwide. Analysis of the strong-motion data is consistent with field observations of damage. Both the recorded accelerations and ground observations provide evidence for topographic amplification of the seismic waves. Magnitude-recurrence curves have been established for El Salvador and adjacent regions. Beta values of 1.96 for Central America and 1.84 for El Salvador (or b values of 0.85 and 0.80) are in the middle range of worldwide averages. The recurrence rate for magnitude 7.0 earthquakes is high—6 yr for Central America and 25 yr for El Salvador. Approximately 50% of large Central American earthquakes originate within the subducting slab.

Keywords: strong motion, topographic amplification, seismic hazard, in-slab earthquake.

INTRODUCTION

On 13 January 2001 a magnitude (M_W) 7.6 earthquake occurred off the coast of El Salvador (Fig. 1) triggering more than 400 landslides, killing ~1000 people (most of whom perished in the tragic landslide in the Las Colinas neighborhood of Santa Tecla), causing widespread building damage, and forcing tens of thousands of people into temporary shelter. The earthquake was felt throughout Central America as far away as Mexico City and Panama. The National Earthquake Information Service (NEIS) of the U.S. Geological Survey (USGS) located the earthquake at 13.05° N and 88.66° W with a hypocentral depth of 60 km, placing the earthquake within the subducting slab of the Cocos plate. The normal-faulting mechanism (Fig. 1A) provides further evidence that the earthquake was an in-slab event. One of the most unusual and, as yet, unexplained aspects of the 13 January

earthquake was the large number of aftershocks. The Centro de Investigaciones Geotécnicas in El Salvador recorded more than 5000 aftershocks, of which more than of 225 were felt. The large number of aftershocks appears to be a unique feature of this earthquake and not characteristic of in-slab earthquakes in El Salvador. For example, the magnitude 7.3 1982 earthquake was also an in-slab event but did not generate many aftershocks.

The destructive earthquake (M_W 6.5) that followed on 13 February was not an aftershock of the January event but may have been triggered by stress changes resulting from it. The 13 February earthquake was a shallow, crustal, strike-slip earthquake (Fig. 1A) occurring within the Caribbean plate in the region of San Vicente. A third earthquake (M_W 5.1) that occurred on 17 February near San Salvador was an aftershock of neither previous event, but possibly triggered by one of them.

We spent a week (both authors 22–29 January; S.E. remained until 3 February) in El Salvador investigating the damage and landslides associated with the 13 January earthquake. A companion

---

**bent@seismo.nrcan.gc.ca

Bent, A.L., and Evans, S.G., 2004, The M_W 7.6 El Salvador Earthquake of 13 January 2001 and Implications for Seismic Hazard in El Salvador, *in* Rose, W.I., Bommer, J.J., López, D.L., Carr, M.J., and Major, J.J., eds., Natural hazards in El Salvador: Boulder, Colorado, Geological Society of America Special Paper 375, p. 397–404. For permission to copy, contact editing@geosociety.org. © 2004 Geological Society of America

Central American Seismicity: 1973–2001

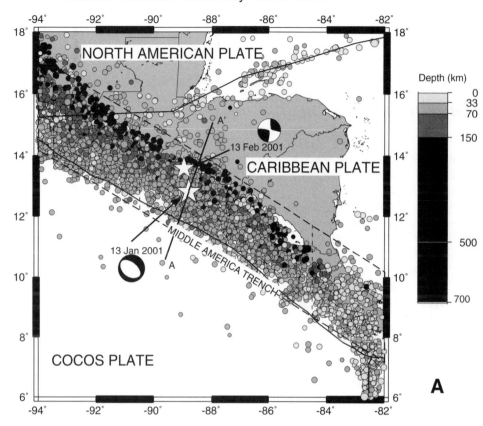

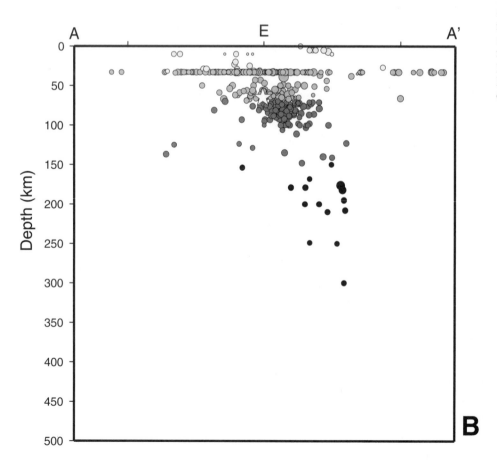

Figure 1. A: Seismicity of Central America from 1973 through 2001. Epicenters from the NEIS database. Symbol size is scaled to magnitude and symbol shade indicates depth. Earthquakes plotted are of magnitude 4.0 and greater. Plate boundaries are indicated by heavy black lines. Epicenters for the 13 January and 13 February 2001 earthquakes are indicated by stars. The Harvard CMT focal mechanisms are shown as lower hemisphere projections with shaded regions representing compressional quadrants.

B: Cross section through line A–A'. Earthquakes occurring with 25 km of either side of the line are plotted.

paper (Evans and Bent, this volume, Chapter 3) discusses the Las Colinas landslide in detail. In the present paper we relate our field observations to recorded strong ground motions and published fault rupture models, discuss evidence for topographic amplification of the seismic waves, and develop magnitude-recurrence relations for in-slab earthquakes in El Salvador and adjacent regions.

SEISMOTECTONIC ENVIRONMENT AND HISTORICAL SEISMICITY

Destructive earthquakes are a frequent occurrence in El Salvador and elsewhere in Central America. Most of the seismicity results from interactions between the Cocos and Caribbean plates where the Cocos plate is being subducted along the Middle America trench (Fig. 1). Earthquakes may occur within either plate as well as along the boundary between them. There are many active volcanoes on the Caribbean plate within El Salvador, but most of the earthquakes occurring in the volcanic regions are tectonic earthquakes and not directly related to volcanic activity (Bommer et al., 1997). An up-to-date summary of the seismicity and seismotectonics may be found in Dewey et al. (this volume, Chapter 27).

The city of San Salvador has sustained significant damage or undergone complete destruction from moderate-sized crustal earthquakes at least twelve times since 1576 (Lomnitz and Schulz, 1966; Harlow et al., 1993), and five times from larger subduction zone earthquakes (Larde, 1960; Harlow et al., 1993). The largest of the crustal earthquakes was an event of estimated magnitude 6.6, comparable to that of 13 February 2001, that occurred in 1854 and caused such complete destruction in San Salvador that the capital was relocated to Santa Tecla for a period of five years. The devastating earthquake of 1986 had a magnitude of only 5.7 but occurred at a shallow depth directly beneath San Salvador. Descriptions of many of these earthquakes may be found in the references cited above and those contained therein, as well as in White and Harlow (1993) and CIG (2001).

Landslides have been triggered by many Salvadoran earthquakes. Lomnitz and Rodríguez Elizarrarás (2001) note that landslides are more often associated with in-slab earthquakes than with crustal events, although they may be triggered by either.

While earthquakes occurring within the subduction zone are often much larger than the shallow crustal earthquakes, they tend to cause less damage principally because they occur farther from heavily populated regions and because they are deeper. Many earthquakes (such as the 13 January 2001 event) that have had only a minor impact on San Salvador have caused moderate to significant damage elsewhere in the country. Because of its small size, El Salvador is also vulnerable to the effect of large earthquakes in the neighboring countries of Guatemala and Honduras.

STRONG GROUND MOTIONS

Two strong-motion accelerograph networks operate within El Salvador. The first is maintained by the Universidad Cen-

troamericana José Siméon Cañas (UCA) and the second by the Centro de Investigaciones Geotécnicas (CIG). The 13 January 2001 earthquake was well recorded by both networks. Data from the UCA network were obtained directly from the university (UCA, 2001), which operates a network of three-component digital SSA-2 accelerographs. Site conditions and instrument characteristics may be found in UCA (2001) and a detailed description of the network also appears in Bommer et al. (1997). Most of these instruments are located in the central region of El Salvador in the vicinity of San Salvador. The CIG network consists of three-component SMA instruments. The data were processed and made available by the U.S. Geological Survey (USGS). The CIG network covers most regions of El Salvador. The radial component and instrument locations of all stations from which data were obtained are shown in Figure 2.

The peak accelerations recorded by the UCA and CIG networks are presented in Table 1. Horizontal components were rotated to their radial and tangential components using the SAC (Tull, 1989) program. Stations with two-letter codes belong to the CIG network and those with four-letter codes to the UCA network. Stations are listed in order of increasing distance from the epicenter.

The ground motion data along with that for 2001 Nisqually and Geiyo in-slab earthquakes will be valuable for refining and verifying existing ground motion relations for in-slab earthquakes. Adams and Halchuk (2002) note that there is debate stemming principally from the size of the data set used over whether the ground motion relations for in-slab earthquakes (Youngs et al., 1997)—currently used by both the Geological Survey of Canada (GSC) and USGS to calculate hazard from in-slab earthquakes in the Pacific Northwest—are valid.

Because large earthquakes cannot generally be treated as point sources, especially at close distances, standard practice for ground motion relations is to define distance as the closest distance from the rupture surface to the station rather than as epicentral distance. The El Salvador earthquake, which occurred offshore and at a depth of 60 km, did not produce a surface rupture. Thus, rupture dimensions must be obtained from waveform modeling. We use the rupture model of Vallée et al. (2003), which defines the landward-dipping nodal plane as the fault plane. Rupture is updip and toward the northwest on a plane ~50 km long and 25 km wide. This model is consistent with our observations of damage (which was more severe in the western than eastern part of the country), with the aftershock locations, and with the strong-motion records that generally show higher accelerations to the west of the epicenter (Fig. 2).

The data are plotted against the Youngs et al. (1997) model for in-slab earthquakes (Fig. 3). The average fit to the data is generally good, although the scatter is large and individual stations deviate significantly (by factors of 3 to 4) from the expected accelerations.

We visited the localities of several of the strong-motion stations and viewed several others by helicopter. Thus, we can compare damage to recorded shaking levels. The higher accel-

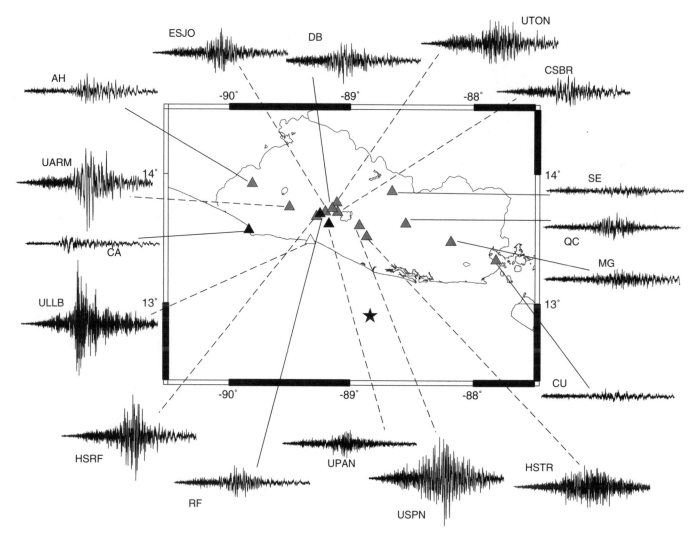

Figure 2. The radial component of the strong-motion records recorded by the UCA (four-letter codes and dashed lines) and CIG (two-letter codes and solid lines) accelerograph networks. The star denotes the epicenter, and the triangles the station sites. Each record shown is 40 seconds in duration. Amplitudes are plotted to scale. The symbol color indicates the soil conditions noted in Table 1 and matches the symbol shading in Figure 3 (white—alluvium, gray—soil, black—rock).

TABLE 1. PEAK ACCELERATIONS

Station Code	Location	Distance (km)	Azimuth (deg)	Site Conditions	Peak acceleration (g)		
					Vertical	Radial	Tangential
QC	Presa 15 de Septiembre	64	166	Soil	0.12	0.16	0.20
MG	San Miguel	70	43	Soil	0.09	0.12	0.12
HSTR	Zacatecoluca	76	354	Soil	0.26	0.25	0.31
USPN	San Pedro Nonualco	87	350	Soil	0.44	0.53	0.55
SE	Sensuntepeque	91	68	Soil	0.05	0.08	0.06
DB	Ciudadela Don Bosco	92	339	Soil	0.16	0.25	0.23
ULLB	La Libertad	93	321	Alluvium	0.66	0.72	1.07
UPAN	Panchimalco	96	334	Rock	0.09	0.16	0.19
RF	Relaciones Exteriores	96	332	Rock	0.19	0.20	0.25
CU	Cutuco	97	62	Soil	0.06	0.08	0.07
CSBR	San Bartolo	103	341	Soil	0.17	0.18	0.18
ESJO	San Salvador	107	335	Soil	0.15	0.28	0.25
HSRF	Santa Tecla	107	330	Soil	0.25	0.50	0.47
UTON	Tonacatepeque	111	342	Soil	0.21	0.27	0.23
UARM	Armenia	127	323	Soil	0.20	0.55	0.45
CA	Acajutla Cepa	139	306	Rock	0.05	0.11	0.14
AH	Ahuachapan	157	318	Soil	0.11	0.19	0.16

erations at Santa Tecla (HSRF) relative to San Salvador (ESJO), which are located at the same distance and almost the same azimuth from the epicenter, are consistent with the higher levels of damage observed in Santa Tecla. The difference may be related to site conditions. Strong motions recorded at Puebla, Mexico, from the magnitude 7.0 Tehuacan, Mexico, earthquake of 15 June 1999 (also an in-slab earthquake) showed a factor of four amplification for a soft soil relative to a hard rock site (EERI, 1999). The Santa Tecla strong-motion instrument is located in the Hospital San Rafael. The older buildings of the hospital suffered considerable damage, whereas the newest building was undamaged and fully operational. The hospital is located only a few blocks from the Las Colinas neighborhood; hence, the recorded accelerations should be a good indication of the shaking at the base of the landslide.

The town of Armenia (UARM), where high accelerations were recorded, was one of the most heavily damaged larger towns we observed. Zacatecoluca (HSTR), which is closer to the epicenter than Armenia, recorded lower accelerations and suffered very little damage. The general observation of higher accelerations in the west is consistent with the observed damage and may be indicative of source directivity. However, the differences could also reflect soil stiffness and depth, and the difference in observed damage might also be related to population density. The one site where the accelerations and damage levels are not consistent is La Libertad (ULLB). While this instrument recorded the highest accelerations, we saw very little building damage in the city despite the fact the many houses were poorly constructed and precariously perched on ledges in the hillside. The only effects of the earthquake we saw were several rockfalls on the road near La Libertad. The lack of damage in La Libertad may be related to the fact that the peak accelerations were at much higher frequencies (~5 Hz) than at the other sites.

Topographic Amplification

Field observations revealed that hilltop towns suffered considerably more damage than their lower-elevation counterparts, suggesting topographic amplification of the seismic signal. This was particularly evident in the Cordillera del Balsamo region, where in addition to high levels of shaking-related building damage, we also observed significant cracking in the ridgetops from Santa Tecla to Comasagua, a hilltop town that sustained extreme damage. Some care must be taken not to overinterpret these observations, as the damage pattern could be related to population distribution or soil conditions, such as the pyroclastic deposits found in many areas. Nevertheless there is a case to be made for topographic amplification. The accelerograph station at San Pedro Nonualco is noted as being located on a ridge, although there is no companion station at the base of the slope. This station recorded relatively high accelerations. Direct evidence for topographic amplification of seismic signals has been observed elsewhere in the past in strong-motion records (Bouchon, 1973; Celebi, 1987; Finn and Ventura, 1994).

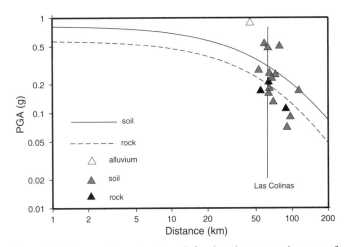

Figure 3. Peak ground acceleration (defined as the geometric mean of the horizontal components) versus distance from the fault rupture. The Youngs et al. (1997) ground motion relations for in-slab earthquakes for rock (dashed line) and soil (solid line) sites are also plotted. The vertical line is plotted at the distance of the Las Colinas landslide.

There are several theoretical studies on the subject of topographic amplification. Bouchon (1973) showed that P and SV waves are amplified near ridgetops and attenuated along the flanks of the ridge and that SH waves are amplified at ridgetops for near-vertical angles of incidence. The relation between incidence angle and amplification may help explain why landslides are more often associated with the deeper offshore earthquakes. The amplification is more pronounced for steeper slopes and when the wavelength is comparable to the size of the topographic anomaly.

A number of other studies undertaken by several researchers (Boore, 1972; Smith, 1975; Bard, 1982; Sánchez-Sesma et al., 1982; Geli et al., 1988) employed different methods of analysis but reached essentially the same conclusions. In a more recent study, Ashford et al. (1997) evaluated steep slopes as well as ridges and included the effects of soil conditions in the deposits behind the ridge crests. Their principal conclusions are that topographic amplification is greatest (50% for SV and 30% for SH) for slopes greater than 60°, when the height to wavelength ratio is 0.2, and when the waveform is dominated by the natural frequency of the area behind the ridge. The frequency at which the peak amplification occurs is proportional to the shear wave velocity. The steep slopes of pyroclastic material ubiquitous throughout El Salvador are particularly vulnerable to topographic amplification. Using Ashford et al.'s (1997) results and the landslide profile of Evans and Bent (this volume, Chapter 3), we would expect a maximum amplification at ~4 Hz at Las Colinas for a hard rock site. However, since the area is underlain by softer pyroclastic material, we would expect maximum amplification at frequencies of ~1 Hz (although there is some uncertainty related to the exact seismic velocities of the material). On the Santa Tecla accelerograms, we see a very strong peak at 2 Hz and slightly smaller peaks at 3 and 0.8 Hz on the radial component, and a peak at 2 Hz on the tangential component. These frequencies are in the range where we would expect to see the strongest additional contribution from topographic amplification.

Hazard

At least four studies have produced seismic hazard maps for El Salvador (Algermissen et al., 1988; Alfaro et al., 1990; Singh et al., 1993; Lindholm et al., 1995). They differ significantly with respect to peak accelerations expected throughout the country. For example, estimates of the 475-year return period accelerations for the San Salvador region range from 0.31 g (Lindholm et al., 1995) to 1.0 g (Singh et al., 1993). A brief description of the methods used and differences between them is found in Bommer et al. (1997) and a more detailed discussion in Bommer et al. (1996). Clearly the issue is unresolved, and a complete resolution is beyond the intended scope of this paper. Recorded data for the January and February 2001 earthquakes should help improve hazard estimates for the region but will not necessarily provide the final word on the subject. Given the accelerations for the January earthquake and the fact that the San Salvador region sustained very little damage relative to previous, recent earthquakes, the lower hazard estimates are probably too low. We use the term seismic hazard specifically to refer to the probability of the ground shaking exceeding a specified level within a specified time period. Seismic risk is used to refer to the negative consequences of the shaking. Bommer et al. (2002) comment that the seismic risk in El Salvador is increasing due to population growth, rapid urbanization in areas of high hazard, and deforestation, which increases the landslide hazard related to earthquakes.

An evaluation of seismic hazard requires knowledge of earthquake distribution, recurrence rates, ground motion, attenuation relations, and site effects. Because these factors as well as proximity to population centers differ from one source zone to another, the hazard from each relevant source should be evaluated separately initially. Obviously, a final hazard assessment for El Salvador must include the contribution from subduction zone (interplate and in-slab) and crustal earthquakes. We have established magnitude-recurrence rates for in-slab earthquakes in El Salvador and adjacent regions.

Earthquakes occurring within the subducting slab have normal focal mechanisms. That is, they represent extensional faulting due to tension in the slab. Those occurring at the plate interface within the subduction zone have thrust mechanisms. Shallow crustal earthquakes in the Caribbean plate in El Salvador generally have strike-slip mechanisms. Thus, focal mechanism is a fairly reliable method for identifying in-slab earthquakes. Depth provides another indication of source zone, as in-slab earthquakes have intermediate depths.

The Harvard CMT catalog (www.seismology.harvard.edu/data) provides a database of large earthquakes occurring since 1977. Comparing the number of earthquakes versus magnitude in the Harvard catalog to the NEIS database, which is complete to smaller magnitudes but which does not contain focal mechanism

information for all earthquakes, it appears that the Harvard catalog is complete for magnitude (M_w) 5.5 and higher, consistent with their goal for global coverage. The principal advantages of using the Harvard catalog for a seismic hazard study are that it provides both depth and focal mechanism information and that the data contained therein were processed in a uniform manner. The chief drawbacks are that the catalog covers a relatively short time period and does not contain information about smaller earthquakes. The latter point is most significant in regions where large earthquakes are infrequent. Ideally, a catalog that includes lower-magnitude earthquakes would be used. These catalogs, however, while generally sufficient to differentiate between shallow crustal and subduction zone earthquakes, rarely have adequate depth control to separate the plate-interface and in-slab earthquakes within the subduction zone.

Earthquakes with normal mechanisms and depths greater than 20 km were selected for further analysis. We define normal to include pure normal and oblique-normal mechanisms. In other words, the normal component of slip must be at least 1/3 the total slip.

The magnitude-recurrence curve for Central America was calculated following the method used by the GSC to calculate recurrence curves for Canada (Weichert, 1980). This technique employs the maximum likelihood method of curve fitting and expresses the slope of the line as β, which is related to the more often used "b" value by β = b ln 10.

In practical terms, a higher β value indicates a higher number of small earthquakes for every large event. Low β values are generally interpreted as indicators of high hazard levels. For earthquakes on a global scale, β values are typically in the range of 1.5–2.5 (b values 0.65–1.10). Not enough is known about in-slab earthquakes to conclude whether they have similar recurrence rates, although preliminary calculations (Bent, 2002) suggest that they do.

Initial calculations considered in-slab earthquakes throughout Central America. The source zone of Astiz (1987) was selected. The zone was defined as part of a global study of intermediate-depth earthquakes, wherein source zone boundaries were based on subduction zone characteristics.

An estimate of the maximum-magnitude earthquake for the source zone is required to calculate the recurrence curve. The lower-bound estimate of 7.7 reflects the largest known in-slab earthquake in the region—the 13 January event. The best (7.9) and upper-bound (8.1) estimates were chosen to allow for slightly larger earthquakes than have occurred in this region. A compilation of intermediate-depth earthquakes worldwide (Astiz, 1987) shows no events larger than 8.1. Tests conducted to determine the effect of the magnitude bounds showed no significant difference in the β values for extremely different lower (7.0) and upper (9.0) bounds but that there was a more significant impact on the expected return period of the very largest earthquakes. The best-fit curve for Central America (Fig. 4) has a β value of 1.96 (b = 0.85) and suggests that an in-slab earthquake of magnitude 7.0 or greater should occur, on average, once every six years. The

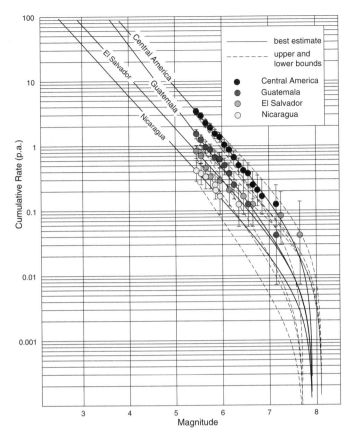

Figure 4. Magnitude-recurrence curves for in-slab earthquakes in Central America and smaller regions therein. Data are from the Harvard CMT catalog. The horizontal axis is moment magnitude. The vertical axis shows the cumulative rate of earthquakes per year. The dots with stochastic error bounds represent the earthquake rates. Note that the two largest Salvadoran events are also the two largest-magnitude points on the Central American curve and that the El Salvador shade has been used to plot them.

TABLE 2. SEISMICITY RATES FOR CENTRAL AMERICA

Region	Beta	b	Average return period for magnitude 7.0 or higher (years)
Central America	1.96	0.85	6
Guatemala	2.19	0.95	23
El Salvador	1.84	0.80	25
Nicaragua	1.84	0.80	50

mala, 25 yr for El Salvador and 50 yr for Nicaragua, suggesting a decrease in the seismicity rate from north to south.

To determine the relative importance of in-slab earthquakes in Central America, we have also calculated magnitude-recurrence curves for the two nearest sources of in-slab earthquakes as defined by Astiz (1987): Colombia and Mexico. We also determined the magnitude-recurrence curves for each source zone using earthquakes from all sources. We must emphasize that the latter curves should not be used directly for seismic hazard assessments because the impact and ground motions of a shallow earthquake directly beneath a city are different from those of a deep offshore earthquake of the same magnitude. Nevertheless, these curves help place in-slab earthquakes in the proper regional context. Since the seismic zones are not of uniform size, the recurrence rates have been normalized to 100 km lengths of the trench axis, assuming uniform seismicity within each zone and the trench lengths of Moore (1982). The results are presented in Table 3. Per 100 km of trench, the recurrence period for in-slab earthquakes of magnitude 7.0 is 90 yr for both Mexico and Central America and 264 yr for Colombia. In-slab earthquakes account for ~50% of Central America's magnitude 7.0 and greater earthquakes. The figures for Mexico and Colombia are 30% and 18% respectively. Thus, in-slab earthquakes are a larger contributor to the seismic hazard of Central America than to neighboring regions.

CONCLUSIONS

A field investigation of damage resulting from the 13 January 2001 El Salvador earthquake was generally consistent with the recorded strong ground motions. That is, areas that sustained heavy damage generally occurred in areas with relatively high recorded accelerations. Strong-motion records, observed damage, and aftershock locations are all consistent with published rupture models, which indicate rupture updip and toward the northwest. Both in the

predicted rate for an earthquake the size of the 13 January 2001 earthquake is about one per hundred years.

Central America was divided into three smaller areas to compare the seismic hazard of El Salvador to that of adjacent regions. Each box measures 3° longitude by 3° latitude. El Salvador is defined as the region between 90.5° W and 93.5° W, corresponding roughly to the national boundaries. The region immediately to the northwest is referred to as Guatemala and that to the southeast as Nicaragua. Note that the boundaries of these regions do not correspond exactly to the borders of each country. The Central American maximum magnitude estimates were assumed for each region. The resulting magnitude-recurrence curves are shown in Figure 4 and Table 2. The β values for El Salvador and Nicaragua are both 1.84 (b = 0.80), and that for Guatemala is 2.19 (b = 0.95). However, the difference between each region and Central America as a whole falls within the uncertainty of the method. The magnitude 7.0 return period is 20–25 yr for Guate-

TABLE 3. MAGNITUDE 7 RETURN PERIODS

Region	All Sources M7+ (years)		In-Slab M7+ (years)	
	Region	Per 100 km	Region	Per 100 km
Mexico	3	27	10	90
Central America	3	45	6	90
Colombia	6	48	33	264

field and in the recorded accelerations we find some evidence for topographic amplification of the seismic waves. Magnitude-recurrence curves for in-slab earthquakes in Central America have been established and indicate that a magnitude 7.0 or greater in-slab event should occur approximately every 6 yr somewhere in Central America and approximately every 25 yr in El Salvador. These results imply that most people will experience several large earthquakes in a lifetime. Although in-slab earthquakes are not the only earthquakes or even the most devastating earthquakes to occur in El Salvador and other regions of Central America, they are a more significant contributor to the overall hazard of this region than to the areas immediately to the north and south.

ACKNOWLEDGMENTS

All maps shown in this paper were produced using the GMT program of Wessel and Smith (1991). We thank John Adams, Z. Lubkowski, and an anonymous reviewer for their constructive comments. Geological Survey of Canada contribution number 2002108.

REFERENCES CITED

Adams, J., and Halchuk, S., 2002, Knowledge of in-slab earthquakes needed to improve seismic hazard estimates for southwestern British Columbia, *in* Kirby, S., et al., eds., The Cascadia subduction zone and related subduction systems—Seismic structure, intraslab earthquakes and processes, and earthquake hazards: U.S. Geological Survey Open-File Report 02-328, p. 149–154.

Alfaro, C.S., Kiremidjian, A.S., and White, R.A., 1990, Seismic zoning and ground motion parameters for El Salvador: The John A. Blume Earthquake Engineering Center Report No. 93, Palo Alto, California, Stanford University.

Algermissen, S.T., Hansen, S.L., and Thenhaus, P.C., 1988, Seismic hazard evaluation for El Salvador, Report for the U.S. Agency for International Development, 21 p.

Ashford, S.A., Sitar, N., Lysmer, J., and Deng, N., 1997, Topographic effects on the seismic response of steep slopes: Bulletin of the Seismological Society of America, v. 87, p. 701–709.

Astiz, L., 1987, 1. Source analysis of large earthquakes in Mexico 2. Study of intermediate-depth earthquakes and interplate seismic coupling [Ph.D. thesis]: Pasadena, California Institute of Technology, 284 p.

Bard, P.-Y., 1982, Diffracted waves and displacement field over two-dimensional elevated topographies: Geophysical Journal of the Royal Astronomical Society, v. 71, p. 731–760.

Bent, A.L., 2002, Seismic hazard from in-slab earthquakes: Seismological Research Letters, v. 73, p. 220.

Bommer, J.J., Hernández, D.A., Navarette, J.A., and Salazar, W.M., 1996, Seismic hazard assessments for El Salvador: Geofisica Internacional, v. 35, p. 227–244.

Bommer, J.J., Udías, A., Cepeda, J.M., Hasbun, J.C., Salazar, W.M., Suárez, A., Ambraseys, N.N., Buforn, E., Cortina, J., Madariaga, R., Méndez, P., Mezuca, J., and Papastamatiou, D., 1997, A new digital accelerograph network for El Salvador: Seismological Research Letters, v. 68, p. 426–437.

Bommer, J.J., Benito, M.B., Ciudad-Real, M., Lemoine, A., López-Menjívar, M.A., Madariaga, R., Menkelow, J., Méndez de Hasbun, P., Murphy, W., Nieto-Lovo, M., Rodríguez-Pineda, C.E., and Rosa, H., 2002, The El Salvador earthquakes of January and February 2001: Context, characteristics, and implications for seismic risk: Soil Dynamics and Earthquake Engineering, v. 22, p. 389–418.

Boore, D.M., 1972, A note on the effect of simple topography on seismic SH waves: Bulletin of the Seismological Society of America, v. 62, p. 275–284.

Bouchon, M., 1973, Effect of topography on seismic SH waves: Bulletin of the Seismological Society of America, v. 63, p. 615–632.

Celebi, M., 1987, Topographic and geological amplification determined from strong-motion and aftershock records of the 3 March 1985 Chile earthquake: Bulletin of the Seismological Society of America, v. 77, p. 1147–1167.

GIG (Centro de Investigaciones Geotécnicas), 2001, Cronología de sismos destructivos en El Salvador: http://www.snet.gob.sv/Geologica/Sismologia/1crono.htm.

Dewey, J.W., White, R.A., and Hernández, D.A., 2004, Seismicity and tectonics of El Salvador, *in* Rose, W.I., et al., eds., Natural hazards in El Salvador: Boulder, Colorado, Geological Society of America Special Paper 375, p. 363–378 (this volume).

EERI (Earthquake Engineering Research Institute), 1999, The Tehuacan, Mexico earthquake of June 15, 1999, Earthquake Engineering Research Institute Special Report: http://www.eeri.org/lfe/mexico_tehuacan.html.

Evans, S.G., and Bent, A.L., 2004, The Las Colinas landslide, Santa Tecla: A highly destructive flowslide triggered by the January 13, 2001, El Salvador earthquake, *in* Rose, W.I., et al., eds., Natural hazards in El Salvador: Boulder, Colorado, Geological Society of America Special Paper 375, p. 25–37 (this volume).

Finn, W.D.L., and Ventura, C.E., 1994, Ground motions, preliminary report on the Northridge, California, Earthquake of January 17, 1984: Vancouver, Canadian Association for Earthquake Engineering, p. 7–81.

Geli, L., Bard, P.-Y., and Jullien, B., 1988, The effect of topography on earthquake ground motion: A review and new results: Bulletin of the Seismological Society of America, v. 78, p. 42–63.

Harlow, D.H., White, R.A., Rymer, M.J., and Alvarez, G.S., 1993, The San Salvador earthquake of 1986 and its historical context: Bulletin of the Seismological Society of America, v. 83, p. 1143–1154.

Larde, J., 1960, Obras Completas, Tomo I, Ministerio de Cultura, Departamento Editorial, San Salvador, 576 p.

Lindholm, C., Rojas, W., Bungum, H., Dahle, A., Camacho, E., Cowan, H., and Laporte, M., 1995, New regional seismic zonation for Central America: Proceedings, 5th International Conference on Seismic Zonation, Nice, v. 1, p. 437–444.

Lomnitz, C., and Rodríguez Elizarrarás, S., 2001, El Salvador 2001: Earthquake disaster and disaster preparedness in a tropical volcanic environment: Seismological Research Letters, v. 72, p. 346–351.

Lomnitz, C., and Schulz, R., 1966, The San Salvador earthquake of May 3, 1965: Bulletin of the Seismological Society of America, v. 56, p. 561–575.

Moore, G.W., 1982, Plate tectonic map of the circum Pacific region, explanatory notes: Tulsa, Oklahoma, American Association of Petroleum Geologists, 14 p.

Sánchez-Sesma, F., Herrera, I., and Aviles, J., 1982, A boundary method for elastic wave diffraction: Application to scattering SH waves by surface irregularities: Bulletin of the Seismological Society of America, v. 72, p. 473–490.

Singh, S.K., Gutiérrez, C., Arboleda, J., and Ordaz, M., 1993, Peligro sismico en El Salvador, Universidad Autónoma de México, Mexico City.

Smith, W.D., 1975, The application of finite element analysis to elastic body wave propagation problems, Geophysical Journal of the Royal Astronomical Society, v. 42, p. 747–768.

Tull, J.E., 1989, SAC—Seismic Analysis Code Users' Guide, Lawrence Livermore National Laboratory, non-paginated.

UCA (Universidad Centroamericana), 2001, Strong-motion data from the January–February 2001 earthquakes in El Salvador: San Salvador, El Salvador, Universidad Centroamericana, digital data set.

Vallée, M., Bouchon, M., and Schwartz, S.Y., 2003, The January 13, 2001 El Salvador earthquake: A multi-data analysis: Journal of Geophysical Research, v. 108, p. ETG 8-1–ETG 8-16.

Weichert, D.H., 1980, Estimation of the earthquake recurrence parameters for unequal observation periods for different magnitudes: Bulletin of the Seismological Society of America, v. 70, p. 1337–1346.

Wessel, P., and Smith, W.H.F., 1991, Free software helps map and display data: Eos (Transactions, American Geophysical Union), v. 72, p. 441, 445–446.

White, R.A., and Harlow, D.H., 1993, Destructive upper crustal earthquakes of Central America since 1900: Bulletin of the Seismological Society of America, v. 83, p. 1115–1142.

Youngs, R.R., Chiou, S.-J., Silva, W.J., and Humphrey, J.R., 1997, Strong ground motion relationships for subduction zone earthquakes: Seismological Research Letters, v. 68, p. 58–73.

MANUSCRIPT ACCEPTED BY THE SOCIETY JUNE 16, 2003

Geological Society of America
Special Paper 375
2004

Strong-motion characteristics of January and February 2001 earthquakes in El Salvador

J.M. Cepeda*
Dpto. de Mecanica Estructural, Universidad Centroamericana "José Simeón Cañas"
(UCA), A.P. (01) 168, San Salvador, El Salvador

M.B. Benito
Universidad Politécnica de Madrid (UPM), Madrid, Spain

E.A. Burgos
Dpto. de Mecanica Estructural, Universidad Centroamericana "José Simeón Cañas"
(UCA), A.P. (01) 168, San Salvador, El Salvador

ABSTRACT

During 2001, very intense seismic activity occurred in El Salvador and concentrated in the in-slab subduction zone and the volcanic chain that runs along the central part of the country from west to east. The ground motions of 188 earthquakes were recorded by 31 accelerographs, which produced a total of 479 records. The purpose of this paper is to present the main characteristics of this strong-motion database and to examine the horizontal ground motion attenuation of earthquakes that occurred in both seismic zones. Even though the database is not complete in terms of magnitude, and site conditions are only known for 2% of the records, an application is carried out by making adjustments of attenuation equations derived from worldwide data. The analysis of the subduction earthquake records shows that the attenuation characteristics of the El Salvador 2001 database seem to have a better agreement with interface-type events rather than with the in-slab type of earthquakes assumed in the present study. The attenuation analysis of the shallow upper crustal events indicates that rupture directivity effects may be relevant and should deserve attention in future assessments of strong ground motion.

Keywords: earthquakes, strong-motion records, peak ground acceleration, response spectral ordinates, attenuation equations.

INTRODUCTION

During January and February 2001, two large earthquakes struck El Salvador, causing major destruction mainly due to widespread landslides and collapse of nonengineered structures. The 7.7 M_w first event occurred on January 13 and had its origin in the in-slab subduction zone that results from the convergence of the subducting Cocos plate under the Caribbean plate. The second event on February 13, 6.6 M_w, was associated with an inland crustal fault. These events were accompanied by intense activity in both source areas. The parameters, focal mechanism, and other source characteristics of these events and their distribution in space and time are detailed in Benito et al. (this volume, Chapter 25). The seismic sources and parameters used in the present study, as well as the magnitude conversions (from M_c to M_w), are based on the above reference. Bommer et al. (2002) have previously presented an overall assessment of the January and February 2001 main shocks, including an assessment of the strong-motion attenuation characteristics of these earthquakes.

*jcepeda@ing.uca.edu.sv

Cepeda, J.M., Benito, M.B., and Burgos, E.A., 2004, Strong-motion characteristics of January and February 2001 earthquakes in El Salvador, *in* Rose, W.I., Bommer, J.J., López, D.L., Carr, M.J., and Major, J.J., eds., Natural hazards in El Salvador: Boulder, Colorado, Geological Society of America Special Paper 375, p. 405–421. For permission to copy, contact editing@geosociety.org. © 2004 Geological Society of America

The seismic activity during 2001 in El Salvador was recorded on 31 accelerographs of three strong-motion networks: the government network operated by SNET (Servicio Nacional de Estudios Territoriales, formerly CIG, Centro de Investigaciones Geotécnicas), and two private networks operated by GESAL (Geotérmica Salvadoreña) and UCA (Universidad Centroamericana "José Simeón Cañas"). Bommer et al. (1997) and Cepeda et al. (1997a) present the design and characteristics of the UCA network. Table 1 gives information on how to obtain the data produced by these three networks.

These strong-motion networks produced records for the main events and the aftershock series. A total of 479 triaxial records from 188 earthquakes in 2001 were collected and analyzed for the present study. Figure 1 shows the epicenters of these 188 earthquakes as well as the locations of the 31 strong-motion stations. The source parameters of the events were taken from the SNET catalog of 2001. The source parameters of the January 13 and February 13

earthquakes were taken from the PDE (Preliminary Determination of Epicenters) catalog of the U.S. Geological Survey (USGS). The geographical coordinates for the strong-motion stations and the number of records in each station are presented in Tables 2 and 3. Prior to the January and February 2001 earthquakes, a major part of the strong-motion records for large earthquakes (M ≥ 6) in Central America were obtained from the March and April 1990 earthquakes in Costa Rica (Cepeda et al., 1997a).

The purpose of this paper is to assess attenuation characteristics of the subduction and volcanic chain (shallow upper crustal) earthquakes respectively, based on the strong-motion records of 2001.

SELECTION OF SEISMIC SOURCES

The definition of the source-site distance of the attenuation models is determined by the type of seismic source. For moderate-

TABLE 1. WEB AND E-MAIL ADDRESSES FOR OBTAINING 2001 STRONG-MOTION DATA

Owner of instruments	Web addresses of institution or link for downloading data	Contact persons	E-mail addresses
SNET	www.snet.gob.sv nsmp.wr.usgs.gov/data_sets/20010113_1.html	Griselda Marroquín Douglas Hernández	gmarroquin@snet.gob.sv dhernandez@snet.gob.sv
GESAL	www.gesal.com.sv	José Rivas	jarivas@gesal.com.sv
UCA	www.uca.edu.sv	Reynaldo Zelaya José Cepeda	rezelaya@ing.uca.edu.sv jcepeda@ing.uca.edu.sv

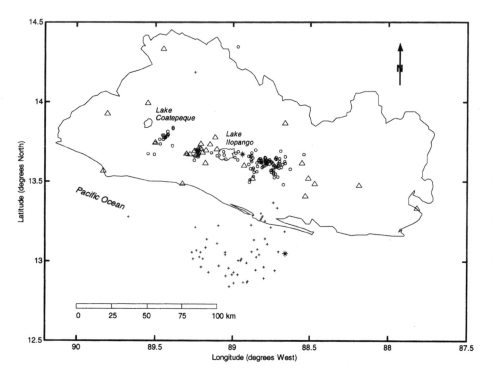

Figure 1. Map of El Salvador showing epicenters of events that produced strong-motion records during 2001. Crosses—subduction earthquake epicenters. Circles—shallow upper crustal earthquake epicenters. Stars—epicenters of the January 13 (southern star) and February 13 (northern star) earthquakes. Triangles—strong-motion stations that recorded at least one earthquake.

TABLE 2. CHARACTERISTICS OF STRONG MOTION RECORDING STATIONS

Code	Description	Owner	Latitude (°N)	Longitude (°W)	Instrument
AH	Ahuachapán	CIG	13.925	89.805	SMA-1
AR	Armenia	UCA	13.744	89.501	SSA-2
BA	San Bartolo	UCA	13.704	89.106	SSA-2
BE	Berlin	GESAL	13.497	88.529	SSA-2
CA	CEPA, Acajutla	CIG	13.567	89.833	SMA-1
CI	Centro de Investigaciones Geotécnicas, San Salvador	CIG	13.698	89.173	SMA-1
CM	CESSA, Metapán	CIG	14.333	89.450	SMA-1
CU	Cutuco	CIG	13.333	87.817	SMA-1
DB	Ciudadela Don Bosco, Soyapango	CIG	13.733	89.150	SMA-1
EX	Externado, San Salvador	UCA	13.707	89.207	SSA-2
LI	La Libertad	UCA	13.486	89.327	SSA-2
MG	San Miguel	CIG	13.475	88.183	SMA-1
NO	San Pedro Nonualco	UCA	13.602	88.927	SSA-2
OB	Observatorio, San Salvador	CIG	13.681	89.198	SMA-1
PA	Panchimalco	UCA	13.614	89.179	SSA-2
QC	"15 de septiembre" dam (zero level)	CIG	13.616	88.550	SMA-1
RF	Relaciones Exteriores (bottom of borehole)	CIG	13.692	89.250	SMA-1
RS	Relaciones Exteriores (ground level)	CIG	13.692	89.250	SMA-1
SA	Santa Ana	CIG	13.992	89.550	SMA-1
SE	Sensuntepeque	CIG	13.867	88.663	SMA-1
SM	Santiago de María	CIG	13.486	88.471	SMA-1
SS	Seminario "San José de La Montaña" (ground level), San Salvador	CIG	13.705	89.225	SMA-1
ST	Santa Tecla	CIG	13.675	89.300	SMA-1
TE	Hospital San Rafael, Santa Tecla	UCA	13.671	89.279	SSA-2
TO	Tonacatepeque	UCA	13.778	89.114	SSA-2
TR	Planta Boca Pozo (TR-9), Berlín	GESAL	13.520	88.512	SSA-2
UC	Universidad Centroamericana, Antiguo Cuscatlán	CIG	13.677	89.236	SMA-1
VF	Viveros de DUA (bottom of borehole), San Salvador	CIG	13.737	89.209	SMA-1
VI	San Vicente	UCA	13.642	88.784	SSA-2
VS	Viveros de DUA (ground level), San Salvador	CIG	13.737	89.209	SMA-1
ZA	Zacatecoluca	UCA	13.517	88.869	SSA-2

size events, the use of hypocentral or epicentral distance, i.e., point modeling of the source, is appropriate when the dimensions of the rupture are small compared to the source-site distance. This is not the case for large-magnitude events. In the following sections, two earthquake cases are discussed: January 13 and February 13.

January 13 Earthquake

The source parameters of this earthquake reported by NEIC (National Earthquake Information Center) are origin time, 17:33:32.38 u.t.; coordinates of epicenter, 13.049° N and 88.660° W; focal depth, 60 km; and reported magnitudes of 6.4 m_b, 7.8 M_S, and 7.7 M_W. The effect of source dimensions on the recorded motions has been examined by Cepeda (2001a). The vertical component was used in order to minimize the effects of amplification (or deamplification) due to topographic or geologic features. A very large scatter was observed when peak

vertical acceleration was plotted against hypocentral distance. This fact can be partially explained if the source dimensions are taken into account. For such a major earthquake, whose rupture extended along a large portion of the Salvadoran coastline, it is clear that source dimensions and source-site distances for the strong-motion station are comparable. Hence, point modeling for the source does not seem to be adequate for this event. In the following paragraphs the use of a rupture plane as the event source is examined.

Consistent with the strike and dip angle of the rupture plane defined by Benito et al. (this volume, Chapter 25), a robust fit was performed for the main event and the aftershocks within two days after the main shock (Lay and Wallace, 1995). The distribution of aftershocks and the USGS solution for the main shock are shown in Figure 2. For the regression, only events in the epicentral area were taken as aftershocks.

The fit yields the following result:

TABLE 3. GEOLOGY OF STATIONS AND NUMBER OF RECORDS

Code	Geology§	Description	Subduction Records	Shallow Upper Crustal Records	NEHRP Site
AH	s3	Acid pyroclastites, volcanic epiclastites ("brown tuffs")*	1	N.D.	D
AR	s3	Acid pyroclastites, volcanic epiclastites ("brown tuffs")*	28	23	D
BA	s4	Acid pyroclastites ("white earth")*	20	17	D
BE	s2	Andesitic and basaltic effusives: piroclastites*	N.D.	1	D
CA	b1	Volcanic epiclastites, pyroclastites, lava flows*	1	N.D.	Rock
CI	s5'a	Basaltic and andesitic lavas, predominantly from San Salvador volcano#	N.D.	2	Rock
CM	Q'f	Alluvium, locally with pyroclastites*	1	N.D.	Rock
CU	c3	Andesitic–basaltic effusives*	1	N.D.	Rock
DB	s4	Volcanic ashes ("white earth"), low consolidated#	1	2	D
EX	s4	Volcanic ashes ("white earth"), low consolidated#	17	16	D
LI	Q'f	Alluvium, locally with pyroclastites*	29	5	C
MG	s3	Acid pyroclastites, volcanic epiclastites ("brown tuffs")*	1	N.D.	C
NO	c1	Acid pyroclastites, volcanic epiclastites, welded tuffs*	33	4	D
OB	s4	Volcanic ashes ("white earth"), low consolidated#	1	2	D
PA	c1	Acid pyroclastites, volcanic epiclastites, welded tuffs*	8	10	Rock
QC	b3	Andesitic–basaltic effusives*	1	1	Rock
RF	s3'a	"Brown tuffs," locally with ashes and scoria#	1	1	C
RS	s3'a	"Brown tuffs," locally with ashes and scoria#	1	1	D
SA	s3	Acid pyroclastites, volcanic epiclastites ("brown tuffs")*	1	N.D.	D
SE	b1	Volcanic epiclastites, pyroclastites, lava flows*	1	N.D.	Rock
SM	s3	Acid pyroclastites, volcanic epiclastites ("brown tuffs")*	1	N.D.	D
SS	s4	Volcanic ashes ("white earth"), low consolidated#	1	1	D
ST	s3	Acid pyroclastites, volcanic epiclastites ("brown tuffs")*	1	1	D
TE	s2	Andesitic and basaltic effusives: piroclastites*	29	13	C
TO	c1	Acid pyroclastites, volcanic epiclastites, welded tuffs *	24	23	D
TR	s2	Andesitic and basaltic effusives: piroclastites*	1	N.D.	D
UC	s3'a	"Brown tuffs," locally with ashes and scoria#	N.D.	2	D
VF	s4	Volcanic ashes ("white earth"), low consolidated#	N.D.	2	C
VI	s4	Acid pyroclastites ("white earth")*	21	65	D
VS	s4	Volcanic ashes ("white earth"), low consolidated#	1	2	D
ZA	b1	Volcanic epiclastites, pyroclastites, lava flows*	28	31	Rock
TOTAL			254	225	

Note: N.D.—no data
N.A.—not applicable
§—symbols used in the geologic map of El Salvador
*—Obtained from El Salvador geologic map 1:500,000
#—Obtained from San Salvador geologic map 1:15,000
C—NEHRP soil type C
D—NEHRP soil type D

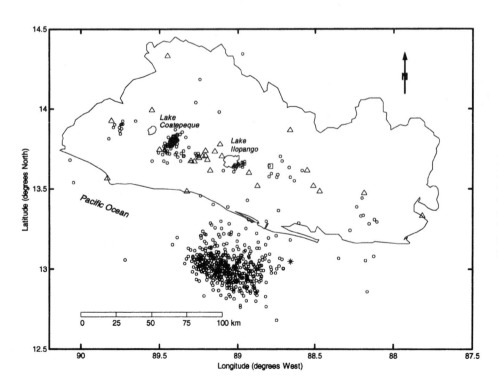

Figure 2. Map of El Salvador showing epicenter of the January 13 main shock (star) and epicenters of earthquakes (circles) that occurred within a two-day period after the main shock. Triangles—strong-motion stations that recorded the January 13 earthquake. Square—station VI.

Depth = 482699 – 0.339032 *Longitude* – 1.69855 *Latitude* (1)

where *Depth*, *Longitude*, and *Latitude* are in meters, and the projection used for these coordinates is Lambert NAD 27. This projection is used in order to be consistent with the convention of the existing geographical data for El Salvador.

The strike angle of the calculated plane is 281.3°, and the dip angle was fixed to 60° according to the fracture plane solution given by Benito et al. (this volume, Chapter 25). Both the strike and dip angles closely agree with the fault plane solution, which is a normal fault due to a rupture of the subducting Cocos plate oceanic crust. The borders of the rupture plane are constrained by the distribution of aftershocks.

In Table 4, the trend of peak vertical acceleration versus distance to rupture surface presents a more consistent trend, showing that this parameter of distance agrees more adequately with the ground-motion attenuation behavior.

February 13 Earthquake

The source parameters of this earthquake reported by NEIC (National Earthquake Information Center) are origin time, 14:22:05.82 u.t.; coordinates of epicenter, 13.671° N and 88.938° W; focal depth, 10 km; and the magnitudes reported are 5.5 m_b, 6.5 M_S, and 6.6 M_W. Using the Wells and Coppersmith (1994) relations, the rupture length was estimated to be 21.6 km, and the range of epicentral distances for the recording stations is between 78.7% and 284.3% of the above estimated length, which indicates that the source-site distances are comparable to the source dimensions, and therefore it is possible to anticipate that a point model of the source may not be adequate for the calculation of distances in the attenuation analysis.

The conditions for a fit to a rupture plane using the distribution of aftershocks are very poor in this case. The reason is that the high density and very close spacing of mapped geological faults in the surroundings of the epicentral area makes the task of assigning every event to a single fault almost impossible.

In order to better determine the fault rupture, the procedure described by Cepeda (2001b) was followed. The alignment and length of local faults were identified on a 1:100,000 geological map of El Salvador and compared with the strike of the focal mechanism presented by Benito et al. (this volume) and the above estimate of rupture length. A good agreement was found with the fault shown in Figure 3B, and this fault has been used as the earthquake source. The length of the surface trace of this fault is 23.7 km, compared to the 21.6 km given by the

TABLE 4. RECORDS FOR JANUARY 13, 2001 EARTHQUAKE:
PEAK ACCELERATION, PEAK VELOCITY AND PSEUDOSPECTRAL ACCELERATIONS FOR 0.3 s AND 1.0 s

| Code | Rupture distance (km) | North-South | | | | Vertical | | East-West | | | |
		PGA (cm/s²)	PGV (cm/s)	PSA T = 0.3 s (cm/s²)	PSA T = 1 s (cm/s²)	PGA (cm/s²)	PGV (cm/s)	PGA (cm/s²)	PGV (cm/s)	PSA T = 0.3 s (cm/s²)	PSA T = 1 s (cm/s²)
LI	61.3	1092	53.2	1290	285	604	15.9	564	35.5	958	237
ZA	72.0	255	12.3	362	140	247	8.6	305	19.1	410	229
PA	75.4	173	9.2	223	174	87	7.3	151	9.4	182	111
SM	77.7	864	27.8	1607	350	432	16.1	702	40.4	2011	415
NO	78.9	569	37.5	1063	402	430	18.2	479	26.4	1789	319
TE	79.2	486	57.0	1103	385	239	18.5	477	34.2	1112	389
TR	79.4	453	18.6	1017	163	235	18.0	364	24.2	1279	220
ST	79.4	588	60.5	1119	514	464	21.6	761	43.3	2570	343
RF	81.0	204	19.5	476	233	184	13.9	205	16.6	470	251
RS	81.0	317	27.6	1207	280	323	15.3	298	22.9	1026	268
OB	81.4	420	38.4	1096	555	301	13.0	372	26.2	1052	507
SS	83.2	267	15.0	544	211	157	11.3	247	20.3	656	330
EX	83.7	295	25.4	962	441	151	11.9	273	17.4	584	394
BA	85.2	154	25.2	615	491	163	15.2	195	31.2	485	454
CA	86.5	106	18.6	209	282	49	4.2	96	14.6	226	183
VS	86.5	301	21.9	N.D.	N.D.	207	12.5	306	37.3	N.D.	N.D.
DB	87.1	221	23.2	473	523	157	11.3	245	19.2	502	183
QC	87.2	149	23.5	365	209	120	10.2	183	16.0	574	163
AR	87.3	589	49.6	751	1050	219	19.6	445	53.3	1183	657
MG	91.9	118	12.1	215	252	88	6.0	133	12.8	204	225
TO	92.0	258	23.1	594	424	201	9.8	230	23.2	611	208
SE	108.7	81	8.5	213	115	57	6.2	60	9.1	190	71
SA	112.1	133	19.5	373	407	50	6.2	84	13.6	169	175
CU	113.6	76	13.8	205	100	62	4.0	78	8.6	179	149
AH	114.8	210	16.6	335	335	121	10.8	143	14.9	318	324
CM	144.1	14	1.7	23	18	N.D.	N.D.	12	2.2	21	25

Note: Stations ordered with increasing rupture distance.
Rupture distance—distance to plane of rupture; PGA—horizontal peak ground acceleration; PGV—horizontal peak ground velocity; PSA—pseudospectral acceleration; T—period; N.D.—No data.

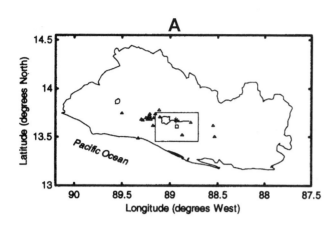
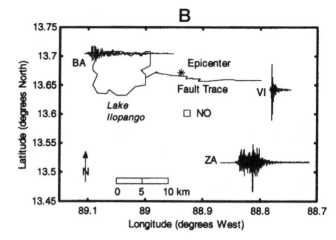

Figure 3. A: Map of El Salvador showing epicenter of the February 13 main shock (star) and strong-motion stations that produced records (triangles). Square—station NO. Rectangle marks the epicentral area, which is expanded in part B of this figure. B: Map of epicentral region of the February 13 earthquake. Star—epicenter. Square—station NO. North-south acceleration time histories are presented for stations BA, VI, and ZA.

empirical estimate. The E-W alignment of the surface trace matches closely the strike angle deduced from the focal mechanism. Since the reported dip angle is steep, the fault trace is approximated by the vertical projection of the rupture surface. In the following sections, for each recording station, the source to site distance reported for this event is the shortest horizontal distance to the fault trace shown in Figure 3.

Cepeda (2001b) presents an additional feature of the February 13 earthquake. This feature is an effect of strong-motion directivity, which is particularly observed in the recorded ground motions in the surroundings of the fault trace. In Figure 3B, if a west-east unilateral direction of the rupture is assumed, the expected acceleration time histories at VI (in the rupture direction) should show very high frequencies and large amplitudes, whereas the signal at BA (opposite to the rupture direction) should have low amplitudes and a low frequency content. It is assumed that the geologic conditions in VI and BA are not very different and hence are not expected to have a strong influence in the ground motions. These trends are confirmed after comparing with the observed ground motions. Figure 3B shows the north-south component for both stations. This figure also shows the trace recorded at ZA. The envelope of this record appears to indicate a multiple rupture of the fault. This same envelope was also observed at PA, which has almost the same latitude as NO and is slightly to the west of the limits of the window shown in Figure 3B.

CHARACTERISTICS OF STRONG-MOTION RECORDING STATIONS

Tables 2 and 3 show the characteristics of the strong-motion stations and instruments. The geology has been obtained from the digital versions of the 1:500,000 geological map of El Salvador

and the 1:15,000 geological map of San Salvador, digitized and geographically referenced by Brizuela and Menjívar (2001). When a site was found in both maps, the classification and description was taken from the 1:15,000 map of San Salvador. Each station was assigned a National Earthquake Hazard Reduction Program (NEHRP) site class (see Dobry et al., 2000), which is listed in the last column of Table 3. The NEHRP site class is originally calculated as the average shear wave velocity within the top 30 m (FEMA, 1997). Rock sites have average velocities of 760 m/s to 1500 m/s. NEHRP C sites are very dense soils or soft rocks with average velocities of 360 m/s to 760 m/s, and NEHRP D sites are defined as stiff soils with average velocities of 180 m/s to 360 m/s.

Very limited geotechnical information was available from the stations in order to make direct assignments of site classes. Standard Penetration Test blow counts were made available in stations RS, SS, and VS by SNET. Shear wave velocity profiles for stations CI and UC are presented by Italtekna-Italconsult (1987). Hence, site classification according to NEHRP was directly made only for the above five stations. For the rest of the stations, which produced 98% of the records, the assignation of a site class was made indirectly based on a combination of the following indicators: the surface geology as indicated by the geologic maps, the shear wave velocities reported by Italtekna-Italconsult (1987) for different types of volcanic materials in El Salvador, and the distribution of residuals at every station in a preliminary strong-motion attenuation analysis assuming uniform site conditions in all stations. The third indicator was considered only in the stations of the UCA network, because the number of records was considered significant and the average values of residuals were assumed to reflect the geologic site conditions.

Figures 2 and 3 show the distribution of recording stations for the January 13 and February 13 earthquakes.

PROCESSING OF RECORDS

The type and number of records used in the present study were 32 analogue records from SMA-1 instruments and 447 digital records from SSA-2 accelerographs.

The SMA-1 film records from the January 13 earthquake were processed by the USGS, and the rest of the SMA-1 records were processed by SNET. The analogue-to-digital conversion was performed by digitizing the original record from film.

The SSA-2 digital records produced by the GESAL and UCA networks were processed for this study using the Strong Motion Analyst software by Kinemetrics. The processing of the SSA-2 records followed these steps:

1. File conversion from SSA format to EVT format.

2. File conversion from EVT format to uncorrected acceleration V1 format.

3. File conversion from uncorrected acceleration V1 format to corrected acceleration V2 format. This step includes an instrument and baseline correction, a high-pass filtering of velocity and displacement, and the computation of peak acceleration, velocity, and displacement. The correction method that was selected is the Shakal and Ragsdale method (Shakal and Ragsdale, 1984). The corner frequency used for high-pass filtering varied in the range from 0.12 to 0.20 Hz. The terminal frequency used for low-pass filtering was 45 Hz.

4. File conversion from corrected acceleration V2 format to Fourier and response spectra V3 format. Response spectra were

computed for 5% damping and for periods of 0.3 s and 1.0 s. Specifically, the values of pseudo spectral velocity (PSV) and pseudo spectral acceleration (PSA) were computed. The above periods were selected because they are of interest for earthquake engineering purposes, since these values are spectral periods suggested by the current NEHRP seismic code design guidelines for the assessment of short- and long-period response to strong ground motion (FEMA, 1997). These spectral parameters will also be used later in the attenuation analysis of the 2001 records.

STRONG-MOTION PARAMETERS

Peak values of acceleration and velocity for each component are listed in Tables 4 and 5 for the January 13 and February 13 earthquakes, respectively. Also pseudo spectral acceleration and pseudo spectral velocity are presented. Stations are sorted by increasing source to site distance. The parameters selected for the distance were described previously.

The San Vicente (VI) station recorded the January 13 earthquake, but the record is cut due to an instrument malfunction during the earthquake. The instrument stopped recording during the earthquake due to a power failure in the electrical supply and in the main internal battery. Peak ground acceleration (PGA) values for this cut record are 154.6, 138.2, and 118.2 cm/s^2 for the N-S, vertical, and E-W components, respectively. These values are likely to be lower than the actual PGA because the record envelope and duration appear not to have reached the strongest

TABLE 5. RECORDS FOR FEBRUARY 13, 2001 EARTHQUAKE:
PEAK ACCELERATION, PEAK VELOCITY, AND PSEUDOSPECTRAL VELOCITIES FOR 0.3 s AND 1.0 s

Code	Rupture distance (km)	North-South				Vertical		East-West			
		PGA (cm/s^2)	PGV (cm/s)	PSV T = 0.3 s (cm/s)	PSV T = 1 s (cm/s)	PGA (cm/s^2)	PGV (cm/s)	PGA (cm/s^2)	PGV (cm/s)	PSV T = 0.3 s (cm/s)	PSV T = 1 s (cm/s)
VI	2.5	425	14.6	25.7	8.2	229	4.0	232	6.2	17.7	4.9
BA	12.5	104	25.6	16.2	41.1	121	6.9	139	22.3	18.7	45.6
ZA	15.5	400	20.0	33.9	44.4	257	9.8	296	20.5	35.9	24.7
TO	18.0	238	30	29.9	77.6	235	10.6	246	24.6	33.2	28.5
DB	18.2	98	14.8	13.0	17.3	54	4.6	92	12.2	14.1	21.4
CI	19.3	135	19.9	8.6	10.1	58	3.8	69	8.4	10.4	25.8
PA	20.2	182	7.5	11.3	7.4	44	2.2	105	4.6	6.1	5.0
OB	21.7	105	6.7	12.5	22.6	67	3.3	102	13.9	13.7	10.1
EX	23.1	121	15.2	9.0	19.1	51	2.7	97	6.1	9.9	8.3
VF	24.3	40	3.1	3.3	11.4	31	2.9	39	7.2	2.4	7.9
VS	24.3	76	8.2	5.8	9.0	45	3.5	58	8.7	7.5	14.5
SS	24.9	64	5.7	5.6	11.5	43	2.6	70	10.8	6.0	4.9
UC	25.7	N.D.	N.D.	N.D.	N.D.	39	2.1	57	8.5	8.4	14.5
RF	26.2	42	3.7	4.2	6.9	26	1.8	42	7.4	4.2	7.3
RS	26.2	57	3.9	6.8	8.3	34	2.2	62	8.1	7.5	6.5
QC	27.4	19	6.4	3.9	4.6	17	2.4	26	5.0	2.9	3.9
TE	30.3	46	6.4	2.8	8.7	22	2.0	40	4.8	3.4	6.1
ST	32.6	38	6.4	2.8	7.0	19	2.2	41	7.4	3.2	11.2
BE	34.3	32	4.4	5.3	5.8	30	2.3	70	6.8	6.4	10.9
LI	40.5	90	4.5	12.5	3.7	35	2.4	92	5.0	5.8	3.4
AR	55.1	28	3.2	3.5	4.4	26	1.3	36	1.9	2.7	4.2

Note: Stations ordered with increasing rupture distance. Rupture distance—horizontal distance to surface projection of rupture; PGA—horizontal peak ground acceleration; PGV—horizontal peak ground velocity; PSV—pseudospectral velocity; T—period; N.D.—No data.

part of the shaking, and it was decided not to include it in any part of this study. In addition, there is a record of the February 13 event in San Pedro Nonualco (NO) that was not included in the analysis. In this station, the floor anchor was found broken in the first maintenance visit after this earthquake, presumably due to the very strong shaking (PGA values are 1105.5, 729.4, and 1360.8 cm/s^2 for N-S, vertical, and E-W components, respectively). Figure 4 shows the acceleration time histories of these two records.

Figures 5 and 6 present the observed PGA, PSA, and PSV values versus distance for the January 13 and February 13 earthquakes.

PREDICTION OF STRONG GROUND MOTION PARAMETERS

Regression analysis of the 479 accelerograms was applied to obtain strong ground motion relationships for horizontal PGA, horizontal PSA, and horizontal PSV, for in-slab subduction and shallow upper crustal earthquakes. The number of records for each station is summarized in Table 3.

The total data set comprised 479 triaxial records from 188 earthquakes, of which 61 have their origin in the subduction zone and 127 in the upper crustal faults systems. Table 6 summarizes the parameters characteristics of each subset.

Subduction Earthquakes

Cepeda et al. (1997b) compared a Central American database of 178 subduction earthquake records with the predictions by Alfaro

et al. (1990), Bommer et al. (1996), Crouse (1991), and Youngs et al. (1997), all recent studies of ground motion, which give attenuation estimates specifically for subduction tectonic regimes.

Figure 7A shows the magnitude-distance distribution for the 254 in-slab subduction earthquake records during 2001. The distribution shows a gap in the magnitude range from 5.9 to 7.6 M$_w$, which is explained by the fact that almost the entire subduction activity of 2001 occurred along the rupture area of the January 13 earthquake, and the energy release associated with this large-magnitude event did not leave conditions for the occurrence of a new large earthquake within the one-year time frame of the present study. This is different from the volcanic chain earthquakes. For example, in Figure 7B, the 6.6 M$_w$ and 5.3 M$_w$ events took place in different fault systems, namely around San Vicente volcano and San Salvador volcano, and in fact, the February 13 earthquake seems to have acted as a trigger for the 5.3 M$_w$ February 17 earthquake (Benito et al., this volume, Chapter 25).

As the magnitude gap in the subduction activity of 2001 includes earthquakes that are important for earthquake engineering purposes, in the present study the attenuation analysis is performed by taking a basic equation that includes earthquakes in the different ranges of magnitudes, and then by making the necessary adjustments to the basic equation for the residuals of the observed strong-motion parameters during 2001.

A recent attenuation equation for in-slab and interface subduction earthquakes is given by Atkinson and Boore (2003), herein referred to as AB03. This equation, which is used here as the basic equation, has been derived by using 1200 records from Japan, Cascadia, Mexico, and Central America, produced

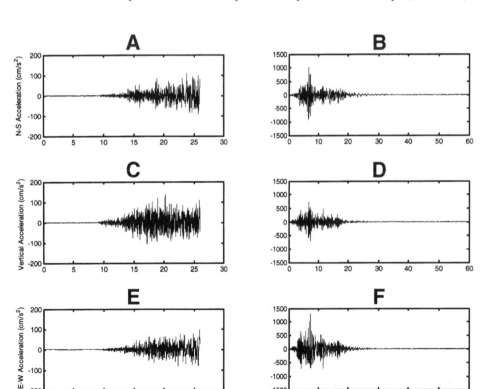

Figure 4. Acceleration time histories of January 13 earthquake record at station VI (left traces, A, C and E) and February 13 earthquake record at station NO (right traces, B, D, and F). A and B: North-south components. C and D: Vertical components. E and F: East-west components.

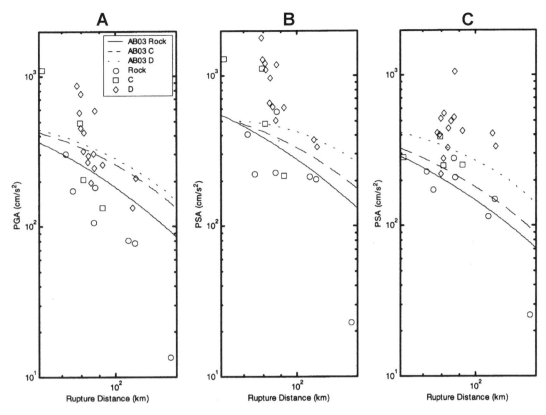

Figure 5. Comparison of observed strong-motion parameters for January 13 earthquake to predicted attenuation curves by Atkinson and Boore (2003). For the observed values, only the largest component is shown. A: Peak ground acceleration. B: Pseudo spectral acceleration for 0.3 s. C: Pseudo spectral acceleration for 1.0 s. AB03—Atkinson and Boore (2003). Circles—rock sites. Squares—NEHRP C sites. Diamonds—NEHRP D sites.

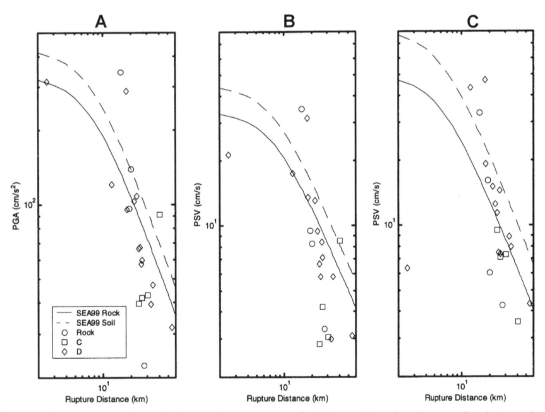

Figure 6. Comparison of observed strong-motion parameters for February 13 earthquake to predicted attenuation curves by Spudich et al. (1999). Observed values are geometric mean. A: Peak ground acceleration. B: Pseudo spectral velocity for 0.3 s. C: Pseudo spectral velocity for 1.0 s. SEA99—Spudich et al. (1999). Circles—rock sites. Squares—NEHRP C sites. Diamonds—NEHRP D sites.

TABLE 6. SUMMARY OF CHARACTERISTICS OF DATABASE USED IN ATTENUATION ANALYSIS

Type of event	Number of earthquakes	NEHRP D site records	NEHRP C site records	Rock site records	Total records	M_{min}	M_{max}	h_{min} (km)	h_{max} (km)	r_{min} (km)	r_{max} (km)
Subduction	61	153	60	41	254	2.8	7.7	26.1	111.8	57.5*	190.2*
Shallow upper crustal	127	160	21	44	225	2.4	6.6	1.9	26.4	0.4#	102.8#
Total	188	313	81	85	479	N.A.	N.A.	N.A.	N.A.	N.A.	N.A.

Note: M_{min}—minimum moment magnitude; M_{max}—maximum moment magnitude; h_{min}—minimum focal depth; h_{max}—maximum focal depth; r_{min}—minimum rupture distance; r_{max}—maximum rupture distance.

*distance to plane of rupture for events with large magnitudes or hypocentral distance for low and moderate magnitude events.

#horizontal distance to surface projection of rupture for events with large magnitudes or epicentral distance for low and moderate magnitude events.

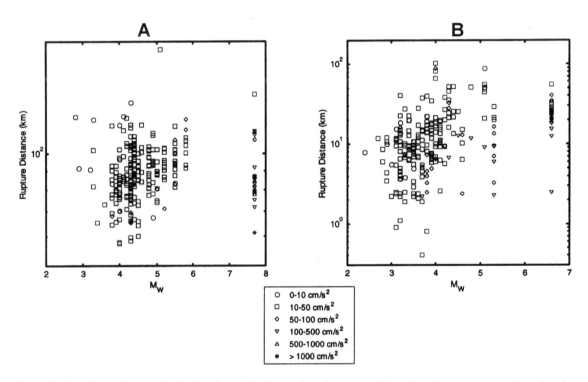

Figure 7. Magnitude-distance distribution for subduction earthquake records (A) and shallow upper crustal earthquake records (B). Marker types are classified by range of horizontal peak ground acceleration.

by more than 500 earthquakes of moment magnitude 5 to 8.3 recorded within 300 km. The equation has the following form:

$$\log Y = C_1 + C_2 M + C_3 h + C_4 R - g \log R + C_5 \, sl \, S_C + C_6 \, sl \, S_D + C_7 \, sl \, S_E \tag{2}$$

in which Y is peak ground acceleration (PGA) or 5% damped pseudo spectral acceleration (PSA) in cm/s², using the random horizontal component; in this study, the larger component will be used. M is moment magnitude; h is focal depth in km; $R = \sqrt{D_{fault}^2 + \Delta^2}$, in which D_{fault} is the closest distance to the

fault rupture surface, in km; in this study, D_{fault} is the hypocentral distance for all events other than the January 13 earthquake; $\Delta = 0.00724 \, (10^{0.507 \, M})$; $S_C = 1$ for NEHRP C soils, or 0 otherwise; $S_D = 1$ for NEHRP D soils, or 0 otherwise; $S_E = 1$ for NEHRP E soils, or 0 otherwise; $g = 10^{(0.301-0.01M)}$ for in-slab events; $sl = 1$ for $PGA_{rx} \leq 100$ cm/s² or frequency $f \leq 1$ Hz; $sl = 1 - (f-1)(PGA_{rx} - 100)/400$ for $100 < PGA_{rx} < 500$ cm/s² (1 Hz $< f < 2$ Hz); $sl = 1 - (f-1)$ for $PGA_{rx} \geq 500$ cm/s² (1 Hz $< f < 2$ Hz); $sl = 1 - (PGA_{rx} - 100)/400$ for $100 < PGA_{rx} < 500$ cm/s² ($f \geq 2$ Hz); $sl = 0$ for $PGA_{rx} \geq 500$ cm/s² ($f \geq 2$ Hz); PGA_{rx} is the predicted PGA on rock (NEHRP B site) in cm/s².

Table 7 presents the regression coefficients and standard deviations in the original AB03 equations and in the adjusted forms of the equations. The standard deviations in the original AB03 equations are the values reported by Atkinson and Boore (2003), whereas the standard deviations in the adjusted equations are calculated from the residuals of the records in 2001. The coefficients for PSA at a period of 0.3 s have been interpolated from the original coefficients given in the AB03 equations at periods of 0.2 s and 0.4 s. The interpolation has been performed assuming that there is a linear variation of log(PSA) versus log(frequency).

Figure 5 shows the predicted curves from AB03 for the January 13 earthquake along with the observed strong-motion parameters. In this plot it is interesting to see the behavior of the La Libertad (LI) station. LI is the closest station to the rupture surface and seems to be in the rupture direction (central station in the coast in Fig. 2). Consistently, this station shows a large underestimation for PGA (Fig. 5A) and for high-frequency PSA (Fig. 5B), whereas for low-frequency PSA (Fig. 5C) the predicted value agrees closely with the observed PSA. This high-frequency response was also observed by Bommer et al. (2002) when they showed the acceleration response spectrum for this station. A sharp high-frequency response seems to be also a local characteristic of LI, as it was presented by Bommer et al. (1997) for a moderate-magnitude subduction earthquake.

Figures 8A and 8B show the distribution of residuals in terms of magnitude and distance for PGA using AB03. The residuals are also classified by distance and magnitude, respectively. Figure 8A shows a clear linear dependence on magnitude, whereas observation of Figure 8B indicates an underestimation of PGA, without any dependence on distance. The underestimation is larger as the magnitude becomes smaller. The adjustment was performed to the constant term C_1 and the magnitude coefficient C_2. Figures 8C and 8D present the distribution of residuals after the adjustment of the coefficients. The fit of the observed 2001 data is similar to the database of the original AB03 equation, which is apparent when the adjusted standard deviation of 0.26 is compared to the slightly higher value of 0.27 reported by Atkinson and Boore (2003).

The distributions of residuals for PSA at periods of 0.3 s and 1.0 s are presented in Figures 9 and 10. Note the similarities of the distribution of residuals in Figures 9A and 10A compared to Figure 8A. The dependence on magnitude follows the same trend, even though the underestimation for low magnitudes is higher in PSA than in PGA. As in the case of the adjustment of the PGA equation, the coefficients that were modified are the constant term C_1 and the magnitude coefficient C_2. The standard deviations increase in the adjusted equations compared with the original AB03 equations.

An interesting observation from Table 7 is the trend of the adjustments in the C_1 and C_2 regression coefficients toward the coefficients for interface events. For example, the constant term in AB03 for PGA is −0.04713, and after the adjustment C_1 is 2.93078. This value is close to the coefficient 2.991 for interface events in the AB03 equation. A similar trend appears in the magnitude coefficient of 0.6909 for AB03 in-slab events and the smaller adjusted coefficient of 0.2877, compared to the AB03 coefficient of 0.03525 for interface events. These trends are also observed in the case of the coefficients for PSA. These observations impose a question about the type of the subduction events in the El Salvador database. Based on the focal mechanisms and the tectonics of the region, in-slab-type subduction earthquakes were assumed in the present analysis. However, it may be interesting in future studies to make an assessment and a revision of the location of earthquakes within the subduction area in the surroundings of El Salvador.

Shallow Upper Crustal Earthquakes

Analysis of attenuation for shallow upper crustal earthquakes in Central America has been also performed by Cepeda et al. (1997b). They compared a database of 116 Central American shallow crustal earthquake records with the estimates of Alfaro et al. (1990), Ambraseys et al. (1996), and Spudich et al. (1997).

TABLE 7. REGRESSION COEFFICIENTS AND STANDARD DEVIATIONS IN ATTENUATION EQUATIONS FOR IN-SLAB SUBDUCTION EVENTS

Strong motion parameter	Attenuation equation	C_1	C_2	C_3	C_4	C_5	C_6	σ
PGA	AB03	−0.04713	0.6909	0.01130	−0.00202	0.19	0.24	0.27
PGA	AB03 after adjustments	2.93078	0.2877	N.A.	N.A.	N.A.	N.A.	0.26
PSA (T = 0.3 s)	AB03*	0.2173	0.73915	0.00339	−0.00184	0.14	0.33	0.28#
PSA (T = 0.3 s)	AB03 after adjustments	3.31445	0.34496	N.A.	N.A.	N.A.	N.A.	0.32
PSA (T = 1 s)	AB03	−1.02133	0.8789	0.00130	−0.00173	0.10	0.30	0.29
PSA (T = 1 s)	AB03 after adjustments	1.85185	0.51846	N.A.	N.A.	N.A.	N.A.	0.33

Note: C_1, C_2, C_3, C_4, C_5 and C_6—regression coefficients; σ—standard deviation of residuals; PGA—horizontal peak ground acceleration (cm/s²); PSA—horizontal pseudo spectral acceleration (cm/s²); T—period; AB03—Atkinson and Boore (2003); N.A.—not applicable, coefficient not adjusted.
* linearly interpolated from Atkinson and Boore (2003) using regression coefficients for T = 0.2 s and T = 0.4 s.
# standard deviation in Atkinson and Boore (2003) for T = 0.2 s and T = 0.4 s.

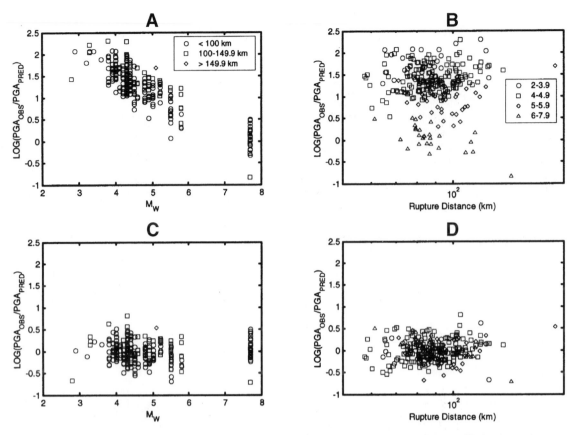

Figure 8. Distribution of residuals of peak ground acceleration versus magnitude (A) and distance (B) for the AB03 equation (Atkinson and Boore, 2003) using subduction earthquake records. C and D: Distributions of adjusted residuals.

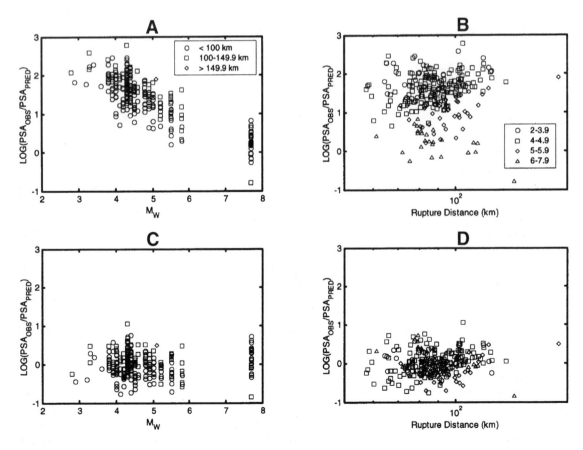

Figure 9. Distribution of residuals of pseudo spectral acceleration (period = 0.3 s) versus magnitude (A) and distance (B) for the AB03 equation (Atkinson and Boore, 2003) using subduction earthquake records. C and D: Distributions of adjusted residuals.

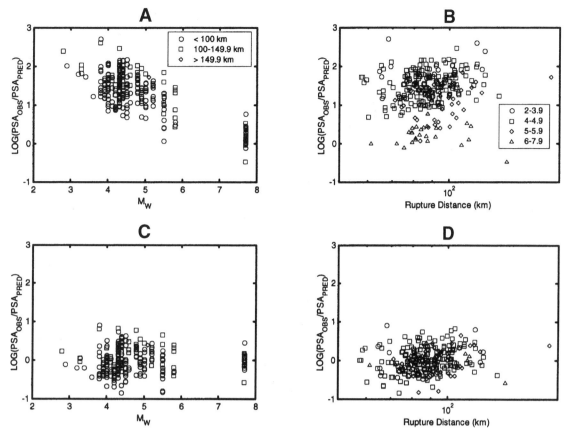

Figure 10. Distribution of residuals of pseudo spectral acceleration (period = 1.0 s) versus magnitude (A) and distance (B) for the AB03 equation (Atkinson and Boore, 2003) using subduction earthquake records. C and D: Distributions of adjusted residuals.

Figure 7B shows the magnitude-distance distribution for the shallow upper crustal records of 2001 classified according to horizontal PGA. The distribution covers most of the magnitude ranges, except in the range from 5.4 to 6.5 M_w.

As the above gap of magnitudes includes damaging earthquakes of interest for earthquake engineering purposes, the database is not complete, and in this study it is not considered appropriate to derive an attenuation equation based solely on these data. The preferred approach is to use an existing attenuation equation and make the necessary adjustments based on the distribution of the residuals of the observed data.

The Spudich et al. (1999) equation, SEA99, is used as the base for the analysis. This relation was developed based on 142 strong-motion records from 39 extensional regime earthquakes in the United States, Central America, Italy, Mexico, Greece, New Zealand, Turkey, and Holland. The original equation is directly applicable to earthquakes with moment magnitudes in the range of 5.0–7.7 and with source-site distances less than 70 km.

The SEA99 equation has this form:

$$\log Y = B_1 + B_2(M-6) + B_3(M-6)^2 + B_5 \log r + B_6 \Gamma \quad (3)$$

in which Y is PGA (in g) or pseudovelocity response (cm/s) at 5% damping for the geometric mean horizontal component of motion; B_1 to B_6 are the regression coefficients; M is moment magnitude; $r = \sqrt{r_{jb}^2 + H^2}$, in which r_{jb} is the closest horizontal distance (km) to the vertical projection of the rupture; in this study, r_{jb} is equal to the epicentral distance for all earthquakes, except in the case of the February 13 earthquake; H is a regression value; $\Gamma = 0$ for rock and $\Gamma = 1$ for soil.

Table 8 lists the regression coefficients and standard deviations in the original SEA99 equations and in the adjusted versions of the equations. The standard deviations in the original SEA99 equations are the values reported by Spudich et al. (1999), whereas the standard deviations in the adjusted forms are calculated from the residuals of the records in 2001.

Figure 6 shows the predicted attenuation curves by SEA99 for the February 13 earthquake along with the observed strong-motion parameters. In this case, the closest station to the fault is the San Vicente, VI, station (leftmost diamond in Fig. 6). The directivity effect at VI that was previously discussed by Cepeda (2001b) is also apparent in the variation of the observed PGA and PSV. It can be seen that PGA (Fig. 6A) and PSV at high

TABLE 8. REGRESSION COEFFICIENTS AND STANDARD DEVIATIONS
IN ATTENUATION EQUATIONS FOR SHALLOW UPPER-CRUSTAL EVENTS

Strong motion parameter	Attenuation equation	B_1	B_2	B_3	B_5	B_6	H	σ
PGA	SEA99	0.299	0.229	0	−1.052	0.112	7.27	0.203
PGA	SEA99 after adjustments	−0.0423	N.A.	N.A.	N.A.	N.A.	N.A.	0.288
PSV (T = 0.3 s)	SEA99	2.263	0.334	−0.070	−1.020	0.121	7.72	0.232
PSV (T = 0.3 s)	SEA99 after adjustments	2.270	N.A.	N.A.	N.A.	N.A.	N.A.	0.347
PSV (T = 1 s)	SEA99	2.276	0.450	−0.014	−1.083	0.210	6.01	0.269
PSV (T = 1 s)	SEA99 after adjustments	0.6949	N.A.	N.A.	−0.0585	N.A.	N.A.	0.370

Notes: B_1, B_2, B_3, B_5 and B_6—regression coefficients; H—regression value used for calculation of the distance term in the attenuation equation; σ—standard deviation of residuals; PGA—peak ground acceleration (cm/s^2) for the geometric mean horizontal component; PSV—pseudo spectral velocity (cm/s) for the geometric mean horizontal component; T—period; SEA99—Spudich et al. (1999); N.A.—not applicable, coefficient not adjusted.

frequencies (Fig. 6B) have a close agreement with the predicted values. However, PSV at low frequencies (Fig. 6C) is significantly overestimated, which can be partially explained by the effects of the rupture directivity at this station. At San Bartolo station, BA, which is the second closest station to the fault (second leftmost diamond in Fig. 6), the observed strong-motion parameters also show an apparent effect of the rupture directivity, but this time for a station that is opposite to the direction of rupture propagation. For PGA (Fig. 6A) a slight overestimation is seen from the SEA99 attenuation curves. This overestimation is reduced for the high-frequency PSV (Fig. 6B). However, for the low-frequency PSV (Fig. 6C), the SEA99 equation predicts an underestimated value. These observations are an indication of the rupture directivity effects at BA, which are producing amplitudes that are large for the low-frequency contents of the signal and small for the high-frequency response.

Figure 11 shows the distributions of residuals for observed PGA in terms of magnitude and distance, and grouped in classes. Figures 11A and 11B show the residuals calculated from the original SEA99 equation. The trend of the data in these figures indicates some overestimation by the SEA99 equation. This overestimation does not have a clear dependence on magnitude (Fig. 11A). There is a slight dependence on distance in Figure 11B. This variation is small compared to the scatter of the residuals. After making an adjustment of the constant term B_1, Figures 11C and 11D present the distribution of residuals in the adjusted version of the SEA99 equation.

In the case of the observed pseudo spectral velocity values for a period of 0.3 s, Figures 12A and 12B show that there is a balanced distribution of overestimations and underestimations. Again, there seems to be no clear dependence on magnitude or distance. After adjusting the constant term B_1, Figures 12C and 12D present the distribution of residuals in the adjusted form of the SEA99 equation. The constant term in Table 8 shows a slight increase from

the original 2.263 to the adjusted 2.270, which is an indication that the original SEA99 is only making a small underestimation.

The observed pseudo spectral velocities for a period of 1.0 s are generally lower than the predicted values, as it is shown in Figures 13A and 13B. Some trend in the distribution in terms of distance is present in Figure 13B. In this case, the regression was adjusted in the constant term B_1 and in the distance coefficient B_5. Figures 13C and 13D present the distribution of residuals. Note the significant increase in the distance coefficient B_5 after making the adjustments: from −1.083 to −0.0585. This is an indication of a low geometrical spreading attenuation of the low-frequency contents of the strong ground motion.

DISCUSSION

Our estimates of attenuation for shallow upper crustal and subduction earthquakes have a number of aspects that deserve further study and review. Firstly, it is necessary to collect more records from moderate-magnitude events in order to complete the magnitude-distance distribution, particularly in the range of M_W 5.9 to 7.6 for subduction earthquakes and M_W 5.4 to 6.5 for shallow upper crustal events. Secondly, the attenuation characteristics of the 2001 subduction earthquakes show a closer agreement with the characteristics of interface-type events rather than with the assumed in-slab type. This observation suggests the necessity of further studies of the type of subduction earthquakes in the surroundings of El Salvador. In a third aspect, the ground motion attenuation for the February 13 earthquake shows some indication of the effects of rupture directivity. Bommer et al. (2001) also identified rupture directivity effects in the strong-motion recordings from an earlier upper crustal earthquake in 1986. These observations indicate that this effect should have proper consideration at least in the development of site-specific earthquake hazard assessments. Finally, the fact that geologic site

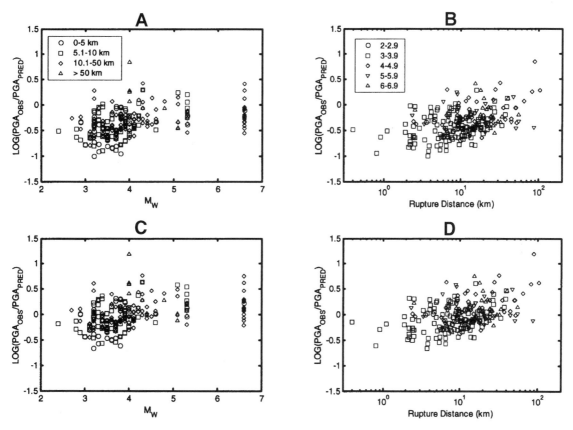

Figure 11. Distribution of residuals of peak ground acceleration versus magnitude (A) and distance (B) for the SEA99 equation (Spudich et al., 1999) using shallow upper crustal earthquake records. C and D: Distributions of adjusted residuals.

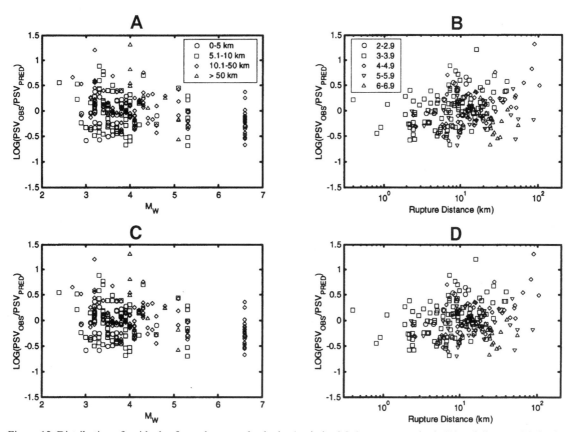

Figure 12. Distribution of residuals of pseudo spectral velocity (period = 0.3 s) versus magnitude (A) and distance (B) for the SEA99 equation (Spudich et al., 1999) using shallow upper crustal records. C and D: Distributions of adjusted residuals.

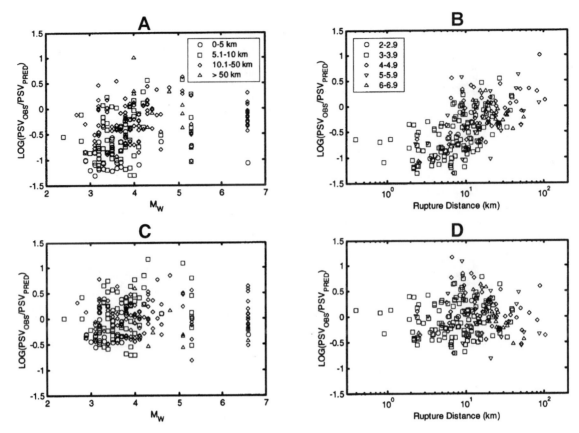

Figure 13. Distribution of residuals of pseudo spectral velocity (period = 1.0 s) versus magnitude (A) and distance (B) for the SEA99 equation (Spudich et al., 1999) using shallow upper crustal earthquake records. C and D: Distributions of adjusted residuals.

conditions are known in detail for only 2% of the strong-motion database is a clear indication that the task of performing detailed site investigations at every station is urgent in order to establish site classes more reliably.

ACKNOWLEDGMENTS

Our colleagues from SNET (Servicio Nacional de Estudios Territoriales) have provided very helpful assistance in the provision of the necessary seismological, strong-motion, and geotechnical data in the SNET stations. Special thanks to Manuel Díaz, Douglas Hernández, and Griselda Marroquín. Their comments and exchange of ideas have also been an invaluable resource for our study. Carlos Pullinger from SNET and Patricia de Hasbun from UCA (Universidad Centroamericana "José Simeón Cañas") have devoted their interest and efforts to strengthen the interinstitutional bonds between SNET, UPM (Universidad Politécnica de Madrid), and UCA.

José Rivas from GESAL (Geotérmica Salvadoreña) kindly provided the records and information from their seismological network in Berlín, Usulután.

David Boore from the U.S. Geological Survey and Conrad Lindholm from NORSAR provided very useful comments after their in-depth and rigorous revisions. The attenuation analysis for the spectral ordinates was suggested by David Boore. Julian Bommer from Imperial College London also provided very detailed and rigorous observations that greatly improved the contents of this paper.

REFERENCES CITED

Alfaro, C.S., Kiremidjian, A.S., and White, R.A., 1990, Seismic zoning and ground motion parameters for El Salvador: The John A. Blume Earthquake Engineering Center Report No. 93, Palo Alto, California, Stanford University, 157 p.

Ambraseys, N.N., Simpson, K.A., and Bommer, J.J., 1996, Prediction of horizontal response spectra in Europe: Earthquake Engineering and Structural Dynamics, v. 25, p. 371–400.

Atkinson, G.M., and Boore, D.M., 2003, Empirical ground-motion relations for subduction-zone earthquakes and their application to Cascadia and other regions: Bulletin of the Seismological Society of America, v. 93, no. 4, p. 1703–1729.

Benito, M.B., Cepeda, J.M., and Martínez Díaz, J.J., 2004, Analysis of the spatial and temporal distribution of the 2001 earthquakes in El Salvador, *in* Rose, W.I., et al., eds., Natural hazards in El Salvador: Boulder, Colorado, Geological Society of America Special Paper 375, p. 339–356 (this volume).

Bommer, J.J., Hernández, D.A., Navarrete, J.A., and Salazar, W.M., 1996, Seismic hazard assessments for El Salvador: Geofísica Internacional, v. 35, p. 227–244.

Bommer, J.J., Udías, A., Cepeda, J.M., Hasbun, J.C., Salazar, W.M., Suárez, A., Ambraseys, N.N., Buforn, E., Cortina, J., Madariaga, R., Méndez, P., Mezcua, J., and Papastamatiou, D., 1997, A new digital accelerograph network for El Salvador: Seismological Research Letters, v. 68, p. 426–437.

Bommer, J.J., Georgallides, G., and Tromans, I.J., 2001, Is there a near-field for small-to-moderate magnitude earthquakes?: Journal of Earthquake Engineering, v. 5, no. 3, p. 395–423.

Bommer, J., Benito, B., Ciudad-Real, M., Lemoine, A., López-Menjívar, M., Madariaga, R., Mankelow, J., Méndez de Hasbun, P., Murphy, W., Nieto-Lovo, M., Rodríguez-Pineda, C.E., and Rosa, H., 2002, The El Salvador earthquakes of January and February 2001: Context, characteristics, and implications for seismic risk: Soil Dynamics and Earthquake Engineering, v. 22, p. 389–418.

Brizuela, B., and Menjívar, L., 2001, Sistemas de información geográfica en el manejo de peligros naturales [B.S. thesis]: Universidad Centroamericana "José Simeón Cañas," El Salvador, 184 p.

Cepeda, J.M., 2001a, Información acelerográfica de los terremotos del 13 de enero y 13 de febrero de 2001: Revista Asociación Salvadoreña de Ingenieros y Arquitectos, March, no. 139, p. 28–33.

Cepeda, J.M., 2001b, Ciento cincuenta días después del 13 de enero y 15 años después del 10 de octubre: algunas consideraciones técnicas sobre los sismos y sus efectos: Revista Estudios Centroamericanos, ECA, v. 631–632, p. 459–472.

Cepeda, J.M., Salazar, W.M., and Bommer, J.J., 1997a, Caracterización de la red acelerográfica "Talulin," *in* Méndez de Hasbun, P., and Hasbun, J.C., eds., Proceedings, Seminar for the Assessment and Mitigation of Seismic Risk in Central America, San Salvador, El Salvador, p. 109–118.

Cepeda, J.M., Salazar, W.M., and Bommer, J.J., 1997b, Ecuaciones de atenuación para Centroamérica: I. Aceleración máxima del terreno, *in* Méndez de Hasbun, P., and Hasbun, J.C., eds., Proceedings, Seminar for the Assessment and Mitigation of Seismic Risk in Central America, San Salvador, El Salvador, p. 119–128.

Crouse, C.B., 1991, Ground-motion attenuation equations for earthquakes on the Cascadian subduction zone: Earthquake Spectra, v. 7, p. 201–236.

Dobry, R., Borcherdt, R., Crouse, C., Idriss, I., Joyner, W., Martin, G., Power, M., Rinne, E., and Seed, R., 2000, New site coefficients and site classification system used in recent building seismic code provisions: Earthquake Spectra, v. 16, p. 41–67.

FEMA, 1997, NEHRP recommended provisions for seismic regulations for new buildings and other structures: FEMA 302, Federal Emergency Management Agency, Washington, D.C., 335 p.

Italtekna-Italconsult, 1987, Valutazione della pericolosita sismica nelle aree del Distretto Sanitario A3 (San Salvador) e del Distretto 7 (Apopa). Parte 4ª—studi di risposta sismica locale ed elaborazione delle carte di Riconstruzione, Republica Italiana, Ministerio degli Affari Esteri, Direzione Generale per la Cooperazione allo Sviluppo, 300 p.

Lay, T., and Wallace, T.C., 1995, Modern global seismology: San Diego, California, Academic Press, 521 p.

Shakal, A.F., and Ragsdale, J.T., 1984, Acceleration, velocity and displacement noise analysis for the CSMIP accelerogram digitization system, *in* Proceedings, 8th World Conference on Earthquake Engineering, San Francisco, v. 2, p. 111–118.

Spudich, P., Fletcher, J.B., Hellweg, M., Boatwright, J., Sullivan, C., Joyner, W.B., Hanks, T.C., Boore, D.M., McGarr, A., Baker, L.M., and Lindh, A.G., 1997, SEA 96—A new predictive relation for earthquake ground motions in extensional tectonic regimes: Seismological Research Letters, v. 68, p. 190–198.

Spudich, P., Joyner, W.B., Lindh, A.G., Boore, D.M., Margaris, B.M., and Fletcher, J.B., 1999, SEA 99—A revised ground motion prediction relation for use in extensional tectonic regimes: Bulletin of the Seismological Society of America, v. 89, no. 5, p. 1156–1170.

Wells, D.L., and Coppersmith, K.J., 1994, New empirical relationships among magnitude, rupture length, rupture width, rupture area and surface displacement: Bulletin of the Seismological Society of America, v. 84, no. 4, p. 974–1002.

Youngs, R.R., Chiou, S.J., Silva, W.J., and Humphrey, J.R., 1997, Strong motion attenuation relationships for subduction zone earthquakes: Seismological Research Letters, v. 68, p. 58–73.

MANUSCRIPT ACCEPTED BY THE SOCIETY JUNE 16, 2003

Geological Society of America
Special Paper 375
2004

Monte Carlo seismic hazard maps for northern Central America, covering El Salvador and surrounding area

Behrooz Tavakoli*
David Monterroso
Uppsala University, Department of Earth Sciences, Seismology, Villavägen 16, S-752 36 Uppsala, Sweden

ABSTRACT

A Monte Carlo approach is utilized to evaluate the ground motion hazard and its uncertainties in northern Central America. In this approach, the seismic source boundary location is only considered to construct different earthquake catalogs. For each catalog, a Monte Carlo simulation is then used to generate numerous synthetic catalogs that have the same properties as the observed catalog but with random occurrences of earthquakes following the Poisson model and with epicentral locations randomly determined from the observed catalog. Selecting models to describe seismic source zones and earthquake recurrence is not required since such considerations are implicit in the synthetic earthquake catalogs. Thus, uncertainties related to gathering and manipulating the appropriate data for seismic hazard parameters and source zone boundaries are not propagated through the analysis into the hazard result. The characteristic earthquake recurrence model is used to describe the occurrence of large-magnitude events from the geological data. A set of new seismic hazard maps, exhibiting probabilistic values of peak ground acceleration (PGA) with 50%, 10%, and 5% probabilities of exceedance (PE) in 50 yr, is presented for a large area of northern Central America, including El Salvador and Guatemala. The hazard is highest in the coastal regions adjacent to the megathrust and the transcurrent fault zones. As a case history, disaggregation of seismic hazard is carried out for cities of San Salvador and Guatemala by using a spatial distribution of epicenters around these sites to select design ground motion for seismic risk decisions.

Keywords: design earthquake, disaggregation, El Salvador, Monte Carlo simulation, northern Central America, seismic hazard, synthetic catalog, uncertainty.

INTRODUCTION

Quantitative estimation of earthquake ground motion at a particular site may be analyzed by deterministic or probabilistic methods. The deterministic seismic hazard approach (DSHA) uses ordinary variables that are assigned to a particular earthquake scenario. It consists of a postulated occurrence of an earthquake of fixed size (controlling earthquake) occurring at a specified location (source-to-site distance). The probabilistic seismic hazard approach (PSHA) includes one or more independent variables that are characterized by a probability distribution from which values are drawn. Probabilistic concepts allow spatial and temporal uncertainties to be explicitly considered in the evaluation of seismic hazard. A deterministic model makes a fixed-point estimate while a probabilistic model estimates a value range, which reflects the uncertainty about the true value. Thus, both approaches have their advantages and disadvantages. To take the advantages of both approaches to obtain a uniform basis for hazard evaluation, the DSHA must be converted into the PSHA in a described seismicity model. This helps to characterize the range of potential values in the hazard evaluation and assess the probability of exceeding specific target values. Results can

*Behrooz.Tavakoli@geo.uu.se

Tavakoli, B., and Monterroso, D., 2004, Monte Carlo seismic hazard maps for northern Central America, covering El Salvador and surrounding area, *in* Rose, W.I., Bommer, J.J., López, D.L., Carr, M.J., and Major, J.J., eds., Natural hazards in El Salvador: Boulder, Colorado, Geological Society of America Special Paper 375, p. 423–433. For permission to copy, contact editing@geosociety.org. © 2004 Geological Society of America

be presented as frequency or cumulative distributions, each of which corresponds clearly to the range and likelihood of possible values. It also illustrates the contributions of different magnitude-distance pairs to the exceedance of the probabilistic ground motion, which is useful to define the design earthquake(s) for the dominant contributor(s). Statistical regression analysis and Monte Carlo simulation are two different techniques to estimate such ranges of values. A disadvantage of statistical regression analysis is that one should follow a prescribed fitting procedure, and hence the systematic control of uncertainty is complicated. One of the main tasks in Monte Carlo simulation is the generation of random numbers from prescribed probability distribution to incorporate the uncertainty. The simulation process for a given set of generated random numbers is completely deterministic. This property is extremely useful for hazard processes. However, a realistic probability distribution is required to fit each uncertain variable. In general, the normal and lognormal probability distributions are used for this purpose. A set of numbers that follow the normal distribution can be obtained by averaging several uniform random numbers ($0\rightarrow1$). For sufficiently large sample size, the central limit theorem ensures the normality of the distribution of the averages. If the resulting distribution does not fit the data, then the uniformly distributed numbers can be converted into a set with a more suitable distribution. For instance, the frequency of earthquakes that occur in the catalog decreases in accordance with an increase in a given earthquake magnitude, hence the probability of selection of each magnitude value is not uniform. Those magnitudes that occur more frequently in the catalog contribute more to the seismic hazard. Thus, if earthquakes are occurring randomly in time and space, a Gutenberg-Richter recurrence relation can be utilized as a formula in conversion.

The approach commonly used to produce seismic hazard maps for Central America is the conventional PSHA developed by Cornell (1968). In this region, hazard assessments have assumed that the area around a given site can be divided into seismic source zones with different activity rates (Algermissen et al., 1988; Alfaro et al., 1990; Singh et al., 1993; Bommer et al., 1998; Cáceres and Kulhánek, 2000). There is much disagreement among experts concerning the demarcation of seismic source zones, the estimation of seismic hazard parameters for those zones, and the hazard results. Comparison of the horizontal PGA seismic hazard maps with 10% probability of exceedance (PE) in 50 yr, from different seismic hazard studies in El Salvador and surrounding areas, shows that the variation in the PGA depends on each of the input models and parameters incorporated into the hazard analysis (Bommer et al., 1996). Previous studies to estimate seismic hazard in other areas show that the Monte Carlo approach has many advantages over conventional PSHA techniques (Musson, 2000; Ebel and Kafka, 1999; Cramer et al., 1996). In this paper, we have chosen the Monte Carlo approach developed by Ebel and Kafka (1999), which eliminates the need to define source zones and hazard parameters to compute ground motion at the site. Using Monte Carlo simulation to convert a deterministic model to a probabilistic model involves three steps:

First, replace uncertain input values with generation of random numbers; second, run a simulation that evaluates the model many times and saves the outcome of each evaluation as one observation; and third, analyze the result of all observations.

The objective of the present paper is to take advantage of these approaches and techniques in preparation of seismic hazard maps in northern Central America, in particular El Salvador and Guatemala. First, we create a catalog using published data, convert all entries to a common magnitude scale (M_W), decluster the catalog using established methods, and create synthetic catalogs to test parameter sensitivities. Then we include the effect of uncertainties on the final hazard estimates and show the contribution of earthquake magnitude, M, distance, R, and other incorporated uncertainty to seismic hazard at a site from disaggregation studies.

SPATIAL DISTRIBUTION OF EPICENTERS

The spatial pattern of observed seismicity represents the expected pattern of future earthquake occurrences. It means that zones of spatially concentrated seismicity are often identified as potential source zones even if they have not generated more than moderate-magnitude earthquakes. When an epicenter and then a magnitude from an existing catalog are randomly selected, the general observed spatial distribution of seismicity remains relatively stable. It is assumed that large earthquakes, with probability of occurrence described by the Gutenberg-Richter relationship, can occur anywhere there is an epicenter in the catalog. The principal justification for this assumption is confirmed when the historical earthquakes are compared with more recent instrumental locations (Ambraseys and Adams, 2001). Figure 1 shows the pattern of seismicity in northern Central America and the western Caribbean. To form a uniform basis for an earthquake catalog, we homogenized all reported magnitudes to moment magnitude through a multistep process, since it provides a measure of the size of faulting events and more reliable estimates of seismic hazard. We have first used the empirical relations described in Wyss and Habermann (1982) to convert m_b into M_S. Relationships of Ekström and Dziewonski (1988) were subsequently applied to determine the seismic moment of each earthquake from M_S magnitude. The Harvard catalog of Centroid-Moment Tensor (CMT) solutions has been considered for the period 1976 through the present to provide most of the seismic moments from large earthquakes. As a final step, a unified moment magnitude was computed by using the relation of Hanks and Kanamori (1979). The first part of the compiled catalog used for the hazard analysis refers to the period from 1900 through 1963 (Rojas et al., 1993; White and Harlow, 1993; Ambraseys and Adams, 2001). We used an updated version of the catalog by Engdahl et al. (1998) for the period from 1964 through 1999, and PDE data (Preliminary Determinations of Epicenters from the National Earthquake Information Center, NEIC) for the period from 2000 to March 2002. In this study, the regional limits of the catalog are $12°$–$16°$ N and $82°$–$92°$ W. The catalog contains 4100 earthquakes of magnitude $M_W \geq 4.0$. Earthquake locations indicate that the spatial pattern

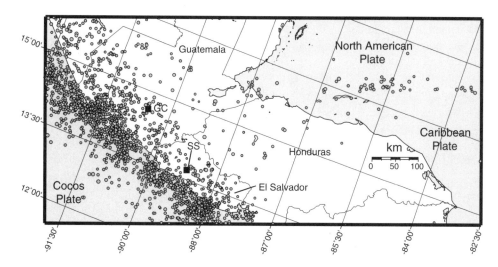

Figure 1. Epicentral map of northern Central America and western Caribbean showing independent earthquakes (circles) with $M_{\mathrm{w}} \geq 4.0$ for the period of 1900–2001. The sites of San Salvador (SS) and Guatemala City (GC) are locations of two major cities (squares) used in the seismic hazard analysis.

of seismicity is not uniform throughout this region. In order to have more uniform seismicity, the region is subdivided into five tectonic domains to construct different earthquake subcatalogs (Fig. 2). The boundaries of the tectonic domains in the region are related to the geometry and character of the documented recent active faults and the pattern of seismicity. The first subcatalog is based on areas that have given rise to large earthquakes (e.g., megathrust subduction zone and intraslab zone); the second, on areas of spatially concentrated seismicity along fault ruptures (e.g., transcurrent fault systems and volcanic chain zone). These two subcatalogs also include the locations of ongoing small- or moderate-magnitude earthquakes. Subsequently, the areas of intraplate seismicity, where earthquake rates are low and the association of earthquakes with faults is poorly understood, are the basis for the third subcatalog. In these areas there are difficulties in demarcating earthquake source zones, therefore a Monte Carlo approach may be most applicable. It does not require the identifi-

cation of seismic sources and their seismicity. This information is replaced by catalogs of past earthquakes in the region.

Spatially Smoothed Seismicity in the Region

We divided the region into small blocks (0.25° by 0.25°) and evaluated seismic hazards on each block. The size of block was set so that it was large enough to accommodate the maximum horizontal epicenter error for nearly all observed earthquakes. If values of X, Y, and Z represent location of events, the cumulative probability that the earthquake will occur around that point is defined as

$$F_{X,Y,Z}(x, y, z) = \sum_{\{x_i \leq x, y_j \leq y, z_k \leq z\}} p_{X;Y;Z}\left(x_i y_j z_k\right) \tag{1}$$

which is simply the sum of probabilities associated with all points (x_i, y_j, z_k) in the subset $\{x_i \leq x, y_j, y, z_k \leq z\}$. The spatially smoothed

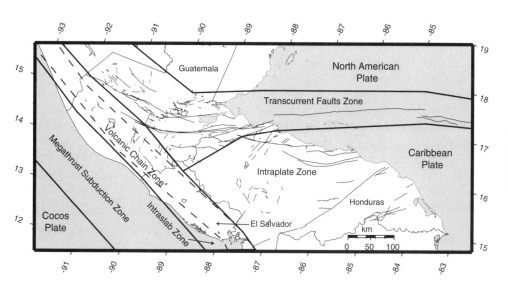

Figure 2. Seismic source boundary locations to construct different earthquake catalogs used to produce the Monte Carlo seismic hazard maps.

seismicity (Frankel, 1995) linked with each point in the region is considered as normally distributed around that point. To calculate seismicity, we choose Monte Carlo techniques. If σ_l is the standard deviation in location, we can just generate configurations for the location of each epicenter randomly and approximate the real seismicity averages by Monte Carlo averages

$$\bar{n}_i = \frac{1}{N}\left(\frac{\sum_{j=1}^{N}\exp\left(-E_{ij}/\sigma_l^2\right)}{\sum_{j=1}^{N}\exp\left(-E_{ij}/\sigma_l^2\right)}\right) \qquad (2)$$

in which E_{ij} is the square of the distance between the ith and jth cells, σ_l^2 is the variance in length units, n_j is the number of earthquakes in each cell, and N is the total number of events within the normal density function. To create the synthetic catalog, we allowed a random distance variation of up to 20 km from the different possible earthquake locations.

TEMPORAL DISTRIBUTION OF EARTHQUAKES

In order to model and predict earthquake recurrence, first the available subcatalogs should demonstrate Poissonian distribution or temporal-spatial independent earthquakes. Then the degree of completeness of the catalog must be assessed. This requires the removal of aftershocks and other dependent events. The algorithm used to identify and successfully eliminate clusters of aftershocks consists of a variant of time and space windows around the main shocks developed by Reasenberg (1985).

Catalog completeness is important in the calculation of recurrence model parameters since one must account for the fact that the detection threshold of some magnitudes may change over the time span of the catalog. The method used to determine the number of independent events consists of the definition of various magnitude ranges and completeness periods for which the earthquake detection probability is homogeneous. To estimate the completeness periods for various magnitude ranges, the magnitude scale is divided into half-magnitude intervals beginning with M_w 4.0 and ending with M_w 8.5. For each magnitude interval, we vary the time by five-year periods from 1900 to 1970, and yearly from 1970 to 2001. The completeness periods are quantified using the time intervals at which a significant change in the cumulative number of events is observed. For instance, completeness period estimates for the subduction source zone show that the observed catalog may be considered complete at five levels: $M_w \geq 7.0$ for 1900, $M_w \geq 6.5$ for 1925, $M_w \geq 6.0$ for 1940, $M_w \geq 5.0$ for 1965, and $M_w \geq 4.0$ for 1974.

Seismicity of each tectonic domain can be obtained by counting all the constrained earthquakes between two certain magnitude levels and by calculating a single b value from the corresponding subcatalog. This is accomplished by randomly generating synthetic catalogs and estimating their respective b values with the maximum likelihood procedure. The resulting b

values are assumed to be normally distributed. The mean value of this distribution is taken as an estimator for the b value of the subcatalog. This estimate is often based on short historical records, and may not be reliable for long time intervals. This constraint provides motivation for geological studies to characterize the size and recurrence rate of large earthquakes that are captured in a longer geological record. Estimating the specific characteristics of future ruptures, and relating those to magnitude, determines maximum magnitudes, which define the characteristic earthquake (Schwartz and Coppersmith, 1984). The size and recurrence rate of characteristic earthquakes can be obtained by a frequency-magnitude relationship with a slope based on geological data. For the Motagua-Polochic fault system, most large earthquakes occur within a rupture area of ~6200 km², at an average slip rate of ~2.1 cm/yr (DeMets et al., 1994). If we assume $\mu = 3.3 \times 10^{11}$ dyne/cm² for crustal rock, then the seismic moment rate will be 4.29×10^{25} dyne-cm/yr. The largest seismic moment measured (1976 Guatemala earthquake) is $M_0 = 2.81 \times 10^{27}$ dyne-cm, and the largest seismic moment expected within the zone is $M_0^{max} = 5.62 \times 10^{27}$ dyne-cm (Wells and Coppersmith, 1994). Introducing the last three values into the recurrence interval relation (Molnar, 1979), we would expect the recurrence interval of the 1976 Guatemala event to be 207 yr. According to the subduction model in El Salvador, the rupture area that generates large interface earthquakes is 23,000 km². Thus, the seismic moment rate will be 5.77×10^{26} dyne-cm/yr, where $\mu = 3.3 \times 10^{11}$ dyne-cm⁻² and the average slip rate of 7.6 cm/yr (DeMets et al., 1994) are used. The largest seismic moment that has been estimated by Geomatrix (1995) and McCaffrey (1997) for large plate-interface earthquakes is about 2.24×10^{28} dyne-cm. Therefore, the recurrence interval for the 2001 El Salvador earthquake with $M_0 = 4.57 \times 10^{27}$ dyne-cm is 40 yr.

As shown in Figure 3, this range of recurrence periods gives a range of $b = 0.22$ to $b = 0.28$ for $M_w \geq 7.5$. Using $b = 0.25$ and moment magnitude ranges of 6.5 to 8 provide log (N) = –0.44– 0.25M_w for the transcurrent fault events and log (N) = 0.27– 0.25M_w for the subduction zone. Note that these two recurrence models, which are governed by the geological data, describe the characteristic earthquakes. Geological evidence indicates that the characteristic earthquakes occur more frequently than would be implied by extrapolation of the Gutenberg-Richter recurrence relation, which leads to very low b values for these models. We allowed such large earthquakes to occur at any epicenter in the respective subcatalogs. The results illustrate a significant reduction in the rate of occurrence of moderate-magnitude earthquakes and an increase in the rate of the largest events. This difference can affect the assessment of seismic hazard at a site, depending on whether the moderate-magnitude events or the large events contribute most to the hazard.

Long-Term Synthetic Catalogs

To generate a large number of synthetic catalogs from the observed catalog, Monte Carlo estimation should be performed.

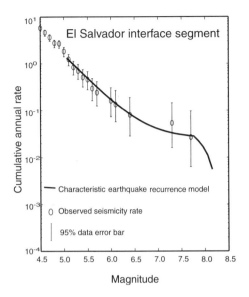

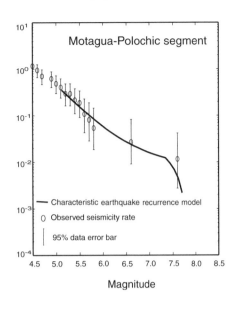

Figure 3. Recurrence relationships used to model subduction (left) and transcurrent fault (right) seismicity. Recurrence rates developed from independent seismicity and from fault slip rates. Predicted recurrence rates are shown for the characteristic earthquake model.

The procedure is based on randomly pulling out earthquakes from the observed catalog and then replacing them. First, we select an epicenter from the observed earthquake subcatalog and allow random variation up to 20 km to include it in the synthetic catalog. A magnitude for that epicenter is then selected according to the observed magnitude-frequency relationship. With such random draws, we construct sufficient data sets with exactly the same number of earthquakes as the actual data set. To construct a long-term synthetic catalog, the annual rate of events ($\lambda = 1/\Delta T$) is first used as a time increment between events, and then we allow the nth earthquake to take place at time $(n-1)\Delta T$ for each synthetic catalog. Computer-generated random numbers are used to simulate many long-term synthetic catalogs in which each possible sample of a given size n has an equal probability of being selected. To select a sample of size n from the original catalog, one thousand synthetic catalogs are generated for different values of duration in the range 5000 to 100,000 yr. Each of the synthetic catalogs is analyzed to obtain its fitted parameters.

GROUND MOTION ATTENUATION CHARACTERIZATION

In order to represent the ground motion at a site, the states Y_{ij} are defined to include information on the site ground motion. For a specified location of earthquake, the ground motion is commonly given by

$$Y_{ij} = f\left(5\%, M_i, R_i, \varepsilon, GC, T\right) \qquad (3)$$

where the Y_{ij} is the ground motion parameter for 5% of critical damping, M_i is a randomly selected magnitude, R_j is the distance from randomly selected epicenter to the site, ε is the scatter parameter for the ground motion, GC represents the geological conditions, and T is the natural period in the attenuation

relationship. To compute from all possible events the probability of exceedance of a certain level of ground motion (y), we define a new space Q. This area can be discretized into the convenient number of intervals such that

$$Q = \left[Y_{ij} = \phi(M_i R_j) > y, Y_{ij} = \phi(M_i R_j) < y\right] \qquad (4)$$

We used the peak ground acceleration (PGA) as a ground motion parameter, and rock sites as the geological conditions. The logarithm of the PGA is considered to be normally distributed with standard deviation σ. The scatter parameter is often defined as the product of a set of errors, η_{ij}, with this standard deviation. If the standard deviation is fixed, the ground motion model (equation 3) can be written as $Y_{ij} = f(M_i, R_j, \varepsilon, GC, T) + \eta_{ij} \sigma$.

Two types of uncertainties, aleatory and epistemic, are distinguishable in seismic hazard studies. To incorporate the aleatory uncertainty ($\eta_{ij} \sigma$) in the set of Monte Carlo simulations, the Y_{ij} for each magnitude is determined, and the corresponding distance to the site is used to correct it (Fig. 4). One of the ground motions from the specific distance is randomly selected as the ground motion for that event at the site. Epistemic uncertainty can be represented by considering different attenuation relations into the analysis (Toro et al., 1997). Predictive relationships for earthquake ground motion and response spectral values are empirically obtained by well-designed regression analyses of a particular strong-motion parameter data set. Three categories of regional ground motion attenuation relationships are used as alternative branches to estimate the ground motions due to all earthquakes in northern Central America and the western Caribbean plate. (1) Empirical crustal relationships used in the transcurrent fault zone, the volcanic, and the megathrust zones with focal depth less than 55 km include those of Abrahamson and Silva (1997), Campbell (1997), and Sadigh et al. (1997). (2) Three attenuation relations are used to represent predictive relationships and their

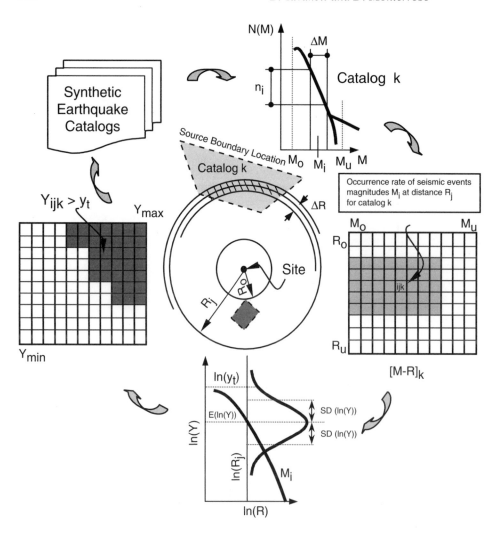

Figure 4. A typical Monte Carlo seismic hazard flowchart to depict the aggregate and disaggregate contribution of sources to the seismicity of the site.

associated uncertainties for the Y_{ij} at various natural periods in the intraplate region. These are given in Atkinson and Boore (1995), Frankel et al. (1996), and Toro et al. (1997). (3) Finally, the subduction zone relationship of Youngs et al. (1997) is applied for intraslab sources. Due to the low seismicity rates in the intraplate region, there are very few strong-motion data available. Therefore, a stochastic point source model to simulate strong ground motions at hard rock sites defined the attenuation relationships used in this case. In the subduction zones and the active crustal regions, attenuation relationships are based on strong ground motion recordings on rock site conditions. We assumed a focal depth range of 55 to 70 km for intraslab earthquakes. The attenuation models are given equal weighting factor.

MONTE CARLO SIMULATION FOR SEISMIC HAZARD ANALYSIS

According to the Monte Carlo method, the area surrounding a given site can be constrained by seismic source boundary locations to construct different earthquake catalogs (Fig. 4). The catalogs will represent earthquakes of different sizes, up to the maximum earthquake at the site, due to the different possible earthquake locations. The earthquakes are usually assumed to have a uniform spatial distribution over the time period of the catalogs, and a Poisson temporal distribution with an activity rate (λ) describing their occurrence. The activity rate is determined empirically from the past seismicity data. In general, $\lambda(t,x,y,z)$ depends not only on a particular location of event in time-space (t, x, y, z) but also on the times and locations of preceding events. When events in the catalog follow a Poisson process, $\lambda(t,x,y,z)$ depends only on time-space coordinates. We subdivide the spatial region into a finite number of subregions and fit the Poisson process models to the data within each subregion. Then the conditional rate may be written

$$\lambda(t,x,y,z) = \sum_i \lambda_1(t) \times \ell_i(x,y,z) \qquad (5)$$

where ℓ_i is an indicator function denoting the relevant region.

For each catalog, the seismic recurrence relationship shows that possible ranges of magnitude and distance will be $M_0 < M < M_u$ and $R_0 < R < R_u$, respectively (Fig. 4). Using a Monte Carlo approach to generate a synthetic earthquake catalog, we can discretize the ranges of M and R into a convenient number of intervals and a grid superimposed over the region of interest, as depicted in Figure 4. As a result, the region is divided into bins or segments of dimensions ΔM and ΔR. For each bin, ij, in the grid of a catalog k, the number of occurrences in the interval ΔM around M_i per year and unit area is determined by the portion of earthquakes in the catalog k that is located at a distance R_j. We enter this value at the location ij of the $[M–R]$ matrix for catalog k and repeat the procedure for all i's and j's in $[M–R]_k$ and all earthquakes in the synthetic catalogs. The information contained in a row, i, of a matrix can be physically interpreted as the contribution of all earthquakes in catalog k located at a distance R_j from the site. The above procedure can be repeated for all catalogs.

The probability $q_{ij}(y)$ that the spectral acceleration, Y_{ij}, of a structure with period T and damping factor 5% on a certain ground condition, GC, at the site under consideration exceeds y as a consequence of the occurrence of an earthquake, E_{ij}, in bin ij could be evaluated by

$$q_{ij}(y) = P\left[Y_{ij} > y \middle| T, GC, E_{ij}\right] \qquad (6)$$

This earthquake (E_{ij}) is usually identified by magnitude, M, distance, R, and standard deviation, ε, away from the median with respect to level y. At this stage, to obtain the rate of occurrence of various levels of ground motion at the site of interest, any attenuation relationship (denoted above as equation 3) can be used. Taking a target level y_t, we can scan the $[M–R]$ matrix and identify all pairs M_i and R_j for which $Y_{ij} = f(5\%, M_i, R_j, \varepsilon, GC, T) > y_t$ as illustrated in Figure 4. Then, the probability of Y_{ij} exceeding y in t years may be expressed as

$$P\left[Y_{ij} > y \middle| t\right] = 1 - \exp\left(-t\lambda_{(y)}\right) \qquad (7)$$

where

$$\lambda_{(y)} = \sum_{i=1}^{N_M} \sum_{j=1}^{N_R} \lambda_{ij} q_{ij}(y) \qquad (8)$$

where N_M and N_R are number of possible values of magnitude and distance, respectively. Comparison of equations 5 and 8 illustrates that the indicator function can be replaced by the probability $q_{ij}(y)$ in a given region. Equation 8 counts the number of times that ground motion exceeds the target level and equation 7 is then used to calculate the probability that ground motion will exceed that level.

The Monte Carlo method can be used to investigate the properties of any proposed estimation parameter in the seismic hazard analysis. The essence of the Monte Carlo method is straightforward, to select earthquakes that would actually generate the ground motion parameters (e.g., PGA) at a specified site.

To simulate a series of ground motion parameters on the computer for each earthquake in the synthetic catalog, we proceed as follows:

1. Define the system by prescribing (*a*) the model equation, also called attenuation function, (*b*) the way in which uncertainties are incorporated in the model of the observations, (*c*) the probability distribution of all the uncertainties, and (*d*) the synthetic catalog, which provides a priori distribution for the analysis.

2. Assign true values to all parameters in the attenuation function and to those in the distribution of uncertainties.

3. Select a set of values of the independent variables in the synthetic catalog and calculate the associated set of true values of the ground motion parameters from the attenuation equations.

4. Use the computer to produce a set of errors η_{ij} drawn from the prescribed probability distribution. The simulated ground motion measurements are obtained by combining the uncertainties with the parameter values. For additive uncertainties, the ith error is simply added to the ith magnitude value at jth distance. This provides simulated measurements for ground motion.

5. Acting as though the real ground motion parameters are unknown, we estimate the ground motion parameters for the particular set of independent values.

6. Repeat the series of simulated measurements N times by repeating steps 4 and 5, each time with a new set of uncertainties.

7. We use appropriate methods to estimate properties of the distribution of parameter estimates. The expected value of our parameter estimator is estimated by the mean of our parameter estimates.

The above simulation procedure is reasonably flexible. We can estimate the sample properties for a model, linear or nonlinear, and its corresponding parameter values. Moreover, the effect of different probability distributions upon different methods (e.g., maximum likelihood method) can be assessed. We do not use seismic source boundary locations and seismic hazard parameters, since these are implicit in the observed earthquake catalog. The simulation is completely deterministic and can be explicitly applied to convert a deterministic hazard model to a probabilistic hazard model. The hazard-consistent earthquake scenarios and the contribution of individual earthquakes to the total hazard are calculated by making a simple disaggregation procedure.

RESULTS FOR NORTHERN CENTRAL AMERICA

Deterministic-probabilistic seismic hazard maps for northern Central America and the western Caribbean plate are shown in Figures 5A–C. A spacing of 0.25° in latitude and longitude is used to create new hazard maps. The maps depict probabilistic values of peak ground acceleration (PGA) with 50%, 10%, and 5% probabilities of exceedance (PE) in 50 yr, corresponding to return periods of 72, 475, and 975 yr, respectively. The reasons for developing the various levels of the catalog completeness and a logical progression from models and parameter values are discussed above. The Ran_Haz program (Ebel and Kafka, 1999) modified by the present authors is used for computing seismic hazard from a synthetic earthquake catalog.

Results of a seismic hazard are usually expressed in terms of the hazard curves, which indicate the annual probability of exceed-

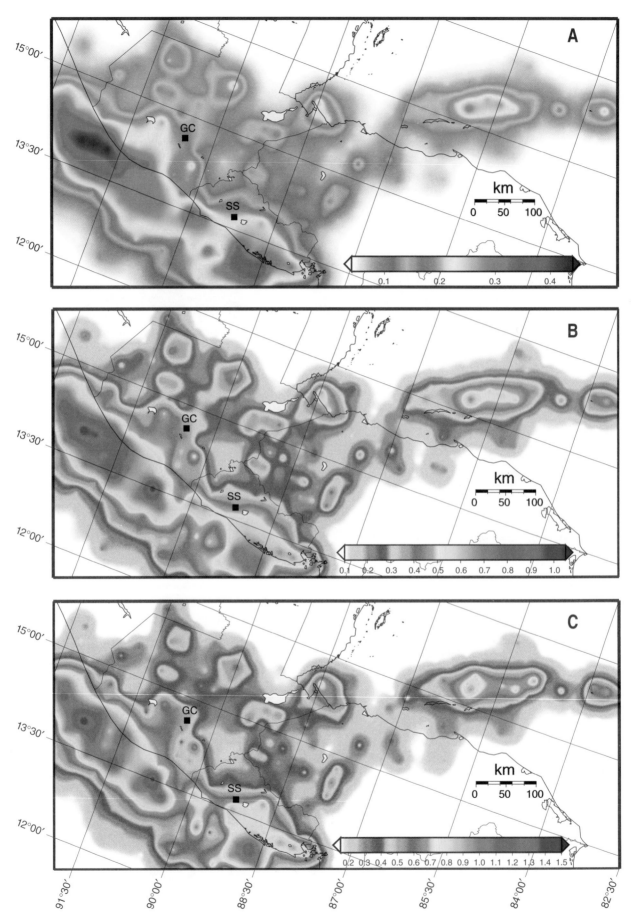

Figure 5. Seismic hazard map of northern Central America and western Caribbean by using a Monte Carlo approach. Peak ground acceleration (%g) is depicted having 50% (A), 10% (B), and 5% (C) probability of being exceeded in 50 yr.

ance (PE) of different values of the PGA. The seismic hazard results for San Salvador (SS) are 0.215 g, 0.407 g, and 0.614 g, and for Guatemala City (GC) are 0.187 g, 0.344 g, and 0.504 g with 50%, 10%, and 5% PE in 50 yr, respectively. These results tell nothing about what type of earthquakes actually produce the PGA values at a site, or whether these accelerations are due to a moderate event at a short distance or a large event at a long distance.

Disaggregating the hazard results, which shows the contributions of different magnitude-distance pairs to the exceedance of the probabilistic ground motion, is a useful approach to see the role of specific earthquakes (Musson, 1999). If, for example, we consider a return period of 475 yr for a specific earthquake, then the annual probability of exceedance of a target spectral acceleration (e.g., 10% PE in 50 yr) can be calculated by disaggregating all of the earthquake contributions. Figure 6 illustrates the spatial distribution of earthquakes around San Salvador and Guatemala City from the synthetic catalogs that produce peak acceleration of at least PGA = 0.40 g at the sites (cities). The distribution of earthquakes shows a cluster of moderate-magnitude events close to the two cities and a large event at a larger distance. The volcanic chain and subduction zone dominate the hazard at the considered sites. This may be explained by the fact that the intervals of time between successive earthquakes are not necessarily uniform. In intraplate environments, such as the volcanic chain zone, there are no persistent segments, and both point-specific slip and earthquake magnitude are variable. Thus, the intervals between events are often highly variable and characterized by clusters of events that are concentrated in small areas. On the other hand, in interplate environments, such as the subduction zone, the rate of cumulative strain is high, and slip along a master

segment repeats periodically. Therefore, in these areas, intervals between events are relatively uniform and clustering of events is not expected. Clustering of the estimated hazard can also be observed in Figure 5.

The total seismic hazard is expressed as the sum of the contributions from each possible combination of the $[M-R]_k$ matrix on each of the earthquake catalogs. The mean values of magnitude and distance are considered to identify the seismic events dominating the hazard at the sites. In the case of San Salvador (Fig. 6), this procedure leads to two *controlling earthquakes* and *source-to-site distances* characterized by $\overline{M} = 5.86$, $\overline{R} = 10.2$ km, and $\overline{\eta} = 0.7$ for the volcanic chain zone and $\overline{M} = 7.5$, $\overline{R} = 41.8$ km, and $\overline{\eta} = 1.8$ for the subduction zone. In Guatemala City (Fig. 6), this approach reports $\overline{M} = 5.47$, $\overline{R} = 6.3$ km, and $\overline{\eta} = 1.1$ near the Motagua fault system and $\overline{M} = 6.96$, $\overline{R} = 63.45$ km, and $\overline{\eta} = 1.26$ for the volcanic chain.

A practical example of two controlling earthquakes to represent a realistic seismic loading for the dynamic analysis of structures in a region can be found in San Salvador. Two accelerograms were recorded with similar soil conditions, one from the 1982 subduction event, M_S 7.3, ~60 km southwest of San Salvador, and the other from the 1986 volcanic event, M_S 5.4, south of San Salvador. Bommer et al. (1998) have pointed out that the plateau of the subduction spectrum occurs at longer periods than the volcanic spectrum.

CONCLUSIONS

The Monte Carlo hazard maps demonstrate that the PGA in El Salvador decreases from the southeast of the country in both

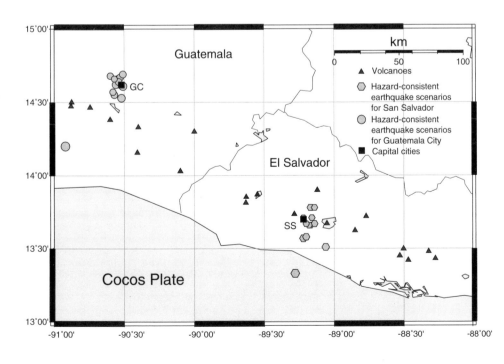

Figure 6. Magnitude-distance pairs distribution from synthetic catalogs that contribute to the hazard at San Salvador (SS) and Guatemala City (GC). The dominant earthquakes are considered to be the mean magnitude and distance of the seismic events that caused the target ground motion at the sites.

northern and western directions. The average PGA with a 10% PE in 50 yr is estimated to be 0.41 g for San Salvador, which is less than the PGA values presented in previous hazard studies. However, the 1994 edition of the El Salvador seismic code introduces a PGA value of 0.4 g for San Salvador (López et al., this volume, Chapter 23), which is comparable with the PGA estimated in this study.

The disaggregation results show the location and the size of the postulated earthquakes to calculate design ground motion at the sites. The effects of coupling along the strike of the interplate interface control the locations and mechanisms of seismicity around San Salvador. Dip-slip movements, which were documented during the 1976 Motagua earthquake, appear responsible for the events that occurred near Guatemala City.

These two clusters of earthquakes can have short duration and control the shorter periods of design spectra. On the other hand, the high PGA values obtained for the subduction zone are due to the occurrence of large-magnitude earthquakes such as the 2001 El Salvador earthquake. Most of these earthquakes have recurrence times of ~40 yr. They also have considerable duration and control the longer periods of design spectra. Thus, for the sites that are exposed to shaking from more than one seismic source, the two controlling earthquakes determined in this study should be first used for the scaling of ground motion records to be consistent with design spectra. The associated spectral shape is then produced by contributions both from the volcanic chain (as small nearby earthquake sources) and from the subduction zone (as large distant earthquake sources).

To determine the corresponding parameters at the surface of the sites, local site effects play an important role in earthquake-resistant design and must be taken into account for the ground surface motions.

ACKNOWLEDGMENTS

The study presented in this paper has been carried out at the Department of Earth Sciences, Seismology, Uppsala University, Uppsala, Sweden. One of us (D.M.) has been supported through the SERCA project financed by the Swedish Agency for Research Cooperation with Developing Countries, SAREC. The authors would like to thank Ota Kulhanek, J. Bommer, and two anonymous reviewers for their comments and useful suggestions.

REFERENCES CITED

Abrahamson, N.A., and Silva, W.J., 1997, Empirical response spectral attenuation relations for shallow crustal earthquakes: Seismological Research Letters, v. 68, p. 94–27.
Alfaro, C.S., Kiremidjian, A.S., and White, R.A., 1990, Seismic zoning and ground motion parameters for El Salvador: The John A. Blume Earthquake Engineering Center Report No. 93, Palo Alto, California, Stanford University.
Algermissen, S.T., Hansen, S.L., and Thenhaus, P.C., 1988, Seismic hazard evaluation for El Salvador: Report for the U.S. Agency for International Development, 21 p.
Ambraseys, N.N., and Adams, R.D., 2001, The seismicity of Central America: A descriptive catalogue 1898–1995: London, Imperial College Press, 309 p.

Atkinson, G.M., and Boore, D.M., 1995, Ground-motion relations for eastern North America: Bulletin of the Seismological Society of America, v. 85, p. 17–30.
Bommer, J.J., Hernández, D.A., Navarrete, J.A., and Salazar, W.M., 1996, Seismic hazard assessment for El Salvador: Geofisica Internacional, v. 35, p. 227–244.
Bommer, J., McQueen, C., Salazar, W., Scott, S., and Woo, G., 1998, A case study of the spatial distribution of seismic hazard (El Salvador): Natural Hazards, v. 18, p. 145–166.
Cáceres, D., and Kulhánek, O., 2000, Seismic hazard of Honduras: Natural Hazards, v. 22, p. 49–69.
Campbell, K.W., 1997, Empirical near-source attenuation relationships for horizontal and vertical components of peak ground acceleration, peak ground velocity, and pseudo-absolute acceleration response spectra: Seismological Research Letters, v. 68, p. 154–179.
Cornell, C.A., 1968, Engineering seismic risk analysis: Bulletin of the Seismological Society of America, v. 58, p. 1583–1606.
Cramer, C.H., Petersen, M.D., and Reichle, M.S., 1996, A Monte Carlo approach in estimating uncertainty for a seismic hazard assessment of Los Angeles, Ventura and Orange counties, California: Bulletin of the Seismological Society of America, v. 86, p. 1681–1691.
DeMets, C., Gordon, R.G., Argus, D.F., and Stein, S., 1994, Effect of recent revisions to the geomagnetic reversal time scale on estimates of current plate motions: Geophysical Research Letter, v. 21, p. 2191–2194.
Ebel, J.E., and Kafka, A.L., 1999, A Monte Carlo approach to seismic hazard Analysis: Bulletin of the Seismological Society of America, v. 89, p. 854–866.
Ekström, G., and Dziewonski, A.M., 1988, Evidence of bias in estimations of earthquakes size: Nature, v. 332, p. 319–323.
Engdahl, E.R., Hilst, V.D., and Buland, R., 1998, Global teleseismic earthquake relocation with improved travel times, and procedures for depth determination: Bulletin of the Seismological Society of America, v. 88, p. 722–743.
Frankel, A., 1995, Mapping seismic hazard in the central and eastern United States: Seismological Research Letters, v. 66, p. 8–21.
Frankel, A., Mueller, C., Barnhard, T., Perkins, D., Leyendecker, E.V., Dickman, N., Hanson, S., and Hopper, M., 1996, National seismic hazard maps: Documentation June 1996: U.S. Geological Survey, Open-File Report 96-532, 110 p.
Geomatrix, 1995, Seismic design mapping, State of Oregon, Consultant's report for Oregon department of transportation: San Francisco, California, Geomatrix Consultants.
Hanks, T.C., and Kanamori, H., 1979, A moment magnitude scale: Journal of Geophysical Research, v. 84, p. 2348–2350.
López, M., Bommer, J.J., and Pinho, R., 2004, Seismic hazard assessments, seismic design codes, and earthquake engineering in El Salvador, in Rose, W.I., et al., eds., Natural hazards in El Salvador: Boulder, Colorado, Geological Society of America Special Paper 375, p. 301–320 (this volume).
McCaffrey, R., 1997, Influences of recurrence times and fault zone temperatures on the age-rate dependence of subduction zone seismicity: Journal of Geophysical Research, v. 102, p. 22,839–22,854.
Molnar, P., 1979, Earthquake recurrence intervals and plate tectonics: Bulletin of the Seismological Society of America, v. 69, p. 115–133.
Musson, R.M.W., 1999, Determination of design earthquakes in seismic hazard analysis through Monte Carlo simulation: Journal of Earthquake Engineering, v. 3, no. 4, p. 463–474.
Musson, R.M.W., 2000, Generalised seismic hazard maps for the Pannonian Basin using probabilistic methods: Pure and Applied Geophysics, v. 157, p. 147–169.
Reasenberg, P.A., 1985, Second-order moment of central California seismicity, 1969–1982: Journal of Geophysical Research, v. 90, p. 5479–5495.
Rojas, W., Bungum, H., and Lindholm, C., 1993, A catalog of historical and recent earthquakes in Central America: NORSAR Report, Reduction of natural disasters in Central America, earthquakes preparedness and hazard mitigation, seismic zonation and earthquake hazard assessment.
Sadigh, K., Chang, C.Y., Egan, J.A., Makdisi, F., and Youngs, R.R., 1997, Attenuation relationships for shallow crustal earthquakes based on California strong motion data: Seismological Research Letters, v. 68, p. 180–189.
Schwartz, D.P., and Coppersmith, K.J., 1984, Fault behavior and characteristic earthquakes: examples from the Wasatch and San Andreas fault zones (USA): Journal of Geophysical Research, v. 89, p. 5681–5698.

Singh, S.K., Gutierrez, C., Arboleda, J., and Ordaz, M., 1993, Peligro sísmico en El Salvador: Universidad Nacional Autónoma de México, México.

Toro, G.R., Abrahamson, N.A., and Schneider, J.F., 1997, Model of strong ground motions from earthquakes in central and eastern North America: Best estimates and uncertainties: Seismological Research Letters, v. 68, p. 41–57.

Wells, D.L., and Coppersmith, K.J., 1994, New empirical relationships among magnitude, rupture length, rupture width, rupture area, and surface displacement: Bulletin of the Seismological Society of America, v. 84, p. 974–1002.

White, R.A., and Harlow, D.H., 1993, Destructive upper-crustal earthquakes of Central America since 1990: Bulletin of the Seismological Society of America, v. 83, p. 1115–1142.

Wyss, M., and Habermann, R.E., 1982, Conversion of m_b to M_S for estimation the recurrence time of large earthquakes: Bulletin of the Seismological Society of America, v. 72, p. 1651–1662.

Youngs, R.R., Silva, W.J., and Humphrey, J.R., 1997, Strong ground motion attenuation relationships for subduction zone earthquakes: Seismological Research Letters, v. 68, p. 58–73.

MANUSCRIPT ACCEPTED BY THE SOCIETY JUNE 16, 2003

Geological Society of America
Special Paper 375
2004

Tsunami hazards in El Salvador

Mario Fernández*

Central America Seismological Center (CASC), Apdo. 214-2060, San José, Costa Rica

Modesto Ortiz-Figueroa

Centro de Investigación Científica y Educación Superior de Ensenada (CICESE), Km 107 Carretera Tijuana-Ensenada, Ensenada, Baja California 22860, México

Raúl Mora

Escuela Centroamericana de Geología, Universidad de Costa Rica, Apdo. 214-2060, San José, Costa Rica

ABSTRACT

A review of historical data for locally and regionally generated tsunamis indicates that 11 events have struck the coast of El Salvador between 1859 and 2002. Two of these tsunamis caused significant death and destruction. Five of the tsunamis were triggered locally, three regionally (from other countries of Central America), and three were triggered by distant earthquakes in the Aleutian, Chilean, and Colombian subduction zones. Nearly 200 fatalities have been caused by local tsunamis. Reported damages range from coastal flooding to the destruction of villages. Earthquakes of magnitude 7 or higher having epicenters offshore or onshore (close to the coastline) could trigger tsunamis that would impact the coastal areas of El Salvador. A preliminary estimation of tsunami hazard indicates that the entire Pacific coast of Central America is at risk from impact by tsunamis. In this paper, we illustrate the potential damage in El Salvador that could be caused by a distant tsunami with a numerical simulation of the 31 January 1906 Colombia tsunami.

Keywords: earthquakes, tsunamis, Central America, simulation, hazard.

INTRODUCTION

Tsunamis are ocean (or sea) waves generated by large and abrupt perturbations of the ocean (or sea) floor (or surface) caused by the sudden offset of a fault, volcanic eruptions, large subaerial or submarine landslides, or meteor impacts. In deep waters, tsunamis can travel at speeds up to 800 km/h because their phase speed is proportional to the water depth (the phase speed equals the square root of the acceleration of gravity times the water depth). Though the amplitude of a tsunami is small (a few centimeters) in deep waters, it becomes large (because of shoaling amplification) as the tsunami approaches the shoreline.

Most tsunamis originate in the Pacific Ocean along the many subduction zones of the rim of fire because of perturbations of the sea floor associated with submarine earthquakes. Some of these perturbations have set large tsunamis in motion, such as that generated by the 1960 Chilean earthquake. Tsunamis can travel thousands of kilometers away from their source, causing death and destruction over large areas far from the earthquake epicenter. As a result, tsunamis are second only to meteorological events in their potential for widespread destruction (Okal, 1994).

Tsunamis are perceived as presenting little natural hazard to Central America when compared to hazards posed by more frequent processes such us earthquakes, volcanic eruptions, and those of meteorological origin. Contrary to this perception, however, historical data show that tsunamis have caused

*mefernan@cariari.ucr.ac.cr

Fernández, M., Ortiz-Figueroa, M., and Mora, R., 2004, Tsunami hazards in El Salvador, *in* Rose, W.I., Bommer, J.J., López, D.L., Carr, M.J., and Major, J.J., eds., Natural hazards in El Salvador: Boulder, Colorado, Geological Society of America Special Paper 375, p. 435–444. For permission to copy, contact editing@geosociety.org. © 2004 Geological Society of America

widespread death and destruction across Central America (Fernández et al., 1999; Fernández et al., 2000). Locally generated tsunamis occur sufficiently frequently that their hazards should not be ignored. Furthermore, the Central American coasts are subject to strong tsunamis generated by distant earthquakes from Alaska to South America. Unfortunately, the most recent tsunami disasters have been largely forgotten, and the consequences of tsunamis are largely unknown to most of the population, including coastal residents.

TSUNAMIS

The Salvadoran section of the Middle American Subduction Zone, located off the Pacific coast of El Salvador (Fig. 1), is an area of high stress and crustal deformation that has generated large compressive earthquakes in the past two centuries. Furthermore, bending of the subducting Cocos plate at depths of 40 km (Burbach et al., 1984) creates tensional stress that causes large intraplate earthquakes such as the M = 7.6 earthquake on 13 January 2001, 40 km offshore of El Salvador (Bommer et al., 2002). This environment of high stress and large earthquakes makes the coast of El Salvador particularly vulnerable to tsunamis.

The main cause of tsunamis that could affect El Salvador is dislocation of the ocean floor associated with moderate- to large-magnitude earthquakes. A plot of the epicenters of six tsunamigenic earthquakes is presented in Figure 1. Four of the epicenters are inland; in the vicinity of shorelines, only two of these earthquakes occurred offshore. Along the coastal segment of Central America, which extends from western Nicaragua to southern Mexico, the strongest recent activity is characterized by earthquakes of M ≥ 6. Although there are no reported depths for several of these events, they likely resulted from the oblique subduction of the Cocos plate beneath the Caribbean plate.

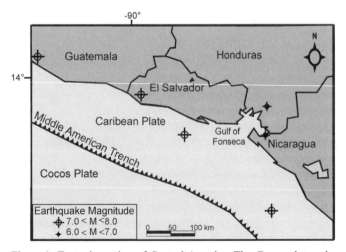

Figure 1. Tectonic setting of Central America. The Cocos plate subducts under the Caribbean plate along the Middle American Trench (MAT) and generates stress and seismicity in El Salvador. Epicenters of earthquakes from Table 1 are shown.

According to historical records, 11 tsunamis have struck the coast of El Salvador in the last 143 yr: two in the nineteenth century and nine in the twentieth century. The 26 August 1859 tsunami is the first one documented in currently available historical records. It is likely that earlier tsunamis in the region went unreported simply because they were too small to cause significant damage or because they struck uninhabited coasts. The increase in reported tsunamis during the twentieth century is due to improved global communications and the growth of the population in coastal communities. The negative aspect of this growth is that more people are now exposed to the risk of tsunamis; consequently, the impacts of future tsunamis will be greater than those of the past.

Local Tsunamis (0–100 km from El Salvador)

Locally generated tsunamis have been more devastating in El Salvador than regionally or distally generated tsunamis. Five locally generated tsunamis hit El Salvador between 1859–2001. Two of them, in 1859 and 1902, were extremely destructive. In fact, these two tsunamis have been the most damaging in the history of El Salvador. Fortunately, three of the local tsunamis did not cause severe damage. Locally generated tsunamis may be triggered within a few kilometers of the shoreline, which means that there may be little time to take action to protect people from a tsunami once it has been triggered.

Earthquakes as small as M 6 have triggered tsunamis in El Salvador. However, the largest and most destructive tsunamis were associated with local earthquakes having magnitudes ≥ 7. An M 7.0 earthquake on 26 February 1902, which took place off the coast of El Salvador, produced the largest and most devastating tsunami to hit El Salvador. This large earthquake produced three tsunami waves of unreported height. Larde y Larín (2000) estimated that the tsunami had a maximum height of 20 m and indicated that the waves penetrated 100 m inland near La Libertad and Acajutla. The waves flooded a 120-km-long segment of coast, damaging houses and trees, and killed almost two hundred people. Receding waters denuded a considerable area of the shoreline.

There are nine important communities exposed to the risk of tsunamis along the coast of El Salvador: La Unión, San Rafael de Tasajera, El Zapote, Marcelino, La Libertad, El Majahual, Acajutla, Barra de Santiago, and Garita Palmera (Fig. 2). La Unión, La Libertad, and Acajutla are the largest ports of El Salvador and the places where the majority of the coastal residents are concentrated. Acajutla is one of the communities that has been most affected by tsunamis in Central America. It was flooded by tsunamis in 1859 and 1957. La Unión is a harbor having a good record of tsunamis owing to the existence of a tide gauge there. There are no reports of extensive damage caused by tsunamis in this port. High grounds along the coast that can provide safe refuge from tsunamis include playa Playita–playa Flor, playa El Cuco–playa El Espino, and La Libertad–Barra El Maguey (Fig. 2). The remaining coastal segments of El Salvador

TABLE 1. PARAMETERS OF EARTHQUAKES AND TSUNAMIS OF EL SALVADOR

No	Date–Time	Elat	Elon	Ed	Em	Tsunami location	Tm	Te
1	1859–0826	13.00	87.50	ND	6.2	Gulf of Fonseca	1.5	CO-CA
2	1859–1209	13.75	89.75	40	7.0	Acajutla	1.5	CO-CA
3	1902–0226	13.00	89.50	30	7.0	Garita Palmera	2.0	CO-CA
4	1906–0131	01.00	81.30	ND	8.1	The entire coast	2.0?	NA-SU
5	1906	ND	ND	ND	ND	Los Negros Beach	–1.0	ND
6	1919–0629	13.50	87.5	>40	6.7	Gulf of Fonseca	–1.0	CO-CA
7	1950–1005	10.00	85.70	<60	7.7	The entire coast	–1.0	CO-CA
8	1950–1023	14.30	91.80	S	7.3	Gulf of Fonseca	–1.0	CO-CA
9	1957–0310	51.63	171.40	ND	8.1	The entire coast	3.0	PA-NO
10	1960–0522	38.20	73.50	32	8.5	The entire coast	4.0	NA-SU
11	1992–0901	11.70	87.40	S	7.2	Gulf of Fonseca	–2.5	CO-CA

Note: S—Shallow earthquake with macroseismic or instrumental evidence for a focus in the upper crust, No.—Number of the event, Elat—Earthquake latitude, Elon—Earthquake longitude, Ed—Earthquake depth, Em—Earthquake magnitude (Ms), Tm—Tsunami Magnitude (Imamura-Iida Tsunami Magnitude Scale, in Molina, 1997), Te—Tectonic environment, CO-CA—Cocos–Caribe Margin, NA-SU—Nazca–South American Margin, PA-NO—Pacific–North American Margin, ND—No Data. From Fernández et al., 2000.

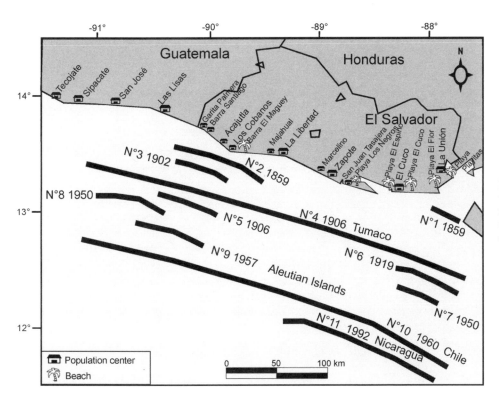

Figure 2. Map of coastal areas in El Salvador affected by tsunamis. Solid lines represent expanse of affected areas. The number of the tsunami in Tables 1 and 2 and year of occurrence is shown.

are flat areas where people must go well inland before escaping the reach of a tsunami.

Regional Tsunamis (100–700 km from El Salvador)

Three regional tsunamis triggered by large (M > 7.0) earthquakes impacted the coast of El Salvador during the last century. Two of these tsunamis were relatively small (Costa Rica, 1950; Guatemala, 1950), but the third (Nicaragua, 1992) was exceptionally large near its source. The Nicaragua tsunami had a small amplitude by the time it reached El Salvador and therefore caused little or no damage. These regional tsunamis were recorded around the Pacific in Hawaii and Japan, leading us to conclude that they affected the whole coast of El Salvador.

TABLE 2. DAMAGE AND REMARKS ON TSUNAMIS

No.	Date–Time	Description
1	1859–0826	The tsunami generated by the earthquake caused severe damage to houses and boats. As a consequence of the furious waves, two vessels and a brigantine sank at La Union, Gulf of Fonseca, El Salvador. In addition, two canoes were damaged. Reports indicate that the situation in the sea was horrible.
2	1859–1209	This tsunami was accompanied by noise. The main observed effects in the water body were a very agitated ocean in Acajutla, and high waves and recession of the water far from the shore, leaving fish floundered on the beach and terraces. Because of the recession, the docks and riverboat yards dried up almost to the breakwater. Standing out among the damage to structures is the destruction of the state warehouses and the flooding of the breakwater and the customhouse. The reports also say that caves and grottos collapsed.
3	1902–0226	There are similarities between this tsunami and that of 1859. First, both struck the western coast of El Salvador. Second, both were noisy tsunamis; in this case, a loud rumble like cannon shots was heard. Finally, both experienced a falling of the water level, exposing the ocean bottom for a considerable distance. A large wave arose from the sea and reached the coast, causing damage to property and washing homes and trees out to the ocean. Three waves were observed. The magnitude of devastation from the tsunami was exceptional. The coast from Garita Palmera to Barra de Santiago and beyond (a distance of about 120 km) was flooded. Barra de Santiago village was heavily damaged. The death toll was over 185 in the affected area.
4	1906–0131	Small tsunami waves were recorded throughout the Pacific Ocean. The tsunami was observed along the entire Pacific coast of Central America.
5	1906–0525	High waves invaded Los Negros Beach.
6	1919–0629	Flooded area in Gulf of Fonseca.
7	1950–1005	Small oscillations of the ocean level were recorded at La Union, El Salvador. Small oscillations of the ocean level were recorded at La Union, El Salvador. Iida et al. (1967) reported a wave of 8.9 meters high at La Libertad (El Salvador) but that is obviously a mistake.
8	1950–1023	A wave about 30 cm high was recorded at San José, Guatemala. A wave of 10 cm was also reported at La Union.
9	1957–0310	This is a remote-source tsunami. The associated earthquake was located at the Aleutian Islands Subduction Zone. A sea wave several meters high hit the Acajutla port and killed an undetermined number of coastal residents.
10	1960–0522	The tsunami from the great Chilean earthquake of 1960 (the largest seismic event in human history). This tsunami affected the entire Pacific coast of Central America. It was recorded at La Union (El Salvador) and San José (Guatemala) with heights of 0.5 m.
11	1992–0902	This is the largest known tsunami of Central America. In El Salvador, the effect was minimal.

Note: from Fernández et al., 2000.

Distant Tsunamis (> 700 km from El Salvador)

Remotely generated large tsunamis may cause damage to a particular region depending on the distance from the tsunami source and on the orientation and geometry of the initial sea surface deformation (e.g., Kowalik and Murthy, 1987). For a circular tsunami source, natural spreading will cause the wave-front amplitude to decrease by approximately equal amounts in all directions as the tsunami propagates over long distances. In contrast, an elongated tsunami source prevents wave-front spreading by focusing flow energy in a narrow beam oriented perpendicular to the elongated source. The wave-front amplitude at the head of the beam will not decrease as rapidly as in the case of the circular source, which can lead to damage at longer distances. These basic concepts regarding source effects on wave-front attenuation may help us explain the effects of distant tsunamis on the coast of El Salvador. For example, the 1957 Alaska tsunami that was generated by an elongated rupture along the Aleutian Islands caused extensive damage and loss of life in Acajutla-Salvador. In contrast, the 1960 Chile tsunami, which was generated by a magnitude M_w 9.5 earthquake,

was barely perceptible in Central America because its energy beam was oriented along the Great Circle that connects Chile and Japan. Despite the long distance between Chile and Japan, this tsunami caused extensive damage in southern Japan and on the Hawaiian Islands, which were located along the tsunami's path.

Among the potential sources for distant tsunamis (Fig. 3), the Colombian Subduction Zone is particularly noteworthy because of its proximity to Central America as well as its orientation. The 31 January 1906 Colombia earthquake (M_w 8.8; Kanamori and McNally, 1982) generated a large tsunami that washed away all homes on the low-lying coast above the rupture area and killed an estimated 500–1500 people (Soloviev and Go, 1975). The rupture length of this earthquake was estimated to be 400–500 km along the Pacific Coast of Colombia (Keleher, 1972).

NUMERICAL SIMULATION OF TSUNAMI PROPAGATION

We use a numerical model of tsunami propagation to illustrate the potential hazard to El Salvador of a large tsunami trig-

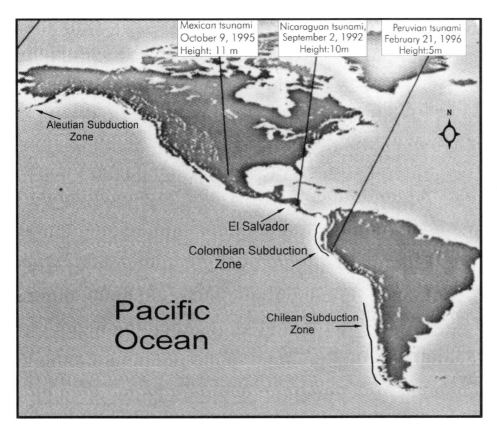

Figure 3. Potential source areas of distant tsunamis. Data indicate Chile, Colombia, and Aleutian subduction zones can generate earthquakes that trigger destructive tsunamis in El Salvador. Data for three regional tsunamis are also shown.

gered by an earthquake along the Colombian Subduction Zone. Our purpose is to emphasize the need for tsunami awareness and educational programs in El Salvador. The numerical simulation is based on the 31 January 1906 Colombia earthquake. The results of the simulation are shown as: amplitude and travel time for the tsunami wave front, which is useful for warning purposes; estimated tsunami height along the coast of Central America; the temporal signature of the tsunami at the coast and in the open ocean; and tsunami runup height along the coast of El Salvador.

The code used to model tsunamis in this study is available through the Tsunami Inundation Modeling for Exchange (TIME) program of the Tohoku University. Goto et al. (1997) provide a description of the model and its use. The theory of shallow-water flows is used to model tsunamis when the ratio of water depth to wavelength is small. The theory assumes that vertical acceleration of water particles is negligible compared to the gravitational acceleration and that pressure distributions can be approximated by hydrostatic profiles. Consequently, the vertical motion of water particles has no effect on the pressure distribution. In addition, the horizontal velocity of water particles is considered vertically uniform.

On the basis of these approximations, the motion of tsunami waves is expressed by the depth-averaged shallow-water equations (Pedlosky, 1979):

$$\frac{\partial \eta}{\partial t} + \nabla \bullet \mathbf{M} = 0 \qquad (1)$$

$$\frac{\partial \mathbf{M}}{\partial t} + gh\nabla \eta = 0 \qquad (2),$$

in which t is time, η is the vertical displacement of the water surface above the equipotential surface, h is the ocean depth, g is the gravitational acceleration, and M represents the vector quantities (U, V) of the horizontal depth–averaged volume flux, in which U and V are the east-west and north-south components, respectively.

Equation (1) describes mass conservation for incompressible fluids, and equation (2) expresses conservation of momentum. Notice that the effect of Earth's rotation on horizontal acceleration (the Coriolis effect) is not included in equation (2). Kowalik and Whitmore (1991) demonstrated that even for tsunami propagation distances encompassing an east-west displacement as large as 30 degrees, the Coriolis effect is not important. Friction and nonlinear terms are also suppressed in equation (2). Shuto (1991) and Kowalik and Murthy (1987) found that for water depths greater than 50 m, both friction and nonlinear terms become negligible relative to linear terms. This finding is important because the numerical simulation of tsunamis in deep waters uses only the linear terms in equation (2), which minimizes the number of computations needed to conduct a simulation. To simulate tsunami runup heights in shallow water, frictional and nonlinear terms must be included in the momentum conservation equations (discussed below). However, inclusion of such terms requires accurate, high-resolution near-shore bathymetry to produce appropriate results. At present, such high-resolution data for the coastal waters of El Salvador, or any of Central America, do not exist. Therefore, to approximate the needed high-resolution bathymetry along coastal El Salvador, we interpolated a 3-second grid spacing from the ETOPO-2 bathymetric database (Smith and Sandwell, 1997).

We determine tsunami characteristics in open water greater than 50 m deep by solving equations (1) and (2) in spherical coordinates, using an explicit finite-difference method that utilizes a staggered leapfrog scheme (Goto et al., 1997). In the computations, the time step was set to 1 s, and the bathymetric grid spacing was set to 2 minutes.

The radiation condition at the offshore boundary is taken from the characteristic solution of the wave equation:

$$(U, V) = \pm \eta \sqrt{gh} ,$$

where the sign is taken such that the wave will propagate outward from the computational domain.

The initial condition for the model consists of a description of sea-surface deformation; it assumes that the instantaneous topography of the sea surface approximates the vertical deformation of the sea floor produced by the earthquake. The time evolution of sea-floor displacement is not included in equation (1). The assumption that the time evolution of deformation is negligible is valid as long as sea-floor deformation is sufficiently large compared to the water depth at the tsunami source and the rupture velocity is short compared to the tsunami-propagation velocity. Kowalik and Whitmore (1991) have shown that the inclusion of a finite rupture velocity in the model has little effect on the energy-flux distribution of the tsunami.

Vertical deformation of the sea floor, produced by an earthquake, is determined from analytical expressions for the internal deformation of a continuous media due to shear and tensile faulting (e.g., Mansinha and Smylie, 1971; Okada, 1992). Fault parameters and rupture mechanisms usually are estimated from inversion of seismic data and assumptions about local geologic properties. The rupture mechanism of the 31 January 1906 earthquake, however, is not well known. Therefore, we computed a vertical deformation of the sea floor by using the dislocation model of Mansinha and Smylie (1971). This model considers a simple rupture geometry and uniform slip distribution on the fault plane. We assumed a uniform reverse-slip distribution of 10 m over an area of 400 km × 100 km on a fault plane dipping 30° SE at a depth of 30 km. These parameters approximately fit the area and seismic moment of the $M_w > 8.4$ earthquake.

The location of the fault plane used for our simulation (see Fig. 4) lies along the Pacific Coast of Colombia and approximates the location of the 31 January 1906 earthquake provided by Kelleher (1972). A profile of the resulting sea floor deformation along the transect AB illustrated in Figure 4, perpendicular to the strike of the fault plane, is shown in Figure 5.

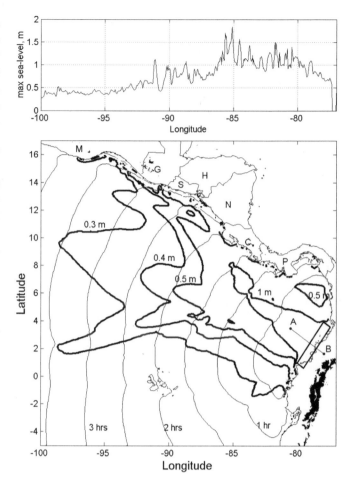

Figure 4. Numerical simulation of a large tsunami generated by an earthquake along the Colombian Subduction Zone. Top—Maximum wave height in water depth of 50 m along the coast of Central America: México (M); Guatemala (G); El Salvador (S); Honduras (H); Nicaragua (N); Costa Rica (C); Panamá (P). Bottom—Tsunami wave fronts (thin lines) at 30 minute time steps. Contours of maximum wave height (thick lines) in meters. The rectangle along the coast of Colombia indicates the surface projection of the assumed fault plane (400 km × 100 km). The transect AB, perpendicular to the fault plane, is drawn as a reference for illustrating the tsunami initial condition in Figure 5.

Results from the numerical simulation of the propagation of a tsunami generated off the coast of Colombia are shown in Figure 4. Tsunami travel times were obtained by tracking the wave front at every time step in the numerical model. Figure 4 shows that tsunami arrival times ranged from 1 hour for Panama to 4 hours for southern Mexico and Guatemala. A distant tsunami triggered by an earthquake off the coast of Colombia comparable to that of January 1906 could be expected to reach coastal El Salvador within 2.5 to 3 hours. Depending on the hour of the day and the day of the week when the earthquake occurred, such an arrival time might or might not allow time to warn coastal residents in El Salvador of an approaching tsunami.

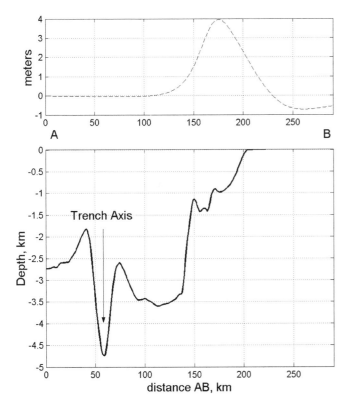

Figure 5. Top—Inferred vertical sea-floor deformation across the transect AB (Fig. 4) generated by a reverse fault dipping 30° southeast at a depth of 30 km. An average slip magnitude of 10 m was considered across the fault area (400 km x 100 km). Bottom—Bathymetric profile across transect AB.

The primary energy gradient of the simulated tsunami is oriented perpendicular to the fault rupture (Fig. 4). Part of the energy in the tsunami is deflected by the sea floor bathymetry toward the coast of Central America, where shoaling amplifies the tsunami wave height at near-shore sites. The pattern of near coastal maximum wave heights in water 50 m deep exhibits three distinct features: from Panama to Costa Rica, the maximum wave height approaches 1 to 1.5 m (arrives within 1 to 2 hours); from Nicaragua to Guatemala, the maximum wave height approaches 0.5 to 1 m (arrives within 2.5 to 3.5 hours); and along southern Mexico, the maximum wave height approaches 0.5 m and remains nearly constant (arrives within 3.5 to 4 hours).

From these offshore wave heights, we can roughly estimate the water heights at near-shore sites by applying an empirical one-dimensional amplification factor. Specific coastal configurations may reduce or increase the amplification factor depending on whether the coastal bathymetry induces convergence or divergence of the wave front. The one-dimensional shoaling amplification factor between a water depth of 50 m and a water depth of 2 m at near shore sites is $(50/2)^{1/4}$, which is based up on Green's Law for one-dimensional shoaling of monochromatic water waves along a constant slope (e.g., Lamb, 1932).

The Tsunami Signature in the Open Ocean and at the Coast

In the simulated tsunami, the first wave of the wave train has the greatest amplitude. However, for some observed tsunamis, the second or the third wave has the greatest amplitude. Therefore, coastal residents should be aware that the hazard has not necessarily diminished after the arrival of first tsunami wave. Figure 6 illustrates the temporal signature of the tsunami in the open-ocean and at a near shore site. The open-ocean signature of the simulated tsunami (at 5° N, 90° W) has an emergent character due to uplift on the oceanward side of the initial deformation (see Figure 5). The open-ocean signature displays a high-amplitude pulse followed by a wave train with decreasing amplitude. In contrast, the near-shore signature of the tsunami along coastal El Salvador exhibits features that reflect a response to local bathymetry. The initial wave arriving at the coast has an emergent character, but the amplitudes of subsequent waves decrease more slowly than those in the open ocean.

Tsunami Runup Along the Coast of El Salvador

We estimated the tsunami runup height along the coast of El Salvador by including nonlinear and friction terms in the momentum conservation equations (2). The resulting expressions are given by:

$$\frac{\partial U}{\partial t}+\frac{\partial}{\partial x}\left(\frac{U^2}{D}\right)+\frac{\partial}{\partial y}\left(\frac{UV}{D}\right)+gD\frac{\partial \eta}{\partial x}+\frac{gm^2}{D^{\frac{7}{3}}}U\sqrt{U^2+V^2}=0$$

$$(3)$$

$$\frac{\partial U}{\partial t}+\frac{\partial}{\partial x}\left(\frac{U^2}{D}\right)+\frac{\partial}{\partial y}\left(\frac{UV}{D}\right)+gD\frac{\partial \eta}{\partial x}+\frac{gm^2}{D^{\frac{7}{3}}}U\sqrt{U^2+V^2}=0$$

Parameters t, η, and g are as defined in equations (1) and (2). In addition, $D = (\eta + h)$ is the instantaneous depth of the water column, where h is the average depth of the water column. $U = u(\eta + h)$ and $V = v(\eta + h)$ are the horizontal depth-averaged volume flux vectors in the longitudinal (x) and latitudinal (y) directions respectively, where u and v are the corresponding velocities of water particles. A Manning's roughness parameter, m, is included to account for friction and is considered to have a constant value $m = 0.025$.

The expression for the bottom friction terms in equations (3) deserves a brief explanation. Prior to depth-averaging, friction terms are expressed by analogy to the quadratic friction law in a uniform flow (e.g., Dronkers, 1964):

$$\frac{\tau_x}{\rho}=\frac{f}{2D}u\sqrt{u^2+v^2}\,, \qquad \frac{\tau_y}{\rho}=\frac{f}{2D}v\sqrt{u^2+v^2}\,, \qquad (4)$$

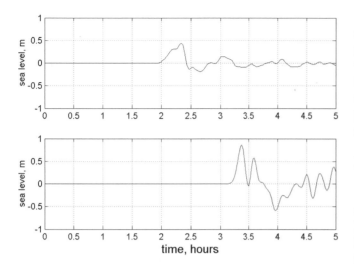

Figure 6. (Top)—Open-ocean temporal signature of simulated tsunami in water 3000 m deep. (Bottom)—Near-shore tsunami signature in water 50 m deep. The origin of the time axis in both figures is taken as the origin time of sea-floor deformation in the numerical model.

where τ_x and τ_y are the bottom stress in the directions x and y, respectively, ρ is fluid density, and the friction coefficient (f) is related to Manning's roughness parameter m by:

$$m = \sqrt{\frac{fD^{\frac{1}{3}}}{2g}} \qquad (5).$$

Consequently, bottom friction terms are expressed as:

$$\frac{\tau_x}{\rho} = \frac{gm^2}{D^{\frac{4}{3}}} u\sqrt{u^2 + v^2}, \qquad \frac{\tau_y}{\rho} = \frac{gm^2}{D^{\frac{4}{3}}} v\sqrt{u^2 + v^2} \qquad (6).$$

By integrating (6) from the sea floor to the water surface, friction terms acquire the aspect presented in equations (3). Notice that water depth is an important factor controlling the magnitude of the friction terms.

We determine near-shore tsunami characteristics by solving equations (1) and (3) using an explicit rectangular staggered leapfrog scheme and an interpolated high-resolution grid for the selected coastal region of El Salvador. A computation cell was judged to be submerged if $D = \eta + h > 0$ or dry if $D = \eta + h < 0$.

For a wave front located between the dry and submerged cells, the discharge across the boundary between the two cells is computed if the ground height in the dry cell is lower than the water level in the submerged cell. Otherwise, discharge is considered zero. The boundary condition on the seaward side of the interpolated high-resolution grid is compatible with the boundary of the open-ocean grid.

Tsunami runup computations along the coast of El Salvador are illustrated in Figure 7. Runup height (above mean sea level) ranges from 0.5 m at El Cuco to 2 m at Acajutla. Because the tsunami in 1906 arrived at El Salvador during the low tide (1.5 m below mean sea level), there is a possibility that the tsunami runup at that time did not overtop the high tide level in most places along the coast. Therefore, tide level plays an important role in the effects of a potential tsunami.

EDUCATION AND HAZARD ASSESSMENT

Education is a fundamental aspect of tsunami hazard mitigation. Politicians, scientists, and the public, especially those who live in coastal areas, must properly perceive and understand the hazards of tsunamis before the danger can be mitigated effectively. People living in areas prone to tsunamis need to recognize the threat and learn how to respond to it. The following case examples illustrate the importance of education in reducing the impacts of tsunamis.

With regard to the Okushiri (Japan) tsunami of 1993, Gonzalez (1999) notes, "The loss of lives in this event was a great tragedy, but it is clear that both warning technology and community education greatly reduced the number of casualties. The Japan Meteorological Agency issued timely and accurate warnings, and many residents saved themselves by fleeing to high ground immediately after the main shock—even before warning. Okushiri clearly demonstrated that the impact of tsunamis can be reduced."

In 1998, the Papua New Guinea tsunami was a great disaster, and more than 2000 people were killed. However, this tragedy helped others one year later when another tsunami (Vanuatu 1999) struck the country. Caminade et al. (2000) state,

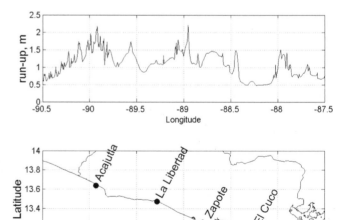

Figure 7. Tsunami runup height along the coast of El Salvador computed using a high-resolution bathymetric grid interpolated from ETOPO-2.

The small number of casualties was due to prior education and a party. Because of a wedding on the day of the earthquake, most everyone was still up celebrating when the earthquake occurred. A lookout was sent to note the condition of the sea. When he reported that the water was receding, villagers concluded that a tsunami was coming, and they ran to a nearby hillside to escape the wave. Villagers credited their response to a video of the 1998 Papua Guinea tsunami, which they had seen a few months before. The only casualties were those too elderly to escape the wave, those who returned for possessions after the passage of the first wave, and a man so drunk on Kava that he ignored people who were directing him to safety. The tsunami also occurred three days after a full moon, so the village was well lit despite a lack of electricity.

It is clear that simple activities such as presentations and watching videos on natural hazards can save lives. For this reason, emergency committees or technical groups should visit coastal communities to explain the nature, cause, and effect of tsunamis. We recently discussed tsunamis with people who work on mitigating natural disasters in El Salvador. After the discussion, participants worked with topographic maps to delineate areas damaged by historical tsunamis and to identify places of refuge along the coast of El Salvador. These activities will be repeated in the future and will include field visits to better understand the potential hazards of tsunamis. Coastal residents need to learn that a tsunami will be destructive if its height reaches a few meters and that when an earthquake occurs they must seek high ground or go inland. Following an earthquake, it is important to observe the ocean level and know what to do if a sudden regress is detected.

CONCLUSIONS

Eleven historical tsunamis have been reported along the coast of El Salvador since 1859, four of which were destructive and flooded villages, destroyed houses, and killed at least 185 people. Of these eleven tsunamis, five were triggered locally, three regionally (Costa Rica, Guatemala, and Nicaragua), and three from distant sources (Chile, Colombia, and Alaska). The most destructive tsunamis that struck El Salvador were associated with large earthquakes triggered along the tectonic margin of the Cocos and Caribbean plates. The most dangerous tsunamigenic earthquakes are those having magnitudes 7 or higher, with epicenters offshore. Those having epicenters onshore or close to the shoreline generally have caused limited damage by tsunamis.

Our study has demonstrated that the entire coast of El Salvador is prone to tsunamis. A distant tsunami triggered in the coastal region of Colombia could reach El Salvador within 3 hours and have waves heights in excess of 1 meter. The safest places to seek refuge along the coast are high ground in La Libertad–Barra El Maguey, Playa El Cuco–Playa El Espino, and Playa Playita–Playa El Flor.

Although few destructive tsunamis have struck El Salvador in historical time, the threat of tsunamis should not be ignored. Population increases and urban expansion along the coast now place considerable populace at risk. Further investigations to accurately define areas at highest risk are needed, and educational plans to mitigate the impact of tsunamis should be instituted.

ACKNOWLEDGMENTS

This research was partially supported by CEPREDENAC (Centro de Coordinación para la Prevención de Desastres Naturales en América Central). We thank the Government of México for its help in developing tsunami research in Central America, two anonymous reviewers for their comments and suggestions, and B. Williams and J. Major for helpful comments that improved the manuscript. M.O. acknowledges support from NIED (National Research Institute for Earth Science and Disaster Prevention-Japan).

REFERENCES CITED

Bommer, J., Benito, M., Ciudad-Real, M., Lemoine, A., López-Menjivar, M., Madariaga, R., Mankelow, J., Méndez de Hasbum, P., Murphy, W., Nieto-Lovo, M., Rodríguez-Pineda, C., and Rosa, H., 2002, The El Salvador earthquakes of January and February 2001: context, characteristics, and implications for seismic risk: Soil Dynamics and Earthquake Engineering, v. 22, p. 389–418.

Burbach, G., Frolich, V., Pennington, W., and Matumoto, T., 1984, Seismicity and tectonics of the subducted Cocos plate: Journal of Geophysical Research, v. 89, p. 7719–7735.

Caminade, P., Charlie, D., Kanoglu, U., Koshimura, S., Matsutomi, H., Moore, A., Ruscher, Ch., Synolakis, C., and Takahashi, T., 2000, Vanuatu Earthquake and Tsunami Cause Much Damage, Few Casualties: Eos (Transactions, American Geophysical Union), v. 81 no. 52, p. 641–646.

Dronkers, J., 1964, Tidal Computations in Rivers and Coastal Waters: Amsterdam, North-Holland Publishing Company, 518 p.

Fernández, M., Havskov, J., and Atakan, K., 1999, Destructive Tsunamis and Tsunami Warning in Central America: The Science of Tsunami Hazard, v. 17, no. 3, p. 173–186.

Fernández, M., Molina, E., Havskov, J., and Atakan, K., 2000, Tsunamis and Tsunami Hazards in Central America: Natural Hazards, v. 22, no. 2, p. 91–116.

Gonzalez, F., 1999, Tsunami: Scientific American, p. 56–67

Goto C., Ogawa, Y., Shuto, N., and Imamura, F., 1997, Numerical Method of Tsunami Simulation with the Leap-Frog Scheme, Intergovernmental Oceanographic Commission of UNESCO: IUGG/IOC TIME Project: Manuals and Guides # 35, Paris, 4 Parts.

Iida, D., Cox, C., and Pararas-Carayannis, G., 1967, Preliminary catalogue of tsunamis occurring in the Pacific Ocean: University of Hawaii, Hawaii Institute of Geophysics, data repository no. 5.

Kanamori, H., and McNally, K.C., 1982, Variable rupture mode of the subduction zone along the Ecuador-Colombia coast: Bulletin of the Seismological Society of America, v. 72, p. 1241–1253.

Kelleher, J., 1972, Rupture Zones of Large South American Earthquakes and Some Predictions: Journal of Geophysical Research, v. 77, no. 11, p. 2087–2103.

Kowalik, Z., and Murthy, T.S., 1987, Influence of the Size, Shape and Orientation of the Earthquake Source Area in the Shumagin Seismic Gap on the Resulting Tsunami: Journal of Physical Oceanography, v. 17, no. 7, p. 1057–1062.

Kowalik, Z., and Whitmore, P., 1991, An investigation of two tsunamis recorded at Adak, Alaska: Science of Tsunami Hazard, v. 9, no. 2: p. 67–83.

Lamb, L., 1932, Hydrodynamics, 6th edition: Cambridge, University Press, 738 p.

Larde y Larín, J., 2000, El Salvador, inundaciones e incendios, erupciones y terremotos, 2nd Edition: San Salvador, El Salvador, Dirección de Publicaciones e Impresos, 397 p.

Mansinha, L., and Smylie, D., 1971, The Displacement Fields of Inclined Faults: Bulletin of the Seismological Society of America, v. 61, no. 5, p. 1433–1440.

Molina, E., 1997, Tsunami Catalogue for Central America: Bergen, Norway, Institute of Solid Earth Physics, University of Bergen, 87 p.

Okada, Y., 1992. Internal deformation due to shear and tensile faults in a half-space: Bulletin of the Seismological Society America, v. 82, p. 1018–1040.

Okal, E., 1994, Tsunami Warning: beating the waves to death and destruction: Endeavour, New Series, v. 18, no. 1, p. 38–43.

Pedlosky, J., 1979, Geophysical Fluid Dynamics: Springer-Verlag, 624 p.

Shuto, N., 1991, Numerical Simulation of Tsunamis—Its Present and Near Future: Natural Hazards, v. 4: p. 171–191.

Smith, W., and Sandwell, D., 1997, Global Seafloor Topography from Satellite Altimetry and Ship Depth Soundings: Science, v. 277, p. 1956–1962.

Soloviev, L., and Go, Ch. N., 1975, A catalogue of tsunamis on the eastern shore of the Pacific Ocean (1513–1968): Moscow, Academy of Sciences of the USSR, Nauka Publishing House, 204 p.

MANUSCRIPT ACCEPTED BY THE SOCIETY JUNE 16, 2003

Geological Society of America
Special Paper 375
2004

The earthquake of January 2001:
Strengths and weaknesses in the social response

Ludvina Colbeau-Justin
Laboratory of Environmental Psychology CNRS UMR 8069, University René Descartes, Paris 5, France

Max Mauriol
Fire Service Department of Martinique, French West Indies

ABSTRACT

Multiple unfavorable conditions have inflicted heavy casualties and damage that severely affect El Salvador; these include the consequences of civil war, systematic delinquency organized into armed gangs, and a strong exposure to natural risks. The lack of earthquake response planning prior to the serious earthquake of January 2001 aggravated the efficiency of the government's response and prolonged the victims' difficulties. The earthquake disorganized families and individuals as well as institutions and communities. Despite the extent of the disaster, the flaws exposed concerns regarding a few aspects of risk management, especially institutional ones. However, the actions of some authorities were effective, such as the quick response of the national government. Furthermore, the social response was quite reactive and positive. The mobilization of NGOs, the involvement of the media, and the capacity of communities and individuals to face the catastrophe should be highlighted. This paper will present both negative and positive aspects of the social response. The lesson to be learned is the great capability of mobilization shown by unofficial actors. The institutional weaknesses were reduced by taking into account the resilience of the civil society.

Keywords: Crisis management, social response to disaster, social vulnerability, resiliency, psychological impact.

INTRODUCTION

This paper presents the results of a post-earthquake mission, organized in February 2001 by the AFPS, which is the French Association for Earthquake Engineering (Colbeau-Justin and Mauriol, 2001). The approach taken by the AFPS intends to adopt a new perspective by demonstrating the complexity underlying the concept of natural risk. The multidisciplinary team, composed of seismologists, geologists, engineers, architects, psychologists, members of emergency service organizations, governmental authorities, and others, tries to change the traditional view of risk assessment as a purely scientific enterprise. Therefore, the different factors (physical, social, cultural,

political, psychological, etc.) leading to specific situations after a destructive earthquake are identified (Guillande et al., 2001). Another purpose of the AFPS is didactic. Actors (members or nonmembers of AFPS), who might take part in disaster management, are trained and sensitized by participation in the assessment of the situation.

Fourteen years after the earthquake of October 1986, which left 1400 dead in the capital, El Salvador experienced, once again, a violent earthquake that left 1000 dead, one million affected, and 200,000 homes totally razed. In this area, where natural catastrophes (hurricanes, earthquakes, landslides, floods, and drought) strike with regularity, the socioeconomic reality of El Salvador and the low efficiency of prevention systems increase the catastrophes' negative impact on society. Massive deforestation, added to the lack of urban planning and coordination, encourages the

*colbeau@psycho.univ-paris5.fr

Colbeau-Justin, L., and Mauriol, M., 2004, The earthquake of January 2001: Strengths and weaknesses in the social response, *in* Rose, W.I., Bommer, J.J., López, D.L., Carr, M.J., and Major, J.J., eds., Natural hazards in El Salvador: Boulder, Colorado, Geological Society of America Special Paper 375, p. 445–452.

proliferation of precarious housing in exposed zones, such as gullies, accesses, and riverbanks. When the armed conflicts ended, El Salvador devoted great energy to rebuilding infrastructure, allowing the country to attract foreign investment. The earthquake happened when the country expected that "dollarization," the local currency set equal to the U.S. currency (January 2001), would help to invigorate economic growth.

Fritz (1968) defines the elements of disaster: "An event concentrated in time and space which inflicts on a community such damages that the result is a social structures breakdown and the obstruction of all or part of social functions." In the El Salvador earthquake of 2001, political interests sometimes interfered with crisis management despite the clear necessity for coordination. The emergence of local groups and of organization on the micro-social level substituted for a lack of political and organizational cohesion. The new participants compensated, with their own resources, for some gaps in the institutional response. The destruction appeared to be so severe that the official assistance teams, who were also victims of the disaster, could not handle it. One of the first decisions made by the government was to split the management of some activities. So, the initial search and rescue activities were carried out by local networks. However, the wounded were relayed to local professionals and professionals coming from numerous international organizations.

AN INSTITUTIONAL APPROACH STILL EVOLVING

For a long time, El Salvador has been known as a country with a strong seismicity. Since the beginning of the Spanish colonial period, religious assistance has been sought and a church named "Virgen de los Terremotos" (Virgin of Earthquakes) was built. However, to this day, the authorities have not committed themselves to a global preventive process. Nevertheless, the earthquake of 1986 alerted the authorities that wanted, at the time, to react. No progress was made until the creation of the national crisis committee, the COEN (Comité de Emergencía Nacional), which was placed under the authority of the president of the republic. This committee is responsible for emergency response planning. Although the members of the committee have been appointed, there is still no real policy in place. The committee's resources are limited, and it does not have the necessary infrastructure to provide assistance.

Preventive Information

Information booklets, edited by the Interior Ministry and distributed by schools or by the national fire and rescue service, provide the population with information about earthquake risk and mitigation. Evacuation training is supposed to be carried out by each school in order to prepare everyone to react quickly to the threat. In theory, teachers are trained in evacuation procedures, and exercises are organized monthly in schools. However, there is no administrative unit in charge of checking that these drills are carried out. Furthermore, the media are not actively associated with the dissemination of preventive information even though they are involved in the distribution of the booklets.

The Operational Level

The existing means of assistance are definitely insufficient compared to the existing risks. For example, among 5.2 million inhabitants, there are fewer than 400 firemen, and there are no volunteers. We must emphasize that the lack of volunteer firefighters contrasts with what can be traditionally observed in Caribbean Hispanic cultures like Colombia or Cuba.

As far as seismic risk is concerned, there are no rescue squads trained to search for buried people or in clearing techniques. There are no "pre-positioned" specific means (that is to say, human and material means) specially assigned to the management and the immediate response following natural catastrophes. Therefore, this lack of resources and policy makes the country totally dependent on the intervention of international assistance.

A new supervisory group for natural hazards, SNET (Servicio Nacional de Estudios Territoriales), composed of people from different ministries or agencies, has been created. The Center for the Prevention of Natural Disaster in Central America (CEPREDENAC), a coordination center for strengthening the capacity of the region to reduce the vulnerability of the population, is involved in this new group. One member will act as consultant and organize this group so that it becomes self-sufficient.

Social Impacts and Social Answers

The heavy casualties and the extent of damage severely affected the country, which was already suffering from several serious difficulties: the consequences of a civil war, systematic delinquency organized into armed gangs, and strong exposure to a variety of natural hazards. Poverty, a rapidly growing population, and the absence of planning for urban development were identified as the main factors of social vulnerability.

As far as building architecture is concerned, inequality is obvious in the distribution of victims and damage. Clay and wood construction were responsible for lethal flat collapses. The damage emphasized that the traditional adobe construction culture has been lost since the Spanish colonization and the introduction of new building techniques. There was no real transmission of the adobe techniques, but rather a copy of existing buildings. Consequently, it was observed that the very old adobe buildings offered a better resistance to earthquakes than the more recent ones. Another aggravating factor was the change in the use of buildings. Old one-family houses were converted to multifamily housing, which involved rearrangement of the interior and degradation of the structure.

Mobilization

The size and extent of the damage generated a real shock within the population. Fortunately, the government's reaction time was very short. That was a positive point, considering that delays can increase the number of victims, as was the case after the earthquake in Kobe, Japan. The president of the republic gathered the COEN within two hours following the impact. The call for international assistance was ordered immediately without waiting for aerial reconnaissance evaluations about important landslides and massive destruction of particular precarious homes. This quick response can be explained by two factors: The president was personally concerned because the earthquake hit the presidential buildings, and, by coincidence, all the ministers were gathered there for a meeting at the time of the earthquake. These elements favored the swift mobilization by the government.

Seeking probity and integrity, the government quickly named two foreign independent organizations to control and manage the distribution of international assistance. Here we see the desire to present an appearance of indisputable integrity toward foreign investors in order to preserve an income source upon which the economic growth of the country is built.

According to a widely held tradition in Latin America, the First Lady of the republic has an important position during the crisis management. In the field, the president's wife brings an emotive and human tone to the government control, personally participating in the distribution of humanitarian assistance. Meeting the population, she enhances the president's image.

The Rescue

The blockage of major traffic arteries by collapses and the limited means to carry out aerial reconnaissance prevented obtaining a quick, comprehensive view of the damages. Both main roads (from north to south and east to west) were obstructed; one month later, one of them was still only open to alternate traffic. Landslides caused many deaths. Huge, dry soil landslides buried hundreds of victims who could not always be located, and carried away crops and houses in the mountains. With the limited number of rescuers and the lack of means adapted to earthquakes (no specialized rescue squads with dogs specially trained for finding buried people), the government turned to the army and to international assistance and declared a state of emergency for the country. The government's main goals focused on the recognition of damaged sites, obtaining international assistance, receiving foreign rescuers, and the repair of collapsed lifelines (electricity, telephone, water, etc.).

Firemen, rescuers, and doctors, reinforced by the army, although they were also affected, were mobilized to answer the primary needs: care of the wounded and water delivery to the still accessible damaged zones that lacked a water supply. At the national level, the resources available only allowed for the rescue of the victims reachable without specific equipment (so-called "surface" victims). As far as the search for deeply buried victims was concerned, foreign teams with specialized equipment responded, but there was an significant intervention delay.

Rescues of buried, uninjured victims who died shortly after their release shows that the "crush syndrome" was not treated. After burial in a collapse, a victim should be approached with a specific methodology. When a victim's limb is compressed by rubble, toxins are produced in that part of his body. Once the limb is released, the toxins spread all through the body, quickly inducing death. An amputation or serious therapeutic management (dialysis, etc.) is necessary in these circumstances. The press reported deaths following rescue that were most likely the result of this syndrome. One cannot say if the reason was a lack of rescue culture or ignorance of the risk, or rather a lack of resources. Furthermore, the extent of lethal landslides made the timely search for victims difficult. Most buried victims were considered deceased (see below).

International Assistance

The call for assistance was made very quickly, within hours following the disaster, to neighboring countries (Colombia, Guatemala, Mexico, and Cuba) with the ability to mobilize specialized forces and means for rescue. With remarkable organization, COEN (nicknamed "Maestro de la Improvisación" [master of improvisation] by the media) evaluated the specific needs of affected sites and distributed the international teams. However, some dysfunctions were observed in the international assistance. Some countries wished to intervene in places where they had previously intervened, even if those sites were less damaged than others. For the countries involved in international assistance, the disaster provided an opportunity for their Civil Defense departments to be trained and to get experience in emergency intervention.

Humanitarian assistance rushed to El Salvador. However, as was also seen in Colombia after the Quindío earthquake, the distribution raised some problems. On the one hand, the goal of equitable distribution involved establishing a rather inflexible bureaucratic system, slowing down the distribution of assistance. On the other hand, controversy quickly occurred over the government's choices in making the distribution. Despite the government's claim of integrity and transparency, the media widely reported that municipalities belonging to the same party as the majority profited by the distribution of assistance. Due to the difficulty of equitably coordinating the aid and managing the priorities at the local level, the distribution process changed. The municipal governments, churches, and other organizations were then involved, which simplified the distribution of donations in the more affected municipalities.

A great number of NGOs (nongovernmental organizations) were active. Their important reaction was due to their previous involvements in the country because of civil war and the consequences of Hurricane Mitch. However, this previous experience resulted in an imbalance in the territorial coverage by NGOs, only partly filled in by the Civil Defense foreign departments being at the service of affected municipalities, who suffered low media coverage.

AN ACTIVE MOBILIZATION OF THE SOCIETY

The Part of the Media: A Constructive Process

Media play an essential part in both short- and long-term social response (Romano, 1997). They create a channel to the outside world for the affected populations. The first thing many people did after the earthquake was to switch on the radio and television for information about the extent of the disaster and for local information about what streets to walk on without being stuck. Later, the media provided pedagogical information about risks.

Media analysis clearly shows a temporal evolution in topics, especially in the newspapers. During the first days, subjects were devoted to the victims and the reasons for fatalities. Furthermore, rescue tasks were explained. After evaluation of the material damages came a phase where opinions and advice were given to the population. The newspapers played an important role as mediator, reinforcing the social link and allowing separated people (in the country or even abroad) to establish contact, for instance, by exchanging e-mails.

Beyond a straight informative function, newspapers took charge of prevention, acting as a substitute for government. For example, through an article extensively illustrated with very explicit graphics validated by scientists, they distributed advice on protecting oneself during the tremors. They also distributed helpful lists of emergency telephone numbers for readers. Health measures were not forgotten, as numerous warnings on epidemic risks were given in shelters. In the same prevention process, they provided advice about how to evaluate damaged houses and how to locate the safest places in case of tremors.

As usually noticed in other contexts of disasters, the newspapers denounced what they considered governmental policy failures in seismic emergency management. The lack of preventive measures prior to the disaster made the front page on Sunday, January 21, particularly insisting on aggravating conditions in the Las Colinas tragedy. This topic focused the journalists' attention, leaving in the background other similarly dramatic situations. Television, particularly channel 12, put pressure on the central government to change its distribution strategy toward the victims.

However, providing information was not one of television's main goals; instead, they were obsessed by the landslide's drama, the dead, and the homeless. Although justified in the first days, according to interviewed people, this concern appeared excessive after a while. Two weeks later, after the emergency period, the local newspapers took a cathartic turn, opening discussion sections in their pages to allow people affected by the drama to tell their stories and to share their experience. The press collected the earthquake perceptions, published children's drawings, victims' stories, and so on.

Populations' Attitudes and Reactions

The First Moments: Stand Up and Protect

Immediate response to sudden disaster has some similarities and some differences according to economic status, stage of the family cycle, and culture. Archea (1990) speculates, quite reasonably considering our results, that after the first second of shock, people move to protect what is most dear to them.

During the post-earthquake mission, a questionnaire was completed by 82 people living in an urban environment (San Salvador). Forty men and 42 women were questioned. This questionnaire was tested previously, after an earthquake in Martinique, by F. Leone (Geode, University of French West Indies). It allowed definition of the main behavioral trends and an evaluation of the authorities' actions.

The results show that there was no visible panic. As reported in our interviews and in the media, people tried to act as calmly, rationally, and effectively as possible under the circumstances, the same as shown in many investigations (Quarantelli, 1976; Laska, 1990).

Half the sample (47%) adopted the appropriate protective behavior when the impact occurred: to go under the table or under a doorframe (28%) or to get out of the building when the person stood on the ground-floor level (19%). Only 17% declared they did not adopt any protection behavior and stayed rooted on the spot or ran away. These results are interesting if we take into account that there was no risk mitigation policy in El Salvador. Furthermore, 31% declared they collect preventive information in the newspapers, 67% were asking for official information, and 74% wanted to participate in exercises in order to improve the efficiency of behavior during an earthquake. Sims and Baumann (1983) showed that a form of preparedness is to engage in self-protective behavior. Self-protective behavior is most likely to occur in persons who know about the risk, believe they have some control over their own health and job, and believe that precautions really work.

Disaster victims do not act as independent individuals; their responses are to a large extent influenced by their families. Family members try to meet as soon as they learn that a disaster has occurred. This point must be underlined, even if it appears obvious, because it may cause interference with the efforts of crisis management. It deals with the importance of the social network and communication needs, because 50% of the people were anxious to know what had happened to their close relatives and tried to call, saturating the already damaged telephone network.

Most of the sample (82%) feared another earthquake, which actually occurred one month after the first one, and 71% adopted protective measures in their home after the earthquake, by adding a rescue light or making reserves of rations or water. It must be emphasized that the majority of people explain in a scientific way the reason why the earthquake occurred, even if 9% assign its origin to divine intervention. Moreover, 20% mention the role of God in their survival.

The evaluation of the efficiency of the public authorities was harsh; 43% judged the efficiency as bad or very bad, 23% judged it fairly efficient, and only 28% judged it good. Nobody evaluated the efficiency of the public authorities as excellent. This point may involve some psychological issues because distress is greater when victims distrust authorities (Goldsteen et al., 1989).

Some elements of the authorities' management, the police presence, the fast mobilization, and very swift call for the international assistance, were viewed as strong points. In contrast, the negative points were the lack of governmental preparation for the threat, even when the country is know to be highly seismic, the lack of risk planning, the lack of information or education about seismic risk, and the absence of previously organized structures to face catastrophes and distribute assistance.

Another element that raised a problem was the treatment of the missing and presumed dead. Just after the earthquake, the number of dead and wounded reported in formal speeches and in the international press did not take into account the missing. This fact is explainable by the authorities' will to put forward reliable and definitive numbers, and, in this way, not to be accused of an excessive catastrophism. People buried by a landslide in which bodies could not be found were not taken into account in the official statistics. However, witness on the spot of a landslide could certify the absence of family and friends. This "double" disappearance (real death and official nonrecognition of death) raises a psychological problem in the acceptance of the death of a close relative. Surveys showed that a funeral for the dead was a necessary ritual for acceptance of the tragedy and a condition of the recovery of the community. The government's desire for transparency was not interpreted favorably by the population.

In this way, a rumor spread of a political manipulation to create a fictitious electorate. All catastrophe situations tend to generate rumor; it is the medium by which a message of a population's anxiety gets to institutions. Its appearance in a crisis corresponds to a classical and necessary phenomenon for social thought: Rumor answers at once psychological and social purposes. In a state of uncertainty, and facing the shock created by the event, it has an explanatory function because it allows people to understand and to justify a version of reality that individuals fear or manage in a wrong way. Allport and Postman (1945) define rumor as a "general statement that we represent as true, without having real facts allowing us to verify its exactitude."

Care of Possible Psychological Troubles

Some psychologists focus on the mental health outcomes of disasters for individuals (Aldwin and Revenson, 1987; Aptekar, 1994). After the earthquake, the Salvadorian health services noted an increase in need for medical attention, especially for mental health attention.

Immediately, many victims show signs of post-traumatic stress disorder (PTSD), but depression and amnesia are also seen. In many ways, victims avoid the overwhelming reality facing them. For example, those who survived the November 1985 volcanic mudflow that killed 22,000 people in Armero, Colombia, often did not want to know the full extent of the tragedy and resorted to a primitive thinking style (Krug et al., 1998), so-called "magical thought." Mental health problems are greater when more people are killed. For example, in the Armero disaster, major depression and post-traumatic stress disorder affected ~50% of the survivors (Lima and Pai, 1992–1993). After a big

Australian bush fire, families experienced more conflict, withdrawal, and irritability, and mothers tended to overprotect their children. After the 1989 earthquake near San Francisco, college students had many more nightmares than students in another (unaffected) town (Wood et al., 1992).

Mental health problems appear to decline if social and health structures are active, or with time after the disaster (Carr et al., 1997; Chahraoui et al., 2003). However, this does not always occur or it may not occur very rapidly. Two years after an Australian earthquake, substantial mental health problems persisted. Among victims of the large Armero lahar, the rate of emotional distress over a four-year period declined from 65 to 31%. One could say that stress declined by 50%, but one could also say that, after four years, almost a third of the population was still suffering. These statements assume that the disaster does subside in a certain way, whereas living in a house located in an area at risk means that it now becomes a semipermanent place of danger and defilement. This point is very important for mitigation and disaster awareness.

COHESION IN THE CRISIS: A FACTOR OF RESILIENCE

The organization of the community to face the disaster is conditional upon the reference system model to which it belongs. The Kobe earthquake management was done in a society characterized by a high bureaucracy level; the Manjil earthquake in Iran (de Vanssay, 1991) came under a religious vision of the earthquake as a divine punishment; survivors saw themselves as the "elected" chosen by God. In the Latin American cultural context, the dominating element is the fact that the group (the community) represents a strong value, and family relationships are of particular importance; we talk about a high "familism" level. The newsletter of the Pan-American Health Organization, *Disasters* (January 2002, No. 86), underlines that local communities have always played a critical role in reducing their vulnerability to disasters. Many of the most successful disaster reduction initiatives have been the result of community participation and ownership of both the activities and the results.

In accordance with the above observations, three social responses to the earthquake appeared.

1. A particularly important one was the convergence of non-victims to the places of the disaster, adding a population looking to take advantage of the situation. The attraction of promises of rebuilding for disadvantaged populations belonging to villages or nearby areas was mentioned by many of our interviewees. This phenomenon occurred after the catastrophe produced by Hurricane Mitch. The government proposed consistent aid to victims, as well as immediate assistance in rebuilding. Considering the difficulty in verifying such dealings, the resort to social control exerted by the district inhabitants, the communities' leaders, or the district committee is very widely used. Furthermore, different researchers (Colbeau-Justin and de Vanssay, 1997) show that, faced with the inability of institutions, social relationships are

reinforced by a mechanism of substitution and become a social cohesion norm.

2. Leaders arose who took charge in small communities at the level of razed districts. Their effectiveness varied according to local circumstances. The disorganization produced by the earthquake was not complete. In this way, communities quickly developed with a "community leader," reflecting the already existing structure in towns and villages. This leader could have proposed himself spontaneously or been chosen after or before the catastrophe. His abilities are in proportion to his goodwill and to his involvement in the community. Security and hygiene relations with the world outside, but also psychological support, maintaining the balance inside the group, and so on, are part of the assignments he assumes on a material and social level.

3. The social support and development of community spirit took the place of official institutions. One frequent immediate post-disaster phenomenon was the therapeutic community, when even unrelated people work hard together during a disaster or very soon afterward. Some victims were helped by family friends; such social support would seem to be a key element in recovery from disaster stress. However, research indicates that the matter is not as simple as one may think. Having social support of friends and family eases the pain, but does not necessarily stop the pain (Colbeau-Justin and Barouch, 1996).

One can see in these initiatives the development of a Latin American cultural characteristic, and perhaps more particularly, that the lack of strong institutions leads to social support within the family or the social group.

THE DIFFICULT RECONSTRUCTION PROCESS

The Physical Reconstruction

A press conference given by the president on January 15 showed there was no post-earthquake reconstruction policy. The same attitude was repeated on a communal and individual level. That is a problem because previous studies show that the longer the reconstruction process, the slower the recovery of the community (Horlick-Jones et al., 1995). Town councils recommended building temporary shelters from sheet metal or plastic to answer the immediate demand for housing. Questioned about long-term rebuilding, it appears that since such shelters' quality lasts about ten years, they already positioned themselves into a long-term phase.

Three weeks after the earthquake, people who lost their homes were not rebuilding. The persistent tremors made the inhabitants anxious, and indeed, one month after the earthquake, another destructive and lethal event occurred.

Concerning aid for reconstruction, it must be noted that private individuals practically do not insure themselves. No specific system of indemnity for natural catastrophe exists. Furthermore, private insurance is still out of reach for the great majority of the population. The few insured people did not find any problem receiving refunds for their damaged goods. Once the insurance company was informed, it sent an expert to examine the concerned building, to authorize work, and to define the amount of the indemnity. In return, a $10.5 million temporary housing program was settled in order to build 63,000 temporary housing units in the 100 most affected municipalities. Every unit was given $167 for construction of 16 m² of floor area. Through town councils, the government granted this amount (1500 colones in the former currency) for financial and material aid for each razed home. Municipalities distributed the amounts to the affected families. However, it is not apparent that specific information about earthquake-resistant building would be provided.

Rebuilding promises are an important attraction to the disadvantaged nonvictim populations. This phenomenon is classical and has already been observed after Hurricane Mitch. The resort to social control, exerted by district inhabitants, the community leaders, or the district committee, has been very widely used. This kind of social regulation could lead to some excess. A previous study (Bisch et al., 1999) showed a predominance of positive aspects such as solidarity and social cohesion.

Reconstruction will also be essential for the economy. Landslides razed many coffee plantations. Another problem, which is less often mentioned, is the impact on fishing. The earthquake occurring at sea, the consequences on the submarine flora, and therefore on the fauna, cannot yet be evaluated. It seems that this fact could have an impact on migratory movements of birds that could not feed themselves as usual. Fishermen (100,000 families, often organized in cooperatives, are involved in this activity) have already seen a negative impact of the earthquake on fish production.

The motivation to rebuild is not only to return as rapidly as possible to a normal situation, as before the earthquake, but is also directed to reinforce individual abilities to cope.

Psychological Reconstruction: A Real Will of Recovering

Evaluations of catastrophes usually concern real figures: number of dead and wounded, material damages, and economic costs. These data are reported by both national and international press, as well as by public institutions, and allow mobilization of material and physical assistance. Much less attention is given to less visible injuries that may still persist for life: psychological traumas.

Connecting physical reconstruction to the reinforcement of the social links is a positive goal. Reinforcing the importance of social support allows everyone to rely on others; social cohesion is strengthened and social life normalized in spite of the institutional disorganization.

Individual identity is based on relationships maintained with others, such as family, friends, and social group, as well as with the physical environment. In this way, a village or a razed district is not a simple physical fact, but also corresponds to the disappearance of a personal and collective identity. The foundation of social life (church, school, shops, town) disappears, throwing into question everyday life and all it involves in terms of stability and sense of control. The trauma is collective and does not affect only those who have lost their close family or their home, but also all those who lived through this situation of emergency.

On all levels, we aid initiatives whose aim is to restore social links, to help the victims cope, to release anguish or fear, to return to reality again. However, most of these initiatives are isolated in time and too rarely planned or institutionalized. Municipalities of the affected towns recruited leaders among professionals and volunteers. Their aim was to bring play activities to the main public squares and most damaged zones, activities like singing and playing games or sports that are primarily directed at the young people and children.

NGOs were quickly aware of the necessity of bringing psychological support, even if informal, to the victims, with teams available in the late evenings and going into shelters to talk and offer a cup of coffee. This is a fraternal process, not only a care process. Volunteers or group leaders recruited by damaged municipalities animated the main public squares and most damaged zones with entertainment activities to encourage the youngest people to express the emotions they felt during the earthquake.

CONCLUSION: REINFORCING INDIVIDUAL AND SOCIAL ABILITIES TO COPE

This paper argues that, despite the objective hazard, risk is socially constructed. Risk assessment is inherently subjective and represents a blending of science and judgment with important psychological, social, cultural, and political factors.

The post-seismic experience feedback showed the importance of integrating psychosocial insights into the crisis management. At the population level, there are strong dynamics that may be used to simplify the management of an emergency. If neglected, these same dynamics may be an obstacle or at the very least a hindrance.

At an assistance level, the lack of specialized means and the penury of the assistance services limited El Salvador's reactive ability, making it dependent on external assistance. The Salvadorian experience therefore confirms that good preparation, as well as anticipatory organization, would allow acceptable assistance. Several experiences around the world can confirm this point (Horlick-Jones et al., 1995).

A culture of prevention was lacking. This is shown by the lack of measures to mitigate effects of natural phenomena, the non-respect of seismic building standards, and feelings that previous tragedies were not taken to be lessons. The attitude spontaneously adopted by the media concerning prevention and pedagogy was to emphasize it, because it tends to oppose this situation. The use of clear maps and explanations, supported and formulated by well-known scientific experts, should be noted and encouraged in other prevention systems. As a matter of fact, instead of information revealed by public authorities (whose credibility may sometimes be diminished), choosing different informational vectors may favor the integration of adaptive knowledge and practices.

It must be emphasized that there is a gap between the population's awareness level (its knowledge about the seismic history of the country, the safety instructions to follow, the characteristics of earthquakes) and the level of preparation by the public authorities. Regardless of difficulties, the community was ready to cope and to take part in the reconstruction process. Many local initiatives proved that damaged communities were not waiting for assistance but took care of themselves. There is a very strong demand for state intervention, and more particularly, for public information and education, reinforcement of institutions, creation of specialized structures, and so on.

At a social dynamics level, the cohesion phenomenon that one observes regularly after a catastrophe was present. It appears even more strongly when precariousness is no longer individual but collective and involves questions about the functioning of the society itself. Cohesion is reinforced when the temporary replaces long-established conditions, and starting over involves a new beginning at the individual, economic, and social level. Sims and Baumann (1983) showed that the same process could be observed in the case of tornado threat. In this particular context of Salvadorian institutional weakness, social cohesion proved to be an essential motive for raising up the society. It is true that the notions of solidarity and of group are strong values in Latin America and that disorganization at the national level might be expected. Nevertheless, the communities' action at a "micro-local" level was particularly efficient and stood on several steps, including rescuing and reconstruction. The limitations of risk science and the importance and difficulty of maintaining trust and confidence in public authorities during a crisis point to the need for a new approach. It focuses upon introducing more public participation into both risk assessment and risk decision making in order to make the decision process more democratic and efficient, to improve the relevance and quality of technical analysis, and to increase the legitimacy and public acceptance of the resulting decisions.

We conclude, noting that recognizing interested and affected citizens as legitimate partners should improve the risk management. We have underlined that the Salvadorian society is based on high family and social values: familism. This can develop effective mitigation as well as strong reconstruction dynamics following a disaster, no matter how severe the destruction and degradation of the economical situation may be.

ACKNOWLEDGMENTS

Grateful thanks to Morena Azucena for her help and support in El Salvador, to Frédéric Leone and Thierry Lesales for their scientific help and participation, and to Professor Michael Carr, who made a great contribution to improving this paper.

REFERENCES CITED

Aldwin, C.M., and Revenson, T.A., 1987, Does coping help? A reexamination of the relation between coping and mental health: Journal of Personality and Social Psychology, v. 53, p. 337–338.

Allport, G.W., and Postman, L.J., 1945, The basic psychology of rumor: Transactions of the New York Academy of Sciences, Series II, VIII, p. 61–81.

Aptekar, L., 1994, The psychology of disaster victims, *in* Aptekar, L., ed., Environmental disasters in global perspective: New York, G.K. Hall, p. 79–126.

Archea, J., 1990, Two earthquakes: Three human conditions, *in* Yoshitake, Y., et al., eds., Current issues in environment-behavior research: University of Tokyo, p. 23.

Bisch, P., Bour, M., Colbeau-Justin, L., and Givry, M., 1999, Le séisme du Quindío (Colombie): Cahiers de l'AFPS, Paris, 93 p.

Chahraoui, K., Laurent, A., Colbeau-Justin, L., Weiss, K., and de Vanssay, B., 2003, Stress psychologique des sinistrés des inondations de la Somme: une étude exploratoire: L'information Psychiatrique, v. 79, no. 4, p. 307–318.

Carr, V.J., Lewin, T.J., Kenard, J.A., Webster, R.A., et al., 1997, Psychological sequelae of the 1989 Newcastle earthquake: II. Exposure and morbidity profile during the first 2 years post-disaster: Psychological Medicine, v. 27, p. 167–178.

Colbeau-Justin, L., and Barouch, G., 1996, La demande de sécurité des victimes d'inondations: Instance d'Evaluation des Politiques Publiques de Prévention des Risques Naturels, LPENV CNRS UMR 8069.

Colbeau-Justin, L., and de Vanssay, B., 1997, Evaluation of the efficiency of networks, Société Française des Risques Majeurs (SFRM), European Union, Euroconferences: Control and management of seismic risk: Spetzes, Greece.

Colbeau-Justin, L., and Mauriol, M., 2001, Gestion de crise et reconstruction, *in* Guillande, R., et al., eds., Le séisme du Salvador: Cahiers de l'AFPS, Paris, p. 8.

de Vanssay, B., 1991, Quelques aspects de sociologie des désastres: Cahier technique No. 5, AFPS, Paris, p. 63.

Fritz, C.E., 1968, Disasters: International Encyclopedia of Social Sciences, v. 4, p. 202–207.

Goldsteen, R., Schorr, J.K., and Goldsteen, K.S., 1989, Longitudinal study of appraisal at Three Mile Island: Implications for the life event research: Social Science and Medicine, v. 28, p. 389–398.

Guillande, R., Audras, F., Balandier, P., Bour, M., Coco, R., Colbeau-Justin, L., Fabriol, H., Garry, P., Gunot, S., Hubert, F., and Mauriol, M., 2001, Gestion de crise et reconstruction, Le séisme du Salvador: Cahiers de l'AFPS, Paris, p. 87.

Horlick-Jones, T., Amendola, A., and Casale, R., editors, 1995, Natural Risk and Civil Protection, European Commission: London, E & FN Spon, p. 554.

Krug, E.G., Kreesnow, M.J., Peddicord, J.P., Dahlberg, L.L., Powell, K.E., Crosby, A.E., and Annest, J.L., 1998, Suicide after natural disasters: New England Journal of Medicine, v. 338, p. 373–378.

Laska, S.B., 1990, Homeowner adaptation to flooding: An application of the general hazards coping theory: Environment and Behavior, v. 22, p. 320–357.

Lima, B.R., and Pai, S., 1992–1993, Responses t the psychological consequences of disasters in Latin America: International Journal of Mental Health, v. 21, p. 59–71.

Quarantelli, E.L., 1976, Human response in stress situations: Laurel, Maryland, Johns Hopkins University Press, 189 p.

Romano, L.E., 1997, Implicaciones sociales de los terremotos en San Salvador, Seminario sobre Evaluacion y Mitigacion del Riesgo Sismico en al Aera Centroamericana, 22–26 Sept: Universitad Centoamericana José Simeon Cañas, San Salvador, El Salvador, 9 p.

Sims, J.H., and Baumann, D.D., 1983, The tornado threat: Coping styles of the North and South: Science, v. 1976, p. 1386–1391.

Wood, J.M., Bootzin, R.R., Rosenhan, D., Nolen-Hoeksema, S., et al., 1992, Effects of the 1989 San Francisco earthquake on frequency and content of nightmares: Journal of Abnormal Psychology, v. 101, p. 219–224.

MANUSCRIPT ACCEPTED BY THE SOCIETY JUNE 16, 2003

Geological Society of America
Special Paper 375
2004

Social perspectives on hazards and disasters in El Salvador

Ana Elizabeth Araniva de Gonzalez

USAID/El Salvador, Boulevard y Urbanización Santa Elena, Antiguo Cuscatlán, La Libertad, El Salvador

ABSTRACT

Nobody can be fully prepared to face a disaster, but in El Salvador, the path to disaster preparedness is just starting. The last three major natural disasters that have affected El Salvador have presented images that show that the people most affected are always the poorest in the country, the least prepared, who live in high-risk areas or under inappropriate housing conditions.

The impact of Hurricane Mitch and of the January and February 2001 earthquakes on the country has been more profound than any other natural disaster in the last 50 years. It has changed government structures, created social consciousness, alerted the population to what a high-risk area is, and it has also promoted new "disaster-prepared" communities. However, El Salvador was not ready to face the disasters. Neither governments at the local and the national level nor the population were concerned with the possibility of a disaster. The majority of the communities were not aware of the risks that they were facing. Regulations to enforce better planning for new housing development had not been applied. The donor community has helped to cope after the disasters with integral activities that promote economic reactivation, improvement of the infrastructure, and call attention to the environmental degradation that is creating more hazardous conditions for the future. El Salvador is trying to rebuild with better disaster resistance, with the assistance of the donor community and the effective participation of its society.

Keywords: El Salvador, flooding, reconstruction, earthquake.

INTRODUCTION

Hurricane Mitch slammed into El Salvador about November 2, 1998. Ironically, this day is observed as a social and religious event, during which the people pay respect to the dead. Mitch was a surprise to everybody, including the Meteorological Division in the Ministry of Agriculture. Although not as hard hit as Honduras or Nicaragua, with major landslides and thousands of deaths (Honduras reported more than 5000 and Nicaragua 3000), El Salvador suffered 374 people dead or missing, 60,000 people displaced, and roughly 65,000 hectares of land flooded (Crone et al., 2001). Estimates of damage ranged from $132.5 million to $1.7 billion. The hardest hit were in the low-lying areas of the flood plains in the low Lempa and Río Grande de San Miguel areas (Fig. 1).

Official reaction to disaster starts when a country declares that it is facing a national disaster, and the government of El Salvador promptly declared a status of national emergency. Days after the rain had stopped, the signs of the flooding were still present in Usulután: The corn crops were damaged, the adobe houses were literally "dissolved," and there were no bridges to cross the Lempa river. Along the Río Lempa, local people described a huge wave (from the extreme water release on the Río Lempa *15 de septiembre* hydroelectric dam) that had come over the bridge and the houses, flooding everything downstream. They told how they had lost all their belongings because they lived on the edge of the river, but not their lives because luckily they were already in a shelter. Why were they living on the edge of the river? Because they had resources close by. Were they aware of the risk of living there? No, nothing like this had ever happened. These were the common answers of people in the areas of San Vicente and La

*Egonzalez@usaid.gov

Araniva de Gonzalez, A.E., 2004, Social perspectives on hazards and disasters in El Salvador, *in* Rose, W.I., Bommer, J.J., López, D.L., Carr, M.J., and Major, J.J., eds., Natural hazards in El Salvador: Boulder, Colorado, Geological Society of America Special Paper 375, p. 453–459. For permission to copy, contact editing@geosociety.org. © 2003 Geological Society of America.

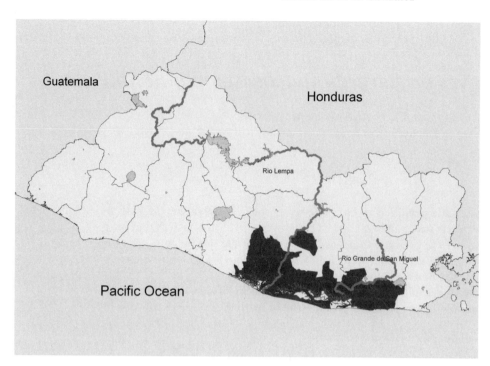

Figure 1. Map of the area most affected by Hurricane Mitch in El Salvador.

Paz. The scene was the same everywhere: adobe houses gone, as well as the loss of basic grain crops and animals.

The shelters, mostly schools, were crowded, a result of the high population density in El Salvador. Only a few people, mainly old settlers, in the lower Río Lempa had suffered from floods in the past; nonetheless, they choose to live in the area because the land is productive and provides them income and agricultural production in good years.

The area around the town of Chilanguera, San Miguel, was also hit by a flash flood, killing 126 people, and many others were missing. The scenes were moving and the effect was direct on the local inhabitants, who lost everything in the disaster. Were they aware of the risk of living there? "No, it is the will of God," one lady told me.

The Road to Recovery

El Salvador received a good international aid response. Most of the direct assistance came as in-kind contributions. Data from the Ministry of Foreign Affairs estimated that cash contribution channeled through the Government of El Salvador (GOES) during the emergency phase was $4.3 million. The U.S. government provided around $37.7 million through USAID (U.S. Agency for International Development), USDA (U.S. Department of Agriculture), and the Department of Defense.

Donor agencies and nongovernmental organizations started working in the post-emergency attention period. USAID designed a special objective (SpO), and it was named "Reduced Vulnerability of the Rural Poor to Natural Disasters in Targeted Areas."

The focus of the assistance was on reducing the vulnerability of the rural poor in the most affected areas.

The impact of the storm was greatest on the rural poor. The highest losses of basic grains and livestock were sustained mainly by the rural communities. These same communities had lost their homes and were drinking contaminated water from shallow wells and nearby streams. Here, the loss was not high in economic costs, but for the people it was all that they had. What was needed in order for these people to rebuild stronger communities?

USAID decided that simply restoring the previous conditions was not a satisfactory solution and that the response from the mission needed to include stimulating economic activity, restoring and expanding access to basic community services, and mitigating the environmental impact of future natural disasters.

Consultations with partners and stakeholders were held, but it was the field visits that proved to be most important. Assessments were conducted by the different offices, particularly the Water and Environment Office, which financed a survey on water contamination issues. Water quality analyses conducted on many wells in the flood plains of the Río Lempa revealed that the primary type of contamination was bacteriological. Solutions, such as cleaning up and disinfections of wells, were implemented to lower the risk to the health of the communities.

RECONSTRUCTION PLANS FOR THE AFFECTED AREA

The focus area had a population of ~153,000 in 150 communities; more than 37,000 people were identified as the most affected. The main considerations that guided reconstruction activity are described in the following sections (USAID, 1999).

Stimulating Economic Activity

It is essential to stimulate the ability of the rural poor to get back on their feet and avoid dependence on donations. Also, by increasing economic activity, the communities will be better poised for future natural disasters. The floods of November 1998 came just a week before the harvest of basic grains. About 80 percent of the farmers in the selected area grow basic grains, and probably most of them lost their crops. With the loss of the crop, the farmers lose their seed grain and the income needed to buy materials for planting in the following year. Without these materials, the farmers are left unable to replant, and they prolong their dependence on assistance, donations, and support from families in the United States. Economic reactivation is thus retarded.

New rural roads and electricity lines are key factors for economic activity, job creation, and overall improvement of the quality of life in rural areas. The use of electricity facilitates the formation of rural micro-enterprises, and in the selected area only 5% of the population had access to electricity compared with 26% in other rural areas and 65% nationwide.

Communities in the affected areas need access to basic social services if they are to realize their development potential and reduce their vulnerability to natural disasters. The floods and landslides exacted a toll on social infrastructure, mainly water systems, community wells, latrines, houses, schools, and local clinics. The selected area was deficient in these services, and the houses were badly constructed with weak materials and were vulnerable to flooding. The lack of access to clean water is a critical aspect of development in El Salvador. In this area, ~55% of the families obtain their water for household consumption from shallow wells from a water table only 2–5 meters below the surface in the dry season; 25% obtain their water from streams or vendors who obtain the water from unknown sources; and only 20% get their water from domiciliary piped water systems.

Most wells in the region were contaminated with dirt, human waste, and dead animals. The wells do not have protective casings and sanitary seals. Most of the work consisted of cleaning up and disinfecting wells to make the water suitable for drinking. The mixing of latrine overflow and sewage with floodwaters exacerbated the contamination of wells and groundwater. Saline intrusion was another problem that became evident during the assessment process, and it is an issue that will not be easily addressed.

Since the risk area in which we were working is susceptible to annual flooding, the houses that were constructed in this area of activity were elevated on concrete platforms to provide additional flood protection. The solution served as an example for other donors on how to improve the social infrastructure in risk areas.

The design of the schools also took into account measures to make them more resistant to flooding, either by elevating them or using water-impermeable blocks.

Mitigating the Environmental Impact of Future Natural Disasters

The devastating impact of Hurricane Mitch in El Salvador and in the region is directly linked to severe degradation of the environment by deforestation, overpopulation, and the lack of effective early warning systems. Reducing the vulnerability of the rural poor to natural disasters must include steps to better manage our watersheds while also improving community, local, and national government disaster preparedness.

Along the coast, the mangroves that prevent seawater intrusion and make the area less susceptible to the effects of high waves and hurricanes are disappearing due to extensive overcutting and development for new houses. In the hills, the lack of vegetation due to deforestation leads to greater runoff and landslides. Lack of tree cover in the upper Lempa watershed also magnifies the problem of siltation in reservoirs. Deforestation of sub-watersheds on both sides of the lower Lempa and Río Grande de San Miguel increases the risk of flooding to the local population.

Different working missions interviewed relevant GOES offices to better ascertain their needs. The Meteorological and Hydrological Division (SMHN) addresses the flooding and weather-related issues, and the Center for Geotechnical Investigations (CIG) in the Ministry of Public Works (MOP) addresses landslides and related problems. In both institutions, there were highly trained professional people, but with low salaries and very low motivation. Even with their low motivation, they were doing their work under very difficult circumstances: They had cars assigned to their offices, but lacked the gasoline; they had a fax machine, but they did not have the paper to receive the faxes; they had computers, but did not have updated software, did not have an uninterrupted power supply or the wiring to protect them in the case of an electrical storm, and they did not have access to the Internet in order to use weather-related information.

The inherent weakness of these vital institutions made the country even more vulnerable to all kinds of natural disasters. There were concerns that all the equipment that the U.S. government was going to provide would be wasted if these institutions were not strengthened, and there were also concerns about the access of this information and of these systems to the rural poor or to the people that needed it the most.

The fears of providing more equipment to these institutions grew as the months passed and more assistance from the National Oceanic and Atmospheric Administration (NOAA), the Environmental Protection Agency (EPA), USDA, and the U.S. Geological Survey (USGS) came to these weak offices. USAID was concerned about the sustainability of the assistance, and it was easy to find reasons. Here are some concrete examples:

NOAA was providing a variety of assistance, mainly to the meteorological service, but the situation of their offices was not appropriate for new servers and high-tech computers. The Ministry of Agriculture had their offices in a brand new building, but the Meteorological and Hydrological Services were in the old Ministry of Agriculture building, without air conditioning and with a lack of much-needed security for all the new equipment. USAID requested meetings with the Minister of Agriculture, but only the Director of Natural Resources was able to meet with us, in spite of repeated attempts. Internally, the USAID special-objective director also asked the mission director to talk about these issues with the Minister of Agriculture. A meeting was set once with the minister, but at the last moment, the minister had to go somewhere else, and the mission director met with the Director of Natural Resources once more. This dissonance was very disappointing and also caused concern about the sustainability of these programs.

Other solutions were sought because time was running out. The security of all the equipment, as well as the possibility of the service to the affected communities, was at stake, and the opportunity for the country to have a working early warning system was delayed. Exploratory meetings were held with academics from the Central American University (UCA) and with the Río Lempa Hydroelectric Commission (CEL) to locate possible sites for the systems. Finally, with the help from the Minister of Environment, the decision was made to situate the systems within the government, and steps were taken to accommodate the equipment in a technically sound location. The financial sustainability was still pending, but the political decision to support the program had been made.

While the sustainability issue on the government side was a concern, the disaster preparedness component was advancing at the local level; both communities and local governments supported the work that was being done. Many donor agencies were working in this aspect, and local communities got organized and received the training needed. Also, communities started recognizing the problems that they were facing and the risks that surrounded them. Towns such as Berlín in Usulután formed local Disaster Management Committees, with support from NGOs (nongovernmental organizations), local government, private enterprises, and international donor agencies. That effort is still working and has served as a model to others.

All these efforts complemented the work of the USGS and NOAA toward the installation of a sophisticated early warning system in the Lempa watershed. The Río Lempa was chosen as an area for regional activity, and the Guatemala mission was managing it through the Central American Integration Secre-

tariat (SICA). This accomplishment exemplifies the regional integration efforts.

THE 2001 EARTHQUAKES

The work was advancing and Mitch was almost out of our minds when the January 13th earthquake hit. And it was almost unthinkable that another earthquake could happen, but we were affected again on February 13th. The greatest impact of these earthquakes was associated with massive landslides (Baum et al., 2001). It was impressive and frightening to see the effects in Las Colinas, San Vicente, Lake Ilopango, and Los Chorros (Jibson and Crone, 2001). Anybody who visited all those sites and still continues to do so is more aware of the risk that we all face and of the fragility of the land that we live in.

Many of the towns affected by Mitch were also affected by the earthquakes, and the disaster preparedness committees were effective, but the effects of both earthquakes overwhelmed all preparation efforts.

After the earthquakes, GOES agencies lacked the equipment to determine their magnitude or obtain accurate information about the aftermaths. The public lost confidence in the information that was being provided by the government, and information was based on hearsay and gossip only, creating confusion.

Efforts Responding to the Earthquakes

USAID is conducting a reconstruction activity with a total investment of $170 million, but many other donors are also conducting reconstruction activities. Table 1 presents a picture of the destruction.

Since most of the investment will be in infrastructure, steps are being taken to avoid new settlements and reconstruction activities in high-risk areas. With technical assistance from the USGS, studies are being conducted to plan for better and safer reconstruction efforts. El Salvador has the highest population density in Central America, and moving people out of high-risk areas is not a practical solution. So, trying to provide mitigation and other alternative solutions is the most needed approach.

TABLE 1. EARTHQUAKE DAMAGE STATISTICS (USAID, 2001)

Damage	Number of dead or complete destruction and losses	Number of injured or damaged infrastructure
Personal losses	1,159	8,122
Houses	149,528	185,338
Schools	465	2,216
Health facilities	141	
Public buildings	990	
Churches	417	
Losses to private infrastructure: coffee, poultry, mills	$50 million	

The earthquakes and Hurricane Mitch had a damaging impact on the socio-cultural infrastructure well beyond economic impacts. Some of the most important are presented below.

Slums Were Established

Those people who lost their homes but did not own the land on which they were built are less likely to qualify for a new house. Title of ownership is a requirement of many housing projects. Also, a person who resides in a high-risk area is excluded from many reconstruction projects due to strict guidelines and an effort to place these people in a safer area. Most people will keep the temporary shelter as their new home until it falls down (Fig. 2).

Destruction of the Adobe Houses

Most people in the rural areas built their homes from adobe because they had the resources available, but most of these homes were old and had never been maintained or renewed. People that do not have funds or are not being supported by a reconstruction project are using adobe again because they do not have another option. That situation is changing the rural and urban appearance of most towns, and the new homes are going to be smaller than the older adobe homes and will not provide the same insulation against the tropical heat. But the reconstruction effort is also creating a sense of teamwork, because the houses are being rebuilt as a result of a community effort (Figs. 3 and 4).

Destruction of Colonial Churches

Old colonial churches were built of adobe, and most of them were already in a precarious state (Fig. 5). Churches in small towns are the gathering place for the community and host other activities besides religious ones. Towns are using parks and basketball courts and are working together to gather funds to build a new church. Because their church was destroyed, people were

Figure 2. Temporary shelter home.

Figure 3. Destroyed adobe house.

Figure 4. Construction of new homes.

Figure 5. Destroyed church, Santa María Ostuma, Cuscatlán.

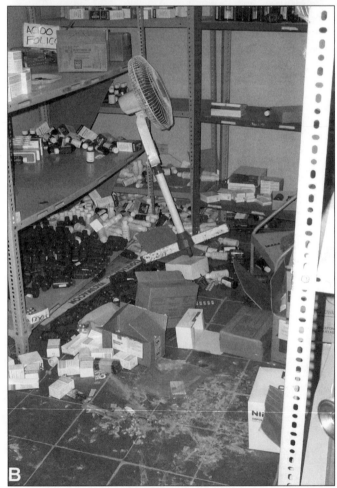

Figure 6. School and clinic damage in San Vicente.

thinking that the earthquakes were a punishment for so many killings in the history of El Salvador (civil war, crime, kidnappings), and priests had to talk about this during masses to explain that these destructions were caused by natural events.

Destruction of Rural Schools and Clinics

For many rural communities the construction of a school is a great achievement. As a result of the 2001 earthquakes, 465 schools were destroyed or suffered major damage (Fig. 6). The GOES is making an effort in this respect, but El Salvador was in a financial deficit even before the earthquakes. The same effort is developing for clinics, which were also damaged. The communities are already trying to obtain funds to rebuild, and some are using camps and other sites as schools or clinics to avoid losing this vital infrastructure for their children and families.

SUMMARY AND CONCLUSIONS

It is difficult to say that positive changes have come from the disasters, but some significant changes are evident in the society after these events.

Social Awareness

Many people are now afraid to live where they have their homes. Other people are paying more attention to where they buy their homes. Some new developments announce that they are "certified to be risk free" or "not damaged during the earthquake."

New Roles in Local Governments

Municipalities have established disaster prevention or disaster management groups in order to work with communities in the case of a disaster; these are also functioning for ongoing threats such as the dengue epidemics. Also, municipalities are conduct-

ing "land planning" or "emergency territorial plans" in order to have manageable information about the risk areas. These plans may then serve as tools to provide or deny construction permits. Municipalities have faced problems due to the issue of construction permits in high-risk areas that have resulted in casualties.

Moreover, different donors are promoting this activity and are providing funding to make it happen.

Changes in Government Institutions

The biggest change that is directly related to Mitch and the earthquakes is the establishment of SNET (Servicio Nacional de Estudios Territoriales). As a result of the lack of trust in GOES institutions to handle technical information during the most recent disasters, the government decided to create SNET (Fig. 7). It was assigned to the Ministry of Environment and Natural Resources, with financial and technical autonomy. It also established that the Meteorological and Hydrological Services from the Ministry of Agriculture and the Seismic and Geological Services within the Ministry of Public Works will be transferred to SNET. SNET has gained public attention: They are providing vital information on weather, seismic, geological, and hydrological aspects. They have published their first investigations, and their technical personnel are on the news often, gaining presence, trust, and

visibility. They have their own Web site (www.snet.gob.sv), on which they present information on weather forecasts and seismic activity, as well as investigative reports.

Among SNET's main objectives is "To develop the scientific research and the studies specialized for uses related to the prevention and reduction of risks, as much in the field of the disasters as in development and territorial planning, and transfer the results of these investigations and studies to the units of government so that each one executes the recommendations according to its capacity." This objective clearly recognizes the importance of disaster prevention and making accurate information available. It also recognizes the responsibility within the government to provide that information to the people who need it most: They should be the ones who benefit most.

ACKNOWLEDGMENTS

The completion of this article relied on the documents and work of my colleagues at USAID/El Salvador. I wish to thank former director Dr. Ken Ellis for all the support during the design of the Mitch strategy. I also want to thank John S. Moore (USDA) for his expert assistance in preparing the final part of this document with the native English review.

Decree of Creation
OFFICIAL NEWSPAPER VOLUME no. 353, 18 of October of 2001
DECREE No. 96
THE PRESIDENT OF THE REPUBLIC OF EL SALVADOR, CONSIDERING:

I. That the obligation of the State is to assure to the inhabitants of the Republic the enjoyment of the constitutional rights, specially the right to life and physical integrity, property and possession;

II. That El Salvador is exposed to the incidence of threats of natural and environmental origin, by its high seismic and volcanic activity, to be subject in addition to the impact of phenomena of hydro-meteorological origin and by those threats that are formed in the social processes of transformation of natural means and in the proliferation of more and more vulnerable establishments;

III. That is necessary to know the constituent factors of risk, threats and vulnerability, as a basis to adopt measures that guarantee suitable levels of security for the population in the face of the events and processes of risks of disasters;

IV. Which for such reasons it is necessary to create a responsible organization to produce opportune information in this matter and that defines measures adapted for the prevention and the reduction of risks before disasters.

THEREFORE,
In use of its constitutional faculties,
IT DECREES:

Creation of the National Service of Territorial Studies

Figure 7. Decree of SNET's creation.

REFERENCES CITED

Baum, R.L., Crone, A.J., Escobar, D., Harp, E.L., Major, J.J., Martinez, M., Pullinger, C., and Smith, M.E., 2001, Assessment of landslides hazards resulting from the February 13, 2001, El Salvador earthquake: U.S. Geological Survey Open-File Report 01-119, 20 p.

Crone, A.J., Baum, R.L., Lidke, D.J., Sather, D., Bradley, L.A., and Tarr, A.C., 2001, Landslides induced by Hurricane Mitch in El Salvador—An inventory of descriptions of selected features: U.S. Geological Survey Open-File Report 01-444, 24 p.

Jibson, R.W., and Crone, A.J., 2001, Observations and recommendations regarding landslides hazards related to the January 2001, M-7.6 El Salvador earthquake: U.S. Geological Survey Open-File Report 01-141, 19 p.

USAID (U.S. Agency for International Development), 1999, USAID/El Salvador, Mitch special objective: Reduced vulnerability of the rural poor to natural disaster, 52 p.

USAID (U.S. Agency for International Development), 2001, USAID/El Salvador, Earthquake special objective: The lives of the rural poor bettered in the aftermath of the earthquake disasters, 25 p.

MANUSCRIPT ACCEPTED BY THE SOCIETY JUNE 16, 2003

Geological Society of America
Special Paper 375
2004

Risks and disasters in El Salvador:
Economic, environmental, and social aspects

L. González

*Centro de Información, Documentación y Apoyo a la Investigación, Universidad Centroamericana José Simeón Cañas,
Autopista Sur, San Salvador, El Salvador*

L. Romano*

*Centro de Información, Documentación y Apoyo a la Investigación, Universidad Centroamericana José Simeón Cañas,
San Salvador, El Salvador*

L. Salamanca

Departamento de Organización del Espacio, Universidad Centroamericana José Simeón Cañas, San Salvador, El Salvador

ABSTRACT

The accumulated environmental impact, population growth, and human settlements in high-risk zones have made the disaster problem more complex, with disasters of small and medium magnitude affecting the most impoverished groups. The impact of disasters has increased due to a vulnerability accumulation process and unsustainable interaction with ecosystems.

With the Spanish conquest and colonization, the situation in the Salvadoran territory changed drastically, due to the new philosophies on production and the relation between natural resources and human beings. Land was used not only for essential subsistence production, but for cultivating products for agro-exportation. As a result, cultivated lands were expanded, vegetal coverage and natural resources were diminished, and an increasing process of unsustainable use of natural resources was underway. Irregular urban growth patterns, the concentration of demographic and socioeconomic activity in the Metropolitan Area of San Salvador (AMSS), and the lack of territorial regulations are other key issues that increase risk factors.

In a country like El Salvador, permanently affected by earthquakes, floods, and droughts, disasters have evidenced the deficiencies in citizen involvement and in the social awareness of different institutions. The social capacity to deal with El Salvador's environment, including its risks, needs to be strengthened in order to reduce the impact of disasters.

Disasters are the result of a risk accumulation process, caused by the intervention of economic factors, territorial characteristics, and social organization. Regarding this, one main conclusion points out that the reduction of risks and the impact of disasters is a matter of social transformation and is not only a technical issue.

Keywords: El Salvador, risk management, disasters, economic development, environment.

*Corresponding author: luis_romano@hotmail.com

González, L. Romano, L., and Salamanca, L., 2004, Risks and disasters in El Salvador: Economic, environmental and social aspects, *in* Rose, W.I., Bommer, J.J., López, D.L., Carr, M.J., and Major, J.J., eds., Natural hazards in El Salvador: Boulder, Colorado, Geological Society of America Special Paper 375, p. 461–470. For permission to copy, contact editing@geosociety.org. © 2003 Geological Society of America.

INTRODUCTION

During the last two decades, Latin America has been exposed to the effects of thousands of disasters of every kind of magnitude. In the past five years, this situation has become so serious that it has led to a deep reflection about the causes and the actions that should be taken in order to surpass a model focused only on emergencies and move toward a model that aims to reduce the impact of the disasters.

One of the most remarkable advances has been the introduction of the concept of "risk" in several institutional approaches, and consequently the introduction of a concept in which disasters are not only a result of physical phenomena, but of the complex interaction of social processes that take effect when disaster happens. However, it cannot be denied that there is still a long way to go before the state, the communities at risk, and society adopt adequate policies and practices to face the defiant reality of disasters.

Disasters provoked by the 2001 earthquakes are the most recent evidence of the interaction between seismic threats and a vulnerable social context: The impact was selective to a considerable degree. With few exceptions, it damaged poor communities without access to basic services, where houses were built with deficient materials and without any criteria for structural safety. The earthquakes caused a notable increase in the percentage of population that lives in poverty, from 47.5 to 51.2% (PNUD, 2001). The earthquakes left more than 164,000 families homeless and more vulnerable to future risks. It also left total losses appraised as high as 12% of the gross national product (GNP) (CEPAL, 2001). Obviously, this impact is a setback in the processes of poverty relief, in the reduction of the residential deficit, in economic growth, and, in general terms, in the pursuit of sustainable development.

In this paper, we want to improve the knowledge of the risk processes and factors that lead to disasters, noting that disasters are not merely a result of natural or physical events. The first section offers a general overview of risks and disasters in El Salvador, focusing not only on geological disasters, but also on hydrometeorological and human-caused disasters. In the second section, we review the economic transformations that have conditioned land use and the location of human settlements, and the way this has led to a society at risk, not only because of these locations, but also because of their high poverty level and the unsustainable use of natural resources.

The third section discusses the aspects related to territorial use and the social construction of risks. In the fourth section, we discuss how other social factors could provoke vulnerability and worsen the effects of disasters. This is important because the social factors determine the conceptions, visions, and options in the presence of disasters. Finally, in the fifth section, the main conclusions and some challenges that should be faced in order to reduce risk and to achieve sustainable development are presented.

RISKS AND DISASTERS IN EL SALVADOR: A GENERAL OVERVIEW

Since pre-Columbian times, the population of El Salvador has been menaced by intense seismic activity. The Salvadoran territory is part of the so-called "Fire Belt" of the Pacific Ocean, where earthquakes and eruptions are just part of daily life. There is evidence of geological activity and earthquakes before the Spanish conquest and colonization. Archaeological evidence points to the fact that, during pre-Hispanic times, indigenous populations were severely affected by disasters (Manzanilla, 1997). The "Joya de Cerén" case is the clearest example: Evidence of the destruction of housing, crops, and belongings in 590 A.D., as a result of the eruption of the Caldera volcano, was found at this location. Some researchers suggest that this eruption forced the migration of indigenous population and the desertion of the Salvadoran territory (Manzanilla, 1997).

This information suggests that geological activity—which includes volcanic eruptions and earthquakes—is not a contemporary phenomenon. It has a long history that has not been adequately studied. It cannot be denied that there have been some efforts to compile a considerable amount of information on earthquakes and the disasters they have produced (Lardé y Larín, 1978; Martínez, 1978; Romano, 1996a). However, the list of threats and disasters in El Salvador is quite extensive and is not limited to earthquakes. Some of the most intense and frequent threats that provoke disasters in El Salvador are described in Table 1.

This quick glance at Salvadoran reality shows how the risk problem has different facets: The geological characteristics of the zone, combined with the hydrometeorological system, types of human interaction with the environment, and the different forms of social organization, can create conditions that generate diverse kinds of disasters. In this sense, it can be said that the disaster grows through a risk construction process. This is the argument in the following two sections, in which we examine the case of San Salvador and the progressive increase of hydrometeorological disasters, especially those provoked by floods and droughts.

San Salvador: A Martyred City

San Salvador, the capital city, has been the location of recurrent disasters provoked by earthquakes since it was founded. The frequency and magnitude of these disasters have increased and have generated processes of social change. It is clear that the intensity of the 1524 earthquake, when San Salvador was a small town or, more accurately, a camping area for Spanish troops, cannot be compared with the 1575 earthquake, in which many important buildings—built during the previous thirty years—were destroyed (Lardé y Larín, 1978). Throughout time, vulnerability and risk in San Salvador have grown in the same way as the city and its economic, social, and political life have grown.

A comparison of economic losses reveals clearly that San Salvador suffers the most severe risk. As stated above, the amount of damage caused by the 2001 earthquakes—neither of

TABLE 1. EL SALVADOR: MAIN HAZARDS AND IMPACTS

Hazard	Description	Examples in El Salvador
Landslides	Human settlements and ecosystems are affected through the volcanic chain in the central region (Campos and Velis, 1991). Earthquakes can cause landslides and deaths.	Panchimalco, 1762. Almost 300 people were killed in the Montebello neighborhood in 1982 (Romano, 1997b). In January and February 2001, 600 people died in La Colinas, a residential zone located in the lower part of Cordillera del Bálsamo.
Floods	There were few floods in the pre-Columbian and colonial period. Nowadays, floods are common and affect mainly the areas and population located on the coastal zone.	San Andrés valley and Tacachico town (XVII century) (Lardé and Larín, 1978). Several cases of floods and disasters during the rainy season (from May to October) are reported annually in local newspapers.
Earthquakes	The oldest and most fatal threat in El Salvador. During five centuries, San Salvador has been destroyed several times and the Spanish conquerors rebuilt the city each time. On some occasions, they tried to move the city to a safer location (Lardé and Larín, 1978). Other cities have become affected as they have grown through the centuries.	The first earthquake that was reported took place in 1524, just a few months after the founding of the city of San Salvador. Between 1524 and 2001, San Salvador has been destroyed 22 times by earthquakes (Acevedo and Romano, 2001). There also have been several disasters induced by earthquakes in other parts of the country. These disasters occurred in 1915, 1932, 1936, 1937, 1951, 1982, and 2001 (Acevedo and Romano, 2001).
Volcanic eruptions	The main cities in Central America and 60% of its population are settled on the Pacific Coast, where better soils and climate are encountered, but, at the same time, the area is characterized by geological threats such as continental and local seismic faults, tsunamis, and volcanic eruptions.	The presence of volcanoes indicates a high level of geological activity. Volcanic eruptions and the accompanying seismic activity have been the cause of the destruction of San Salvador. The most recent destructive event occurred in 1917 (Campos and Castillo, 1991).
Droughts	Droughts have side effects, such as damage to agriculture, increase in electric energy generation costs, and reduction in water supply (CEPRODE, 1997a).	Droughts became more recurrent during the second part of twentieth century, although historical and archeological evidence show that these phenomena are characteristic of the climate in Latin America (García Acosta, 1997).
Forest fires	Forest fires are caused by high temperatures. They take place on land that becomes dry due to droughts. An agricultural practice which consists of burning the land to prepare it for the next crop sowing is another cause of fires. They destroy forests, vegetation, and habitat for biodiversity. Fires are an unattended threat in El Salvador.	During 1998, after the worst episode of El Niño, forest fires increased by 77% (CEPRODE, 1998) and it is estimated that every year 3,967 forested areas are affected by fires (CCAD and others, 1998).
Building of infrastructure without environmental and risk reduction measures	Infrastructure and building construction without any environmental criteria. In several cases, the infrastructure construction has provoked floods and other apparently natural phenomena.	In San Salvador, a large number of the floods that occur every year are a result of the obstruction of creeks and rivers by basements of bridges. This is the case for the floods in communities such as Nuevo Israel and La Vega, located near the Acelhuate river.
Inappropriate use of technology	In spite of El Salvador being a country where industrial development is scarce—limited to food and chemical products, aluminum, asbestos, cement, and pharmaceutical industries— massive poisonings due to improper handling of highly polluting substances (e.g., chlorine, cobalt, and others) are reported.	In 1989, four workers were affected by radioactivity in a small industry in Soyapango (Romano, 1997b). In 1994, a large oil spill in the Acajutla port revealed a long history of pollution (CEPRODE, 1994).
Excessive pollution of natural resources	Other pollution risks are caused by the lack of plans for processing industrial effluents (PRISMA, 1997); The sewage in cities with sewage systems is not processed before it is discharged into nearby water sources.	According to official data, about 90% of the water available in the country is contaminated (MARN, 2001).

which had its epicenter in San Salvador—equaled 12% of the GNP of the year 2002 (CEPAL, 2001), while the damage caused by the 1986 earthquake—which had its epicenter near the city—equaled 27% of the 1985 GNP (CEPAL, 1986). It is obvious that the most severe risk—understanding *risk* as the probability and magnitude of losses and damages (Wilchez-Chaux, 1998)—is located in San Salvador, where the main economic, social, and political activities take place. The possibility of a more widespread disaster is still latent.

Disasters provoked by earthquakes have affected social organization and building techniques. Changes in construction materials and building techniques were made after the seventeenth-century disasters and at the end of nineteenth century (Romano, 1996a). The 1854 earthquake made it necessary to temporarily move the capital city (Browning, 1982). The 1986 and 2001 earthquakes forced the government to change the building and seismic engineering codes (Bommer, 1996).

As mentioned before, earthquakes are not the only cause of recurrent disasters. The danger during rainy season has become, little by little, the main basis for disasters. Although the damage is lower than those provoked by earthquakes, the accumulated impact may be equal or greater.

Disasters Caused by Floods and Droughts

There are some reports of disasters linked to the rainy season during the nineteenth century, but those were sporadic. The evidence suggests that the rainy season was not a threat during the colonial times. The primary disasters during this period were the flooding of the Zapotitán valley in 1653—caused by the eruption of Playón volcano and the obstruction of the Sucio riverbed (Lardé y Larín, 1978)—and the flooding of Ateos and nearby towns, caused by the San Dionisio flood of 1762 (Lardé y Larín, 1978), which also caused a landslide in Panchimalco (Lardé y Larín, 1978).

During the twentieth century, disasters during the rainy season became part of normal life. In the 1990s, at least thirty cases of flooding were registered every year in the creeks and rivers of San Salvador and in the lower parts of the Paz, Sensunapán, Lempa, Grande, and Goascorán Rivers (Campos and Velis, 1991; CEPRODE, 1997b; Acevedo and Romano, 2001). Additionally, floods now take place in locations where this phenomenon did not happen before, i.e., some small villages on the banks of the San Francisco River in Morazán, a province located in the northeastern zone of the country.

These incidents of flooding are accompanied by secondary disasters due to rain, such as triggered collapses and landslides, which were already frequent before 2001 and became more widespread after that year's earthquakes (see map on www.snet.gob.sv). The marginalized urban communities located near brooks, creeks, and on sandy ground in San Salvador, or in middle-class neighborhoods and semirural settlements in the mountains and on the volcanic slopes, are amongst the most highly affected areas.

Droughts have also become a cause of several disasters, especially among peasants who have to face loss in their harvests every two or three years (Acevedo and Romano, 2001; CEPRODE, 1997a; Romano, 1997a). Over the last thirteen years, there were droughts in 1991, 1994, 1997, 2000, and 2001.

The accumulated environmental impact, population growth, and human settlements in high risk zones have made the disaster problem more complex, with disasters of small and medium magnitude affecting the most impoverished groups.

Some Key Processes of the Disaster Dynamics

The situations presented above show that disasters had been increasing due to a vulnerability accumulation process and unsustainable interaction with ecosystems. Earthquakes in San Salvador (and in the rest of the country) are not something new, but their impact is growing. Rains and cyclones are not new either, but the proliferation of disasters caused by floods is a fact. The increasing number of disasters caused by floods can be explained by the unsustainable use of natural resources since the Spanish conquest: the environmental deterioration and the annihilation of the basic functions of the ecosystems (the control of water overflows, floods, and water capture), the populating of zones exposed to floods, and riverbed obstruction (Romano, 1996b).

These factors are grounded in processes that have generated the present risk conditions: economic transformations, urbanization, and the organization of a society based on exclusion, inept in the face of the natural and physical realities of the Salvadoran territory. Due to the fact that these three processes are key to explaining the reality of risks and disasters, each process is explained briefly in the following sections.

ECONOMIC TRANSFORMATION AND RISK GENERATION

With the Spanish conquest and colonization, the situation of the Salvadoran territory changed drastically, due to new philosophies on production and relations between natural resources and human beings. Land was not only used for cultivating essential subsistence production, but for cultivating products for agro-exportation, such as balsam, cocoa, and, above all, indigo. As a result, cultivated lands were expanded, vegetal coverage and natural resources were diminished, and an increasing process of unsustainable use of natural resources was underway. In the following, we will show how the most important Salvadoran socioeconomic changes are linked with the increase of risk factors and processes.

From Agro-exporting Economy in the Colony to Agro-exporting Economy in the Republic

During the first years of the colony, the environment was not drastically manipulated, nor were the properties and production systems of the indigenous communities. These communities had a magical vision of nature; they were grateful for its nurture and, therefore, respected it. This vision coexisted during three centuries with the commercial agriculture that the Spaniards oriented to agro-exportation. Agro-exportation was built not only upon balsam, cocoa, indigo, coffee, sugar, cotton production, and cattle farming, but also on the depletion of vegetation (Browning, 1982).

This period was marked by low population density, as the result of a drastic reduction of indigenous peoples caused by diseases the Spaniards transmitted to them, such as chicken pox, smallpox, measles, and tuberculosis. Low demand for natural resources and the apparent limitations on the further expansion of the agro-exporting–oriented model explain how, during the first centuries of the colonial age, the scarce indigenous communities could keep the larger part of their lands.

During the Republican Era, there was a transition toward coffee growing, which led to the expropriation of the indigenous community lands (Menjívar, 1980) that had been respected and even favored by colonial authorities up to this point (Browning, 1982). The expropriation forced the rural population to migrate to cities along the volcanic chain threatened by earthquakes and left landless the peasants who came to constitute the social group most vulnerable to floods and drought.

Economic Modernization in the Second Part of the Twentieth Century

During the second part of the twentieth century, there were some important efforts oriented toward agricultural diversification and the introduction of industries. This process is known as the "modernization" of the Salvadoran economy. In fact, the favorable prices of coffee after the end of World War II produced important income, which promoted investment in new areas by coffee producers in order to increase their income (Dada Hirezi, 1985). The lands fit for coffee growing were already in use. The only available lands were located in the northern zone of the country—used previously for indigo growing—and in the southern part of El Salvador, which were uninhabitable due to the presence of diseases such as malaria and yellow fever (Quezada, 1989).

Unlike coffee, cotton requires the total depletion of vegetal coverage, which implied a massive deforestation along the coast, an environmentally weak zone. The proliferation of plagues that affected cotton crops led to the massive use of pesticides that eventually made this cultivation environmentally and economically unviable.

Industrialization—"limited to chemical products, food, and asbestos and cement products" (Dada Hirezi, 1985)—was introduced steadily in towns but, in actuality, aggravated the population growth of the city of San Salvador, already consolidated as the main urban center in the country. During the second part of the twentieth century, in El Salvador and other Latin American countries, the industrialization and urbanization processes that favored population concentration in cities such as San Salvador began (López, 1986). These processes led to the expansion of San Salvador as nearby municipalities were added to its territory, leading to the formation of the San Salvador Metropolitan Area (AMSS is the Spanish acronym), which includes 13 municipalities: San Salvador, Mejicanos, Ciudad Delgado, Cuscatancingo, Ayutuxtepeque, San Marcos, Nueva San Salvador, Antiguo Cuscatlán, Soyapango, Ilopango, San Martín, Apopa, Nejapa, and Tonacatepeque.

These processes set the conditions for risk due to floods in the coastal zone of the country—the main cotton-growing zone—and earthquakes in San Salvador.

The Rise of the Twenty-First Century

The twentieth-century modernization process was followed by social convulsions that led to the civil war (1979–1992) and to relative economical stagnancy or, worse yet, recession. During the 70s and 80s, there were no important changes in productive activities, but there was an important transformation in the property regime due to land reform and the nationalization of the bank. Coffee production-related activities and the incipient industrial activities were the motor of the economy in this period.

The situation began to change in 1989 with the first government of the right-wing Alianza Republicana Nacionalista (ARENA) party, which reversed the nationalization of banking and international trade and combined this with an economic liberalization process that included the opening of the Salvadoran economy to foreign trade, public services, privatizations, and State reductions.

The old agro-exporting model was substituted for a model based on two main elements: textile assembly plants (*maquila*), created with foreign investments, and the economic support that many Salvadoran families received from their relatives in the United States.

These changes favored the growth of urban economic activities, such as construction, trading, and services, which replaced agriculture as the main economic activity. Economic reforms entailed a drastic tax reduction for importations, including agricultural products such as corn, beans, and sorghum.

Under these conditions—high poverty levels, unsatisfied basic needs, scarce investment in agricultural technology, and economic politics that affect income—almost any external event can easily cause a crisis or disaster. A ten-day rain in a deforested zone can turn into a disaster due to the frailty of ecosystems and the productive activities and limited adaptability of its inhabitants. Equally, erosion, obstruction in riverbeds, and floods appear in the same zones where precipitation levels reach more than 50 mm. This phenomenon is recorded in newspapers, which describe disasters caused by floods every year and droughts that take place approximately every two and a half years (Campos and Velis, 1991; Romano, 1997a; Romano, 1997b).

Risks and Disasters Are Caused by a Specific Development or Underdevelopment Process

Risk processes and disasters are determined by economic factors. Changes in the modalities or zones of production have led to certain uses of the soil and natural resources and to social inequalities and poverty.

The initial focus on agro-exporting, the later focus on coffee growing, and the demographic concentration in the higher seismic activity zones has configured a society that is highly exposed to geological phenomena. The changes in land tenure created more pressure on frail soils (the hills in the northern zone of the country and coastal zones), inadequate use of natural resources, and environmental deterioration. Maps of land use (see www.snet.gob.sv) show that most of the territory is being cultivated for basic grains and pastures, even in places were this is not recommended.

In this period, the "anti-agricultural bias" of the Salvadoran economic system damaged the rural way of life by causing the decline in the prices of agricultural produce, which became more severe with the fall of international coffee prices since 2001. The negative impact of State reduction diminished the economic and technological support for agriculture, as can be verified by the budget and personnel cuts in the Agriculture Ministry (PRISMA, 1997; Romano, 1999). In January 2002, most of the employees of the National Center for Technological Transference (CENTA is the Spanish acronym), a key part of the ministry, were laid off. Obviously, this worsens the technical deficit in agriculture.

Equally, the concentration of population in urban centers increased with the conversion of old coffee farm land into residential zones, shopping centers, universities, or industrial plants. This concentrates more population and investments in high seismic activity zones and along high-risk zones such as riverbanks and brooks and on sandy grounds or zones vulnerable to landslides. With this change, disasters not only affect poor urban communities, but middle- and high-class neighborhoods and industrial zones as well.

THE CONSTITUTION OF RISK IN THE SALVADORAN URBANIZATION PROCESS

The irregular urban growth patterns, the concentration of demographic and socioeconomic activity in the AMSS, and the lack of territorial regulations, linked to the Salvadoran urbanization process, amplify risk factors.

According to the 1992 census, more than 30% of the country's population and more than 50% of the urban population lived in the AMSS, an area that equals 3% of the national territory (Barry and Rosa, 1995), while 10% of the country's urban population lived in Santa Ana and San Miguel, the two other main urban zones.

This drastic imbalance has caused increasing environmental pressure, which is visible in three aspects: potable water availability, solid waste handling, and urban land availability (Lungo, 1995). These factors have reached a critical level that may impede the medium- and long-term sustainability of the country.

Characteristics of the Main Urban Centers

The rapid expansion of the AMSS is characteristic of the Salvadoran urbanization process (Baires and Lungo, 1989). This process accelerated disproportionately in the 1980s, when the migra-

tions from rural zones to urban centers increased dramatically. The AMSS's population increased 5.5-fold between 1950 and 1992, from 606 inhabitants per square kilometer in 1950, to 3322 inhabitants per square kilometer in 1992 (Chinchilla, 1999).

This rapid urbanization and demographical concentration has taken over important forested zones, mainly coffee zones, on the slopes of San Salvador volcano, the hills of San Jacinto, and the Bálsamo mountain chains in the southern zone of the AMSS. Population pressure in these zones and the expansion into the northern, northeastern, and southwestern zones cause damage to agricultural and forest soils and the last subterranean water reserves of the country (Lungo, 1995).

In 1993, 27% of the AMSS population lived in illegal neighborhoods, cottages, and shelters; 38% of the housing in these settlements was built on already occupied lands, 40% of this housing was built with waste material, and 13% of the people dwelled in settlements in high-risk locations, such as along brooks, riverbanks, and on unstable slopes (Acevedo, 1994).

The unbalanced growth of the AMSS and the lack of soil treatment policies to ensure the protection of the main natural resources are the causes of environmental degradation.

Territorial Management

In spite of the rapid urban growth in El Salvador, appropriate planning and management instruments to facilitate land use policies are virtually nonexistent. Up to now, few instruments have been designed to regulate the use of territory in El Salvador. Those that have been designed have not effectively contributed to land use regulation. The most recent was the Urban Development Master Plan, the application of which is scarce because ongoing urban processes are tainted by marginality and illegality.

The lack of formality in the urbanization processes, an obvious consequence of poverty, is reinforced by the current vision of the regulatory framework, in which there is a clear lack of mechanisms to regulate that informality. In fact, what prevails is an excess of regulations, a lack of flexibility, and inadequate criteria that do not respond to the current reality (Chinchilla, 1999).

DISASTERS AND SOCIETY

It is a fact that disasters triggered by floods, droughts, or earthquakes make clear to what level the different social sectors are unified or not. Disasters test the level of democratic development of a society, how citizens are involved in key issues, and how institutions attend to social demands. In a country like El Salvador, permanently shaken by earthquakes, floods, and droughts, disasters have evidenced the deficiencies in citizen involvement and in the social awareness of the different institutions.

State and Disasters

The most recent disasters in El Salvador—Hurricane Mitch in 1998 and the 2001 earthquakes—made it evident that the institutions of the Salvadoran State are not ready to confront or prevent disasters. Neither do they have a coherent plan to mitigate, prevent, and attend to disasters. The 1986 earthquake showed much of the institutional deficiencies in disaster prevention and attention (Alvarado, 1986; Goitia, 1986); however, the war made it impossible to discuss these institutional deficiencies.

What is evident currently is the fact that the State is not able to create national systems to prevent, mitigate, and respond to disasters, as have other countries such as Colombia (Wilchez-Chaux, 1998). National prevention systems help to distinguish what sectors are involved in these situations and make their responsibilities clear, but the institutions in El Salvador have two characteristics that restrain their effectiveness: an excessive focus on emergency attention and their centralized nature, which makes it difficult to make decisions (Romano, 1999).

This is exemplified by the fact that the National Emergency Committee (COEN is its Spanish acronym) has never made any effort to authorize a seismic resistance code, even in the aftermath of disaster situations. Legally, its actions are restricted to preparing for and attending to emergencies, with few resources and a small budget. During the last decade, its budget was less than $60,000 (Romano, 1999). That explains why COEN was not able to give attention to the entire situation and yielded its functions to a commission of business people: the National Solidarity Commission (CONASOL).

On the other hand, the institutional frailty of the government is evidenced in the scarcity of initiatives to involve the populations affected by the disaster and other local representatives. The government must assist with the most difficult aspects of emergencies, but emergency attention is a social problem that should also be assumed by affected and vulnerable populations, private enterprise, municipalities, and development organizations.

Eighteen months after the disaster of 2001, no institutional changes were made, in spite of the fact that civilian organizations proposed the reformation of the "Civilian Defense Law," and no long-term plan on how to respond to forthcoming emergencies in the country was written.

Governmental institutions are still improvising in order to face emergencies, without a development strategy to reduce risks and to promote security. An example is the way that the Agriculture Ministry reacts to droughts: Instead of pursuing better technological and productive conditions for rural areas, it only commits itself to distribute food and seeds to the peasants.

Limits of Democracy

The fact that the different social sectors, especially those least protected, lack significant channels by which they can transmit their needs and concerns to higher government circles is an example of the weakness of the State apparatus and society in general. This can be explained by the centralization of the decision-making process and the carrying out of those decisions. In this context, it is clear why mechanisms to mediate the discussion between the government and society do not exist.

The democratization of El Salvador cannot continue while public participation is insufficient and does not permeate the areas of risk and disaster intervention. This participation cannot be real if State institutions are unable to respond to public demands. Disasters during the past decades have made this evident.

Adaptation and the Response of the Educational System

The education of the population of a country at risk like El Salvador is as much a government responsibility as the management of foreign aid, and the limitations of the Salvadoran institutions are once again evidenced in this area. The educational system has serious voids regarding issues such as risk prevention and relief, although these have been improved after the 2001 earthquakes by the creation of higher-level courses on risks and disasters and by some educational programs organized by the government and several international and social organizations.

Themes related to risks and disasters should be incorporated into university education, taking into account the characteristics of each field. For example, in economics, this can be studied from the point of view of economic incentives for risk control; in engineering, it would be suitable to study the norms of earthquake-resistant construction or the sustainable use of natural resources. It will be necessary to adapt the curriculum to include new courses and course content.

Up until the 2001 earthquakes, the Education Ministry had made few efforts to give public schools emergency plans (Cruz, 2001). That is why floods, drought, and earthquakes damaged the educational system severely, forcing its officials to improvise responses in order to face every new crisis.

On the other hand, there are processes that aim to reduce educational vulnerability. Since 1999, in the aftermath of Hurricane Mitch, non-governmental organizations have been educating those communities that frequently suffered flood damage. A good example is the case of the communities in the lower part of the Lempa River, which have a solid social organization. Its settlers are being trained in risk and disaster intervention issues.

Communities and Disasters

The most recent disasters in El Salvador showed not only how incapable the government was to attend to them, but also the important role of the population in disaster protection and attention. Risk is socially generated, and its causes can be avoided through a social process of sensitization and through social, economic, and environmental changes.

From a community approach, it is possible to note how disasters are an opportunity for altruism and for strengthening a sense of belonging that makes it easy to educate communities on the causes of disasters and to organize communities for reconstruction and protection against disasters. Although it cannot be denied that in the first moments of disasters "the injury syndrome" (Wilchez-Chaux, 1998) is prevalent—the fact that those people affected by a disaster feel impotent and believe that the only way to survive crisis is with external help—the spontaneous efforts to rescue victims and develop more organized reconstruction efforts is also remarkable.

However, it is equally undeniable that those efforts rely more on goodwill rather than technique and knowledge, which makes it clear that it is necessary to improve the community proficiency in risk and disaster intervention. The current interest of nongovernmental and international organizations to raise the proficiency level in high-risk communities has been demonstrated in dozens of recent risk intervention projects. As a key point, it is important to remember that by April 2003 there were two entities in place for dialogue and effort coordination of risk intervention projects: the Initiative for Risk Intervention and the Permanent Dialogue for Risk Intervention, which gather some 30 organizations, most of them nongovernmental and international.

The government's inability to properly respond to disaster, and the scarce participation of the population at risk in public policy decision making, has left room for private initiative to take charge of the problems generated by disasters. During the 1986 and 2001 earthquakes, this was more than clear: The business sector not only supported the government, but took charge of channeling foreign aid for affected communities. In the presence of the institutional deficiencies, the business sector attended to those people the government was supposed to help. This is not a negative fact in itself, but it makes clear the governmental inclinations toward centralization and toward an elitist emergency planning and attention scheme. Wijkman and Timberlake (1985) even point out the fact that sometimes populations affected by disasters have to cope with situations in which aid is centralized by armed forces and health assistance organizations, which act like an "invading army."

The Role of Aid

The way in which assistance has been organized assures that it passes through the hands of private enterprise before reaching the affected population, making the process more bureaucratic and centralized. Protests regarding the delay in the delivery of aid were published abroad by local and international media. Government institutions are relegated to a secondary level when high-scale disasters occur, as happened in 2001, while the population—without mechanisms for participation—remains passive in the face of the decisions that business leadership makes with government. The positive impact the aid could have on the population dilutes itself in an endless sea of intermediaries who are not in touch with the needs of those affected by disasters.

The 2001 earthquakes are especially illustrative of this reality: Private enterprise, acting through a committee, centralized the aid in San Salvador, while the affected population—mostly from other cities like Usulután, San Vicente, Cuscatlán, and La Paz—did not have access to medicine, water, or food. These "goods" were classified and taken to warehouses in San Salvador for later delivery. According to the schemes of an ad hoc committee created by the central government, neither the municipalities

nor the population could decide anything about the reception and distribution of the aid. Thus, those organizations or sectors that could guarantee a better distribution were excluded (CIDAI, 2001). There was a remarkable improvement when municipalities were allowed to participate in the aid distribution.

Distortions of Reality: An Obstacle to Disaster Prevention and Relief

The educational situation is a key factor to risk prevention and relief in El Salvador: Traditional perceptions that disasters are a result of nature's anger and the belief that the only thing that can be done is to respond to emergencies affect the capacity for risk mitigation.

Governmental analyses of the earthquake disaster of 2001 agree with the traditional viewpoint; they defend that the causes of disasters were exclusively natural. Moreover, disasters tend to provoke mythical interpretations of geological phenomena. For instance, the rumors about the imminent collapse of Central America caused by the earthquakes and the supposed predictions of another and yet more severe earthquake by a newborn child were believed to be true by a significant part of public.

It cannot be denied, however, that since 2001, government officials and civilian society have evolved, and current visions tend to incorporate the perspective that social processes and factors condition the evolution of disasters. This can be seen in the creation of the National Center of Territorial Studies (SNET, by its Spanish acronym), in charge of doing research and drawing up proposals to support risk attention nationwide. SNET also works to incorporate disaster intervention issues in municipal programs of the "Social Investment for Local Development Fund," and provides some courses about disaster intervention.

PERSPECTIVES ON DISASTER PREVENTION AND RELIEF FOR THE TWENTY-FIRST CENTURY

Throughout this paper, we have maintained that disasters are the result of a risk accumulation process, caused by the intervention of economic factors, territorial characteristics, and social organization. Regarding this, some conclusions can be stated:

1. The proliferation of disasters caused by floods, drought, and landslides has become a permanent phenomenon, recorded in every rainy season.

2. Disasters caused by geological phenomena are not that frequent, but their comparative impact is so severe that recovery takes years or even decades.

3. Economic development models have had serious implications for risk management, because they require a reorientation in economic growth patterns, in the use of the territory, and in the distribution of benefits.

4. The territorial use patterns in El Salvador, characterized by high population concentration in the AMSS and in the coastal zone, exceed environmental capacities—above all, access to urban land and water. This increases the social exclusion factors

and vulnerability. At the same time, this process excludes other parts of the country that also require investment.

5. The government is not sufficiently adapted to the reality of risks and disasters. There are some ambiguous signs in the measures taken. An example of this ambiguity is the creation of SNET while, on the other hand, there has been too much delay in the discussion, formulation, and promulgation of new legislation to build up governmental institutions to meet the needs of risk intervention.

6. The almost null participation of the vulnerable population in decision making is not only circumscribed to decisions linked with emergency attention, but with disaster intervention in general. This is a case of accumulated risks by previous economic development models.

7. Progress in official and unofficial education programs is clear and points to sensitization, consciousness raising, and analysis of the risk factors and processes. The next challenge is how to turn this sensitivity and consciousness into policies, plans, and strategies that help confront the causes of risks and disasters.

8. The reevaluation of the role of local population and institutions is a positive consequence of the earthquakes; it shows conceptual progress by both the government and the population. The increase in local organizations and projects oriented toward risk attention will undoubtedly contribute to reducing the organizational vulnerability of the population.

9. More important is the possibility that the population could influence public policymaking in order to include issues such as the support of economic growth with social equity and environmental sustainability, territorial regulations, education of the public in risk, and disaster intervention, and the support of processes for organizational enhancement and the empowerment of communities at risk.

ACKNOWLEDGMENTS

The authors wish to acknowledge Dr. Julian Bommer for his worthy contribution to the publication of this paper.

REFERENCES CITED

Acevedo, C., 1994, Raíces económicas de la crisis ecológica en Centroamérica: Realidad magazine, no. 37, p. 7–32.

Acevedo, C., and Romano, L., 2001, Economía, desastre y desarrollo sostenible: San Salvador, Facultad Latinoamericana de Ciencias Sociales (FLACSO), 157 p.

Alvarado, J.A., 1986, Lineamientos generales para una estrategia de reconstrucción: Boletín de Ciencias Económicas y Sociales, no. 6, p. 357–366.

Baires, S., and Lungo, M., 1989, San Salvador (1880–1930): la lenta consolidación de la capital, *in* Dutrénit, S., ed., El Salvador: México, Nueva Imagen, p. 338–363.

Barry, D., and Rosa, H., 1995, Población, Territorio y Medio Ambiente en El Salvador: Programa Salvadoreño de Investigación sobre Desarrollo y Medio Ambiente (PRISMA) bulletin, no. 11, p. 1–15.

Bommer, J., 1996, Terremotos, urbanización y riesgo sísmico en San Salvador: Programa Salvadoreño de Investigación sobre Desarrollo y Medio Ambiente (PRISMA) bulletin, no. 18, p. 1–11.

Browning, D., 1982, El Salvador, la tierra y el hombre (2nd edition): San Salvador, Ministerio de Educación, Dirección de Publicaciones, 525 p.

Campos, N., and Castillo, L., 1991, Los desastres en El Salvador. Una visión histórico social. Desastres por actividad sísmica y vulcanológica: San Salvador, CEPRODE (Centro de Protección para Desastres), 109 p.

Campos, N., and Velis, L., 1991, Los desastres en El Salvador. Una visión histórico social. Desastres por actividad hidrometeorológica: San Salvador, CEPRODE (Centro de Protección para Desastres), 115 p.

CCAD (Comisión Centroamericana de Ambiente y Desarrollo), et al., 1998, Estado del ambiente y los recursos naturales en Centroamérica: San José, CCAD, 179 p.

CEPAL (Comisión Económica para América Latina y el Caribe), 1986, El terremoto de 1986 en San Salvador: daños, repercusiones y ayuda requerida: San Salvador, CEPAL, 71 p.

CEPAL (Comisión Económica para América Latina y el Caribe), 2001, El Salvador. Evaluación del terremoto del martes 13 de febrero de 2001: San Salvador, CEPAL, 53 p.

CEPRODE (Centro de Protección para Desastres), 1994, Noticias sobre desastres: Actualidades sobre Desastres bulletin, no. 9, p. 7.

CEPRODE (Centro de Protección para Desastres), 1997a, Efectos de El Niño en El Salvador: Actualidades sobre Desastres bulletin, no. 19, p. 11–16.

CEPRODE (Centro de Protección para Desastres), 1997b, Cronología de desastres, accidentes y ecología (1997): Actualidades sobre Desastres bulletin, no. 20, p. 6–11.

CEPRODE (Centro de Protección para Desastres), 1998, Los incendios forestales en el primer cuatrimestre de 1998: Actualidades sobre Desastres bulletin, no. 21, p. 4–7.

Chinchilla, R., 1999, La Gestión del suelo urbano y su relación con la generación de riesgos ambientales: El caso del Área Metropolitana de San Salvador: Canada, Universidad de Montreal, 42 p.

CIDAI (Centro de Información Documentación y Apoyo a la Investigación), 2001, Consideraciones económicas, sociales y políticas del terremoto del 13 de enero: ECA Estudios Centroamericanos, no. 627–628, p. 19–57.

Cruz, J., 2001, Terremotos y salud psicosocial: ECA Estudios Centroamericanos, no. 626–628, p. 279–282.

Dada Hirezi, H., 1985, La economía de El Salvador y la integración centroamericana 1945–1960: San Salvador, UCA Editores, 55 p.

García Acosta, V., editor, 1997, Historia y Desastres en América Latina: volumen II: Lima, LA RED, CIESAS, ITDG, 385 p.

Goitia, A., 1986, El impacto socio-económico del terremoto: Boletín de Ciencias Económicas y Sociales, no. 6, p. 343–356.

Lardé y Larín, J., 1978, El Salvador. Terremotos, Incendios e Inundaciones: San Salvador, Academia Salvadoreña de Historia, 227 p.

López, C., 1986, Industrialización y urbanización en El Salvador 1969–1979: San Salvador, UCA Editores, 74 p.

Lungo, M., 1995, Problemas ambientales, gestión urbana y sustentabilidad del AMSS: Programa Salvadoreño de Investigación sobre Desarrollo y Medio Ambiente (PRISMA) bulletin, no. 12, p. 1–11.

Manzanilla, L., 1997, Indicadores arqueológicos de desastres: Mesoamérica, Los Andes y otros casos, *in* García Acosta, V., ed., Historia y Desastres en América Latina: volumen II: Lima, LA RED, CIESAS, ITDG, p. 33–58.

MARN (Ministerio del Ambiente y Recursos Naturales), 2001, Informe Nacional Estado del Medio Ambiente 2000: El Salvador, MARN, 46 p.

Martínez, M., 1978, Cronología sísmica y eruptiva de la República de El Salvador a partir de 1520: San Salvador, Centro de Investigaciones Geotécnicas, 60 p.

Menjívar, R., 1980, Acumulación originaria y desarrollo del capitalismo en El Salvador: Costa Rica, Editorial Universitaria Centroamericana, 57 p.

PNUD (Programa de las Naciones Unidas para el Desarrollo), 2001, Informe sobre Desarrollo Humano: El Salvador 2001: San Salvador, PNUD, 576 p.

PRISMA (Programa Salvadoreño de Investigación sobre Desarrollo y Medio Ambiente), 1997, Dinámica de la degradación ambiental: San Salvador, PRISMA, 29 p.

Quezada, J., 1989, Base ecológica de la violencia en El Salvador: una propuesta de restauración ambiental: Presencia, v. 1, no. 4, p. 106–123.

Romano, L., 1996a, Implicaciones sociales de los terremotos en San Salvador (1524–1919), *in* García Acosta, V., ed., Historia y Desastres en América Latina: volumen I: Lima, LA RED, CIESAS, ITDG, p. 71–96.

Romano, L., 1996b, El Salvador: historización de los desastres naturales y de la degradación ambiental: San Salvador, CEPRODE (Centro de Protección para Desastres), 32 p.

Romano, L., 1997a, Efectos económicos y sociales de la sequía en El Salvador: San Salvador, CEPRODE (Centro de Protección para Desastres), 56 p.

Romano, L., 1997b, Catálogo de desastres, accidentes y ecología (1915–1990): San Salvador, CEPRODE (Centro de Protección para Desastres), 61 p.

Romano, L., 1999, La protección civil en El Salvador. Propuestas de redefinición: San Salvador, CEPRODE (Centro de Protección para Desastres), 57 p.

Wijkman, A., and Timberlake, L., 1985, Desastres naturales ¿fuerza mayor u obra del hombre?: Nottingham, UK, Earthscan, 175 p.

Wilchez-Chaux, G., 1998, Auge, Caída y Levantada de Felipe Pinillo, Mecánico y Soldador o Yo voy a correr el riesgo. Guía de LA RED para la gestión local del riesgo: Quito, ITDG/LA RED, 154 p.

MANUSCRIPT ACCEPTED BY THE SOCIETY JUNE 16, 2003

Geological Society of America
Special Paper 375
2004

Damages and losses of the 2001 earthquakes in El Salvador

Ricardo Zapata Martí
Roberto Jovel
United Nations Economic Commission for Latin America and the Caribbean, Presidente Masaryk 29, piso 12,
Mexico D.F. 11570, Apartado Postal 6-718, México

ABSTRACT

This paper describes the results of the United Nations Economic Commission for Latin America and the Caribbean (ECLAC) appraisal of direct and indirect damages caused by the January and February 2001 El Salvador earthquakes. The appraisal provided the basis for a meeting held in Madrid in March 2001 at which the international donor and financial community learned about the damages and losses sustained and decided the types and amounts of cooperation to be provided for the reconstruction of El Salvador.

Total damages and losses caused by the 2001 January and February earthquakes in El Salvador were estimated by ECLAC at U.S.$1.604 billion. Of this amount, $939 million (59%) reflect direct damages, and the remaining $665 million (41%) reflect indirect losses. Thus, most of the economic impact of the earthquakes resulted from direct losses of capital or assets. Indirect losses are estimates of changes in future economic flows.

The ECLAC appraisal led to an intense reconstruction effort from within El Salvador and provided a basis for a flow of international cooperation in the form of donations, credits, and restructuring of available external resources. Those external resources facilitated a process of recovery that, although still incomplete, would have been more difficult without them.

Keywords: Socio-economic impact, environmental impact, disaster assessment, methodology.

INTRODUCTION

The United Nations Economic Commission for Latin America and the Caribbean (ECLAC) appraised damages and losses in El Salvador following the January and February 2001 earthquakes. A comprehensive damage and loss assessment methodology, developed and refined by ECLAC (2002a) and used for the past 30 years in the Latin America and Caribbean region, was used to conduct the appraisal.

Some basic definitions concerning damages and losses are in order to assist the reader unfamiliar with the ECLAC methodology in clearly understanding the terms used in this paper. Direct damages refer to total or partial destruction of assets; indirect losses are the modifications to future economic flows (i.e., lower production and higher expenditures, or lower revenues in prod-

ucts and services that will arise as a result of direct damages). Reconstruction costs represent the amount of funds required to repair or rebuild the lost assets, in fact replacing direct damages. They normally include technological improvements, mitigation measures, reengineering or reinforcement, or risk reduction and mitigation costs. Reconstruction costs are normally higher than direct damages. Finally, macroeconomic effects are the results of direct damages and indirect losses on the overall national economic performance and social well-being of the population. Readers interested in details of the damage and loss assessment methodology (ECLAC, 2002a) are advised to examine its contents at the ECLAC Web page (www.eclac.cl/mex) as well as the Web page of the World Bank's Disaster Management Facility and the Provention Consortium (www. proventionconsortium.org/toolkit.htm).

Zapata Martí, R., and Jovel, R., 2004, Damages and losses of the 2001 earthquakes in El Salvador, *in* Rose, W.I., Bommer, J.J., López, D.L., Carr, M.J., and Major, J.J., eds., Natural hazards in El Salvador: Boulder, Colorado, Geological Society of America Special Paper 375, p. 471–480. For permission to copy, contact editing@geosociety.org. © 2003 Geological Society of America.

DAMAGE AND LOSS ASSESSMENT

Total damages and losses caused by the 2001 January and February earthquakes in El Salvador were estimated at U.S.$1,604 million (ECLAC, 2001). Of this amount, $939 million (59%) reflects direct damages, and the remaining $665 million (41%) reflects indirect losses. Thus, most of the economic impact of the earthquakes resulted from direct losses of capital or assets. Indirect losses are estimates of changes in future economic flows. Table 1 shows details on these figures.

Figure 1 classifies the total impact from various types of damages. It reveals that the largest proportion of damages and losses was sustained by physical infrastructure and equipment (64%), followed by an increase in costs and reduced income in the provision of goods and services, mainly transport (31%), and by future production (5%). This breakdown of damage is consistent with the expected effects of disasters caused by geological phenomena. In contrast, when a disaster is caused by hydrometeorological phenomena, future production is most heavily affected (Jovel, 1986).

Of special relevance in the aforementioned breakdown is that two-thirds of the total damages and losses were sustained by the private sector, compared to one-third by the public sector. However, because the government assumes the risk of damages sustained by the poor, in addition to public sector assets, the reconstruction process will require a sizable contribution by the public sector and the international cooperating community. Furthermore, the great amount of asset damages and their impact on future economic flows will translate into increased fiscal pressures on the government and on the country's external sector accounts.

SECTOR DAMAGE AND LOSS BREAKDOWN

The distribution of damages among the main economic sectors affected by the earthquakes is shown in Figure 2. Social

sectors (39% of total damages and losses), infrastructure (29%), productive sectors (21%), and the environment (6%) were affected the most. The most affected individual sectors included transportation ($433 million), housing and human settlements ($334 million), industry and commerce ($246 million), and education and culture ($211 million) (Table 1).

The total estimated damages and losses ($1604 million) represents a high absolute figure. However, the real impacts of damages and losses are realized when they are compared to the country's economy and living conditions. In this regard, we note that total damages and losses are equivalent to (1) 12% of El Salvador's 2000 gross domestic product (GDP); (2) more than 40% of the amount of national exports in 2000; and (3) 42% of the annual rate of gross capital formation. In another comparison, the total losses amount to four times the annual GDP in the construction sector. Therefore, the impact of the earthquakes should not be underestimated. For comparison with other disasters, Hurricane Mitch inflicted regional losses equivalent to 13% of the five Central American countries' combined GDP. If we assume that El Salvador's construction sector capacity were solely devoted to the replacement of lost assets, temporarily setting aside any other type of construction, reconstruction of damages caused by the 2001 earthquakes would require a period of at least four years.

The great percentages of damage that occurred in the social sectors (housing, education, and health), and in the production capacity of industry and commerce, underscore the real dimension of the earthquake tragedy, because small entrepreneurs and people in the lower income strata were affected the most.

GEOGRAPHICAL DISTRIBUTION OF DAMAGE AND LOSSES

An analysis of the geographical distribution of damages illustrates the impacts of the earthquakes on the population. Table 2 and Figures 3 and 4 show the results of an analysis

TABLE 1. SUMMARY OF DAMAGES CAUSED BY THE 2001 EARTHQUAKES IN EL SALVADOR*

Sector and sub sector	Type of loss			Type of property	
	Total	Direct	Indirect	Public	Private
Total	1,604	939	665	567	1,037
Social	617	496	120	238	379
Education and culture	211	190	20	69	142
Health	72	56	16	72	–
Housing	334	250	84	97	237
Infrastructure	472	97	375	171	301
Electricity	16	3	13	3	13
Water and sanitation	23	19	4	13	10
Transport	433	75	358	155	278
Productive	339	244	96	15	324
Agriculture and fisheries	93	39	55	13	80
Industry and commerce	246	205	41	2	244
Environment	103	102	1	103	–
Other damages and expenditures	73	–	73	40	33

Note: Source: ECLAC estimates.
*In millions of U.S. dollars at 2001 prices.

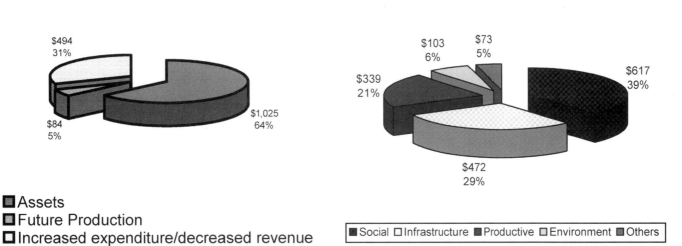

US$ million

US$ million

$494
31%

$1,025
64%

$84
5%

$103
6%

$73
5%

$339
21%

$617
39%

$472
29%

- ■ Assets
- ☐ Future Production
- ☐ Increased expenditure/decreased revenue

☐ Social ☐ Infrastructure ■ Productive ☐ Environment ■ Others

Figure 1. Total impact by type of damage.

Figure 2. Breakdown of damages by sectors.

TABLE 2. GEOGRAPHICAL DISTRIBUTION OF DAMAGES CAUSED BY THE JANUARY
AND FEBRUARY 2001 EARTHQUAKES IN EL SALVADOR

Department	Total damage million U.S.$	Per capita damage U.S.$ per person	Per capita GDP U.S.$ per person*	Damage/GDP %
Ahuachapán	20.3	64	2,242	2.9
Cabañas	3.5	23	2,191	1.1
Chalatenango	1.4	7	2,578	0.3
Cuscatlán	147.1	735	3,335	22.1
La Libertad	263.6	399	5,121	7.8
La Paz	270.5	943	3,020	31.2
La Unión	4.1	14	2,803	0.5
Morazán	0.8	5	2,475	0.2
San Miguel	47.5	101	3,526	2.9
San Salvador	199.5	103	4,142	2.5
San Vicente	243.7	1,533	2,671	57.4
Santa Ana	94.7	175	3,356	5.2
Sonsonate	127.0	289	3,252	8.9
Usulután	180.4	534	2,789	19.1

Note: Source: ECLAC estimates.
* United Nations Development Programme (UNDP, 2001).

conducted for each of the country's departments or provinces. Those data indicate total damages and losses in absolute terms and as well as on a per capita basis, and they also compare total damages to each department's productive capacity as expressed by its GDP.

The effects of both earthquakes were geographically concentrated in the departments of San Vicente, La Paz, and Cuscatlán. Populations in those departments suffered losses that ranged from $700 to $1500 per capita. After those cities, the departments most affected were Usulután, La Libertad, and Sonsonate, listed in decreasing order of damages (Table 2; Fig. 3).

The geographical distribution of per capita damages has both positive and negative implications. Damages and losses were concentrated in relatively well developed departments whose populations have a greater probability of self-recovery than those of poorer departments (e.g., Cabañas, Morazán, Ahuachapán, and La Unión), where poverty was concentrated prior to the disasters. Therefore, losses in human development were less severe in the poorest areas of the country.

Figure 4 shows the prevailing human development index (HDI) by department prior to the earthquakes. The HDI is a composite indicator developed on the basis of income, living conditions, education, health services, etc., by the United Nations Development Programme (UNDP, 2001).

On the positive side, the earthquake disasters afford El Salvador an opportunity to undertake mitigation measures as part

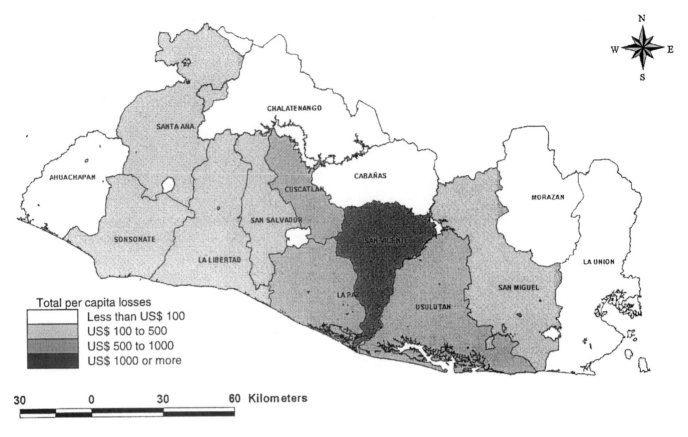

Figure 3. Geographical distribution of damages and losses caused by the January and February 2001 earthquakes in El Salvador (U.S.$ per capita).

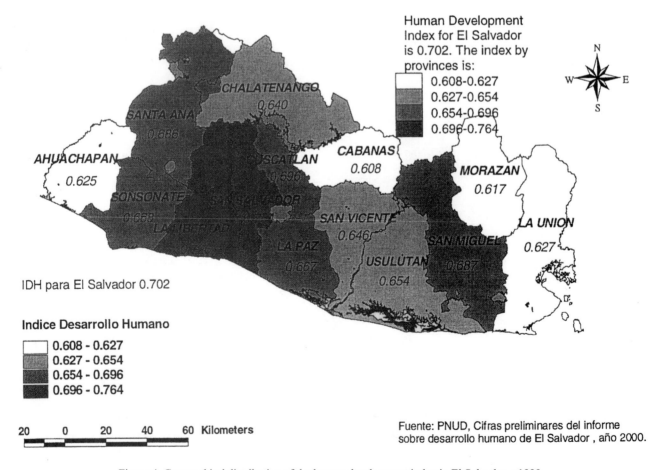

Figure 4. Geographical distribution of the human development index in El Salvador, ~1999.

of the reconstruction of lost or damaged assets. These disasters provide an opportunity to improve housing and productive conditions, and to reduce the vulnerability and fragility of social and physical infrastructure that prevailed prior to the earthquakes.

On the negative side, the modest human development improvements that had been achieved during recent years were set back in those departments that were most affected by the earthquakes. Thus, the poverty distribution was adversely affected and modified by the earthquakes. Figure 5 shows the revised HDIs in all departments.

As a consequence of the earthquakes, the departments of San Vicente, La Paz, and Usulután changed from some of the better-developed to the least-developed departments in the country, since they now have the lowest human development indexes.

Reconstruction of infrastructure damaged in the earthquakes required that financial resources be assigned to the departments that were most affected, where, coincidentally, major development investments were already under way prior to the disaster. This will no doubt result in the postponement of poverty reduction investments in those departments that were less developed and poorer prior to the earthquakes.

THE MAGNITUDE OF THE DISASTER

With regard to the magnitude of damage, expressed as a percentage of GDP of the affected areas, the highest impacts occurred in San Vicente (57%), La Paz (31%), Cuscatlán (22%), and Usulután (19%) (Table 2; Fig. 6). In these departments, the earthquakes caused the loss of a considerable proportion of the annual GDP in the brief two or three minutes that they lasted.

The dimension of the damages caused by the earthquakes can be better understood by comparing the magnitudes of damages in the affected departments with those of other recent disasters that have occurred in the Latin American and Caribbean region. As shown in Figure 7, the magnitude of damages of these earthquakes in the most affected departments in El Salvador is greater than those caused by torrential storms and debris flows in Venezuela in 1999 or by Hurricane Georges in the Dominican Republic in 1998. Only Hurricanes Mitch in Honduras (1998) and Keith in Belize (2000) had higher damage magnitudes. This comparison shows that the disaster in El Salvador was of a very high magnitude, especially in San Vicente.

SUMMARY OF DAMAGE AND LOSSES ANALYSIS

The analysis of damages and losses, in both absolute and relative terms, caused by the 2001 earthquakes enables us to define the main characteristics of the disasters as follows:

- A relatively large amount of damages and losses, two-thirds of which were sustained by the private sector.
- Interruption and destruction of road transport, which considerably increased operational costs.
- Major destruction and effects on housing and human settlements, particularly in smaller urban areas, which aggravated preexisting housing deficits.
- Destruction and losses in education and health services, which increased existing gaps between wealthy and less wealthy departments and slowed social development efforts.

Figure 5. Geographical distribution of the human development index in El Salvador following the January and February 2001 earthquakes.

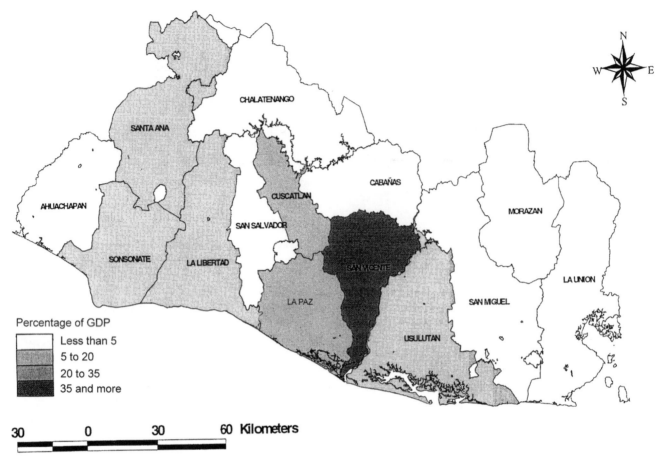

Figure 6. Geographical distribution of damages caused by the January and February 2001 earthquakes in El Salvador as a percentage of GDP.

- Damage and destruction to the production capacities of micro-, small-, and medium-sized enterprises in agriculture, industry, and commerce, but with only limited negative effects on large enterprises.
- Considerable environmental damage owing to extensive losses in agricultural land due to landslides, slope destabilization, and aggravation of already severely fragile soil conditions.
- Significant local concentration of damages in the central area of the country.
- Major economic losses in various departments, both on a per capita basis and in relation to GDP.

The described damages and losses that resulted from the two earthquakes must be considered within a broader context. Destroyed assets represent more than 40% of the country's gross annual capital formation, which provides a measure of the efforts that will be required to replace them. Replacement of assets will have to be made at considerably higher costs than the estimated value of losses at the time of the disaster. We estimate a necessary investment effort of no less than $2 billion, in sharp contrast to the limited capacity of the local construction sector. It may take up to five years to replace lost assets, and communities and their inhabitants will sustain losses in their living conditions during this time. Unexpected emergency expenditures and reconstruction investment pressures, even taking into consideration the expected contributions from the international community, will bear on public sector finances and the ongoing macroeconomic stabilization efforts.

Another major economic effect of the earthquakes concerns damages to transportation infrastructure. Those damages impose not only increased costs, but also long delays and longer routes for the transport of cargo and people. Direct and indirect damages are estimated at nearly $358 million. All users and consumers in El Salvador will feel the effects of damages to transportation infrastructure. Those damages will further reduce social well-being and impact prices in the country and beyond, because of the significant damage to the Pan-American Highway and the regional nature of the road network.

Production losses, estimated to have affected only 3% of the country's exports, were heavily concentrated on micro and small enterprises. Affected activities, mainly oriented to the domestic markets, will result in income loss for those entrepreneurs and will also cause supply shortages of several products and services, requiring the importation of substitutes to meet internal demands.

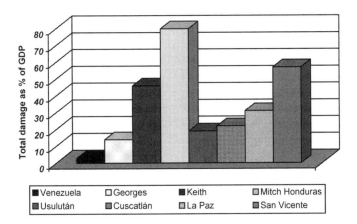

Figure 7. Comparative effect of damages in the most affected departments in El Salvador with other recent disasters in Latin America and the Caribbean.

Finally, economic impacts of the El Salvador earthquakes reached beyond the country, giving these disasters a regional scope. For example, disruption of the Pan-American Highway led to the use of longer and costlier alternate routes that affected intraregional trade. Furthermore, nearby countries sustained economic losses when tourists canceled reservations even though the infrastructure and tourist attractions in those countries sustained no direct damages by the earthquakes.

SOME CROSSCUTTING ISSUES OF THE DISASTER

Both men and women sustain the ravages of disasters, but women face special consequences related to their gender. Unique negative effects of disasters suffered by women can be broken into two categories: (1) impact on women's "reproductive work," which includes renewal of the labor force (child care and education), provision of household chores (labor and time in house care, food preparation, water supply, personal care, etc.), and care of nonworking family members (the sick, the old, the physically and mentally challenged); and (2) impact on the so-called "cottage" or "backyard" economy, in which women undertake income-generating and food-supplementing activities (growing vegetables and raising small animal species). Table 3 shows the impact of the 2001 earthquakes on women. Data were estimated from special field surveys of affected women conducted by the ECLAC field mission. Public sector impacts distributed by gender, which for women is estimated at $289 million, must be added to the figures shown in Table 3. Thus, the total economic impact of the 2001 earthquakes on women is estimated at more than $1 billion.

Our analysis of economic impact of the earthquakes also considered impact on the environment. The earthquakes triggered numerous landslides. A rough estimate of environmental damage indicated a loss of $35 million (Table 4). This estimate does not include restoration costs of the affected areas and is based only on stabilization costs in terms of the engineering

work required on sloping terrain. A more complete estimate of environmental damages would appraise biodiversity losses and damages to ecosystems, habitat losses, watershed reduction through sedimentation inland and on the coastal plains, increased instability of slopes, loss of fertile land, changes in the patterns of replenishing of groundwater reservoirs, and landscape loss for recreation. Permanent loss in environmental services and losses caused by temporary provision of those services until recuperation works are completed were not estimated because there is no baseline valuation available for those services in El Salvador.

MACROECONOMIC EFFECTS

The impacts described above had effects on the Salvadorian economy and society as a whole, and not just on the departments affected. Macroeconomic impacts of the earthquakes were estimated on the basis of an analysis of their effects on the economy in 2001 as well as in future years. The analysis permitted an estimation of potential effects on economic growth, inflation, and gaps in the balance of payments and public sector finances. We estimate a possible post-earthquake fiscal gap associated with

TABLE 3. DAMAGE BY THE 2001 EARTHQUAKES TO SALVADORIAN WOMEN

Type of damage	Million U.S. dollars, at 2001 prices
Direct damages	300.8
Housing, furnishing	146.1
Industry, commerce and services	117.0
"Backyard economy" assets	37.7
Indirect effects	414.4
Loss of jobs and income outside of the household*	(34.7)
Production losses in the household	116.8
"Backyard economy" production	25.0
Micro and small informal production	24.0
Production activities	91.8
Increase in reproductive work	276.5
Other losses	21.1
Total damages and losses	715.2

Note: Source: ECLAC estimates.
* This amount must be deducted from total damage to avoid duplication in the valuation of increased reproductive work.

TABLE 4. ENVIRONMENTAL DAMAGES ASSOCIATED TO RESTORATION COSTS AFTER THE JANUARY AND FEBRUARY 2001 EARTHQUAKES IN EL SALVADOR

Type of damage	Million U.S.$ at 2001 prices
Stabilization of landslides and avalanches in the Chinchontepec (San Vicente) volcano	9,500
Stabilization of landslides and avalanches of the Jiboa river watershed	14,000
Stabilization of landslides and avalanches in San Pedro Perulapán	11,500
Total	35,000

Note: Source: ECLAC estimates.

new reconstruction loans from international financial institutions to average $390 million for five years. This increased fiscal pressure on the Government of El Salvador (GOES) will have an impact on domestic savings and on investment capacity. The feasibility of GOES undertaking reconstruction expenditures is considered possible if fresh, "soft" (i.e., long repayment terms, lower interest rates, and extended grace periods) external resources are made available. The El Salvador Central Reserve Bank (BCR) and the International Monetary Fund (IMF) both highlighted the required concessionary nature of external loans: 20-year maturity, 5-year grace period, and a preferential annual rate of 7.5% (London Interbank Offering Rate). If such loans are made available, the short-term indebtedness would not increase in the three years immediately following the disasters.

Three alternative scenarios of possible post-disaster economic evolution were prepared by ECLAC on the basis of damages and consequences caused by the earthquakes. These scenarios estimate economic evolution over a five-year period and are based respectively on the historical investment ratio for the economy. The scenarios are (1) $150 million of investment for the first year and subsequent annual average investments of more than $400 million for another four years to complete reconstruction; (2) $380 million annual average investments for five years; and (3) $400 million of investment in the first year and subsequent annual average investments of $375 million for another four years. In each scenario, the level of expenditure and public investment is associated with estimated debt conditions, and expenditure feasibility depends on the productive and growth capacity of the country both in terms of infrastructure and organization. The time frame for the reconstruction phase affects the economic evolution in these scenarios and is a factor that can be modified depending on the nature of the reconstruction loans. Ensuing conditions faced by the country, such as severe drought and depressive world-coffee-market condition (ECLAC, 2002b), also affect projected economic evolution and could extend the macroeconomic effects of these earthquakes beyond five years. Changes in interest rates and conditions of loans for reconstruction will alter the cost and service of fresh debt. Any concessionary components of attained loans could favor faster reconstruction and reduce pressure on the basic macroeconomic fundamentals. The results for each economic-evolution scenario are described below and summarized in Table 5.

Pessimistic Scenario

A pessimistic scenario is based on resource availability of $150 million for 2001 and $1,750 million during 2002–2005. It results in real GDP growth greater that of the year 2000, but because of the reduced flow of resources for reconstruction in 2001, it does not allow a timely recovery of the productive structure and causes a downturn of most major macroeconomic indexes.

Probable Scenario

The most probable scenario is based on resource availability of $380 million in 2001 and $1,520 million during 2002–2005. This scenario allows for GDP growth to double the 2000 rate, and reduces annual price increases. The fiscal deficit and the current account gap would expand due to the reconstruction efforts and import increases. The underlying deficit grows to 2.7% of GDP, while the reconstruction costs level at 2.1% of GDP, increasing the global deficit for 2001 to 4.8% of GDP.

Optimistic Scenario

An optimistic scenario requires reconstruction funds of $400 million for 2001 and $1,500 million for 2002–2005. This scenario produces a higher GDP growth, reduced inflation with respect to that of 2000, and fiscal and external balances that remain within reasonable levels.

The preceding scenarios are intended to provide a measure of the full dimension of the potential impact of the earthquakes as perceived immediately after the disaster. At that time, there was no certainty with regard to the amount of foreign cooperation, the volume of credits effectively obtained or renegotiated to be used in 2001, required reconstruction execution periods and the sequence of the resources' flows, or whether funds obtained would be under soft term conditions.

The assessments presented here do not consider the possible existence of alternative sources to partially finance reconstruction, such as bond emissions, sale of public assets, privatization of goods and services provided by government, or the use of private contractors. Other possible sources of funding include an increased internal savings effort and enhanced fiscal revenue to reduce the pressure of increased public expenditure in both cur-

TABLE 5. SUMMARY OF MAIN ECONOMIC INDICATORS UNDER THREE RECONSTRUCTION SCENARIOS FOR 2001 AFTER THE JANUARY AND FEBRUARY 13 EARTHQUAKES IN EL SALVADOR

	Scenario 1 (pessimistic)	Scenario 2 (probable)	Scenario 3 (optimistic)
Real GDP growth	3.0%	3.5–4.0%	4.0–5.0%
Inflation	4.3%	3.0%	3.0%
Fiscal deficit	5.0–5.5%	4.8–5.0%	2.7–3.0%
Current account deficit/GDP	4.0%	3.5%	2.5%
Public debt/GDP	35%	33%	32.3%

Note: Source: ECLAC estimates.

rent (during the emergency and immediate rehabilitation phases) and capital accounts (investments in the reconstruction process for five or more years). Figure 8 shows possible GDP performance under each of the aforementioned scenarios.

We recommend a prudent and consistent macroeconomic policy to handle the necessary resources to finance the national reconstruction plan. The policy should consist of handling public finances in such a way as to not generate additional risks of macroeconomic instability, keep an adequate level of international reserves and debt services costs under control, and not negatively affect productive capacity and employment already weakened by the earthquakes. A possible basket of resources for reconstruction includes concessionary loans by multilateral institutions, bond emissions, government resources, and fiscal measures to enhance revenues such as widening the tax base and fiscal efficiency. A recent change in El Salvador's tax law was oriented to promote tax efficiency and to address traditional loopholes and tax evasion normally faced by developing countries.

We do not deem it possible to apply a government revenue enhancement policy through increased taxes, given the entrepreneurial expectations that were expressed directly to the ECLAC mission and in polls that were conducted after the earthquakes. Those polls showed a negative response to change in tax policy. The prevailing perception was that domestic demand would diminish if income and employment did not recover promptly. In addition, the possible increase in demand in the construction sector might occur at the expense of reduced demand in other sectors. Reconstruction was expected to increase government's current expenditure and to negatively affect programmed capital and social investment. A so-called "winter plan" (emergency measures of temporary housing, slope stabilization, and road repairs before the rainy season) was also expected to further diminish resource availability.

Under any possible reconstruction scenario, public sector deficit would have to be financed through new loans, even under the assumption that the Central Reserve Bank would continue to accumulate international assets associated with an increase in foreign remittances. This situation would be aggravated if loans constituted a major fraction of reconstruction funding sources in the medium and long term. Under the most probable scenario, debt servicing costs might reach a level equal to about one third of GDP and still be considered adequate.

CONCLUSIONS

The January and February 2001 earthquakes in El Salvador caused total damages and losses estimated to exceed $1.6 billion. Of this total, 59% ($939 million) reflected direct damages and 41% ($665 million) reflected indirect losses. Physical infrastructure and equipment sustained the largest proportion of damages and losses. Our preliminary estimates of damages and losses, by geographical distribution and economic and social sectors, provided a timely framework for quantifying resources needed for reconstruction programming. This information proved valuable for mobilizing external cooperation and international assistance.

ACKNOWLEDGMENTS

We thank two anonymous reviewers for comments that improved our presentation. This paper does not constitute an official document of ECLAC, and the opinions expressed herewith do not necessarily reflect ECLAC's official position.

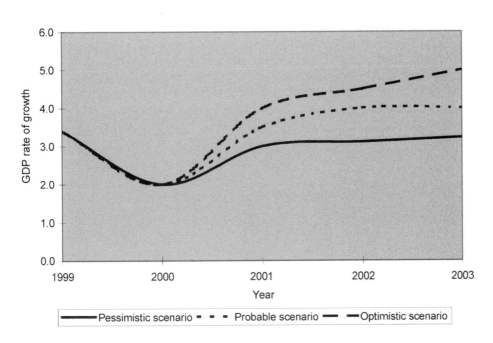

Figure 8. Possible reconstruction scenarios for the period immediately following the 2001 earthquakes.

REFERENCES CITED

ECLAC (Economic Commission for Latin America and the Caribbean), 2001, El terremoto del 13 de enero de 2001 en El Salvador; Impacto socio-económico y ambiental: Mexico, 21 February 2001 (LC/MEX/L.417), 115 p., and El Salvador, Evaluación del terremoto del martes 13 de febrero de 2001: Mexico, 28 February 2001, 53 p.

ECLAC (Economic Commission for Latin America and the Caribbean), 2002a, Handbook for estimating the socio-economic and environmental effects of disasters (LC/MEX/L.519): Mexico City, 386 p.

ECLAC (Economic Commission for Latin America and the Caribbean), 2002b, El impacto socioeconómico y ambiental de la sequía de 2001 en Centroamérica (LC/MEX/L.510/Rev.1/E): February 2002, 65 p.

Jovel, R., 1986, Natural disasters and their economic and social effects, *in* ECLAC Review, No. 38, Santiago, Chile, p. 139–146.

UNDP (United Nations Development Programme), 2001, Informe sobre el desarrollo humano en El Salvador: San Salvador, United Nations Development Programme, CD-ROM.

MANUSCRIPT ACCEPTED BY THE SOCIETY JUNE 16, 2003

Index

Volcán Santa Ana
  vs. Ilopango caldera: 197
  vs. Kelud volcano: 122
  vs. La Caldera de Ilopango: 200
  studies of: 138–145
  vs. Volcán Chaparrastique: 211
diffusivity value in water: 200
Fickian efflux equation for: 196
pH value link: 199
transport mechanisms: 231
carbon, isotopic analysis of
  biogenic material: 199, 210
  crust/mantle structure: 197–199, 210
  La Caldera de Ilopango: 197–199
  soil, forest: 199, 210
  Volcán Chaparrastique: 210
  Volcán San Salvador: 230, 233–234
carbonate: 197–199, 234
Caribbean plate
  boundary with NA plate: 373, 381
  Chortis block: 6, 21
  crust/mantle structure
    earthquake trigger: 28, 397
    thickness of: 381
  fault systems
    earthquake trigger: 70
    Motagua-Polochic fault: 6, 264, 426
  motion of
    in earthquake modeling: 369
    forearc region: 373
    velocity of: 302, 373
  subduction zone. *see also* Cocos plate
    asperities in: 370, 390
    geotectonic style of: 369
    location of
      at convergent margin: 260, 302
      thrust interface: 368
      in trench: 302, 380
    oblique convergence in
      components of: 341–342
      direction of: 322
      right-lateral shear zone: 302
      stress direction, rotation of: 360
    slip azimuth: 368
    volcanic front formation: 2, 70
    Wadati-Benioff characteristics: 368–370
  volcanic slope failures: 21
  vs. West Indies plate: 21
Caribbean Sea: 264, 266
Cascade Mountains: 156
Casita volcano: 90
catalogs, earthquake: 257–258, 261–266, 357–362
cattle: 465
Cauta: 259
Cecropia: 239
Celtis: 239
cement: 282, 295–296, 465
cementation: 65
Cenozoic, photosynthesis in: 199
Central America
  Belize: 475
  Caribbean plate. *see* Caribbean plate
  Cocos plate. *see* Cocos plate
  Costa Rica. *see* Costa Rica
  earthquakes in

database of: 357–362
  recurrence interval: 403–404
El Salvador. *see* El Salvador
Guatemala. *see* Guatemala
Honduras. *see* Honduras
Nicaragua. *see* Nicaragua
Panamá. *see* Panamá
plate tectonics: 2. *see also* Central American arc
pre-Columbian societies. *see* indigenous people
seismic monitoring in: 258
volcanic edifices in: 18
volcanic slope failures: 5
Central American arc
  crust/mantle structure
    Chortis block: 6, 21
    gas, release/movement of
      Volcán Santa Ana: 142
      volcano surveillance by: 194
    isotopic analysis of: 197–199
    magma mixtures in: 176–177, 181
    Motagua-Polochic fault: 6, 264, 426
    perturbation of: 140
    stress field changes: 194
    subduction zone
      annual rate of subduction: 191
      dip angle: 214
    thickness of
      Caribbean plate: 381
      Cocos plate: 383
      El Salvador: 214
      Guatemala: 214
      Nicaragua: 214
      variation in: 6
    volcanic edifice elevation factor: 18
  location of graben: 137
  Middle America Trench. *see* Middle America
    Trench
  segmentation of: 6, 214, 390
  volcanic front formation: 2, 149
Central America Seismological Center (CASC):
    258, 261–262
Centroid Moment Tensor (CMT)
  vs. Coulomb stress model: 352
  database, Central American: 357–359
  depth default: 367
  magnitude coverage: 402
  vs. USGS/NEIC: 367
  of Wadati-Benioff zone events: 368–369
Ceren: 113–116, 154, 157, 462
Cerro Chambala: 219
Cerro Chino: 118
Cerro Colorado: 21
Cerro La Hoya: 231–232, 234
Cerro Jícaro: 11, 13
Cerro El Limbo: 214. *See also* Ojo de Agua
Cerro Negro
  correlations to: 197, 210, 232
  gas, release/movement of
    carbon dioxide: 232
    radon: 197, 206, 210
Cerro El Pacayal: 214. *See also* Volcán Chinameca
Cerro Pacho: 138, 140–144
Cerro Quemado: 21
Cerro San Jacinto: 57, 322, 466
Cerro Verde: 137–144

Cerro Zapote: 113
Cerros de Mariona: 322
Chalatenango: 258, 274. *See also* La Palma
Chalchuapa
  eruptions, impact on: 239–240
  graben, trend of: 19
  indigenous people
    economy, basis of: 239
    settlement patterns: 238–240, 243
    stelae from: 112
  Laguna Cuzcachapa: 238–240
  pollen analysis: 239, 240
  settlement dates: 238, 240, 243
charcoal: 162, 241
Chenopodiaceae: 239, 241
Chiapas: 390, 392–395
chicken pox: 465
Chilanguera: 454
Chile
  earthquakes in: 435, 438
  hospital design codes: 272, 279
  tsunamis in: 435, 438
  Volcán Petero: 18
  Volcán Socompa: 16
Chilean-type subduction zone: 369
China: 305
Chinameca: 258, 302. *See also* Volcán Chinameca
Chiquihuat River. *See* Río Chiquihuat
Chiriqui River: 111
chlorine: 122, 127–128, 185
Cholula: 114
Chorti people: 115
Chortis block: 6, 21
churches: 394, 457–458
cinder cones: 18
Ciudad Delgado: 465
Ciudad Vieja: 381
civil war
  casualties from: 2, 118, 316
  churches, destruction of: 457–458
  land, ownership of: 291, 465
  schools, impact on: 316
clay
  building material
    in adobe blocks: 291, 446
    in mortar: 292
  cementation of: 65
  fracture zones, sealing of: 253
  in lahars: 17–18, 90
  petrology of
    La Caldera de Ilopango: 163, 169
    Tierra Blanca: 57
climate
  carbon dioxide, impact on: 197
  crater lake impact on: 130–131, 143
  precipitation. *see* precipitation
  soil tests, impact on: 142–144, 194, 197
  wind conditions
    radiosonde studies: 154
    velocity of: 200
clinopyroxene
  geochemical analysis of: 221
  petrology of
    La Caldera de Ilopango: 178–181
    Volcán Chaparrastique: 204, 220